Topics in Applied Physics

Volume 137

Topics in Applied Physics is a well-established series of review books, each of which presents a comprehensive survey of a selected topic within the domain of applied physics. Since 1973 it has served a broad readership across academia and industry, providing both newcomers and seasoned scholars easy but comprehensive access to the state of the art of a number of diverse research topics.

Edited and written by leading international scientists, each volume contains high-quality review contributions, extending from an introduction to the subject right up to the frontiers of contemporary research.

Topics in Applied Physics strives to provide its readership with a diverse and interdisciplinary collection of some of the most current topics across the full spectrum of applied physics research, including but not limited to:

- Quantum computation and information
- Photonics, optoelectronics and device physics
- Nanoscale science and technology
- Ultrafast physics
- Microscopy and advanced imaging
- Biomaterials and biophysics
- Liquids and soft matter
- Materials for energy
- Geophysics
- Computational physics and numerical methods
- Interdisciplinary physics and engineering

We welcome any suggestions for topics coming from the community of applied physicists, no matter what the field, and encourage prospective book editors to approach us with ideas. Potential authors who wish to submit a book proposal should contact Zach Evenson, Publishing Editor:

zachary.evenson@springer.com

Topics in Applied Physics is indexed by SCOPUS and books in this series are submitted for indexing to Web of Science.

2018 Impact Factor: 0.746

More information about this series at http://www.springer.com/series/560

Yutaka Yoshida · Guido Langouche
Editors

Modern Mössbauer Spectroscopy

New Challenges Based on Cutting-Edge Techniques

With Contributions by A. I. Chumakov, J. Evers, N. Kojima, P.-E. Lippens, R. Masuda, C. McCammon, A. Okazawa, M. Reissner, R. Röhlsberger, R. Rüffer, M. Saito, V. Schünemann, M. Seto, Y. Yoshida

Springer

Editors
Yutaka Yoshida
Shizuoka Institute of Science and Technology
Fukuroi, Shizuoka, Japan

Guido Langouche
University of Leuven
Leuven, Belgium

ISSN 0303-4216 ISSN 1437-0859 (electronic)
Topics in Applied Physics
ISBN 978-981-15-9421-2 ISBN 978-981-15-9422-9 (eBook)
https://doi.org/10.1007/978-981-15-9422-9

This Springer imprint is published by the registered company Springer Nature Singapore Pte Ltd.
The registered company address is: 152 Beach Road, #21-01/04 Gateway East, Singapore 189721, Singapore

Preface

Mössbauer spectroscopy is one of the most powerful methods providing deep insight into the atomistic nature of matter. Properties such as valence state, chemical bond, spin state, internal magnetic field, electric field gradient, lattice vibrations, etc., were studied for various kinds of materials. Use is made of nuclear probes, meaning that materials can be studied nondestructively. Therefore, the method has widely been applied to scientific research since the discovery of the Mössbauer effect in 1958. Moreover, since the first demonstration of nuclear resonance scattering of synchrotron radiation (SR) in Hamburg in 1985, SR-based Mössbauer spectroscopy has extensively been developed, allowing us to measure various kinds of Mössbauer spectra whose observation is impossible by using ordinary radioactive sources. Moreover, in addition to Mössbauer spectra, the measurement of element-specific vibrations has become possible using nuclear resonant inelastic scattering.

These cutting-edge techniques have attracted much attention in many scientific fields such as materials science, physics, chemistry, biology, earth science, etc. Today, we witness an interdisciplinary research community of Mössbauer spectroscopists all over the world. Extensive exchange of experience exists in this Mössbauer community by means of at least one international conference every year. We mention the International Conference on the Applications of the Mössbauer Effect (ICAME) organized by the International Board on the Applications of the Mössbauer Effect (IBAME) which, e.g., also guards the nomenclature used in Mössbauer spectroscopy, the International Symposium on Industrial Applications of the Mössbauer Effect (ISIAME) and regional conference series such as LACAME in Latin America and MECAME in the Mediterranean countries, among others. We should also mention the Mössbauer Effect Data Center (MEDC), Dalian, China, which is a fairly unique institute which essentially keeps track of publications involving Mössbauer spectroscopy.

Today, we are facing a generation change with new researchers following in the footsteps of the early pioneers of Mössbauer spectroscopy. This creates a need to preserve and clearly present previously acquired knowledge in Mössbauer spectroscopy for new generations. We are happy with the success of our first tutorial book on Mössbauer spectroscopy (Mössbauer Spectroscopy—Tutorial Book,

Springer 2013) meant for final-year bachelor students and available also as e-book, leading to numerous downloads from all over the world.

One of the main purposes of this new tutorial book is to transfer the central knowledge of Mössbauer spectroscopy not only to the next generations, but also to scientists working in other research fields. Accordingly, we present an overview of new methodologies developed at different synchrotron facilities, we shed light on some advanced materials such as batteries and other energy materials, we focus on a wide variety of applications of Mössbauer spectroscopy in physics, chemistry, biology, geoscience, and we discuss the emergence of imaging techniques in Mössbauer spectroscopy, which we expect to attract new users in the community.

The book begins in its first three chapters with new efforts in modern Mössbauer spectroscopy using synchrotron radiation. The authors of these chapters and their research groups lead nuclear resonant scattering experiments and related technique development at synchrotron accelerator facilities like the European Synchrotron Radiation Facility (ESRF), the Super Photon Ring-8 Gev (SPring-8) and the Deutsches Elektronen-Synchrotron (DESY). In Chaps. 4 and 5, the book demonstrates the applications of such cutting-edge techniques to chemistry–biology and geoscience, respectively. In Chaps. 6 and 7, it further describes the applications of Mössbauer spectroscopy to new functional materials, i.e., metal complexes and Li- and Na-ion batteries, respectively. The last two chapters, Chaps. 8 and 9, are devoted to two important measuring techniques: Mössbauer spectroscopy under external magnetic fields in Chap. 8 and microscopic Mössbauer techniques on diffusion in solids in Chap. 9. Those techniques are expected to play an essential role in investigating and characterizing magnetic structures and microstructures in materials.

This second tutorial book project originated at the occasion of the 9th TOYOTA RIKEN International Workshop on New Developments and Prospects for the Future of Mössbauer Spectroscopy (IWMS2018) at Toyota Commemorative Museum of Industry and Technology in Nagoya city from November 15 to 17, 2018. The workshop was organized by Toyota Physical and Chemical Research Institute (TOYOTA RIKEN) and co-organized by Shizuoka Institute of Science and Technology and Institute for Integrated Radiation and Nuclear Science, Kyoto University. We would like to thank Dr. Akiyuki Tokuno, Springer, for his intensive supports.

Editors

Fukuroi, Japan

Yutaka Yoshida
Guido Langouche

Contents

Contributors

Aleksandr I. Chumakov ESRF, Grenoble, France

Jörg Evers Max-Planck-Institut für Kernphysik, Heidelberg, Germany

Norimichi Kojima Toyota Physical and Chemical Research Institute, Nagakute, Aichi, Japan

Pierre-Emmanuel Lippens Institut Charles Gerhardt, UMR 5253 CNRS Université de Montpellier, 34095 Montpellier Cedex 5, France

Ryo Masuda Institute for Integrated Radiation and Nuclear Science, Kyoto University, Osaka, Japan

Catherine McCammon Bayerisches Geoinstitut, Universität Bayreuth, Bayreuth, Germany

Atsushi Okazawa Division of Chemistry, Institute of Liberal Education, Nihon University School of Medicine, Tokyo, Japan

Michael Reissner Institute of Solid State Physics, TU Wien, Vienna, Austria

Ralf Röhlsberger Deutsches Elektronen-Synchrotron DESY, Hamburg, Germany; Helmholtz Centre for Heavy Ion Research (GSI), Darmstadt, Germany

Rudolf Rüffer ESRF, Grenoble, France

Makina Saito Institute for Integrated Radiation and Nuclear Science, Kyoto University, Osaka, Japan

Volker Schünemann Department of Physics, Technische Universität Kaiserslautern, Kaiserslautern, Germany

Makoto Seto Institute for Integrated Radiation and Nuclear Science, Kyoto University, Osaka, Japan

Yutaka Yoshida Shizuoka Institute of Science and Technology, Fukuroi, Japan

Chapter 1
Historical Developments and Future Perspectives in Nuclear Resonance Scattering

Rudolf Rüffer and Aleksandr I. Chumakov

Abstract In few decades, Nuclear Resonance Scattering of synchrotron radiation developed from a dream to an advanced suite of powerful methods, gathering a wide range of applications from general relativity to nanoscience, combining unprecedented properties of nuclear resonance and synchrotron light, and expanding studies to multiply extreme conditions. This article reviews fundamentals of nuclear resonance physics and properties of synchrotron radiation, provides a short historical overview of the fascinating development, major techniques and instrumentation of the method, and gives a brief snapshot of modern applications and yet coming opportunities.

List of Acronyms

ADC	Analog-to-Digital Converter
APD	Avalanche Photo Diode
APS	Advanced Photon Source
BLS	Brillouin Light Scattering
CFD	Constant Fraction Discriminator
CONUSS	COherent NUclear resonant Scattering by Single crystals
CRL	Compound Refracrive Lens
CZ	Cloudy Zone
DAC	Diamond Anvil Cell
DESY	Deutsches ElektronenSYnchrotron
DFT	Density Functional Theory
DORIS	Doppel-Ring-Speicher
DOS	(phonon) Density Of States

R. Rüffer (✉) · A. I. Chumakov
ESRF, 38043 Grenoble, France
e-mail: rueffer@esrf.fr

A. I. Chumakov
e-mail: chumakov@esrf.fr

Y. Yoshida and G. Langouche (eds.), *Modern Mössbauer Spectroscopy*,
Topics in Applied Physics 137, https://doi.org/10.1007/978-981-15-9422-9_1

EBS	Extremely Brilliant Source
ESRF	European Synchrotron Radiation Facility
FZP	Fresnel Zone Plates
GGG	Gadolinium Gallium Garnet
GINRS	Grazing Incidence Nuclear Resonance Scattering
HASYLAB	HAmburger SYnchrotronstrahlungsLABor
HRM	High Resolution Monochromator
INS	Inelastic Neutron Scattering
IR	Infra-Red (spectroscopy)
IUVS	Inelastic UltraViolet Scattering
IXS	Inelastic X-ray Scattering
IXSNRA	Inelastic X-ray Scattering with Nuclear Resonance Analysis
KB	Kirkpatrick-Baez
MB	Mössbauer
MCA	Multi-Channel Analyzer
ML	Monolayer
MOKE	Magnetic Optical Kerr Effect
NBD	Nuclear Bragg Diffraction
NCD	Nano-Crystalline Diamond
NFS	Nuclear Forward Scattering
NIS	Nuclear Inelastic Scattering
NQES	Nuclear Quasi-Elastic Scattering
NRS	Nuclear Resonance Scattering
NSAS	Nuclear Small Angle Scattering
PETRA III	Positron Electron Tandem Ring Accelerator
QCP	Quantum Critical Point
RAMAN	RAMAN spectroscopy
RSMR	Rayleigh Scattering of Mössbauer Radiation
SMS	Synchrotron Mössbauer Source
SPring-8	Super Photon Ring 8 GeV
SQUID	Superconducting Quantum Interference Device
SR	Synchrotron Radiation
SRMS	Synchrotron Radiation Mössbauer Spectroscopy
SRPAC	Synchrotron Radiation based Perturbed Angular Correlation
TAC	Time-to-Amplitude Converter
TDI	Time Domain Interferometry
TDPAC	Time Differential Perturbed Angular Correlation
UHRIXS	Ultra-High-Resolution Inelastic X-ray Scattering
UHV	Ultra High Vacuum
XFEL	X-ray Free Electron Laser
XRD	X-Ray Diffraction
YIG	Yttrium Iron Garnet

1.1 Introduction

"Nuclear resonance fluorescence of γ-radiation" without recoil was discovered and explained by R. L. Mößbauer in 1958 ("Kernresonanzfluoreszenz von Gammastrahlung in Ir^{191}" [1, 2]) and later named "Mössbauer effect" and the related spectroscopy consequently Mössbauer spectroscopy. Already shortly after this earth-breaking discovery thoughts went from absorption to scattering and diffraction experiments. Scattering experiments were reported by Black and Moon early as 1960 [3], followed by grazing incidence experiments by Bernstein and Campbell [4], and diffraction experiments by Black et al. [5]. New phenomena such as interference between electronic and nuclear scattering, enhancement of the radiative channel (speed-up), and suppression of the incoherent channels were discovered and clarified. This early period has been reviewed by Smirnov (experiments) [6], and by van Bürck (theory) [7]. At that time synchrotron radiation has already been known since 1947, however, mainly as a nuisance for high energy accelerator experiments.

In the remaining part of the chapter, nuclear resonance and synchrotron radiation will be introduced. The following historical chapter shall in general highlight, on the example of nuclear resonance scattering, the challenges in scientific developments, which are not only scientificly but also technically and politically driven.

The main part is devoted to an overview of the richness of nuclear resonance techniques and spectroscopies, which span from investigations of magnetic and electronic properties, static and dynamic, and structural dynamics on various energy and time scales, to γ-optics and other fundamental research. Eventually, selected examples of applications will not only showcase the unique fields of research accessible with nuclear resonance techniques but also look in the bright future with the new light sources at the horizon.

1.1.1 Nuclear Resonance

Nuclear Resonance Scattering (NRS) with synchrotron radiation (SR) combines the outstanding properties of the Mössbauer effect with those of synchrotron radiation. Since its first convincing observation in 1984 [8] a rapid development of the technique with many facets followed. Thanks to the outstanding properties of 3rd generation synchrotron radiation sources nuclear resonance techniques are nowadays known for their extreme energy resolution and timing properties offering a wide range of applications.

Nuclear resonance techniques including Mössbauer spectroscopy are related to the recoiless resonant scattering, absorption, and emission of x-rays and γ-rays[1] by atomic nuclei. This effect is the same as in the atomic shell and well known e.g. from the yellow emission lines of sodium where light is absorbed and re-emitted by a transition of an electron between the 3p and the 3s atomic levels. Though the basics

[1] we use the term γ-ray for x-rays coming from a nucleus.

are the same nuclear resonance scattering was only discovered by Mößbauer in 1958 [1]. What are the reasons for this late discovery?

There are some important differences in the scattering of "optical" x-rays and γ-rays. First of all the transition energies E_0 are in the range of 1–10 eV and 10–100 keV, respectively. Further, the relative energy width $\Delta E/E_0$ of the involved atomic and nuclear levels are quite different partly due to the different excitation energies. As a consequence the recoil connected with the absorption and emission of a photon is a special problem for γ-rays because their recoil energy is so large that there is virtually no overlap of the absorption and emission lines in single atoms.

The recoil energy E_R is given as

$$E_R = \frac{E_\gamma^2}{2Mc^2}\,, \tag{1.1}$$

with $E_\gamma \cong E_0$ the photon energy, M the atomic mass, and c the velocity of light. For a typical Mössbauer isotope, e.g. ^{57}Fe with $E_0 = 14.4\,\text{keV}$ and $M \approx 57\,\text{u}$,[2] the resulting recoil energy is 1.956 meV, which is about six orders of magnitude bigger than the natural line width Γ_0 of its nuclear level ($\Gamma_0 = 4.66 \cdot 10^{-9}\,\text{eV}$). On the other hand for the abovementiond sodium case ($E_0 \approx 2\,\text{eV}$ and $M \approx 22\,\text{u}$) the loss in energy due to recoil is virtually zero ($E_R \approx 10^{-11}\,\text{eV}$). Finally, the thermal motion and the resulting Doppler broadening of the absorption and emission lines is another important aspect. Consequently, there exists full overlap in case of optical x-rays and no overlap for γ-rays. The achievement of R. L. Mößbauer was the understanding that, when an atom is bound in a solid, the entire solid ($M \to \infty$) will take the recoil and then the energy loss for the γ-ray is negligible. However, also in a solid vibrations of the atoms exist around their equilibrium position. R. L. Mößbauer showed that with a certain probability, expressed by the Lamb-Mössbauer factor (f_{LM}), no vibrations will be involved in the scattering, absorption, and emission process, respectively.

In order to shed some more light on this discovery and the f-factor we have to discuss some basic features of scattering. Generally for resonant scattering, the coherent elastic scattering amplitude f_{res} is given by [9]

$$f_{res} = \left(\frac{\Gamma_\gamma}{2ik}\right) \int_0^\infty dt\, e^{i(\omega-\omega_0)t}\, e^{-(\Gamma/2\hbar)t} \langle e^{-i\mathbf{k}_f\,\mathbf{r}(t)}\, e^{i\mathbf{k}_0\,\mathbf{r}(0)}\rangle, \tag{1.2}$$

where Γ_γ and Γ is the radiative and total resonance linewidth, respectively, $\hbar\omega_0 = E_0$ the resonance energy, $\mathbf{k}_0$ and $\mathbf{k}_f$ the wave vectors of the incident and reflected wave, respectively, and $\mathbf{r}(0)$ and $\mathbf{r}$(t) are the displacement from the equilibrium position of the interacting particle at times zero and t, respectively. The $\langle\rangle$ represents the time average over the characteristic interaction times.

[2]Unified atomic mass unit u $\hat{=}$ 931.494 MeV/c^2.

Zero point and temperature motion of a scatterer (e.g. atom or nucleus) are characterized by vibration times $\omega_m^{-1} \approx 10^{-14}$–$10^{-13}$ s. They have to be compared with the characteristic scattering times.

For resonant x-ray scattering, the characteristic scattering times are about 10^{-16}–10^{-15} s and hence fast compared to ω_m^{-1}, i.e., $t \to 0$ and the displacements from the equilibrium position can effectively be taken at $t \approx 0$, hence

$$\langle e^{-i\mathbf{k}_f \mathbf{r}(t\approx 0)} e^{i\mathbf{k}_0 \mathbf{r}(0)} \rangle \approx \langle e^{-i[(\mathbf{k}_f - \mathbf{k}_0)\mathbf{r}]} \rangle = e^{-\frac{1}{2}\langle [(\mathbf{k}_f - \mathbf{k}_0)\mathbf{r}]^2 \rangle} . \tag{1.3}$$

This is the well known expression for the Debye-Waller factor f_D.

On the contrary, for Mössbauer resonances the scattering is slow, about 10^{-9}–10^{-6} s, compared to ω_m^{-1}, i.e., $t \to \infty$ and the displacements from the equilibrium position can effectively be considered as uncorrelated, hence

$$\langle e^{-i\mathbf{k}_f \mathbf{r}(t\approx \infty)} e^{i\mathbf{k}_0 \mathbf{r}(0)} \rangle \approx \langle e^{-i\mathbf{k}_f \mathbf{r}} \rangle \langle e^{i\mathbf{k}_0 \mathbf{r}} \rangle = e^{-\langle x^2 \rangle E_\gamma^2/(\hbar c)^2} , \tag{1.4}$$

with $\langle x^2 \rangle$ the expectation value of the squared vibrational amplitude in the direction of the γ-ray propagation, the so-called mean-square displacement [10]. This expression is called the Lamb-Mössbauer factor f_{LM}. Equation 1.4 immediately shows that f_{LM} gets very small at higher energies, i.e., for practical reasons only nuclei with low lying nuclear levels ($\lesssim 100$ keV) are considered as Mössbauer nuclei.

In order to describe the f_{LM} one would need a detailed and comprehensive description of the phonon spectrum of the solid lattice. In general that is not available and for most cases the simpler Einstein or Debye model is sufficient to describe for example the temperature dependence of the Lamb-Mössbauer factor [10].

$$f_{LM}(T) = \exp\left[\frac{-3E_\lambda^2}{k_B \Theta_D M c^2} \left\{ \frac{1}{4} + \left(\frac{T}{\Theta_D} \right)^2 \int_0^{\Theta_D/T} \frac{x}{e^x - 1} dx \right\} \right] , \tag{1.5}$$

with Θ_D the Debye temperature and k_B the Boltzmann constant. Inspection of Eq. 1.5 reveals that a large f_{LM} is expected for low transition energies, low temperatures, and high Debye temperatures.

Due to the sharp width of the nuclear levels and their large separation, compared to atomic levels, the linewidth of the radiation for nuclear resonant scattering, absorption, and emission is a Lorentzian curve as given by the so-called Breit-Wigner equation:

$$I(E) = \frac{\Gamma_0/(2\pi)}{(E - E_0)^2 + (\Gamma_0/2)^2} , \tag{1.6}$$

with the natural linewidth $\Gamma_0 = \hbar/\tau_0$ (τ_0 the natural lifetime) of the nuclear level. At resonance the scattering amplitude may peak much higher than for atomic scattering, e.g. for ^{57}Fe as $440\, r_0$ with $r_0 = 2.810^{-15}$ m the classical electron radius [9].

The discovery by R. L. Mößbauer made available a hard γ-ray source with unprecedented energy resolution $\Delta E/E_0 \approx 10^{-13}$ and longitudinal coherence length $l = c\,\tau_0 \approx 40\,\mathrm{m}$ in case of ^{57}Fe even nowadays only comparable with highest precision laser systems. Those properties made applications such as wavelength standard [11], Hanbury Brown-Twiss [12], interferometry [13, 14], and Lamb shift [15] feasible. Furthermore, they established spectroscopies on an atomistic scale, which are element and even isotope sensitive and non-destructive. Applications comprise two main fields, *hyperfine spectroscopy* and *structural dynamics*. In hyperfine spectroscopy NRS is complementary to other nuclear techniques and yields useful information on atomic, magnetic, and electric properties. Those fields of applications benefit most, which exploit the specific properties of synchrotron radiation and therefore allow for applications to high pressure, to grazing incidence geometry (surfaces and multilayers), to single crystals, and to very small samples. Structural dynamics on a ps to μs time scale as free or jump diffusion as well as rotational motions can directly be measured in the time domain by nuclear quasi-elastic scattering techniques. On the fast time scale the (partial) phonon density of states is directly accessible by (nuclear) inelastic scattering techniques.

For the application two cases may be distinguished, (i) excitation of the nuclear levels by 'white' SR with sharp pulses in time (ps) and the successive spectroscopy in the time domain and (ii) excitation of the nuclear levels by highly monochromatized radiation (energy width neV to peV defined by the nuclear level width Γ_0) with 'long' pulses in time (ns to μs) and the successive spectroscopy in the energy domain. Both are, generally speaking, connected by the Heisenberg uncertainty principle.

A comprehensive overview on the technique, experimental and theoretical, as well as on applications is given in the review book by Gerdau and de Ward [16]. The theoretical background was laid by Hannon and Trammell [17, 18] and Afanas'ev and Kagan [19, 20]. A detailed overview on optics may be found by Shvyd'ko [21] and on Nuclear Condensed Matter Physics by Röhlsberger [22]. An introduction to Mössbauer spectroscopy with applications is given by Gütlich, Bill, and Trautwein [10].

1.1.2 Synchrotron Radiation

Synchrotron radiation became a synonym for all electromagnetic radiation generated by transverse acceleration of relativistic charged particles. The name dates back to its first observation in the General Electric 70 MeV synchrotron. In contemporary synchrotron radiation facilities the particles are accelerated to their nominal energy by linear accelerators (linacs) and circular accelerators (synchrotrons). Eventually, the particles are fed into the storage ring where they travel on a stable, closed orbit and with fixed energy. This procedure allows for defined and optimum properties of the generated SR for the various applications.

The transverse acceleration in the storage ring is achieved by static transverse magnetic fields acting on charged particles such as electrons and positrons (see Fig. 1.1). For an electron, i.e., a particle of charge $-$e and of momentum $\mathbf{p} = m \cdot \mathbf{v}$,

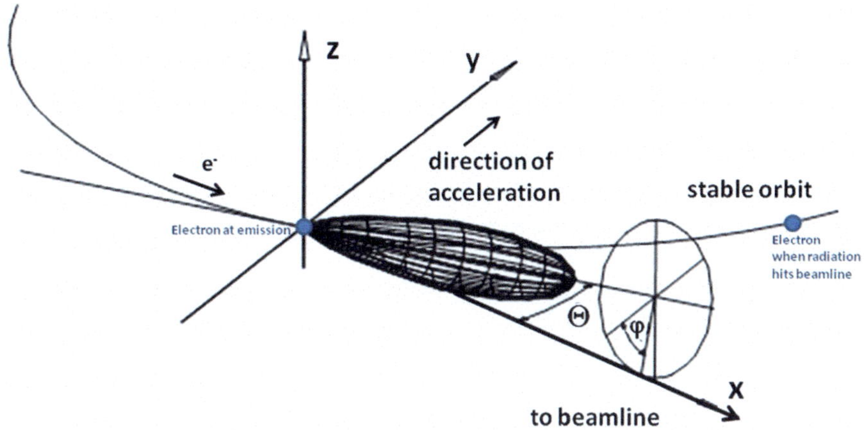

Fig. 1.1 Geometry of the synchrotron radiation. The acceleration points to the center of the storage ring. The radiation peaks in direction of the beamline in the $d\Omega = d\Theta\, d\phi$ cone. Note: The electron emits SR at the origin position and will be at the indicated position when that SR arrives at the beamline

with **v** its velocity and m its mass, the acceleration $\dot{\boldsymbol{v}}$ by a transverse magnetic field **B** is given by the Lorentz force $\mathbf{F}_L$:

$$\mathbf{F}_L = \frac{\mathrm{d}\mathbf{p}}{\mathrm{d}t} = -e \cdot [\mathbf{v} \times \mathbf{B}]\,. \tag{1.7}$$

Because the acceleration $\dot{\boldsymbol{v}}$ fully determines all characteristics of the emitted radiation, the synchrotron radiation can be purpose-engineered by selecting the type of particle, the particle current, the particle energy, and the spatial configuration of the magnetic fields. With that the resulting synchrotron radiation can be tailored in frequency, polarization, time structure, source sizes, and emission angles.

In case that the accelerating particles are at rest in the frame of the observer, like the electrons in an antenna, the radiated power P in the solid angle $\mathrm{d}\Omega = \mathrm{d}\Theta \mathrm{d}\phi$ is given by classical non-relativistic electrodynamics:

$$\frac{\mathrm{d}P}{\mathrm{d}\Omega} = \frac{e^2}{4\pi c^3} \cdot \dot{v}^2 \cdot \sin^2\Theta, \tag{1.8}$$

and the total radiated power is given by

$$P = \frac{2}{3}\frac{e^2\dot{v}^2}{c^3} \qquad (Larmor formula). \tag{1.9}$$

The emission is characterized by:

(i) maximum power is radiated perpendicular to the direction of the acceleration,
(ii) there is zero radiation in the direction of acceleration, and

(iii) the polarization is along the direction of acceleration.

The for us interesting situation is the case where the velocity **v**, close to the velocity of light, is perpendicular to the acceleration $\dot{\boldsymbol{v}}$ of the emitting charge, i.e., relativistic charged particles in a transverse magnetic field in a circular particle accelerator: here the relativistic contraction of the emitted radiation is around the direction of maximum emission of the "antenna", while in the case where velocity **v** and acceleration $\dot{\boldsymbol{v}}$ are parallel (e.g. in a linear accelerator) the emitted radiation in forward direction would be zero.

The Lorentz transformation of the non-relativistic emission [23] leads to a contraction of the emission parallel to the velocity of the charge, which can be approximated for highly relativistic particles of total energy E by

$$\frac{\mathrm{d}P}{\mathrm{d}\Omega} \simeq \frac{2e^2\dot{v}^2}{\pi c^3}\gamma^6 \cdot \frac{1}{(1+\gamma^2\Theta^2)^3}\left[1 - \frac{4\gamma^2\Theta^2\cos^2\phi}{(1+\gamma^2\Theta^2)^2}\right] \tag{1.10}$$

and to an enhancement by a factor γ^4 ($\gamma = E/mc^2$ with m the rest-mass of the charged particle) of the emitted power P compared to the non-relativistic case as in Eq. 1.9:

$$P = \frac{2}{3}\frac{e^2\dot{v}^2}{c^3} \cdot \gamma^4. \tag{1.11}$$

Furthermore, the radiation appears in the laboratory frame highly collimated in a forward cone with an angle

$$\Delta\Theta = \pm\frac{1}{\gamma}. \tag{1.12}$$

This opening angle $\Delta\Theta$ defines for an observer the length L and duration Δt of the synchrotron light pulse, respectively:

$$L = \frac{4}{3}\frac{\rho}{\gamma^3} \quad \text{and} \quad \Delta \mathrm{t} = \mathrm{L/c}, \tag{1.13}$$

with ρ the radius of the curved path of electrons.

1.1.2.1 Emittance and Brilliance

The emittance and the brilliance are the two key parameters, which are nowadays used, to compare synchrotron radiation sources.

The emittance of a beam is defined as the area, enclosed by the one σ line,[3] of the particle density distribution. In case of zero emittance, i.e., an ideal beam, all particles travel on the closed orbit, which is given by the magnetic lattice and the nominal energy of the particles. The cross-section of this beam is pointlike and its divergence is zero. In this ideal case the synchrotron radiation from a given point shows the characteristics of single particle synchrotron light emission. This situation is called diffraction limited (dl).

Point size σ_{dl} and minimum divergence σ'_{dl} for light of wavelength λ are given under Gaussian approximation by

$$\sigma_{\mathrm{dl}} = \frac{\sqrt{L\lambda}}{2\pi} \quad \text{and} \quad \sigma'_{\mathrm{dl}} = \sqrt{\frac{\lambda}{L}}, \tag{1.14}$$

with L the apparent axial extension of the source [24]. This apparent axial extension could be either the undulator length in the case of an undulator source (see Sect. 1.1.2.3) or the particle's path length needed for a deflection of $\Delta\Theta = 2/\gamma$ in a bending magnet source (Eq. 1.13). In such a case of a "diffraction limited synchrotron light source" each phase space occupied by the synchrotron light beam (photon emittance) would be with $i = x, y$

$$\varepsilon_{\mathrm{dl,i}} = \sigma_{\mathrm{dl,i}} \cdot \sigma'_{\mathrm{dl,i}} = \frac{\lambda}{4\pi}. \tag{1.15}$$

For undulator radiation these values are first approximations and may need modifications accounting e.g. for non-Gaussian profiles (see e.g. [25]).

The dimensions for the electron beam may be calculated in first approximation—without considering dispersion and energy spread—by the following expressions, with ε_i the emittance and β_i the β-function of the lattice ($i = x, y$):

$$\sigma_i = \sqrt{\varepsilon_i \cdot \beta_i} \quad \text{and} \quad \sigma'_i = \sqrt{\frac{\varepsilon_i}{\beta_i}}. \tag{1.16}$$

Eventually, the effective x-ray beam dimensions are then given by the convolution of these quantities with the diffraction limited values $\sigma_{\mathrm{dl,i}}$, $\sigma'_{\mathrm{dl,i}}$ of the X-ray beam:

$$\sigma_{\mathrm{T_i}} = \sqrt{\sigma_i^2 + \sigma_{\mathrm{dl,i}}^2} \quad \text{and} \quad \sigma'_{\mathrm{T_i}} = \sqrt{\sigma_i'^2 + \sigma'^{\,2}_{\mathrm{dl,i}}}. \tag{1.17}$$

The brilliance B is the peak flux density in phase space

$$B = \frac{\text{photons/s}}{\sigma_x \sigma_y \cdot \sigma'_x \sigma'_y \cdot d\varepsilon/\varepsilon}, \tag{1.18}$$

[3]All 'σ' values in this paragraph are 'root mean square' (rms) values assuming Gaussian distributions. They have to be multiplied by $2\sqrt{2\ln 2} \approx 2.355$ in order to get the corresponding 'full width at half maximum' values.

with σ_i in mm, σ_i' in mrad, and $d\varepsilon/\varepsilon$ in 0.1% energy bandwidth. In Anglo-American publications very often the brilliance is named "(spectral) brightness".

1.1.2.2 Time Properties

The energy loss of the particles due to the radiated power (synchrotron radiation) (see Eq. 1.11) has to be supplied back to keep the particles on their stable orbit. That is done by rf-transmitters and so-called cavities in the storage ring. The involved frequencies are in the UHF regime (352 MHz for the ESRF) and define together with the circumference of the ring the stable positions, called buckets, for the particles. In case of the ESRF with a circumference of 844 m 992 buckets are available to be filled with electrons separated by 2.8 ns. In principle any pattern of buckets, out of the 992, may be filled to allow for the necessary flexibility to the demand at the experimental stations. Those filled buckets are called bunches and are about 100 ps in length. Typical filling pattern at the ESRF are currently the multi-bunch mode (7/8 + 1: meaning that 7/8th of the ring is closely filled with 868 bunches plus a single bunch just in the center of the remaining gap leaving 176 ns empty space for timing experiments), 16 bunch mode (16 buckets are filled with a separation of 176 ns each), and 4 bunch mode (4 buckets are filled with a separation of 704 ns each).

Most of the NRS experiments rely on those two timing modes, 4- and 16-bunch mode. A special signal (bunch clock) is available for synchronizing this timing with the experimental needs. An important parameter is the "purity" of the filling. It is defined as the ratio between the number of photons emitted by electrons in accidentally filled buckets and the number of photons emitted by the electrons of the nominal bunches. Routinely a purity better than 10^{-9} up to 10^{-11} is reached by special cleaning procedures at the ESRF.

1.1.2.3 Insertion Devices

So far we have only considered the bending magnets, responsible for the transverse acceleration and for keeping the particles on a closed orbit, as a source for synchrotron radiation.

Third generation SR sources are however characterized by their additional synchrotron radiation sources, so called "insertion devices", special magnet structures inserted in dedicated "straight sections" of the storage ring. Their properties may be tailored to the needs of the experimental stations. The most simple case is a planar magnet structure of alternating short dipole magnets, see Fig. 1.2. They are arranged such that the field varies sinusoidally along the particle trajectory

$$B(x, y = 0, z = 0) = B_0 \cdot \cos\left(\frac{2\pi}{\lambda_u} \cdot x\right), \tag{1.19}$$

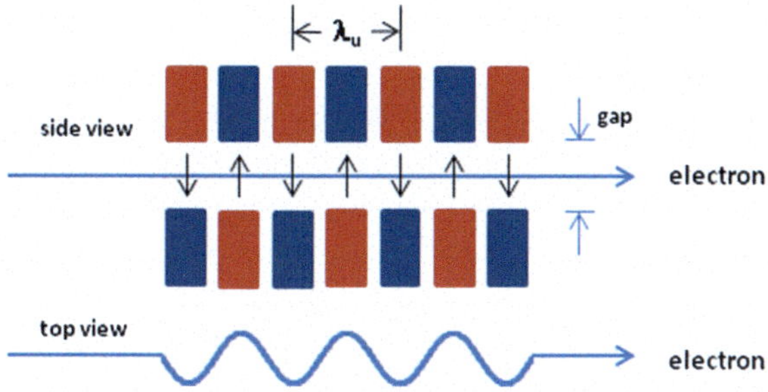

Fig. 1.2 Schematics of an insertion device. Top panel: side view of the alternating dipole magnets (red: north pole, blue: south pole) with the electron beam traveling in the center with λ_u the magnetic period and gap the magnetic gap of the undulator. Bottom panel: top view of the sinusoidal electron movement due to the alternating magnetic field. The amplitude is not to scale and is in the order of microns

with B_0 the peak magnetic field and λ_u the magnetic period. The particle will then oscillate with the amplitude

$$y(x) = \frac{K}{\gamma\, 2\pi/\lambda_u} \cos\left(\frac{2\pi}{\lambda_u} \cdot x\right), \tag{1.20}$$

with $K = \frac{ecB_0\lambda_u}{2\pi mc^2}$ the so called deflection or strength parameter. For $K \gg 1$ the insertion device is called a wiggler and for $K \ll 1$ an undulator. The (magnetic field) strength can be varied by changing the magnetic gap height, which in turn will change the K value and eventually the wavelengths (energies) of the SR spectrum.

Nowadays, mainly undulators will be used, which allows one to optimize the photon beam quality such as its brilliance, energy spectrum, signal-to-noise ratio or to provide special chracteristics such as dedicated (elliptical, circular) polarizations.

For the various harmonics k of an undulator the dependence on K and λ_u can be expressed in photon wavelengths λ as:

$$\lambda_k = \frac{\lambda_u}{2\gamma^2 k}\left(1 + \frac{1}{2}K^2 + \gamma^2\Theta^2\right), \tag{1.21}$$

and in photon energies ε as:

$$\varepsilon_k = \frac{4\pi\hbar c\gamma^2 k}{\lambda_u\left(1 + \frac{1}{2}K^2 + \gamma^2\Theta^2\right)}. \tag{1.22}$$

In this paragraph we could only mention those quantities, which are most important with respect to NRS experiments. There exist a waste amount of literature, which treats all aspects of SR, such as the review work by Wiedemann [24] with references therein.

1.2 Historical Development

In this chapter we want—on the example of NRS with SR—to point out how research may evolve from first ideas and discussions to the hopefully successful finish. What are the scientific dreams, at the beginning and at the end of such an endeavour? What are the technical challenges? Are there other aspects, which were crucial for the development and the science of such a project? Eventually, we can report on the successful achievement of this endeavour: Nuclear Resonance Scattering with Synchrotron Radiation.

Even when there were already some discussions going on, the "kick-off" moment for nuclear resonance with synchrotron radiation was certainly the Mössbauer conference 1974 in Bendor (France). In his contribution Stan Ruby discussed synchrotron radiation as a new source for Mössbauer spectroscopy instead of radioactive sources [26]. It was at a time when SR was mainly a nuisance for the researchers doing high-energy physics with the new colliders using highly energetic electrons and positrons. This unavoidable SR prevented them to increase the particle energy to higher and higher values. Anyway, first attempts had already started to use that radiation in a parasitic manner e.g. for atomic and solid state physics. Now, also the Mössbauer community started with several groups all over the world trying to excite Mössbauer isotopes with SR.

1.2.1 Scientific Dreams

Of course the idea itself—creating a resonant Mössbauer beam out of the SR—is highly appealing and warrants to be tackled. That was for sure one of the important reasons to start the endeavour. On the other hand the discussions at that time show that the combination of the outstanding features of both, the SR and the MB effect, would make new experiments feasible either out of reach or deemed extremely difficult.

Diffraction experiments were named very early with emphasis on the "phase problem" [27]. The temporal coherence of the new source and superradiance has been mentioned as well [28]. In the course of the preparation for a dedicated SR source in Europe, a first set of experiments has been proposed [29]:

- **Analysis of the magnetic structure of complicated compounds** which would benefit from the defined polarization of the SR.
- **Time experiments** such as the observation of relaxation processes between hyperfine levels, the time dependence of the Debye-Waller factor, and the interference between the Mössbauer effect and the conversion or other competing processes. Even laser activation of the environment of the MB nuclei synchronous to the SR flash, nowadays known as pump-and-probe, has been proposed.
- **Nuclear γ-optics** which includes dichroism, magnetooptic effects such as double refraction and Faraday rotation. Further, the interference of the nuclear and the

electronic scattering, the nuclear Borrmann effect, and the suppression of the inelastic channel has to be mentioned.
- **Phase sensitive optics** will allow interference experiments with a large coherence length.
- **Gravitation experiment** in tribute of the very difficult and famous red-shift experiment by Pound and Rebka [30].
- **Small angle scattering**, known as small angle x-ray scattering (SAXS), may be expanded to the Mössbauer radiation to become sensitive as well to magnetic and electric domains/distributions.

This is an impressive list of topics, which was updated for the final technical design report of the ESRF [31]. Remarkable from a present-day perspective is the focus on investigations of "basic science" effects, spectroscopy is nearly absent. Looking back, the explanation might be simply the background of the involved people. They were mainly grown up in nuclear physics.

As we will witness in the Sects. 1.7 and 1.8 the development over the years went after all mainly in a different direction.

1.2.2 Technical Challenges

The main challenge was the unfavorable signal-to-noise-ratio. SR is thought to be a "white" radiation source, i.e., the energy band width of the radiation coming from the bending magnets of the synchrotron spans from literally "zero" to several tens keV. The low energy part could easily be absorbed, however, still a band width of let's say 20 keV was left. On the other side, in case of the 14.4 keV nuclear level in ^{57}Fe, the nuclear level width is 4.66 neV. That means a signal-to-noise-ratio of about 10^{-13}. How to find those resonant quanta in this haystack?

In the following we will discuss various ideas and avenues.

1.2.2.1 Time and Polarization Properties

As discussed above, SR is a pulsed radiation with pulse widths Δt in the 100 ps regime and—in principle—selectable repetition rate. That means that one may accordingly adopt the time structure to the life time τ_0 of the excited nuclear levels. Then it is possible to excite the nuclear levels instantaneously ($\Delta t < \tau_0$) and to record their de-excitation products such as γ-rays and conversion electrons inbetween the synchrotron radiation pulses. Due to the fact that electronic scattering processes are very fast one should see after very intense "prompt" scattering processes only the "slow" scattered γ-radiation from the scattering via the nuclear resonances. That means one would be able to generate out of the SR a pure beam of γ-radiation through time discrimination. Unfortunately, even with todays fast detector systems, it is vrtually impossible for the detectors to sustain the extreme count rates beyond 10^9 photons/s.

However, the time discrimination technique has widely been used as an additional option in the quest of the resonant quanta.

Similary, one might use the different polarization behaviour of electronic and nuclear resonance scattering. As discussed above, SR is linearly polarized in the plane of the synchrotron, i.e., the electric wave vector is parallel to the acceleration and perpendicular to the velocity of the particles. For horizontal (in-plane) 90°-scattering the electronic scattering (electric dipole (E1)) would be "zero", whereas e.g. in case of ^{57}Fe with its 1/2–3/2 magnetic dipole transition (M1) the nuclear scattering could stay due to optical activity. In the early days this could not efficiently be accomplished due to technical challenges. However, meanwhile with state-of-the-art SR sources and crystal optics this endeavour was successful reaching polarization purities in the 10^{-9} to 10^{-10} regime, see e.g. [32].

1.2.2.2 Thin Films and Total Reflection

Another method to suppress the electronic towards the nuclear scattering lies in the different strength of their scattering amplitudes. In case of ^{57}Fe the intensity of nuclear scattering is about 400 times bigger than the electronic one.

Trammell et al. [33] have calculated that in case of a 10^3 layer thick iron crystal only about three successive (332)-reflections would be sufficient to filter a pure γ-ray beam out of the SR. They assumed 90°-scattering in order to benefit as well from the different scattering properties, E1 and M1. However, to our knowledge no such experiment has been successfully conducted.

Staying with thin films, however now in total reflection geometry, one may even omit the need for single crystals. Then the key parameter is the different index of refraction for electronic and nuclear scattering. Further, one may coat the surface with a $\lambda/4$ anti-reflection layer, such as in optical applications, for an additional suppression of the electronic scattering. These are the so-called "Grazing Incidence Anti-Reflection" (GIAR) films [34]. Theory predicts suppression factors of 10^3–10^4. Various solutions have been discussed and laid down in several publications [35–38].

In practice those suppression values could not be reached at that time due to the rather big divergence and beam size of the SR and the non perfect big mirrors needed for total reflection. Nowadays with the improved and dedicated SR sources those ideas flourish e.g. in γ-optical applications, see Röhlsberger and Evers [39].

1.2.2.3 Pure Nuclear Reflections

The most promising avenue seemed to be nuclear Bragg diffraction (NBD) and the use of so-called pure nuclear reflections.

In contrast to the well-known Bragg diffraction at the electrons, electric charge scattering, now the scattering proceeds resonantly via the interaction of the radiation with the excited nuclear levels (Mössbauer levels). In consequence nuclear Bragg diffraction is only possible at distinct energies, the Mössbauer transition energies, and therefore shows a very pronounced and complex resonance character.

The theoretical treatment of nuclear diffraction has independently been developed by two groups well before any thoughts went to SR. Trammell [40] mentioned first the possibility; later Kagan et al. [41], Afanas'ev and Kagan [19, 20], and Hannon and Trammell [17, 18] laid down the theory in detail.

Pure nuclear reflections are named such reflections, which are forbidden due to crystal symmetry, i.e., the electronic scattering amplitude is zero, but allowed for nuclear resonance diffraction due to a different symmetry caused by the properties of the Mössbauer nuclei. Intuitively this might be understood for a hypothetical crystal, which has alternative layers of ^{56}Fe- and ^{57}Fe-atoms with lattice constant d. For electronic scattering the lattice constant is d, because for both isotopes the number of electrons is the same, whereas for nuclear resonance scattering via the ^{57}Fe nuclear exited level the "Mössbauer lattice" constant is 2d. Choosing an appropriate Bragg angle the electronic scattering interferes destructively whereas the nuclear resonance scattering constructively giving rise to the pure nuclear reflection.

Already shortly after the discovery of the Mössbauer effect first diffraction experiments have been conducted with radioactive sources and pure nuclear reflections reported. Those may be classified in reflections due to accidential extinction and magnetic, electric, and combined magnetic and electric superstructures, respectively. Early examples may be found e.g. for an accidental reflection in the [080]-reflection of $K_4Fe(CN)_6{\cdot}3H_2O$ [42], for "magnetic" reflections in α-Fe_2O_3 [43] and $FeBO_3$ [44], for "electric" reflections in the (003)-reflection of $Na_2Fe(CN)_5NO{\cdot}2H_2O$ [45] and for tellur [46], and finally for "combined" reflections in $KFeF_4$ [47], yttrium-iron-garnet (YIG) [48], and Fe_3BO_6 [49].

1.2.3 Political Facts

It might be surprising to find here as well a paragraph on this issue, politics. However, we are very sure, looking back with our todays knowledge and the current scientific situation, that already at that time "politics" played a major role, in one or another way.

It was at a time were "impact factors" and "high-impact" journals have not yet played that important role as today. However, the budgetary situation became more crucial and more and more big scientific facilities competed with each other and with other projects. In the beginning, the endeavour of NRS with SR was thought to be on a short timescale. However, when it took longer and longer the funding was no longer granted on that time scale and most of the "beginners" had to stop the endeavour. Eventually, it was only the Hamburg group, which could convince the funding agencies to invest in "no results" for a long period of nine years, and eventually, it paid off. Would that still be possbile nowadays?

Another long-term issue was the discussion on dedicated SR sources. It became very early clear that the common use of those accelerators for high-energy physics and synchrotron radiation based science, respectively, will never result in optimum conditions for both communities. Therefore, already shortly after first SR experiments in the 70th, discussions started e.g. in Europe for a dedicated high-energy SR source. This was as well an endeavour, which should finally last for more than a decade. The Mössbauer project was one of the very well adapted cases for such a high-energy source especially due to the high transition energies of nuclear levels. It really neededd a vision and courage to make the case for those early proposals of an European SR facility [29] without having seen any "resonant quanta"! However, it was important and essential and was rewarded with a first dedicated high-energy SR facility in the world, the ESRF, and a Mössbauer beamline, called later Nuclear Resonance beamline.

During the discussion of that facility and also later during the start-up and continuous operation a strong community was mandatory for a successful and promising future. This is nowadays even more important with all the frequent evaluations, comparisions between projects, and short term goals. Without such a community, the proper networking, and the endless fights even the best ideas are damned to fail. So far the nuclear resonance community is small and has to make up in the synchrotron radiation scene.

1.2.4 First Results—The Needle in the Haystack

In the course to find the needle in the haystack nearly every avenue described above, and may be more, has been tried over nearly a decade of unsuccessful attempts. However, only very few of those attempts have been reported and may be traced back [50].

Eventually, the pure nuclear reflection strategy brought the break-through in a more than expected and convincing clearness [8].

The measurements were conducted at the Mössbauer test station at HASYLAB located at the storage ring DORIS (Desy, Hamburg). DORIS was running in 2- and 4-bunch mode with 480 ns and 240 ns spacing, respectively, and provided SR from a bending magnet. The original experimental set-up is shown in Fig. 1.3. 22 m downstream of the source point the high-heat-load monochromator, two Ge (111) single crystals, monochromatized the SR at 14.4 keV with $\Delta E = 13$ eV. After the monochromator a beam size of 2 mm × 20 mm (v × h) was defined by a slit system and an intensity of about 10^9photons/(mA s) measured. Eventually, (3–6) × 10^{10} 14.4 keV photons/s were available for the experiment.

The nuclear monochromator consisted of two 15-μm thick single-crystalline films of YIG (88% enriched in ^{57}Fe) epitaxially grown on the [100] surface of a gadolinium gallium garnet (GGG) single crystal with 30 mm in diameter. An amount of about 10 g ^{57}Fe, purchased over several years, and several years of discussion were necessary for the conviction and production of those films by the Philips research laboratory in Hamburg. The two crystals were aligned in non-dispersive, and with respect to the high-heat-load monochromator, in dispersive geometry. Both YIG crystals were set

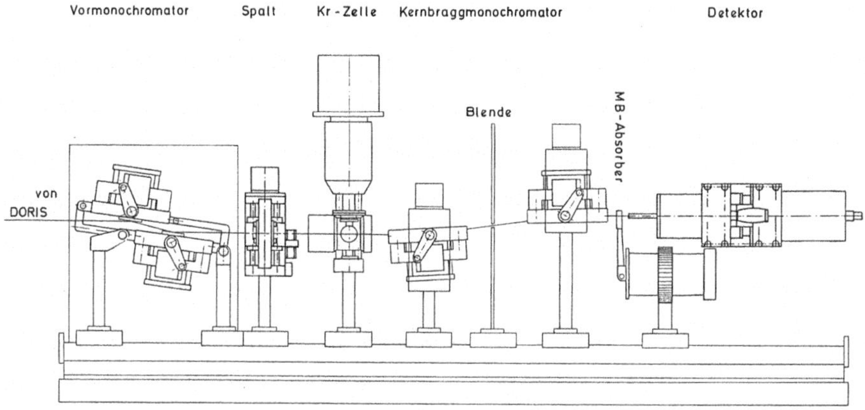

Fig. 1.3 Experimental set-up at the "Mössbauer Test station" F4 at DORIS, Desy (Hamburg) (from [51]). The entire set-up is about 2 m in length. Vormonochromator: high-heat-load monochromator, Spalt: slit system, Kr-Zelle: energy calibration with Kr K-edge, Blende: aperature, Kernbraggmonochromator: nuclear monochromator, MB-Absorber: MB-absorber on MB-driving system, Detektor: high-purity germanium solid state detector (cooled)

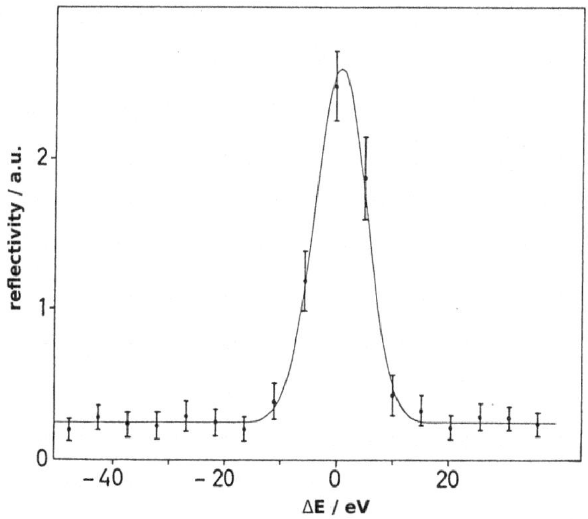

Fig. 1.4 Resonance curve of the diffracted γ-quanta after the nuclear monochromator as function of the energy of the high-heat-load monochromator. The solid line is a fit with a Gaussian distribution and the energy scale is centred to its maximum (from [51]). (Reprinted figure with permission from [8], Copyright (1985) by the American Physical Society)

to the [002] pure nuclear reflection. A small external magnetic field (30 mT) assured a defined and single domain magnetic state of the crystals.

As detector system a high-purity germanium solid state detector with about 1 keV energy resolution and a conventional fast-slow coincidence system (see Fig. 1.14) with an overall 18 ns time resolution has been used.

The performance of the system was verified by measuring the so-called resonance curve of the nuclear monochromator, i.e., the intensity of the resonant quanta, as integrated over a time window between 32 and 137 ns after the prompt pulse, versus the energy of the system. Figure 1.4 displays the reflectivity of the nuclear monochro-

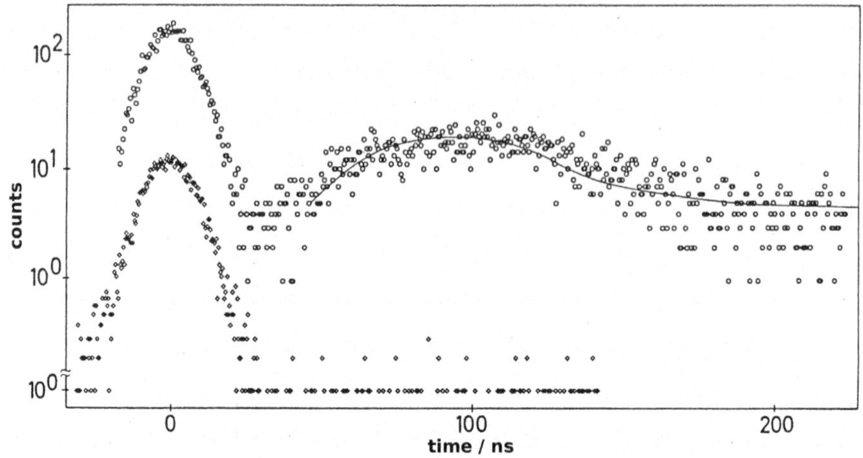

Fig. 1.5 Time spectra of the diffracted γ-rays after the nuclear monochromator on resonance (upper curve) and off resonance (lower curve). The solid line is a fit with the dynamical theory of NBD (from [51]). (Reprinted figure with permission from [8], Copyright (1985) by the American Physical Society)

mator versus the energy defined by the high-heat-load monochromator. The energy width of 10.13(5) eV is determined by the widths of the germanium and YIG crystal reflections and their dispersive geometry.

The time spectra after the nuclear monochromator consist of a "prompt" component, certainly due to residual background radiation and Umwegreflections, and, on resonance, of the delayed nuclear scattering response, see Fig. 1.5. The solid line through the delayed component has been calculated with CONUSS applying the dynamical theory of NBD [53] accounting for a two-crystal arrangement.

Finally, a Mössbauer spectrum proved in another way the existence of resonant quanta (Mössbauer quanta). A stainless steel single-line absorber (1 mg $^{57}Fe/cm^2$) on a conventional MB-driving system was installed downstream of the nuclear monochromator, see Fig. 1.3. The resulting spectrum after about 30 h of effective measuring time is shown in Fig. 1.6. The remarkable high effect ($\sim$40%) allowed even with the low count rate of 0.5 Hz for a clear spectrum. The theory (solid line) resembles nicely the measured spectrum.

With these three proofs, undoubtedly, a 1-Hz beam of nearly pure resonant γ-quanta was now available for further experiments. And indeed, these results again triggered a new rush not only to use but also to improve and develop this technique as it is shown in the following sections.

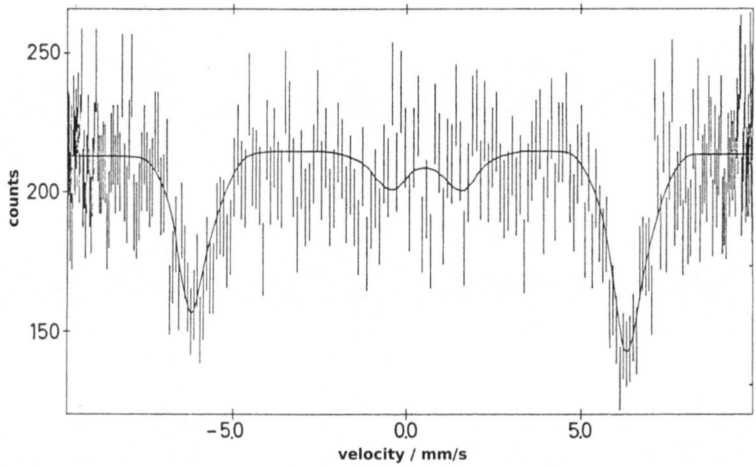

Fig. 1.6 Mössbauer absorption spectrum after the nuclear monochromator with a thick stainless steel absorber. The solid line is a fit with the dynamical theory (from [51]). (Reprinted figure with permission from [8], Copyright (1985) by the American Physical Society)

1.3 Techniques in NRS

After these first experiments several groups with different expertise in more technical fields such as high resolution crystal optics, fast detector systems, synchrotron radiation, nuclear resonance scattering as well as in various fields of applications newly started with nuclear resonance scattering with synchrotron radiation. A very fruitful and successful synergy emerged from this combined expertise and effort, which are summarized, not in historical order, in this chapter.

1.3.1 Synchrotron Mössbauer Source

The Synchrotron Mössbauer Source (SMS), see Fig. 1.7 SMS, provides a 100% resonant, polarized, and collimated γ-ray beam for high energy-resolution applications and energy domain Mössbauer spectroscopy, furtheron called SMS spectroscopy. The high energy resolution ($\Delta E \sim$ neV) is achieved by nuclear monochromators based on nuclear Bragg diffraction from pure nuclear reflections rather than by radioactive sources. In both cases the resolution is governed by the natural line width Γ_0 of the concerned nuclear level. First attempts towards a SMS were already published [8, 54, 55] in the early days of NRS. The breakthrough came with dedicated SMSs for the Mössbauer isotope of ^{57}Fe, utilizing $^{57}FeBO_3$ single crystals as nuclear monochromator, at various beamlines [56, 57]. Since that time SMS spectroscopy flourish with numerous applications.

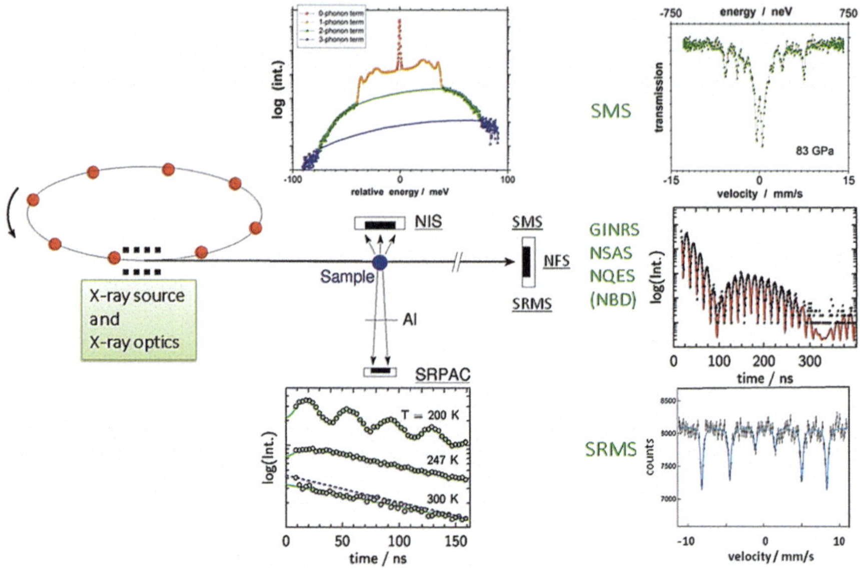

Fig. 1.7 Scheme of the set-up for some techniques in NRS: SMS, SRMS, NFS, NIS, and SRPAC. Furthermore, the corresponding time and energy spectra are schematically shown measured by the detectors for SMS-NFS-SRMS, SRPAC and NIS. For time resolved measurements and SRMS the storage ring is operated in *few bunch* mode with e.g. 176 ns spacing between adjacent buckets at the ESRF (red bullets). For measurements with the SMS timing mode is not an issue and any filling mode may be used. The x-ray source, nowadays an undulator, produces the well collimated synchrotron radiation, which is monochromatized at the nuclear resonances by dedicated x-ray and γ-ray optics with bandwidths ranging from eV to neV. Finally the radiation impinges on to the sample. Depending on the scattering process and the scattering geometry different techniques are exploited. Variable sample environments allow for the combination of high/low temperature, high pressure, and external magnetic field

In contrast to traditional Mössbauer spectroscopy, the energy spectrum of the radiation emitted by the $^{57}FeBO_3$ single crystal of the SMS is the result of the interference of two spectral lines (each with a Lorentzian distribution, Eq. 1.6) with almost equal resonance energies. Therefore, the energy spectrum of the radiation emitted by the SMS is a squared Lorentzian distribution rather than a single Lorentzian [58]:

$$I(E) = \left[\frac{\Gamma_0/(2\pi)}{(E-E_0)^2+(\Gamma/2)^2}\right]^2, \tag{1.23}$$

where Γ is the width of each of the interfering lines, E is the energy of radiation, and E_0 is the resonance energy. The knowledge of the exact shape of the energy distribution, i.e., the instrumental function, is mandatory for proper data evaluation. Furthermore, the radiation is fully polarized with the electric wave field in the vertical plane.

Normally, spectroscopy is carried out in transmission geometry as in conventional Mössbauer spectroscopy with radioactive sources. In both cases the energy of the γ-quanta is varied via Doppler shift in the neV to μeV regime by moving the source and after the sample the energy spectrum is recorded. The excellent collimation of the γ-ray beam from the SMS allows as well for scattering experiments, which were very challenging with radioactive sources due to their isotropic emission spectrum and only exceptionally done.

Sofar the SMS has been developed for the case of ^{57}Fe, which was favoured due to the availability of highly perfect single crystals enriched in the Mössbauer isotope. However, generally speaking, this may be achieved for any Mössbauer isotope provided one is able to grow perfect single crystals or structures, which show pure nuclear reflections and preferably a single line.

1.3.2 Synchrotron Radiation Based Mössbauer Spectroscopy

Another technique in the energy domain was recently presented by Seto et al. [59]: Synchrotron radiation based Mössbauer spectroscopy (SRMS), see Fig. 1.7. The idea is as follows: The sample under investigation (in the beam) "modulates" the synchrotron radiation, which is then analyzed (demodulated)—by time integration and an efficient resonance detector—in the energy domain utilizing a conventional Mössbauer driving system. In order to improve the signal-to-noise-ratio timing mode is still required for the suppression of the prompt radiation by time-gating. That implies that the time integration can only be carried out over a finite time window, which in turn influences the energy spectrum. A detailed description will be given elsewhere in this book by Seto et al. [60].

1.3.3 Nuclear Forward Scattering

Nuclear forward scattering [61, 62] may be considered as the time analog to Mössbauer spectroscopy; in fact it is its scattering variant. The 'white' synchrotron radiation excites all Mössbauer levels in the sample and creates a *coherent collective nuclear state*. In the static case this nuclear state will decay in the forward direction giving rise to an excess of intensity at delayed times (see Fig. 1.7 NFS). The time scale is determined by the lifetime τ_0 of the involved nuclear level. Multiple scattering may influence the measured time response giving rise to *dynamical beats*. Furthermore, in case of split nuclear levels due to hyperfine interaction (electric, magnetic) an additional interference pattern from all involved nuclear levels, the so-called *quantum beat* structure is superimposed.

An important variant of NFS is reflectometry or grazing incidence nuclear resonance scattering (GINRS) for the investigation of thin films, surfaces etc. Even when the scattering angle in GINRS is very small (mrad) one has to account for it in the evaluation of the spectra due to an enhancement of the effective thickness (see Sect. 1.4.2) by interference effects and the grazing incidence geometry.

Small structures such as magnetic and electric domains may give rise to nuclear small-angle scattering (NSAS) [63, 64]. Recording both, the prompt (charge) scattered signal and the delayed (nuclear) one, one easily distinguishes between structural and magnetic/electric structures.

1.3.4 Synchrotron Radiation Based Perturbed Angular Correlation

For the techniques described above, SMS spectroscopy, SRMS, and NFS, the Lamb-Mössbauer factor has to be greater than zero in order to apply those spectroscopies. As discussed in Sect. 1.1 that might not be the case for samples at high temperature, for liquids and gases, or for nuclei with high energetic transitions.

Then the spatially incoherent channel of nuclear resonant scattering, which is independent of the Lamb-Mössbauer factor, offers a way out of these difficulties via Synchrotron Radiation based Perturbed Angular Correlation (SRPAC) [65, 66] (see Fig. 1.7 SRPAC). It can be considered as a scattering variant of time differential perturbed angular correlation (TDPAC) [67]. In SRPAC, on the contrary to NFS, the nuclear levels are excited from the ground state during incoherent, single-nucleus resonant scattering of SR. The interference of indistinguishable paths via these intermediate nuclear levels, split by magnetic dipole and/or electric quadrupole interaction, allows one to investigate hyperfine interactions $\hbar\Omega$ and spin dynamics.

The scattering intensity can be written as

$$I_{\mathrm{SRPAC}}(t) = I_0\, \mathrm{e}^{-t/\tau_0}\{1 - A_{22}G_{22}(t)\} \tag{1.24}$$

with A_{22} the anisotropy and angular term and G_{22} the perturbation factor.

SRPAC can be applied to all nuclei with an isomeric state with energies attainable by synchrotron radiation, so far, for practical reasons, with energies up to 150 keV. In addition, also the contrast of the SRPAC signals might be much larger than for TDPAC since unfavourable transitions to the excited states from above can be avoided. Furthermore, the SRPAC technique can be used in a site-selective option [68].

1.3.5 Time Interferometry and Rayleigh Scattering

For samples without Mössbauer isotopes, in addition to fast (phonon) dynamics accessible with IXSNRA (Sects. 1.3.6.2 and 1.5.2) one may also retrieve slow relaxation (ps to ms time scale) dynamics. This is done in the time domain with time domain interferometry (TDI) [69] and in the energy domain with Rayleigh scattering with Mössbauer radiation (RSMR) [70, 71].

The TDI setup includes the sample under investigation and two resonance absorbers. The first one is placed in the path of the incident radiation upstream of the sample and the second one downstream of the sample in the path of the scattered radiation at the defined momentum transfer under investigation.

In the default scheme [69], these are single-line absorbers. One of those is at rest, the other one is driven by a MB transducer with constant velocity, defining the difference in the energies of the two nuclear resonance transitions. A partial probability of nuclear forward scattering of each photon by the upstream and downstream absorbers gives the probe and the reference wave, respectively.

In absence of slow dynamics, nuclear forward scattering from both samples proceeds elastically, and an interference of the probe and reference waves displays quantum beats, defined by the difference in the resonance energies of the two absorbers.

Under relaxation conditions, nuclear forward scattering of the probe wave acquires with time a phase shift, which damps the beats. The envelope of the fading contrast of the quantum beats provides the momentum-transfer dependent auto-correlation function of the density fluctuations (intermediate scattering function) [69].

For the RSMR set-up a radioactive source or a nuclear monochromator, the SMS, prepares a high-resolution γ-ray beam and after the sample a resonance detector analyzes the scattered radiaton at the defined momentum transfer under investigation. In absence of dynamics the MB spectrum resembles the source characteristics whereas in case of dynamics the line width will broaden according to the dynamics.

A detailed description of RSMR with SR will be given elsewhere in this book by Seto et al. [60].

1.3.6 Inelastic Scattering

1.3.6.1 Nuclear Inelastic Scattering

Nuclear inelastic scattering [72] measures the *(partial) phonon density of states* (DOS) of the Mössbauer element in the sample [72–74]. The principle set-up is shown in Fig. 1.7 NIS.

The synchrotron radiation is monochromatized by a high resolution monochromator (HRM) with (sub-) meV energy resolution. At resonance the x-ray pulse creates a coherent collective nuclear state as in NFS, which decays either in forward direction or incoherently, due to internal conversion, spin flop etc., in the entire solid angle of 4π. This feature gives a simple and effective method at hand to measure the instrumental function (detector NFS) in parallel with the inelastic spectrum (detector NIS). While scanning the HRM the nuclear resonance can be excited when at the same time a phonon is created or annihilated. In NIS this is a purely incoherent process with a perfect averaging over the momentum $\mathbf{q}$. The successive nuclear decay proceeds as a single exponential in the entire solid angle covered by the '4π'-detector (detector NIS). In practice only a solid angle of about π can be achieved.

1.3.6.2 Inelastic X-Ray Scattering with Nuclear Resonance Analysis

NIS needs the Mössbauer isotope in the sample. Phonons in samples not containing Mössbauer isotopes may be studied with "Nuclear Resonance Energy Analysis of Inelastic X-Ray Scattering" (IXSNRA) [75]. In this case the set-up resembles the classical three-axes-spectrometer for inelastic scattering experiments consisting of the monochromator, sample, and analyzer. Nevertheless, the set-up looks like that for NIS (see Fig. 1.7 NIS). However, the (energy) analysis is now carried out with a "resonance detector" containing the Mössbauer isotope as a high-resolution analyzer.

1.3.7 Other Techniques

There exist some other techniques and extensions related to NRS. They can not be discussed here in detail, however, for further reading they are cited (the first publication) in the following.

Everything started with Nuclear Bragg Diffraction [8], however, little has been done during the past decades [76–78].

Several techniques have been developed in order to overcome the need for high-resolution x-ray optics and for dedicated timing modes such as the lighthouse effect [79, 80], stroboscopic detection of nuclear resonance [81], polarimetry [82–84], μeV-resolved spectroscopy [85].

1.4 Hyperfine Spectroscopy

Hyperfine spectroscopy, i.e., the investigation of electric and magnetic properties, can be carried out with the SMS, SRMS, NFS, and SRPAC. The SMS and (partly) SRMS, like the conventional Mössbauer source, probes with its narrow energy bandwidth the nuclear hyperfine splitting of the levels one by one in the energy domain whereas NFS and SRPAC are spectroscopies in the time domain. NFS, as a coherent scattering process, gives access to the splitting of all nuclear levels, ground and excited, due to interference in the time spectra. In contrast SRPAC, as an incoherent single nucleus scattering process, only gives access to the splitting of the intermediate nuclear level, i.e., the excited one.

1.4.1 Absorption

The cases of the SMS and SRMS resemble closely the "classical" Mössbauer spectroscopy (see e.g. [10] and as well in various chapters of this book). Set-up, experimental details as well as the data evaluation are the same as in classical Mössbauer spectroscopy. The only difference is the source with its special properties such as lineshape, 100% linear polarization, and 100% resonant quanta, which is discussed in Sects. 1.3.1 and 1.6.1.3.

1.4.2 Dynamical Beats

For samples with a large effective thickness, $t_{\mathrm{eff}} = n_{\mathrm{MB}}\, d\, \sigma_0\, f_{\mathrm{LM}}$, with n_{MB} the density of Mössbauer nuclei, d the geometrical thickness, σ_0 the resonant Mössbauer cross section, and f_{LM} the Lamb-Mössbauer factor, a *speed-up* effect is observed and dynamical beats (*Bessel beats*) show up in the time spectra of NFS.

In Fig. 1.8 the situation is shown for Mössbauer and NFS spectra. For a thin sample, $t_{\mathrm{eff}} = 1$, we have a single Lorentzian line in the Mössbauer spectrum, which corresponds to an exponential decay in the NFS spectrum (Fig. 1.8a, b, dotted lines). Increasing the effective thickness ($t_{\mathrm{eff}} = 25$) the Mössbauer line becomes much wider and non-Lorentzian (Fig. 1.8a, solid line). In the NFS spectrum (Fig. 1.8b, solid line) we observe two features: (1) a speed-up, showing up as an acceleration of the initial decay rate accompanied by an increase in intensity at early times (from 10^7 to 10^9 units in the present example) and (2) a dynamical beat structure is superimposed to the decay. From this beat structure the effective thickness and correspondingly f_{LM} can very precisely be determined.

1.4.3 Quantum Beats

As it is well known, hyperfine interactions might shift and split the nuclear levels. This is described by the Hamiltonian $\mathscr{H}$ as:

$$\mathscr{H} = \mathscr{H}_0 + \mathscr{E}_0 + \mathscr{M}_1 + \mathscr{E}_2 + \ldots \tag{1.25}$$

with $\mathscr{H}_0$ representing all terms, which do not include hyperfine interactions and cause no shift or splitting, $\mathscr{E}_0$ the Coulombic interactions, i.e., the electric monopole term (isomer shift), $\mathscr{M}_1$ the magnetic dipole and $\mathscr{E}_2$ the electric quadrupole interaction. A detailed discussion might be found in [10] and will be mentioned in more detail as well in other chapters of this book. In the energy domain such as Mössbauer spectroscopy several absorption lines reveal this splitting whereas in the time domain such as NFS and SRPAC an interference pattern, the quantum beat structure, shows up.

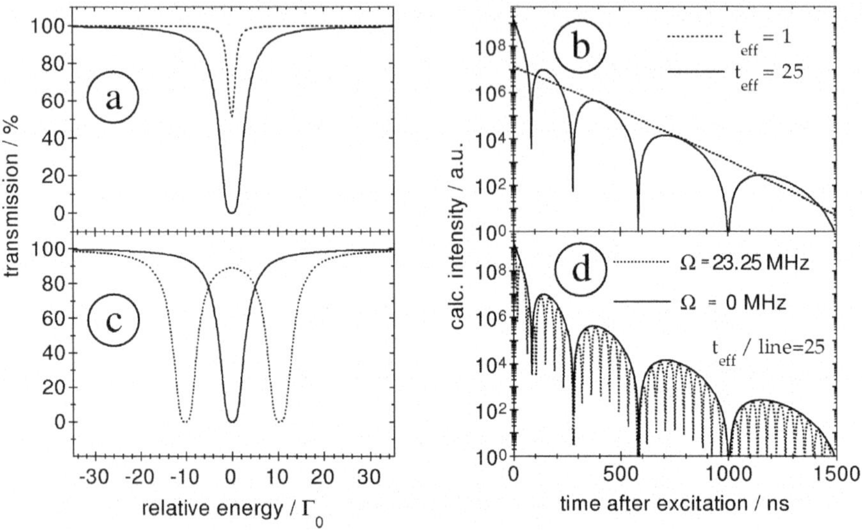

Fig. 1.8 Simulations of corresponding spectra of Mössbauer spectroscopy (left panel, with $\Gamma_0=\hbar/\tau_0$) and NFS (right panel) for thin ($t_{eff}=1$) and thick ($t_{eff}=25$) samples. For the single lines (panel a and b) the thin sample leads to an exponential decay whereas the spectrum from the thick sample is further modulated by the dynamical beats. In panel c the energy spectra of a thick sample ($t_{eff}=25$) with one (solid line) and two (dotted line) transition lines, respectively, are shown. In panel d the corresponding time spectra are shown. The sample with the two lines shows in addition to the dynamical beats the fast, equidistant quantum beat structure superimposed (from [86])

In case of electric quadrupole interaction ($\mathscr{E}_2$) the excited nuclear state, I = 3/2 in case of ^{57}Fe, splits to the $\pm 3/2$ and $\pm 1/2$ levels. This gives rise to two absorption lines in Mössbauer spectroscopy (see dotted line in Fig. 1.8 c). The corresponding spectrum in NFS shows up as an interference pattern of these two transitions with a single frequency Ω of the quantum beats (see dotted line in Fig. 1.8 d). Due to the thick sample with an effective thickness $t_{eff} = 25$, the dynamical beat structure is strongly modulating the quantum beats as an envelope. It is clearly seen that the quantum beat structure is equidistantly spaced whereas for the dynamical beat structure the distance of the minima increases with time. This can be described for NFS in a good approximation by

$$I_{NFS}(t) \propto \frac{t_{eff}}{t/\tau_0} \cos^2\left(\frac{\Omega t}{2}\right) \cdot e^{-\frac{t}{\tau_0}} \cdot J_1\left(\sqrt{4t_{eff}\, t/\tau_0}\right)^2 \tag{1.26}$$

and for SRPAC by

$$I_{SRPAC}(t) \propto e^{-\frac{t}{\tau_0}}\{1 - A_{22}\cos^2\left(\frac{\Omega t}{2}\right)\}. \tag{1.27}$$

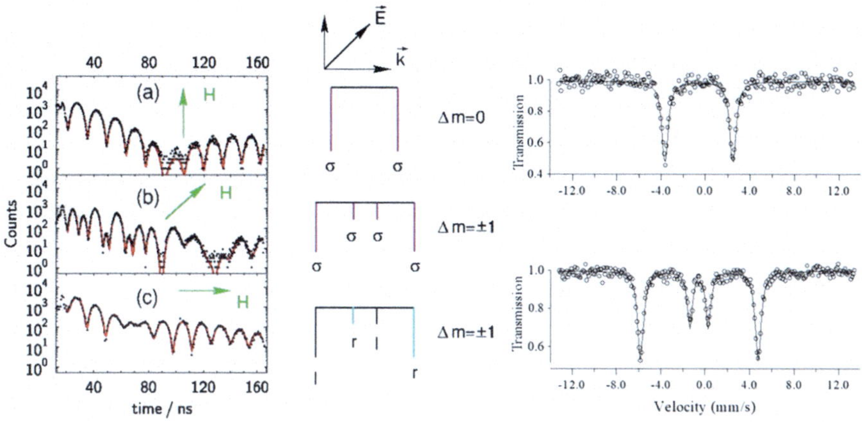

Fig. 1.9 Left panel: measured time spectra of iron (NFS) in case of magnetic hyperfine interaction for various alignments of the hyperfine field **H** with respect to the wave vector **k** and the polarization **E**. Solid lines are fits according to the full theory. middle panel: nuclear transition lines with their polarization state, σ—σ linearly polarized, l and r—left and right hand circular polarized, respectively. Δm - change of magnetic quantum number. (Reprinted by permission from Nature Springer: ref. [87], copyright 2003) right panel: Measured energy spectra of iron (SMS) in case of magnetic hyperfine interaction for various alignments of the hyperfine field **H** with respect to the wave vector **k** and the polarization **E** (note: **E** points now up). For the absorption spectra the two cases for $\Delta m = \pm 1$ are indistinguishable. Solid lines are fits with the transmission integral. (Reprinted figure with permission from [54], Copyright (1997) by the American Physical Society)

The term $\cos^2(\frac{1}{2}\Omega \cdot t)$ describes in both cases the quantum beats. The Bessel function J_1 describes the dynamical beats and A_{22} the anisotropy and angular dependence. For comparison the unsplit case is shown as solid lines in the same figures. Changing the strength of the hyperfine interaction will result in a different splitting and correspondingly in a different quantum beat frequency Ω.

In case of magnetic hyperfine interaction full splitting of the nuclear levels occurs giving rise to six nuclear transitions and correspondingly to six absorption lines in Mössbauer spectroscopy with ^{57}Fe. In NFS spectroscopy a more detailed interference pattern will result. Contrary to conventional Mössbauer spectroscopy where the γ-rays from the radioactive source are normally unpolarized now the x-rays from the synchrotron radiation source and the γ-rays from the SMS are highly linearly polarized. This feature strongly modifies the spectra. Corresponding time and energy spectra are displayed in Fig. 1.9. The important parameters are the orientation of the three vectors with respect to each other, the wave vector **k**, the polarization vector **E**, and the hyperfine field vector **H**.

If all three vectors are perpendicular to each other only the two $\Delta m = 0$ transitions contribute to the spectrum, resulting in a two-line spectrum for the SMS (Fig. 1.9 upper right), and consequently in a simple quantum beat pattern with one single frequency and high contrast (Fig. 1.9 upper left). In case of NFS a similar spectrum with only one frequency, however, with less contrast appears when **H** $\parallel$ **k**. In this case the two $\Delta m = +1$ and the two $\Delta m = -1$ transitions interfere independently giving now rise to left and right hand circular polarization (Fig. 1.9c). Finally, for **H** ∥ **E** all $\Delta m = \pm 1$ transitions interfere giving rise to a more complicated spectrum which is sigma polarized (Fig. 1.9b). The slow overall modulation is caused by dynamical beats due to the finite effective thickness. In case of the SMS the two cases with $\Delta m = \pm 1$ transitions are indistinguishable in a simple absorption experiment and the resulting spectra resemble four-line absorption spectra (Fig. 1.9 lower right).

1.5 Structural Dynamics

Structural dynamics is accessible via quasi-elastic scattering techniques in the energy domain by RSMR or directly in the time domain by TDI and NFS/SRPAC measuring translational and rotational dynamics and via inelastic scattering techniques, NIS and IXSNRA, in the energy domain giving access to the phonon density of states.

1.5.1 Quasi-elastic Dynamics

Nuclear quasi-elastic scattering measures structural dynamics on a ps to μs time scale. The coherent and the incoherent channel can be utilized. In the first case, the coherent channel, the Lamb-Mössbauer factor has to be greater than zero ($f_{LM} > 0$). The set-up is the same as in NFS (see Fig. 1.7). The incoming x-ray pulse creates a coherent collective nuclear state which decays in the static case in forward direction. Dynamics, e.g. the jump of a Mössbauer nucleus from one atomic site to another (in space and angle, see e.g. [88]), destroys this state. As a consequence no x-ray is scattered in forward direction, i.e., the measured intensity in the NFS detector is decreased at later times. We will get an 'accelerated decay' or a 'damping' of the NFS intensity $I_{NFS}(t)$, which might be described in a simplified picture by:

$$I(t) \propto I_{\text{NFS}}(t)\, \mathrm{e}^{-2\lambda_t t}\, \mathrm{e}^{-\lambda_r t}. \tag{1.28}$$

The first exponential is related to the van Hove self-intermediate function [89] with λ_t being the translational relaxation rate and the second one to the rotational correlation function with λ_r being the rotational relaxation rate.

In the second case, the incoherent channel, the scattering is independent of the Lamb-Mössbauer factor. The set-up is the same as in SRPAC (see Fig. 1.7). The incoming x-ray pulse selectively excites a single nucleus, which decays in the static case with an angular distribution according to the anisotropy parameter, A_{22}. Dynamics, e.g. rotational motion monitored by the electric hyperfine interaction $\hbar\Omega$, changes this distribution and gives rise to a damping of the intensity signal.

The perturbation factor G_{22} in the scattering intensity I_{SRPAC} (see Eq. 1.24) reduces in the slow approximation to

$$G_{22}(t) \propto \mathrm{e}^{-\lambda_r t} \cos^2\left(\frac{\Omega t}{2}\right) \tag{1.29}$$

and in the fast approximation to

$$G_{22}(t) \propto \mathrm{e}^{-(\Omega^2/\lambda_r)t}. \tag{1.30}$$

Combining both techniques the translational and rotational relaxation rates can be separately extracted.

In both cases described above the sample has to contain the Mössbauer isotope. For samples without Mössbauer isotopes a variant—time domain interferometry [69] or Rayleigh scattering with Mössbauer radiation—may be utilized, see Sect. 1.3.5.

In the default TDI setup [69], with two identical single line resonance absorbers, the interference of the probe and reference waves is described by

$$I(\mathbf{q}, t) \propto I_{\mathrm{NFS}}(t)(1 + f_{qe}(\mathbf{q}) \cdot \cos \Omega t \cdot S^n(\mathbf{q}, t)), \tag{1.31}$$

where $I_{\mathrm{NFS}}(t)$ is the single line response of the two resonance absorbers, Ω is the frequency difference in the response of the two absorbers, $f_{eq}(\mathbf{q})$ is the elastic (including the quasi-elastic part) fraction of scattering, and $S^n(\mathbf{q}, t)$ is the normalized intermediate scattering function, i.e., the normalized Fourier transform of the dynamic structure factor.

As described by Eq. 1.31 the TDI pattern displays an oscillating dependence, where an envelope of the fading contrast is given by the intermediate scattering function.

For more details on the TDI technique see in this book Seto et al. [60].

1.5.2 Phonon Density of States

Nuclear inelastic scattering measures the *(partial) phonon density of states* of the Mössbauer element in the sample. As an example the intensity of nuclear inelastic scattering of synchrotron radiation in a polycrystalline α-iron sample at room temperature is shown as a function of energy of the incident radiation (Fig. 1.10).

The central peak corresponds to elastic scattering. The structure beyond the central peak shows the energy dependence of inelastic scattering (furtheron called "energy spectrum"), accompanied either by creation ($E > 0$) or by annihilation ($E < 0$) of phonons. At ambient temperature one may recognize various contributions to the energy spectrum, which correspond to inelastic scattering accompanied by creation or annihilation of a different number of phonons.

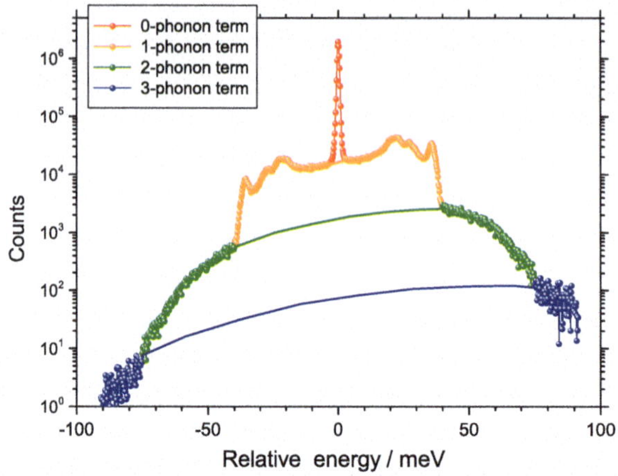

Fig. 1.10 Expansion of the energy spectrum of nuclear inelastic scattering of synchrotron radiation in α-iron in multi-phonon terms. The data were taken at room temperature. Different symbols show the regions of the spectra, where the corresponding contributions are dominant. The lines are the calculations according to Eqs. 1.32–1.34 and convoluted with the instrumental function of the monochromator

The normalized probability of nuclear inelastic scattering $W(E)$ can be decomposed in terms of a multiphonon expansion [90]

$$W(E) = f_{\mathrm{LM}} \left(\delta(E) + \sum_{n=1}^{\infty} S_n(E) \right) . \tag{1.32}$$

The Dirac δ-function $\delta(E)$ describes the elastic part of scattering (zero-phonon term), and the n-th term of the series $S_n(E)$ represents the inelastic scattering accompanied by creation (annihilation) of n phonons. The one-phonon term is given by

$$S_1(E) = \frac{E_R \cdot g(|E|)}{E(1 - \mathrm{e}^{-\beta E})}, \tag{1.33}$$

and the subsequent terms under harmonic approximation may be found through the recursive relation:

$$S_n(E) = \frac{1}{n} \int_{-\infty}^{\infty} S_1(E') \cdot S_{n-1}(E - E')\,\mathrm{d}E' . \tag{1.34}$$

Here $\beta = (k_{\mathrm{B}}T)^{-1}$ with k_{B} the Boltzmann constant, T the temperature; $E_{\mathrm{R}} = \hbar^2 k^2/2M$ the recoil energy of a free nucleus; k the wave vector of the x-ray quantum; M the mass of the atom. The function $g(E)$ is the normalized phonon density of states

$$g(E) = V_0 \frac{1}{(2\pi)^3} \sum_j \int \mathrm{d}\mathbf{q}\, \delta\left[E - \hbar\omega_j(\mathbf{q})\right], \tag{1.35}$$

where V_0 is the volume of unit cell, index j numerates the branches of the dispersion relation $\hbar\omega_j(\mathbf{q})$, $\mathbf{q}$ is the phonon momentum, and the integral is taken within the first Brillouin zone. The detailed theory of nuclear inelastic scattering has been published by several authors [90–92].

Using the sum rules [93, 94], from the nuclear inelastic scattering spectra and from the phonon density of states other (thermo)dynamic quantities can be derived such as the Lamb-Mössbauer factor, the mean square displacement, the mean kinetic energy, the mean force constant, the mean force, and the second order Doppler shift. In addition, from the density of states the lattice specific heat at constant volume and pressure, and the vibrational entropy can be determined.

In summary, NIS gives direct access to the partial phonon density of states and various (thermo)dynamic quantities. It is complementary to methods as inelastic neutron, x-ray, and light scattering. In those techniques mainly dispersion relations are measured, which are fitted to a model and in a final step the phonon density of states can be derived. For more details see Seto et al. [60].

As mentioned above, IXSNRA measures an "x-ray generalized" phonon density of states. The data evaluation procedure is along the same route as for NIS outlined above. In the data treatment, the x-ray generalized phonon density of states can be reduced to the true phonon density of states using a so-called "correction function", when it is available from theory or computer simulation [95].

1.6 Experimental Details

NRS relies very much on the outstanding brilliance and timing properties of 3rd generation synchrotron radiation sources such as APS, ESRF, PETRA III, and SPring-8. Dedicated insertion devices, perfect high-resolution and focusing/collimating x-ray optics, and fast detector systems assure optimal conditions for NRS applications.

As an example the layout of the Nuclear Resonance beamline at the ESRF is shown in Fig. 1.11. The undulators define the maximum available photon flux. One set of the magnet structures, U20, is optimized for 14.4 keV, the resonance energy of the most utilized MB isotope, ^{57}Fe, and the other magnet structure (U27) for the transition energies of other MB isotopes starting with 21.5 keV (^{151}Eu). The first optics hutch (OH1) contains the cryogenically cooled high-heat-load mocnochromator (6–80 keV). The second optics hutch (OH2) contains all high-resolution optical elements. Three experimental hutches are available to the users for their experiments.

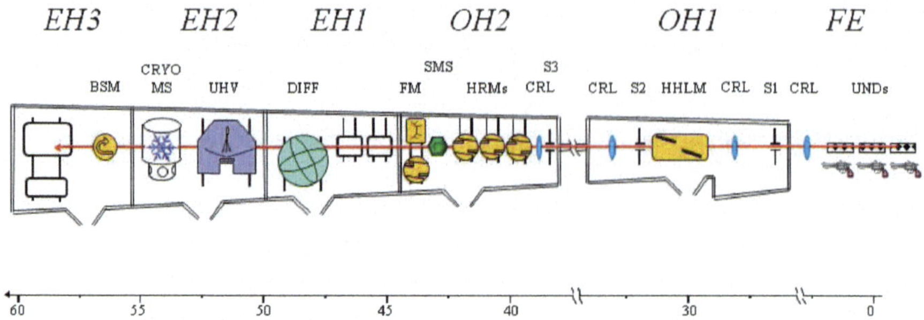

Fig. 1.11 Schematic layout of the nuclear resonance beamline (ID 18) at the ESRF. FE—front end; UNDs—three revolver type undulators with periods λ_u of 20 and 27 mm; OH1, OH2—optics hutches, EH1, EH2, and EH3—experimental hutches, CRL—focusing and collimating compound refractive lenses, S1, S2, and S3—slit systems, HHLM—Si (111) high heat-load monochromator, HRMs—high-resolution monochromators, SMS—synchrotron Mössbauer source, FM—focusing monochromator, DIFF—DIFFractometer, UHV—set-up for in-situ preparation, characterization and NRS measurements of surfaces, interfaces, and nano-structures, CRYO—CRYOmagnet system, BSM—backscattering monochromator

1.6.1 X-Ray Optics

Nuclear level widths are in the range of neV to μeV (e.g. 4.66 neV for ^{57}Fe), whereas energy bandwidths of SR from undulators are in the range of 100 eV. That immediately makes clear that dedicated and optimized SR sources and excellent x-ray optics are needed to achieve not only highest brilliance but also highest flux. Nowadays, x-ray optics for NRS applications ranges from eV resolution, via meV resolution, and eventually to neV resolution, i.e., spanning more than eleven orders of magnitude. The needed resolution is on the one side determined by the spectroscopy (e.g. NIS) and on the other side by the demand not to overload the detector system (e.g. NFS).

1.6.1.1 High Resolution Monochromator

For sub-meV resolution, mandatory e.g. for phonon spectroscopy, in-line silicon high-resolution monochromators based on the 'nested design' [96] or the flux-optimized design [97] are common practice. They may be tailored with high efficiency for energies below about 30 keV.

Figure 1.12 left panel displays the nested-design consisting of two channel-cut high-quality Si crystals, which are mounted in a nested geometry. Asymmetric cuts and a proper choice of reflections assure high throughput and fixed and high energy resolution.

The adopted scheme of the flux-optimized HRM is shown in Fig. 1.12 right panel. Actually, at the ESRF, we have implemented, as shown in the figure, two of those HRMs next to each other on one mechanics, each consisting of four flat Si single crystals, which offer 0.5 and 2 meV resolution just by a horizontal translation of the

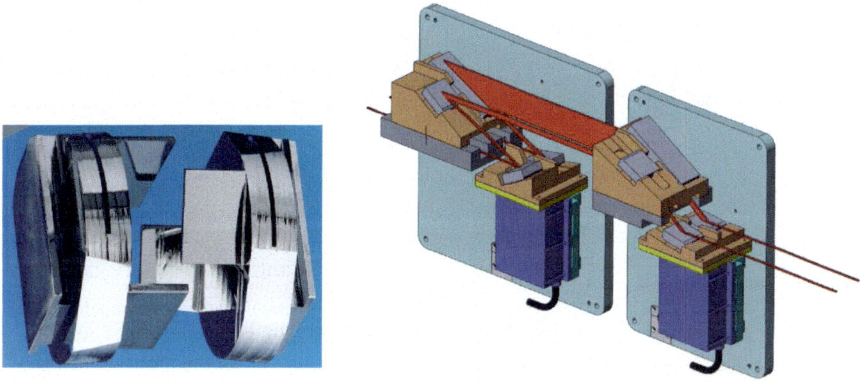

Fig. 1.12 Left panel: nested design of a four-bounce channel-cut HRM. Right panel: double-resolution monochromator with high-energy resolution and optimized intensity: the two first crystals in asymmetric reflections provide in two steps extreme angular collimation of the incident radiation. The two subsequent reflections with the opposite indexes of asymmetry serve as two-step energy analyzer

mechanics. This solution allows especially in NIS for a trade-off between energy resolution and flux.

1.6.1.2 Backscattering Monochromator

For higher energies these silicon HRMs are no longer efficient and backscattering monochromators may be an alternative. The classical approach with a silicon crystal [98] does not work for NRS due to the fixed energies given by the Mössbauer levels. Other materials have to be exploited such as sapphire [21]. However, the needed crystal's quality is still lacking [99].

1.6.1.3 Synchrotron Mössbauer Source

For extreme monochromatization, the SMS provides a fully resonant, polarized, and collimated γ-ray beam with an energy resolution in the neV regime at 14.4 keV [54]. Currently, SMSs are installed at SPring-8 [56] and the ESRF [57] (see Fig. 1.13 as an example at the ESRF). Details of the SMS, theoretical and technical, are laid down in several articles [52, 54, 57, 58, 100–102].

The key element of the SMS is an iron borate $^{57}FeBO_3$ single crystal enriched in the ^{57}Fe isotope up to 95%. Iron borate is a canted antiferromagnet with a Néel temperature of 348.35 K. The crystal is set to electronically forbidden but nuclear allowed [111]- or [333]-reflections to extract the purely scattered nuclear signal. At room temperature $^{57}FeBO_3$ shows in this diffraction geometry with polarized SR a four line spectrum due to hyperfine splitting of the nuclear levels with a flux

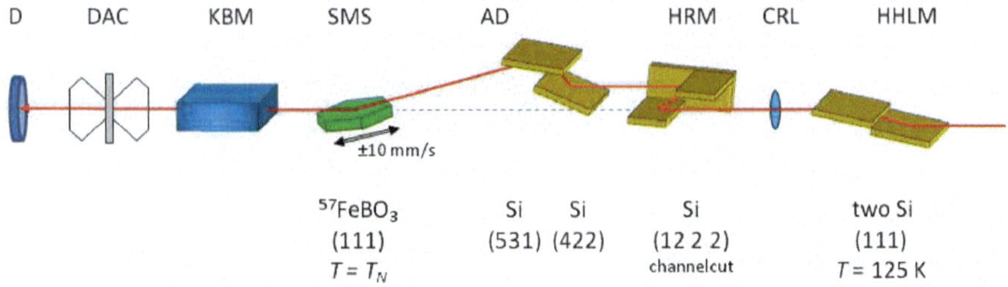

Fig. 1.13 Optical scheme for a high-pressure experiment at the ESRF with the DAC using the SMS based on the $FeBO_3$ (111) pure nuclear reflection. HHLM, high-heat-load monochromator; CRL, compound refractive lens; HRM, Si (12 2 2) high-resolution monochromator; AD, Si (422)—Si(531) angular deflector; SMS, iron borate crystal inside the furnace with magnets and mounted on the Mössbauer transducer and two-circle element; KBM, Kirkpatrick-Baez multilayer mirrors (focusing optics); DAC, diamond anvil cell; D, detector

of about 10^6 photons/s. However, a special case of combined magnetic dipole and electric quadrupole interactions can be realised close to its Néel temperature in the presence of a weak external magnetic field, where a single-line spectrum of the emitted radiation can be obtained. The energy width of the emitted line is close to the natural width of the Mössbauer resonance (see Eq. 1.23). It is very sensitive to the temperature and magnetic field applied across the crystal. Therefore, the temperature of the borate crystal has to be stabilized with milli-Kelvin accuracy. Depending on the exact temperature the linewidth may be chosen according the experimental needs. However, narrower linewidths correlate with lower flux. For example, for a linewidth of $3\,\Gamma_0$ the flux is about $2\,10^4$ photons/s whereas, for a linewidth of $2\,\Gamma_0$ it is about 10^4 photons/s. The energy variation of the γ-ray beam is achieved as in classical MB spectroscopy by Doppler shift of the "source", i.e., in this case the borate crystal. Due to technical issues it is done by a sinusoidal acceleration.

The heat-load of the incoming x-ray beam on the crystal is reduced by the upstream HRM. Further, the angular deflector (AD, see Fig. 1.13) together with the borate crystal keeps the γ-ray beam horizontal.

Furthermore, as discussed in Sect. 1.3.1, the $^{57}FeBO_3$ crystal rotates the sigma polarization by 90°, i.e., the γ-ray beam is π-polarized. Other polarization options may be achieved with phase plates [103].

1.6.1.4 Focusing

A big asset in nuclear resonance techniques compared to classical MB spectroscopy are the focusing capabilities of the radiation. Several options may be employed such as bent monochromators [104], KB-optics [105], compound refractive lenses (CRL) [106], and Fresnel zone plates (FZP) [107]. A general requirement to focusing optics for nuclear resonance applications is that it has to keep the high spectral flux, i.e., to accept the entire SR beam. In case of bulky sample environment such as cryo-magnet or UHV systems a rather large focal distance (~1 m) is an asset. The best trade-off is achieved with graded multilayer optics in Kirkpatrick-Baez geometry [108]. At the Nuclear Resonance beamline at the ESRF a spot size of about $5 \times 10\,\mu m^2$ (v × h) at 1 m focal distance and with high flux is available.

1.6.2 Detectors

For nuclear resonance scattering experiments detectors with (sub-) ns time resolution, high dynamical range, and fast recovery time are mandatory. Beam intensities of 10^9 photons/s in the prompt pulse (i.e. 200 photons per bunch in 16-bunch mode at the ESRF) are common conditions. The detector must survive this intense prompt flash and be able to count few nanoseconds later a single photon event of the delayed nuclear radiation. State-of-the-art detectors are nowadays avalanche photo diode (APD) detectors [109, 110].

The performance can be summarized as:

- dynamic range and linearity is assured over nine decades of intensity
- efficiency: 40% at 14.4 keV
- background: 0.02 photon/s
- time resolution 100 ps to about 1 ns.

Depending on applications and energy regime stacked (up to 24) and inclined detectors are common practice in order to improve the efficiency.

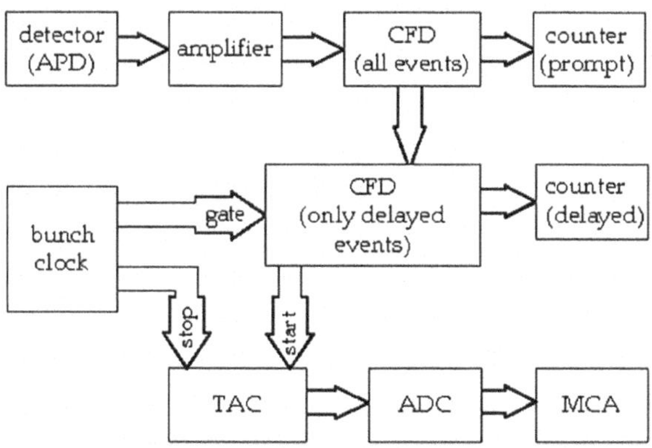

Fig. 1.14 Timing electronics for NRS experiments: it allows one to count both all events, which are pratically the overhelming charge scattered events, and the *delayed* nuclear scattered events. The time spectra are recorded with a MCA. CFD—constant fraction discriminator, TAC—time-to-amplitude converter, ADC—analog-to-digital converter, MCA—multichannel analyzer, bunch clock—reference rf-signals from the SR source

Time spectra are collected in the traditional 'start-stop' technique (see Fig. 1.14). Normally, the detector signal is the "start" signal whereas the "stop" signal is derived from the bunch clock, i.e., the reference signal of the rf-frequency of the SR source.

Fast time-to-digital converter and digitizer techniques are under development in order to improve (i) the response time at early times (< 5 ns) needed especially for Mössbauer isotopes with short life times, (ii) the throughput, and (iii) to allow for "multi-hit" operation. The multi-hit option together with high throughput avoids time gating of the prompt pulse and is mandatory for experiments at the new XFEL sources [111]. Furthermore, fast position sensitive detectors [112–114] received attention for applications such as NSAS, TDI, SRMS.

1.6.2.1 Resonance Detector

For some applications such as IXSNRA, SRMS, and RSMR the energy analysis is carried out with a so-called resonance detector. In principle it consits of an x-ray detector like an APD with its timing electronics and an absorber or a scatterer, which contains the Mössbauer isotope for the resonance analysis.

In the simplest case like IXSNRA the APD is covered e.g. by an $\alpha{-}^{57}$Fe foil. When the energy of the incident radiation matches the energy of the nuclear transition, the scattering leads to nuclear forward scattering in the foil of the resonance isotope. By time discrimination only the (delayed) NFS signal of the time spectrum is taken and time integrated. This resulting signal measures only "resonant quanta" and is used for the energy analysis.

For SRMS the details of the detector system are more sophisticated and will be described in detail in this book by Seto et al. [60]. In case of RSMR (and the SMS) normally time discrimination is not necessary due to the "100%" resonant beam.

1.6.3 Sample Environment

Versatile sample environment is a prerequisite for a holistic approach of the investigation of novel systems. Nuclear resonance techniques by themselves may contribute in the study of electric, magnetic, and structural properties, static and dynamic. Combining those techniques with other on-line and off-line techniques such as XRD, MOKE, transport measurements, etc., will foster a holistic picture. Especially the small beam size and divergence of synchrotron radiation favours sophisticated experimental environments not feasible in the home laboratories with radioactive sources. Standard equipment allows for applying e.g. a combination of low temperature, high pressure, and high magnetic field. Other examples are UHV systems for in situ investigation of surface, interfaces, and nano-structured materials [115]; laser heating system combined with high pressure for in situ investigation of the Earth's interior and new materials [116]; pulsed high-magnetic field [117]; combined RAMAN and IR environment for the investigation of cross-over systems and protein dynamics [118].

1.7 Applications

In this chapter, on the basis of recent applications and developments of NRS at the Nuclear Resonance beamline at the ESRF, we attempt to foresee future applications of NRS, focusing on most challenging scientific cases. Driven by the expected instrumental development of Nuclear Resonance Scattering at Extremely Brilliant Source (EBS) at the ESRF, we analyse possible future applications in two conceptual frames: *Nano-Scale Science* and *Micro-eV Atomic Dynamics*.

1.7.1 Nano-Scale Science

With the EBS Upgrade Programme [119], the Nuclear Resonance beamline at the ESRF expects to offer users with the beam size of about 200 nm. This development is expected to be most beneficial for studies at ultra-high pressure and in fields such as magnetism, superconductivity, geoscience, and nano-paleomagnetism.

1.7.1.1 Ultra-High Pressure

High-pressure studies are one of the evident highlights of nuclear resonance scattering experiments with synchroton radiation sources. This is driven by the intrinsic small size and collimation of synchrotron radiation and by the straightforward focusing capabilities using compound refractive lenses and Kirkpatrick-Baez optics, respectively.

At present, the available beam size at nuclear resonance beamlines is about 10 μm. This is sufficient to use most elaborated single-stage diamond anvil cells (DACs), allowing to reach pressures as high as a few hundreds GPa. Figure 1.15 outlines the studies of magnetism in Ni metal, conducted at pressures up to 260 GPa [120]. Though the statistical accuracy of the data at highest pressures is relatively moderate, it nevertheless enables the reliable determination of the magnetic hyperfine field. The data provide a solid evidence that Ni stays ferromagnetic up to 260 GPa, the highest pressure where magnetism has been observed so far (Fig. 1.15, left panel). They also reveal the importance of accounting for relativistic effects in the theory of magnetic interactions in Ni (Fig. 1.15, right panel).

The above example showcases the frontiers of high-pressure research at existing nuclear resonance beamlines. However, they are still far away from the limits of available static-pressure equipment, determined by recently developed double-stage diamond anvil cells [121].

In the double-stage DAC, a sample is pressurized between two micro-anvils (semi-balls made of nanocrystalline diamonds), which are attached to the culets of the opposed gem quality diamonds (Fig. 1.16). Due to the small diameter of the

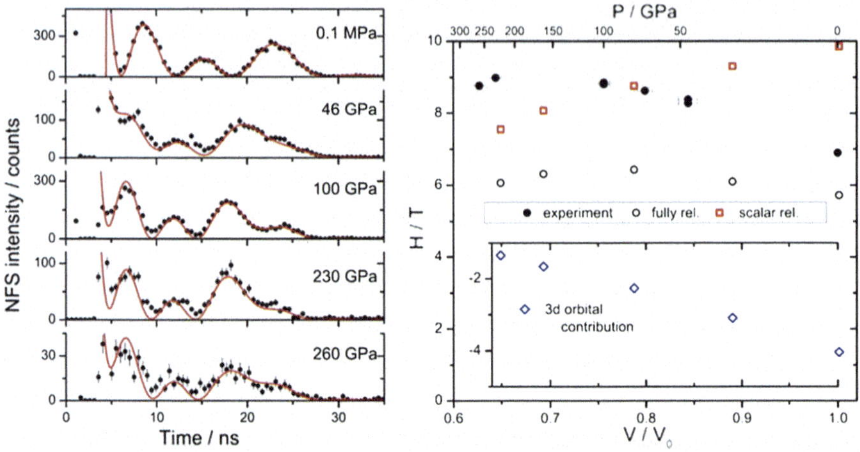

Fig. 1.15 Left: time evolution of nuclear forward scattering for Ni at various pressures. The *(red) solid lines* show the fit. The period of oscillations of the signal is inversely proportional to the magnetic hyperfine splitting. Right: pressure dependence of the magnetic hyperfine field in Ni from the experiment (*solid black circles*) and from fully relativistic ab initio calculations (*open black circles*). Positive and negative contributions to the calculated field are shown as scalar relativistic calculations (*red open squares*) and 3d orbital contribution (*blue open diamonds*). (Reprinted figure with permission from [120], Copyright (2013) by the American Physical Society)

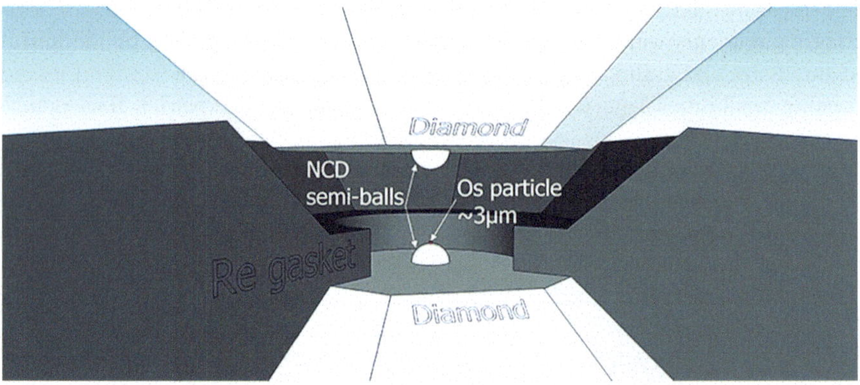

Fig. 1.16 The double-stage diamond anvil cell for ultra-high pressure generation. Semi-balls made of nanocrystalline diamond (NCD) of extraordinary strength are attached to the culets of the opposed gem quality diamonds of the DAC. A sample of osmium (Os), shown as a small red dot on the top of the lower semi-ball, has a size of ca. 3 microns [122] (Image courtesy Elena Bykova, University of Bayreuth)

micro-anvils (10–50 μm), the static pressure limits of these cells approach the TPa range [122].

The typical size of samples in the double-stage DAC is about few microns (Fig. 1.16). In order to study a system under reasonably homogeneous-pressure conditions, the size of the synchrotron radiation beam should be on the (sub-)micron scale. This is not yet available at present nuclear resonance beamlines, but should be enabled by coming Extremely Brilliant Sources of synchrotron radiation.

Furthermore, the sub-micron beam size will greatly improve the quality of high-pressure and high-temperature experiments with laser heating, allowing one to focus the probe beam on the centre of the most heated spot of the sample. With such a beam, the available pressure and temperature range will entirely cover the conditions of the Earth interior, and will open access to studies of matter under conditions of Super-Earth planets interiors.

1.7.1.2 Magnetism

The intrinsic sensitivity of Nuclear Resonance Scattering to the value and direction of magnetic hyperfine fields, combined with a small beam size, allows for imaging of magnetic structures. One of the most impressive example of such studies is the imaging of the spin structure of a magnetic spring [123].

Utilizing the probe layer technique, i.e., inserting ^{57}Fe in various depths of the thin iron film (see Fig. 1.17 right panel), NRS directly probes the actual spin structure in various depths by selectively exciting the ^{57}Fe layer at various lateral positions. The sample investigated here is a bilayer system consisting of 11 nm Fe on 30 nm $Fe_{55}Pt_{45}$ in the hard-magnetic tetragonal $L1_0$ phase. A wedge-shaped 0.7 nm thick ^{57}Fe film with a slope of 0.5 nm/mm has been produced. Different depths D in the sample can thus be probed by adjusting the displacement Δx of the sample transversely to the incident beam [123]. Evaluation of the spectra reveals the depth dependence of the rotation of the magnetization in the iron film with atomic resolution (Fig. 1.17 left panel).

Improving the beam size to the sub-micron scale will enable magnetic imaging of more numerous systems like magnetic domains and domain walls, vortexes, skyrmions, etc. In this sense, the abilities of nuclear resonance scattering will approach the resolution of magnetic imaging with tunnelling microscopy (Fig. 1.18), with the additional option of imaging the chemical and oxidation states of systems. Furthermore, this approach will enable imaging under external stimuli such as extreme temperature and pressure conditions and external electric/magnetic fields and give access to burried and interface layers, not easily accessible by other methods.

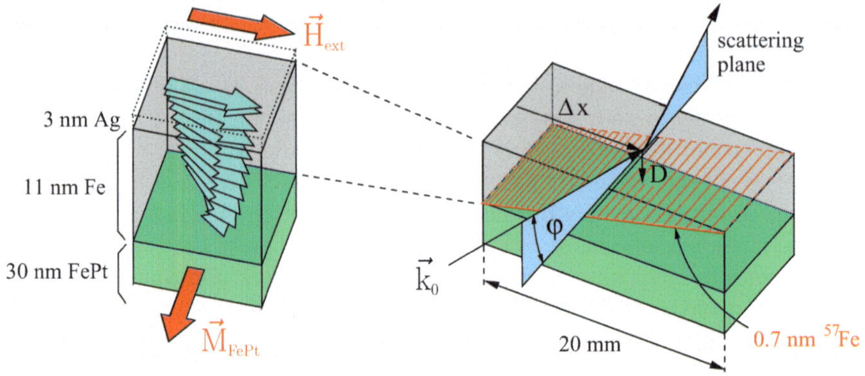

Fig. 1.17 Right panel: scattering geometry of the sample (11 nm Fe on FePt). The incoming x-ray beam with wavevector $\mathbf{k_0}$ impinges the sample at a lateral position Δx probing the spin structure via the wedge-shaped ^{57}Fe probe layer (0.7 nm) at depth D. Left panel: image of the derived spin structure (blue arrows) of the iron layer (image courtesy R. Röhlsberger, Desy, 2002)

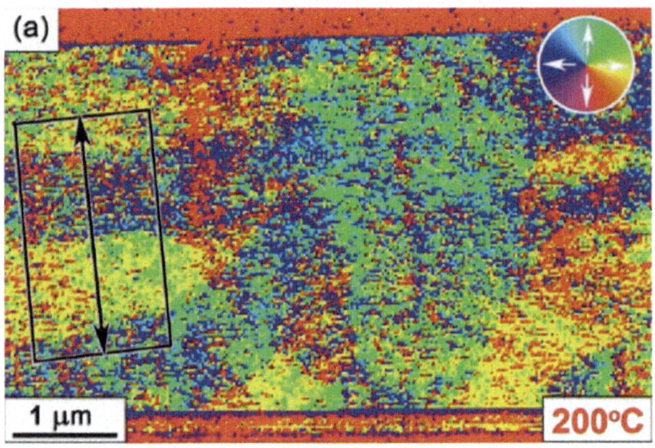

Fig. 1.18 Differential phase contrast magnetic imaging of the planar FeRh thin films. The direction of magnetisation is depicted in the colour wheels (inset). Reprinted through Creative Commons CC BY licence: Ref. [124], copyright 2017

1.7.1.3 Superconductivity

Nuclear resonance scattering provides an elegant and general way to study superconductivity using the Meissner-Ochsenfeld effect [125]. Similar to other applications, this allows one to deal with superconductivity at extreme conditions, which is not easily accessible with standard methods.

A non-magnetic Mössbauer probe is placed inside a sample, which becomes a superconductor (Fig. 1.19). When the NFS spectrum is measured without external magnetic field, one gets a simple exponential decay from a non-magnetic sample. By applying an external magnetic field a splitting of the nuclear levels will appear and, consequently, will cause a quantum beat structure in the NFS spectrum with frequencies corresponding to the strength of the external field [126].

If now the sample is cooled below the critical temperature T_c, the sample becomes superconducting, and the magnetic field is expelled from the sample (Fig. 1.19). Then

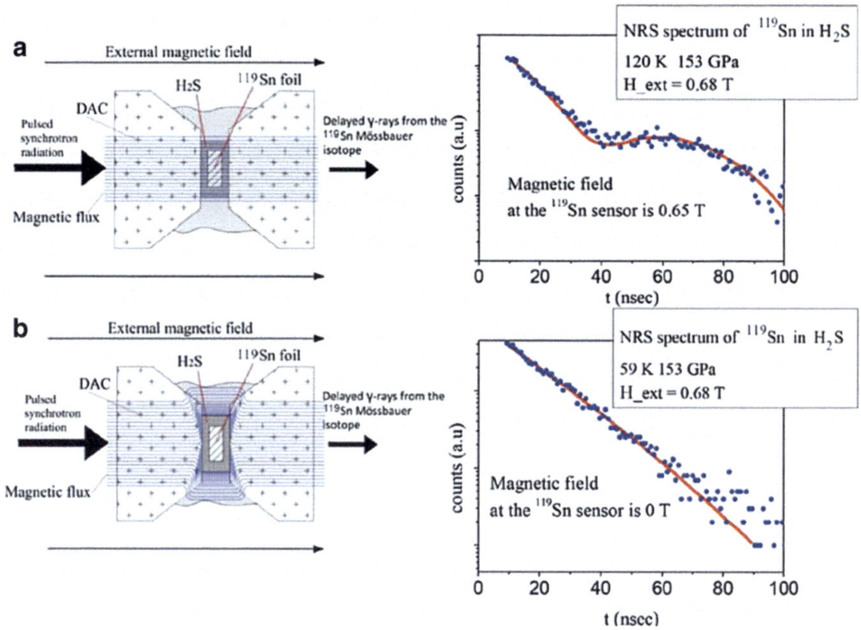

Fig. 1.19 Tin foil, surrounded by compressed H_2S, is located in a DAC at a pressure of about 153 GPa. Synchrotron radiation excites the nuclei of the Mössbauer isotope ^{119}Sn. **a** At high temperature, nuclear forward scattering shows quantum beats due to magnetic splitting of the nuclear levels. **b** At temperatures below the superconducting transition in H_2S, the tin foil is screened from the external magnetic field and, consequently, there is no splitting of the nuclear levels and no magnetic quantum beats in the time spectrum (From [126]. Reprinted with permission from AAAS)

the Mössbauer probe does no longer experience the field and one will again get a simple exponential decay.

In essence, this approach is an alternative to conventional monitoring superconductivity in measurements of magnetic susceptibility with SQUIDs. Instead of putting a sample inside a SQUID, here a small sensor is placed inside a sample. The feasibility of this approach is allowed by the small, micron-scale, size of the synchrotron radiation beam [126].

A further reduction of the beam size to the sub-micron scale will open wide perspectives to study spatial details of superconducting transitions and to image the structures of vortexes and superconducting lamellas. It is well established that the disappearance of a superconducting state proceeds in a spatially-inhomogeneous manner, with creation of an elaborated relief of avalanches of conducting domains penetrating into the residual islands of superconductivity (Fig. 1.20). Studies of the temperature and external field evolutions of these structures at extreme pressure is the next challenge for Nuclear Resonance Scattering.

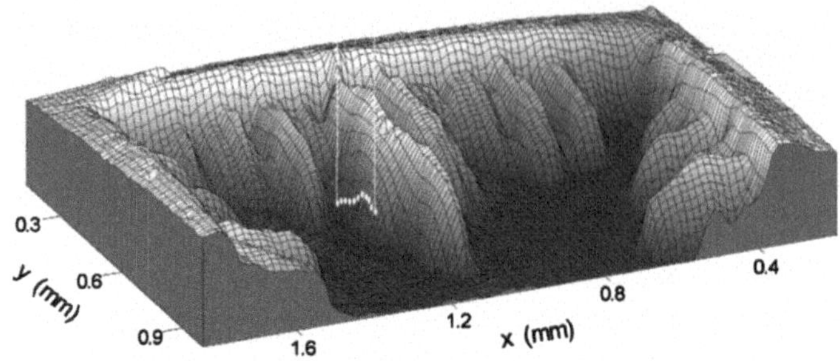

Fig. 1.20 Magnetic landscape in a Nb foil where an array of micro Hall probes (white dots) detect avalanches coming down the slope of the largest flux ridge. The intensity along the vertical axis is proportional to the magnetic field inside the sample (Reprinted figure with permission from [127], Copyright (2004) by the American Physical Society)

1.7.1.4 Nano-imaging

Besides its sensitivity to magnetic structures, nuclear resonance scattering is an established tool to monitor chemical and oxidation states of solids. This enables imaging of chemical and valence structures in real space.

In imaging chemical and oxidation profiles, the mission of nuclear resonance beamlines is to study systems with highly complex micro-size spatial composition. One of the most challenging scientific cases in this aspect are studies of meteorites. Meteorites are slowly cooled (2–9 K per million years) systems, most often consisting of olivine crystals in an Fe-Ni matrix originating from the mantle of a 200-km-radius asteroid [129]. They carry the time-resolved geological, thermal, and magnetic history of the Universe. One of the most intriguing items of these systems is the so-called cloudy zone (CZ), a nanoscale intergrowth of tetrataenite (ordered FeNi) islands of 10–200 nm in size embedded in an Fe-rich matrix, which reveals the time-series record of magnetic activity on an asteroid body. With the current available 10 μm spatial resolution in SMS spectroscopy our understanding of meteoritic metal could already be improved by allowing the cloudy zone to be measured separately from the surrounding metal phases [128]. Scanning along the line crossing the cloudy zone (Fig. 1.21, left panel) is accompanied by drastic changes of the measured Mössbauer spectra, developing from schreibersite to kamacite and further to plessite contributions (Fig. 1.21, right panel). The available 10 μm spatial resolution reveals that the plessite region, previously considered to be a pure-chemical state, actually contains an essential contribution of antitaenite. This finding indicates that the saturation of magnetization in this regions occurs at much lower temperatures than previously thought [128].

However, the most intriguing knowledge on the magnetic history of the Universe is stored in the fine structure of the cloudy zone (Fig. 1.22). The age of the cloudy zone decreases with the distance from the tetrataenite rim [129]. Thus, performing nano-scale imaging of the magnetic, chemical, and coordination states of the

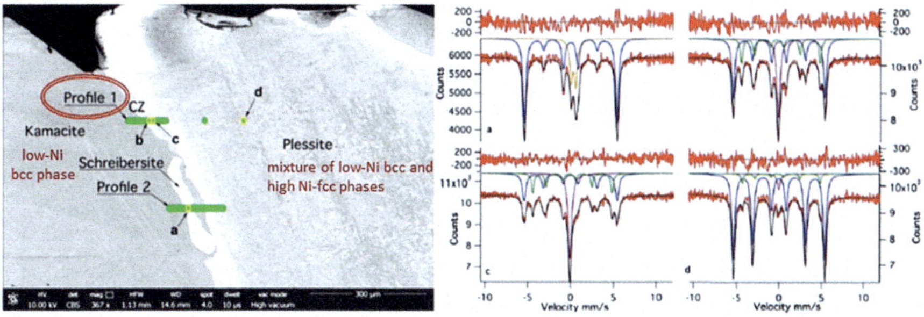

Fig. 1.21 Let panel: scanning electron microscopy image of the Esquel meteorite sample. Green bars mark areas where spectra were obtained. Yellow ovals mark sites where spectra shown on right panel were taken. The size of the ovals represents the size of the beam. Right panel: A selection of representative Mössbauer spectra from the Esquel meteorite: **a** spectrum containing schreibersite; **b** spectrum acquired in the coarse cloudy zone (CZ), closer to the large kamacite lamella; **c** spectrum acquired in the fine cloudy zone, closer to plessite; and **d** spectrum acquired deep into plessite. Reprinted through Creative Commons CC BY 4.0 licence: [128], copyright 2017

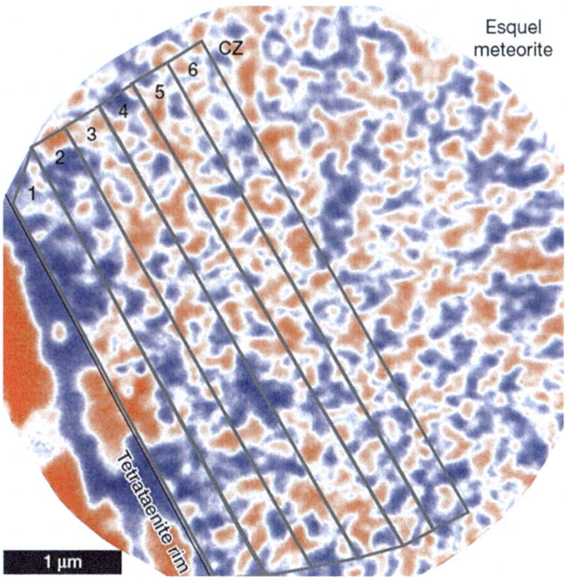

Fig. 1.22 Representative X-ray photoemission electron microscope image of the kamacite, tetrataenite rim and cloudy one (CZ) in the Esquel pallasites. Blue and red colours correspond to positive and negative projections of the magnetization along the X-ray beam direction in the meteorite. The CZ displays a complex interlocking pattern of positive and negative domains. The age of the CZ decreases with distance from the tetrataenite rim. (Reprinted by permission from Nature Springer: [129], copyright 2015)

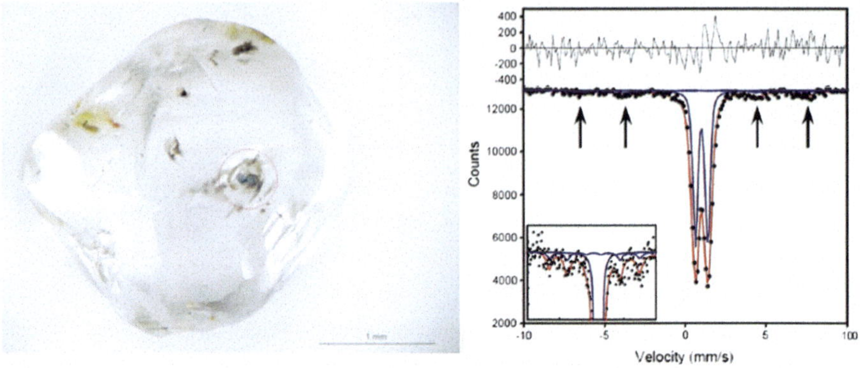

Fig. 1.23 Left: diamond containing ferropericlase inclusion AZ2. Right: Mössbauer spectrum of the inclusion. Solid circles: experimental data; red line: full transmission integral fit assuming one quadrupole doublet and one magnetic sextet. The arrows indicate the positions of four peaks of the magnetic sextet, which can be seen more clearly in the inset that shows a magnified view near the baseline (Reprinted from [130]. Copyright (2016), with permission from Elsevier)

cloudy zone will allow one to recover magnetic, temperature, and thermodynamic conditions of slowly cooled asteroid bodies in their million-years traveling through the Universe. This is what will be possible in nuclear resonance scattering with the expected improvement of the spatial resolution to 100–200 nm.

Another challenging system for nano-imaging of magnetic, chemical, and oxidation states with nuclear resonance techniques are iron-bearing inclusions in diamonds. Syngenetic mineral inclusions in diamonds are pristine witnesses of the chemical and mineralogical environment during diamond formation and thus represent a direct window into the Earth interior. They were for a long period the only source to get information about the interior of the Earth such as the spin and valence states, and the Fe^{3+}/Fe_{tot} ratio of its constituents. Still nowadays they are the only source to get any material at hand from the interior [130].

Already with the presently available resolution of about 10 μm, one can perform spatially-resolved studies of the biggest inclusions. Figure 1.23 shows an example of an iron-bearing inclusion with the size of $192 \times 85 \times 105\,\mu m^3$ (left panel) and the Mössbauer spectrum of one of the parts of the inclusion (right panel), where the presence of a magnetic state is indicated [130].

The expected improvement of the spatial resolution to 100–200 nm will allow for more detailed mapping of the iron-bearing inclusions. Such investigations of the neighbour co-existing chemical phases will enable accessing a wealth of thermodinamical properties of formation of these systems.

Furthermore, the smaller beam size will enable investigations of much smaller inclusions, which accounts for about 95% of the available samples. This will for the first time allow for statistically representative studies.

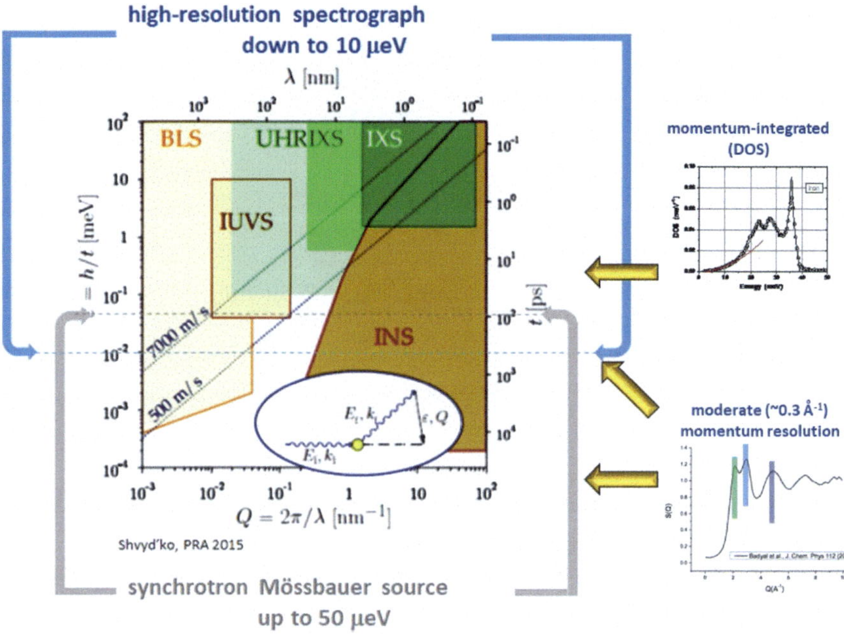

Fig. 1.24 Left panel: Energy-momentum (ε-Q) and relevant time-length (t-λ) space of excitations in condensed matter and how it is accessed by different inelastic scattering probes. The Extremely Brilliant Source Upgrade Programme suggests to improve the energy resolution to about 50 μeV. In addition, the Synchrotron Mössbauer Source available at the Nuclear Resonance beamline at ESRF will allow for scattering experiments with an energy transfer from 1 neV up to 50 μeV for either momentum integrated or moderate momentum resolution conditions (right panels). This will allow one to cover the ~ neV to ~ meV energy range entirely. (Reprinted left figure with permission from [131], Copyright (2015) by the American Physical Society)

1.7.2 Micro-eV Atomic Dynamics

With the Extremely Brilliant Source Upgrade Programme, the Nuclear Resonance beamline at the ESRF expects to offer users with the energy resolution of about 50 μeV. This development will be possibly most beneficial for soft atomic dynamics, geoscience, and glass physics.

1.7.2.1 No-Man's-Land

Figure 1.24 (left panel) shows how the energy-momentum space (or the relevant time-length space) of excitations in condensed matter is accessed by different inelastic scattering probes such as neutrons (INS), x-rays (IXS), ultra-high-resolution IXS (UHRIXS) [131], and photons (ultraviolet (IUVS) and Brillouin (BLS)). It also shows the remaining gap, not accessible by any technique. The expected improvement in

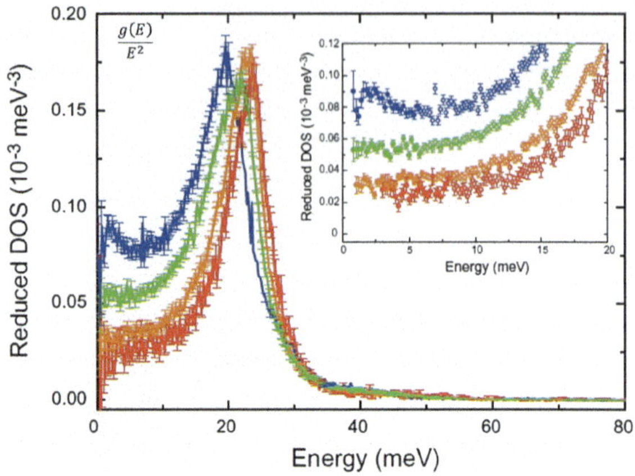

Fig. 1.25 Iron partial reduced phonon density of states (DOS) for four samples of ferropericlase. The inset shows an enlarged reduced DOS (Reprinted from [132] Copyright (2014), with permission from Elsevier)

the energy resolution to about 50 μeV is supposed to close the gap and thereby to enhance capabilities of x-ray techniques.

In addition to the proposed closing of the remaining gap in the energy transfer region from above, we also anticipate to close the gap from below. In particular, the Synchrotron Mössbauer Source available at the Nuclear Resonance beamline at ESRF allows for scattering experiments with an energy transfer from 1 neV to 10 μeV. With an upgraded setup, we plan to reach the energy transfer up to 50 μeV. Thus, both instruments together allow one to cover entirely the ∼neV to ∼meV energy regime.

Even though the proposed development is related either to momentum-integrated studies (DOS) (right top panel) or to measurements with only a moderate momentum resolution ($0.3\,\text{Å}^{-1}$) (right bottom panel), this can be clearly qualified as entering no-man's-land, which will not only help to solve known scientific questions but also will inevitably lead to new fascinating discoveries in fields of geoscience and glass physics.

1.7.2.2 Geoscience

Inelastic x-ray scattering and nuclear inelastic scattering are nearly the only tools to access sound velocity data at extreme conditions such as high pressure and high temperature. While these measurements are relatively easy for model samples, they become progressively complicated for systems with large unit cells and lower mean sound velocities, because the Debye-like parabolic approximation of the DOS is then only valid for much lower energies.

For example, for some samples of ferropericlase [132], a proper determination of sound velocities can be achieved within the energy range below 5 meV (Fig. 1.25, red, orange, and green curves), whereas for other samples this requires precise measurements of the DOS well below 1 meV (Fig. 1.25, blue curve).

Other geophysically meaningful systems like perovskites and wustite require even better energy resolution. Therefore, an achievement of 50 μeV resolution will

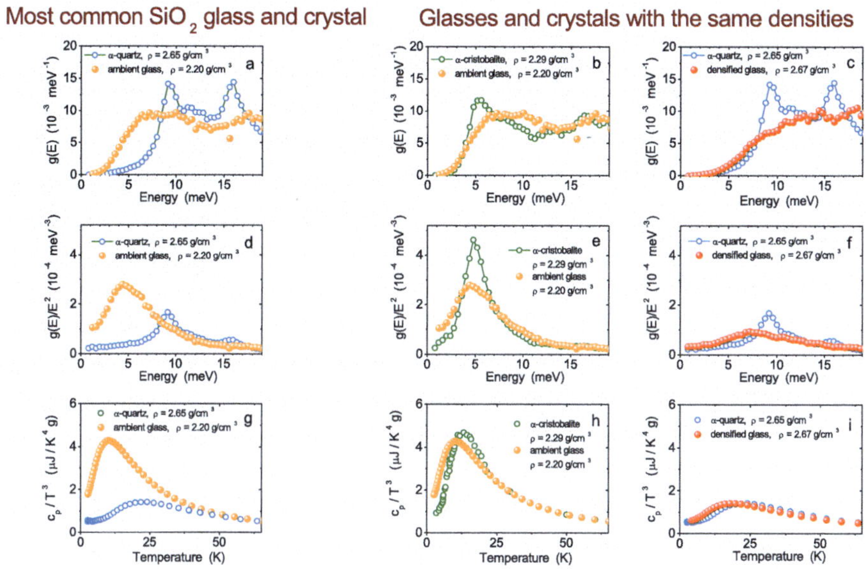

Fig. 1.26 The phonon density of states $g(E)$ (**a**–**c**), the reduced density of states $g(E)/E^2$ (**d**–**f**), and the heat capacity c_P/T^3 (**g**–**i**) for various glassy and crystalline polymorphs of SiO_2. The left panels (**a**, **d**, **g**) compare the most common glassy and crystalline polymorphs and reveal a noticeable difference in the displayed properties. The middle and the right panels compare the glassy and crystalline polymorphs with matched densities, namely, the low-density (**b**, **e**, **h**) and the high-density (**c**, **f**, **i**) polymorphs. They demonstrate that the atomic dynamics and thermodynamics of the glassy and crystalline polymorphs with matched densities do not differ much from each other (Reprinted figure with permission from [95], Copyright (2014) by the American Physical Society)

greatly improve the accuracy of sound velocity measurements, especially for extreme pressure-temperature conditions.

1.7.2.3 Glass Physics

Measurements of the phonon density of states with the presently available energy resolution of about 0.5 meV were already decisive to answer some important scientific questions in glass physics such as the puzzle of the so-called Boson peak (Fig. 1.26). Reliable accessing the atomic dynamics in the 2–5 meV energy range and an ideal integration over the entire range of allowed momentum transfer provide the data, which clearly reveal that the atomic dynamics and thermodynamics of the glassy and crystalline polymorphs with matched densities, contrary to common believes, do not differ much from each other [95].

With the energy resolution improved to about 50 μeV, nuclear resonance scattering may proceed further, to tackle one of the most important scientific cases in glass physics, the nature of the glass transition.

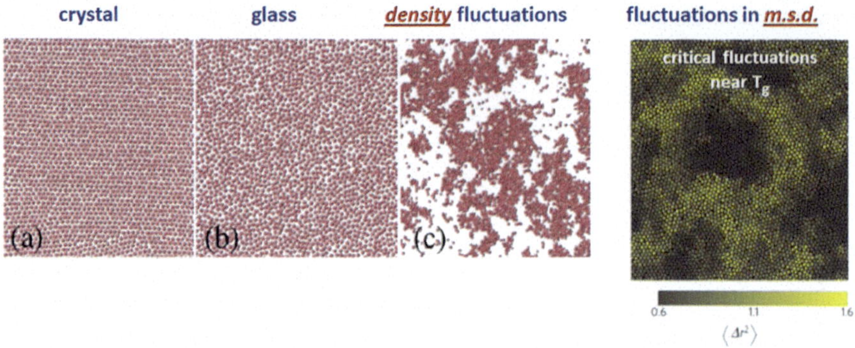

Fig. 1.27 **a** A periodic crystalline structure does not flow because preserving the crystalline order requires moving an extensive set of particles. **b** A mechanically rigid glassy structure exhibits neither the long-range order of a crystal nor the large-scale density fluctuations observed at an ordinary critical point. **c** Large-scale critical density fluctuations near the critical point. (Reprinted figure with permission from [133], Copyright (2011) by the American Physical Society) Right panel: spatial distribution of the mean-square displacement (m.s.d.) of a glass-forming liquid. (Reprinted by permission from Nature Springer: [134], copyright 2010)

Understanding the glass-liquid transition is a challenge, which has resisted the everlasting research efforts of soft condensed matter physics over centuries. Evidence has mounted in recent years that the viscous slowing down of super-cooled liquids might be related to the existence of genuine phase transitions, but of very peculiar nature [133]. One of the most interesting consequences of these ideas is the existence of dynamic heterogeneities (also known as correlated relaxation or correlated diffusion), which have been discovered to be (in the space-time domain) the counterpart of critical fluctuations in standard phase transitions [134].

Dynamic heterogeneity refers to the existence of transient spatial fluctuations in the local dynamical behaviour. The domains of different mobility have no counterpart in the density fluctuations and only appear when dynamics is considered (Fig. 1.27).

Although conceptually of crucial importance, only recently a rather direct evidence for this cooperative motion became available with a suggested cooperative length-scale of about 5–20 molecular diameter at the glass transition [134]. On momentum- and energy-transfer scales, this corresponds to $\sim$1 nm^{-1} and $\sim$100 µeV, respectively. Thus, the expected energy resolution of about 50 µm will undoubtedly contribute to the understanding of the nature of glass-liquid transitions.

1.8 NRS with X-Ray Free Electron Lasers

Impressive achievements and a bright future of NRS studies with synchrotron radiation sources can yet be augmented by emerging opportunities of Nuclear Resonance Scattering with X-Ray Free Electron Lasers (XFELs). In comparison to synchrotrons,

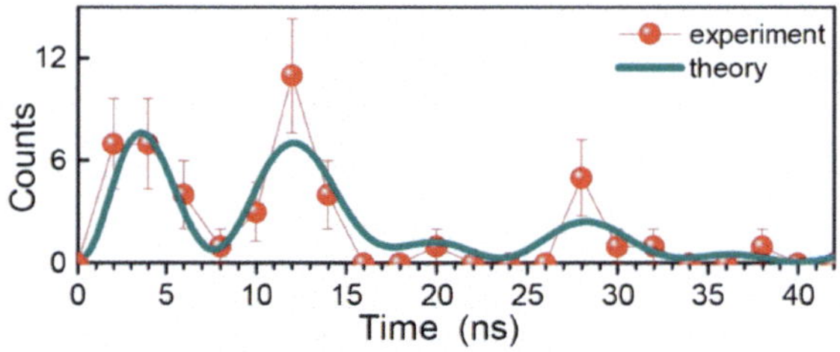

Fig. 1.28 The time evolution of the decay for one (circles) out of twelve registered 50-photon excitations. The solid line is theory fit. The error bars show the standard deviations related to the numbers of counts per channel. (From [111])

these facilities promise to increase the count rate by several orders of magnitude and to allow for NRS pump-probe experiments on the femto-second time scale.

With XFEL radiation, a time spectrum of Nuclear Resonance Scattering can be recorded in a single-shot of a laser pulse. In the first NRS experiment at an XFEL [111], the most probable number of nuclear resonance photons in a single shot was ~ 8, the mean number of photons 16, and the maximum number of recorded photons 68. Single-pulse time spectra recorded after a shot with highest photon numbers provide already sufficient statistical accuracy for single-shot hyperfine spectroscopy.

Figure 1.28 shows the time spectrum measured for a single pulse of XFEL radiation with 50 detected photons. The data were fit with the theory using the nuclear magnetic hyperfine splitting as adjustable parameter. The obtained value of the splitting 513(30) neV is in the excellent agreement with the previously reported value of 512.6 neV [135]. The proper determination of the hyperfine parameters is the first demonstration of hyperfine spectroscopy with a single shot of the XFEL radiation.

This result opens access to nuclear resonance pump-probe experiments with femtosecond time resolution at the presently available XFEL facilities and with APD detectors. Indeed, coherent nuclear scattering relies on the phasing of scattering over a nuclear ensemble and occurs only if the phasing/position of atoms are well defined, relative to each other at the excitation and emission time. This condition is valid for solids, but not for liquids. This allows one to use nuclear forward scattering to study, for example, heat transfer or melting (or even magnetic order) on $\sim$ femto-second time scales, only limited by the XFEL pulse length.

Consider the scheme shown in Figure 1.13, where a resonant sample is placed just after the double-crystal monochromator (HHLM). Because the sample is located before high-resolution optics, it is still illuminated by the short pulses of the XFEL, with $\sim$ fs pulse width. Suppose that the sample is melted by a short pulse of a pump laser.

If the sample remains liquid at the arrival of the probe XFEL radiation, the measured spectrum will be the one prepared only by nuclear diffraction in the iron borate crystal. On the contrary, if the sample cools down to a solid state before the arrival of the probe XFEL radiation, the conditions for nuclear forward scattering are fulfilled, and the time spectrum measured after the sample and the iron borate crystal will be modified. The same approach can be also applied to study dynamics of magnetism by e.g. heating the sample above the Curie point.

1.9 Summary

The article provided a short summary of ideological, instrumental, and scientific developments of Nuclear Resonance Scattering of synchrotron radiation. Comparison of the first pioneering steps in the development of the technique with the gained ability to solve most challenging scientific cases evidences the mature status of this still young field. Moreover, it reveals the permanently increasing rate of the development, opening access to new directions of studies, new types of systems, and new environmental conditions. Thus, there are all reasons to believe in flourishing future of this fascinating way to discover the most intriguing aspects of nature.

Acknowledgements The development of the techniques as well as the waste amount of applications in various scientific fields—only a glimpse could be presented in this article—was only possible with the great involvement of all our colleagues over the years at the beamline and with the users coming to the beamline with incredible ideas for new challenging science. We would as well acknowledge the fruitful collaboration and friendship of our colleagues from the other Nuclear Resonance beamlines at APS, PETRA III, and SPring-8. However, nothing would have started without our mentors, when we were students, E. Gerdau and G. V. Smirnov.

References

1. R.L. Mössbauer, Zeitschrift für Physik **151**(2), 124 (1958). https://doi.org/10.1007/BF01344210
2. R.L. Mössbauer, Naturwissenschaften **45**(22), 538 (1958). https://doi.org/10.1007/BF00632050
3. P.J. Black, P.B. Moon, Nature **188**, 481 (1960)
4. S. Bernstein, E.C. Campbell, Phys. Rev. **132**, 1625 (1963). https://doi.org/10.1103/PhysRev.132.1625
5. P.J. Black, G. Longworth, D.A. O'Connor, Proc. Phys. Soc. **83**(6), 937 (1964). http://stacks.iop.org/0370-1328/83/i=6/a=305
6. G.V. Smirnov, Hyperfine Interactions **27**, 203 (1986)
7. U. van Bürck, Hyperfine Interactions **27**, 219 (1986)
8. E. Gerdau, R. Rüffer, H. Winkler, W. Tolkdorf, C.P. Klages, J.P. Hannon, Phys. Rev. Lett. **54**, 835 (1985)
9. J.P. Hannon, G.T. Trammell, Hyperfine Interactions **123–124**(1–4), 127 (1999). https://doi.org/10.1023/A:1017011621007
10. P. Gütlich, E. Bill, A.X. Trautwein, *Mössbauer Spectroscopy and Transition Metal Chemistry* (Springer, 2011)
11. Yu.V. Shvyd'ko, M. Lerche, J. Jäschke, M. Lucht, E. Gerdau, M. Gerken, H.D. Rüter, H.-C. Wille, P. Becker, E.E. Alp, W. Sturhahn, J. Sutter, T.S. Toellner, Phys. Rev. Lett. **85**, 495 (2000). https://doi.org/10.1103/PhysRevLett.85.495
12. E. Gluskin, E.E. Alp, I. McNulty, W. Sturhahn, J. Sutter, J. Synchrotron Rad. **6**, 1065 (1999)

13. M. Yabashi, K. Tamasaku, T. Ishikawa, Phys. Rev. Lett. **88**, 244801 (2002). https://doi.org/10.1103/PhysRevLett.88.244801
14. Yu.V. Shvyd'ko, M. Lerche, H.-C. Wille, E. Gerdau, M. Lucht, H.D. Rüter, E.E. Alp, R. Khachatryan, Phys. Rev. Lett. **90**, 013904 (2003). https://doi.org/10.1103/PhysRevLett.90.013904
15. R. Röhlsberger, K. Schlage, B. Sahoo, S. Couet, R. Rüffer, Science **328**(5983), 1248 (2010). https://doi.org/10.1126/science.1187770. http://www.sciencemag.org/content/328/5983/1248.abstract
16. E. Gerdau, H. de Waard (eds.), *Nuclear Resonant Scattering of Synchrotron Radiation* (Baltzer Science Publishers, 1999/2000)
17. J.P. Hannon, G.T. Trammell, Phys. Rev. **169**, 315 (1968). https://doi.org/10.1103/PhysRev.169.315
18. J.P. Hannon, G.T. Trammell, Phys. Rev. **186**, 306 (1969). https://doi.org/10.1103/PhysRev.186.306
19. A.M. Afanas'ev, Yu. Kagan, JETP **21**, 215 (1965)
20. A.M. Afanas'ev, Yu. Kagan, JETP **37**, 987 (1973)
21. Yu. Shvyd'ko, *X-Ray Optics—High-Energy-Resolution Applications, Optical Science*, vol. 98 (Springer Publishers, 2004)
22. R. Röhlsberger, Nuclear condensed matter physics with synchrotron radiation, in *Springer Tracts in Modern Physics*, vol. 208 (Springer Publishers, 2004)
23. J.D. Jackson, *Classical Electrodynamics* (Wiley, New York, 1975)
24. H. Wiedemann, *Particle Accelerator Physics I + II* (Springer, Berlin, 1993)
25. T. Tanaka, H. Kitamura, J. Sync. Radiation **16**, 380 (2009). https://doi.org/10.1107/S0909049509009479
26. S.L. Ruby, J. Phys. **35**(C6), 209 (1974)
27. R.L. Mössbauer, Naturwissenschaften **60**, 493 (1973)
28. R.L. Cohen, P.A. Flinn, E. Gerdau, J.P. Hannon, S.L. Ruby, G.T. Trammell, Workshop on new directions in Mössbauer spectroscopy, in *AIP Conference Proceedings*, vol. 38, ed. by G. Perlow (AIP, New York, 1977), pp. 140–148. https://doi.org/10.1063/1.31071
29. E. Gerdau, M. Mueller, H. Winkler, R. Rüffer, European Synchrotron Radiation Facility, in *vol. Supplement I—The Scientific Case*, ed. by Y. Farge, P. Duke (European Science Foundation (ESF), Strasbourg, 1979), pp. 69–75
30. R.V. Pound, G.A. Rebka, Phys. Rev. Lett. **4**, 274 (1960). https://doi.org/10.1103/PhysRevLett.4.274
31. E.P. Team (ed.), *ESRF Foundation Phase Report* (ESRF, Grenoble, 1987)
32. B. Marx, K.S. Schulze, I. Uschmann, T. Kämpfer, R. Lötzsch, O. Wehrhan, W. Wagner, C. Detlefs, T. Roth, J. Härtwig, E. Förster, T. Stöhlker, G.G. Paulus, Phys. Rev. Lett. **110**, 254801 (2013)
33. G.T. Trammell, J.P. Hannon, S.L. Ruby, P. Flinn, R.L. Mössbauer, F. Parak, Workshop on new directions in Mössbauer spectroscopy, in *AIP Conference Proceedings*, vol. 38, ed. by G. Perlow (AIP, New York, 1977), p. 46. https://doi.org/10.1063/1.31071
34. J.P. Hannon, G.T. Trammell, M. Mueller, E. Gerdau, H. Winkler, R. Rüffer, Phys. Rev. Lett. **43**, 636 (1979). https://doi.org/10.1103/PhysRevLett.43.636. URL https://link.aps.org/doi/10.1103/PhysRevLett.43.636
35. J.P. Hannon, N.V. Hung, G.T. Trammell, E. Gerdau, M. Mueller, R. Rüffer, H. Winkler, Phys. Rev. B **32**, 5068 (1985). https://doi.org/10.1103/PhysRevB.32.5068
36. J.P. Hannon, N.V. Hung, G.T. Trammell, E. Gerdau, M. Mueller, R. Rüffer, H. Winkler, Phys. Rev. B **32**, 5081 (1985). https://doi.org/10.1103/PhysRevB.32.5081
37. J.P. Hannon, G.T. Trammell, M. Mueller, E. Gerdau, R. Rüffer, H. Winkler, Phys. Rev. B **32**, 6363 (1985). https://doi.org/10.1103/PhysRevB.32.6363
38. J.P. Hannon, G.T. Trammell, M. Mueller, E. Gerdau, R. Rüffer, H. Winkler, Phys. Rev. B **32**, 6374 (1985). https://doi.org/10.1103/PhysRevB.32.6374
39. R. Röhlsberger, J. Evers, Quantum Optical phenomena in Nuclear Resonant Scattering, in This book, Chapter 3

40. G.T. Trammell, Gamma-ray diffraction by resonant nuclei, in *Chemical Effects of Nuclear Transformations*, vol. I (International Atomic Energy Agency, Vienna, 1961)
41. Yu. Kagan, A.M. Afanas'ev, I.P. Perstnev, JETP **27**, 819 (1968)
42. P.J. Black, I.P. Duerdoth, Proc. Phys. Soc. **84**, 169 (1964)
43. G.V. Smirnov, V.V. Sklyarevskii, R.A. Voskanyan, A.N. Artem'ev, JETP Lett. **9**, 70 (1969)
44. P.P. Kovalenko, V.G. Labushkin, V.V. Rudenko, V.A. Sarkisyan, V.N. Seleznev, JETP Lett. **26**, 85 (1977)
45. R.M. Mirzababaev, G.V. Smirnov, V.V. Sklyarevskii, A.N. Artem'ev, A.N. Izrailenko, A.V. Babkov, Phys. Lett. A **37**, 441 (1971)
46. V.S. Zasimov, R.N. Kuz'min, A.Yu. Aleksandrov, A.I. Firov, JETP Lett. **15**, 277 (1972)
47. A.V. Kolpakov, E.N. Ovchinnikova, R.N. Kuz'min, Phys. Stat. Sol. (b) **93**, 511 (1979)
48. H. Winkler, R. Eisberg, E. Alp, R. Rüffer, E. Gerdau, S. Lauer, A.X. Trautwein, M. Grodzicki, A. Vera, Z. Phys. B **49**, 331 (1983)
49. P.P. Kovalenko, V.G. Labushkin, A.K. Ovsepyan, E.R. Sarkisov, E.V. Smirnov, JETP Lett. **39**, 573 (1984)
50. E. Gerdau, U. van Bürck, R. Rüffer, Hyperfine Interactions **123/124**, 3 (1999)
51. R. Rüffer, Reine Kernbraggstreuung am Yttrium-Eisen-Granat—Eine Methode zur Monochromatisierung von Synchrotronstrahlung. Ph.D. Thesis, Universität Hamburg, Hamburg (1985)
52. T. Mitsui, M. Seto, S. Kikuta, N. Hirao, Y. Ohishi, H. Takei, Y. Kobayashi, S. Kitao, S. Higashitaniguchi, R. Masuda, Japan. J. Appl. Phys. **46**(2R), 821 (2007). http://stacks.iop.org/1347-4065/46/i=2R/a=821
53. W. Sturhahn, E. Gerdau, Phys. Rev. B **49**, 9285 (1994). https://doi.org/10.1103/PhysRevB.49.9285
54. G.V. Smirnov, U. van Bürck, A.I. Chumakov, A.Q.R. Baron, R. Rüffer, Phys. Rev. B **55**(9), 5811 (1997)
55. Q.A. Pankhurst, N.S. Cohen, L.F. Barquín, M.R.J. Gibbs, G.V. Smirnov, J. Non-Cryst. Solids **287**(1–3), 81 (2001). https://doi.org/10.1016/S0022-3093(01)00545-2. http://www.sciencedirect.com/science/article/pii/S0022309301005452
56. T. Mitsui, N. Hirao, Y. Ohishi, R. Masuda, Y. Nakamura, H. Enoki, K. Sakaki, M. Seto, J. Sync. Rad. **16**, 723 (2009)
57. V. Potapkin, A.I. Chumakov, G.V. Smirnov, J.P. Celse, R. Rüffer, C. McCammon, L. Dubrovinsky, J. Sync. Radiation **19**(4), 559 (2012). http://dx.doi.org/10.1107/S0909049512015579
58. G.V. Smirnov, A.I. Chumakov, V.B. Potapkin, R. Rüffer, S.L. Popov, Phys. Rev. A **84**, 053851 (2011). https://doi.org/10.1103/PhysRevA.84.053851
59. M. Seto, R. Masuda, S. Higashitaniguchi, S. Kitao, Y. Kobayashi, C. Inaba, T. Mitsui, Y. Yoda, Phys. Rev. Lett. **102**, 217602 (2009). https://doi.org/10.1103/PhysRevLett.102.217602
60. M. Seto, R. Masuda, M. Saito, Synchrotron-radiation-based energy domain Mössbauer spectroscopy, nuclear resonant inelastic-scattering, and quasi-elastic scattering using Mössbauer gamma rays, in This book, Chapter 2
61. J.B. Hastings, D.P. Siddons, U. van Bürck, R. Hollatz, U. Bergmann, Phys. Rev. Lett. **66**, 770 (1991). https://doi.org/10.1103/PhysRevLett.66.770
62. U. van Bürck, D.P. Siddons, J.B. Hastings, U. Bergmann, R. Hollatz, Phys. Rev. B **46**, 6207 (1992). https://doi.org/10.1103/PhysRevB.46.6207
63. Yu.V. Shvyd'ko, A.I. Chumakov, A.Q.R. Baron, E. Gerdau, R. Rüffer, A. Bernhard, J. Metge, Phys. Rev. B **54**, 14942 (1996). https://doi.org/10.1103/PhysRevB.54.14942
64. D.L. Nagy, L. Bottyán, B. Croonenborghs, L. Deák, B. Degroote, J. Dekoster, H.J. Lauter, V. Lauter-Pasyuk, O. Leupold, M. Major, J. Meersschaut, O. Nikonov, A. Petrenko, R. Rüffer, H. Spiering, E. Szilágyi, Phys. Rev. Lett. **88**, 157202 (2002)
65. A.Q.R. Baron, A.I. Chumakov, R. Rüffer, H. Grünsteudel, H.F. Grünsteudel, O. Leupold, Europhys. Lett. **34**(5), 331 (1996). https://doi.org/10.1209/epl/i1996-00460-0
66. I. Sergueev, U. van Bürck, A.I. Chumakov, T. Asthalter, G.V. Smirnov, H. Franz, R. Rüffer, W. Petry, Phys. Rev. B (Condensed Matter and Materials Physics) **73**(2), 024203 (2006). https://doi.org/10.1103/PhysRevB.73.024203. http://link.aps.org/abstract/PRB/v73/e024203

67. G. Schatz, A. Weidinger, *Nuclear Condensed Matter Physics: Nuclear Methods and Applications* (Wiley, New York, 1996)
68. M. Seto, S. Kitao, Y. Kobayashi, R. Haruki, Y. Yoda, T. Mitsui, T. Ishikawa, Phys. Rev. Lett. **91**, 185505 (2003). https://doi.org/10.1103/PhysRevLett.91.185505
69. A.Q.R. Baron, H. Franz, A. Meyer, R. Rüffer, A.I. Chumakov, E. Burkel, W. Petry, Phys. Rev. Lett. **79**, 2823 (1997)
70. D.C. Champeney, Rep. Prog. Phys. **42**, 1017 (1979)
71. R. Masuda, T. Mitsui, S. Kitao, S. Higashitaniguchi, Y. Yoda, M. Seto, Japan. J. Appl. Phys. **47**(10, Part 1), 8087 (2008). https://doi.org/10.1143/JJAP.47.8087
72. M. Seto, Y. Yoda, S. Kikuta, X.W. Zhang, M. Ando, Phys. Rev. Lett. **74**, 3828 (1995)
73. W. Sturhahn, T.S. Toellner, E.E. Alp, X. Zhang, M. Ando, Y. Yoda, S. Kikuta, M. Seto, C.W. Kimball, B. Dabrowski, Phys. Rev. Lett. **74**, 3832 (1995). https://doi.org/10.1103/PhysRevLett.74.3832
74. A.I. Chumakov, R. Rüffer, H. Grünsteudel, H.F. Grünsteudel, G. Grübel, J. Metge, O. Leupold, H.A. Goodwin, EPL (Europhys. Lett.) **30**(7), 427 (1995). http://stacks.iop.org/0295-5075/30/i=7/a=009
75. A.I. Chumakov, A.Q.R. Baron, R. Rüffer, H. Grünsteudel, H.F. Grünsteudel, A. Meyer, Phys. Rev. Lett. **76**, 4258 (1996). https://doi.org/10.1103/PhysRevLett.76.4258
76. H. Thieß, M. Kaisermayr, B. Sepiol, M. Sladecek, R. Rüffer, G. Vogl, Phys. Rev. B **64**, 104305 (2001)
77. R.A. Brand, Private communication
78. B. Fultz, Private communication
79. R. Röhlsberger, T.S. Toellner, W. Sturhahn, K.W. Quast, E.E. Alp, A. Bernhard, E. Burkel, O. Leupold, E. Gerdau, Phys. Rev. Lett. **84**, 1007 (2000). https://doi.org/10.1103/PhysRevLett.84.1007
80. T. Roth, O. Leupold, H.C. Wille, R. Rüffer, K.W. Quast, R. Röhlsberger, E. Burkel, Phys. Rev. B **71**(14), 140401 (2005). https://doi.org/10.1103/PhysRevB.71.140401
81. R. Callens, R. Coussement, C. L'abbé, S. Nasu, K. Vyvey, T. Yamada, Y. Yoda, J. Odeurs, Phys. Rev. B **65**, 180404 (2002). https://doi.org/10.1103/PhysRevB.65.180404
82. D.P. Siddons, J.B. Hastings, U. Bergmann, F. Sette, M. Krisch, Nuclear Instrum. Methods Phys. Res. Sect. B Beam Interactions Mater. Atoms **103**(3), 371 (1995). https://doi.org/10.1016/0168-583X(95)00654-0. http://www.sciencedirect.com/science/article/pii/0168583X95006540
83. T.S. Toellner, E.E. Alp, W. Sturhahn, T.M. Mooney, X. Zhang, M. Ando, Y. Yoda, S. Kikuta, Appl. Phys. Lett. **67**(14), 1993 (1995). https://doi.org/10.1063/1.114764
84. C. L'abbé, R. Coussement, J. Odeurs, E.E. Alp, W. Sturhahn, T.S. Toellner, C. Johnson, Phys. Rev. B **61**, 4181 (2000). https://doi.org/10.1103/PhysRevB.61.4181
85. R. Röhlsberger, E. Gerdau, R. Rüffer, W. Sturhahn, T. Toellner, A. Chumakov, E. Alp, Nuclear Instrum. Methods Phys. Res. A **394**, 251 (1997)
86. H. Grünsteudel, Nuclear resonant scattering of synchrotron radiation on iron containing biomimetic compounds. Ph.D. Thesis, Medizinische Universiät zu Lübeck, Lübeck (1998)
87. O. Leupold, K. Rupprecht, G. Wortmann, Struct. Chem. **14**, 97 (2003)
88. S. Dattagupta, Phys. Rev. B **14**, 1329 (1976)
89. L. Van Hove, Phys. Rev. **95**, 249 (1954). https://doi.org/10.1103/PhysRev.95.249
90. K.S. Singwi, A. Sjölander, Phys. Rev. **120**, 1093 (1960)
91. V.G. Kohn, A.I. Chumakov, R. Rüffer, Phys. Rev. B **58**, 8437 (1998)
92. W. Sturhahn, V.G. Kohn, Hyperfine Interactions **123/124**, 367 (1999)
93. H.J. Lipkin, Hyperfine Interactions **123/124**, 349 (1999)
94. H.J. Lipkin, Phys. Rev. B **52**(14), 10073 (1995)
95. A.I. Chumakov, G. Monaco, A. Fontana, A. Bosak, R.P. Hermann, D. Bessas, B. Wehinger, W.A. Crichton, M. Krisch, R. Rüffer, G. Baldi, G. Carini Jr., G. Carini, G. D'Angelo, E. Gilioli, G. Tripodo, M. Zanatta, B. Winkler, V. Milman, K. Refson, M.T. Dove, N. Dubrovinskaia, L. Dubrovinsky, R. Keding, Y.Z. Yue, Phys. Rev. Lett. **112**, 025502 (2014). https://doi.org/10.1103/PhysRevLett.112.025502

96. T. Ishikawa, Y. Yoda, K. Izumi, C.K. Suzuki, X.W. Zhang, M. Ando, S. Kikuta, Rev. Sci. Instrum. **63**, 1015 (1992)
97. M. Yabashi, K. Tamasaku, S. Kikuta, T. Ishikawa, Rev. Sci. Instrum. **72**(11), 4080 (2001). https://doi.org/10.1063/1.1406925. http://scitation.aip.org/content/aip/journal/rsi/72/11/10.1063/1.1406925
98. E. Burkel, J. Peisl, B. Dorner, EPL (Europhys. Lett.) **3**(8), 957 (1987). http://stacks.iop.org/0295-5075/3/i=8/a=015
99. I. Sergueev, H.C. Wille, R.P. Hermann, D. Bessas, Yu.V. Shvyd'ko, M. Zając, R. Rüffer, J. Sync. Radiation **18**(5), 802 (2011). https://doi.org/10.1107/S090904951102485X
100. A.I. Chumakov, M.V. Zelepukhin, G.V. Smirnov, U. van Bürck, R. Rüffer, R. Hollatz, H.D. Rüter, E. Gerdau, Phys. Rev. B **41**, 9545 (1990). https://doi.org/10.1103/PhysRevB.41.9545
101. T. Mitsui, M. Seto, R. Masuda, Japan. J. Appl. Phys. **46**(10L), L930 (2007). http://stacks.iop.org/1347-4065/46/i=10L/a=L930
102. T. Mitsui, M. Seto, R. Masuda, K. Kiriyama, Y. Kobayashi, Japan. J. Appl. Phys. **46**(7L), L703 (2007). http://stacks.iop.org/1347-4065/46/i=7L/a=L703
103. T. Mitsui, Y. Imai, R. Masuda, M. Seto, K. Mibu, J. Sync. Radiation **22**(2), 427 (2015). https://doi.org/10.1107/S1600577514028306
104. M. Krisch, A. Freund, G. Marot, L. Zhang, Nuclear Instrum. Methods Phys. Res. Sect. A Accelerators. Detect. Assoc. Equip. **305**(1), 208 (1991). https://doi.org/10.1016/0168-9002(91)90536-Y. http://www.sciencedirect.com/science/article/pii/016890029190536Y
105. R. Barrett, R. Baker, P. Cloetens, Y. Dabin, C. Morawe, H. Suhonen, R. Tucoulou, A. Vivo, L. Zhang, Proc. SPIE **8139**, 813904 (2011). https://doi.org/10.1117/12.894735
106. A. Snigirev, V.G. Kohn, I. Snigireva, B. Lengeler, Nature **384**, 49 (1996)
107. T.M. Mooney, E.E. Alp, W.B. Yun, J. Appl. Phys. **71**(11), 5709 (1992). https://doi.org/10.1063/1.350506
108. O. Hignette, G. Rostaing, P. Cloetens, A. Rommeveaux, W. Ludwig, A.K. Freund, Proc. SPIE **4499**, (2001). https://doi.org/10.1117/12.450227
109. S. Kishimoto, Nuclear Instrum. Methods Phys. Res. A **309**, 603 (1991)
110. A.Q.R. Baron, Hyperfine Interactions **125**, 29 (2000)
111. A.I. Chumakov, A.Q.R. Baron, I. Sergueev, C. Strohm, O. Leupold, Yu. Shvyd'ko, G.V. Smirnov, R. Rüffer, Y. Inubushi, M. Yabashi, K. Tono, T. Kudo, T. Ishikawa, Nature Phys. **14**, 261 (2018). https://doi.org/10.1038/s41567-017-0001-z
112. Multiscalers (2019). https://www.fastcomtec.com/products/ufm/
113. S. Kishimoto, T. Mitsui, R. Haruki, Y. Yoda, T. Taniguchi, S. Shimazaki, M. Ikeno, M. Saito, M. Tanaka, Rev. Sci. Instrum. **85**(11), 113102 (2014). https://doi.org/10.1063/1.4900862
114. P. Fajardo, A.Q.R. Baron, H. Dautet, M. Davies, P. Fischer, P. Göttlicher, H. Graafsma, C. Hervé, R. Rüffer, C. Thil, J. Phys. Conf. Ser. **425**(6), 062005 (2013). http://stacks.iop.org/1742-6596/425/i=6/a=062005
115. S. Stankov, R. Rüffer, M. Sladecek, M. Rennhofer, B. Sepiol, G. Vogl, N. Spiridis, T. Ślęzak, J. Korecki, Rev. Sci. Instrum. **79**(4), 045108 (2008). https://doi.org/10.1063/1.2906321. http://link.aip.org/link/?RSI/79/045108/1
116. I. Kupenko, L. Dubrovinsky, N. Dubrovinskaia, C. McCammon, K. Glazyrin, E. Bykova, T.B. Ballaran, R. Sinmyo, A.I. Chumakov, V. Potapkin, A. Kantor, R. Rüffer, M. Hanfland, W. Crichton, M. Merlini, Rev. Sci. Instrum. **83**(12), 124501 (2012). https://doi.org/10.1063/1.4772458
117. C. Strohm, P. Van der Linden, R. Rüffer, Phys. Rev. Lett. **104**, 087601 (2010). https://doi.org/10.1103/PhysRevLett.104.087601
118. K. Muffler, J.A. Wolny, H.P. Hersleth, K.K. Andersson, K. Achterhold, R. Rüffer, V. Schünemann, J. Phys. Conf. Ser. **217**(1), 012004 (2010). http://stacks.iop.org/1742-6596/217/i=1/a=012004
119. ESRF Upgrade Programme Phase II (2015-2019)—White Paper; ESRF Upgrade Programme Phase II (2015–2022)—Technical Design Study (2013, 2014). www.esrf.eu
120. I. Sergueev, L. Dubrovinsky, M. Ekholm, O.Y. Vekilova, A.I. Chumakov, M. Zając, V. Potapkin, I. Kantor, S. Bornemann, H. Ebert, S.I. Simak, I.A. Abrikosov, R. Rüffer, Phys. Rev. Lett. **111**, 157601 (2013). https://doi.org/10.1103/PhysRevLett.111.157601

121. L. Dubrovinsky, N. Dubrovinskaia, V.B. Prakapenka, A.M. Abakumov, Nature Commun. **3**, 1163 (2012)
122. L. Dubrovinsky, N. Dubrovinskaia, E. Bykova, M. Bykov, V. Prakapenka, C. Prescher, K. Glazyrin, H.P. Liermann, M. Hanfland, M. Ekholm, Q. Feng, L.V. Pourovskii, M.I. Katsnelson, J.M. Wills, I.A. Abrikosov, Nature **525**, 226 (2015)
123. R. Röhlsberger, H. Thomas, K. Schlage, E. Burkel, O. Leupold, R. Rüffer, Phys. Rev. Lett. **89**, 237201 (2002)
124. T.P. Almeida, R. Temple, J. Massey, K. Fallon, D. McGrouther, T. Moore, C.H. Marrows, S. McVitie, Sci. Rep. **7**, 17835 (2017)
125. W. Meissner, R. Ochsenfeld, Naturwissenschaften **21**, 787 (1933). https://doi.org/10.1007/BF01504252
126. I. Troyan, A. Gavriliuk, R. Rüffer, A. Chumakov, A. Mironovich, I. Lyubutin, D. Perekalin, A.P. Drozdov, M.I. Eremets, Science **351**, 1260 (2016)
127. E. Altshuler, T.H. Johansen, Rev. Modern Phys. **76**, 471 (2004)
128. R. Blukis, R. Rüffer, A.I. Chumakov, R.J. Harrison, Meteoritics Planet. Sci. **52**, 925 (2017)
129. J.F.J. Bryson, C.I.O. Nichols, J. Herrero-Albillos, F. Kronast, T. Kasama, H. Alimadad, G. van der Laan, F. Nimmo, R.J. Harrison, Nature **472**, 517 (2015)
130. F. Nestola, V. Cerantola, S. Milani, C. Anzolini, C. McCammon, D. Novella, I. Kupenko, A. Chumakov, R. Rüffer, J.W. Harris, Lithos **265**, 328 (2016)
131. Yu. Shvyd'ko, Phys. Rev. A **91**, 053817 (2015)
132. R. Sinmyo, K. Glazyrin, C. McCammon, I. Kupenko, A. Kantor, V. Potapkin, A.I. Chumakov, R. Rüffer, L. Dubrovinsky, Phys. Earth Planet. Interiors **229**, 16 (2014). https://doi.org/10.1016/j.pepi.2013.12.002. http://www.sciencedirect.com/science/article/pii/S0031920113001921
133. L. Berthier, Physics **4**, 42 (2011)
134. H. Tanaka, T. Kawasaki, H. Shintani, K. Watanabe, Nature Mater. **9**, 324 (2010)
135. U. van Bürck, R.L. Mössbauer, E. Gerdau, R. Rüffer, R. Hollatz, G.V. Smirnov, J.P. Hannon, Phys. Rev. Lett. **59**, 355 (1987)

Chapter 2
Synchrotron-Radiation-Based Energy-Domain Mössbauer Spectroscopy, Nuclear Resonant Inelastic Scattering, and Quasielastic Scattering Using Mössbauer Gamma Rays

Makoto Seto, Ryo Masuda, and Makina Saito

Abstract Nuclear resonant scattering spectroscopy using synchrotron radiation (SR) has been applied to a wide variety of scientific applications. An excellent feature of this method is that element (isotope)-specific information on the electronic and phonon states can be obtained using the energy selectivity of SR. The use of high-brilliance SR as an excitation source for Mössbauer spectroscopy allows imaging measurement under extreme conditions, such as high pressures, very high or low temperatures, and strong external magnetic fields. Additionally, diffusion and fluctuation of atoms can be observed by taking advantage of the ultranarrow width of the nuclear excited states. We introduced the concepts of the methods with an emphasis on these excellent features. Furthermore, the unique features involved in the measurements are highlighted and discussed.

List of Abbreviations and Symbols

SR	Synchrotron radiation
NRIS	Nuclear resonant inelastic scattering
NIS	Nuclear inelastic scattering
NRIXS	Nuclear resonant inelastic X-ray scattering
NRVS	Nuclear resonant vibrational spectroscopy
PDOS	Phonon density of states
DFT	Density functional theory
RSMR	Rayleigh scattering of Mössbauer radiation
TDI	Time-domain interferometry
APD	Avalanche photo diode
NFS	Nuclear (resonant) forward scattering

M. Seto (✉) · R. Masuda · M. Saito
Institute for Integrated Radiation and Nuclear Science, Kyoto University, Kumatori-Cho, Sennan-Gun, Osaka 590-0494, Japan
e-mail: seto@rri.kyoto-u.ac.jp

Y. Yoshida and G. Langouche (eds.), *Modern Mössbauer Spectroscopy*,
Topics in Applied Physics 137, https://doi.org/10.1007/978-981-15-9422-9_2

SMS Synchrotron Mössbauer source
RI Radioisotope
FWHM Full width at half maximum
DAC Diamond anvil cell
QEGS Quasielastic γ-ray scattering
RC Radiative coupling

2.1 Introduction

The use of nuclear probes is advantageous for the study of condensed matter because it enables measurement that would be difficult with electron systems; it gives isotope-specific information for complex compounds. It is possible to attain atomic position resolution if nuclear probes are implanted at specific positions with atomic resolution such as a special isotope monolayer in an artificial multilayer film made by molecular beam epitaxy method. The hyperfine interactions between nuclei and electrons are useful for information on the electronic states. Mössbauer spectroscopy, which uses the recoilless nuclear resonant absorption effect known as the "Mössbauer effect," is a famous and effective method for the electronic state measurement using the hyperfine interactions [1]. Mössbauer spectroscopy is a powerful and well-established method used in a wide variety of research areas, such as materials, chemical, biological, earth, and fundamental physical sciences. The recoilless fraction gives lattice dynamics information in addition to electronic state information. Since nuclear probes are element (isotope)-specific, measurement with very dilute probes is sometimes possible. Besides, they may even attain atomic layer resolution. Moreover, states of the specific atoms of interest in complex compounds can be elucidated. Another outstanding feature to be addressed is the narrow width of a nuclear excited state compared with the resonant excitation energy (e.g., the width of the first nuclear excited state of ^{57}Fe is 4.66 neV, while the resonant excitation energy is 14.4 keV). This feature can be used for the slow dynamics study of resonant atoms in macromolecules, viscous liquids, etc., by observing the broadening of Mössbauer spectra. With this feature, we can use ultranarrow-width γ-rays as spectroscopic probes for the slow dynamics study of the sample containing no resonant atoms. This method is known as "Rayleigh scattering of Mössbauer radiation (RSMR)" [2].

Radioisotope (RI) sources are used in performing Mössbauer spectroscopy, and the preparation of an RI source for each nuclide to be studied is required. In particular, the measurement laboratory should be placed near to a reactor or an accelerator that generates the RI sources with short lives. Mössbauer effect measurement without preparing the RI sources was achieved using energy-tunable SR as a nuclear excitation source [3]. This achievement made Mössbauer measurement easily accessible for nuclides other than the limited ones (such as ^{57}Fe, ^{119}Sn, and ^{151}Eu) with long-lifetime parent RI sources. However, the photon energy range depends on the storage ring and insertion devices installed (the maximum energy with enough intensity is usually limited to below 10^5 eV). The energy selectivity of SR allows the measurement of ^{40}K Mössbauer spectra, of which the observation is impossible using ordinary radioactive sources because the first excited state of ^{40}K is not populated by any radioactive parent nuclides [4]. Additionally, a prominent advantage of Mössbauer spectroscopy using the energy-tunable SR is the measurement of element (isotope)-specific phonon energy spectra, which is difficult with RI sources [5]. Furthermore, the combination of electronic and phonon-state measurements generates site-specific phonon energy spectra [6]. Moreover, high-brilliance SR enables imaging measurement, high-pressure measurement, the measurement of tiny samples, etc. In particular, this is advantageous for RSMR measurement, which can measure angle-dependent energy transfer and, therefore, requires small angular divergence of the probe γ-rays.

As shown, the use of nuclear excitation process and SR has realized unique and effective spectroscopic methods applicable to vast research areas. Note that nuclear resonant scattering spectroscopy using SR has two aspects. One is that it enables the element (isotope)-specific measurement of electronic states and dynamics with advanced features. The other is the ultranarrow width of the emitted γ-rays. In Sect. 2.2, as an element (isotope)-specific method for electronic state (and slow dynamics) measurement, we discuss synchrotron-radiation-based Mössbauer spectroscopy that yields absorption-type Mössbauer spectra [7]. This method enables advanced measurement using the excellent features of SR, such as Mössbauer spectroscopic measurement under high pressures. Furthermore, this method gives absorption-type spectra similar to those obtained by Mössbauer spectroscopy with RI sources. Since this method was developed in 2009 and the efficiency was much improved recently [8], we explain the details of this method precisely based on some recent results obtained using this method, including the comparison of the other similar methods. In Sect. 2.3, we discuss the nuclear resonant inelastic scattering (NRIS) spectroscopy that gives element (isotope)- and site-specific phonon densities of states. This method is sometimes called "nuclear inelastic scattering (NIS)," "nuclear resonant inelastic X-ray scattering (NRIXS)," and "nuclear resonant vibrational spectroscopy (NRVS)," depending on the scientific field. This method has been very actively applied to research areas, such as condensed matter physics, earth sciences, and biosciences. Recently, the local structure of active sites in enzymes has been studied, and these results are introduced. Moreover, site-specific phonon measurement is explained in addition to element (isotope)-specific phonon measurement. In Sect. 2.4, RSMR using SR is introduced, which uses the ultranarrow width of the emitted γ-rays with high directivity. As discussed, ultranarrow-width γ-rays are produced with this method. It means this method uses only a small part (approximately 10^{-9} eV) out of the much wider width (approximately in the electron volt range) of the SR. Even though the small angular divergence of SR is efficient for the angle-dependent measurement compared with RI sources that emit γ-rays in all directions, much improvement in the efficacy is possible. We recently achieved improvement by developing multiline methods and obtained new results on slow dynamics. The newly developed method is precisely explained, and recent results obtained using this method are introduced.

2.2 Synchrotron Radiation-Based Mössbauer Spectroscopy

SR-based Mössbauer absorption spectroscopy is a method to obtain the Mössbauer energy absorption spectra of various nuclides. As you can see, for instance, in many other chapters in this book, Mössbauer spectroscopy is mostly performed with ^{57}Fe as its isotope probe. However, Mössbauer spectroscopy is also possible with many isotopes other than ^{57}Fe. Mössbauer experiments with these isotopes have been performed using RI even before the development of methods using SR. This situation can be shown in the periodic table of Mössbauer elements, as shown in Fig. 2.1. This table and further information are also shown in the Mössbauer Effect Data Center website [9]. Nevertheless, Mössbauer experiments with isotopes other than ^{57}Fe and ^{119}Sn are by far fewer than those with the two isotopes. One major difficulty is in the preparation of parent nuclides. The γ-ray sources for ^{57}Fe and ^{119}Sn can be purchased, but not for the other nuclides. Hence, nuclear reactors or particle accelerators are required. Some isotopes, for example ^{40}K, have no appropriate parent RI; Mössbauer spectroscopy using this isotope has been performed using the "in-beam" method, where the Mössbauer experimental instruments (often including the cryostat for cooling) have been combined with particle beam irradiation system [10]. This problem is fairly solved if SR is applied as the source. SR-based Mössbauer spectroscopy is one of such methods. In this section, its property, measurement system, and analysis method are described.

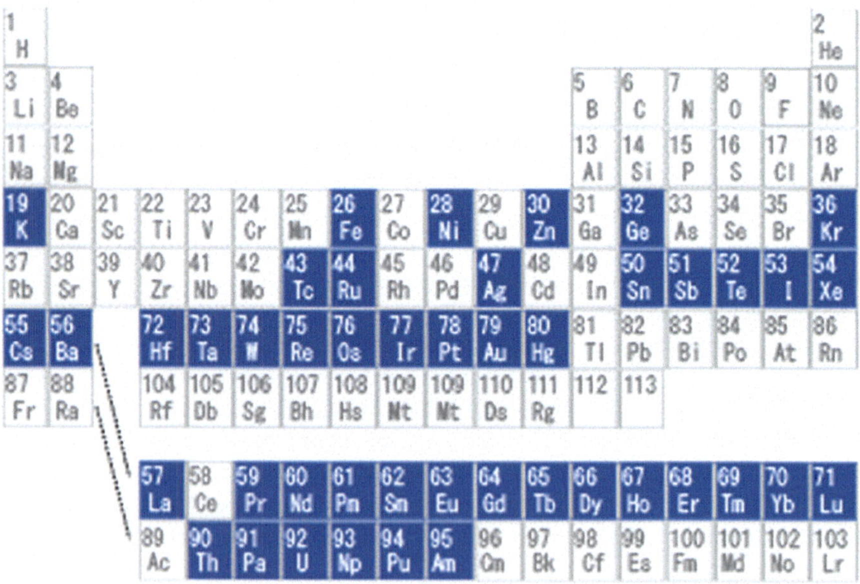

Fig. 2.1 Periodic table of elements in which Mössbauer effect has been observed (reproduced from Seto (2013) [11]

2.2.1 Special Features

This method was originally proposed in 1974 [12] and thus is the oldest method of Mössbauer spectroscopy using SR. However, it was developed in 2009 [7] after the development of other methods, such as the synchrotron Mössbauer source (SMS) [13] and nuclear forward scattering (NFS) [14]. The comparison of these methods is mentioned in a later section. Here, we focus on the properties of SR-based Mössbauer spectroscopy. First, as already described, this method is suitable for Mössbauer measurement using various isotopes, especially those with a relatively short half-life of a few nanoseconds. As shown in Fig. 2.2, many isotopes show their half-life in this range. Although iron and tin are present in various materials, new diverse frontier materials are also composed of many elements. Hence, Mössbauer spectroscopy with these isotopes plays a critical role in elucidating the electronic states of their composition elements. For example, isotope selectivity has yielded unique information on samples in monoatomic position resolution in ^{57}Fe Mössbauer spectroscopy [15]. This property has become more important to understand the mechanism of the novel function of frontier materials if it is applied to various resonant isotopes. Until now, SR-based Mössbauer spectroscopy of ^{40}K [17], ^{61}Ni [18], ^{73}Ge [7], ^{119}Sn [19], ^{125}Te [20], ^{127}I, ^{149}Sm [21], ^{151}Eu [22], ^{174}Yb [8], and ^{189}Os [23] has already been performed. This multi-isotope property is due to the whiteness of SR in the energy domain and the measurement mechanism of this method, described in the following section. Second, this method is suitable for tiny sample amounts, often seen in the frontier materials, such as nanoparticles and thin films, and materials under extreme conditions, such as low temperature, high pressure, and gas environment for an in situ experiment, which is due to the high brilliance of SR. Third,

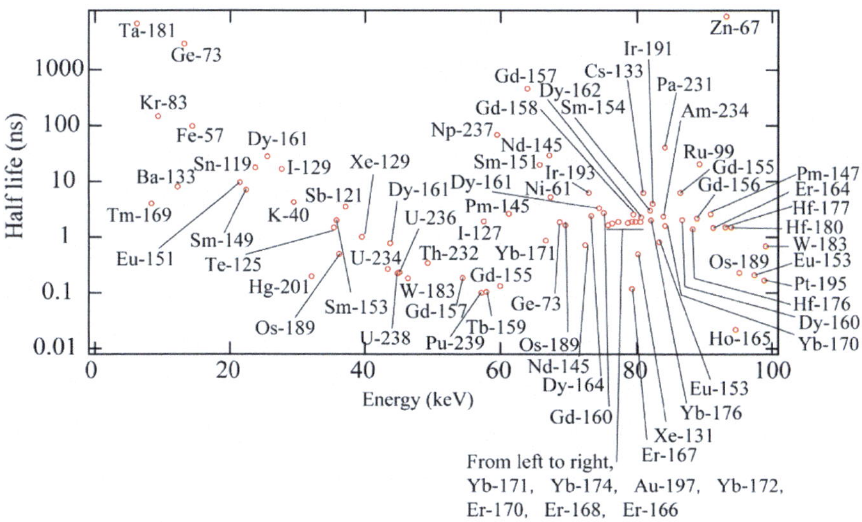

Fig. 2.2 Energy and half-life of potential Mössbauer isotopes

this method can be combined with X-ray optical techniques because of the very low angular divergence of SR and its highly polarized property. For example, the total reflection technique, which occurs in the condition of typically some milliradians, can be effectively combined and applied to Mössbauer spectroscopy of thin films.

2.2.2 *Instrumentation*

In principle, the components other than SR are as follows: monochromator(s), transmitter, scatterer, a velocity-controlling instrument (Mössbauer transducer), and a timing detector. A schematic diagram is shown in Fig. 2.3. First, SR from a beamline undulator passes through a pair monochromator Si crystal. The normal beamline monochromator usually determines the energy of SR with the bandwidth of some electron volts. This energy width is much wider than the hyperfine structure to analyze in the transmitter or scatterer in the figure. The energy modulation by the hyperfine structure is typically below some 10^{-6} eV. Thus, the energy profile of SR from the monochromator is virtually white, as shown in Fig. 2.3b. After the monochromator, the SR penetrates the transmitter in Fig. 2.3a. A sample under study or an energy reference substance at SR-based Mössbauer spectra is positioned as the transmitter. In fact, one of these two should be arranged as the transmitter, and the other should be as the scatterer. The difference between them is described later in this section.

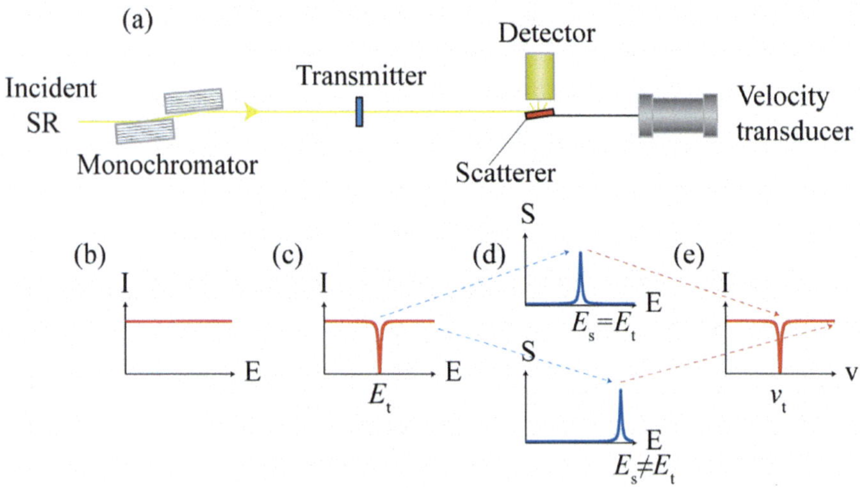

Fig. 2.3 A schematic diagram of the instrumentation and mechanism of SR-based Mössbauer spectroscopy. **a** The arrangement of instruments. **b** The energy profile of SR from the monochromator; **c** The profile of SR behind the transmitter, which has a resonant energy E; **d** The profile of the scattering from the scatterer, which has a resonant energy E_S controlled by the velocity transducer in the case of $E_S = E_t$ in the upper panel and $E_S \neq E_t$ in the lower panel; **e** The detected intensity dependence on the scatterer's velocity. At the velocity v_t, $E_S = E_t$ is satisfied

After the transmitter, SR has an energy profile reflecting the hyperfine structure of the transmitter. This situation is shown in Fig. 2.3c. Subsequently, SR is incident on the scatterer, and the scattering from the scatterer is detected by the detector, arranged just above (or below) the scatterer to cover a large solid angle from the scatterer. Here, either the transmitter or the scatterer is connected with a velocity transducer to control the nuclear resonant energy by the Doppler effect of light. Figure 2.3a shows the case in which the transducer is connected with the scatterer, and Fig. 2.3d shows the energy profile of the scattering from the scatterer, in which the resonant energy is controlled by the velocity transducer. The scatterer is usually in the shape of a plate or film and inclined to minimize the self-absorption of the scattering. At the detector, both high time response of the sub-nanosecond order and high dynamic range are required to distinguish the delayed nuclear resonant scattering from the prompt electronic scattering by setting an appropriate time window. This is because the incoherent nuclear resonant scattering was emitted with some delay corresponding to the lifetime of the excited state of the probe nuclide. Thus, we usually use an avalanche photodiode (APD) detector [24] (See 1.6.2 in this book).

A simple description of the instrumentation was presented in the previous text. In advanced cases, we have some additional components and some desirable conditions. As for monochromator, we can arrange other sets of monochromator(s) to obtain higher energy resolution. A resolution of some milli electron volts is possible by a Si crystal monochromator using higher index. Such a high-resolution monochromator drastically improves the efficiency in this method because it yields the deeper absorption spectra by suppressing non-Mössbauer nuclear resonant absorption, that is, nuclear resonant absorption with recoil. Furthermore, it also alleviates the damage to the sample. The high-resolution monochromators are available for X-rays with the energy below typically 40 keV. However, the reflectivity of such a high-resolution monochromator becomes low in higher-energy range, especially over 60 keV. Thus, we cannot always use such a monochromator. As for the transmitter and scatterer, there are some conditions. One condition is that they should not undesirably vibrate. In the measurement system, the precise control of the relative velocity between the transmitter and scatterer corresponds to the energy scan, and such undesirable vibration spoils the control. Nevertheless, the actual measurement system often includes the sources of vibration, such as vacuum pumps, refrigerators, and air compressors. Vibration from them increases the apparent energy width of the spectra and reduces their absorption depth because the spectra obtained is the convolution of the vibrational profile and the real Mössbauer spectra. When the hyperfine structure to be measured is sufficiently larger than the effect of vibration, the former effect on width might not be quite problematic, but the latter on the absorption depth is still very problematic. Therefore, the vibration should be sufficiently less than the bandwidth of nuclear levels in the velocity expression: $v_\Gamma = c\Gamma/E_{res}$. Here, c denotes the speed of light, Γ denotes the natural width of the nuclear level, and E_{res} denotes the nuclear resonant energy of the nuclide. Next, we discuss the arrangement of the sample under study. It is usually more convenient to arrange the sample at the transmitter because the limitation of the transmitter in addition to the vibration control is milder than that of the scatterer. The important limitation of the transmitter is that SR penetrates

Table 2.1 List of standard references for some isotopes

Nuclides	Energy reference	Nuclide	Energy reference
^{40}K	KCl	^{125}Te	Mg_3TeO_6
^{61}Ni	$Ni_{0.86}V_{0.14}$	^{149}Sm	Sm_2O_3 or SmB_6
^{73}Ge	Li_2GeO_3 or GeO_2	^{151}Eu	EuF_3
^{119}Sn	$CaSnO_3$	^{174}Yb	YbB_{12}

it. This limitation is quite mild because the incident SR is hard X-rays. Consequently, we can arrange any environmental cells, such as a low-temperature cryostat, high-pressure cell, and gas chamber. Conversely, the limitation of the scatterer is a little complicated. The scattering from the scatterer should be efficiently detected by the detector. This limits the shape of the scatterer, as described. Nevertheless, thin film samples are sometimes good as the scatterer. As for the energy reference substance, it should be a chemical compound showing no hyperfine splitting in their resonant nuclear levels. Table 2.1 shows the chemical compounds used as suitable energy references. Some of them show small hyperfine splitting, which is not as large as the natural line width of the nuclear resonance and is not a big problem in actual use. In addition, when we perform the SR-based Mössbauer spectroscopy with isotopes whose resonant energy is high, typically above 40 keV, both the transmitter and scatterer should be cooled down to obtain a reasonable recoilless fraction. For example, the recoilless fraction of pure metal is quite different between the 14.4-keV nuclear resonance of ^{57}Fe and the 67.4-keV nuclear resonance of ^{61}Ni, although they occupy a similar position in the periodic table of elements. The recoilless fraction of nickel metal in ^{61}Ni Mössbauer spectroscopy is miserably low at room temperature, although that of iron metal in ^{57}Fe Mössbauer spectroscopy is still high at that temperature, as shown in Fig. 2.4. However, it is sometimes difficult to support both velocity control and low temperature simultaneously. One solution is to use a cryostat, where helium gas is used as a heat transfer medium. The similar effective thickness of the transmitter and scatterer is another preferable condition to obtain the spectra

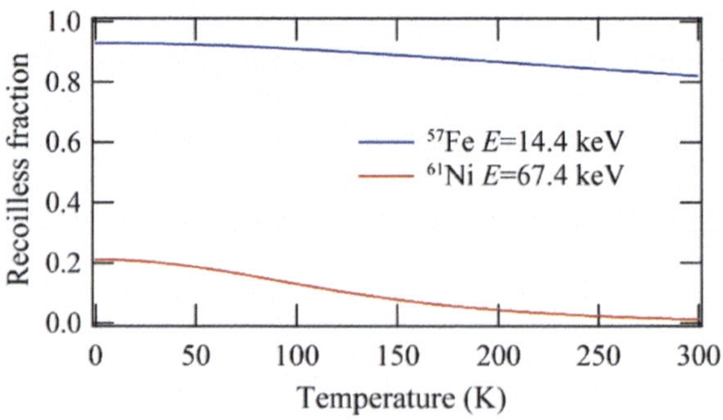

Fig. 2.4 Recoilless fraction of pure metal in ^{57}Fe and ^{61}Ni Mössbauer spectroscopy, calculated based on the Debye vibrational model. See Eq. (1.5) in Chap. 1

with reasonable absorption depth. If the scatterer is too thick, the absorption depth becomes shallow. Another condition concerns the detector. The detector usually detects scattered γ-rays, whose energy is the same as the nuclear resonant energy, and fluorescent X-rays following the internal conversion process, whose energy is lower than that of the γ-rays, as delayed nuclear resonant scattering. The detection of fluorescent X-rays becomes important in SR-based Mössbauer spectroscopy using high-resonant-energy isotopes. This is because the detection efficiency of APD becomes very low when detecting high-energy X-rays. For example, the energy of γ-rays at ^{61}Ni Mössbauer spectroscopy is 67.4 keV, while that of fluorescent Kα X-rays of Ni atom is 7.5 keV. The detection efficiency of APD, whose depletion layer is 150 μm, is 93% for the 7.5-keV X-rays and 0.9% for the 67.4-keV γ-rays. In the latest measurement system, the scatterer and detector are packed in a vacuum chamber to detect internal conversion electrons [8]. The internal conversion coefficients of many Mössbauer nuclides are higher than 1. Furthermore, the detection efficiency of the APD detector is usually 100% against electrons with energies below 100 keV. Therefore, the detection rate of the detector becomes many times higher with the electron detection in Mössbauer spectroscopy using those nuclides. For example, when a sample under an extreme condition is studied by SR-based Mössbauer spectroscopy using high-resonant-energy isotopes, the sample is often arranged in a large environmental cell and cannot be moved by a velocity transducer. Additionally, this cell blocks the internal conversion electrons. Even in that case, arranging the sample as the transmitter and the energy reference substance as the scatterer should support the following three conditions: velocity control, low temperature, and vacuum condition for the electron detection. In those cases, heat transfer by thin copper wires or springs satisfies the conditions to some extent. However, the efficiency of heat transfer by this method is usually inferior to that by the helium exchange gas technique.

2.2.3 *Analysis of the Spectra*

The spectra obtained are the velocity-dependent intensities and are thus very similar to the conventional Mössbauer spectra using RI. One example is shown in Fig. 2.5. A clear difference is observed between the spectra, and a simple Lorentzian evaluation was enough to see the transition in valence. In many simple cases, this kind of Lorentzian evaluation is sufficient. For this kind of rough evaluation, programs for analyzing the conventional Mössbauer spectra using RI can be applied.

However, the exact line shape of SR-based Mössbauer spectra is not Lorentzian. A deviation between the Lorentzian evaluation lines and experimental data is observed around the shoulders of absorption profiles in Fig. 2.5. Furthermore, you can see a zigzag shape at the background in the experimental data. They do not correspond to the minor component in the sample and are characteristics of SR-based Mössbauer spectra, caused by the measurement method described above. They depend on some factors, such as the effective thickness of the transmitter and the scatterer, chemical composition, and time window at the APD detector. Now, we consider the detailed

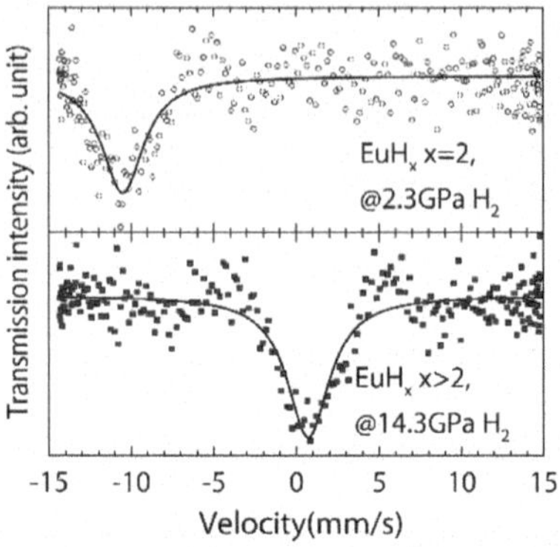

Fig. 2.5 ^{151}Eu SR-based Mössbauer spectra of Eu metal under high *hydrogen* pressure. The energy reference substance was EuF_3 as the scatterer. Rectangles represent experimental data, and lines represent fitting curves by Lorentzian shape. The upper panel shows the Eu^{2+} valence under 2.3 GPa hydrogen, and the lower panel shows Eu^{3+} valence under 14.3 GPa hydrogen (reproduced from Matusoka et al. (2011) [25])

expression on these characteristics [26]. For simplicity, we assume the following conditions: (1) the transmitter is a sample under study and includes only one chemical compound with one site and (2) the scatterer is an energy reference substance and connected to the velocity transducer, as shown in Fig. 2.3. Our aim here is to express the detected intensity as a function of the velocity of the scatterer. First, the propagating coherent field amplitude of the transmitter E_{t} as a function of the dimensionless frequency of w, that is, in the unit of natural linewidth, is considered. It corresponds to the absorption by electrons and nuclei in the transmitter and is expressed as follows:

$$E_{\mathrm{t}}(w) = E_{0\mathrm{t}}\exp\left(-\frac{\mu_{\mathrm{et}}z_{\mathrm{t}}}{2}\right)\exp\left(-i\sum_{m}\alpha_{\mathrm{t}m}\frac{\mu_{\mathrm{nt}}z_{\mathrm{t}}}{2(2(w-w_{\mathrm{t}m})+i)}\right). \tag{2.1}$$

Here, the amplitude of the radiation field at the entrance of the transmitter is denoted as $E_{0\mathrm{t}}$. The electronic absorption coefficient of the transmitter is denoted as μ_{et}. The thickness of the transmitter is denoted as z_{t}. The linear absorption coefficient of radiation by the nuclei at resonance in the transmitter is denoted as μ_{nt}. The index for a nuclear transition is denoted as m. The probability and the nuclear resonant energy of the mth transition at the transmitter are denoted as $\alpha_{\mathrm{t}m}$ and $w_{\mathrm{t}m}$, respectively. The hyperfine structure of the transmitter, which is described by isomer shift, quadrupole splitting, and magnetic hyperfine field, is expressed as $w_{\mathrm{t}m}$. The purpose of Mössbauer spectroscopy is to evaluate them. Here, $\mu_{\mathrm{nt}}z_{\mathrm{t}}$ corresponds to Mössbauer effective thickness T_{t}. Next, the propagating coherent field amplitude of the scatterer E_{s} as a function w is considered. It also corresponds to the absorption by electrons and nuclei in the scatterer and is expressed as follows:

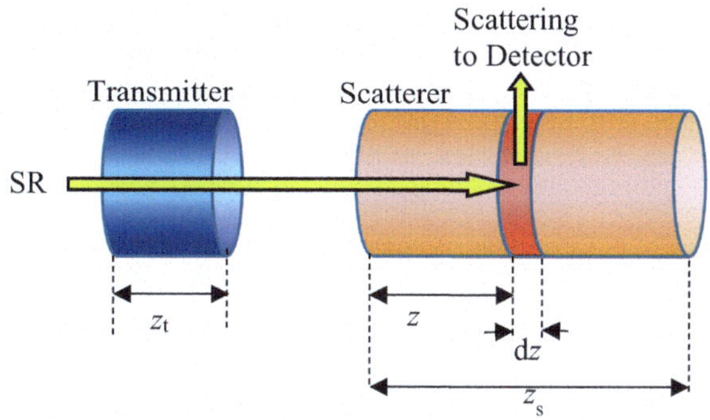

Fig. 2.6 A schematic diagram of parameters used for the detailed expression of SR-based Mössbauer spectra

$$E_s(w, w_s, z) = E_{0s}\exp\left(-\frac{\mu_{es}z}{2}\right)\exp\left(-i\frac{\mu_{ns}z}{2(2(w-w_s)+i)}\right). \tag{2.2}$$

Here, w_s denotes the nuclear resonant energy of the scatterer, and z denotes the depth to which SR penetrates as indicated in Fig. 2.6. The other variables denote similarly to those of the transmitter, although the suffix "t" should be changed to "s," which denotes the scatterer. Equation (2.2) is expressed as a function of w, w_s, and z because w_s is controlled by the velocity transducer and w and z are used for the calculation in the following equations. This field was absorbed by the scatterer at the depth z, as described in the following step, and thus, z is less than the thickness of the scatterer z_s. Because the energy reference substance is assumed at the scatterer, hyperfine splitting is absent, and the summation of the nuclear transition index m is not described. Subsequently, we consider the scattering process at the scatterer. Here, it is enough to consider the w_s-dependent processes, which are as follows: (1) the emission following the recoilless nuclear resonant absorption of radiation and (2) the scattering of the radiation due to the photoelectric absorption. Here, we call the former process as channel A and the latter as channel C according to [26]. In channel A, SR penetrates the transmitter and scatterer until the depth z and is then resonantly absorbed. In this channel, the scattering field $E_A(w, w_s, z)$ satisfies

$$E_A(w, w_s, z) \propto \frac{1}{w - w_s + i/2}E_t(w)E_s(w, w_s, z).$$

Now, we can obtain the detected intensity $I_A(w_s)$ as follows:

$$\begin{aligned} I_A(w_s) &= C_A\int_{\tau_1}^{\tau_2}d\tau\int_0^{z_s}dz\left|\int\frac{dw}{2\pi}\exp(-iw\tau)E_A(w, w_s, z)\right|^2 \\ &= C_A\int_{\tau_1}^{\tau_2}d\tau\int_0^{z_s}dz\left|\int\frac{dw}{2\pi}\frac{\exp(-iw\tau)}{w-w_s+i/2}E_t(w)E_s(w, w_s, z)\right|^2, \end{aligned} \tag{2.3}$$

where C_A denotes a proportionality factor. This expression includes (1) the Fourier transformation of $E_A(w, w_s, z)$ to obtain the time dependence of E_A and the square of its magnitude to obtain the intensity from the field and (2) the integrations on the depth z and τ; the τ integration corresponds to the time window at APD. In channel C, NFS by the transmitter or scatterer is scattered by electrons at depth z. Thus, the incident field at the depth satisfies

$$E_C(w, w_s, z) \propto 1 - E_t(w)E_s(w, w_s, z).$$

Considering cross section of electrons independent of w_s, the detected intensity $I_C(w_s)$ is obtained as

$$\begin{aligned} I_C(w_s) &= C_C \int_{\tau_1}^{\tau_2} d\tau \int_0^{z_s} dz \left| \int \frac{dw}{2\pi} \exp(-iw\tau) E_C(w, w_s, z) \right|^2 \\ &= C_C \int_{\tau_1}^{\tau_2} d\tau \int_0^{z_s} dz \left| \int \frac{dw}{2\pi} \exp(-iw\tau)(1 - E_t(w)E_s(w, w_s, z)) \right|^2, \end{aligned} \tag{2.4}$$

where C_C denotes another proportionality factor. Now, we have the expression of the detected intensity $I(w_s)$ as

$$I(w_s) = I_A(w_s) + I_C(w_s) + I_B, \tag{2.5}$$

where I_B denotes for other processes independent of w_s, such as the process with nuclear resonant scattering with recoil at the scatterer (called channel B).

The narrowing of the energy width and the wavy pattern in the background are now discussed based on Eqs. (2.1)–(2.5). The narrowing effect depends on the time window, as shown in Fig. 2.7. Here, a narrower energy width was obtained compared to the ideal linewidth in conventional Mössbauer spectroscopy, which is twice the natural linewidth. For example, although the ideal linewidth is 2.0 mm/s in conventional ^{174}Yb Mössbauer spectroscopy, a full-width at half maximum (FWHM) of 1.3 mm/s was achieved [8]. This property is advantageous when the nuclear hyperfine structure is small and precise analysis is required. In contrast, the wavy pattern in the background is usually a drawback because it may conceal small components in the spectra, at least in the initial guess for analysis. Proper analysis using the equations is required. Note that if the time window is $[0, \infty]$, both the narrowing effects and the wavy background vanish. In addition to the $[0, \infty]$ time window, in the case of thin transmitter and scatterer, the spectra show Lorentzian shape with the "ideal linewidth," similar to the conventional Mössbauer spectroscopy using RI. Therefore, the Lorentzian approximation described at the beginning of this section corresponds to the case. These situations are similar to the delayed coincidence Mössbauer spectroscopy using RI [27, 28]. Furthermore, the equations also show better conditions for the measurement system, described previously. For example, the high-resolution monochromator enhances the absorption depth of the spectra because it suppresses the channel B process. Furthermore, the effective thickness of the transmitter and that of scatterer (i.e., the sample and energy reference) should be similar.

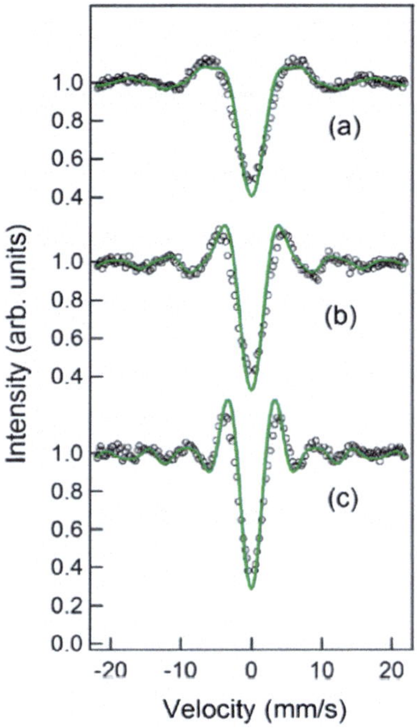

Fig. 2.7 SR-based Mössbauer absorption spectra of ^{151}Eu with the following time windows: **a** 5.7–17.0 ns, **b** 8.1–17.0 ns, and **c** 10.5–17.0 ns. The transmitter and scatterer were EuF_3. Lines are the spectra calculated by Eqs. (2.1)–(2.5) (reproduced from Seto et al. (2010)) [22]. Note that the lifetime $\tau_{151Eu} = 14$ ns

To calculate the above equations, parameters on the isotopes and elements are necessary. Fortunately, we have many databases of them. One good source is the Mössbauer Effect Data Center [9]. For nuclear data, the Table of Isotopes [29] and the National Nuclear Data Center website [30] are also excellent sources. Electronic absorption coefficients are shown in the web database by Sasaki [31] and the National Institute of Standards and Technology (NIST) website [32] based on [33, 34].

We should also consider the type of scattering that the APD detects because the absorption rate of γ-rays, internal conversion electrons, and fluorescent X-rays following the internal conversion processes are different. Particularly, the stopping power of electrons is much higher than those of the other two. Even 100-keV electrons cannot penetrate an aluminum plate of 70 μm. The penetration length can be calculated according to [35] and seen in the web database by NIST [36] based on [37].

Another subject in the analysis of SR-based Mössbauer spectra of various isotopes is the interpretation of the hyperfine structure parameters: the sets of isomer shift, quadrupole splitting, and magnetic hyperfine field. Although many chemical compounds have been studied using ^{57}Fe and ^{119}Sn Mössbauer spectroscopy, few compounds have been studied using other isotopes. Even so, the preceding studies, including experiments by NFS and conventional Mössbauer spectroscopy will help the analysis. A substantial description on various isotopes by Greenwood and Gibb [2], Shenoy and Wagner [38], and Gütlich et al. [39] can be very useful.

2.2.4 Comparison with Other Methods Using SR

Here, the properties of the SR-based Mössbauer spectroscopy are compared with those of the SMS and NFS, the two major methods to observe the hyperfine structure of samples using SR. These methods are in a complementary relation. The differences in these methods described here will help in selecting the best method for a study.

As for the SMS, energy-domain spectra are obtained similarly to SR-based Mössbauer spectroscopy. One large difference between the two is the nuclide availability. In the SMS, a unique component produces Mössbauer radiation, that is, SR whose energy width is as narrow as the natural linewidth of a Mössbauer isotope. This component is a nuclear monochromator at which the nuclear diffraction is allowed, although the usual diffraction by electrons is prohibited [40]. Nuclear monochromator is successfully developed only for ^{57}Fe, and thus, the SMS is limited to ^{57}Fe Mössbauer spectroscopy, which is different from SR-based Mössbauer spectroscopy, in which many isotopes can be used. Furthermore, the nuclear monochromator produces Mössbauer radiation without a time window, and thus, the SMS method has no limit on the time structure on SR in principle. SR-based Mössbauer spectroscopy requires at least some kind of pulsed time structure of SR with the time period for the time window by which the delayed nuclear resonant scattering is selected. Although this difference is only a technical point, it is important to the actual experiment because various bunch modes exist in SR facilities. Neither width-narrowing effect nor the wavy background is seen in the SMS, and thus, the analysis of the SMS spectra is usually more intuitive.

As for the NFS, the hyperfine structure of various isotopes can be observed by both NFS and SR-based Mössbauer spectroscopy. A large difference is noticeable in the appearance of the spectra. The time spectra showing beat patterns due to the interference of nuclear hyperfine splitting at the sample are obtained in NFS. Therefore, NFS is suitable in tracing a reaction depending on the time in the scale of nanoseconds, such as pump-probe spectroscopy. For stable samples, this difference in appearance affects the analyzing model at an initial guess: the energy spectra of SR-based Mössbauer spectroscopy are intuitive. When there are two or more components in the sample, this intuition becomes important. In the experimental viewpoint, NFS requires an appropriate time structure of SR to obtain the time spectra. This limitation is stricter in NFS than in SR-based Mössbauer spectroscopy. The required period between SR pulses in NFS depends on the magnitude of hyperfine splitting to be evaluated. However, the magnitude is not usually obvious. In actual experiments, the period with more than the lifetime of the nuclear excited state is usually enough (the longer the period, the better). This limitation sometimes restricts isotope availability. When we have the suitable timing bunch mode of SR facility, the measuring time tends to be shorter in NFS because the interference pattern of all nuclear levels by hyperfine splitting is measured in the time region without any energy-scanning mechanism. In contrast, the energy dependence of the scattering intensity is measured with the mechanical motion of the energy reference substance in SR-based Mössbauer spectroscopy. The efficiency of the measurement system in NFS is somewhat small in experiments with high-resonant-energy isotopes owing to the detection efficiency of the APD. Only high-energy γ-rays from the sample should be detected in NFS, while low-energy fluorescent X-rays and electrons after the internal conversion processes from the scatterer could be also detected in SR-based Mössbauer spectroscopy. Additionally, it is sometimes difficult to obtain the interference pattern in the time spectra of NFS when the lifetime of the nuclear excited level of the probe isotope is short. In those cases, SR-based Mössbauer spectroscopy is preferable. Thus, SR-based Mössbauer spectroscopy suits Mössbauer spectroscopy using nuclear levels, whose energy is high and whose lifetime is short.

2.3 Nuclear Resonant Inelastic Scattering

While Mössbauer spectroscopy uses the recoilless nuclear resonant excitation effect, the energy tunability of SR enables the measurement of the recoil part. Therefore, nuclear resonant excitation accompanied by phonon excitation can be measured using SR. Hence, the first experiment was performed in 1994 [5], and phonon measurement was conducted [41, 42]. The NRIS method has distinct features favorable for studies concerning the microscopic dynamics (e.g., phonons and molecular vibration) of materials because it provides the element (isotope)-specific dynamics due to the resonant excitation of the specific isotope. In solids, partial phonon densities of states are measured. Furthermore, measurements under extreme conditions, such as high pressures, small samples, and thin films, are possible because of the high brilliance of SR. (For the definition of brilliance, see Chap. 1.) Note that the element (isotope)-specific nature enables the measurement of very dilute resonant atoms in complicated materials. Recently, this method has been actively used to study the local structure of active sites in enzymes along with first-principles density functional theory (DFT).

2.3.1 Instrumentation and Analysis of the Basic Method of NRIS

The measurement is performed through inelastic processes with photons. The energy of each photon is equal to the sum of the energy of the nuclear resonant excited state (based on the ground state) and a phonon (or phonons). Since the typical energy of phonon is 1–100 meV, spectroscopic resolution required for the measurement is approximately several mill electron volts or less. Usually, SR produced by an undulator installed in a recent hard X-ray storage ring is approximately several electron volts, and it is possible to generate X-rays with a bandwidth of approximately several milli electron volts (in special cases, approximately 0.1 meV can be achieved [43]) using Si monochromators. Sapphire (α-Al_2O_3), a material for the monochromator, in addition to Si, has been used particularly for high-energy nuclides [44, 45]. The incident energy can be scanned at approximately the nuclear resonant excitation energy by changing the Bragg angle of the monochromator. Note that the respective nuclear resonant energies of the available isotopes are much more than the usual scan range (100 meV). Therefore, pure element (isotope)-specific measurement is assured.

To observe the phonon energy spectrum, we irradiate a sample containing the resonance isotope as a function of the energy of the generated X-rays with a bandwidth of approximately several milli electron volts. From the irradiated sample, a strong scattering is emitted due to photoelectron effects and so on. Therefore, to discern the relatively weak nuclear scattering from the strong scattering, we observe only delayed photons (e.g., fluorescence X-rays and γ-rays) and electrons (e.g., conversion electrons) emitted at the de-excitation in the time domain because scattering due to electronic processes is promptly emitted at the irradiation. The schematics of the measurement of nuclear resonant excitation accompanied by phonon creation or annihilation are shown in Fig. 2.8. As shown in the figure, if the energy of the incident photon is equal to the sum of the energy of the nuclear resonant excited state and a phonon, one phonon is created. Alternatively, one-phonon annihilation occurs if the sum of the energy of the incident photon and that of a phonon is equal to the sum of the energy of the nuclear resonant excited state. The detection system with a closed-cycle refrigerator cryostat for NRIS measurement using an eight-element APD detector is shown in Fig. 2.9.

We can obtain the phonon density of states (PDOS) weighted by the projection of the polarization vectors on the direction of the incident X-ray radiation from the NRIS measurement. Based on [46], the following expression is relevant:

$$g(E, \boldsymbol{\kappa}) = \frac{V}{(2\pi)^3} \sum_j \int d\mathbf{q} \left|\boldsymbol{\kappa} \cdot \mathbf{e}_j(\mathbf{q})\right|^2 \delta\big(E - \hbar\omega_j(\mathbf{q})\big), \tag{2.6}$$

where V is the volume of the unit cell, $\mathbf{q}$ is the phonon wave vector, $\omega_j(\mathbf{q})$ is the phonon dispersion relation for the branch j, $\boldsymbol{\kappa}$ is the normalized wave vector of the incident X-ray ($\boldsymbol{\kappa} = \mathbf{k}/|\mathbf{k}|$, $\mathbf{k}$: wave vector of incident X-ray), and $\mathbf{e}_j(\mathbf{q})$ is the polarization vector of the vibrations of the resonant atom. For single crystals with a cubic Bravais lattice and polycrystalline materials composed of resonant atoms only, $g(E, \boldsymbol{\kappa})$ is the exact PDOS. In the general case of a polycrystalline material, that is, a material composed of not only resonant atoms but also other nonresonant atoms, averaging over all directions of the incident radiation results in $g(E)$, representing a PDOS weighted by the square amplitude of the resonant atoms. Therefore, we can obtain a partial PDOS for a specific element (isotope). The nuclear resonant inelastic absorption cross section can be expressed using the following weighted PDOS:

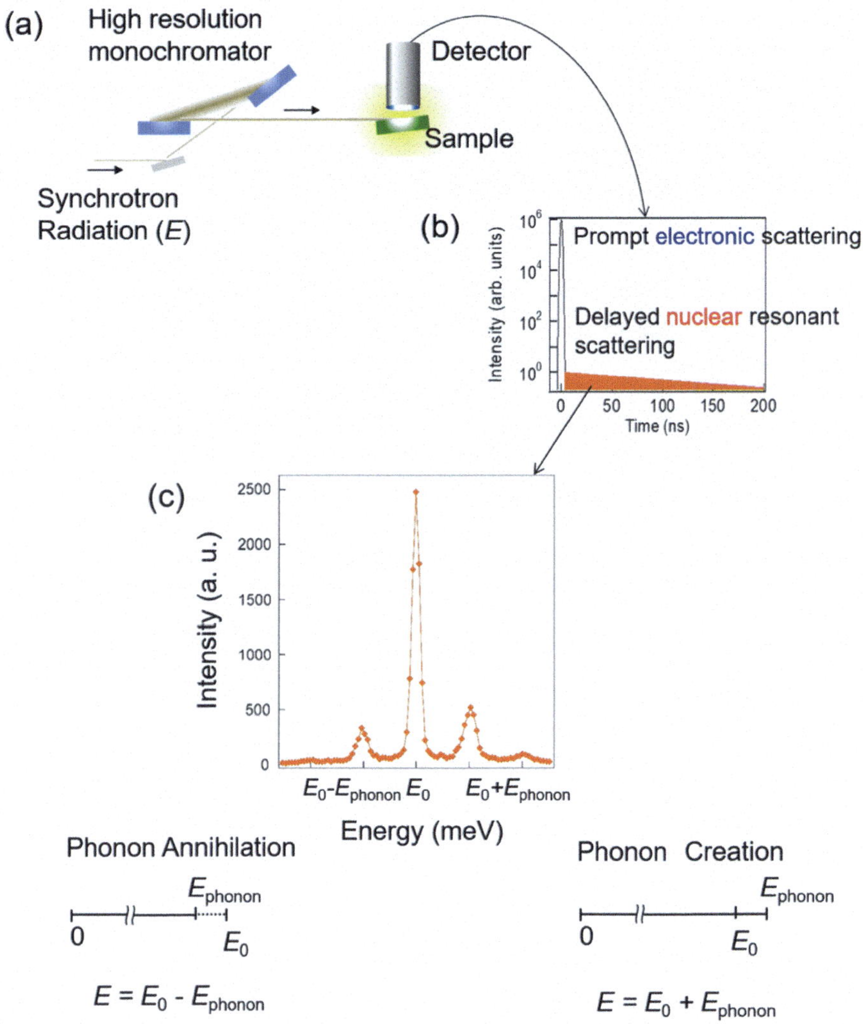

Fig. 2.8 **a** Schematic diagram for the measurement of nuclear resonant inelastic scattering. **b** Nuclear resonant scattering is discerned from prompt electronic scattering in the time domain as delayed scattering because of the finite lifetimes. **c** As a function of the incident photon energy E, the phonon energy spectrum can be obtained by counting the intensity of delayed scattering. In the phonon energy spectrum, the left-side scattering is due to the excitation accompanied by phonon annihilation [$E = E_0$ (nuclear resonant excitation energy) - E_{phonon} (phonon energy)], while the right-side scattering is due to excitation accompanied by phonon creation [$E = E_0 + E_{\text{phonon}}$]

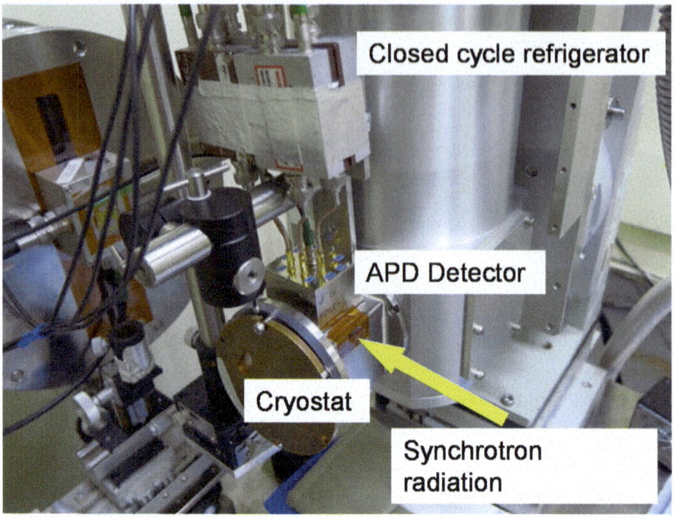

Fig. 2.9 Photograph of the detection system for nuclear resonant inelastic scattering measurement using an eight-element avalanche photodiode (APD) detector. The sample in a cryostat chamber can be cooled with the closed cycle refrigerator

$$\sigma_a(E) = f_{LM} \int \frac{d\tau}{2\pi} \exp(-\mathrm{i}E\tau)\left[\exp(M(\mathbf{k},\tau)) - 1\right], \tag{2.7}$$

where

$$M(\mathbf{k},\tau) = \int dE \ \exp(\mathrm{i}E\tau)S(E,\mathbf{k}), \tag{2.8}$$

$$S(E,\mathbf{k}) = E_R \ g(E,\boldsymbol{\kappa})\frac{\overline{n}+1}{E}, \tag{2.9}$$

$$f_{LM} = \exp(-M(\mathbf{k},0)), \tag{2.10}$$

$$E_R = \frac{(\hbar\mathbf{k})^2}{2M}, \tag{2.11}$$

$$\bar{n} = (\exp(E/k_\mathrm{B}T) - 1)^{-1}. \tag{2.12}$$

In the above expressions, M is the mass of the resonant nucleus and k_B is the Boltzmann constant. Using Fourier transformation of the cross section corresponding to the measured spectrum, $g(E, \boldsymbol{\kappa})$ can be obtained through $M(\mathbf{k},\tau)$ and $S(E,\mathbf{k})$, as shown in Eqs. (2.7) and (2.8). However, before performing these procedures, the subtraction of the elastic peak and the deconvolution of the monochromator widths are required. Since the perfect deconvolution is impossible for the real measured spectrum containing statistical and other errors, the obtained $g(E, \boldsymbol{\kappa})$ sometimes contains the apparent broadening due to the monochromator width. Moreover, special care should be given to this behavior at low-energy region within the monochromator width due to the subtraction. In Fig. 2.10, the obtained PDOS from measured NRIS spectrum is shown using these processes. These procedures are precisely described in [47, 48]. The area ratio of the elastic peak to the total part in the NRIS spectrum as shown in the upper spectrum of Fig. 2.10 may be considered as the recoilless fraction, which is known as the Lamb–Mössbauer factor in Mössbauer spectroscopy. However, it may be incorrect because coherent forward scattering occurs at the exact nuclear resonant energy, while such coherent scattering does not, besides the resonant energy, as discussed in Chap. 1. Although the recoilless fractions are sometimes required to compare the Mössbauer results, PDOS has much better information and gives the recoilless fraction value [47].

In eq. (2.12), $\bar{n}$ indicates Bose factor ruling the temperature dependence of the spectrum. Since phonon is boson, phonon creation is proportional to $\bar{n} + 1$, which is shown in eq. (2.12), while phonon annihilation is proportional to $\bar{n}$. As $\bar{n}$ approaches

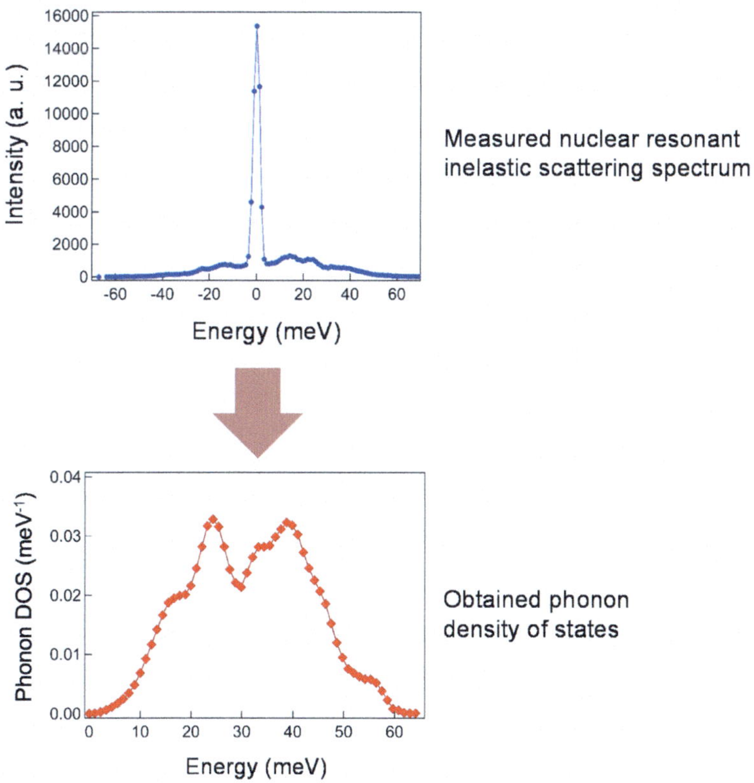

Fig. 2.10 Phonon density of states shown in the spectrum below is obtained from the measured nuclear resonant inelastic scattering spectrum shown above [49]

to zero by lowering the temperature to zero, phonon annihilation is difficult to observe at low temperature as expected. In contrast, phonon creation is possible even at zero temperature as seen in eq. (2.9). Therefore, the asymmetry of the phonon energy spectrum is observed at low temperatures. The NRIS spectra of ^{57}Fe in superconductor $LaFeAsO_{0.89}F_{0.11}$ at 298 K and 15 K are shown in Fig. 2.11 [50]. Note that this method does not provide phonon dispersion relations, which can be obtained from other relevant methods, such as inelastic neutron and X-ray scattering methods, but does provide measurements of polycrystalline, disordered, and amorphous materials. Furthermore, an ideal partial PDOS averaged over phonon momenta can be obtained by observing the emissions due to the nonradiative channel of nuclear de-excitation.

This element-specific phonon information is important because the dynamics of certain atoms in a compound sometimes influence the characteristics of the compound. Therefore, the NRIS of SR that provides element (isotope)-specific phonon energy spectra is advantageous. Moreover, in this method, it is possible to study the dynamics of highly diluted impurities or doped atoms in metals and semiconductors. In Fig. 2.12a, the local PDOS of Fe (0.017 at.%) in Al metal measured by NRIS of SR is shown [51], and the PDOS of Al metal, obtained from a neutron inelastic scattering experiment [52], is also shown in Fig. 2.12c for comparison. A clear difference between the PDOS of ^{57}Fe in Al and that of the Al metal host can

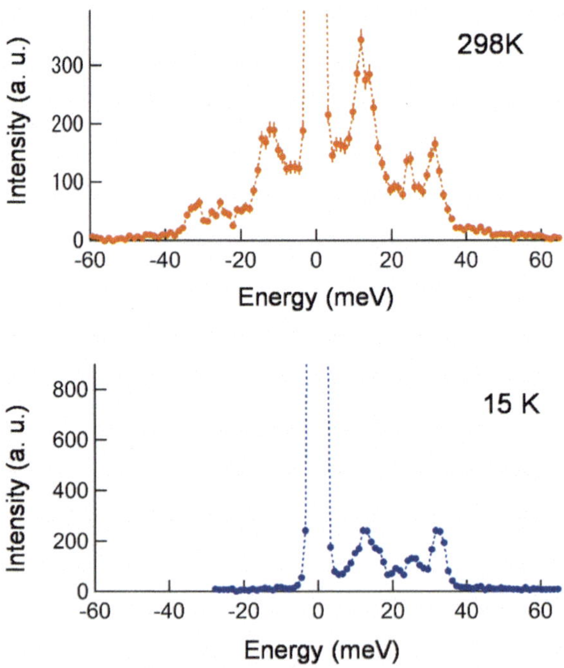

Fig. 2.11 Asymmetry of the phonon energy spectrum is observed at low temperature as shown in the nuclear resonant inelastic scattering spectra of ^{57}Fe in superconductor $LaFeAsO_{0.89}F_{0.11}$ at 298 K and 15 K [50]

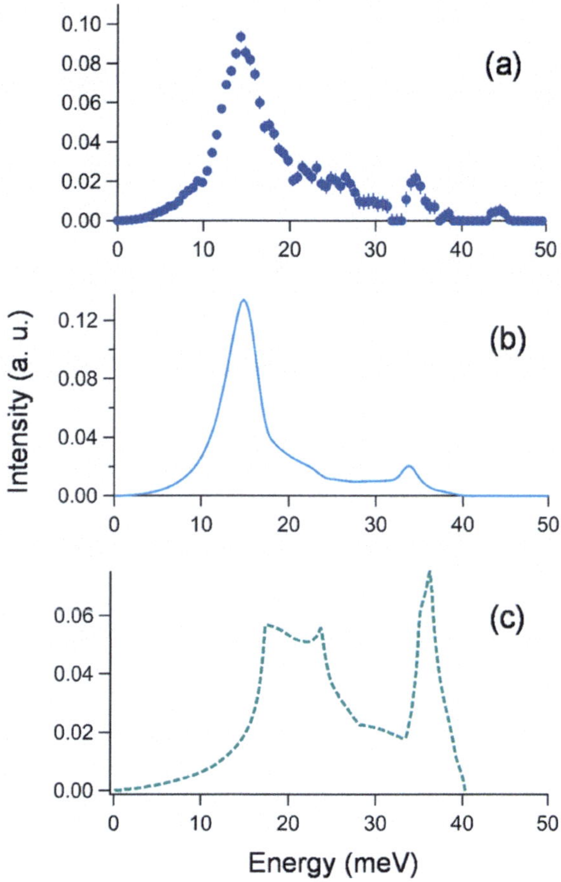

Fig. 2.12 **a** Phonon density of states (PDOS) of ^{57}Fe in Al-0.017 at.% Fe [51]. **b** Response function of the impurity Fe atom to the PDOS of Al metal calculated on the basis of Mannheim's impurity theory [53]. **c** PDOS of Al [52]

be seen by comparing Fig. 2.12a and c. The reliability of the observed spectrum was confirmed using a Green function method that can reproduce the local PDOS from that of the host metal and the coupling between the impurity and host atoms [53]. The phonon spectrum calculated using this method is shown in Fig. 2.12b. It is shown that the calculation well reproduces the measured PDOS of Fe (Fig. 2.12a). The characteristic vibrational modes of Fe in Al were found to be modes that vibrate resonantly with the host Al phonons. Although the measurement of the PDOS of diluted atoms is quite difficult, this method gives clear and reliable results.

2.3.2 Examples of Frontier Science, Especially Biological Application

Element (isotope) selectivity is very advantageous when the function and local structure of a specific atom in a complex compound are of interest. Resonant Raman spectroscopy provides valuable information on the vibrational properties of samples, such as metalloprotein samples. Furthermore, it has been used effectively. However, the selection rule sometimes hampers the observation of vibrational modes. In contrast, NRIS method gives all modes that involve motion of the resonant nuclei. Therefore, it provides the opportunity to study the ligation and oxidation state of a specific site. Many studies have been conducted to determine the vibrational amplitudes and frequencies for specific atoms in a complex biological macromolecule without selection rules using the isotope selectivity (for example [54–76]). In particular, this method has been used to study specific sites in large molecules, such as metalloprotein samples, involving thousands of other atoms. The investigation of the structure around the specific site of interest is conducted by comparing the obtained phonon energy spectrum with the vibrational spectrum obtained from DFT calculations under the assumption of the presumable structural model. This method is effective for samples that cannot be crystallized, such as the intermediates in the catalytic cycles of enzymes. Furthermore, the structural characterization of the reactive Fe(IV) = O intermediate in the catalytic cycles of a mononuclear non-heme iron (NHFe) enzyme (the halogenase SyrB2 from the bacterium *Pseudomonas syringae* pv. *syringae*) was studied (Fig. 2.13) [72]. The intermediate reacts through an initial hydrogen-atom abstraction step and performs subsequent halogenation of the native substrate or hydroxylation of nonnative substrates, Therefore, the revelation of its local structure and mechanism is quite essential and important. In this study, it was indicated that the orientation of the Fe(IV) = O intermediate depends on the substrate, presenting specific frontier molecular orbitals responsible for hydrogen-atom abstraction that can selectively lead to halogenation or hydroxylation. Moreover, NRVS was used in studying the catalytic mechanism of hydrogenases, which catalyze the reversible conversion of molecular hydrogen to protons and electrons. Understanding the mechanism is quite significant because it leads to the development of clean energy sources in producing hydrogen. NRVS was applied to [FeFe] hydrogenase variant lacking the amine proton shuttle, which is stabilizing a putative hydride state [75]. [FeFe] hydrogenases are metalloenzymes that reversibly reduce protons to molecular hydrogen with extremely high efficiency. NRVS spectra clearly showed the bending modes of the terminal Fe–H species that is consistent with the widely accepted models of the catalytic cycle.

The high brilliance of SR allows the PDOS even for small samples to be measured. For example, measurements under extreme conditions in which the accessible sample space is severely limited are possible. The PDOS under high pressures using a diamond anvil cell (DAC), where the sample size is typically less than 1 mm^2, is measured to study the core of the earth [77–79]. Additionally, many important studies using the features of NRIS spectroscopy have been conducted, for example,

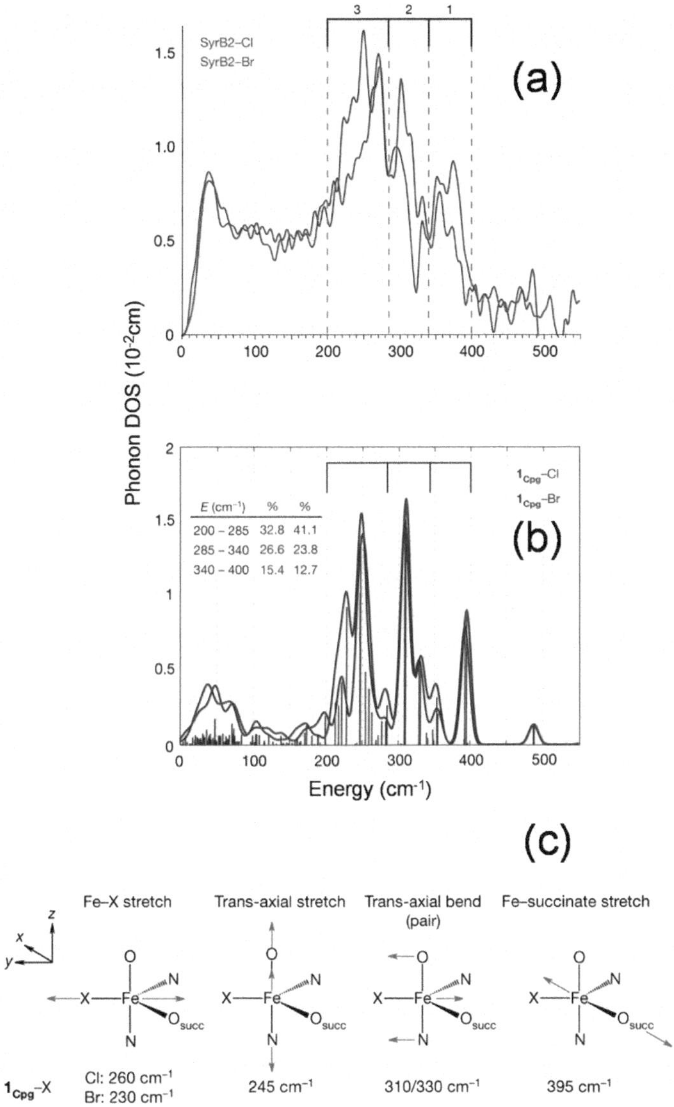

Fig. 2.13 **a** Phonon densities of states of SyrB2–Cl and SyrB2–Br. **b** DFT-predicted the phonon densities of states. **c** DFT-predicted four normal modes of the five-coordinate Fe(IV) = O structure [72]

on nanoparticles [80–85], thin films [86–92], quasicrystals [93], clathrates [94, 95], superconductors [50, 96–98], filled skutterudites [99, 100], and glass [101, 102]. The nuclear resonant scattering method can be applied to not only solids but also liquids. The element-specific diffusion constant of Fe ions in HCl solution [103] and the dynamics of Fe ions in Nafion membranes [104, 105] were measured using the aforementioned method, which are used as ion-exchange membranes.

This method gives element (isotope)-specific phonons but there are compounds containing two or more different atomic states of a particular element; for example, magnetite, which is a mixed-valence Fe compound, is well-known [106]. Although this method cannot distinguish the "site-specific" vibrational properties, information on the difference of the sites is sometimes required because the properties of the individual atomic motion in nonequivalent positions in a compound are not necessarily equivalent; in fact, mixed valence systems, such as iron oxides with a phase transition accompanied by charge splits [107], and spin crossover materials [108], are well known. Furthermore, even in an ideal material containing only one atomic site, there may be atoms with different environments due to imperfections or impurities. Materials with different sites occupied by the same element are not unusual. A method capable of distinguishing the vibrational properties of a specific site is valuable and crucial.

2.3.3 Advanced NRIS Method

Thus far, the measurement of electronic and phonon states has been studied independently, except for the information on the recoilless fraction of the Mössbauer effect. The recoilless fraction sometimes gives important insights for the lattice dynamics study. However, PDOS, which allows us to calculate the recoilless fraction as discussed above, has quite rich information. The combination of the measurement methods of phonon energy spectra and incoherent time spectra through the hyperfine interactions permits the observation of the site-specific PDOS [6]. Since the observed NRIS consists of scattering from individual excited nuclei, the PDOS obtained from the scattering spectrum is the sum of the partial PDOS of individual atoms. Therefore, a partial PDOS of each atom is obtained from the measured NRIS spectrum by discerning the contribution of each atom. The hyperfine interactions between the nuclei and the surrounding electronic states, which split the nuclear energy level, allow this procedure. In that case, quantum beats whose oscillating cycles reflect the splitting energies are observed in the time-domain measurement of incoherently emitted γ-rays during the decay of an excited nucleus. The incoherently emitted γ-rays obey an exponential law with a finite lifetime. In the case of multiple electronic states of a certain element in a compound, the hyperfine interactions for those states are different, and the beat patterns in the time spectrum are expected to be different. The measurement of the quantum beat patterns enables the identification of the respective electronic states and gives the component ratios of the sites. If the nuclear resonant excitation accompanied by phonon excitation occurs at a certain incident radiation energy, the analysis of the quantum beat pattern in the time spectrum reveals the respective phonon contributions from those sites with different electronic states. Performing this procedure at different energies allows us to observe the partial PDOS as distinguished by electronic states.

Using this method, the site-specific PDOS in magnetite (Fe_3O_4) was measured. In magnetite, which is a well-known mixed-valence compound, the iron atoms are located in two nonequivalent positions in the unit cell. One-third of the Fe ions (Fe^{3+}) occupy the *A* sites and are tetrahedrally surrounded by four oxygen ions. The remaining two-thirds of the Fe ions (Fe^{3+} and Fe^{2+}) occupy the *B* sites and are octahedrally surrounded by six oxygen ions. Magnetite is ferrimagnetic, and the magnetic moments of the *A* sites are aligned antiparallel to the magnetic moments of the *B* sites below T_N=858 K. Nuclear magnetic resonance spectroscopy and Mössbauer spectroscopy cannot distinguish the Fe^{3+} or Fe^{2+} ions on the *B* sites, indicating the delocalized nature of the charge carriers with a formal average valence of $Fe^{2.5+}$ at room temperature [106]. The sample measured in this experiment was prepared such that oxidization is prevented and was doped with 7 mol% Ni to replace Fe because some B sites are probably oxidized to Fe^{3+} in air [109]. The enrichment of ^{57}Fe in the used sample was 95.5%. The energy spectrum of NRIS of ^{57}Fe in Fe_3O_4 is shown in Fig. 2.14a. Examples of the incoherent time spectra (measured at 16 meV and 35 meV) are shown in Fig. 2.14b [6]. From these time spectra, the ratio of A site and B site at each phonon energy was obtained by least-squares fitting each time spectrum with two exponential functions accompanied by sinusoidal quantum beats corresponding to A and B sites. These ratios and the PDOS of all Fe atoms give the partial PDOSs of A site and B site. The obtained spectra of the partial PDOS for all Fe sites and the site-specific PDOSs of the A and B sites are shown in Fig. 2.15a. The difference between the partial PDOS s of the A and B sites is clearly observed. In Fig. 2.15b, the PDOSs of Fe in the ideal Fe_3O_4 obtained from ab initio band calculations are shown. The PDOS of all Fe is shown as a black line, and the calculated PDOSs of states of the A and B sites are shown as dashed green and dashed-and-dotted dark yellow lines, respectively. The overall character of the calculated PDOS of all Fe sites in Fe_3O_4 agrees well with the PDOS of all Fe sites measured by NRIS spectroscopy. Neutron inelastic scattering, X-ray inelastic scattering, Raman scattering, and infrared absorption methods are well-known and very useful for the study of atomic dynamics. It is, however, generally impossible to discern the atomic motions of the same element in different environments. Therefore, this method is considered to be unique.

2.3.4 Summary

NRIS spectroscopy is a method used for investigating the vibrational states in substances. Since NRIS spectroscopy uses the nuclear resonant excitation accompanied by phonon creation and annihilation, unique and very effective measurement is possible as shown above. The development of spectroscopic methods is ongoing, and remarkable progress has been achieved in relation to the optics, detectors, and methodologies used. Because this spectroscopy covers a broad range of scientific areas, such as physical, chemical, biological, and earth sciences, recent developments and further improvements of nuclear resonant scattering spectroscopy that will solve

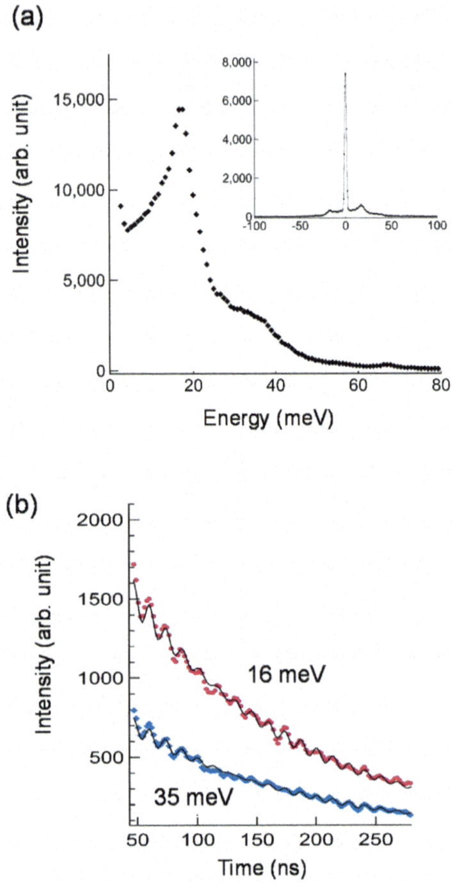

Fig. 2.14 **a** Phonon energy spectrum of nuclear resonant inelastic scattering of ^{57}Fe atoms in Fe_3O_4. Inset is the same spectrum with the elastic peak. **b** Time spectra measured at different incident photon energies referenced from nuclear resonant excitation energy (red circles, 16 meV; blue diamonds, 35 meV). Lines are least-squares fitted spectra with two exponential functions accompanied by sinusoidal quantum beats [6]

problems encountered in electron systems will result in spectroscopy becoming a powerful technique for the study in these areas.

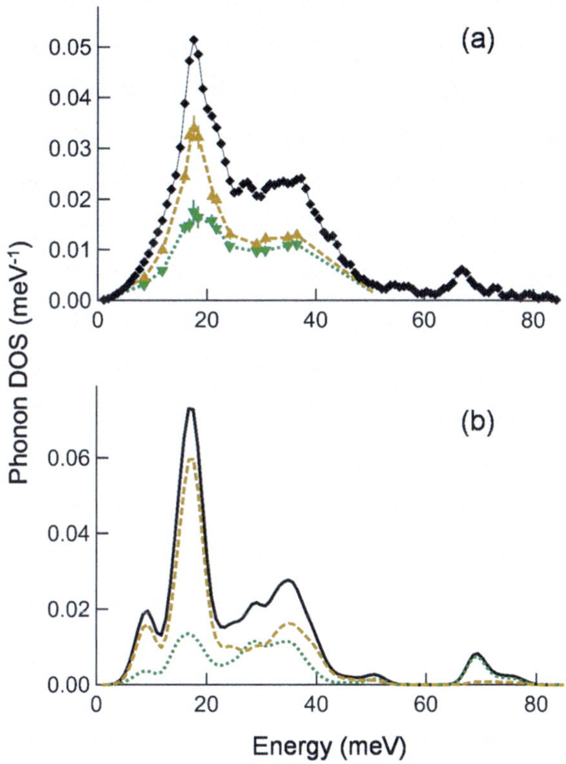

Fig. 2.15 Phonon densities of states (PDOSs) of ^{57}Fe in Fe_3O_4. **a** The PDOSs of Fe are shown by closed black diamonds, and the partial PDOSs of the A and B sites are shown as downward green and upward dark yellow triangles, respectively. Lines are inserted to guide the eye. **b** PDOSs of Fe in the ideal Fe_3O_4 obtained by ab initio calculations. The PDOSs of all Fe are shown as a black line. The calculated partial PDOSs of the A and B sites are shown as dashed green and dashed-and-dotted dark yellow lines, respectively [6]

2.4 Quasielastic Scattering Using Mössbauer γ-Rays

2.4.1 Introduction

Timescale and spatial scale of microscopic density fluctuations in condensed matters are decided by the so-called quasielastic scattering measurements. In Fig. 2.16, we show time and length regions of the fluctuations that can be studied by various quasielastic scattering techniques. In this chapter, we introduce the quasielastic scattering technique using Mössbauer γ-rays as a probe beam. Most of the techniques based on the Mössbauer effect are used to study samples containing nuclear resonant species. In contrast, this technique is used to decide timescale and length scale of the electron density fluctuations in samples that do not contain nuclear resonant species

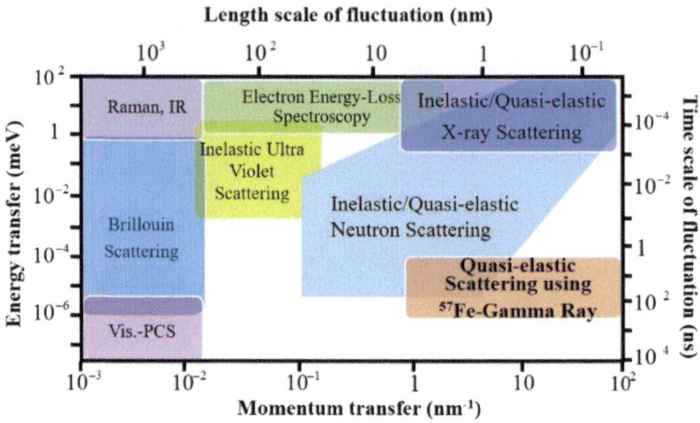

Fig. 2.16 Time and length regions of fluctuations covered by various quasielastic scattering techniques. Quasielastic scattering technique using 14.4-keV γ-rays from ^{57}Fe nuclei covers a unique timescale and length scale

because the Mössbauer effect is used to generate the γ-rays with very high energy resolution and the γ-rays are used for the probe of nonresonant quasielastic scattering study. In Fig. 2.16, we show established time and length regions of the fluctuations that can be studied by quasielastic scattering with 14.4-keV γ-rays from ^{57}Fe nuclei.

This section is composed of the following subsections: In Sect. 2.4.2, basic concepts of quasielastic scattering by nonresonant samples are introduced. In Sect. 2.4.3, the conventional time-domain measurement technique of quasielastic scattering using time-domain interferometry (TDI) with single-line Mössbauer γ-rays is discussed. In Sect. 2.4.4, a finite energy width of incident SR is considered, and the effect on the time spectrum of single-line γ-ray quasielastic scattering is introduced. In Sect. 2.4.5, quasielastic scattering using TDI with multiline Mössbauer γ-rays is described, and its advantage is summarized. In Sect. 2.4.6, results using γ-ray quasielastic scattering are presented. In Sect. 2.4.7, summary and perspective of γ-ray quasielastic scattering are discussed.

2.4.2 Basic Concept of Quasielastic Scattering by Nonresonant Samples

We consider the Rayleigh scattering process of the Mössbauer γ-rays, whose wave vector is $\boldsymbol{k}$, by electrons in a sample, such as liquids. Hereafter, we mainly consider the 14.4-keV Mössbauer γ-rays with 4.66-neV energy width from the first nuclear excited state of ^{57}Fe. In the scattering geometry shown in Fig. 2.17, the γ-rays transfer a momentum $q = |\boldsymbol{k}' - \boldsymbol{k}| = 2k \sin\theta$ to the sample, where $\boldsymbol{k}'$ is the wave vector of the scatted γ-rays and 2θ is the scattering angle. In the elastic Rayleigh scattering case, it follows that $|\boldsymbol{k}| \sim |\boldsymbol{k}'|$. In a simple mono-atom liquid, an electron density

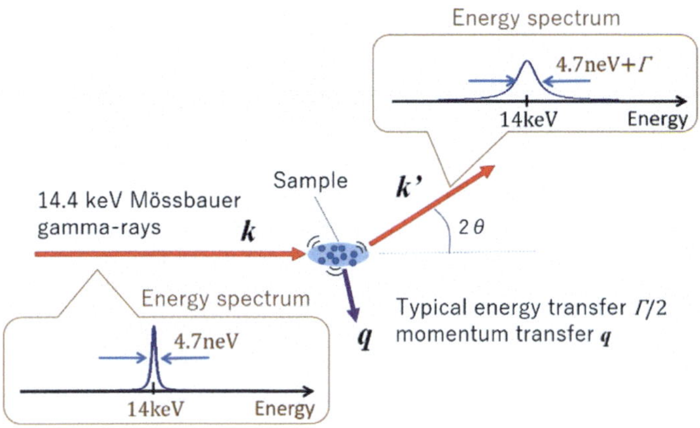

Fig. 2.17 Schematic diagram of quasielastic scattering of Mössbauer γ-rays by a sample

field is written as $\rho(\boldsymbol{r}, t) = \sum_{i=1}^{N} \delta(\boldsymbol{r} - \boldsymbol{r}_i(t))$, where $\boldsymbol{r}$ and t are coordinate and time, respectively; N is a molecular number in the sample; and $\boldsymbol{r}_i$ is a center position of atom i. In the reciprocal $\boldsymbol{q}$ space, the density field is written as $g(\boldsymbol{q}, t) = \int \rho(\boldsymbol{r}, t)\mathrm{e}^{-\mathrm{i}\boldsymbol{q}\cdot\boldsymbol{r}} d\boldsymbol{r}$.

We consider scaler component $q = |\boldsymbol{q}|$, assuming that the sample is isotropic, such as liquids and amorphous solids. When the q dependence of the elastic scattering intensity $I(q)$ is measured, it depends on the static structure factor $S(q)$ as $I(q) = N\,S(q) \equiv \langle g(q, t)g(-q, t)\rangle$, where $\langle\cdots\rangle$ denotes an ensemble averaging over a long time t. $S(q)$ is related to the space correlation function $G(r) = \langle\rho(r, t)\rho(0, t)\rangle$, which shows the spatial correlation of the electron density with a space r, as $S(q) = \int G(r)\exp(\mathrm{i}qr)dr$.

In the presence of atomic/molecular motions in a sample, the energy of the γ-rays is transferred to the sample and vise visa. We consider the microscopic dynamics in timescales between nanosecond and microsecond. Here, we consider quasielastic scattering process, which broadens the width of the energy spectrum, as shown in Fig. 2.17. The energy transfer is very small (approximately 10^{-9} eV) compared to the γ-ray energy (approximately 10 keV). Therefore, we can still assume that $|\boldsymbol{k}| \sim |\boldsymbol{k}'|$.

When we analyze the energy of the scattered γ-rays at q by standard Mössbauer absorption spectroscopy, we can observe that $I(q, E) = NS(q, E)$, where $S(q, E)$ is called the dynamics structure factor. $S(q, E)$ is related to the time and space correlation function $G(r, t)$: $S(q, E) = \int G(r, t)\exp[\mathrm{i}(qr - tE/\hbar)]dtdr$. Here, $\hbar$ is the reduced Planck constant. Inelastic/quasi-elastic X-ray and neutron scattering measures $S(q, E)$. Similarly, in the time domain, the corresponding $S(q, t) = \int S(q, E)\exp(\mathrm{i}tE/\hbar)dE$ is measured by neutron spin echo spectroscopy. Both $S(q, E)$ and $S(q, t)$ show equivalent information on microscopic time–space picture $G(r, t)$.

Quasielastic scattering experiments using Mössbauer γ-rays from RI sources have been performed soon after the discovery of the Mössbauer effect in1960s [110]. The method is called as RSMR [111]. However, because γ-rays from an RI source do not have enough brilliance as a parallel beam required for quasielastic scattering experiments, the method requires much measuring time (e.g., weeks). In this section, we call the quasielastic scattering spectroscopy using the γ-rays as quasielastic γ-ray scattering (QEGS) spectroscopy. Recently, high-brilliance SR is available and widely used for Mössbauer spectroscopy [112]. Using SR, QEGS was demonstrated by some techniques. First, QEGS using ^{57}Fe-nuclear Bragg monochromator, which measures $S(q, E)$, was attempted and demonstrated [113, 114]. Then, QEGS using TDI of ^{57}Fe γ-rays, which observes $S(q, t)$, was demonstrated [115]. These methods allowed a much quicker measurement of the atomic/molecular dynamics than the RSMR method owing to high brilliance and directivity of the SR sources and technological development of the high-resolution monochromator [116] and APD detector [24].

2.4.3 Time-Domain Measurement of Quasielastic Scattering of Mössbauer Gamma Rays Using Synchrotron Radiation

In this subsection, we introduce quasielastic scattering method using TDI with single-line Mössbauer γ-rays.

In Sect. 2.4.3.1, we introduce NFS experiment with single-line γ-ray emitters. This experiment corresponds to the QEGS experiment without a sample. In Sect. 2.4.3.2, the corresponding QEGS using TDI with single-line γ-rays is considered. In Sect. 2.4.3.3, we introduce an interpretation of the time spectrum from space-time diagram. In Sect. 2.4.3.4, we discuss the selectivity of nuclear species for TDI.

2.4.3.1 Introduction of Nuclear Forward Scattering Using Time-Domain Interferometry Setup of Single-Line Mössbauer Gamma Rays

Before the discussion of the QEGS experiment, it is valuable to consider corresponding NFS with two single-line γ-ray emitters because the NFS experiment corresponds to QEGS experiment without a sample and is a basis of QEGS using TDI. The NFS experimental setup is shown in the upper figure of Fig. 2.18a. The incident SR is introduced to two identical materials containing ^{57}Fe with single-line nuclear excitation profile. After transmitting them, the time spectrum of the SR and γ-rays from the materials are detected by a detector, such as an APD detector, with a time resolution of ~1 ns. Hereafter, we call the upstream and downstream emitters as γ-ray emitters 1 and 2, respectively. Here, we assume that emitter 1 is driven with a constant velocity v in the direction of the incident SR wave vector $\boldsymbol{k}$ to change the γ-ray energy from that of downstream emitter 2 by the Doppler effect. The relation between applied

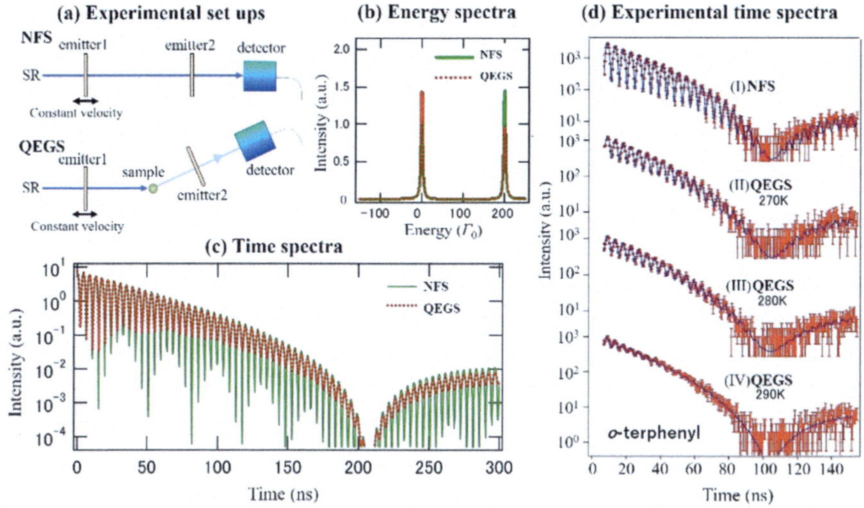

Fig. 2.18 Experimental setups and spectra of NFS and QEGS using the single-line TDI. **a** Experimental setups, **b** theoretical energy spectra, **c** theoretical time spectra, and **d** experimentally obtained time spectra using (I) NFS and QEGS studies on *o*-terphenyl at (II) 270 K, (III) 280 K, and (IV) 290 K. In panel **d**, points represent experimentally obtained counts of γ-rays, and vertical bars represent the statistic errors (standard deviations). The lines are fitting curves using eq. (2.3)

velocity v and the energy shift of γ-rays δE is expressed as $\delta E = E_\gamma v/c$, where c is the speed of light and E_γ is the energy of the γ-rays. For example, for the case of v~10 mm/s, which is easily realized by the velocity transducer produced by e.g. the WissEl GmbH, we obtain $\delta E \sim 100\Gamma_0$, which is much larger than the natural energy width of γ-rays, Γ_0. Simultaneously, this δE value is much smaller than the energy width of the incident SR typically approximately meV (~$10^6\Gamma_0$). Therefore, the nuclear resonant excitation process occurs in the same condition in two emitters. We can detect the interference of the γ-rays in the directive forward scattering component. [14]

In such a case, we consider the energy spectrum of γ-rays from two emitters at the forward detector position. The spectrum shows two peaks sufficiently separated, as shown in Fig. 2.18b. Here, the horizontal axis is a relative energy to the γ-ray energy from emitter 2 scaled by the energy unit of Γ_0. On the NFS time spectrum, we see a beating pattern called a quantum beat, which originates from the interference of γ-rays with different energies from two emitters, as shown in Fig. 2.18c.

First, we describe the property of the incident SR. The amplitude of the incident SR electric field SR in the angular frequency ω domain is expressed as $\hat{E}_0(\omega)$. The incident SR is usually monochromatized around the nuclear excitation energy to reduce the unused radiation for preventing system damage. In such a case, the bandwidth ΔE of $\left|\hat{E}_0(\omega)\right|^2$ is in the order of meV. The phase of each frequency component is assumed to be the same [117].

We write the electric field amplitude of the incident SR in the time t domain as $E_0(t)$, which is related to $\hat{E}_0(\omega)$ by Fourier transformation. The time response functions of emitters 1 and 2 are defined to be as follows [117, 118]:

$$R_1(t) = \delta(t) + G(t)e^{i\delta Et/\hbar}$$

and

$$R_2(t) = \delta(t) + G(t), \tag{2.13}$$

where $\delta(t)$ represents a transmission part without nuclear excitation process and the second term including $G(t)$ represents the nuclear excitation and de-excitation processes with the Mössbauer effect. The energy shift of γ-rays from the upstream emitter is considered by a term that includes the corresponding angular frequency $\delta E/\hbar$. Here, we ignored a transmittance factor because it does not affect the shape of the time spectrum. When the energy shift is sufficiently large ($\delta E \gg \Gamma_0$), unfavorable radiative coupling (RC; photons experience nuclear excitation processes in both emitters) can be neglected [119]. We consider the $\delta E \gg \Gamma_0$ case in the following discussion.

The electric field amplitude after emitter 1 is written as $R_1(t) \otimes E_0(t)$, where $\otimes$ denotes the convolution integral [117, 119]. After transmitting emitter 2, the electric field amplitude is written as $R_2(t) \otimes \{R_1(t) \otimes E_0(t)\}$. When $\left|\hat{E}_0(\omega)\right|^2$ shows a finite bandwidth $\Delta E/\hbar$, $|E_0(t)|^2$ also shows a finite width $\Delta T \sim h/\Delta E$. In the timescale of ΔT, the incident SR partly shows time coherence. ΔT is much smaller than the nuclear response as shown below. Therefore, we assume $E_0(t) \propto \delta(t)$. Here, the electric field amplitude $E_{tot}(t)$ at the detector position can be written as

$$E_{tot}(t) \propto \int_0^\infty R_1(t')R_2(t-t')dt' \approx \delta(t) + G(t)\left(1 + e^{i\delta Et/\hbar}\right). \tag{2.14}$$

Here, the origin of the time is the detection time of the SR pulse. The electric field amplitude of γ-rays $E(t)$ is written as $E(t) \propto G(t)\left(1 + e^{i\delta Et/\hbar}\right)$. The intensity is given by

$$\begin{aligned} I(t) &= |E(t)|^2 \\ &= 2|G(t)|^2[1 + \cos(\delta Et/\hbar)]. \end{aligned} \tag{2.15}$$

Here, the factor $|G(t)|^2$ represents the NFS time spectrum from one emitter. In a thin limit of the emitter thickness, it follows that $|G(t)|^2 \propto e^{-t/\tau_0}$, where τ_0 is the lifetime of the nuclear excited state. Otherwise, $|G(t)|^2$ shows more complex time dependence known as a dynamical beat [120]. In the ^{57}Fe case, τ_0 is ~141 ns. A factor $1 + \cos(\delta Et/\hbar)$ represents a quantum beat modifying $|G(t)|^2$.

Here, the velocity transducer used to drive emitter 1 brings a constant velocity with positive and negative sign, alternatively. Note that the sign of the velocity of the driven emitter 1 does not affect the obtained time spectrum $I(t)$, because only an absolute value of the energy difference between γ-rays from the two emitters is important for the beating pattern on the time spectrum owing to the factor $\cos(\delta Et/\hbar) = \cos(-\delta Et/\hbar)$ in eq. (2.15). In actual experiments, we do not detect signals in the period, where the velocity is not constant owing to the change of sign of the velocity. We also note that following time distributions have to be treated as an incoherent broadening of the time spectrum: (1) the time resolution of the detector (in this study ~1 ns) and (2) the distribution of the arrival time of SR owing to the spatial distribution of the electrons in one bunch (approximately 50 ps (FWHM) in BL09XU of SPring-8). As an experimental time spectrum, we obtain $\bar{I}_{\mathrm{exp}}(t) = I(t) \otimes D(t) + B$, where $D(t)$ is the total incoherent distribution function and B is a background constant noise.

2.4.3.2 Quasielastic Scattering Using Time-Domain Interferometry of Single-Line Mössbauer Gamma Rays

Next, we consider QEGS experimental setup shown in the lower figure of Fig. 2.18a and derive the expression of the QEGS time spectrum. Here, we introduce the sample response function $g(q,t)$. It cannot be immediately assumed that $E_0(t) \propto \delta(t)$ because we do not know the timescale of the sample response [121]. Only the case that the sample shows longer time response than ΔT that we can assume $E_0(t) \propto \delta(t)$. In this subsection, we assume this specific case of $E_0(t) \propto \delta(t)$ because it makes the discussion simpler and instructive. The effect of the finite time width of the incident SR on the QEGS time spectrum is discussed in Sect. 2.4.4 based on the discussion of this subsection.

The electric field amplitude after emitter 1 is $R_1(t) \otimes E_0(t)$. We define t_s as a time when the Rayleigh scattering process of the prompt SR pulse occurs in the sample. The sample response generally depends on both t_s and t. Conversely, the nuclear response is independent of t_s [119, 122]. The electric field amplitude after scattering by the sample is written as $g(q, t_s + t)E_A(t)$, where we used the time response function of the sample $g(q, t_s + t)$ defined in Sect. 2.4.2 [115]. After transmitting emitter 2, the total electric field amplitude $E_{tot}(q, t_s + t)$ at the angle corresponding to q is

$$E_{tot}(q, t_s + t) \propto R_2(t) \otimes \{g(q, t_s + t)[R_1(t) \otimes E_0(t)]\}. \tag{2.16}$$

By assuming $E_0(t) \propto \delta(t)$, $E_{tot}(q, t_s + t)$ is written as

$$E_{tot}(q, t_s + t) \propto \int_0^{\infty} R_1(t')g(q, t_s + t')R_2(t - t')\mathrm{d}t'. \tag{2.17}$$

Neglecting the RC effect, we obtain the electric field amplitude of γ-rays $E(q, t_s + t)$ as

$$E(q, t_s + t) \propto g(q, t_s + t)G(t)\mathrm{e}^{\mathrm{i}\delta Et/\hbar} + g(q, t_s)G(t). \tag{2.18}$$

Similarly, to the NFS case, the theoretical equation of the γ-ray time spectrum is written as $I(q,t_s+t) \propto |G(t)|^2[|g(q,t_s+t)|^2 + |g(q,t_s)|^2 + g(q,t_s+t)g^*(q,t_s)e^{i\delta Et/\hbar} + g^*(q,t_s+t)g(q,t_s)e^{-i\delta Et/\hbar}]$. The experimentally observed $\bar{I}(q,t)$ is obtained by averaging $I(q,t_s+t)$ by t_s over a long measurement time as

$$\begin{aligned}\bar{I}(q,t) \propto |G(t)|^2\big[&\langle|g(q,t_s+t)|^2\rangle + \langle|g(q,t_s)|^2\rangle \\ &+ \langle g(q,t_s+t)g^*(q,t_s)\rangle e^{i\delta Et/\hbar} + \langle g^*(q,t_s+t)g(q,t_s)\rangle e^{-i\delta Et/\hbar}\big] \end{aligned} \tag{2.19}$$

where $\langle\cdots\rangle$ indicates averaging by t_s over a long measurement time. It follows that $S(q,0) = \langle|g(q,t_s+t)|^2\rangle = \langle|g(q,t_s)|^2\rangle$. In classic mechanical cases, it follows that $S(q,t) = \langle g^*(q,t_s+t)g(q,t_s)\rangle = \langle g(q,t_s+t)g^*(q,t_s)\rangle$ suggesting $S(q,t)$ can be treated as a real number [115, 119, 122, 123]. We define $S'(q,t)$ as the intermediate scattering function normalized by $S(q,0)$ as

$$S'(q,t) = \langle g(q,t_s+t)g(q,t_s)\rangle / \langle|g(q,t_s+t)|^2\rangle, \tag{2.20}$$

we obtain

$$\bar{I}(q,t) \propto |G(t)|^2 S(q)\big[1 + S'(q,t)\cos(\delta Et/\hbar)\big]. \tag{2.21}$$

When a single exponential relaxation is assumed for $S'(q,t)$, we can write $S'(q,t) \propto \exp\{-t/\tau\}$, where τ is a relaxation time. Often, there is an intrinsic relaxation in $S'(q,t)$ even when standard samples with no detectable dynamics are measured [115]. To express it, we introduce a relaxation function $F^{\mathrm{int}}(t)$. By using this factor, the time spectrum is given by

$$\bar{I}(q,t) \propto |G(t)|^2\big[1 + F^{\mathrm{int}}(t)S'(q,t)\cos(\delta Et/\hbar)\big]. \tag{2.22}$$

The measured time spectrum $\bar{I}_{\mathrm{exp}}(q,t)$ is written as $\bar{I}_{\mathrm{exp}}(q,t) = \bar{I}(q,t) \otimes D(t) + B$ as described in Sect. 2.4.3.1.

We calculated NFS and QEGS time spectra, as shown in Fig. 2.18c, where we used the following conditions: the effective thickness of each emitter $T_e = 10$, $v = 20$ mm/s, and $\tau = 0.5\tau_0$. The corresponding γ-ray energy spectra are shown in Fig. 2.18b. The figure shows that the time-spectrum shape changes, reflecting dynamics as the disappearance of the quantum beat.

In Fig. 2.18d, experimentally obtained time spectra by (I) NFS and QEGS studies on *o*-terphenyl at (II) 270 K, (III) 280 K, and (IV) 290 K are shown. It can be confirmed that the quantum beat disappears by heating the sample. Least squares fittings using eq. (2.22) were successfully performed considering the time resolution $D(t)$ and background. The obtained relaxation time was confirmed to be consistent with previous results [124].

2.4.3.3 Interpretation of Time Spectrum of Quasielastic Scattering from Space–Time Diagram

We consider why the interference of γ-rays disappears with a time using the space–time diagram of photon paths for QEGS using TDI. We show the diagram in Fig. 2.19. We took the position of the photons on the beam path as a bottom axis of the figure. The vertical axis represents the time t_s. This figure shows that paths I and II constitute an interferometer, in which one arm of the interferometer is in the time domain and the other arm is in the space domain, compared to the usual interferometers with two arms in the space domain [112]. This is the reason why this method is named the "time-domain interferometry." We consider paths I and II of the γ-rays detected at delayed time compared to the SR pulse in the diagram. In Fig. 2.19, filled/empty circle symbols on the beam path express nuclear excitation/de-excitation events in emitters, and the star symbols show the Rayleigh scattering event by the sample. At the filled circle points, nuclear excitation event occurs as if the γ-rays are trapped until the excited nuclei decay. We do not consider the radiative coupling path that γ-rays experience the nuclear excitation event in both emitters because the excitation energies in the two emitters are sufficiently different from each other.

As we can see in eq. (2.21), the amplitude of the beating pattern of the time spectra decays following $S'(q,t)$. This suggests that time coherence of γ-ray photons from the two emitters decreases with time due to diffusion in samples. In other words, the time spectrum shape changes with time from the time spectrum shape calculated from the coherent sum of the electric fields from both emitters to the time spectrum shape calculated from the incoherent sum of the electric fields from each emitter due to the loss of the coherency of γ-rays. This explanation suggests that the measurement efficiency of QEGS using TDI strongly depends on the degree of difference between the coherent and incoherent time spectra, and the efficiency of QEGS using TDI can be improved by selecting the emitters.

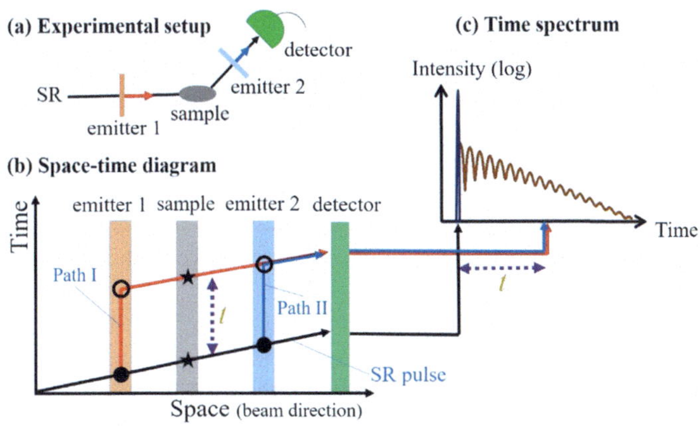

Fig. 2.19 **a** Experimental setup, **b** space–time diagram, and **c** time spectrum of the conventional TDI for QEGS

Table 2.2 Examples of Mössbauer isotopes [112]

Isotopes	Gamma ray energy (keV)	Gamma ray energy width (neV)	Nuclear resonant cross section (barn)
^{57}Fe*	14.4125	4.66	2464.0
^{67}Zn	93.312	0.0498	50.0
^{119}Sn	23.871	25.70	1380.5
^{149}Sm	22.494	64.08	120.1
^{151}Eu*	21.532	47.03	242.6
^{181}Ta	6.238	0.0754	1099.2

*Symbol indicates nuclear species already demonstrated to be available for the QEGS study

2.4.3.4 Selectivity of Isotopes for Quasielastic Scattering Experiment Using Time-Domain Interferometry

In quasielastic scattering measurements using γ-rays, the time (energy) and space (momentum) regions of the measurement depend on the characteristics (lifetime and excitation energy) of the nuclear resonance of isotopes used and other experimental conditions. In QEGS experiments using TDI, the short limit of the accessible time range depends on, for example, the time resolution of the detector, while the long limit depends on the lifetime of the nuclear excited state. The γ-ray energy, energy width, and nuclear resonant cross section for some Mössbauer isotopes are shown in Table 2.2. So far, QEGS experiments using TDI with ^{57}Fe and ^{151}Eu were demonstrated [125].

2.4.4 Effect of Energy Width of Incident Synchrotron Radiation

Here, we consider the general case that the time profile of the incident SR cannot be treated as a delta function [121]. When the high-resolution monochromator with meV energy resolution is used for the monochromatization of incident beam, the corresponding coherent time width is sub-picoseconds, In the timescale, usual condensed matter shows vibration motions. Therefore, normally, the incident SR cannot be treated as a delta function; instead, the sample response in the timescale must be considered by the exact calculation of interaction between the electric field and the sample.

We define a period δT_{12} for quantum beats caused by the interference between γ-rays from emitters 1 and 2. In the general case, the energy spectrum of the γ-rays from each emitter may show multi-peaks. Therefore, there are various quantum beats with various periods. For any quantum beats, we assume $\tau_0 \gg \delta T_{12} \gg \Delta T$. In such a case, the RC effect is negligible. From eq. (2.16), the electric field amplitude of the γ-rays is written as

$$E(q,t_s+t)=\int_{-\infty}^{\infty}dt'\left[g(q,t_s+t)G_1(t-t')E_0(t')+G_2(t-t')g(q,t_s+t')E_0(t')\right].\tag{2.23}$$

The first and second terms of eq. (2.23) describe the electric field amplitudes of the γ-rays passing paths I and II in Fig. 2.19, respectively.

In this subsection, we consider the experimental setup with single-line γ-ray emitters shown in Fig. 2.18a. Here, we assume $G_1(t)=G_2(t)\mathrm{e}^{i\delta Et/\hbar}$ and $G_2(t)=G(t)$. The electric field amplitude of the γ-rays is as follows:

$$\begin{aligned}E(q,t_s+t)=&\int_{-\infty}^{\infty}dt'g(q,t_s+t)G(t-t')\mathrm{e}^{i\delta E(t-t')\hbar}E_0(t')\\&+G(t-t')g(q,t_s+t')E_0(t').\end{aligned}\tag{2.24}$$

The time variation of $|G(t)|$ is much slower than that of $|E_0(t)|$ from the relation $\tau_0\gg\delta T_{12}\gg\Delta T$. Therefore, the first term in eq. (2.24) follows that $g(q,t_s+t)\int_{-\infty}^{\infty}dt'G(t-t')\mathrm{e}^{i\delta E(t-t')\hbar}E_0(t')\cong g(q,t_s+t)G(t)\mathrm{e}^{i\delta Et/\hbar}\int_{-\infty}^{\infty}dt'E_0(t')$. Similarly, the second term follows that $\int_{-\infty}^{\infty}dt'G(t-t')g(q,t_s+t')E_0(t')\cong G(t)\int_{-\infty}^{\infty}dt'g(q,t_s+t')E_0(t')$. Here, we define $g_c(q,t_s)$ as $g_c(q,t_s)\equiv\int_{-\infty}^{\infty}dt'g(q,t_s+t')E_0(t')/\int_{-\infty}^{\infty}dt'E_0(t')$. In case of $E_0(t)=\delta(t)$, it follows that $g_c(q,t_s)=g(q,t_s)$. Using g_c, eq. (2.24) can be written as

$$E(q,t_s+t)\cong\hat{E}_0(0)G(t)\left[g(q,t_s+t)\mathrm{e}^{i\delta Et/\hbar}+g_c(q,t_s)\right],\tag{2.25}$$

where we used the general relation $\int_{-\infty}^{\infty}dt'E_0(t')=\hat{E}_0(0)\equiv\hat{E}_0(\omega=0)$. Here, $\hat{E}_0(\omega)$ is the angular frequency-domain representation of $E_0(t)$.

Here, we consider the meaning of $g_c(q,t_s)$, which is an integration of the product of $g(q,t_s+t')$ and $E_0(t')$ by t'. We show examples of paths I and II of the γ-rays detected at t with an incident time $t'=0$ (long dashed line) and $t'\neq0$ (short dashed line) in the time–space diagrams of Figs. 2.20a and b, respectively. Gamma rays with different incident times t' interfere at the detector owing to the finite coherent width of $E_0(t)$ in both γ-ray paths I and II. The integration in $g_c(q,t_s)$ originates from a characteristic of path II: γ-rays scattered by the sample at various times t_s+t' ($|t'|\lesssim\Delta T$) interfere with each other at the detector position at t. Alternatively, in path I, the γ-rays scattered by the sample at unique time t_s+t interfere with each other at the detector position at t. Therefore, the γ-rays passing path I are not affected by the time width ΔT of the incident radiation. This is the interpretation of eq. (2.25).

From eq. (2.16), we obtain the detected γ-ray intensity:

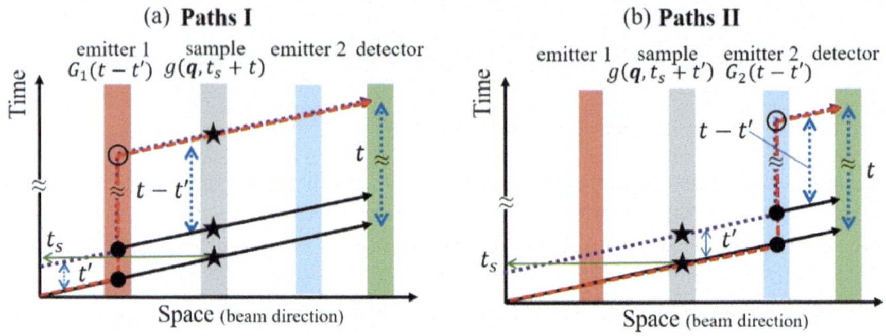

Fig. 2.20 Examples of paths **a** I and **b** II of the γ-rays detected at t with an incident time $t' = 0$ (long dashed line) and $t' \neq 0$ (short dashed line) for the time–space diagrams

$$I(q, t_s + t) \propto |G(t)|^2\big[|g(q, t_s + t)|^2 + |g_c(q, t_s)|^2 + \{g^*(q, t_s + t)g_c(q, t_s) + g(q, t_s + t)g_c^*(q, t_s)\}\cos(\delta Et/\hbar). \tag{2.26}$$

The experimental spectrum $\bar{I}(q, t)$ is obtained by averaging $I(q, t_s + t)$ by t_s over a long measurement time. As we discussed in Sect. 2.4.3.2, it follows $S(q, t) = \langle g^*(q, t_s + t)g(q, t_s)\rangle = \langle g(q, t_s + t)g^*(q, t_s)\rangle$. We define correlation functions $S_{cc}(q, t)$ and $S_c(q, t)$ as $\langle g_c(q, t_s + t)g_c^*(q, t_s)\rangle$ and $\langle g(q, t_s + t)g_c^*(q, t_s)\rangle$ averaged by t_s, respectively. It can be assumed that $S_{cc}(q, t)$ is also a real number. Using these values, the observed time-averaging intensity $\bar{I}(q, t)$ can be written as

$$\bar{I}(q, t) \propto |G(t)|^2\big\{S(q, 0) + S_{cc}(q, 0) + \big[S_c(q, t) + S_c^*(q, t)\big]\cos(\delta Et/\hbar)\big\}. \tag{2.27}$$

Here, $S_c(q, t)$ is written as

$$S_c(q, t) = \langle g^*(q, t_s + t)g_c(q, t_s)\rangle = \frac{1}{\hat{E}_0(0)} \int_{-\infty}^{\infty} dt' E_0(t') S(q, t - t'). \tag{2.28}$$

The γ-ray time spectra are observed in the timescale much longer than ΔT. In the timescale, it can be assumed that $\int_{-\infty}^{\infty} dt' E_0(t') S(q, t - t') \cong S(q, t) \int_{-\infty}^{\infty} dt' E_0(t')$ because, in the measurement time window, the variation of $S(q, t)$ in the timescale ΔT is usually negligible. Therefore, it follows that $S_c(q, t) \cong S(q, t) \int_{-\infty}^{\infty} dt' E_0(t')/\hat{E}_0(0) \cong S(q, t)$ and $S_c(q, t) \cong S_c^*(q, t)$. $\bar{I}(q, t)$ is written as follows:

$$\bar{I}(q, t) \propto |G(t)|^2[S(q, 0) + S_{cc}(q, 0) + 2S(q, t)\cos(\delta Et/\hbar)](\text{at}\, t \gg \Delta T). \tag{2.29}$$

The experimental time spectrum is written as $\bar{I}_{\exp}(q,t) = \bar{I}(q,t) \otimes D + B$ discussed above.

From the definition of $S_{cc}(q,t)$, it follows that

$$S_{cc}(q,t) = \frac{1}{\left|\hat{E}_0(0)\right|^2} \int_{-\infty}^{\infty}\int_{-\infty}^{\infty} dt'dt'' E_0(t') E_0^*(t'') \langle g(q,t_s+t') g^*(q,t_s+t+t'') \rangle$$
$$\equiv \int_{-\infty}^{\infty} dt_d I_0(t_d) S(q,t+t_d) \tag{2.30}$$

where $t_d \equiv t'' - t'$. We define $I_0(t_d) \equiv \int_{-\infty}^{\infty} dt' E_0^*(t') E_0(t_d+t') / \left|\hat{E}_0(0)\right|^2$. The typical timescale of the decay of $I_0(t_d)$ is ΔT as defined in Sect. 2.4.3.1. We show a schematic diagram as example of $S'_{cc}(q,t)$ in Fig. 2.21. When $t \gg \Delta T$, it follows that $S_{cc}(q,t) \sim S(q,t)$ by neglecting the time variation of $S(q,t)$ in the timescale ΔT, as assumed in the above discussion of $S_c(q,t)$. Conversely, when $t = 0 \ll \Delta T$, we obtain $S_{cc}(q,0) = \int_{-\infty}^{\infty} dt_d I_0(t_d) S(q,t_d)$, suggesting that $S_{cc}(q,0) \sim S(q,\Delta T)$. We define the $S_{cc}(q,0)$ value as $S_{cc}(q,0) \equiv f_{\Delta E}(q)$, as shown in Fig. 2.21. When ΔE is sufficiently large and $f_{\Delta E}(q) = 1$ can be assumed, we obtain $\bar{I}(q,t) \propto |G(t)|^2 \left[1 + S'(q,t)\cos(\delta E t/\hbar)\right]$, which is equivalent to eq. (2.21).

$S'(q,t)$ generally shows a form with vibrations and multistep relaxations, which spread over very wide time ranges. See Fig. 2.21 for an example of $S'(q,t)$ and its relation to $S'_{cc}(q,t)$. Here, $\bar{I}(q,t)$ is usually fitted by assuming a relaxation function $F(q,t)$, which represents relaxations of $S'(q,t)$ in the time window of the measurement approximately the timescale of τ_0. In Fig. 2.21, we show an example of $F(q,t)$. Here, we define $f_{\Gamma_0}(q) \equiv \lim_{t \to 0} F(q,t)$. $f_{\Gamma_0}(q)$ is a plateau value of $S'(q,t)$ decided by the fitting of the time spectrum, as shown in Fig. 2.21. Both $f_{\Delta E}(q)$ and $f_{\Gamma_0}(q)$ give unique information on microscopic dynamics. Hence, special attention must be given for these definitions.

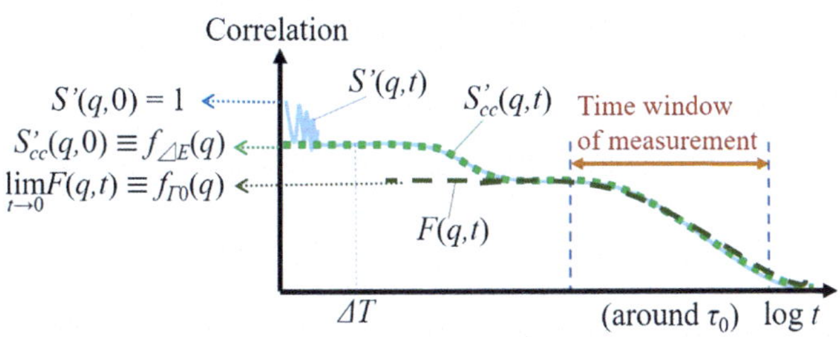

Fig. 2.21 Example of the intermediate scattering function normalized by the static structure factor $S'(q,t)$ with its relation to $S'_{cc}(q,t)$ and assumed $F(q,t)$

Here, we show that neither $f_{\Gamma_0}(q)$ nor $f_{\Delta E}(q)$ cannot be determined by the conventional single-line TDI in the case of finite ΔE width of approximately several meV using a high-resolution monochromator. From eq. (2.29), we obtain

$$\bar{I}(q,t) \propto |G(t)|^2\big\{1 + 2S'(q,t)/[1 + f_{\Delta E}(q)]\cos(\delta Et/\hbar)\big\}. \tag{2.31}$$

In this equation, the cosine term has an additional factor $2/[1 + f_{\Delta E}(q)]$ compared to eq. (2.21). When this expression is used for the fitting, a function $F'(q,t)$ is assumed for $2S'(q,t)/[1 + f_{\Delta E}(q)]$. Using $F(q,t)$, which represents the form of $S'(q,t)$ in the time window of the measurement, $F'(q,t)$ can be expressed as $F'(q,t) = 2F(q,t)/[1 + f_{\Delta E}(q)]$. It follows that $\lim_{t\to 0} F'(q,t) = 2\lim_{t\to 0} F(q,t)/[1 + f_{\Delta E}(q)] = 2f_{\Gamma_0}(q)/[1 + f_{\Delta E}(q)]$. This equation suggests that both free fitting parameters $f_{\Gamma_0}(q)$ and $f_{\Delta E}(q)$ relate to $\lim_{t\to 0} F'(q,t)$. Therefore, in principle, neither $f_{\Delta E}(q)$ nor $f_{\Gamma_0}(q)$ can be determined by the single-line TDI when an identical pair of emitters are used in the incident SR condition with meV energy width. The conventional TDI suffers this uncertainty of the physical meaning of $\lim_{t\to 0} F'(q,t)$.

In contrast, for the case of a multiline emitter case with $|G_1(t)|^2 \neq |G_2(t)|^2$, both $f_{\Delta E}(q)$ and $f_{\Gamma_0}(q)$ can be determined based on the difference between $|G_1(t)|^2$ and $|G_2(t)|^2$ as we discuss in Sect. 2.4.5 [121].

2.4.5 *Time-Domain Interferometry Using Multiline Mössbauer Gamma Rays*

In the case where multiline γ-rays with $|G_1(t)|^2 \neq |G_2(t)|^2$ are used for TDI, the intensity of the γ-rays from eq. (2.23) is written as

$$\begin{aligned} I(q,t_s+t) \propto\ & |G_1(t)|^2|g(q,t_s+t)|^2 + |G_2(t)|^2|g_c(q,t_s)|^2 \\ & + G_1^*(t)G_2(t)g^*(q,t_s+t)g_c(q,t_s) \\ & + G_1(t)G_2^*(t)g(q,t_s+t)g_c^*(q,t_s). \end{aligned} \tag{2.32}$$

Similarly, in Sect. 2.4.4, the observed time-averaging intensity $\bar{I}(q,t)$ can be written as

$$\begin{aligned} \bar{I}(q,t) \propto\ & S(q,0)|G_1(t)|^2 + S_{cc}(q,0)|G_2(t)|^2 \\ & + S(q,t)\big[G_1^*(t)G_2(t) + G_2^*(t)G_1(t)\big](\text{at}\, t \gg \Delta T). \end{aligned} \tag{2.33}$$

Using $S'(q,t)$ and $f_{\Delta E}(q)$, $\bar{I}(q,t)$ can be rewritten as

$$\begin{aligned} \bar{I}(q,t) \propto\ & \big[1 - S'(q,t)\big]\big[|G_1(t)|^2 + |G_2(t)|^2\big] + S'(q,t)|G_1(t) + G_2(t)|^2 \\ & - [1 - f_{\Delta E}(q)]|G_2(t)|^2 \quad (\text{at}\, t \gg \Delta T). \end{aligned} \tag{2.34}$$

The experimental time spectrum is $\bar{I}_{\exp}(q,t) = \bar{I}(q,t) \otimes D(t) + B$. The γ-rays from α-iron foils show energy spectra with a multiline structure. In Fig. 2.22a, we show an example of the experimental setup using α-iron foils as emitters 1 and 2. To satisfy the condition $|G_1(t)|^2 \neq |G_2(t)|^2$, an external magnetic field $\boldsymbol{H}$ is applied to the foils different from each other to selectively allow the excitation between energy levels split by the hyperfine interaction: The direction is horizontal to the electric field of the incident radiation and a direction $\boldsymbol{H} \perp \boldsymbol{k}$ for emitter 1. For emitter 2, the direction is vertical to the electric field of the incident radiation an $\boldsymbol{H} \perp \boldsymbol{k}'$ [126]. The energy spectra of γ-rays in Fig. 2.22b are shown for the case without relaxation and with a relaxation of $\tau = 100$ ns. Here, $\delta E_{12} = \mathrm{h}/\delta T_{12} \sim 20\Gamma_0 \gg \Gamma_0$. When $20\Gamma_0 \gg \Gamma$, the RC effect is negligible. The corresponding time spectra of γ-rays are shown in Fig. 2.22c. The shape of the time spectrum changes following the decay of $S'(q,t)$.

In Fig. 2.22d, the experimentally obtained time spectra by QEGS studies on glycerol are shown. The QEGS spectra were obtained (I) at 40 K at 14 nm^{-1}, (II) at 237.5 K at 14 nm^{-1}, and at 237.5 K at 31 nm^{-1}. We note that the experimental condition is different from the theoretical one. Least squares fitting using eq. (2.34) was successfully performed, and the obtained relaxation time was confirmed to be consistent with previous results [121]. It was shown that the measurement efficiency of the multiline TDI system is much higher than the single-line system one [121, 126, 127].

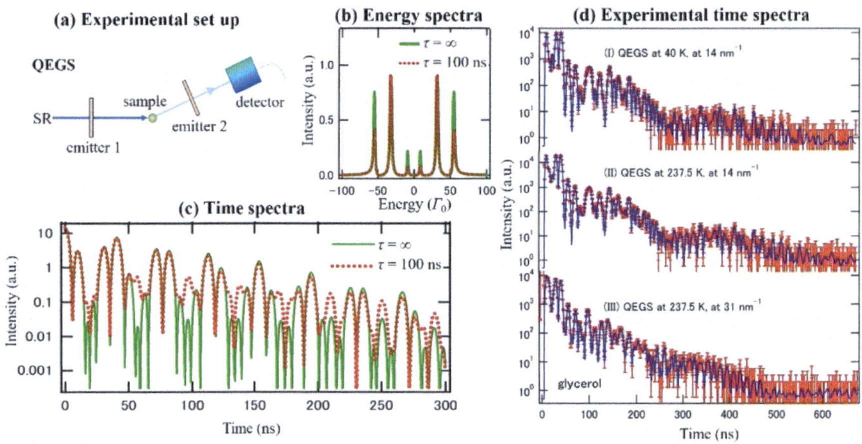

Fig. 2.22 **a** Experimental setup, **b** theoretical energy spectra and **c** theoretical time spectra, and **d** experimentally obtained time spectra from QEGS studies on glycerol. The spectra were obtained (I) at 40 K at 14 nm^{-1}, (II) at 237.5 K at 14 nm^{-1}, and at 237.5 K at 31 nm^{-1}. In panel **d**, the points and vertical bars represent experimentally obtained counts of γ-rays and statistic errors (standard deviations), respectively. The lines represent fitting curves using eq. (2.23)

2.4.6 Results Obtained by Quasielastic Scattering Experiment Using Time-Domain Interferometry of Mössbauer Gamma Rays

So far, QEGS measurements using TDI were performed for alloys, liquid crystals, molecular liquids, ionic liquids, super-ionic conducting glasses, polymers, and polymer nanocomposites. For supercooled glass formers, van der Waals molecular liquid *o*-terphenyl, hydrogen bonding liquid glycerol, ionic liquids, and polymer polybutadiene were measured to study the microscopic dynamics toward glass transition. In glycerol and polybutadiene, the so-called de Gennes narrowing was observed by the q-dependent studies of the structural relaxation process called as the α-process [128]. In addition to the α-process, a local activation process called as the Johari-Goldstein (JG) β-process was observed in *o*-terphenyl and polybutadiene [124, 128]. The branching temperature of the JG β-process from the α-process, which could not be determined by conventional methods, such as dielectric spectroscopy, was accurately obtained. The study of ionic liquid 1-butyl-3-methylimidazolium iodide revealed that the ionic liquids are classified as so-called fragile glass formers [125]. For superionic conducting Na_3PS_4 glass, it was revealed that angstrom-scale translational motion of a part of Na ions occurs at picosecond to nanosecond time scale, while the movement of PS_4 ions rarely occur [129].

To understand the microscopic origin of the shear viscosity, a higher alcohol 3,7-dimethyl-1-octanol with nanometric domain structures was studied using the multiline TDI system [130]. The dynamics of both nanometric- and molecular-scale structures were found to relate to the slower and faster relaxation modes of the shear viscosity. The effect of the presence of Si nanoparticles on bulk polymer (polybutadiene) dynamics was studied for a polymer nanocomposite system [131]. B2 alloy CoGa was studied using the single-line TDI, and the possibility of the atomic diffusion study using diffuse scattering was shown [132].

For soft materials, QEGS was applied for dynamics study of liquid crystal molecules 4-cyano-4'-octylbiphenyl and 11-(4'-cyanobiphenyl-4-yloxy) undecyl pentadecafluorooctanoate under layered structure in the smectic phase [133]. Both the interlayer and intralayer molecular relaxation times were successfully observed using the multiline TDI system. The layer order parameter of the smectic phase was determined, and the anomalous diffusion coefficient could be obtained using the multiline TDI system [134]. The study of cholesteric blue phase revealed that the emergence of the mesostructure is irrelevant to the microscopic molecular structure and dynamics in this system [135]. TDI allows the study of molecular dynamics in smectic, nematic, and cholesteric phases.

2.4.7 Summary and Perspective of Quasielastic Scattering of Mössbauer Gamma Rays

Quasielastic scattering technique using SR-based Mössbauer γ-rays is a promising technique to directly reveal the microscopic dynamics in unique timescales between nanosecond and microsecond.

The quasielastic scattering technique using TDI will be improved by further developments, such as an increase in detection detector efficiency, an increase in the solid angle of γ-ray detection by introducing more detectors, and an increase in γ-ray count rate using more γ-ray lines for measurement. These improvements greatly help in extracting the intermediate scattering function from the spectrum and directly visualizing the decay of the intermediate scattering function.

QEGS using ^{57}Fe nuclear Bragg monochromator is expected to measure the dynamics faster than the timescale covered by current QEGS using TDI, for example, up to 100 ps. Therefore, the development of the energy-domain QEGS system and the combination study with the TDI system expand the accessible timescales by QEGS using the Mössbauer γ-rays. In addition, a further combination study with other techniques, such as quasielastic neutron scattering, allows us to further understand microscopic dynamics in complex systems. These improvements and combination studies are important to understand the macroscopic properties and functions from a microscopic level for both basic science and industrial applications. Furthermore, the TDI system can be applied for studies on dynamical correlations in quantum systems [136].

The fourth-generation SR shows much higher spatial coherency and higher condensing properties than the third-generation SR used for the studies introduced here. Microscopic dynamics of each spatial region in complex systems can be measured selectively by focusing on the γ-rays. Such QEGS system with focused γ-rays is useful, for example, to understand dynamical heterogeneity of glass formers.

Acknowledgements The author is grateful to all collaborators and would like to thank all of the staff at the Institute for Integrated Radiation and Nuclear Science, Kyoto University, National Institutes for Quantum and Radiological Science and Technology, Japan Atomic Energy Agency, SPring-8, and the Photon factory of KEK for their support.

References

1. For example, N.N. Greenwood, T.C. Gibb, *Mössbauer Spectroscopy*. (Chapman and Hall, London, 1971)
2. D.C. Champeney, Rep. Prog. Phys. **42**, 1017 (1979). and references therein
3. Please see Chap. 1 in this book
4. M. Seto, S. Kitao, Y. Kobayashi, R. Haruki, T. Mitsui, Y. Yoda, X.W. Zhang, Yu. Maeda, Phys. Rev. Lett. **84**, 566 (2000)
5. M. Seto, Y. Yoda, S. Kikuta, X.W. Zhang, M. Ando, Phys. Rev. Lett. **74**, 3828 (1995)

6. M. Seto, S. Kitao, Y. Kobayashi, R. Haruki, Y. Yoda, T. Mitsui, T. Ishikawa, Phys. Rev. Lett. **91**, 185505 (2003)
7. M. Seto, R. Masuda, S. Higashitaniguchi, S. Kitao, Y. Kobayashi, C. Inaba, T. Mitsui, Y. Yoda, Phys. Rev. Lett. **102**, 217602 (2009)
8. R. Masuda, Y. Kobayashi, S. Kitao, M. Kurokuzu, M. Saito, Y. Yoda, T. Mitsui, F. Iga, M. Seto, Appl. Phys. Lett. **104**, 082411 (2014)
9. Mössbauer Spectroscopy Periodic Table and Isotopes, Mössbauer Effect Data Center, Dalian, China. http://www.medc.dicp.ac.cn/Resources.php. Accessed 30 Apr 2019
10. D.W. Hafemeister, Phys. Rev. Lett. **14**, 593 (1965)
11. M. Seto, J. Phys. Soc. Jpn. **82**, 021016 (2013)
12. S.L. Ruby, J. Phys. (Paris), Colloq. **35**, C6-209 (1974)
13. E. Gerdau, R. Rüffer, H. Winkler, W. Tolksdorf, C.P. Klages, J.P. Hannon, Phys. Rev. Lett. **54**, 835 (1985)
14. J.B. Hastings, D.P. Siddons, U. van Brück, R. Hollatz, U. Bergmann, Phys. Rev. Lett. **66**, 770 (1991)
15. T. Shinjo, M. Kiyama, K. Sugita, K. Watanabe, T. Takada, J. Magn. Magn. Mater. **35**, 133 (1983)
16. T. Mitsui, R. Masuda, M. Seto, E. Suharyadi, K. Mibu, J. Synchrotron Radiat. **19**, 198 (2011)
17. T. Nakano, N. Fukuda, M. Seto, Y. Kobayashi, R. Masuda, Y. Yoda, M. Mihara, Y. Nozoe, Phys. Rev. B **91**, 140101(R) (2015)
18. S. Kishimoto, F. Nishikido, R. Haruki, K. Shibuya, M. Koshimizu, Hyperfine Interact. **204**, 101 (2012)
19. K. Mibu, M. Seto, T. Mitsui, Y. Yoda, R. Masuda, S. Kitao, Y. Kobayashi, E. Suharyadi, M. Tanaka, M. Tsunoda, H. Yanagihara, E. Kita, Hyperfine Interact. **217**, 127 (2013)
20. M. Kurokuzu, S. Kitao, Y. Kobayashi, M. Saito, R. Masuda, T. Mitsui, Y. Yoda, M. Seto, Hyperfine Interact. **226**, 687 (2014)
21. S. Tsutsui, R. Masuda, Y. Kobayahi Yoda, K. Mizuuchi, Y. Shimizu, H. Hidaka, T. Yanagisawa, H. Amitsuka, F. Iga, M. Seto, J. Phys. Soc. Jpn. **85**, 083704 (2016)
22. M. Seto, R. Masuda, S.S. Higashitaniguchi Kitao, Y. Kobayashi, C. Inaba, T. Mitsui, Y. Yoda, J. Phys: Conf. Ser. **217**, 012002 (2010)
23. J. Yamaura, H. Ohsumi, K. Sugimoto, S. Tsutsui, Y. Yoda, S. Takeshita, A. Tokuda, S. Kitao, M. Kurokuzu, M. Seto, I. Yamauchi, K. Ohgushi, M. Takigawa, T. Arima, Z. Hiroi, J. Phys: Conf. Ser. **391**, 012112 (2012)
24. S. Kishimoto, Nucl. Inst. Meth. Phys. Res. A **309**, 603 (1991)
25. T. Matsuoka, H. Fujihisa, N. Hirao, Y. Ohishi, T. Mitsui, R. Masuda, M. Seto, Y. Yoda, K. Shimizu, A. Machida, K. Aoki, Phys. Rev. Lett. **107**, 025501 (2011)
26. G.V. Smirnov, U. van Brück, J. Arther, G.S. Brown, A.I. Chumakov, A.Q.R. Baron, W. Petry, S.L. Ruby, Phys. Rev. A. **76** (2007)
27. F.J. Lynch, R.E. Holland, M. Hamermesh, Phys. Rev. **120**, 513 (1960)
28. D.W. Hamill, G.R. Hoy, Phys. Rev. Lett. **21**, 724 (1968)
29. R.B. Firestone, V.S. Shirley (eds.), *Table of Isotopes*, 8th edn. (Willey, New York, 1996)
30. E.A. McCutchan, Evaluated nuclear structure data file search and retrieval, Last updated 2019-03-15. National Nuclear Data Center, Brookhaven National Laboratory, New York. (2019) https://www.nndc.bnl.gov/ensdf/ensdf/ensdf.jsp. Accessed 30 Apr 2019
31. S. Sasaki, KEK report **90–16** (1990) 1—143. This report is also in database form at http://www.sasakiken.net/abcoeff/abcoeff2.html
32. M.J. Berger, J.H. Hubbell, S.M. Seltzer, J. Chang, J.S. Coursey, R. Sukumar, D.S. Zucker, K. Olsen, XCOM: photon cross sections database (version 1.5), NIST Standard Reference Database 8 (XGAM), Online Mar 1998 – Last Update Nov 2010. National Institute of Standards and Technology, Maryland, (2010) http://physics.nist.gov/xcom. Accessed in 30 Apr 2019
33. M.J. Berger, J.H. Hubbell, NBSIR **87–3597**, 1 (1987)
34. M.J. Berger, J.H. Hubbell, NIST x-ray and gamma-ray attenuation coefficients and cross sections database. NIST Standard Reference Database 8, Version 2.0, National Institute of Standards and Technology, Gaithersburg, MD, (1990)

35. K. Kanaya, S. Okayama, J. Phys. D Appl. Phys. **5**, 43 (1972)
36. ESTAR (1984) National Institute of Standards and Technology, Maryland, USA. https://physics.nist.gov/PhysRefData/Star/Text/ESTAR.html. Accessed 30 Apr 2019
37. M.J. Berger, M. Inokuti, H.H. Anderson, H. Bichsel, J.A. Dennis, D. Powers, S.M. Seltzer, J.E. Turner, J. Int. Comm. Radiat. Units Meas. os19 (1984) Report 37
38. G.K. Shenoy, F.E. Wagner (eds.), *Mössbauer Isomer shifts* (North-Holland, Amsterdam, 1978)
39. P. Gütlich, E. Bill, A.X. Trautwein, *Mössbauer spectroscopy and transition metal chemistry: fundamentals and applications* (Springer, Berlin, 2011)
40. G.V. Smirnov, V.V. Sklyarevskii, R.A. Voskanyan, A.N. Artem'ev, JETP Lett. **9**, 70 (1969)
41. W. Sturhahn, T.S. Toellner, E.E. Alp, X. Zhang, M. Ando, Y. Yoda, S. Kikuta, M. Seto, C.W. Kimball, B. Dabrowski, Phys. Rev. Lett. **74**, 3832 (1995)
42. A.I. Chumakov, R. Rüffer, H. Grünsteudel, H.F. Grünsteudel, G. Grübel, J. Metge, O. Leupold, H.A. Goodwin, Europhys. Lett. **30**, 427 (1995)
43. M. Yabashi, K. Tamasaku, S. Kikuta, T. Ishikawa, Rev. Sci. Instrum. **72**, 4080 (2001)
44. H.C. Wille, R.P. Hermann, I. Sergueev, O. Leupold, P. Van der Linden, B.C. Sales, F. Grandjean, G.J. Long, R. Rüffer, V. Shvyd'ko Yu, Phys. Rev. B **76**, 140301(R) (2007)
45. Yu.V. Shvyd'ko, *X-Ray Optics*, Springer Series in Optical Sciences **98**, (Berlin, Springer, 2004)
46. K.S. Singwi, A. Sjölander, Phys. Rev. **120**, 1093 (1960)
47. V.G. Kohn, A.I. Chumakov, R. Rüffer, Phys. Rev. B **58**, 8437 (1998)
48. V.G. Kohn, A.I. Chumakov, Hyperfine Interact. **125**, 205 (2000)
49. J. Matsuno, M. Seto, S. Kitao, Y. Kobayashi, R. Haruki, T. Mitsui, A. Fujimoril, Y. Takeda, S. Kawasaki, M. Takano, J. Phys. Soc. Jpn. **73**, 2768 (2004)
50. S. Higashitaniguchi, M. Seto, S. Kitao, Y. Kobayashi, M. Saito, R. Masuda, T. Mitsui, Y.Y. Yoda Kamihara, M. Hirano, H. Hosono, Phys. Rev. B **78**, 174507 (2008)
51. M. Seto, Y. Kobayashi, S. Kitao, R. Haruki, T. Mitsui, Y. Yoda, S. Nasu, S. Kikuta, Phys. Rev. B **61**, 11420 (2000)
52. G. Gila, R.M. Nicklow, Phys. Rev. **143**, 487 (1966)
53. P.D. Mannheim, Phys. Rev. **165**, 1011 (1968)
54. C. Keppler, K. Achterhold, A. Ostermann, U. Van Bürck, W. Potzel, A.I. Chumakov, A.Q.R. Baron, R. Rüffer, F. Parak, European. Biophys. J. **25**, 221 (1997)
55. F. Parak, K. Achterhold, Hyperfine Interact. **123/124** Part I, 825 (1999)
56. J.T.S. Sage, M. Durbin, W. Sturhahn, D.C. Wharton, P.M. Champion, P. Hession, J. Sutter, E.E. Alp, Phys. Rev. Lett. **86**, 4966 (2001)
57. J.T. Sage, C. Paxson, G.R.A. Wyllie, W. Sturhahn, S. Durbin M, P.M. champion, E.E. Alp, W.R. Scheidt, J. Phys.: Condens. Matter **13**, 7707 (2001)
58. Y. Xiao, H. Wang, S.J. George, M.C. Smith, M.W. Adams, F.E. Jenney Jr., W. Sturhahn, E.E. Alp, J. Zhao, Y. Yoda, A. Dey, E.I. Solomon, S.P. Cramer, J. Am. Chem. Soc. **127**, 14596 (2005)
59. T. Petrenko, S.D. George, N. Aliaga-Alcalde, E. Bill, B. Mienert, Y. Xiao, Y. Guo, W. Sturhahn, S.P. Cramer, K. Wieghardt, F. Neese, J. Am. Chem. Soc. **129**, 11053 (2007)
60. S.J. George, R.Y. Igarashi, Y. Xiao, J.A. Hernandez, M. Demuez, D. Zhao, Y. Yoda, P.W. Ludden, L.M. Rubio, S.P. Cramer, J. Am. Chem. **130**, 5673 (2008)
61. C.B. Bell III, S.D. Wong, Y. Xiao, E.J. Klinker, A.L. Tenderholt, M.C. Smith, J.U. Rohde, L. Que Jr., S.P. Cramer, E.I. Solomon, Angew. Chem. Int. Ed. **47**, 9071 (2008)
62. F. Paulat, T.C. Berto, S.D. George, L. Goodrich, V.K. Praneeth, C.D. Sulok, N. Lehnert, Inorg. Chem. **47**, 11449 (2008)
63. W. Zeng, N.J. Silvernail, W.R. Scheidt, J.T. Sage, Nuclear Resonance Vibrational Spectroscopy (NRVS), *Encyclopedia of Inorganic Chemistry* (2008)
64. W.R. Scheidt, A. Barabanschikov, J.W. Pavlik, N.J. Silvernail, J.T. Sage, Inorg. Chem. **49**, 6240 (2010)
65. Z.J. Tonzetich, H. Wang, D. Mitra, C.E. Tinberg, L.H. Do, F.E. Jenney, M.W.W. Adams, S.P. Cramer, S.J. Lippard, J. Am. Chem. Soc. **132**, 6914 (2010)

66. C.E. Tinberg, Z.J. Tonzetich, H. Wang, L.H. Do, Y. Yoda, S.P. Cramer, S.J. Lippard, J. Am. Chem. Soc. **132**, 18168 (2010)
67. W. Zeng, N.J. Silvernail, W.R. Scheidt, J.T. Sage, Nuclear Resonance Vibrational Spectroscopy (NRVS). In *Encyclopedia of Inorganic Chemistry*, 2008 (published online)
68. S. Kamali, H. Wang, D. Mitra, H. Ogata, W. Lubitz, B.C. Manor, T.B. Rauchfuss, D. Byrne, V. Bonnefoy, F.E. Jenney Jr, M.W. Adams, Y. Yoda, E. Alp, J. Zhao, S.P. Cramer, Angew. Chem. Int. Ed. **52**, 724 (2013)
69. D. Mitra, S.J. George, Y. Guo, S.S. Kamali Keable, J.W. Peters, V. Pelmenschikov, D.A. Case, S.P. Cramer, J. Am. Chem. Soc. **135**, 2530 (2013)
70. T. Ohta, J.G. Liu, M. Saito, Y. Kobayashi, Y. Yoda, M. Seto, Y. Naruta. J. Phys. Chem. B **116**, 13831 (2012)
71. K. Park, C.B. Bell III, L.V. Liu, D. Wang, G. Xue, K. Kwak, S.D. Wong, K.M. Light, J. Zhao, E.E. Alp, Y. Yoda, M. Saito, Y. Kobayashi, T. Ohta, M. Seto, L. Que Jr., E.I. Solomon, Proc. Natl. Acad. Sci. U.S.A. **110**, 6275 (2013)
72. S.D. Wong et al., Nature **499**, 320 (2013)
73. H. Ogata et al., Nat. Commun. **6**, 7890 (2015)
74. K. Sutherlin et al., J. Am. Chem. Soc. **138**, 14294 (2016)
75. E.J. Reijerse et al., J. Am. Chem. Soc. **139**, 4306 (2017)
76. V. Pelmenschikov et al., J. Am. Chem. Soc. **139**, 16894 (2017)
77. R. Lübbers, H.F. Grünsteudel, A.I. Chumakov, G. Wortmann, Science **287**, 1250 (2000)
78. H.K. Mao, J. Xu, V.V. Struzhkin, J. Shu, R.J. Hemley, W. Sturhahn, M.Y. Hu, E.E. Alp, L. Vocadlo, D. Alfè, G.D. Price, M.J. Gillan, M. Schwoerer-Böhning, D. Häusermann, P. Eng, G. Shen, H. Giefers, R. Lübbers, G. Wortmann, Science **292**, 914 (2001)
79. J.F. Lin, W. Sturhahn, J. Zhao, G. Shen, H.K. Mao, R.J. Hemley, Science **308**, 1892 (2005)
80. B. Fultz B, C.C. Ahn, E.E. Alp, W. Sturhahn, T.S. Toellner, Phys. Rev. Lett., **79**, 937 (1997)
81. L. Pasquini, A. Barla, A.I. Chumakov, O. Leupold, R. Rüffer, A. Deriu, E. Bonetti, Phys. Rev. B **66**, 73410 (2002)
82. S. Stankov, Y.Z. Yue, M. Miglierini, B. Sepiol, I. Sergueev, A.I. Chumakov, L. Hu, P. Svec, R. Rüffer, Phys. Rev. Lett. **100**, 235503 (2008)
83. Y. Tamada, R. Masuda, A. Togo, S. Yamamoto, Y. Yoda, I. Tanaka, M. Seto, S. Nasu, T. Ono, Phys. Rev. B **81**, 132302 (2010)
84. S. Couet, M. Sternik, B. Laenens, A. Siegel, K. Parlinski, N. Planckaert, F. Gröstlinger, A.I. Chumakov, R. Rüffer, B. Sepiol, K. Temst, A. Vantomme, Phys. Rev. B **82**, 094109 (2010)
85. S. Stankov, M. Miglierini, A.I. Chumakov, I. Sergueev, Y.Z. Yue, B. Sepiol, P. Svec, L. Hu, R. Rüffer, Phys. Rev. B **82**, 144301 (2010)
86. W. Keune, W. Sturhahn, Hyperfine Interact., **123/124** Part I, 847 (1999)
87. R. Röhlsberger, W. Sturhahn, T.S. Toellner, K.W. Quast, E.E. Alp, A. Bernhard, J. Metge, R. Rüffer, E. Burkel, Phys. B **263–264**, 581 (1999)
88. W. Sturhahn, R. Röhlsberger, E.E. Alp, T.T. Ruckert Schrör, W. Keune, J. Magn. Magn. Mater. **198–199**, 590 (1999)
89. T. Ruckert, W. Keune, W. Sturhahn, M.Y. Hu, J.P. Sutter, T.S. Toellner, E.E. Alp, Hyperfine Interact. **126**, 363 (2000)
90. B.R. Cuenya, W. Keune, W. Sturhahn, T.S. Toellner, M.Y. Hu, Phys. Rev. B **64**, 235321 (2001)
91. S. Couet, K. Schlage, R. Rüffer, S. Stankov, Th Diederich, B. Laenens, R. Röhlsberger, Phys. Rev. Lett. **103**, 097201 (2009)
92. S. Stankov, R. Röhlsberger, T. Ślęzak, M. Sladecek, B. Sepiol, G. Vogl, A.I. Chumakov, R. Rüffer, N. Spiridis, J. Łażewski, K. Parliński, J. Korecki, Phys. Rev. Lett. **99**, 185501 (2007)
93. R.A. Brand, G. Coddens, A.I. Chumakov, Y. Calvayrac, Phys. Rev. B **59**, 14145 (1999)
94. R.P. Hermann, W. Schweika, O. Leupold, R. Rüffer, G.S. Nolas, F. Grandjean, G.J. Long, Phys. Rev. B **71**, 174301 (2005)
95. J.S. Tse, D.D. Klug, J.Y. Zhao, W. Sturhahn, E.E. Alp, J. Baumert, C. Gutt, M.R. Johnson, W. Press, Nat. Mater. **4**, 917 (2005)

96. V. Ksenofontov, G. Wortmann, A.I. Chumakov, T. Gasi, S. Medvedev, T.M. McQueen, R.J. Cava, C. Felser, Phys. Rev. B **81**, 184510 (2010)
97. S. Tsutsui, C.H. Lee, C. Tassel, Y. Yoshida, Y. Yoda, K. Kihou, A. Iyo, H. Eisaki, J. Phys. Soc. Jpn. **79**, 013706 (2010)
98. H. Kobayashi, S. Ikeda, Y. Yoda, H. Nakamura, Machida: M Phys. Rev. B **84**, 184304 (2011)
99. G.J. Long, R.P. Hermann, F. Grandjean, E.E. Alp, W. Sturhahn, C.E. Johnson, D.E. Brown, O. Leupold, R. Rüffer, Phys. Rev. B **71**, 140302 (2005) (R)
100. S. Tsutsui, H. Kobayashi, D. Ishikawa, J.P. Sutter, A.Q.R. Baron, T. Hasegawa, N. Ogita, M. Udagawa, Y. Yoda, H. Onodera, D. Kikuchi, H. Sugawara, C. Sekine, I. Shirotani, H. Sato, J. Phys. Soc. Jpn. **77**, 033601 (2008)
101. A.I. Chumakov, G. Monaco, A. Monaco, W.A. Crichton, A. Bosak, R. Rüffer, A. Meyer, F. Kargl, L. Comez, D. Fioretto, H. Giefers, S. Roitsch, G. Wortmann, M.H. Manghnani, A. Hushur, Q. Williams, J. Balogh, K. Parliński, P. Jochym, P. Piekarz, Phys. Rev. Lett. **106**, 225501 (2011)
102. A.I. Chumakov, Hyperfine Interact. **207**, 107 (2012)
103. X.W. Zhang, Y. Yoda, M. Seto, Yu. Maeda, M. Ando, S. Kikuta, Jpn. J. Appl. Phys. **34**, L330 (1995)
104. R. Haruki, M. Seto, S. Kitao, Y. Yoda, Maeda Yu. J. Phys. Soc. Jpn. **69**, 4049 (2000)
105. R. Haruki, M. Seto, S. Kitao, Y. Kobayashi, Y. Yoda, T. Mitsui, Maeda Yu. J. Phys. Soc. Jpn. **70**, 445 (2001)
106. For a review, see N. Tsuda, K. Nasu, A. Fujimori, K. Shiratori, *Electronic Conduction in Oxides*, 2nd edn. (Springer-Verlag, Berlin, 2000), M. Imada, A. Fujimori, Y. Tokura, Rev. Mod. Phys. **70**, 1039, Sec. E (1998)
107. M. Takano, N. Nakanishi, Y. Takeda, S. Naka, T. Takada, Mater. Res. Bull. **12**, 923 (1977), M. Takano, J. Kawachi, N. Nakanishi, Y. Takeda, J. Solid State Chem. **39**, 75 (1981)
108. For a review, see P. Gutlich: in "*Mössbauer Spectroscopy Applied to Inorganic Chemistry*" vol 1, ed. by G.J. Long (Plenum, New York, 1984) pp. 287–337
109. T. Ishikawa, H. Nakazaki, A. Yasukawa, K. Kandori, M. Seto, Mater. Res. Bull. **33**, 1609 (1998)
110. J.A. Elliott, H.E. Hall, Bunbury D St P: Proc. Phys. Soc. **89**, 595 (1966)
111. D.C. Champeney, W.D. Woodhams, J. Phys. B **1**, 620 (1968)
112. R. Röhlsberger, *Nuclear Condensed Matter Physics with Synchrotron Radiation* (Springer-Verlag, Berlin, 2005)
113. J.Z. Tischler, B.C. Larson, L.A. Boatner, E.E. Alp, T. Mooney, J. Appl. Phys. **79**, 3686 (1996)
114. R. Masuda, T. Mitsui, Y. Kobayashi, S. Higashitaniguchi, M. Seto, Jpn. J. Appl. Phys. **48**, 120221 (2009)
115. A.Q.R. Baron et. al., Phys. Rev. Lett. **79**, 2823 (1997)
116. T. Ishikawa et. al., Rev. Sci. Instrum. **63**, 1015 (1992)
117. G.V. Smirov, Hyperfine Interact. **123**(124), 31 (1999)
118. Yu. Kagan, A.M. Afanas'ev, V.G. Kohn, Phys. C **12**, 615 (1979)
119. G.V. Smirnov, V.G. Kohn, W. Petry, Phys. Rev. B **63**, 144303 (2001)
120. G.T. Tramell, Phys. Rev. Lett. **126**, 1045 (1962)
121. M. Saito, R. Masuda, Y. Yoda, M. Seto, Sci. Rep. **7**, 12558 (2017)
122. G.V. Smirnov et. al., Phys. Rev. B **73**, 184126 (2006)
123. U. Balucani, M. Zoppi, *Dynamics of the Liquid State* (Oxford University Press, UK, 1994)
124. M. Saito et. al., Phys. Rev. Lett. **109**, 115705 (2012)
125. M. Saito et. al., Appl. Phys. Express **2**, 026502 (2009)
126. M. Saito et. al., Hyperfine Interact. **206**, 87 (2012)
127. F. Caporaletti et. al., Rev. Sci. Instrum. **88**, 105114 (2017)
128. T. Kanaya, R. Inoue, M. Saito, M. Seto, Y. Yoda, J. Chem. Phys. **140**, 144906 (2014)
129. M. Saito et. al., Phys. Status Solidi B **257**, 2000113 (2020)
130. T. Yamaguchi et. al., J. Phys. Chem. Lett. **9**, 298 (2018)
131. M. Saito et al., Hyperfine Interact. **238**, 99 (2017)
132. M. Kaisermayr et. al., Eur. Phys. J. B **20**, 335 (2001)

133. M. Saito et. al., J. Phys. Soc. Jpn. **81**, 023001 (2012)
134. M. Saito et. al., Phys. Rev. Res. **1**, 012008(R) (2019)
135. M. Saito et. al., Hyperfine Interact. **241**, 14 (2020)
136. S. Castrignano, J. Evers, Phys. Rev. Lett. **122**, 025301 (2019)

Chapter 3
Quantum Optical Phenomena in Nuclear Resonant Scattering

Ralf Röhlsberger and Jörg Evers

Abstract With the advent of high-brilliance, accelerator-driven light sources such as modern synchrotron radiation sources or x-ray lasers, it has become possible to extend quantum optical concepts into the x-ray regime. Owing to the availability of single photon x-ray detectors with quantum efficiencies close to unity and photon-number resolving capabilities, fundamental phenomena of quantum optics can now also be studied at Angstrom wavelengths. A key role in the emerging field of x-ray quantum optics is taken by the nuclear resonances of Mössbauer isotopes. Their narrow resonance bandwidth facilitates high-precision studies of fundamental aspects of the light-matter interaction. A very accurate tuning of this interaction is possible via a controlled placement of Mössbauer nuclei in planar thin-film waveguides that act as cavities for x-rays. A decisive aspect in contrast to conventional forward scattering is that the cavity geometry facilitates the excitation of cooperative radiative eigenstates of the embedded nuclei. The multiple interaction of real and virtual photons with a nuclear ensemble in a cavity leads to a strong superradiant enhancement of the resonant emission and a strong radiative level shift, known as collective Lamb shift. Meanwhile, thin-film x-ray cavities and multilayers have evolved into an enabling technology for nuclear quantum optics. The radiative coupling of such ensembles in the cavity field can be employed to generate atomic coherences between different nuclear levels, resulting in phenomena including electromagnetically induced transparency, spontaneously generated coherences, Fano resonances and others. Enhancing the interaction strength between nuclei in photonic structures like superlattices and coupled cavities facilitates to reach the regime of collective

R. Röhlsberger (✉)
Deutsches Elektronen-Synchrotron DESY, Notkestr. 85, 22607 Hamburg, Germany
e-mail: ralf.roehlsberger@desy.de

Helmholtz Centre for Heavy Ion Research (GSI), Planckstr. 1, 64291 Darmstadt, Germany

J. Evers
Max-Planck-Institut für Kernphysik, Saupfercheckweg 1, 69117 Heidelberg, Germany
e-mail: joerg.evers@mpi-hd.mpg.de

Y. Yoshida and G. Langouche (eds.), *Modern Mössbauer Spectroscopy*,
Topics in Applied Physics 137, https://doi.org/10.1007/978-981-15-9422-9_3

strong coupling of light and matter where phenomena like normal mode splitting and Rabi oscillations appear. These developments establish Mössbauer nuclei as a promising platform to study quantum optical effects at x-ray energies. In turn, these effects bear potential to advance the instrumentation and applications of Mössbauer science as a whole.

3.1 Introduction

The study and applications of light-matter interactions in the optical regime have undergone a revolutionary development over the last decades, to the point where now quantum technologies become a reality. Quantum mechanical phenomena in this interaction are the domain of quantum optics, which encompasses semiclassical setups exploiting the quantum-mechanical nature of the matter, as well as cases in which the quantum character of the light has to be taken into account [1–4]. A key driver for the advancement continues to be the progress in laser source technology, also beyond the visible light regime.

3.1.1 Light Sources for X-Ray Quantum Optics

X-ray quantum optics has not been very prominent in the early phase of x-ray science, not least because of source limitations. For instance, unlike a laser source, typical x-ray sources emit photons into a large number of electromagnetic-field modes, severly restricting the control possibilities offered by the light. This is no longer the case for experimental conditions that can be realized with modern synchrotron radiation sources and x-ray free-electron lasers, together with increasing source brilliance and advances in x-ray optical elements and detection techniques (for a view on the evolution of the brilliance of x-ray sources, see Fig. 3.1). As a result, the study of quantum optical effects in the interaction of light and matter moves within reach at hard x-ray energies, and is becoming increasingly relevant for new enabling experimental possibilities and for the interpretation of data obtained at these radiation sources. Broadly speaking, the long-term goals of this approach are to fully exploit the capabilities offered by the new x-ray sources, and to continue the success story of quantum optics at hard x-ray energies.

3.1.2 X-Ray Quantum Optics with Atomic Resonances

Two key concepts of quantum optics are coherence and interference. Sharp resonances are favorable in this regard, since the narrow linewidth translates into comparably long lifetimes of coherent superpositions of the involved atomic states. At

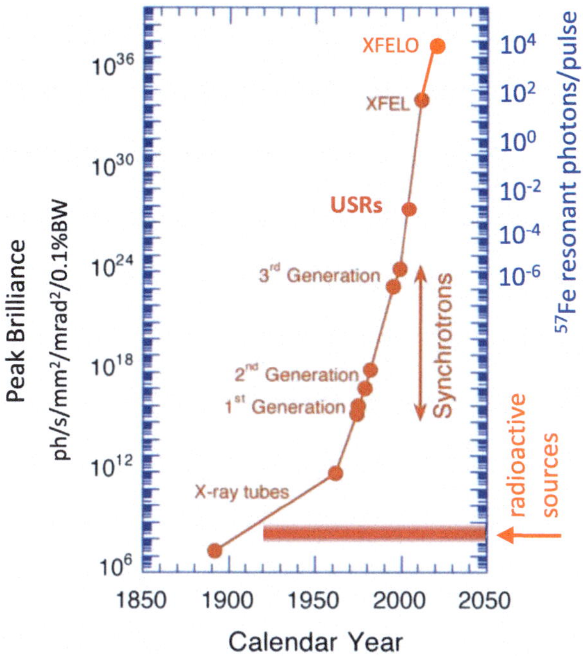

Fig. 3.1 Evolution of brilliance (a.k.a brightness) of x-ray sources since the discovery of x-rays. The advent of synchrotron radiation sources enabled the first accelerator-based nuclear resonant scattering experiments following the proposal by Ruby [5]. The further increase in brilliance resulting from the improvement of storage-ring technology facilitated a multitude of unique applications throughout the natural sciences [6]. The ultimate limit in storage ring technology is reached when the diffraction limit of electron and photon beams is encountered (USR = ultimate storage ring). A further increase in brilliance is possible with x-ray free electron lasers (XFEL) based on the SASE process (SASE = self-amplified spontaneous emission). At these levels, the x-ray pulses may contain several photons within the resonance bandwidth of the nuclear transition, which enables the realization of coherent multiphoton excitations for experiments in quantum and nonlinear optics. Ultimate brilliance values are expected when the SASE process is amplified in a cavity as proposed in the XFEL-oscillator (XFELO) concept [7, 8]

hard x-ray energies, however, it becomes increasingly difficult to find sharp electronic resonances in atoms because they are intrinsically lifetime-broadened due to strong competing interactions within the inner electron shell, see Fig. 3.2. A fortunate exception from this rule are nuclear resonances. If they are of sufficiently low energy (< 100 keV) and if the nucleus is bound in a solid, we observe the Mössbauer effect of recoilless absorption and emission of photons. This leads to the immediate consequence of coherence in the scattering of radiation from nuclear resonances because the interaction is completely elastic (the final state and the initial state are identical). As a result, nuclear resonances of Mössbauer isotopes are particularly promising in terms of coherence and interference effects. On the other hand, the Mössbauer resonances are much more narrow than the spectra of the pulses delivered by mod-

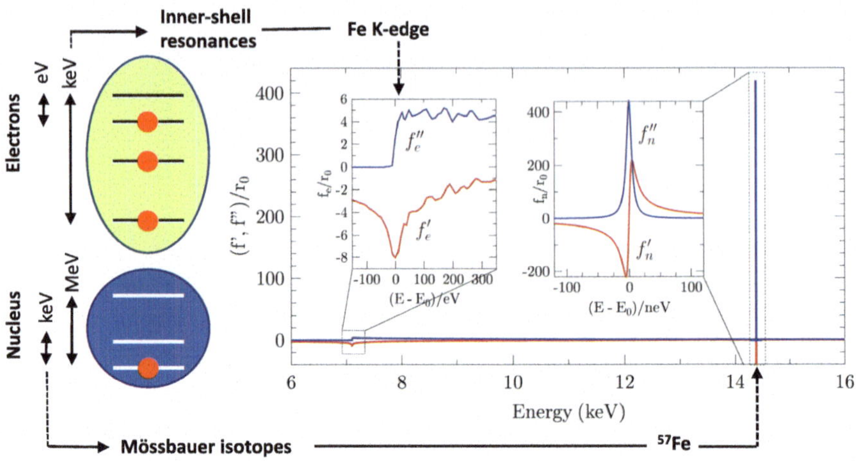

Fig. 3.2 Bound states of electrons or nucleons in atoms are the origin of electromagnetic resonances in matter. While the keV excited states of inner-shell electrons are affected by several competing decay channels and thus are strongly lifetime broadened, the isolated keV nuclear resonances can be observed with their ultranarrow natural linewidth if the nuclei are bound in a solid. This is due to the Mössbauer effect, in which the whole solid with its large mass acts as a recoil partner so that the recoil energy exchanged with the solid during absorption or emission is negligibly small. The right graph shows the real and imaginary parts of the atomic scattering amplitudes f' and f'', respectively, in the vicinities of the Fe K-edge at 7.1 keV and the 14.4 keV nuclear resonance of ^{57}Fe. Please note the relative amplitudes of the electronic and nuclear scattering amplitudes as well as their largely different energy scales and spectral shapes. While the Fe K-edge absorption proceeds from a bound state into the continuum, the 14.4 keV transition can be considered as an almost ideal two-level system connecting two discrete nuclear levels

ern x-ray sources, such that it is challenging to strongly drive nuclear resonances as compared to corresponding electronic resonances. Already these general observations separate electronic and nuclear resonances into complementary platforms to establish quantum optical concepts at x-ray energies. This review will focus on how quantum optical phenomena can be realized in the regime of hard x-rays via the nuclear resonances of Mössbauer isotopes.

3.1.3 Collective and Virtual Effects in Quantum Optics

Next to coherence and interference, the structure and control of field modes, photon correlations, entanglement, vacuum fluctuations and virtual processes, spontaneous and stimulated emission, and nonlinear optical interactions are important elements of quantum optics. They are fundamentally affected if many identical atoms are interacting with the same radiation field. This has led to the development of the research area of cooperative emission in quantum optics, which has been mostly developed on theoretical grounds during many decades because the preparation of ensembles of identical emitters was by far not trivial for a long time. This situation has changed significantly in recent years, e.g., due to the development of storing and manipulating of atoms in electromagnetic traps, but also by the possibility to prepare resonant atoms in solid state environments in a very controlled fashion. It turns out that such a controlled preparation of identical atoms is naturally realized for certain experimental settings involving Mössbauer isotopes, and that the engineering of cooperative effects in turn is indispensable for implementing advanced quantum optical schemes with nuclei.

Collective and virtual effects in the interaction of identical atoms with single photons are the source of intriguing phenomena in atomic physics and quantum optics [9–22] extending into the regime of hard x-rays [23–29]. The cooperative character of the interaction modifies the decay rate [30] (see Fig. 3.3) and shifts the resonance energy of the atomic ensemble as compared to a single atom [31], also known as the collective Lamb shift. Nowadays, these effects are becoming increasingly attractive to create entangled atomic ensembles [32] for applications ranging from quantum memories [33], quantum information processing [14] to radiative transport of energy in light-harvesting systems [34]. In particular, as discussed in this review, they also allow for the design of cooperative nuclear level schemes [28, 29, 35–40].

The collective decay rate of an ensemble of identical resonant atoms was introduced by Dicke in his pioneering work on superradiance [30]. In contrast to the atomic Lamb shift, the collective Lamb shift emerges when a virtual photon emitted from one atom is not absorbed by the same atom but by another atom within the ensemble [31, 41]. The investigation of the collective Lamb shift induced by virtual processes has received stimulated theoretical interest [15, 18, 19, 31, 41–44] that has been accompanied by recent experimental studies [27, 45–48]. Virtual transitions not only lead to a shift of the transition energy, but have an interesting effect on the collective decay rate as well [19–21]: They partially transfer population from the initially superradiant state into slowly decaying states, resulting in a trapping of the atomic excitation. On the other hand, virtual transitions open additional decay channels for otherwise trapped states. It lies at the heart of superradiance that the presence of many identical atoms opens a large number of potential decay channels for collective excitations. From that perspective such systems are appealing examples for open and marginally stable quantum many-body systems [49].

3.1.4 X-Ray Cavities as Enabling Tool for Nuclear Quantum Optics

Today it is possible to experimentally access collections of identical resonators in a controlled fashion, ranging from atomic Bose-Einstein condensates to quantum dots in solid state systems. Moreover, laser technology has reached a level of advancement that allows to control the light-matter interaction down to timescales of attoseconds. Currently this field of research progresses to shorter and shorter wavelengths into the regime of hard x-rays. However, it is not only the sharpness of the nuclear resonances and their favorable coherence properties which render Mössbauer nuclei an ideal candidate to experimentally explore cooperative phenomena at x-ray energies. Also, the possibility to engineer the interaction of x-rays with nuclei and the coupling between nuclei via their geometric arrangement or by embedding them into photonic nanostructures opens many fascinating routes to realize quantum optical concepts with nuclei. In this respect, thin-film x-ray cavities and multilayers have become an enabling technology for nuclear quantum optics. These cavities transform the propagating x-ray field delivered by the source into a standing wave field structure, and the precise placement of the nuclei within this standing wave allows for an accurate tuning of the interaction of the nuclei with the x-rays. Another decisive aspect is that the cavity geometry facilitates the excitation of single cooperative radiative eigenstates of the embedded nuclei, and to tailor the superradiant enhancement of the resonant emission as well as the collective Lamb shift. The possibilities are further enriched if the magnetic substructure of the nuclei is exploited, or if different nuclear ensembles are embedded within a single thin film structure. Then, the cavity fields can be employed to generate atomic coherences between different nuclear states, and to induce couplings between nuclear states up to the regime of strong collective coupling, opening additional new possibilities. This enabled the implementation of archetype quantum optical phenomena such as electromagnetically induced transparency, spontaneously generated coherences, Fano resonances and others. While much progress has already been achieved on the level of single excitations, we anticipate further enrichment of this fascinating field of physics facilitated by the ongoing development of modern x-ray sources like high-brilliance synchrotrons and x-ray lasers, see Fig. 3.1. These sources are capable of delivering many resonant photons in each single radiation pulse, providing a direct route towards multiphoton x-ray optics, and opening perspectives for associated effects like stimulated emission, x-ray lasing, nonlinear optics and more.

3.1.5 Outline of this Review

This review is organized as follows. In Sect. 3.2 of this chapter we review the properties of nuclear resonances as almost ideal two-level systems that can be prepared as identical emitters in various structural arrangements. This leads us then in Sect. 3.3 to discuss general properties of ensembles of Mössbauer isotopes forming a cooperative atomic environment concerning their radiative properties. Specifically, in Sect. 3.4 we will discuss the properties of the nuclear exciton, i.e., the state that is formed after impulsive excitation of a nuclear ensemble by a radiation pulse, the duration of which is much shorter than the collective nuclear lifetime. Section 3.5 describes the most fundamental effect of cooperative emission, the collective Lamb shift, the observation of which was enabled via the application of planar x-ray cavities. While this section contains a semiclassical description of the underlying physics to illustrate the basic concepts of x-ray cavities as 'enabling technology' for this field, the following Sect.

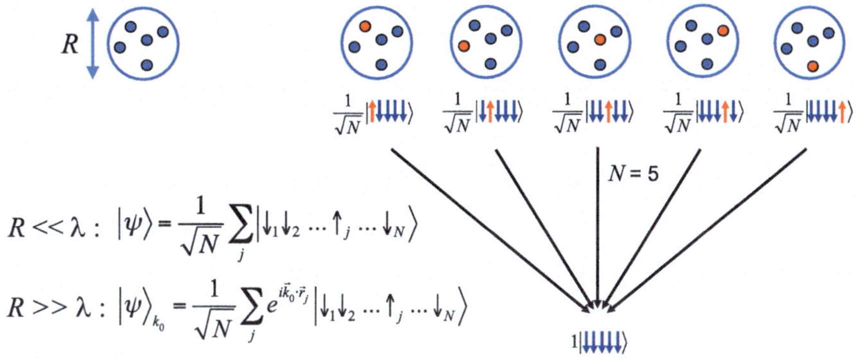

Fig. 3.3 Illustration of single-photon superradiance according to Dicke [30]. If a sample consisting of N identical resonant atoms is excited by a single photon, each of the atoms can be excited (red dots), but we do not know which one. Therefore, each Fock state of the system with one atom exited (red arrow up) contributes with the same probability to the state vector $|\psi\rangle$ of the whole sample. Since all the singly excited states decay to the same ground state, the sample can radiate its energy via N different pathways, so the decay proceeds N times faster than the decay of a single atom. This applies for the case that the linear dimensions of the sample are smaller than the wavelength. In the opposite case, the relative spatial phases of the atoms have to be taken into account which leads to a complex non-exponential temporal evolution of the collective decay that is strongly directional [26, 50]. In experiments with x-rays, samples are typically much larger than the radiation wavelength, so this is the most frequently encountered case. For the description of collective nuclear resonant scattering the states $|\psi\rangle_{k_0}$ have been coined 'nuclear excitons' [26], in a more general perspective they are referred to as 'timed Dicke states' [9]

3.6 provides a fully quantum optical description of the Mössbauer nuclei in x-ray cavities, setting the stage for inclusion of multiphoton excitation conditions. Section 3.7 is then devoted to quantum optical effects in x-ray cavities that result from the formation of coherences in this particular environment, like Fano resonance control, electromagnetically induced transparency, spontaneously generated coherences, and slow light. Further engineering of the atomic environment to form superlattices or coupled cavities allows one to reach the regime of collective strong coupling. This is discussed in Sect. 3.8, illustrated by the observation of normal-mode splitting and Rabi oscillations between nuclear ensembles. Finally, Sect. 3.9 provides an outlook on the ongoing development of modern high-brilliance x-ray sources and how they will contribute to further development of this exciting research field.

3.2 Nuclear Resonances of Mössbauer Isotopes as Two-Level Systems

In the x-ray regime, the nuclear resonances of Mössbauer isotopes provide almost ideal two-level systems to study the effects of cooperative emission. After being proposed by Ruby in [5], the use of synchrotron radiation for nuclear resonant scattering was demonstrated first by Gerdau et al. in 1985 for nuclear Bragg diffraction [51] and by Hastings et al. in [52] for nuclear resonant forward scattering. Since then the technique became an established method at many synchrotron radiation sources around the world with a multitude of applications in various fields of the natural sciences [6, 53, 54]. The most widely used isotope in this field is ^{57}Fe with a transition energy of E_0 = 14.4125 keV, a natural linewidth of Γ_0 = 4.7 neV, corresponding to a lifetime of τ_0 = 141 ns. Beamlines at present-day 3rd generation synchrotron radiation sources like ESRF, APS, SPring8 and PETRA III deliver a spectral flux of about 10^5 photons/s/Γ_0. The radiation comes typically in pulses with a duration of a few 10 ps, so that excitation and subsequent emission can be treated as independent processes.

Before discussing cooperative effects in the resonant interaction of many identical nuclei with a common radiation field, it is instructive to discuss first the scattering behavior of a single atom. The scattered field of an atom in momentum-frequency space is given by [55]:

$$\mathbf{A}(\mathbf{k},\omega) = -c\frac{\delta_+(k,\omega)}{(2\pi)^4}\int \langle\Psi_f|\mathbf{M}(\mathbf{k},\omega,\mathbf{k}',\omega')|\Psi_i\rangle\, \mathbf{A}_0(\mathbf{k}',\omega')d^3k'\,d\omega' \qquad (3.1)$$

The scattering process described by this equation can be read from right to left: The incoming field is represented by $\mathbf{A}_0(\mathbf{k}',\omega')$, which may be understood as the wave function of a photon. In fact, $|\mathbf{A}_0(\mathbf{k},\omega)|^2 d^3k\,d\omega$ is the probability of finding the incoming photon in the mode characterized by the wave vector $\mathbf{k}$ and energy ω. $\mathbf{M}$ is the scattering operator of the atom for scattering an incident photon with $\mathbf{k}',\omega'$ into an outgoing photon with $\mathbf{k},\omega$ and $\delta_+(k,\omega) = -4\pi c/(\omega^2 - k^2c^2 + i\epsilon)$ is the propagator of the outgoing photon. The scattering operator $\mathbf{M}$ depends on the electromagnetic current $\mathbf{b}$, on the Hamiltonian $\mathbf{H}$ and on the propagator $\mathbf{G}_0$ of the atom:

$$\begin{aligned}&\mathbf{M}(\mathbf{k},\omega,\mathbf{k}',\omega') \qquad (3.2)\\ &= \frac{i}{c}\int e^{-i(\mathbf{kx}-\omega t)}e^{i\mathbf{H}t}\,\mathbf{b}(x)\,\mathbf{G}_0(t-t')\,\mathbf{b}(x')\,e^{-i\mathbf{H}t'}\,e^{i(\mathbf{k}'\mathbf{x}'-\omega' t')}d^3x\,d^3x'\,dt\,dt'\end{aligned}$$

The propagator $\mathbf{G}_0$ itself can be expressed in terms of the Hamiltonian $\mathbf{H}$ and the level-shift operator Δ:

$$\mathbf{G}_0(t-t') = \frac{i}{2\pi}\int \frac{e^{-i\omega(t-t')}}{\omega - \mathbf{H} - \Delta(\omega)}\,d\omega \qquad (3.3)$$

The level-shift operator is in general non-Hermitian. Δ consists of a radiative contribution Δ_γ resulting from the perturbation of the atom by its own photon field (the self energy) and of a non-radiative contribution Δ_α that originates from internal conversion. The real part of Δ gives the single-atom Lamb shift [56] while the imaginary part is the decay width of the transition. The smaller this imaginary contribution becomes, the more pronounced is the resonance behavior of the propagator. This leads to a strong enhancement of the scattering in the vicinity of sharp nuclear resonances.

Since we are interested here in coherent elastic scattering ($\omega = \omega'$) we consider only the diagonal elements of the operator **M**. The currents **b** of the atom are split into the nuclear and the electronic part. This gives three contributions to the scattering operator: the pure electronic part **E**, the pure nuclear part **N**, and an interference term between the nuclear and electronic currents that can be neglected in most cases. The nuclear contribution to the atomic scattering operator for an unsplit (single-line) nuclear resonance is given by:

$$\mathbf{N}(\mathbf{k},\omega,\mathbf{k}',\omega) = e^{i(\mathbf{k}-\mathbf{k}')\cdot\mathbf{R}}\,\frac{f_0\,(\Gamma_0/2)}{\omega-\omega_0-i\Delta} \quad \text{with} \quad f_0 = \frac{f_{LM}}{2k_0}\frac{2I_e+1}{2I_g+1}\frac{1}{1+\alpha} \tag{3.4}$$

where f_{LM} is the Lamb-Mössbauer factor, I_g and I_e are the spins of the ground and excited nuclear states, respectively, and α is the coefficient of internal conversion. Effectively, the situation of an unsplit ground and excited state justifies a scalar approach to the scattering problem. As we will see in the next section, Δ is not a property of the single atom only, but can be greatly affected by cooperative effects, i.e., by the radiative coupling of many identical atoms.

3.3 The Nuclear Level Width in a Cooperative Atomic Environment

In an ensemble of many identical atoms a radiated photon may interact not only with the same atom but also with identical atoms within the same ensemble. To describe this interaction a diagrammatical approach was introduced by Friedberg et al. in [31] that is illustrated in Fig. 3.4.

This leads to the complex-valued self-energy correction $\Delta_C = L_C + i\,\Gamma_C$ of the collective resonance energy of the atomic ensemble. To sum all these repeated dia-

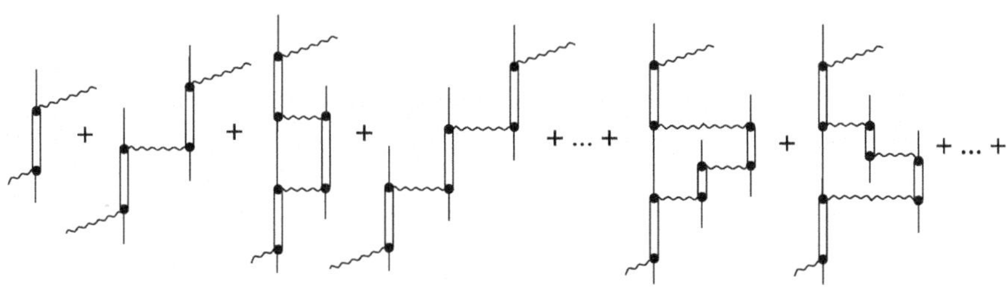

Fig. 3.4 Photon scattering from a resonant atom (vertical double line: excited state) involving the exchange of virtual photons (horizontal wavy lines) with other atoms in the ensemble. The total amplitude is given by the sum over all possible diagrams

grams in Fig. 3.4 we note that the Fourier transform $\tilde{\mathbf{G}}(\omega)$ of the excited-state propagator $\mathbf{G}(t - t')$ satisfies the Dyson equation

$$\tilde{\mathbf{G}}(\omega) = \tilde{\mathbf{G}}_0(\omega) - i\tilde{\mathbf{G}}_0(\omega)\,\tilde{\Delta}_C(\omega)\,\tilde{\mathbf{G}}(\omega) \tag{3.5}$$

where $\tilde{\mathbf{G}}_0(\omega) = 1/(\omega - \mathbf{H} - \Delta(\omega))$ is the uncorrected propagator of the single atom. Solving Eq. (3.5) for the corrected propagator yields:

$$\tilde{\mathbf{G}}(\omega) = \frac{1}{\omega - \mathbf{H} - \Delta(\omega) - \Delta_C(\omega)}\,. \tag{3.6}$$

It should be noted that the Dyson equation above only provides the proper summing of the repeated diagrams in Fig. 3.4. The amplitudes of the individual diagrams have to be calculated before. The diagrammatical technique has been applied in a pioneering paper [31] to calculate the collective Lambshift. Since then the CLS has been calculated for various geometries (sphere, cylinder, slab) and models for the electromagnetic field (scalar/vector) [11, 31, 41, 57, 58]. The result of these calculations in the large-sample limit, i.e., for $k_0\,R \gg 1$ with R being the size of the sample can be summarized as follows:

$$\Delta_C \approx i\Gamma_C\left(1 - \frac{i\,S}{k_0 R}\right) \quad \text{with} \quad \Gamma_C = \frac{3}{2}\frac{N}{(k_0 R)^2}\,\Gamma_0 = \frac{\rho\lambda^2 R}{2\pi}\,\Gamma_0\,, \tag{3.7}$$

where ρ is the number density of resonant atoms in the sample. S is a factor that depends on the shape of the sample and on the scalar/vector model of the field. Thus, for the CLS to be observable, the quantity $\rho\,\lambda^3$ has to be sufficiently high. In gaseous samples, however, an increase of the density goes along with the increase of interactions between atoms, leading to collisional broadening of the resonance line. In condensed matter systems significantly higher number densities than in gases can be reached without these perturbing effects. In this case a detrimental effect that could quench cooperative emission is the inhomogeneous broadening of atomic and nuclear resonances due to interactions of the resonators with their environment. While atomic resonances are most susceptible to the interaction with their surrounding, nuclear resonances are much less affected. In fact, by controling the environment of the Mössbauer isotopes in solids it is possible to prepare ensembles of identical resonators with high number density while still keeping the natural linewidth of the transition. It appears that the narrow nuclear resonances of Mössbauer isotopes provide an almost ideal two-level system for the study of cooperative effects in the interaction of x-rays with matter.

In the following we will describe a procedure how to calculate the eigenmodes of an ensemble of resonant atoms that yields the eigenfrequencies together with the complex self-energy correction Δ_C. The derivation follows in great parts the treatment given in [26], p. 234ff.

3.4 The Nuclear Exciton, Radiative Eigenstates and Single-Photon Superradiance

Soon after the discovery of the Mössbauer effect it became clear that ensembles of nuclei collectively excited by single photons bear a number of fascinating properties. These states have been called 'nuclear excitons' and the physics of them was explored theoretically by Afanas'ev and Kagan [23] as well as Hannon and Trammell [24, 25]; for an extensive review see [26]. With the advent of high-brilliance synchrotron radiation it became possible to prepare such states and study their properties systematically. Due to the small number of photons per mode of the radiation field at these sources, however, there is in most cases only one photon interacting with the resonant ensemble at a time. In the following we investigate collectively excited atomic (nuclear) states that have been created by short-pulse excitation containing one photon at most. Since we do not know which nucleus is excited, all possible Fock states $|b_1\, b_2 \ldots a_j \ldots b_N\rangle$ containing one excited nucleus (a_j) while the others (b_i) are in the ground state, contribute with equal weight to the state vector of the whole system. In this sense, the superradiant excitonic states are those of maximum delocalization of the excitation energy.

For the case that the sample extension R is much smaller than the wavelength of the radiation, $k_0 R \ll 1$, the exciton state is written as

$$|\Psi_e\rangle = \frac{1}{\sqrt{N}} |b_1\, b_2 \ldots a_j \ldots b_N\rangle. \tag{3.8}$$

This state is fully symmetric with respect to exchange of any two atoms, therefore it is often called the symmetric Dicke state. In most cases of optical physics up into the x-ray regime, however, the opposite limit is encountered where $k_0 R \gg 1$, so that the spatial position of the atoms within the ensemble has to be taken into account:

$$|\Psi_e(\mathbf{k}_0)\rangle = \frac{1}{\sqrt{N}} \sum_j e^{\mathrm{i}\, \mathbf{k}_0 \cdot \mathbf{R}_j} |b_1\, b_2 \ldots a_j \ldots b_N\rangle, \tag{3.9}$$

where $\mathbf{k}_0$ is the wave vector of the incident photon and $\mathbf{R}_j$ denotes the position of the jth atom. This state was introduced to describe coherent nuclear resonant scattering as 'nuclear exciton' [23, 26] or more recently as 'timed Dicke state' [41], because atoms at various locations within the extended sample are excited at different times.

3.4.1 Radiative Normal Modes

The collective spectral response of a given ensemble of emitters and the temporal evolution of its decay can be obtained by determination of the radiative normal modes. The scattering of an external wave proceeds via virtual excitation of these modes as intermediate states. The decay of a collectively excited state is then a superposition of exponentially decaying normal modes of the system. In the following we will summarize how to obtain the Hamiltonian equation of motion of the system that determines the complex normal mode frequencies ω_n, based on the formalism layed out in Ref. [26]. Due to retardation effects in the 'timed Dicke state' of Eq. (3.9), the Hamiltonian is symmetric rather than Hermitian. For that reason the eigenmodes $|\Psi_n\rangle$ are transpose orthogonal rather than Hermitian orthogonal. A general superposition exciton state $|\Psi_e\rangle = \sum a_n |\Psi_n\rangle$, prepared by pulsed excitation, will develop dynamical beats in the time evolution of its decay, resulting from destructive interference effects between the light emitted from the normal modes. Under certain conditions, however, a single superradiant eigenmode $|\Psi_e(\mathbf{k}_0)\rangle$ can be excited that exhibits a simple enhanced exponential decay. This is the case for single crystalline samples if the wavevector $\mathbf{k}_0$ of the incident photons satisfies a symmetric Bragg condition or if $\mathbf{k}_0$ excites a single mode in a cavity [59]. If $\mathbf{k}_0$ is off-Bragg (i.e. transmission in forward direction) then $|\Psi_e(\mathbf{k}_0)\rangle$ is a superposition of normal modes. The spread of frequencies of these modes and their Hermitian nonorthogonality determine the superradiant decay at early times and the emergence of dynamical beats thereafter. Because the energy bandwidth of the synchrotron radiation pulses (meV - eV, depending on the degree of monochromatization) is much larger than the natural linewidth of the nuclear transition (4.7 neV for ^{57}Fe), the incident pulse covers the energies of all radiative eigenmodes of the sample, such that their excitation only depends on arrangement of the nuclei.

In a classical system of resonators with oscillating dipole moments, the coupled equations of motion lead to an eigenvalue equation from which the eigenfrequencies and the eigenvectors of the semi-stationary (decaying) normal modes can be determined:

$$\tilde{h}\,\mathbf{X} = \omega\,\mathbf{X} \tag{3.10}$$

with $\mathbf{X}$ being an N-component vector that contains the amplitudes of all N oscillators. $\tilde{h}$ is the Hamitonian of the system. In a quantum mechanical description one obtains the equations of motion by taking the Fourier transform of the decaying exciton $\mathbf{G}_0(t-t')\,|\Psi_e(\mathbf{k}_0)\rangle$ with $\mathbf{G}_0(t-t')$ given by Eq. (3.3) [26]. The Hamiltonian equation of motion has the same shape as Eq. (3.10) where the state vector is now the nuclear exciton

$$\mathbf{X} = |\Psi_e\rangle = \sum_j c_j\,|b_1\,b_2\ldots a_j\ldots b_N\rangle = \begin{pmatrix} c_1 \\ c_2 \\ \vdots \\ c_N \end{pmatrix} \tag{3.11}$$

where here and in the following for notational simplicity we identify the quantum mechanical states with their vector representation in the basis of Fock states $|b_1\,b_2\ldots a_j\ldots b_N\rangle$. The Hamiltonian is given by

$$h_{ii} = \omega_0 - \frac{i\Gamma}{2} \quad \text{and} \quad h_{ij}(i \neq j) = -\frac{\Gamma_\gamma}{2}\kappa_{ij}\frac{e^{ik_0 R_{ij}}}{k_0 R_{ij}} \tag{3.12}$$

with

$$\kappa_{ij} \approx \begin{cases} \frac{3}{2}\left(3\cos^2\Theta - 1\right)\left[\frac{1}{(k_0 R_{ij})^2} - \frac{i}{k_0 R_{ij}}\right] & \text{(near zone } R_{ij} \ll \lambda) \\ \frac{3}{2}\sin^2\Theta & \text{(far zone } R_{ij} \gg \lambda) \end{cases} \tag{3.13}$$

where Θ is the angle between the wavevector of the outgoing photon and the polarization direction of the oscillator as determined by the polarization of the incident photon. The complex frequencies $\omega_m = \omega'_m - i\Gamma_m/2$ of the normal modes are obtained via the determinant equation

$$\text{Det}[\tilde{h} - \omega\,\tilde{\mathbf{1}}] = 0 \tag{3.14}$$

where $\tilde{\mathbf{1}}$ is the $N \times N$ unity matrix. The resulting frequencies ω'_m and the decay widths Γ_m will generally be different from the corresponding values of an isolated nucleus. After determination of the eigenvectors $\mathbf{X}_m$ we obtain the $N \times N$ matrix U that diagonalizes the Hamiltonian $\tilde{h}$ (the rows of U are the transpose eigenvectors $\mathbf{X}_m^T$):

$$U\,\tilde{h}\,U^{-1} = \tilde{\omega} \tag{3.15}$$

with $U^{-1} = U^T$ and $\tilde{\omega}$ being the diagonal eigenvalue matrix $[\tilde{\omega}]_{mn} = \omega_m\,\delta_{mn}$. Since the trace of a matrix is an invariant under a similarity transformation, we have $\text{Tr}(\tilde{h}) = \text{Tr}(\tilde{\omega})$, which is equivalent to:

$$\sum_m \omega_m = N\left(\omega_0 - \frac{i\Gamma}{2}\right) \tag{3.16}$$

From this equation two important sum rules for the real and imaginary part follow:

$$\sum_m \delta\omega = \sum_m (\omega'_m - \omega_0) = 0 \tag{3.17}$$

$$\sum_m \Gamma_m = N\,\Gamma_0 \tag{3.18}$$

The frequency shift sum rule, Eq. (3.17), means that the frequency shifts of all modes average to zero. If some modes are selectively excited or unequally populated, one may nevertheless observe an overall net shift. The decay width sum rule, Eq. (3.18), states that the decay width averaged over all modes equals that of a single resonator. With the normal mode state vectors $|\Psi_m\rangle$ and their complex frequencies ω_m now at hand, we can calculate the time evolution of any single-exciton state $|\Psi_e\rangle$ via

$$|\Psi_e(t)\rangle = \sum_m a_m\, e^{-i\omega_m t}|\Psi_m\rangle \qquad \text{with} \qquad a_m = \langle\Psi_m^T|\Psi_e\rangle \tag{3.19}$$

In a synchrotron experiment with broadband excitation, we observe the decay of the excitation probability that is given by

$$I(t) = \langle \Psi_e(t) | \Psi_e(t) \rangle = \sum_{m,n} a_n^* a_m \, e^{i(\omega_n^* - \omega_m)t} \langle \Psi_n | \Psi_m \rangle \tag{3.20}$$

Since the normal modes $|\Psi_m\rangle$ are transpose orthogonal rather than Hermitian orthogonal, we have in general $\langle \Psi_n | \Psi_m \rangle \neq 0$. This gives rise to dynamical beats between the modes in the temporal evolution $I(t)$ of the decay. In the following subsections we discuss the most frequently encountered cases, i.e., forward scattering and Bragg scattering with particular emphasis on cooperative effects encountered in these geometries.

3.4.2 Forward Scattering

The state vector in Eq. (3.8) corresponds to the small sample limit, also called the simple Dicke limit. The time evolution of the decay of this state to the ground state is strictly exponential, but due to the lack of spatial phasing there is no directionality involved. On the other hand, for extended samples ($kR \gg 1$) the spatial phasing in Eq. (3.9) leads to directional emission that is the situation most frequently encountered in experiments, especially in the regime of hard x-rays.

The exciton $|\Psi_e(\mathbf{k}_0)\rangle$ created by the synchrotron pulse can be considered a Bloch wave given by Eq. (3.9). However, the Bloch waves are generally not the true radiative normal modes in a crystal. In general, the Bloch state $|\Psi_e(\mathbf{k}_0)\rangle$ is a superposition of radiative eigenmodes, which exhibit a distribution of eigenfrequencies and decay rates. In all cases, the initial decay is always superradiant but the decay at delayed times is drastically different, depending on whether the exciton $|\Psi_e(\mathbf{k}_0)\rangle$ is an eigenmode or not.[1] In the case of an eigenmode, the scattered signal $I(t)$ exhibits a pure exponential decay with an enhanced decay rate. On the other hand, if $|\Psi_e(\mathbf{k}_0)\rangle$ is a superposition of eigenmodes, the superradiant components die out quickly, leaving a superposition of slowly decaying components with a distribution of eigenmode frequencies. This leads to a slowly decaying beating signal at delayed times, referred to as dynamical beats or propagation quantum beats, as illustrated in Fig. 3.5. They have been observed not only for nuclear resonant scattering [50], but also for coherent forward scattering from excitons in the optical domain [60]. Quantitatively, the response function of the sample, characterizing the amplitude of the scattered light for an incident field $\delta(t)$, is given by

$$A(t) = \delta(t) - e^{-\Gamma_0 t/2\hbar} \, \frac{\Gamma_C}{\hbar} \, \frac{J_1(\sqrt{4\Gamma_C t/\hbar})}{\sqrt{\Gamma_C t/\hbar}}, \tag{3.21}$$

---

[1] In a great part of the literature about the collective Lamb shift the values given are valid only for the initial phase of the temporal evolution where the decay can be considered superradiant.

where Γ_C depends on the thickness of the sample, see Eq. (3.7) with $R = L_{\parallel}$. At early times the decay is essentially superradiant with an enhanced decay width given by $\Gamma_0 + \Gamma_C$. At delayed times the decay of the exciton proceeds with an envelope given by $1/\sqrt{t^3}$ and an onset of dynamical beats $\sim \cos^2(\sqrt{t})$. The width Γ_C for the initial radiative decay is given by [26]

$$\Gamma_C(\mathbf{k}_0) = \frac{\Gamma_\gamma}{4\pi N} \int d\Omega \, \sin^2\Theta \, |S(\mathbf{k} - \mathbf{k}_0)|^2 \tag{3.22}$$

with

$$S(\mathbf{k} - \mathbf{k}_0) = \sum_{j=1}^{N} \exp[-i(\mathbf{k} - \mathbf{k}_0) \cdot \mathbf{R}_j] \tag{3.23}$$

where $\mathbf{k}$ is the wavevector of the outgoing photon, and the sum runs over all N atoms in the sample. $|S(\mathbf{k} - \mathbf{k}_0)|^2 = N^2$ in those directions $\mathbf{k}$ where constructive interference takes place for the amplitudes emitted from all nuclei, as it applies for forward scattering. In this case the decay width is given by

$$\Gamma_C(\mathbf{k}_0) = \Gamma_\gamma \frac{N}{4\pi} \Delta\Omega \tag{3.24}$$

where $\Delta\Omega$ is the solid angle around $\mathbf{k}_0$ for which $(\mathbf{k} - \mathbf{k}_0) \cdot (\mathbf{R}_i - \mathbf{R}_j) < 1$ for all interatomic distances. Thus, due to the phasing, the emission preferentially proceeds into the direction of the incident photon wave vector.[2]

Equation (3.24) implies that $\Delta\Omega$ strongly depends on the dimensionality and the shape of the sample. For a 3-dimensional sample we find that $\Delta\Omega \approx (\lambda/L_\perp)^2$, where $L_\perp$ is the dimension of the sample transverse to $\mathbf{k}_0$. In this case we obtain

$$\Gamma_C(\mathbf{k}_0) = \frac{1}{4\pi} \rho \lambda^2 L_{\parallel}(\mathbf{k}_0) \Gamma_\gamma, \tag{3.25}$$

where $L_{\parallel}(\mathbf{k}_0)$ is the dimension of the sample along the direction of $\mathbf{k}_0$ and $\rho = N/(L_\perp^2 L_{\parallel})$ is the number density of resonant atoms in the sample. The product $\rho \lambda^2 L_{\parallel}$ has an interesting interpretation: It is the number of resonant atoms in a column with cross section λ^2 and length $L_{\parallel}$, as illustrated in Fig. 3.5.

It is instructive to take another view on forward scattering by dividing the sample into M thin layers (see Fig. 3.5) and solve for the time dependent response of the oscillators in each layer as they act under the influence of the radiation fields from all the other layers after pulsed excitation. The initial phasing of the emitters in each layer is assumed to be symmetrical, i.e., they radiate equally in both forward and

[2]The directional emission of single photons has been called counterintuitive [9] since no macroscopic dipole moment is involved in the single-excitation timed Dicke state Eq. (3.9) like it is the case, e.g., for a fully inverted system. Therefore one might expect that a weakly excited system radiates with an undirected emission pattern as a single atom does. The directionality, however, is just another consequence of the coherence involved in the scattering process.

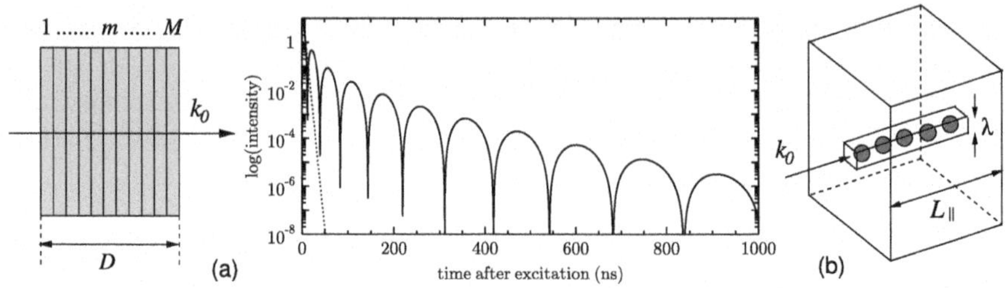

Fig. 3.5 **a** Dynamical beats in nuclear resonant forward scattering through an optically thick foil of $D = 20\,\mu$m stainless steel, where the ^{57}Fe nuclei act as single-line resonators. For times later than $t \sim \hbar/\Gamma_C$ after excitation the temporal evolution is dominated by dynamical beats described by Eq. (3.21). The dashed line in the graph illustrates the initial superradiant part of the temporal evolution that proceeds as $I(t) = I_0\,\exp[-(1+\chi)\,t/\tau_0]$ with a speedup factor of $\chi = 60$. **b** $\chi = \Gamma_C/\Gamma_0 \sim \rho\,\lambda^2\,L_{\|}$ is the number of resonant atoms N in the column of cross section λ^2. Reprinted from [61], Copyright 2012, with permission from Wiley

backward directions. In forward scattering, however, there is no radiative coupling with another scattering channel (as it is in Bragg geometry), and this leads to an asymmetry: While the mth layer acts under the influence of the $(m-1)$ upstream layers, the downstream $(M-m)$ layers have no effects on the mth layer. As a result, the emitters in the first layer radiate at their natural resonance energy ω_0 and decay rate Γ_0 while the emitters in the Mth layer are driven by the fields from all upstream layers. This strong driving eventually forces the downstream layers out of phase with the upstream layers resulting in dynamical beats and a nonexponential decay at late times. On the other hand, if the incident wave vector satisfies the symmetrical Bragg condition, then the radiated waves are constructive in both transmitted and reflected directions. As a result, the driving forces on each oscillator in the sample are equal, leading to a normal mode oscillation with superradiant decay width Γ_C at the natural resonance frequency ω_0.

It should be noted that the formalism outlined so far is valid only in the local or Markov approximation, i.e., for slowly evolving systems that do not change much while the signal propagates through the sample. In case of large samples that violate the local approximation the dynamics becomes nonlocal in time and one expects collective oscillations in the atomic population resulting from subsequent emission and reabsorption of radiation within the sample [15, 62, 63].

3.4.3 Bragg Scattering

In contrast to forward scattering, the Bragg exciton is an eigenmode that radiates at the natural resonance frequency ω_0 with an exponential accelerated decay. This can be understood via the normal mode analysis presented above. For a crystal consisting of M resonant layers, each layer separately has two-dimensional normal modes corresponding to Bloch waves. The layers scatter into the same outgoing field modes, which at the same time couples the different layers. As a result, the crystal of M layers with a spacing of d will give rise to M different linear combinations of the single-layer solutions [26]. Factorizing out the two-dimensional component corresponding to the single-layer Bloch waves, the part of the excitonic state characterizing the superposition of the different layers is given by

$$|\Psi_e(\mathbf{k}_0)\rangle = \frac{1}{\sqrt{M}}\begin{pmatrix} 1 \\ e^{ig_0 d} \\ \vdots \\ e^{ig_0 d(M-1)} \end{pmatrix}, \tag{3.26}$$

with $g_0 = k_0 \sin\Theta_B$. At the Bragg angle Θ_B we have $n\lambda = 2d \sin\Theta_B$ with a natural number n, so that $g_0 d = n\pi$ if the Bragg condition is fulfilled. Thus, at the exact Bragg angle for a symmetric Bragg reflection, the phasing is such that only the superradiant eigenmode state is excited. With increasing deviation from the Bragg angle, in addition to the superradiant eigenmode, various other normal modes are virtually excited. As a result, the weighted resonance energy is shifted from ω_0, and the effective decay width is reduced. Quantitatively, the reflectivity of a thin crystal consisting of M resonant layers in the vicinity of the resonance energy ω_0 is given by [26]:

$$R(\omega,\delta) = \frac{1}{1 - iM\delta}\,\frac{i\Gamma_C/2}{[\omega_0 - \omega - i\Gamma_0/2 - i(\Gamma_C/2)/(1 - iM/\delta)]} \tag{3.27}$$

with $\delta = k_0 d \cos\Theta_B\,\delta\Theta$. d is the spacing of the lattice planes, $\delta\Theta = \Theta - \Theta_B$ is the deviation of the incidence angle Θ from the exact Bragg angle Θ_B, and Γ_C is given by Eq. (3.25). Thus, in Bragg geometry, the effect of the collective resonant scattering is an enhancement of the decay width to

$$\Gamma = \Gamma_0 + \frac{\Gamma_C}{1 + (M\delta)^2} \tag{3.28}$$

while the resonance frequency ω_0' is shifted relative to ω_0 by the amount

$$\Delta\omega_c(\delta) = \omega_0' - \omega_0 = \frac{M\delta}{1 + (M\delta)^2}\,\frac{\Gamma_c}{2} \tag{3.29}$$

Accordingly, in Bragg geometry the collective Lamb shift $\Delta\omega_c$ changes from negative to positive values if the angle of incidence crosses the Bragg angle from below. Such a behaviour has been experimentally observed in nuclear Bragg diffraction from perfect single crystals of $FeBO_3$ [64]. It was also reported in a theoretical study for a density modulated slab of material [65].

3.5 Cooperative Emission and the Collective Lamb Shift in a Cavity

It has been shown in the previous Sect. 3.4.3 that radiative eigenstates of a resonant collection of identical atoms can be selectively excited by proper phasing of the resonators. This is the case, for example, if the atoms are arranged in a crystal and the incident wavevector matches a symmetric Bragg reflection. Here we discuss another phasing scheme for a superradiant eigenstate that leads to large cooperative effects and exhibits a high degree of experimental tunability. This is the case if the resonant atoms are embedded in a planar cavity that is excited in its first-order mode, as sketched in Fig. 3.6. An ultrathin layer of ^{57}Fe atoms is located in the plane at $z = 0$ the center of the cavity. The layer system that forms the planar cavity consists of a material of low electron density (e.g., carbon) as a guiding layer that is sandwiched between two layers of high electron density (e.g., Pt) acting as total reflecting mirrors.

The two phasing schemes are in fact closely related. In the Bragg case, the phasing leads to constructive interference if the condition $n\lambda = 2d \sin \Theta_B$ is satisfied. Then, different scattering pathways through the crystal add up in phase. We can relate this expression to the cavity case by rewriting $\lambda = 2\pi/k_\lambda$ with the wave number k_λ, and evaluating the corresponding wave number normal to the cavity surface via $\sin \Theta_B = k_\perp/k_\lambda$. Using $\lambda_\perp = 2\pi/k_\perp$, the Bragg condition becomes $d = n\lambda_\perp/2$, which is the usual resonance condition for the nth mode of a perfect resonator with length d. One may therefore interpret a cavity as a Bragg setting "folded" into one layer via the action of the mirrors. This way, also the scattering pathways shown in Fig. 3.6 can be related to the corresponding pathways in the Bragg case.

To find the complex eigenfrequencies of the system we reverse the solution procedure outlined above, first obtaining the eigenmodes by symmetry and then solving for the eigenfrequencies. We find for the electric field in the regions above and below the resonant layer at $z = 0$

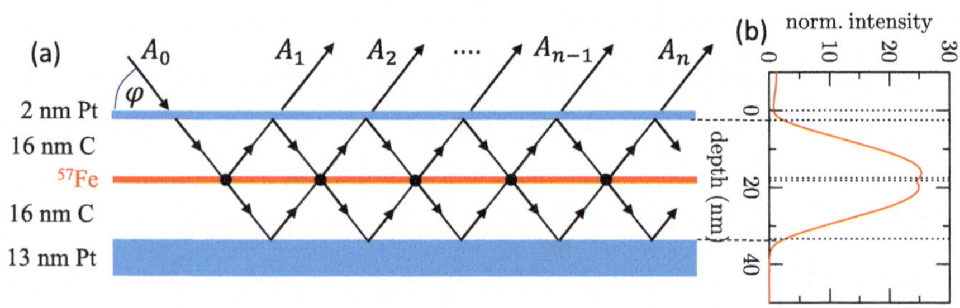

Fig. 3.6 **a** Structure of the planar cavity and scattering geometry used for calculation of the CLS for an ensemble of resonant ^{57}Fe nuclei embedded in the center of its guiding layer. **b** Depth dependence of the normalized radiation field intensity in the first-order guided mode of the cavity. Dashed lines mark the interfaces between layers

$$E(z) = \begin{cases} A\,(e^{ik_z z} + r_m\, e^{-ik_z z}), & (z > 0) \\ B\,(e^{-ik_z z} + r_m\, e^{ik_z z}), & (z < 0) \end{cases} \tag{3.30}$$

where k_z is the wavevector in z direction and r_m is the reflection coefficient of the mirrors. The reflection and transmission coefficients of the ultrathin resonant layer (thickness d) are given by

$$r_r = \frac{ifd}{1 - ifd}, \quad t_r = 1 + r_r, \quad \text{with} \quad f_n = f_n(\omega) = \frac{2\pi\rho}{k_0^2}\,\frac{f_0(\Gamma_0/2)}{\omega - \omega_0 - i(\Gamma_0/2)} \tag{3.31}$$

where f_n is the nuclear scattering amplitude with f_0 defined in Eq. (3.4). Matching the fields in Eq. (3.30) above and below the resonant layer under conditions of resonant transmission and reflection leads to

$$\begin{pmatrix} A \\ B \end{pmatrix} = \begin{pmatrix} r_r r_m & t_r r_m \\ t_r r_m & r_r r_m \end{pmatrix} \begin{pmatrix} A \\ B \end{pmatrix} \tag{3.32}$$

The eigenfrequencies are determined from the coresponding determinant equation:

$$r_m(r_r \pm t_r) = 1 \tag{3.33}$$

where the sign distinguishes between the odd and even solutions for the field in the cavity. Odd modes are those with a minus sign; they have a node at $z = 0$ and thus do not interact with the resonant layer. For the even modes Eq. (3.33) turns into $r_m(2r_r + 1) = 1$ from which we derive that $fd = i\,(1 - r_m)/(1 + r_m)$ which yields the complex eigenfrequency

$$\omega = \omega_0 - \frac{i\Gamma_0}{2}\left[1 - \frac{2\pi\rho f_0 d}{k_0^2}\left(\frac{1 + r_m}{1 - r_m}\right)\right]. \tag{3.34}$$

From this expression we obtain the frequency shift

$$L_C = \frac{2\pi\rho f_0 d}{k_0^2}\,\Gamma_0\,\mathrm{Im}\left(\frac{1 + r_m}{1 - r_m}\right). \tag{3.35}$$

For highly reflecting mirrors with $|r_m| \approx 1$ the expression on the right can become quite large. Effectively, the cavity promotes the exchange of real and virtual photons between the resonant atoms within the ensemble, leading to large values for the cooperative decay width and the collective Lamb shift.

For a more rigorous description we treat the propagation of x-rays in stratified media within a transfer matrix formalism [66]. Owing to the high energies of x-rays compared with electronic binding energies in atoms, the refractive index n of any material is slightly below unity. Thus, n is commonly written as $n(E) = 1 - \delta$. Accordingly, in the X-ray regime, every material is optically thinner than vacuum, thus total reflection occurs for angles of incidence (measured relative to the surface) below the critical angle $\phi_c = \sqrt{2\delta}$. Since $\delta = 10^{-6} \ldots 10^{-5}$ for hard X-rays with energies between about 10 and 20 keV, the critical angle φ_C is typically a few mrad. In the regime of total reflection, the radiation penetrates only a few nm into the material via the evanescent wave. In the example shown in Fig. 3.6, the top Pt layer is thin enough (2.2 nm) so that x-rays impinging under grazing angles can evanescently couple into the cavity.

Constructive superposition of the partial waves inside the cavity occurs at certain angles when the thickness of the guiding layer equals an integer multiple of the standing wave period that is given by $(\lambda/2)/\sqrt{\varphi^2 - \varphi_C^2}$, where φ_C is the critical angle of total reflection of the guiding layer material. This leads to a strong amplification of the local photonic density of states, limited only by the photoabsorption in the guiding layer material. In the first-order mode excited at about $\varphi = 2.5$ mrad, illustrated in Fig. 3.6, one obtains a 25-fold enhancement of the normalized intensity in the center of the cavity.

In the following we calculate the spectral response of this system around the nuclear resonance energy to determine the collective decay width and the collective Lamb shift of the nuclei in the cavity. This can be accomplished via a perturbation expansion of the resonant reflectivity R of the cavity in powers of the nuclear scattering amplitude f_n at the angular position $\varphi = \varphi_1$ of the first-order mode [27]. Each order of the perturbation series of R corresponds to one of the outgoing partial waves A_i that are emitted from the nuclear ensemble at the 'vertices' (denoted by the black dots) in the diagram. In order to sum up all the partial waves A_i, we note that the scattered amplitude in the nth outgoing wave is related to the $(n-1)$th amplitude via

$$A_n = (\mathrm{i}\,d\,f_n)\,p\,q\,A_{n-1} \tag{3.36}$$

Here d is the thickness of the ^{57}Fe layer and p and q are the amplitudes of the wavefields (at the position of the resonant nuclei) propagating in the directions of the incident and the reflected beams, respectively. The depth dependence of the relevant product $p\,q$ for the first-order mode of the cavity used here is shown in Fig. 3.6b. For the first vertex we have $A_1 = (\mathrm{i}\,d\,f_n)\,p^2\,A_0$ that also includes the coupling of the radiation into the cavity. Finally, the sum over all orders results in

$$R = \mathrm{i}\,d\,f_n p^2 \sum_{k=0}^{\infty} (\mathrm{i}\,d\,p\,q\,f_n)^k = \frac{\mathrm{i}\,dp^2 f_n}{1 - \mathrm{i}\,d\,p\,q\,f_n}. \tag{3.37}$$

Inserting $f_n(\omega)$ as defined in Eq. (3.31) we obtain a spectral response that is again a Lorentzian resonance line

$$R(\omega) = \frac{C\,d\,p^2\,(\Gamma_0/2)}{\omega - \omega_0 + L_C + \mathrm{i}(\Gamma_0 + \Gamma_C)/2} \quad \text{with} \quad C = \frac{2\pi\rho\,c_n}{k_0\,k_{0z}} \tag{3.38}$$

that exhibits a decay width of $\Gamma_C = C\,d\,|\mathrm{Re}(p\,q)|\,\Gamma_0 =: \chi\,\Gamma_0$ and an energy shift of $L_C = -C\,d\,\mathrm{Im}(p\,q)\,\Gamma_0/2$. Combining these results into one expression for the complex-valued frequency shift Δ_C, we obtain

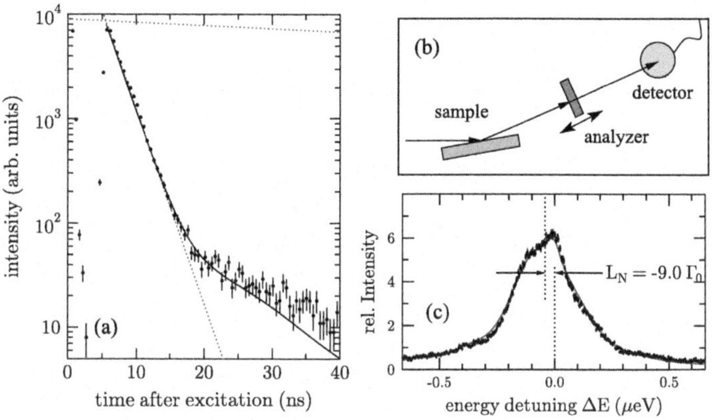

Fig. 3.7 **a** Measured time response of a 1.2 nm thick layer of ^{57}Fe atoms embedded in the center of the planar cavity (Fig. 3.6), excited in the first-order mode. The decay proceeds exponentially over two orders of magnitude with a speedup of $\chi = 65$ compared to the natural decay (upper dashed line). At later times the decay levels off into a curve with a much smaller slope, resulting from residual hyperfine interactions of the nuclei in the C matrix. **b** Experimental setup to record the energy spectrum of the radiation reflected from the cavity. The analyzer is a 6 μm thick foil of stainless steel $^{57}Fe_{0.55}Cr_{0.25}Ni_{0.20}$ where the ^{57}Fe exhibits a single-line nuclear resonance. It is mounted on a Doppler drive in order to obtain the spectrum by recording resonantly scattered photons as function of the drive's velocity. **c** The measured energy spectrum is strongly broadened due to the superradiant enhancement. Its center is shifted by about -9 Γ_0 which is the collective Lamb shift for this sample [27]. Reprinted from [67], Copyright 2015, with permission from Springer Nature

$$\Delta_C = \mathrm{i}\,\Gamma_C \left(1 + \mathrm{i}\,\frac{\mathrm{Im}(p\,q)}{|\mathrm{Re}(p\,q)|}\right) \tag{3.39}$$

The collective Lamb shift in single-photon γ-ray superradiance has been experimentally confirmed in an experiment at the European Synchrotron Radiation Facility (ESRF) [27], see Fig. 3.7.

While for a spherical atomic cloud the collective Lamb shift scales with the quantity $\rho\lambda^3$ [31, 41], in this setting it scales with $\rho_A\lambda^2$, where ρ_A is the areal density of the resonant nuclei in the sample. Here the ensemble of resonant nuclei effectively appears to be two-dimensional because all nuclei within the thin layer are confined to a dimension that is small compared to the period of the standing wave in the cavity. As a result, the cooperative emission from the nuclei in the cavity takes place in the limit $k_{0z}d \ll 1$, so that essentially the small-sample limit of Dicke superradiance is realized here, while the directionality of the emission is kept because the resonant nuclei interact only with one guided mode of the cavity with a well defined wavevector. One may speculate that if the resonant atoms are confined in a 1-dimensional structure like a fiber, the collective Lamb shift might scale as $\rho_L\lambda$, where ρ_L is the linear density of the atoms [61, 68]. This could lead to relatively large values of the CLS. The preparation of corresponding samples is certainly more demanding, although x-ray waveguides with a 2-dimensional confinement of the photon field have already be demonstrated [69]. Even more interesting it is to consider a 0-dimensional confinement of the atoms, e.g., in a 3D cavity. This will be practically impossible for x-rays, but has been demonstrated with microcavities in the optical regime [70].

The cavity is an ideal laboratory to study features of cooperative emission. In the following we exploit this to study the dependence of the CLS on the size of the sample. For that purpose we increased the thickness of the resonant layer while keeping the areal density of the resonant atoms constant. Calculations of the cavity reflectivity for an extended ensemble of atoms distributed over the standing wave within a 3rd order guided mode are shown in Fig. 3.8a. A close inspection reveals two prominent features: First, one observes a sharp dip in the reflectivity spectrum at the exact resonance energy ($\Delta = 0$). This structure is very reminiscent of the transparency dips that appear in the phenomenon of electromagnetically induced transparency (EIT) in quantum optics [71, 72]. As will be discussed in Sect. 3.7.2, there is indeed a mechanism which leads to EIT in the case of nuclear resonant scattering from a cavity that contains resonant atoms. Second, the CLS (determined from the center of gravity of the curve), is a non-monotonous function of the thickness of the atomic ensemble within the cavity. This behavior can be studied particularly well if higher-order modes are employed where the resonant atoms can be distributed over a large range of kR values within the cavity, see Fig. 3.8b. The results are displayed in Fig. 3.8c. For comparison, we have used the function $a + b(\sin 2kR)/(2kR)$ (dashed lines) to pinpoint the functional dependence of the oscillations in the CLS with increasing sample size as it was predicted first in [31] and recently experimentally verified [45]. A rigorous theory to describe this behaviour on the basis of the cavity geometry used here, however, still has to be developed.

3.6 Quantum Optics of Mössbauer Nuclei in X-Ray Cavities

The reflectance and spectral response of an x-ray cavity can be calculated using different techniques (see Fig. 2 in [40]). One approach is Parratt's formalism [73], in which all possible scattering pathways arising from the material boundaries are summed up. A generalization of this technique which enables one to include resonant multipole scattering with its polarization dependence has been formulated in a transfer-matrix formalism (for an overview, see [6]). A numerical implementation of this formalism is provided via the CONUSS software package [74]. An alternative approach involves the direct numerical integration of Maxwell's equations, via a finite-difference time-domain method [75]. The analysis so far, however, focused to the case of linear light-matter interaction with classical light fields. Moreover, these methods do not enable one to interpret the obtained spectra in terms of the underlying nuclear dynamics. In the following, we therefore focus on a recently developed quantum optical framework for the description of ensembles of nuclei in x-ray cavities [35, 37]. The key idea of this approach is to relate the entire system comprising the x-ray cavity and the large ensemble of multi-level nuclei to that

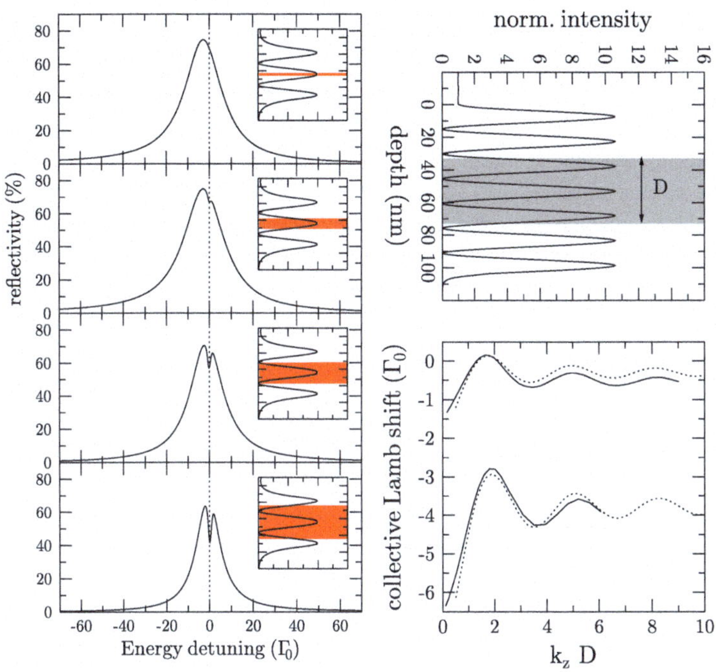

Fig. 3.8 The collective Lamb shift (CLS) for an extended layer of resonant atoms within the cavity. Left column: cavity reflectivity in the third-order guided mode for increasing thickness D of the resonant layer. The insets show the cavity cross section with the resonant layer (red) and the standing wave intensity pattern (solid line). Right column: CLS as function of $k_z D$ for the 7th and 11th order mode in a Pt/C/Pt cavity with a 100 nm thick guiding layer. Similar curves have been observed recently in an experiment involving thin layers of atomic vapor [45]. Reprinted from [61], Copyright 2012, with permission from Wiley

of a corresponding "artificial atom", that is, a quantum optical few-level structure which for low probing fields gives rise to the same response. This method has the advantage that it treats the x-ray field as quantized, enables one to explore non-linear and quantum effects, allows for a full interpretation, and for a quantitative modeling of experiments. On the other hand, designing the cavity geometry and the nuclear level structure in a suitable way enables one to design artificial atoms with properties which reach beyond what is available in natural atoms [29, 68]. We note that the quantum optical model presented here recently was promoted to an ab-initio theory, in which the model parameters and the realized artificial quantum system can directly be calculated from a given cavity structure [40, 76]. As compared to previous models, this ab-initio theory further predicts qualitatively new phenomena, e.g., related to the effect of off-resonant cavity modes on the nuclear dynamics.

3.6.1 Quantum Optics of the Empty Cavity

To illustrate the photonic environment in a cavity without nuclei, we restrict the discussion to a single cavity mode and neglect the light polarization for the moment. The probing x-rays have frequency ω and a wave vector $\mathbf{k}$ that defines the incidence angle θ. The cavity modes are characterized by discrete wave number components perpendicular to the cavity surface, but continuous wave number components along the surface. The boundary conditions impose that the wave vector $\mathbf{k}_C$ inside the cavity has a component $|\mathbf{k}_C|\cos(\theta)$ along the cavity surface. The component transverse to the cavity surface, however, is fixed by the guided mode standing wave condition to $|\mathbf{k}|\sin(\theta_0)$, if θ_0 is the incidence angle under which this mode is driven resonantly for incident wave number $|\mathbf{k}|$. Thus, $|\mathbf{k}_C| = \sqrt{|\mathbf{k}|^2\cos^2(\theta) + |\mathbf{k}|^2\sin^2(\theta_0)}$, and a detuning $\Delta_C = \omega_C - \omega \approx -\omega\theta_0\Delta\theta$ between the cavity resonance frequency and the frequency of the incident light can be defined, which can be tuned via small variations in the incidence angle $\Delta\theta = \theta - \theta_0$ from the resonance condition. With this detuning, the Heisenberg equation of motion for the cavity mode in the absence of nuclear resonances characterized by annihilation [creation] operators a $[a^\dagger]$ is given by

$$\frac{d}{dt}a = -(\kappa + i\Delta_C)a + \sqrt{2\kappa_R}\, a_{\text{in}}\,, \tag{3.40}$$

where κ is the overall damping rate of the cavity mode, κ_R characterizes the evanescent coupling into and out of the cavity mode, and a_{in} the applied x-ray field. In practice, κ_R can be adjusted, e.g., by choosing the thickness of the cavity top layer through which the x-rays evanescently couple into the cavity mode. From the cavity field operators, the empty cavity reflectance $|R_c|^2$ can be obtained via the input-output relations [77] $a_{\text{out}} = -a_{\text{in}} + \sqrt{2\kappa_R}\, a$, using $R_c = \langle a_{\text{out}}\rangle / \langle a_{\text{in}}\rangle$. In the stationary state (SS) $\dot{a}^{(SS)} = 0$, such that

$$a^{(SS)} = \frac{\sqrt{2\kappa_R}\, a_{\text{in}}}{\kappa + i\Delta_C} \tag{3.41}$$

and

$$R_c = \frac{2\kappa_R}{\kappa + i\Delta_C} - 1. \tag{3.42}$$

At the so-called critical coupling condition $2\kappa_R = \kappa$, the reflectance $|R_c|^2$ vanishes on resonance $\Delta_C = 0$, which can be interpreted as destructive interference between light reflected from the outside of the cavity with that coupling out of the cavity mode. If operated in this regime, the cavity can be employed to suppress a significant part of the background photons, facilitating the detection.

3.6.2 Quantum Optics of a Cavity Containing Resonant Nuclei

Next we consider the effect of the nuclei on the cavity, restricting the discussion to the single-excitation subspace spanned by the two collective states $|G\rangle$ and $|E\rangle$. The nuclei effectively act as a source term for x-ray photons in the empty cavity equation of motion (3.40), which is modified to

$$\frac{d}{dt}a = -(\kappa + i\Delta_C)a + \sqrt{2\kappa_R}\, a_{\text{in}} - ig^*\sqrt{N}|G\rangle\langle E|, \tag{3.43}$$

where g is the x-ray-nucleus coupling constant. Thus the reflectance Eq. (3.42) becomes

$$R = R_{\text{c}} - \frac{i}{a_{\text{in}}}\frac{2\kappa_R}{\kappa + i\Delta_C}g^*\sqrt{N}\langle E|\hat{\rho}|G\rangle, \tag{3.44}$$

where $\hat{\rho}$ is the density operator characterizing the nuclei. If the empty cavity reflectance R_{c} vanishes on resonance in critical coupling, then the observable reflectance originates from the nuclei alone, and therefore ideally forms a signal without any background.

The result Eq. (3.44) can be generalized in a straightforward way to accomodate for arbitrary input and output photon polarizations, as well as the magnetic substructure of the nuclear levels, as it may result from nuclear Zeeman splitting in the presence of a magnetic hyperfine interaction in a ferromagnetic environment. We denote the input [output] polarization unit vectors as $\hat{a}_{\text{in}}$ [$\hat{a}_{\text{out}}$], the two cavity mode polarization unit vectors as $\hat{a}_1$ and $\hat{a}_2$, and define $\mathbf{1}_\perp = \hat{a}_1\hat{a}_1^* + \hat{a}_2\hat{a}_2^*$. The different transitions from the ground state manifold to the excited state manifold within each nucleus are labeled with index μ, and have a dipole moment $\mathbf{d}_\mu$ and a Clebsch-Gordan coefficient c_μ. Since the nuclei initially are distributed over the different ground states, we further define the number of nuclei in the ground state of transition μ as N_μ, and generalize the exciton Eq. (3.9) to $|E_\mu\rangle$ as the exciton created upon excitation on transition μ. Then,

$$R = R_{\text{c}}\, \hat{a}_{\text{out}}^*\hat{a}_{\text{in}} - \frac{i}{a_{\text{in}}}\frac{2\kappa_R}{\kappa + i\Delta_C}g^*\sum_\mu(\hat{a}_{\text{out}}^* \cdot \mathbf{1}_\perp \cdot \hat{d}_\mu)\, c_\mu\sqrt{N_\mu}\langle E_\mu|\hat{\rho}|G\rangle. \tag{3.45}$$

It can be seen that the empty cavity response R_{c} can be filtered out using orthogonal input and output polarizations, as expected. The nuclei, however, can scatter between these two orthogonal modes, such that this crossed polarization setting again is a method to detect the nuclear response without background via a high-purity polarimetry setup [78]. Further, the different transitions μ can be interpreted as a collective few-level system, with number of relevant states determined by the input and output polarization, as well as the nuclear quantization axis defining the magnetic substates. This setting with magnetic sublevels therefore enables one to realize quantum optical few-level systems [35, 37]. As evidenced by the coherent addition of the scattering channels, the responses of the different transitions within this few-level system may interfere, providing access to a rich variety of quantum optical phenomena.

3.6.3 Nuclear Dynamics in the Cavity

It remains to determine the nuclear dynamics, i.e., its evolution under the action of an x-ray pulse, in order to determine the density matrix ρ entering the reflectance in Eqs. (3.44) and (3.45). Here we want to illustrate this for the simplest case of two-level nuclei and a single cavity mode a. The original Hamiltonian is of Jaynes-Cummings-type [2, 29], and contains interaction terms of the form $S_+^{(n)}a$, describing the annihilation of a cavity photon (a) together with an excitation of nucleus n ($S_+^{(n)}$), as well as the reverse process. The problem can be simplified considerably be exploiting that the fastest timescale in the problem typically is given by the cavity lifetime $1/\kappa$. In this "bad cavity" limit characterized by short photon trapping times, the cavity modes adiabatically follow the much slower evolution of the nuclei. As a consequence, the cavity operators can approximately be replaced by their steady-state values Eq. (3.43), which results in an effective Hamiltonian for the nuclei alone. In the radiative eigenmode basis it is given by [35, 37]

$$H = -\hbar\Delta|E\rangle\langle E| + \hbar(\Omega_{\text{eff}}\,|E\rangle\langle G| + \text{H.c.}) + \hbar\Delta_{LS}\,|E\rangle\langle E|. \tag{3.46}$$

The interpretation of this Hamiltonian is straightforward. It is equivalent to the Rabi model for a driven two-level system [2]. However, the effective Rabi coupling constant $\Omega_{\text{eff}} = g\sqrt{N}\,a^{(SS)}$ in H_Ω is not given by the bare nucleus-cavity coupling g, but modified by cooperative effects as indicated by the superradiant enhancement factor $\sqrt{N}$, as well as by the cavity field as indicated by the presence of the steady-state value of the field operator $a^{(SS)}$. Furthermore, the usual detuning Δ between x-ray frequency and bare nuclear transition frequency is augmented by an additional contribution

$$\Delta_{\text{LS}} = |g|^2 N\ \text{Im}[(\kappa + i\Delta_C)^{-1}], \tag{3.47}$$

which arises due to the radiative coupling between the nuclei and which can be interpreted as the cooperative Lamb shift. Similar to the Hamiltonian parts, also the incoherent spontaneous emission of the individual nuclei γ is modified by

$$\Gamma = 2|g|^2 N\ \text{Re}[(\kappa + i\Delta_C)^{-1}]. \tag{3.48}$$

In linear response, the desired nuclear polarization $\langle E|\rho|G\rangle$ is governed by

$$\langle E|\dot{\rho}|G\rangle = -i\Omega_{\text{eff}} + i\left[(\Delta - \Delta_{LS}) + \frac{i}{2}(\gamma + \Gamma)\right]\langle E|\rho|G\rangle. \tag{3.49}$$

Inserting the steady state solution obtained from $\langle E|\dot{\rho}|G\rangle = 0$ into Eq. (3.44) finally yields

$$R = R_c - i\left(\frac{\kappa - i\Delta_C}{\kappa + i\Delta_C}\frac{2\kappa_R}{\kappa}\right)\frac{\Gamma/2}{(\Delta - \Delta_{LS}) + \frac{i}{2}(\gamma + \Gamma)}, \tag{3.50}$$

with the empty cavity reflectance R_c defined in Eq. (3.42). The nuclear response therefore comprises a Lorentzian shifted with respect to the bare nuclear resonance frequency by the collective Lamb shift Δ_{LS}, and with superradiant broadening Γ of the natural line width γ, as already found using a different formalism in Eq. (3.38).

3.7 Quantum Optical Effects in Cavities

3.7.1 *Interferometric Phase Detection via Fano Resonance Control*

In Sect. 3.5 it was found that a resonantly driven cavity containing resonant two-level nuclei features a Lorentzian spectral response, broadened by superradiance, and shifted by the cooperative Lamb shift. Having the expression for the cavity reflectance Eq. (3.50) at hand, we can start by exploring the cavity response off-resonance with the cavity mode [79]. Close to the resonance, $\Delta_C = \delta_C(\theta - \theta_{\text{min}})$, such that the detuning between x-rays and cavity mode can experimentally be tuned by varying the x-ray incidence angle θ around the resonance angle θ_{min}.

To simplify the analysis, we specialize to strongly superradiant cavities ($\gamma \ll \Gamma$) in critical coupling ($\kappa = 2\kappa_R$) and rewrite Eq. (3.50) using Eq. (3.48) to give

$$|R|^2 = \frac{|\varepsilon + q|^2}{1 + \varepsilon^2}\sigma_0\,, \tag{3.51}$$

where we defined the dimensionless energy $\varepsilon = (\Delta - \Delta_{LS})/(\Gamma/2)$, the prefactor $\sigma_0 = [1 + \kappa^2/\Delta_C^2]^{-1}$, and the so-called q-factor $q = \kappa/\Delta_C$. The cavity response thus takes the form of a Fano resonance [80], which is an ubiquitous spectroscopic signature in light-matter interactions [81, 82]. The Fano resonance arises, because there are two interfering pathways for the light to propagate through the sample. First, the spectrally broad cavity response, which is of relevance it the light does not interact with the nuclei. Second, a spectrally narrow bound-state contribution arising from the scattering on the nuclei. The relative phase between the two contributions is given by $\phi = -\arg(q - i)$, and it turns out that this phase determines the line shape, which may range from Lorentzian absorption features via dispersion-like asymmetric structures up inverted Lorentzian lines [83]. Conversely, external control over this relative phase can be used to manipulate the lineshape [83, 84]. Since close to resonance, $q = \kappa/[\delta_C(\theta - \theta_{\text{min}})]$, we find that changing the incidence angle allows

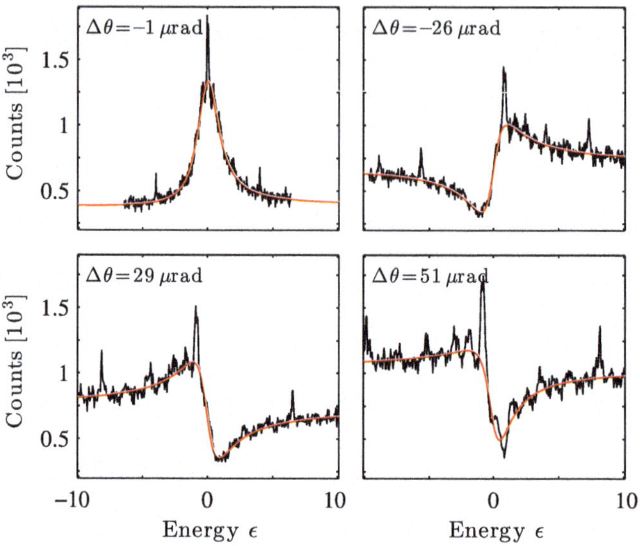

Fig. 3.9 Fano lineshape control with nuclei embedded in a thin film cavity. The panels show the reflectance recorded at incidence angles deviating from the resonance angle by $\Delta\theta$ indicated in the panels. The figure shows raw data without baseline subtraction such that the absolute scaling cannot directly be compared. Red lines show fits with a generic Fano line shape. The cavity is a Pd(4 nm)/C(36 nm)/Pd(14 nm) structure with a 1.2 nm thick layer of ^{57}Fe nuclei in the middle of the guiding C-layer. Reprinted from [79], Copyright 2015, with permission from the American Physical Society

one to tune the Fano parameter and thus the spectral lineshape. Example lineshapes are shown in Fig. 3.9, clearly demonstrating the control mechanism.

The analytic expression Eq. (3.51) enables one to interpret the cavity spectra on- and off-resonance as Fano resonances. But more importantly, the phase-sensitivity of the spectra together with the possibility to control the relative phase of the two interfering channels open the possibility to exploit the setup as an interferometer. Note that the cavity setup discussed here enables one to control the phase in a static manner via the x-ray incidence angle, but it has been demonstrated that also a dynamical control over the phase is possible [83, 84]. This approach enables one to measure tiny phase shifts via the asymmetry of the line shape. On the other hand, manipulating the phase can be used to control the light-matter interaction. As an example, the complex nuclear dipole moment induced by the x-rays could be measured using this interferometric approach [79]. From a broader perspective, the line shape control discussed here provides a route towards the implementation of a diverse range of applications relying on Fano interference [81, 82] at x-ray energies. One example, electromagnetically induced transparency, will be discussed in the next Sect. 3.7.2.

3.7.2 Electromagnetically Induced Transparency

As shown in Sect. 3.6, an ensemble of nuclei in a cavity can be effectively considered as an artificial atom with an enhanced decay width and a self-energy correction of its resonance energy given by the collective Lamb shift. The strong spatial variation of the photonic density of states inside the cavity opens the possibility to tune these properties via placement of the atoms within the wavefield of the cavity. An ensemble in an antinode of the cavity field displays a strong superradiant enhancement, which defines a new evolution time scale which is much faster than the usual single-nucleus single decay. Relative to this accelerated time scale, ensembles in a node will remain subradiant, since their dynamics is not accelerated. Thus, a cavity with two atomic ensembles at positions with such markedly different photonic densities of states can be considered as an artificial atom with three levels: one belonging the common ground state and two excited state levels that correspond to the superradient and subradiant ensembles, respectively. Effectively, the subradiant ensemble represents a metastable level because its radiative lifetime is much longer than that of the superradiant ensemble. In this three-level system, all levels are radiatively coupled through the vacuum field of the cavity. Such a coupling gives rise to a key phenomenon of quantum optics, electromagnetically induced transparency (EIT) [72].

The EIT effect arises from the cancellation of resonant absorption due to quantum interference between atomic levels with significantly different radiative lifetimes. In the original version of EIT, the quantum interference is induced by an external laser field tuned to the transition between a metastable level and a shortlived excited state in a three-level system. The basic ingredients for EIT are illustrated in Fig. 3.11a, where a three-level system is shown, represented by the ground state, $|1\rangle$, and two upper states, $|2\rangle$ and $|3\rangle$ with respective energies E_2 and E_3. A strong laser field with Rabi frequency Ω_C induces an atomic coherence between states $|2\rangle$ and $|3\rangle$. Tuning a (weak) probe laser field across the resonant transition $1 \rightarrow 3$ leads to a Fano-type quantum interference [80] that renders the medium almost transparent in a narrow window around the exact resonance frequency. The transparency arises since due to the coherent superposition of the two states $|1\rangle$ and $|2\rangle$, the two excitation pathways $|1\rangle \rightarrow |3\rangle$ and $|2\rangle \rightarrow |3\rangle$ interfere destructively, such that no excitation takes place. The degree of transparency is limited by the dephasing of the atomic coherence resulting from the decay of state $|2\rangle$ or external perturbations. Thus, maximum transparency is observed if $|2\rangle$ can be considered metastable, i.e., if it has a decay width, γ_2, that is negligibly small relative to the radiative decay width, γ_3, of the state $|3\rangle$.

To investigate the possibility of EIT in the ^{57}Fe containing cavity, we replace the spatially extended ^{57}Fe layer in Fig. 3.8 by two layers with a separation of half the period of the standing wave in the cavity, as shown in Fig. 3.10. Energy spectra of the cavity reflectivity are calculated via the transfer matrix formalism already employed in Sect. 3.4.2. Quite remarkably, the appearance of the dip in the reflectivity very sensitively depends on the location of this pair of layers in the cavity. The dip is most pronounced (Fig. 3.10a) if the first of the resonant layers (seen from the top) is

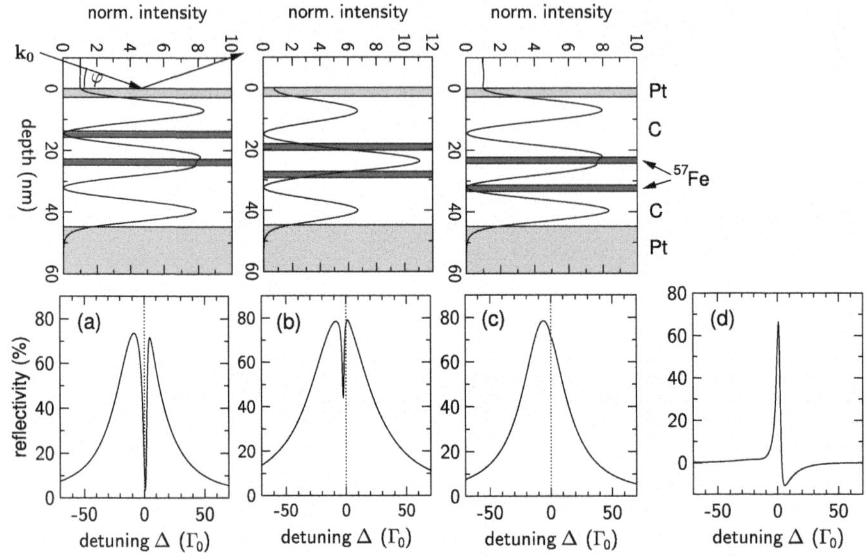

Fig. 3.10 Top row: sample geometry of planar cavities for X-rays containing two 2-nm-thick layers of ^{57}Fe (red) together with the normalized field intensity (solid line) in the 3rd-order guided mode, excited at an angle of incidence of $\varphi = 3.5$ mrad. Bottom row: calculated energy spectra of the cavities reflectivity around the nuclear resonance, together with the difference of the spectra in (**a**) and (**c**), displayed in (**d**). The fundamental difference between the spectra in (**a**) and (**c**) results from the asymmetry of the boundary conditions for the electromagnetic field in the cavity, see supplementary material for Ref. [28]. Reprinted from [61], Copyright 2012, with permission from Wiley

located in a node of the wavefield and the second one is located in an antinode of the wavefield. The dip gradually vanishes if the two layers are displaced by half a period of the standing wave (Fig. 3.10b, c). To determine the spectral shape of the transparency dip we subtract the two spectra in Fig. 3.10a, c. The resulting difference spectrum (Fig. 3.10d) exhibits an asymmetric shape corresponding to a Fano profile [80], thus providing clear evidence for the type of quantum interference that is typical for EIT [85].

In order to analyse the analogy with EIT more closely we expand the cavity reflectivity around the nuclear resonance (details of the derivation are given in the supplementary information of Ref. [28]), resulting in:

$$R(\Delta) = \frac{d_2\, f_0\, \gamma_0 E_{2-+}(i\Delta + \gamma_0)}{(i\Delta + \gamma_0)(i\Delta + \gamma_0[1 + d_2\, f_0\; E_{2--}]) + d_1\, d_2\, f_0^2\, \gamma_0^2\, E_{2-+} E_{1+-}} \tag{3.52}$$

The quantities E_{2-+}, E_{2--}, and E_{1+-} are elements of the transfer matrices that describe the propagation of the photon field in the unperturbed cavity. Equation (3.52) is basically identical to the standard expression for the complex susceptibility in case of EIT [72] if one identifies (see Fig. 3.11)

$$\begin{aligned}\gamma_2 &= \gamma_0 \\ \gamma_3 &= \gamma_0[1 + d_2\, f_0\; E_{2--}] \\ \Omega_c^2 &= d_1\, d_2\, (f_0\, \gamma_0)^2 E_{2-+} E_{1+-}\end{aligned} \tag{3.53}$$

This result prompts an obvious interpretation, supported by the illustration in Fig. 3.11b, c: The two ensembles of nuclei in the node and the antinode of the standing wave field experience two significantly different photonic densities of states, leading to two different collective decay rates γ_2 and γ_3. This effectively converts the nuclei in the cavity into three-level systems with two degenerate upper levels represented by the states $|2\rangle$ and $|3\rangle$, as illustrated in the level scheme of Fig. 3.11c. The expression for Ω_C^2 is proportional to the two transfer matrix elements E_{2-+} and E_{1+-} that describe the transition amplitudes between the two counterpropagating fields in the cavity at the position of the two resonant layers. This indicates that the coupling field arises from the radiative coupling of the two resonant layers via the cavity field: An excited atom in the antinode $|3\rangle$ decays back to the ground state $|1\rangle$ and releases a photon into the cavity. This photon can promote an atom in the node from the ground state into state $|2\rangle$ that eventually decays and again releases a photon into the cavity, and so on. As a result, the two excited states $|2\rangle$ and $|3\rangle$ are coupled through their common ground state $|1\rangle$ via the vacuum field of the cavity, which effectively establishes a control field between the two upper states, represented by the horizontal arrow in Fig. 3.11c. The resulting arrangement of levels in Fig. 3.11c and their coupling resembles closely a Λ-type level scheme as in Fig. 3.11a. It should be noted that the control field Rabi frequency of Eq. (3.53) enters Eq. (3.52) as a complex-valued quantity Ω_C^2 rather than a real number $|\Omega_C|^2$ in the usual expression for an EIT susceptibility. A closer inspection reveals that the imaginary part of Ω_C^2 is small compared to its real part for the cavity configurations employed here. It remains to be investigated in which way the imaginary part of Ω_C^2 affects EIT in these systems.

We want to emphasize that cooperative emission is critical to EIT in this system. While one of the atomic ensembles undergoes single-photon superradiant enhancement leading to a decay width of $\Gamma_C = 2\gamma_3 = d_2\, f_0\, \mathrm{Re}[E_{2--}]\, \Gamma_0$ and a collective Lamb shift of $L_C = -d_2\, f_0\, \mathrm{Im}[E_{2--}]\, \Gamma_0/2$, the decay width $2\gamma_2$ of the other'subradiant' ensemble is given by just the natural line width Γ_0, so that $\gamma_3 \approx 50\, \gamma_2$ in the example shown in Fig. 3.10. Thus, in the presence of a strong superradiant enhancement of state $|3\rangle$, the state $|2\rangle$ is relatively long-lived and thus can be considered as metastable. This is an important condition for a pronounced EIT effect. The superradiantly broadened transition of the nuclei in the antinode provides the continuum of states relative to which the Fano interference in this system takes place. The collective Lamb shift of this level introduces an asymmetry that leads to the characteristic Fano profile of the transparency window. For vanishing CLS, the profile would simply be a Lorentzian line [79, 82, 83].

For an experimental verification of EIT in the x-ray regime, we have prepared an x-ray cavity, shown in Fig. 3.11b, that consists of a Pt(3 nm)/C(38 nm)/Pt(10nm) sandwich structure containing two 3 nm ^{57}Fe layers that occupy a node and an antinode of the cavity field. These two layers represent the subradiant state $|2\rangle$ and the superradiant state $|3\rangle$ that are radiatively coupled via the vacuum field of the cavity, and thereby realize the EIT scheme. Drawing these states and their coupling in a level diagram with the decay width as vertical axis, one obtains a Λ type level system as shown in Fig. 3.11c. Using a spectroscopic detection scheme similar to that employed for measurement of the collective Lamb shift [27], we could record the energy spectrum of one of the hyperfine-split resonances of the Fe in this system, shown in Fig. 3.11d, clearly displaying a pronounced EIT transparency dip at the exact resonance where the system would be completely opaque otherwise [28].

The modulation of the photonic density of states in the cavity facilitates the preparation of ensembles of resonant atoms with greatly different radiative lifetimes. In other words, it lifts the radiative degeneracy of the atoms in the cavity, effectively

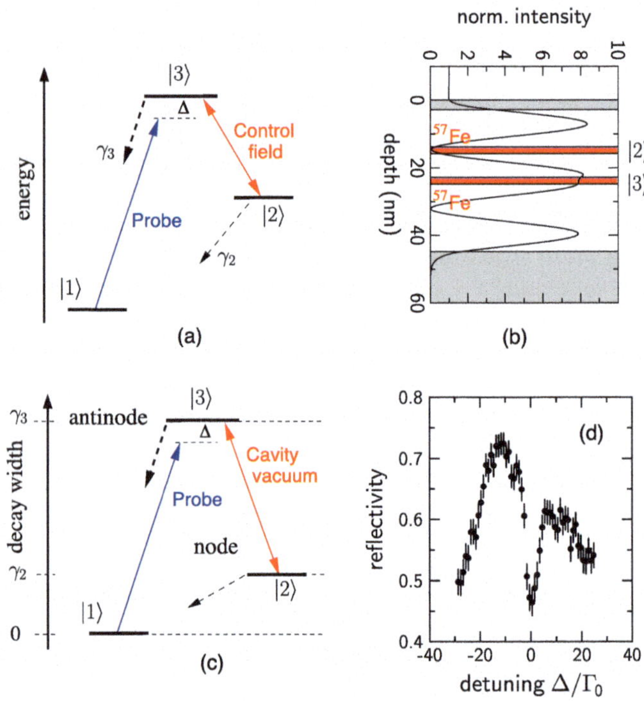

Fig. 3.11 **a** Typical, Λ-shaped level scheme of EIT in quantum optics: a strong laser field with Rabi frequency Ω_C induces an atomic coherence between the metastable level $|2\rangle$ and the upper state $|3\rangle$. The decoherence rate γ_2 can be neglected compared to the decay rate γ_3. The system appears to be transparent for the probe field at resonance ($\Delta = 0$) with the transition $|1\rangle \rightarrow |3\rangle$. **b** Cavity geometry with two layers of ^{57}Fe that can be translated into a Λ-shaped level scheme, if plotted with the decay width as vertical axis, shown in (**c**). While level $|3\rangle$ in the antinode is superradiant, level $|2\rangle$ in the node is subradiant so that $\gamma_2 \ll \gamma_3$. **d** Measured reflectivity spectrum of the cavity shown in (**b**) that clearly shows the EIT transparency dip at the exact resonance energy where the system would be completely opaque without the ^{57}Fe layer in the node [28]. Reprinted from [67], Copyright 2015, with permission from Springer Nature

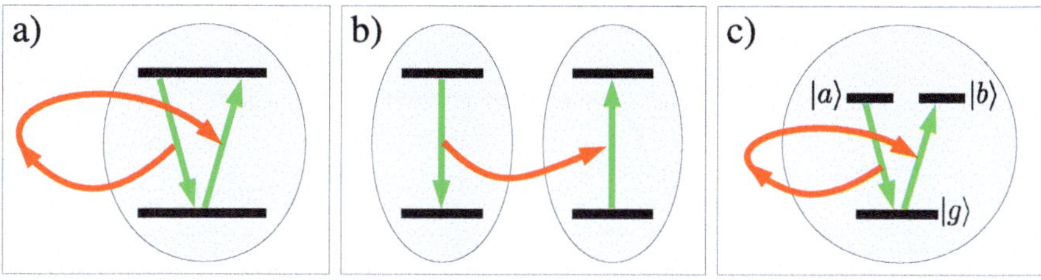

Fig. 3.12 Elementary processes in the interaction of the nuclei with the electromagnetic vacuum. Green straight arrows indicate the nuclear dynamics. Red curved arrows depict the corresponding emission and re-absorbtion of a virtual photon. **a** Emission and absorption of the virtual photon within a single nucleus on a single transition leads to a complex energy correction, contributing to Lamb shift and spontaneous emission. **b** The photon exchange between two nuclei induces dipole-dipole couplings. **c** If the virtual photon couples different transitions within a single nucleus, spontaneously generated coherences arise. Reprinted from [67], Copyright 2015, with permission from Springer Nature

creating extra excited levels that are radiatively coupled amongst each other via the cavity field. Note that this approach is general, as illustrated by the fact that the EIT mechanism discussed here has been adapted to qubits operating at microwave frequencies [86]. The degeneracy of the excited-state levels can also be lifted energy-wise by a magnetic hyperfine interaction (nuclear Zeeman effect). In this case the vacuum field of the cavity then leads to coherences between the excited states that are subject of the following section.

3.7.3 Spontaneously Generated Coherences

So far, we have focused on nuclei without magnetic substructure. In an environment with magnetic fields, e.g., ^{57}Fe exhibits two ground ($I = 1/2$) and four excited ($I = 3/2$) states, which result in a splitting of the Mössbauer spectra into six lines. Superradiant broadening can overcome this splitting, such that the different spectral lines overlap. Naively, one might expect an incoherent addition of the individual spectral contributions of the different transitions. However, deep minima can arise in the spectrum, which suggest the presence of destructive interference. These will turn out to be a consequence of so-called spontaneously generated coherences (SGC) [87–89]. While SGC are linked to numerous potential applications including the suppression of spontaneous decay, stringent conditions on their presence so far have limited the experimental exploration.

The origin of SGC can be understood as follows. In second order perturbation theory, the interaction of the nuclei with the surrounding electromagnetic vacuum field leads to processes as visualized in Fig. 3.12. In (a), an initially excited nucleus deexcites and emits a virtual photon (red line), which subsequently is reabsorbed on the same transition. This process results in a complex correction of the transition energy,

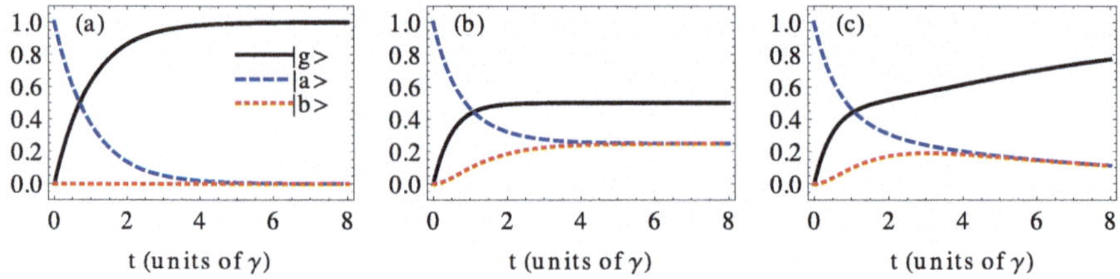

Fig. 3.13 Spontaneous emission dynamics for nuclei initially in state $|a\rangle$ of Fig. 3.12c. **a** Regular spontanous decay without the presence SGC, as it could be observed, e.g., in two-level nuclei. **b** As in (**a**), but with SGC in the three-level system shown in Fig. 3.12c. Part of the nuclei evolve into state $|b\rangle$, and half of the nuclei remain trapped in the excited states. **c** Similar to (**b**), but with energy splitting $\gamma/2$ between $|a\rangle$ and $|b\rangle$, which limits the time over which population can be trapped in the excited states. Reprinted from [67], Copyright 2015, with permission from Springer Nature

the real and imaginary part of which can be interpreted as single-particle Lamb shift and spontaneous decay rate. Similarly, the virtual photon can be reabsorbed by another particle (b), giving rise to dipole-dipole energy exchange between nuclei. This corresponds to the exchange of virtual photons already discussed in Sect. 3.3. To be added is the process shown in (c), where the virtual photon is re-absorbed within the same particle, but on another transition. This state transfer establishes SGC between the two excited states, arising from the interaction with the vacuum only. As a consequence of this coherence, the spontaneous emission from a superposition of the excited states $|-\rangle = (|a\rangle - |b\rangle)/\sqrt{2}$ is suppressed, since the two decay channels $|a\rangle \rightarrow |g\rangle$ and $|b\rangle \rightarrow |g\rangle$ destructively interfere. In contrast, $|+\rangle = (|a\rangle + |b\rangle)/\sqrt{2}$ decays with double decay rate due to constructive interference. Figure 3.13 shows the corresponding temporal evolution. Initially, the nuclei are in state $|a\rangle$. Without SGC, the excited state exponentially decays, and the ground state population grows accordingly (a). With SGC in (b), population is transfered from $|a\rangle$ to $|b\rangle$, establishing a coherence. As a result, half of the nuclei remain trapped in $|-\rangle$, which corresponds to the contribution to the initial state $|\langle -|a\rangle|^2 = 1/2$. In the optical spectra, such trapping states translate into dark lines.

However, the generation of SGC is limited by stringent conditions, which usually are not met for atoms in free space. First, the dipole moments of emitting and absorbing transitions must be non-orthogonal. Second, the energy difference between the upper states should be small compared to the natural line width, since for non-degenerate upper states the free time evolution converts the trapping state $|-\rangle$ into the decaying state $|+\rangle$. Thus, with increasing energy difference the time over which population can be trapped in the excitated states becomes smaller (see Fig. 3.13c), until it eventually can be neglected compared to the natural lifetime. A final condition is that the two involved transitions should share a common ground state $|g\rangle$ to enable the re-absorption, even though there are also effects in the spectrum of the emitted light associated to SGC on transitions with different ground states [90].

With large ensembles of nuclei in x-ray cavities, these limitations can be overcome. One reason is that the coupling between the different transitions is mediated via the cavity rather than free space. Then the condition of non-orthogonal dipole moments $\mathbf{d}_\mu^* \cdot \mathbf{d}_\nu \neq 0$ is relaxed to $(\hat{d}_\mu^* \cdot \mathbf{1}_\perp \cdot \hat{d}_\nu) \neq 0$, where $\mathbf{1}_\perp$ is a projector onto the cavity polarization space [see Eq. (3.45)]. To illustrate the consequence of this, we introduce a coordinate system with cavity surface normal unit vector $\hat{\pi} = \hat{x}$, wave vector $\mathbf{k} = \hat{z}$, and $\hat{\sigma} = \hat{k} \times \hat{\pi} = \hat{y}$. In this case, $\mathbf{1}_\perp = \hat{x}\hat{x}^* + \hat{y}\hat{y}^*$, where the vectors are multiplied with the outer product to form a matrix. If the nuclear magnetization $\mathbf{B}_{\rm hf} \propto \hat{\pi}$, the two circularly polarized transitions have dipole moments $\hat{d}_1 = (\hat{k} + i\hat{\sigma})/\sqrt{2}$ and $\hat{d}_2 = (\hat{k} - i\hat{\sigma})/\sqrt{2}$. Then, $(\hat{d}_\mu^* \cdot \mathbf{1}_\perp \cdot \hat{d}_\nu) = 1/2$ for $\mu, \nu \in \{1, 2\}$. Since the cross terms $\mu \neq \nu$ have the same weight as the diagonal terms $\mu = \nu$, maximum SGC arises. This is possible since the contribution due to $\hat{x}\hat{x}^*$ vanishes in this particular cavity geometry. In contrast, in free space, contributions of different polarizations would cancel each other and $\hat{d}_1^* \cdot \hat{d}_2 = 0$. Thus, the spatial anisotropy of the cavity vacuum leads to the formation of SGC [91, 92]. A second mechanism for SGC in x-ray cavities involves the coupling between different nuclei. In our approach, the combined system of the large ensemble of nuclei and the cavity is described as a single effective nucleus. Within this model, the probing x-ray beam does not resolve the microscopic structure of the system. As a consequence, the dipole-dipole coupling between different nuclei appears as a radiative coupling between different states of the single effective nucleus. Therefore, the nuclear many-body system enables one to engineer an effective single nucleus with properties going beyond those naturally found in nuclei. Finally, in both cases, the superradiant enhancement of spontaneous emission enables one to reduce the perturbing effect of the energy splitting between the different states.

We implemented SGC in a Pd(5 nm)/C(20 nm)/^{57}Fe(2.5 nm)/C(20 nm)/Pd(20 nm) layer system, in which the magnetization of the ferromagnetically ordered Fe layer can be controlled via a weak external magnetic field. An example is shown in Fig. 3.14 for the half-Faraday geometry, in which the magnetization $\mathbf{B}_{\rm hf} \parallel (\hat{k} + \hat{\sigma})$. The input and detection polarization were chosen along $\hat{\sigma}$ and $\hat{\pi}$, respectively. It can be seen that the quantum optical model predicts deep interference minima at around $\Delta = \pm 30\gamma$, which disappear if the mechanism leading to SGC is artificially omitted

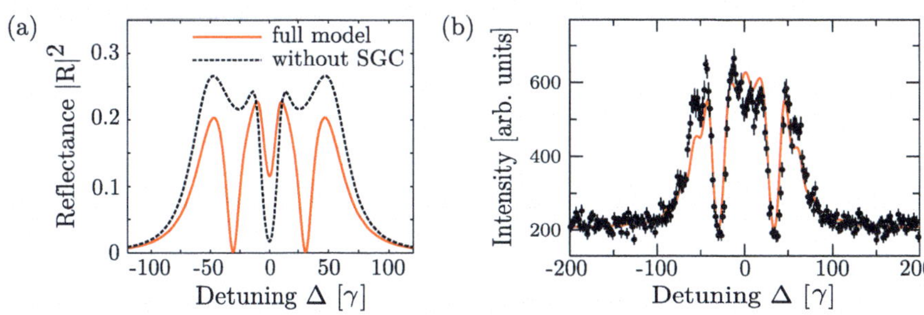

Fig. 3.14 Experimental realization of SGC. **a** Theoretical predictions of the quantum optical model, as well as corresponding results obtained by artificially omitting the SGC contributions. **b** Experimental data from [29], together with a theoretical fit using CONUSS including details of the detecion procedure. The deep minima indicating the presence of SGC can clearly be seen. Reprinted from [67], Copyright 2015, with permission from Springer Nature

in the analysis. The corresponding experimental results clearly verify the presence of these minima, and thus of SGC [29]. The experimental data is overlayed by a theoretical calculation obtained using CONUSS [55, 74] which in addition takes into account the details of the detection method. Interestingly, the intensity drops down to the background baseline, which indicates full interference visibility. This indicates a nuclear quantum system essentially free of perturbations.

From a broader perspecitive, the large ensembles of nuclei with magnetic sublevels in x-ray cavities thus enable one to engineer a variety of tunable quantum optical level schemes, including the possibility to implement SGC.

3.7.4 Tunable Subluminal Propagation of Resonant X-Rays

As discussed in Sect. 3.7.2, the key signature of electromagnetically induced transparency is the vanishing of the linear absorption of a probe beam within a narrow spectral transparency window. However, next to the transparency, EIT is also accompanied by characteristic modifications to the medium's dispersion [72]. In particular, within the transparency window, a steep linear dispersion appears, which can be facilitated to control the group velocity of a light pulse passing through the medium. To see this, we consider the propagation of an electromagnetic wave packet through a medium, given in one dimension by

$$E(x,t) = \frac{1}{2\pi} \int_{-\infty}^{\infty} d\omega E(\omega)\, e^{i(\omega t - kx)}. \tag{3.54}$$

We assume that the spectral width of the wave packet is narrow as compared to the EIT window, and expand the wave number $k = k_R + ik_I$ in leading order of a Taylor series around the center of the EIT window at ω_0 to give $k_R(\omega) \approx k_R(\omega_0) + \frac{\partial k_R}{\partial \omega}|_{\omega_0} (\omega - \omega_0)$ and $k_I(\omega) \approx k_I(\omega_0)$. Note that the linear order of k_I is zero since the absorption has a minimum at ω_0. Inserting this into Eq. (3.54) gives

$$E(x,t) \approx \frac{1}{\sqrt{2\pi}} e^{-k_I(\omega_0)x} \times e^{k_R(\omega_0)\,(x - v_{ph}t)} \times \int_{-\infty}^{\infty} d\omega E(\omega) e^{i\frac{\omega-\omega_0}{v_{gr}}(x - v_{gr}t)}. \tag{3.55}$$

The three parts separated by "×" have a clear interpretation. The first part is the linear attenuation because of the imaginary part of the resonant refractive index $n_I(\omega_0) \propto k_I(\omega_0)$, following the Lambert-Beer law. The second part describes the propagation of the carrier frequency plane wave through the medium. It moves with the phase velocity $v_{ph} = \omega_0/k_R(\omega_0) = c/n_R(\omega_0)$, that is, with the vacuum speed of light c divided by the real part of the index of refraction. The third part shows that the wave packet propagates with the group velocity

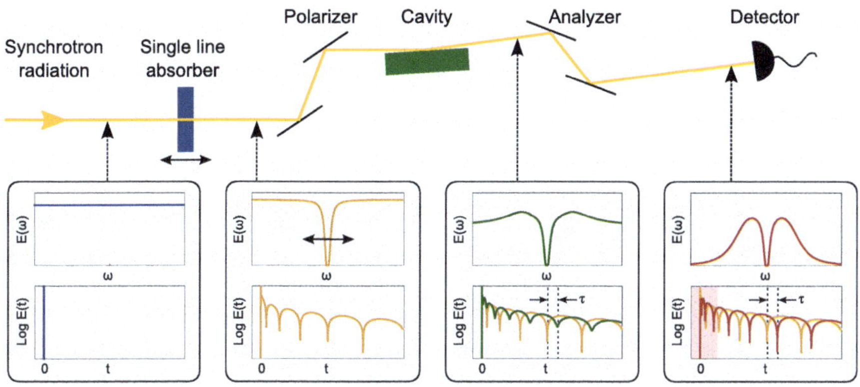

Fig. 3.15 Schematic setup of the experiment. The lower panel shows the temporal and the spectral structure of the x-ray pulse at different points throughout the propagation through the setup. Reprinted from [36], Copyright 2015, with permission from the American Physical Society

$$v_{gr} = \left(\frac{\partial k_R}{\partial \omega} \bigg|_{\omega_0} \right)^{-1} = \frac{c}{n_R(\omega_0) + \omega_0 \frac{\partial n_R}{\partial \omega}|_{\omega_0}}. \tag{3.56}$$

This expression illustrates that without dispersion $\partial n_R / \partial \omega$, the group velocity of the wave packet is equal to the phase velocity. Otherwise, depending on the sign and the magnitude of the dispersion, the group velocity can be much lower than the vacuum speed of light c ("slow light", sub-luminal propagation), larger than c ("fast light", super-luminal propagation), or even negative. All cases have been experimentally implemented with atomic gases, and the control of the group velocity has found numerous applications [72].

Recently, group velocity control and slow light has also been achieved at x-ray energies using Mössbauer nuclei [36]. The concept of the experiment is shown in Fig. 3.15. The setup is motivated by two main experimental challenges. First, a resonant medium with steep positive linear dispersion has to be implemented in order to achieve $v_{gr} \ll c$. For this, a suitably prepared cavity containing ^{57}Fe nuclei was used. Second, a spectrally narrow x-ray pulse must be generated, whose spectrum lies entirely within the linear dispersion part of the medium, i.e., within a bandwidth of about 10–100 neV. This is impossible with conventional monochromators, but can be realized, e.g., using pure nuclear Bragg reflections [93–95] or mechanical choppers [96]. For the experiment, instead another method was developed, based on a single line absorber together with a high-purity polarimetry setup [78].

To explain the generation of the spectrally narrow pulse, we follow the propagation of the x-rays through the setup in Fig. 3.15. In the time domain, the incident spectrally broad synchrotron pulse is well approximated as a Dirac delta function at time $t = 0$, see the lower left panel. Upon passing through the single line absorber, the x-ray pulse is split into two parts. The part which did not interact remains a delta function. The other part which did interact with the nuclei in the single line analyzer leads to a temporally long response of the nuclei, and correspondingly to a narrow spectral absorption resonance which will become the desired spectrally narrow x-ray pulse. Note that because of the thickness of the single line absorber, dynamical beats appear in the time domain, which will turn out to be crucial for the analysis of the experiment. Afterwards, the x-rays pulse is polarized, which does not lead to notable changes because of the natural polarization of the synchrotron radiation, but improves the purity of the polarimetry setup. Next, the x-rays interact with the cavity. As explained below, the nuclei in the cavity are operated in such a way that they slow down and delay the narrow pulse component by a time τ, and at the same time rotate the polarization of part of the scattered light. The analyzer is operated in crossed setting, such that only light with rotated polarization may pass. As a consequence, only light which interacted with the nuclei in the cavity can pass the analyzer. The light seen by the detector thus contains two parts. The first part interacted with the cavity, but not with the single line analyzer. The second part interacted with the cavity and the single line analyzer. At late times, the latter signal dominates, because it is delayed by both the single line analyzer and the cavity, and it comprises the desired signal of a spectrally narrow x-ray pulse which interacted with the cavity. As the lower right panel of Fig. 3.15 shows, this signal in the time domain approximately is a copy of the input pulse, but delayed by τ. This delay is easily visualized independent of the total count rate by the position of the dynamical beat minima.

In order to implement the steep dispersion, we used a variety of the cavity featuring spontaneously generated coherences (see Sect. 3.7.3) optimized for the linear dispersion. Both EIT (see Sect. 3.7.2) and SGC may feature transparency windows and steep linear dispersion, and in fact are related [35, 37]. The advantage of the SGC cavities is the simpler design, and that previous experiments [29] had already demonstrated the possibility to reach almost perfect transparency, see Fig. 3.14. Furthermore, the SGC cavities rely on the magnetic substructure of the nuclei in a way which allows to rotate the polarization of the scattered light as required for the polarimetry method to generate spectrally narrow x-ray pulses.

To analyze light propagation in the cavity setting, in analogy to the procedure leading to Eq. (3.55), we expand the cavity response $R(\omega)$ around the transparency resonance ω_0 in linear order. We find $R(\omega) \approx R(\omega_0)e^{i(\omega-\omega_0)\tau}$, where $\tau = \frac{\partial}{\partial\omega}\arg[R(\omega_0)]$. The action of the cavity on an input pulse $E_{in}(t)$ with spectrum $E_{in}(\omega)$ is therefore given by

$$\begin{aligned} E_{out}(t) &\propto \int E_{in}(\omega)R(\omega)e^{-i\omega t}d\omega \\ &\approx R(\omega_0)e^{-i\omega_0\tau}\int E_{in}(\omega)e^{-i\omega(t-\tau)}d\omega \\ &\propto E_{in}(t-\tau)\,. \end{aligned} \tag{3.57}$$

We thus find that the input pulse preserves its shape, but is delayed by the time τ, as desired for slow light. The expression of τ can be understood by noting that in a cavity setting, the real and imaginary parts of the medium susceptibility are in fact related to the phase $\arg(R)$ and the logarithm of the modulus $\log(|R|)$ of the cavity reflectance R [36].

In the experiment, a cavity consisting of a Pd(2 nm)/C(20 nm)/^{57}Fe(3 nm)/C(21 nm)/Pd(10 nm)/Si layer system was probed in grazing incidence, coupling the x-rays into the fundamental mode. A magnetic field was applied along the beam direction in order to align the magnetization of the nuclei in the so-called Faraday configuration. In this setting, spontaneously generated coherences occur, which lead to a steep linear dispersion [29]. Since full transparency of the medium corresponds to zero intensity of the scattered x-ray, the cavity was designed to avoid full transparency in order to measure the delay of the x-rays also on resonance. The single line analyzer comprised a 10μm thick stainless steel foil ($^{57}Fe_{55}Cr_{25}Ni_{20}$) mounted on a Doppler drive, which generates a spectrally narrow pulse of about 10 ns duration with tunable central frequency. The results of the experiment are shown in Fig. 3.16. (a) presents

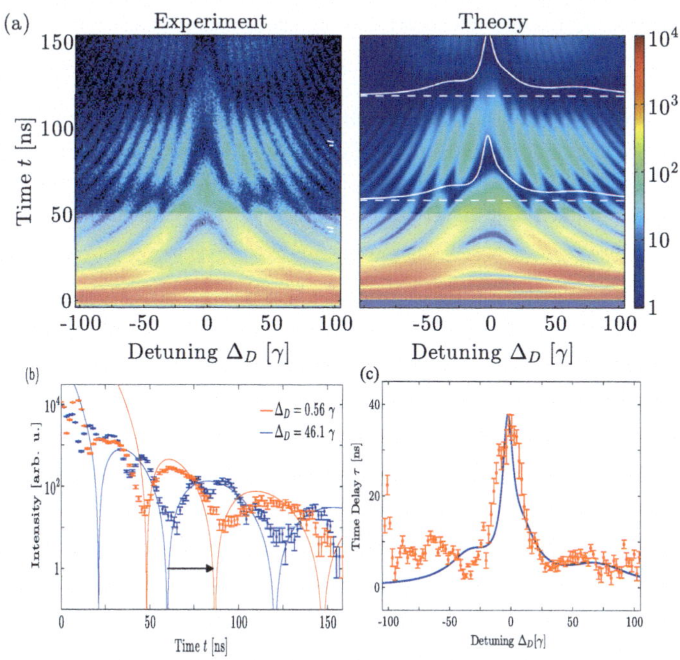

Fig. 3.16 Results of the experiment. **a** Shows the recorded intensity as function of time and the detuning of the single line absorber, as well as the corresponding theory result. Dashed lines in the theory plot indicate the position of the undelayed dynamical beat minima. White lines indicate the actual positions of the minima predicted from theory. **b** Shows two cuts through (**a**) at detunings $\Delta_D = 0.56\gamma$ and $\Delta_D = 46.1\gamma$, respectively. The two curves coincide at late times except for a delay τ, indicated by the black arrow. **c** Shows the observed delay as function of the single line analyzer detuning, i.e., the center frequency of the spectrally narrow x-ray pulse. The corresponding theoretical result is shown in blue. Delays exceeding 35 ns are observed, corresponding to group velocities below 10^{-4} c. Reprinted from [36], Copyright 2015, with permission from the American Physical Society

the experimentally recorded intensity as function of the time after excitation and the detuning of the single line absorber. (b) shows the corresponding theoretical fit, which agrees very well. The white dashed lines indicate the position of the dynamical beats expected without pulse delay. It can be seen that close to resonance $\Delta_D = 0$, the dynamical beat minima observed in the experiment are systematically shifted to later times, as expected for slow light. To illustrate this further, (c) shows cuts through (a) and (b) for $\Delta_D = 0.56\gamma$ and $\Delta_D = 46.1\gamma$, respectively. For the lower detuning, the spectrally narrow x-ray pulse experiences the steep linear dispersion, and thus is slowed down. The higher detuning is outside the steep linear dispersion region and does not lead to slow light. As a result, at late times, the two data sets are essentially identical, except for a temporal delay indicated by the black arrow. Note that at early times the two curves differ, because of the residual contribution of light which did not interact with the single line absorber. Finally, panel (d) shows the experimentally observed delay τ as function of detuning Δ_D. The solid line shows the corresponding theoretical prediction. We find that our cavity allows to induce delays exceeding 35 ns, which corresponds to group velocities below 10^{-4} c [36].

These results constitute another proof of the possibility of manipulating the x-ray optical response of nuclei embedded in cavities to one's desire. Further theoretical calculations predict that with a suitable time-dependent manipulation of the applied magnetic fields, also a complete stopping of the x-ray pulse could be achieved [97]. Possible applications of such techniques include the delay and synchronization of x-ray photons, and the coherence-based enhanced of nonlinear interactions between x-rays and nuclei [72].

3.8 Collective Strong Coupling of Nuclei in Coupled Cavities and Superlattices

A central subject of quantum optics is to manipulate the interaction of light and matter. To achieve this, two important parameters must be controlled. One is the strength of the light-matter interaction. It has to be strong enough that emitted photons have a chance to act back on their emitters. This is the so-called strong-coupling limit in quantum optics [98]. It can be achieved in special environments into which the emitters are embedded. Strong coupling is used in the optical and infrared regimes, for instance, to produce non-classical states of light, enhance optical nonlinearities even at relatively low intensities [99] and control quantum states [100]. The other parameter is the number of modes of the electromagnetic field that the resonant system interacts within this environment. If the number of these modes is too large, the emitted photons might get irreversibly lost when they are distributed over these modes. Strong coupling has been achieved for a variety of systems and energy ranges, but until now not with X-rays. Here we report about the first observation of collective strong coupling of hard X-rays at the nuclear resonance of ^{57}Fe.

3.8.1 Strong Coupling of X-Rays and Nuclei in Photonic Lattices: Normal-Mode Splitting

The usual route to strong coupling between light and resonant atoms is to insert the atoms into a cavity, the Q-factor of which is on the order or larger than that of the atomic resonance [98]. A good cavity restricts the interaction with the light to the one mode allowed by the cavity; furthermore, the intensity of the light within the cavity is large, leading to an enhanced interaction. The strong interaction leads to the coupling of two degrees of freedom; two normal modes form, which are superpositions of the otherwise uncoupled components of the system. If the cavity is probed, for instance by monitoring its reflectivity or transmissivity, its spectral signature shows two dips, which are detuned from the sample resonance and the cavity mode. Upon varying the detuning between cavity mode and resonance, it turns out that the dips undergo a very well-known anti-crossing dispersion; the minimal distance between the branches of the dispersion relation is given by the interaction strength.

While X-ray cavities have been successfully applied to realize quantum optical concepts in the linear regime, fabricating a cavity of sufficient quality to reach the strong coupling regime is not possible yet in the X-ray range. Even in the angular range of grazing incidence, the reflectivity of the cladding mirrors is $\sim$95% only. Compared to the reflectivity achieved for microwave and visible light cavities of 99.999% this is not sufficient. The problem is rooted in the fact that the spectral width of the cavity is much larger than the coupling strength. Essentially, the two dispersive dips mentioned above cannot be resolved [27, 79].

This points the way to another method of coherent control. Since our nuclear exciton interacts with only one mode anyway, we can focus on enlarging the interaction without making use of a cavity, but by enlarging the number of nuclei that contribute to the nuclear exciton, as the collective interaction strength depends on that number. In the following, we will focus our attention on periodic multilayers or periodic resonant systems. These, often referred to as resonant photonic crystals or resonant optical lattices, also restrict the number of modes the resonant matter interacts with.

The propagation of light through periodic arrays of resonant media such as optical lattices or multiple quantum wells has opened intriguing possibilities to control the interaction of light and matter [102–105]. One of the most interesting features of these systems is the appearance and dispersion of bandgaps [106–109], see Fig. 3.17.

Here we describe an optical lattice-like structure with bandgaps in the regime of hard X-rays, consisting of a multilayer with alternating layers of non-resonant ^{56}Fe and nuclear-resonant ^{57}Fe. The electronic part of the index of refraction is identical for both isotopes. The system can thus be modelled as having a uniform background refractive index, with a periodic resonant refractive index superimposed. The resulting bandgap can be observed by measuring the spectrally resolved X-ray reflectivity of the sample close to the Bragg angle of the multilayer. Although similar isotopic multilayer structures have been discussed before [110–113], mainly the angular dependence of the reflectivity has been studied instead of its spectral

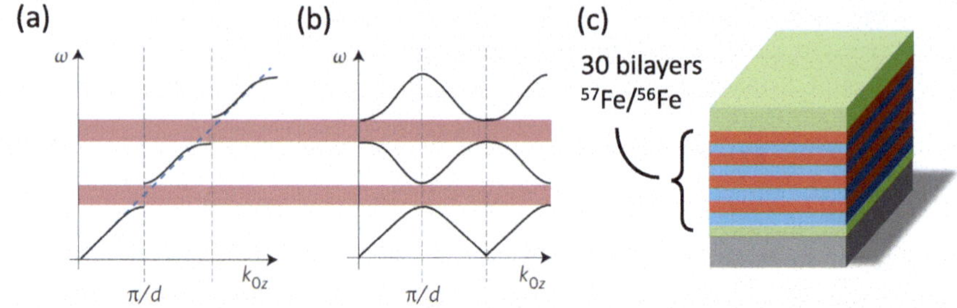

Fig. 3.17 **a** Dispersion relation (plotted as frequency ω against wavevector k_{0z}) for photons in different media: in a medium with uniform refractive index (blue dashed line) and in a photonic crystal with an interlayer spacing d (black solid curves). **b** Brillouin diagram for a material with a modulated refractive index, showing the allowed and forbidden regions for the real wavevector, implying propagation and reflection, respectively. Pink bands indicate photonic bandgap formation. A periodic nuclear resonance structure can result in the formation of an excitonic Bragg reflection feature. **c** A stack of 30 [(1.64 nm ^{56}Fe)/(1.12 nm ^{57}Fe)] bilayers was used to modulate the refractive index of the material for light at the nuclear resonance. Figures **a, b** reprinted from [101], Copyright 2016, with permission from Springer Nature

properties. Here we report about high-resolution spectroscopic studies, and reveal the strong collective interaction between resonant X-rays and nuclei, which leads to the formation of photonic bandgaps. In the following we discuss the connection between the photonic dispersion relation and the reflectivity of the multilayer, the latter being the observable that allows us to experimentally assess the signatures of strong coupling.

Figure 3.18 shows the real and imaginary parts of the dispersion relation around the nuclear resonance and the Bragg peak. There are three distinct contributions to the imaginary part: (1) the uniform electronic absorption, (2) the nuclear absorption, which exhibits a Lorentzian energy dependence around the resonance and (3) the extinction, which determines how deeply the radiation penetrates into the multilayer. The higher the extinction, the fewer periods contribute to the reflection. Nuclear absorption dominates around the resonance, absorbing most incoming radiation. Off resonance, the extinction becomes stronger relative to the absorption, indicating enhanced reflection. Far off resonance and off-Bragg, the electronic absorption suppresses the reflection. The shape of the real part results from a polaritonic effect: X-rays of suitable energy impinging on resonant matter undergo nuclear-resonant forward scattering, which leads to an energy-dependent phase shift of the scattered photons. In periodic media, polaritons excited at certain angles and with certain energies radiate in phase, such that the outgoing radiation has a different direction from the incoming one because the electromagnetic waves interfere destructively in all other directions. In other words, the Bragg condition $k_{0z} = \pi/d$ is fulfilled. Tuning the angle across the Bragg peak, one can observe an angularly resolved polaritonic dispersion relation and the corresponding anticrossing behaviour, as well as a narrow, almost dispersionless contribution around resonance that is visible in Fig. 3.18a, b. This interpretation is supported by a simple quantum-optical model

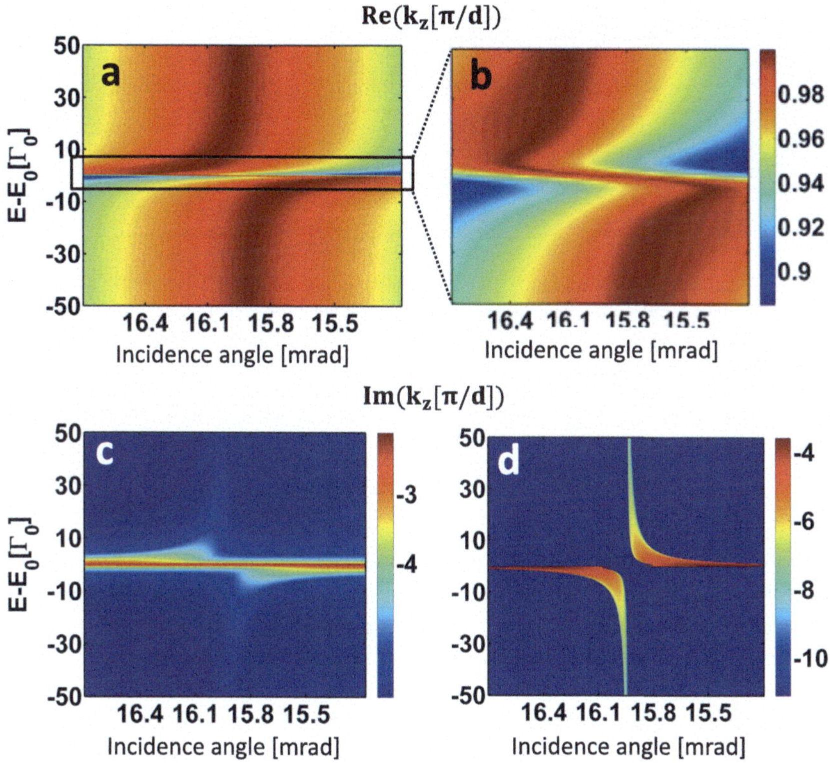

Fig. 3.18 Angular-dependent dispersion relation of a $^{56}Fe/^{57}Fe$ multilayer. The dispersion of the out-of-plane component k_z of an infinite stack of 1.64 nm ^{56}Fe/1.12 nm ^{57}Fe bilayers around the Bragg position is shown. **a** The real part of k_z is encoded in the colour bar in units of π/d. **b** Magnification of the small, dispersionless gap in (**a**). **c** Imaginary part of k_z, logarithmically encoded in the colour bar. **d** The extinction coefficient contribution to the imaginary part, encoded logarithmically. It characterizes how well a sample reflects light and is due to dispersion, not absorption of the materials. An anticrossing is visible at the Bragg position. In comparison with (**c**), it is obvious that the extinction coefficient is a weak contribution to the imaginary part both at the energetic resonance and very far from it, but is dominant in the intermediate range. Figures reprinted from [38]

[38] to explain the dispersion relation in microscopic terms [114, 115]. To account for the finite thickness of the resonant layers, we model our system as a so-called bichromatic optical lattice (containing two atoms per unit cell) and simplify the Hamiltonian until it can be numerically diagonalized for a range of k-vectors at a particular angle. Figure 3.19 compares the resulting quantum mechanical dispersion relation to the reflectivities of two multilayers with a finite number of periods. In Fig. 3.19a (30 periods) a splitting is readily observable, but in Fig. 3.19b (100 periods) the almost fully formed bands seem to diverge at the Bragg angle, because the collective coupling-enhanced splitting is too large for the displayed energy range.

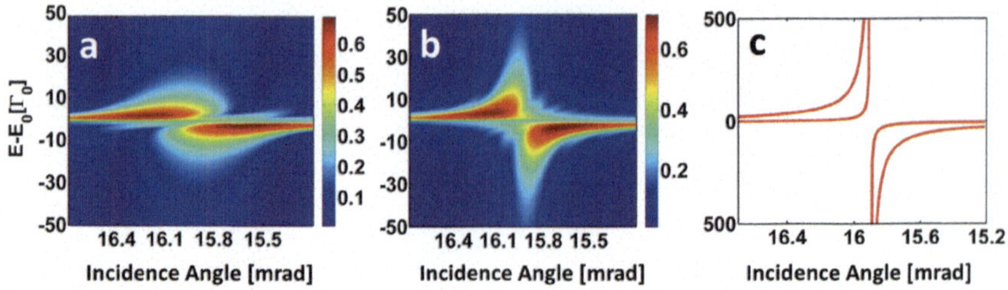

Fig. 3.19 Calculated reflectivity of the $^{56}Fe/^{57}Fe$ multilayer. **a** Calculated reflectivity of a 30-period multilayer and **b** of a 100-period multilayer. **c** Quantum mechanically calculated dispersion relation. The reflectivity of a many-period multilayer is well-described by the model of the multilayer with an infinite number of layers. For 30 periods, the bandgaps are not fully formed and the splitting is smaller. This is due to the collectivity of the light-matter interaction. More layers result in a stronger interaction and a larger splitting. In (**b**) and (**c**), the interaction and splitting are so large that the bandgaps are not observed together in an experimentally accessible energy range. Figures reprinted from [38]

We also see that the reflectivity of the 30-period structure exhibits peaks rather than fully formed bandgaps. The peaks correspond to the lower edges of the bandgaps; here the extinction coefficient is largest and even for a few-period structure there is a sizable reflection. Alternatively, the peaks can be described as superradiant modes, which, on adding more periods to the structure, turn into bandgaps [116]. Note that the descriptions by the transfer matrix formalism and the quantum optical model are qualitatively identical for infinite systems, thus supporting our interpretation of the observed phenomena. The low-dispersion bandgap that appears close to resonance is probably caused by nuclei weakly coupling to the electromagnetic field, for example at the layer interfaces. Although the bandgaps we observe are photonic bandgaps, their dispersion is due to the (strong) collective interaction of light and the nuclei.

In the case where the wavevector of the incoming light approximately fulfills the Bragg condition, the quantum-optical model yields an analytic dispersion relation given by

$$\omega_{j,\pm}(q) = \frac{\omega_{k_{0z}} + \omega}{2} \pm \sqrt{(\frac{\omega_{k_{0z}} - \omega}{2})^2 + 2Ng^2\left[1 - (-1)^j \cos(q\rho)\right]} \qquad (3.58)$$

where index $j = 1,2$, g is the coupling constant, N is the number of unit cells, ρ is the distance between atoms of the same unit cell and q is the reciprocal lattice vector. The similarity to a standard strong coupling dispersion relation [98, 117] with the corresponding Rabi splitting is obvious. In our case, multiple bands undergo this splitting, leading to two bandgaps. The splitting between them is the signature of collective strong coupling. This microscopic model does not include dissipation. However, it has the advantage of giving our results an intuitive and qualitative explanation. The dispersion relation derived from the transfer matrix model is quantitatively more reliable because it includes dissipation and absorption.

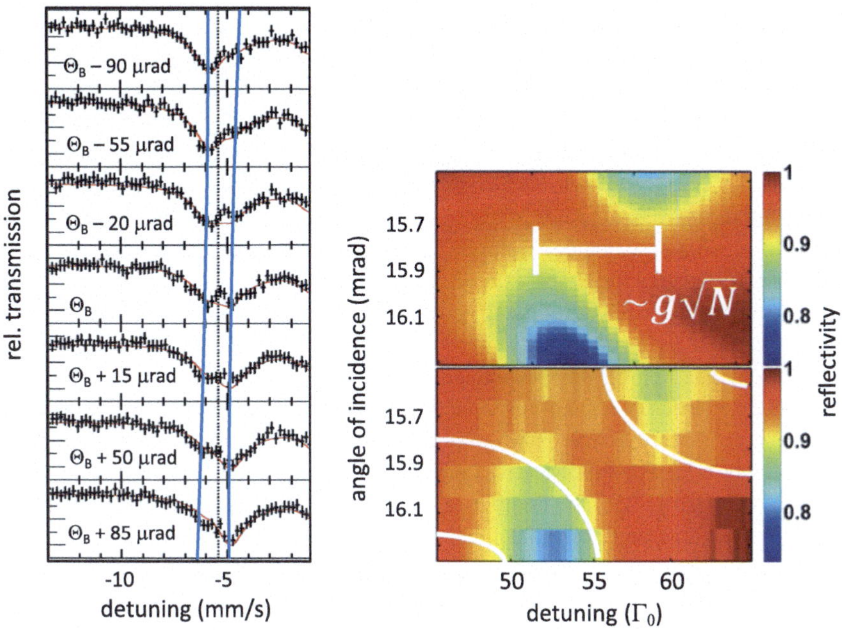

Fig. 3.20 Left: measured energy spectra of the multilayer reflectivity for different angular positions around the Bragg angle θ_B. Displayed is the energy region around the outer resonance line of ^{57}Fe in ferromagnetic Fe. The spectra reveal a splitting of the line, the angular dependence of which displays the characteristic anticrossing behavior (see blue lines as guide to the eye) that is indicative of strong coupling. On the the right hand side the measured intensity is plotted as a colour map; the background baseline is normalized to one. The upper right panel shows simulations, the lower shows the data. The white bar in the upper panel indicates the dip distance which gives the collective coupling strength. The white lines in the lower bar indicate the edges of the photonic band gaps. Right figure reprinted from [38]

To experimentally verify the particular shape of the dispersion relation for a nuclear optical lattice, we prepared a multilayer sample consisting of 30 periods of (1.64 nm ^{56}Fe)/(1.12 nm ^{57}Fe) (average thicknesses) sandwiched between two 4 nm Ta layers, altogether deposited on a Si substrate. This facilitated to measure the splitting in a suitable energy range, but did not permit an accurate quantitative comparison to the model. However, the reflectivity can be simulated by the transfer matrix model. Measurements were performed at the Nuclear Resonance Beamline ID18 of the European Synchrotron Radiation Facility (ESRF). Reflectivity spectra of the sample are shown in Fig. 3.20 together with simulations using the program package CONUSS [55, 74]. Owing to the magnetic hyperfine interaction in the sample, the nuclear resonance of ^{57}Fe is split into four well-separated lines, each of which can be treated as a single-line nuclear resonance (displayed in Fig. 3.18). The outer lines in the measured spectra exhibit the strongest collective coupling, manifesting in a clearly resolved Rabi splitting at the Bragg position. The bands at zero detuning, i.e., at the Bragg angle, have the frequencies $\pm\sqrt{2Ng^2(1-(-1)^j\cos(q\rho))}$ around the nuclear resonance. The splitting is then $4g\sqrt{N}f_j(q\rho)$, where $f_1(q\rho) = \cos(q\rho/2)$ and $f_2(q\rho) = \sin(q\rho/2)$. Depending on whether ρ is smaller or larger than $0.25a$, the bands marked by $j = 1$ or $j = 2$ form the inner bands. We assume that ρ is smaller, giving a splitting of $4g\sqrt{N}\cos(q\rho/2) \approx 8\Gamma_0 \approx 57.3$ MHz. Again, this is the result for 30 layers and should not uncritically be equated with the formula resulting from the infinite model. Nevertheless, the agreement between transfer matrix model simulations and measurements is excellent. An exception is the almost dispersionless gap visible in Fig. 3.18a which could not be resolved experimentally since the energetic width of the analyzer foil smears out sharp spectral features.

To summarize, we have simulated and measured the energy-resolved reflectivities of an isotopic 1.64 nm ^{56}Fe/1.14 nm ^{57}Fe multilayer around the nuclear resonant Bragg peak. The results were explained in terms of the polaritonic propagation of light and excitations of nuclear resonant matter. Within a quantum optical model, we could connect the dispersion to the coupling of x-rays with nuclear excitons and have given a lower bound to the collective coupling strength. Dissipation and the interface roughness that grow with the number of layers thwart any attempt to observe fully-formed band-gaps, and therefore precise values cannot be determined. However, the distinctly observable dispersion and splitting of bands is the first unambiguous evidence of collective strong coupling in the hard x-ray energy range.

Nuclear optical lattices display several unique features absent in other systems, such as extremely high number densities (on the order of 10^{28}m^{-3}) and stability over a wide temperature range. This work could be extended to other isotopes with higher energies and less electronic interaction, such as ^{119}Sn (23.9 keV) or ^{61}Ni (67.4 keV). We therefore anticipate that nuclear resonant periodic multilayers will stimulate x-ray quantum optics research and bring it closer to coherent control of the x-ray-matter interaction. Even further, the concept of polaritons in bichromatic optical lattices or other periodic systems itself is a subject that holds great interest far beyond the x-ray range.

To put our results into perspective we briefly summarize previous work in similar systems, as given in [118]. To the best of our knowledge, two physical systems have yielded phenomena and observations similar to those described here: genuine optical lattices [104, 106, 107] and gratings of excitonic quantum wells. In excitonic quantum wells, a semiconductor is doped periodically; that way, the background refractive index is identical throughout, but there are periodically spaced regions where quantum well excitations are possible [119]. This medium is particular interesting, since it suffers from a similar drawback as ours: too few layers result in an unclear or incomplete formation of band gaps. Since the early 1990s, the results obtained from excitonic quantum wells have been described in a different framework. Instead of assuming an infinite structure, researchers calculated the eigenmodes of these systems for a small number ($\sim$10) of layers [120, 121]. In that case, the eigenmodes are one superradiant Bragg mode, which reflects the radiation in a band much wider than the exciton resonance, and a number of dark modes. In a sense, this is the incipient Bragg band gap. However, experiments [108, 122] showed that with an increasing number of layers, dips in the superradiant mode and a saturation of its width appeared; researchers explained this later in terms of band gaps and standing waves within the band gaps [116, 123], much as the quantum optical model of this chapter. A system resembling a bichromatic array of quantum wells was examined in [102]; although that paper has a different focus, the observed phenomena resemble those presented here.

3.8.2 *Rabi Oscillations via Strong Coupling of Two Nuclear Cavities*

In the regime of strong coupling, the reversible exchange of photons between a cavity mode and an electromagnetic resonance leads to an oscillatory energy transfer between the two systems, the so-called Rabi oscillations. Collective strong coupling of X-rays and nuclei has recently been demonstrated in a periodic multilayer [39], as described in the previous section. The mode splitting and anticrossing dispersion typical of strong coupling was observed in energy-resolved reflectivity measurements. However, the Zeeman splitting of the Mössbauer resonance into several resonance lines together with dissipation and structural imperfections of the layer system prevented a clear detection of Rabi oscillations. Moreover, a conclusive proof of strong coupling requires that the splitting of the resonance is solely due to the interaction between the ensembles and is not affected by Lamb shift contributions. This requires a particular arrangement of the resonant ensembles, as explained in the following.

The central requirement for the strong coupling regime is that the coupling of the mode and the resonant layer be larger than their decay rates given by the the spectral width of the cavity (that is, the inverse of the time a photon is stored in the cavity) and the decay width of the nuclear ensemble. For a single thin-film cavity system, the coupling strength, although larger than the nuclear decay width, is still much smaller than the cavity linewidth. We circumvent this difficulty by introducing a novel double cavity setup (Fig. 3.21), which is described by an effective Hamiltonian fulfilling the desired strong coupling conditions. This ansatz follows the general approach of simulating a complex physical system that mimics a simple Hamiltonian that cannot be implemented straightforwardly. Similar approaches are used extensively in contemporary research, for example to observe the Dicke phase transition [124, 125] or in the use of ultracold quantum gases to simulate magnetism [126] and correlated materials [127, 128].

In the following we discuss a quantum optical description of the double cavity setup in order to show that it indeed fulfills the conditions to observe Rabi oscillations between the two nuclear layers. The interaction between the X-ray field and the two nuclear layers embedded in the double cavity can be described by means of a recently developed quantum-optical model [29, 79] adapted to the particular sample geometry. The Hamiltonian of this interaction is given by

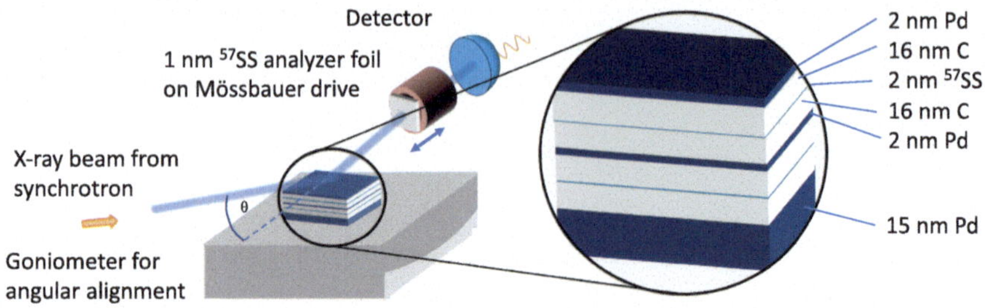

Fig. 3.21 Sketch of the coupled cavity system and experimental geometry to observe Rabi oscillations between two ensembles of ^{57}Fe nuclei. Each cavity consists of a Pd/C/^{57}SS/C/Pd layer stack, where ^{57}SS denotes stainless steel ($Fe_{0.55}Cr_{0.25}Ni_{0.2}$) with its iron content enriched to 95% with ^{57}Fe. Stainless steel does not exhibit ferromagnetic order, so in this alloy the ^{57}Fe isotope presents a single-line resonance at 14.413 keV with no Zeeman splitting. This layer structure realizes two almost identical cavities coupled via a thin Pd interlayer, which also constitutes the top cladding of the lower cavity and bottom cladding of the upper cavity. The layer system was positioned on a goniometer that permitted to control the cavity detunings via adjustment of the angle of incidence θ and to perform $(\theta - 2\theta)$ reflectivity measurements

$$\begin{aligned} H = {} & \Delta_1 a_1^{+} a_1 + \Delta_2 a_2^{+} a_2 + J(a_1^{+} a_2 + a_2^{+} a_1) \\ & - \Delta(|E_1\rangle\langle E_1| + |E_2\rangle\langle E_2|) \\ & + g_1\sqrt{N_1}\left(a_1|E_1\rangle\langle G| + a_1^{+}|G\rangle\langle E_1|\right) \\ & + g_2\sqrt{N_2}\left(a_2|E_2\rangle\langle G| + a_2^{+}|G\rangle\langle E_2|\right), \end{aligned} \tag{3.59}$$

where $|G\rangle = |g_1 g_2 0_1 0_2\rangle$ denotes the nuclear ground state and $|E_1\rangle = |e_1 g_2 0_1 0_2\rangle$, $|E_2\rangle = |g_1 e_2 0_1 0_2\rangle$ denote the states with a single excitation in either of the nuclear layers, respectively. The first line in Eq. (3.59) gives the energies of the cavity modes and their interaction term; the second one likewise the energies of the nuclear ensembles. The third line describes the interaction between the first mode and its nuclear ensemble, while the fourth one depicts the interaction between the second mode and its ensemble. The creation (annihilation) operators of the cavity mode in the first cavity are $a_1^{\dagger}(a_1)$. A second cavity is coupled to this first one by a strength J. The second cavity mode's creation (annihilation) operators are $a_2^{\dagger}(a_2)$. The detunings of these two cavities are denoted by Δ_1 and Δ_2, respectively.

In each cavity, there is a nuclear ensemble coupled to it. In our experiments with synchrotron radiation there is at most one photon in the system, i.e., we work in the one-excitation limit. This allows us to truncate the Hamiltonian and to take into account only the first symmetric timed Dicke state [29] of these ensembles that is coupled to these modes [79]. The Dicke states are denoted as $|E_1\rangle$ and $|E_2\rangle$, and their excitation and deexcitation operators are $|E_{1(2)}\rangle\langle G|$ and $|G\rangle\langle E_{1(2)}|$, respectively. They are coupled to their respective cavity modes with a collective coupling strength $g_1\sqrt{N_1}$, $g_2\sqrt{N_2}$, where the $N_{1(2)}$ is the number of nuclei per ensemble, and $g_{1(2)}$ the coupling strength of an individual nucleus to the respective cavity mode. An additional term Δ describes the energy of the nuclei and their detuning from the incoming beam.

We then continue by prediagonalizing the interaction between the cavity modes, resulting in two cavity supermodes $a_{\pm}$, both coupling to the two nuclear ensembles (see Methods in [39]). We then heuristically add driving terms for both of these cavity modes, which are $i\sqrt{2\kappa_{R\pm}}(a_{in}a_{\pm}^{\dagger} + a_{in}^{*}a_{\pm})$. The energy bandwidth of the empty cavities is very large, on the order of 100 eV, which renders them highly dissipative, i.e., their interactions and the temporal evolution of their modes take place on extremely fast timescales, orders of magnitude faster than the evolution of the nuclear decay. Therefore, the common practice is to adiabatically eliminate the cavity modes and solely follow the temporal evolution of the three states $|g_1 g_2 0_1 0_2\rangle$, $|g_1 g_2 0_1 0_2\rangle$ and $|g_1 e_2 0_1 0_2\rangle$, corresponding to states with no nuclear excitation or one delocalised, collective excitation in layer 1 or 2, respectively, as depicted in Fig. 3.22d. Thus we reduce the problem to that of two interacting low-excitation nuclear ensembles similar to Ref. [79]. The effective Hamiltonian then reads (see Methods in [39]):

$$H = \begin{pmatrix} 0 & \Omega_1 & \Omega_2 \\ \Omega_1^* & \Delta - \delta_1 + i\gamma_1 & g_{12} - i\gamma_{12} \\ \Omega_2^* & g_{12} - i\gamma_{12} & \Delta - \delta_2 + i\gamma_2 \end{pmatrix}, \tag{3.60}$$

where the Hamiltonian is made up of a series of effective parameters given by:

(i) the effective driving strengths of both layers:

$$\Omega_1 = g_1\sqrt{N}\left(\frac{\sqrt{2\kappa_{R+}}}{\kappa_+ + i\Delta_+} + \frac{\sqrt{2\kappa_{R-}}}{\kappa_- + i\Delta_-}\right)a_{in}, \tag{3.61}$$

$$\Omega_2 = g_2\sqrt{N}\left(\frac{\sqrt{2\kappa_{R+}}}{\kappa_+ + i\Delta_+} - \frac{\sqrt{2\kappa_{R-}}}{\kappa_- + i\Delta_-}\right)a_{in}, \tag{3.62}$$

(ii) The effective decay widths, consisting of real and imaginary part, δ and γ, respectively:

$$\delta_i = -\frac{g_i^2 N}{2}\left(\frac{\Delta_+}{\kappa_+^2 + \Delta_+^2} + \frac{\Delta_-}{\kappa_-^2 + \Delta_-^2}\right), \tag{3.63}$$

$$\gamma_i = \frac{g_i^2 N}{2}\left(\frac{\kappa_+}{\kappa_+^2 + \Delta_+^2} + \frac{\kappa_-}{\kappa_-^2 + \Delta_-^2}\right), \tag{3.64}$$

(iii) the effective cavity-mediated coupling strength (real and imaginary part):

$$g_{12} = -\frac{g_1 g_2 N}{2}\left(\frac{\Delta_+}{\kappa_+^2 + \Delta_+^2} - \frac{\Delta_-}{\kappa_-^2 + \Delta_-^2}\right), \tag{3.65}$$

$$\gamma_{12} = \frac{g_1 g_2 N}{2}\left(\frac{\kappa_+}{\kappa_+^2 + \Delta_+^2} - \frac{\kappa_-}{\kappa_-^2 + \Delta_-^2}\right). \tag{3.66}$$

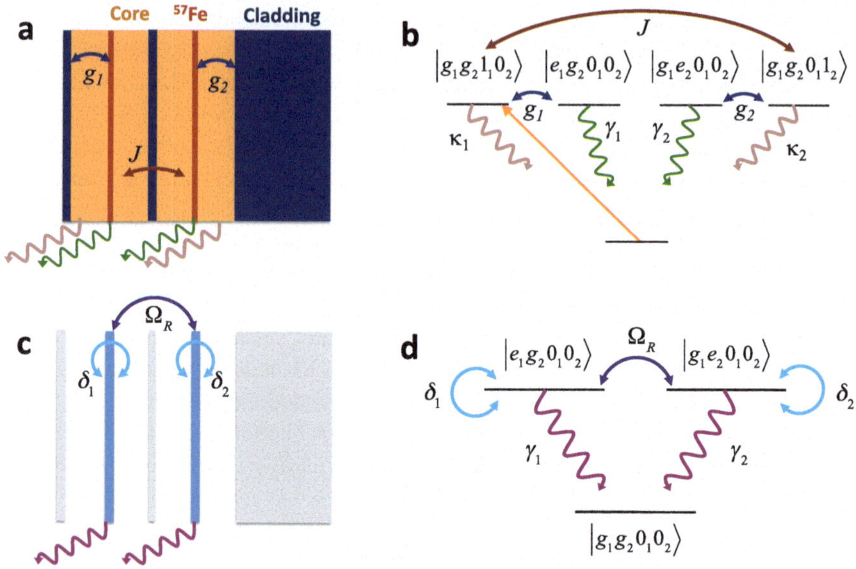

Fig. 3.22 Illustration of the quantum optical models for description of the two coupled cavities. **a** Each cavity contains a thin ^{57}Fe layer located in the antinode of the first mode of each cavity so that the coupling constants g_1 and g_2 are roughly equal. The first cavity is illuminated by a classical driving field. Both cavities are at the same time divided and coupled via a thin cladding layer, the thickness of which must be carefully adjusted to ensure a large coupling constant J. **b** Level scheme of the setup in (**a**). There are four states that are labeled by g or e, describing the states of the nuclear ensemble in layer 1 and layer 2, and by 0 or 1, which denote the number of photons in the first guided modes of cavity 1 and cavity 2. g_i describes the situation when all atoms in layer i are in the ground state, and e_i stands for the first symmetric Dicke state after the delocalized excitation of one nucleus in layer i. **c**, **d** Sketches of the effective set-up achieved after adiabatic elimination of the cavity modes. The original energies of the layers are replaced by those of the Dicke states in individual layers and undergo collective Lamb shifts, denoted δ_1 and δ_2. Their decay is superradiantly enhanced, with decay widths γ_1 and γ_2, and there is an effective interaction Ω_R between the layers. The latter is stronger than the respective superradiant decay rates, leading to strong coupling between the upper two levels. The decay of these dressed states shows Rabi oscillations. Figures b, d reprinted from [39]

Originally, the cavity modes were driven by an outside X-ray beam. Because we have adiabatically eliminated the cavity modes, in our new picture that driving term is now applied to both nuclear ensembles. In this picture the beam effectively drives both ensembles directly, but with an extra dispersion inherited from the cavity mode. Δ is the X-ray detuning to the nuclear resonance energy, considered to be the same in both layers. $\Delta_\pm$ is the detuning of the supermodes from the incoming beam energy.

The new effective Hamiltonian in Eq. (3.60) can be interpreted as follows: The new energies of the states representing collective excitations in layer i are given by $\Delta - \delta_i + i\gamma_i$. γ_i is the cavity-enhanced superradiant decay rate of the Dicke state, and δ_i is the cavity-induced collective Lamb shift [27]. The two states are connected by an effective interaction given by the real part g_{12} and an imaginary part γ_{12}. All parameters in the effective Hamiltonian can be derived from $g_{1(2)}\sqrt{N_{1(2)}}$, $\kappa_\pm$ and $\Delta_\pm$, and $\kappa_{R\pm}$. While the latter three can be extracted from an off-resonant reflectivity curve, the quantity $g_i\sqrt{N_i}$ must be fitted from the experimental data. Note that these new, effective parameters can be tuned via the angle of incidence θ through a wide range of values, as they all depend on the detuning, which is given by $\Delta_\pm = (\sin(\theta_\pm)/\sin(\theta) - 1)\,\omega_0$. Here, $\theta_\pm$ is the angle for which one cavity supermode is exactly at resonance at the nuclear transition energy ω_0.

We can essentially regard the new, effective Hamiltonian as a three-level system (Fig. 3.22d). The first level, corresponding to the ground state (no excited ensemble) serves only to probe the properties of the other two through the (very weak) driving terms. The energies of these two states, in turn, can be tuned via the collective Lamb shifts, as can their superradiant decay terms and their mutual interaction. These upper two levels form a two-level system with tunable energies, decay constants and a tunable mutual interaction. Just as for a regular two-level system interacting with a cavity mode, strong coupling and Rabi oscillations between the two levels can be achieved if the interaction term is larger than the two respective decay terms, in this case $g_{12} > \gamma_1, \gamma_2$. If this can be reached by appropriate tuning, the two upper states will exchange a photon multiple times before the whole system decays, giving rise to Rabi oscillations [129]. This is an equivalent of the standard strong-coupling set-up, where the coupling strength between an atom and a cavity must exceed the latter's decay constants [130–133]. This effective three-level system may also be regarded as the implementation of an artificial Autler-Townes set-up, which also consists of two upper levels whose interaction is probed from a lower ground state. A crucial difference is that the interaction here is cavity-induced, whereas in the Autler-Townes case, a laser beam tuned to the energy difference between the two levels couples them.

The equivalent of the classical Rabi frequency in our case is given by

$$\Omega_R = \sqrt{g_{12}^2 + \gamma_{12}^2 + (\delta_1 - \delta_2)^2}, \tag{3.67}$$

which includes the cavity-mediated coupling strength g_{12} between the two ensembles/states, as well as their relative energy detuning $\delta_1 - \delta_2$. The behaviour of Ω_R in comparison with the superradiant decay rates, and the collective Lamb shifts as a function of the incident angle, are shown in Figs. 3.23b,c. Any change in the incident angle θ leads to modifications of all parameters δ_i, γ_i and g_{12}, which makes it difficult to reach the desired behaviour of the coupling and the decay properties of the system. This raises the question of whether we can find an angle θ at which the strong coupling conditions ($g_{12} > \gamma_i$) are fulfilled. This is indeed the case: In between both supermodes, at large detunings from them, the superradiant decay widths γ_1, γ_2 are insignificant compared to the interaction strength between the two layers. This can be explained because the real part g_{12} of the interaction strength is suppressed by $1/\Delta_\mp^{-1}$ at angles sufficiently far away from the cavity angle of zero detuning, while the imaginary part γ_{12} describing the superradiant decay is suppressed by $1/\Delta_\mp^{-2}$. However, the collective Lamb shifts do not go to zero as quickly, and, according to Eq. (3.67), they are also involved in determining the oscillation rate Ω_R. To observe

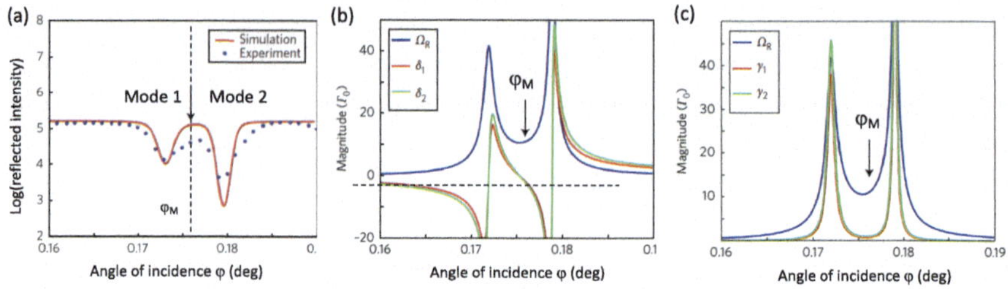

Fig. 3.23 **a** Reflectivity and **b**, **c** effective parameters (given by Eqs. (3.61)–(3.66)) of the Hamiltonian that describes the coupled cavities. φ_M is the optimum coupling angle at which the conditions for the observation of pure Rabi oscillations between the two resonant layers are fulfilled. φ_m lies in the middle between the two cavity modes that show up as dips in the reflectivity displayed in (**a**). At φ_m, the collective Lamb shifts (**b**) in the two layers cancel, and the superradiant enhancements (**c**) are minimal, meaning that according to Eq. (3.67) the splitting is exclusively due to the interaction Ω_R. Figures reprinted from [39]

pure Rabi oscillation - that is, the case when the oscillations are solely induced by the coupling constant, i.e. $\Omega_R = g_{12}$, it is necessary to have zero Lamb shift detuning between the layers. This, in turn, requires that both layers either have the same collective Lamb shift, or none at all.

A brief inspection of the Hamiltonian reveals that this is possible if the Lamb shifts caused by the two cavity modes cancel exactly for both layers. This can be accomplished if the two cavity modes both couple to the nuclear layers with identical strength. The double cavity is precisely the set-up that matches all these requirements: the field modes couple equally strong to both layers, so the Lamb shifts of both layers will always be equal. There are, however, always imperfections in real samples that may lead to slightly different coupling strengths. An easy way to ameliorate this, relies on the fact that the Lamb shift is highly dependent on the sign of the relative detuning. Between two modes, the corresponding contributions will cancel and the Lamb shifts of both layers will be exactly zero. At that particular angular position, illustrated in Fig. 3.23a, we perform our experiment.

The experiment was performed at the nuclear resonance beamline ID18 of the European Synchrotron Radiation Facility (ESRF). The beam was pre-monochromatized to a bandwidth of 1 meV around the nuclear resonance energy by the successive use of high-heat load and high-resolution monochromators.

The sample system for the experiment consists of two thin-film cavities stacked on top of each other, coupled through a thin interlayer as sketched in Fig. 3.21. A resonant ^{57}Fe nuclear ensemble is embedded into each cavity. Specifically, the sample was a (15 nm Pd)/(19 nm C)/2 nm ^{57}SS)/(19 nm C)/(2 nm Pd)/(16 nm C)/(2 nm ^{57}SS)/(19 nm C)/(2 nm Pd) multilayer (Fig. 3.21) fabricated by sputter deposition on a superpolished Si wafer. Here, ^{57}SS indicates stainless steel ($Fe_{0.55}Cr_{0.25}Ni_{0.2}$) with its iron content enriched to 95% with the resonant ^{57}Fe isotope. Stainless steel does not display ferromagnetic order, so the ^{57}Fe isotope presents a single-line resonance at 14.4125 keV with no Zeeman splitting. A weak hyperfine magnetic field distribution

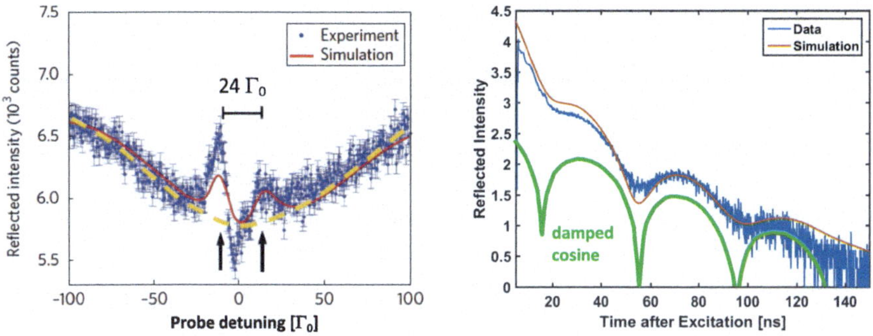

Fig. 3.24 Left: measured energy-resolved reflectivity at the optimum coupling angle φ_m. The two peaks indicated by the arrows are due to the normal mode splitting that translates to the frequency of the Rabi oscillations in (**b**). The modulation of the background (yellow dashed line is a guide to the eye) is due to the detection method involving time gating and has been taken into account in the numerical simulation. Right: temporal response of the double cavity, showing pronounced Rabi oscillations. The theoretical curves are a Fourier transform of the energy-resolved reflectivity, derived from the quantum optical model, and an exponentially damped cosine with a period that corresponds to the Rabi frequency, respectively. Note that these temporal oscillations are evenly spaced, in contrast to dynamical beats [50, 134], the period of which increases with increasing time after excitation. Figures reprinted from [39]

in the sample, however, effectively broadens the ^{57}Fe-resonance linewidth to ~2–3 Γ_0 [135]. This design effectively implements two almost identical cavities coupled via a thin Pd interlayer, which also constitutes the top cladding of the lower cavity and bottom cladding of the upper cavity.

For an energy analysis of the reflected radiation, a 1-μm-thick ^{57}Fe-enriched stainless steel foil was inserted into the beampath. In this foil the ^{57}Fe nuclei exhibited a single-line resonance with a FWHM width of 1.5 Γ_0. This stainless-steel foil was mounted on a Mössbauer drive, which was then moved at velocities of ± 10 mm/s. The ensuing Doppler shifts of $\pm 100\ \Gamma_0$ covered the spectral range of interest. To obtain the energy spectrum of the reflectivity at the optimum angle of incidence $\varphi = \varphi_m$ (Fig.3.24a) the delayed, resonantly reflected photons at that angular setting were detected together with the actual velocity of the Mössbauer drive. The nonresonantly scattered radiation was rejected from being recorded by a temporal gating from 0 to 7 ns after excitation. The temporal gating ensured that the huge background of nonresonantly scattered radiation was rejected from the detection, but the associated time-window effect introduced an unphysical background modulation that is visible in the spectrum. Moreover, because the non-resonant background is already subtracted in the detection process, dips in the reflectivity appear as peaks in the spectrum measured at the APD detector. For the time-resolved measurement, the Mössbauer drive with the analyzer foil was removed. The measured temporal beat pattern in Fig. 3.24b shows a clear, exponentially damped cosine pattern with a period that corresponds the energy splitting of the line in (a), the typical fingerprint of Rabi oscillations.

To summarize, we have succeeded in observing Rabi oscillations in the X-ray regime by using a double cavity set-up, and confirmed the collective strong coupling regime via a quantum optical model. The temporal evolution of the system is marked by the coherent exchange of a photon between two different collective excited nuclear states. The result also points the way to coherent control of X-rays over matter. Already minor changes in the layer structure, such as the positioning of the layers, can create new effective multilevel systems with completely different dynamics, and facilitate artificial quantum optical systems in the X-ray range, with tunable dispersions and interactions.

3.9 Nuclear Quantum Optics with Advanced Sources of X-Rays

A further boost for the field of nuclear quantum optics will be enabled not only by an increase of the spectral flux provided by the x-ray source, but also by qualitatively new properties of the radiation source. One of those is the so-called photon degeneracy η, which is a parameter that gives the number of resonant photons per mode of the radiation field. In the case of a pulsed radiation source where the pulse length is much shorter than the nuclear lifetime, η is simply given by the number of photons per pulse within the nuclear resonance bandwidth. If η is significantly larger than 1, one obviously enters the regime of multiphoton excitation of nuclear ensembles which could open new avenues for nuclear quantum optics, ranging from stimulated emission to nonlinear optics. At most present-day sources, however, values for η are typically much lower than 1, implying that in the majority of the pulses that interact with the sample there is no resonant photon. This means that there is typically only one nuclear resonant photon at a time in the setup. Under favorable conditions, e.g., when large bunch charges like in few-bunch filling mode of the storage ring are realized, there can be a sufficiently large fraction of bunches containing two or more resonant photons. The situation changes drastically with the advent of x-ray laser sources operating in the regime of hard x-rays like the LCLS in Stanford (USA) [136], SACLA in Himeji (Japan) [137], and the European XFEL in Hamburg (Germany) [138–141]. At these sources, values for η larger than 10^2 at 14.4 keV can be expected. In fact, a first experiment devoted to NRS has been recently been performed at SACLA, demonstrating superradiant emission from nuclear excited states with close to 10^2 photons per pulse [142]. In this section we are going to discuss two new types of future sources of x-rays that have great potential to further stimulate the field of nuclear quantum optics. These are the ultimate realizations of synchrotron radiation sources, i.e., diffraction limited storage rings (DLSR) and x-ray free electron laser oscillators (XFELO).

3.9.1 Diffraction-Limited Storage Rings

A new generation of accelerator-based x-ray sources is presently emerging that relies on a novel type of electron optics for the storage-ring, the so-called multi-bend achromat (MBA) [143–145]. The new concept allows for a significant reduction of the horizontal emittance ϵ_x compared to existing facilities. The emittance is a property of a radiation source and is given by the product of beam diameter and beam divergence. The smaller the emittance, the higher is the degree of spatial coherence of the source, which is the most desirable property of a radiation source for focusing or imaging applications. The emittance cannot be arbitrarily decreased, but is limited by the fundamental effect of diffraction of the beam by the shape of its own beam cross section. For that reason there are worldwide great efforts undertaken to reduce the emittance of synchrotron radiation sources as much as possible to reach the diffraction limit. The synchrotron radiation source MAX IV in Lund, Sweden, is the first light source that was successfully commissioned with this new lattice type [146], and SIRIUS in Campinas, Brazil, is currently under construction. Amongst the high-energy synchrotron radiation sources (electron energy equal or larger than 6 GeV), the European Synchrotron Radiation Source (ESRF) in Grenoble, France, will undergo an upgrade starting in winter 2018 to the fourth generation light source ESRF-EBS with an emittance as low as $\epsilon_x \approx 130$ picometer-radian (pmrad) at an electron beam energy of 6 GeV [147]. Many other sources worldwide have upgrade plans along these lines, among which are also the high-energy storage-ring sources Advanced Photon Source (APS) at Argonne National Laboratory (USA) [148], SPring-8 in Harima (Japan), and PETRA IV at DESY in Hamburg (Germany) [149]. The latter facility will be the first source reaching the diffraction limit for hard x-rays at 10 keV photon energy. The implementation of the new electron optics in the storage ring allows for an increase of the spectral brightness by one to two orders of magnitude. This will dramatically change the landscape of synchrotron radiation facilities in the next decade, see Fig. 3.25.

The benefit of diffraction-limited storage rings for the field of nuclear quantum optics will not be an increase in the photon degeneracy parameter η, but primarily the concentration of more resonant photons in a given scattering volume as compared to present-day sources. This is enabled by the high degree of lateral coherence which facilitates a very efficient collimation and focusing of x-rays, leading to an enhanced coupling of x-rays to nuclear ensembles, e.g., in cavities and photonic nanostructures.

3.9.2 X-Ray Free-Electron Lasers: SASE-XFEL and XFELO

A tremendous increase of the photon degeneracy η will be reached by x-ray sources that rely on a fundamentally different mechanism of radiation generation. This is the principle of self-amplified spontaneous emission (SASE) which forms the basis for free-electron lasers (FEL). The SASE principle leads to a increase in the brilliance

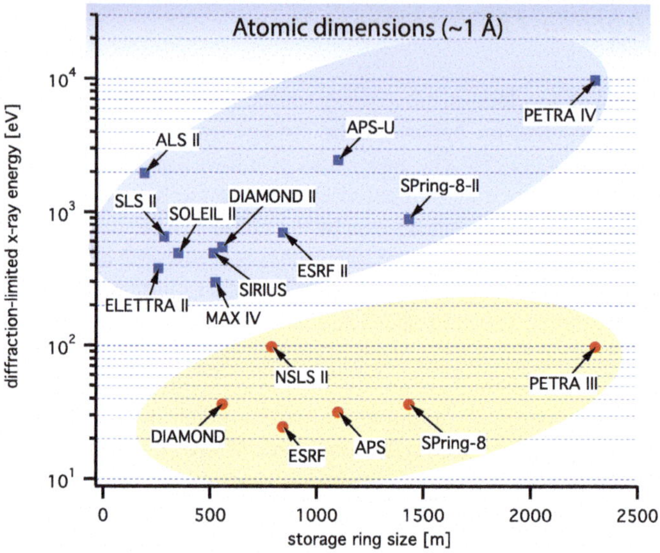

Fig. 3.25 Diffraction limit for selected synchrotron radiation sources and their future upgrades. Reprinted from [149], the authors licensed under CC-BY 4.0

of the radiation by several orders of magnitude compared to synchrotron radiation sources, going along with a corresponding increase in η. The ultimate performance for nuclear resonant scattering experiments, however, will be achieved if the SASE process takes place in a specially designed cavity where the relativistic electrons interact with the x-rays that are circulating in the cavity. Such a device has been termed x-ray free-electron laser oscillator (XFELO), conceptually introduced about ten years ago [7, 150, 151].

An XFELO is a low-gain device, in which an X-ray pulse that circulates in a cavity formed by diamond crystal Bragg mirrors is amplified every time it overlaps with an electron bunch in the undulator, illustrated in Fig. 3.26 . Due to its high reflectivity and excellent thermo-mechanical properties, diamond is the preferred material for the Bragg crystals employed to form the X-ray cavity [152]. An XFELO will work at any photon energy for which the Bragg reflectivity of diamond is sufficiently high and the bandwidth is sufficiently broad so that the initial exponential gain of the intra-cavity pulse energy can be sustained for a reasonable set of electron beam and undulator parameters. This range is expected to extend from 5 to 25 keV [151]. The photon energy can be continuously tuned for a given setting within a range of about 5% by changing the Bragg angle and adjusting the crystal positions so that the cavity roundtrip time remains fixed.

An XFELO will open up completely new possibilities in the field of nuclear resonance scattering (NRS) for isotopes with resonance energies between 5 and 25 keV. Due to the narrow resonance linewidths of Mössbauer transitions, NRS will benefit from the extremely intense, narrow-bandwidth radiation from the XFELO

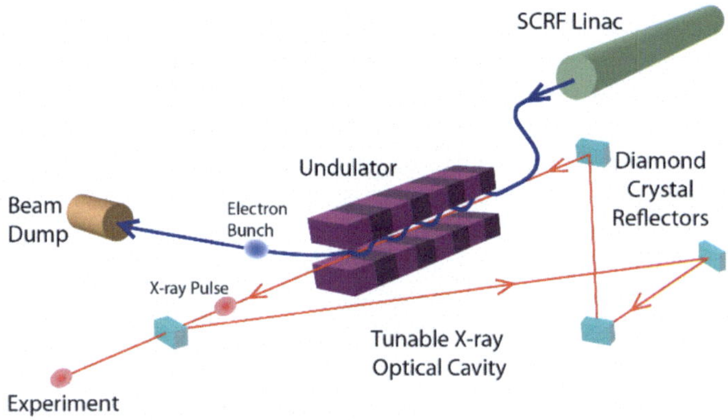

Fig. 3.26 A schematic illustration of an XFELO. Four crystals form a closed X-ray cavity via Bragg reflection. Figure adopted from [8]

in several ways: The anticipated hard X-ray spectral flux of 3×10^9 ph/sec/neV is more than 4 orders of magnitude larger than at existing 3rd generation synchrotron radiation sources [8].

As an example, at 14.4 keV (the transition energy of the ^{57}Fe Mössbauer resonance) the number of photons per pulse is expected to be 1×10^{10} within a pulse length of 680 fs (FWHM) and a spectral bandwidth of 3 meV (FWHM). This amounts to an average spectral flux of about 3×10^{15} photons/sec/meV or 1.5×10^{10} photons/sec/Γ_0, which corresponds to $\eta = 1.5\times10^4$ photons/pulse/Γ_0 where $\Gamma_0 = 4.7$ neV is the natural linewidth of the ^{57}Fe Mössbauer resonance. These numbers are four orders of magnitude larger than those observed at the best third-generation synchrotron radiation sources to date. This allows one to push Mössbauer science far beyond the single photon regime, opening new perspectives for X-ray quantum optics and nonlinear science with nuclear resonances. The full transverse coherence of the radiation will allow for efficient focusing to extremely small spot sizes in the range of 10 nm, enabling one to combine NRS with high-resolution imaging techniques. Moreover, a frequency stabilized XFELO would enable a hard X-ray frequency comb with pulse-to-pulse coherence for unique applications in X-ray coherent control and extreme metrology.

The longitudinal coherence of optical fields is the core requisite to induce and control interference between different quantum pathways in atoms. In nuclei, similar developments so far were restricted to single photons interfering with themselves, due to the lack of sufficiently coherent photon sources. With an XFELO this situation will fundamentally change. Its full coherence and spectral brightness provides new avenues for studying the interaction between X-rays and nuclei under multiphoton excitation conditions, thus offering unique possibilities for nonlinear spectroscopy of the nucleus, as well as for novel approaches to nuclear state preparation and detection.

For example, at low orders of nonlinearity quantum aspects involving X-ray photons could naturally be explored with two or more correlated photons. Potential approaches encompass both the generation of X-ray photon entanglement and its applications, and the exploration of quantum states in the nuclei by subsequent spectroscopic detection of scattered X-ray photons [153]. The availability of multiple coherent photons per pulse in turn enables detection of multiple correlated X-ray photons, providing access to higher-order correlation functions characterizing, e.g., density fluctuations, phonons or similar excitations. This will fuel the development of a broad class of new detection and analysis techniques. With multiple potentially phase-locked driving fields, multi-dimensional spectroscopy techniques come within reach, providing further insight into the dynamics. 2D nuclear spectroscopy might reveal couplings among nuclear transitions that could provide fundamental insight into intra-nuclear interactions, analogous to what is revealed in two-dimensional spectroscopy throughout radio frequency to optical spectral ranges [154, 155]. This and the other 2D measurements mentioned above will require at least phase-related X-ray pulse pairs, which could be generated by splitting one XFELO output pulse by an X-ray split-and-delay line or by applying temporal control of resonantly scattered photons via ultrafast piezo modulation [84].

Further progress is anticipated in the engineering of advanced nuclear level schemes. First steps have recently been demonstrated at 3rd generation light sources, by designing suitable target structures utilizing Mössbauer nuclei embedded in superlattices [38] and planar X-ray cavities [27, 29, 39]. The XFELO will enhance these capabilities by its unique source properties, which, aside from the spectral brightness also includes coherent multi-pulse or multi-color excitation. The XFELO could also facilitate novel nuclear resonance excitation processes, such as non-linear two-photon excitation [156] or four-wave mixing.

The pulse-to-pulse coherence of an energy stabilized XFELO enables one to realize a hard X-ray frequency comb (see Sect. 3.6), facilitating ultrahigh-resolution X-ray spectroscopy of nuclear transitions. Examples include multi-level nuclear transition measurements, probing ultra-narrow X-ray Mössbauer resonances, dynamics of X-ray driven nuclear—electronic transitions, and X-ray + laser double resonance experiments. Facilitated by X-ray comb spectroscopy, fascinating possibilities come into reach: X-ray frequency and wavelength metrology would be enabled by extending the optical frequency comb technologies and techniques to X-ray wavelengths. In addition to probing nuclear physics with unprecedented precision, linking nuclear transitions to the Cs standard can be used to search for the variation of fundamental constants with improved sensitivity [158, 159]. Nonlinear phase-coherent driving and probing at X-ray wavelengths will be possible over long times >10 s. High-quality-factor nuclear transitions like the 12.4 keV level of ^{45}Sc with a lifetime of $\sim$300 ms and $\Gamma_0/E_0 \sim 10^{-19}$ (see Fig. 3.27) can be established as new and improved frequency standards. Importantly, the pulse-to-pulse coherence allows to excite these narrow resonances using a sequence of pulses, offering the possibility of orders of magnitude higher excitation fraction than expected from SASE [153].

With sufficient temporal coherence and high intensity, coherent processes including nuclear coherent population transfer in the stimulated Raman adiabatic passage (STIRAP) technique [160, 161] or nuclear Rabi oscillations [162] are rendered pos-

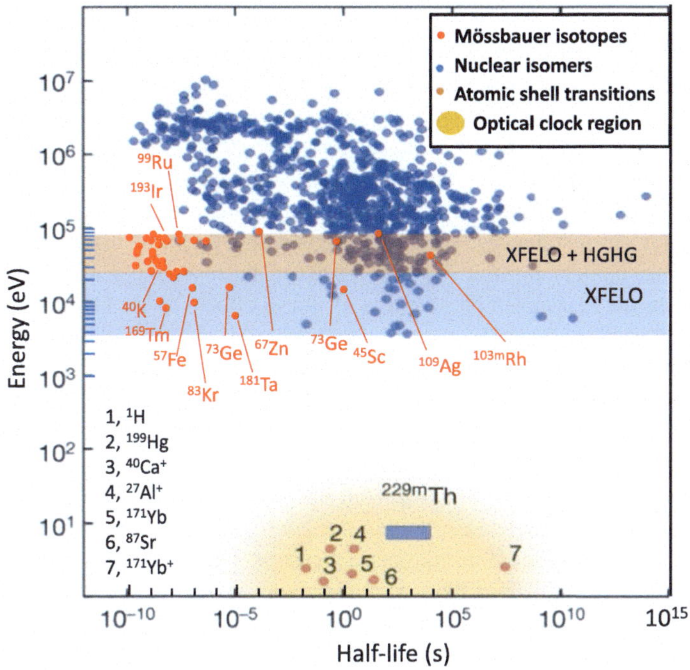

Fig. 3.27 Diagram of Mössbauer isotopes and nuclear isomers together with atomic and nuclear clock transitions in the parameter space of transition energy and half-life. The shaded regions covers the energy range of the XFELO (light blue) and the XFELO with a high-gain high-harmonic (HGHG) extension to reach photon energies above 25 keV (light brown). Modified from [157], Copyright 2016, with permission from Springer Nature

sible. Coherent population transfer would enable controlled pumping, storage and release of energy stored in long-lived nuclear excited states. In addition, also nuclear reactions starting from excited nuclear states driven by the XFELO can be envisaged.

3.10 Concluding Remarks

We have shown in this review that planar cavities and photonic nanostructures like multilayers and superlattices containing Mössbauer nuclei constitute interesting systems to explore quantum optical effects in the x-ray regime. Since the period of the x-ray standing waves in these structures is typically much larger than the thickness of ultrathin layers of Mössbauer atoms embedded in them, one is able to realize the small-sample limit of Dicke superradiance. This is instrumental for the preparation of single radiative eigenmodes of these nuclear ensembles. Accurately controlling their placement within the standing wave field facilitates to tailor their radiative decay width. In the vacuum field of the cavity this leads to several possibilities to prepare coherences between nuclear levels. One of those coupling schemes leads to electromagnetically induced transparency (EIT), another one to spontaneously generated coherences (SGC), and many others are still to be explored, opening exciting perspectives for future research in the field of cooperative emission and quantum optics with hard x-rays. Further engineering of photonic structures in one, two, or even three dimensions with modern thin-film deposition methods and lateral structuring techniques enables one to reach the regime of collective strong coupling between x-rays and nuclear resonances. We have demonstrated the basic effects encountered in this regime, namely normal mode splitting of the resonances and Rabi oscillations between nuclear ensembles. While all these effects could be explored in the limit of single-photon excitation, new avenues could open up under conditions of multiphoton excitation. This includes stimulated emission, nonlinear x-ray optics, multiphoton cooperative emission, novel hybrid light-matter states in cavities, to name a few. Especially intriguing in this area would be the simultaneous coupling of different types of photonic structures like cavities and multilayers to further enhance the multiphoton light-matter coupling with nuclear resonances. Moreover, the concepts presented in this review even open the possibility to transfer them into other wavelength regimes of the electromagnetic spectrum. One could explore these effects also with narrow resonance of inner-shell resonances, where first measurements on the collective Lamb shift have been performed [163]. Further applications could be to realize concepts of quantum memory with nuclear resonances which could even form a bridge into the area of quantum computing implementations.

Acknowledgements The results presented in this review would not have been possible without the eminent contributions of numerous coworkers and colleagues over the past ten years, We are gratefully indebted for their various contributions. Amongst those, there are few who we would like to mention in particular. First and foremost, this is Marlan O. Scully who, during the PQE conference in Snowbird in 2009, inspired our work on the collective Lamb shift, which was the initial spark that ignited the field of cavity-based nuclear quantum optics. It was then during the following years where the lively discussions with Bernhard Adams, Olga Kocharovskaya, Anatoly Svidzinsky and others, constituted a constant source of inspiration for the development of this field. In the following we would like to list—in alphabetical order—those colleagues who contributed in many ways to the development of the field, ranging from stimulating discussions to hands-on work in the laboratories and during the experiments at the synchrotron radiation sources ESRF and PETRA III: Hendrik Bernhardt, Lars Bocklage, Alexander I. Chumakov, Sebastien Couet, Frank-Uwe Dill, Jakob Gollwitzer, Tatyana Gurieva, Johann Haber, Kilian Heeg, Andreas Kaldun, Christoph H. Keitel, Dominik Lentrodt, Robert Loetzsch, Olaf Leupold, Berit Marx, Xiangjin Kong, Christian Ott, Adriana Palffy, Gerhard Paulus, Thomas Pfeifer, Sasha Poddubbny, André Rothkirch, Rudolf Rüffer, Balaram Sahoo, Kai Schlage, Kai-Sven Schulze, Daniel Schumacher, Cornelius Strohm, Ingo Uschmann, Hans-Christian Wille, and Svenja Willing.

References

1. R. Loudon, *The Quantum Theory of Light* (Oxford University Press, 1983)
2. M.O. Scully, M.S. Zubairy, *Quantum Optics* (Cambridge University Press, 1997)
3. M. Fox, *Quantum Optics* (Oxford University Press, 2006)

4. G.S. Agarwal, *Quantum Optics* (Cambridge University Press, 2013)
5. S.L. Ruby, Mössbauer experiments without conventional sources. J. Phys. (Paris) Colloq. **35**, C6–209 (1974)
6. R. Röhlsberger, *Nuclear Condensed Matter Physics with Synchrotron Radiation. Basic Principles, Methodology and Applications.* Number 208 in Springer Tracts in Modern Physics (Springer, Berlin, Heidelberg, 2004)
7. Kwang-Je Kim, Yuri Shvyd'ko, Sven Reiche, A proposal for an x-ray free-electron laser oscillator with an energy-recovery linac. Phys. Rev. Lett. **100**, 244802 (2008)
8. B.W. Adams et al., Scientific opportunities with an x-ray free-electron laser oscillator. arXiv:1903.09317 [physics.ins-det] (2019)
9. M.O. Scully, E.S. Fry, C.H. Raymond Ooi, K. Wódkiewicz, Directed spontaneous emission from an extended ensemble of N atoms: timing is everything. Phys. Rev. Lett. **96**, 010501 (2006)
10. J.H. Eberly, Emission of one photon in an electric dipole transition of one among N atoms. J. Phys. B Atom. Molec. Opt. Phys. **39**, S599 (2006)
11. A. Svidzinsky, J.-T. Chang, Cooperative spontaneous emission as a many-body eigenvalue problem. Phys. Rev. A **77**, 043833 (2008)
12. R. Friedberg, J.T. Manassah, Effects of including the counterrotating term and virtual photons on the eigenfunctions and eigenvalues of a scalar photon collective emission theory. Phys. Lett. A **372**, 2514–2521 (2008)
13. R. Friedberg, J.T. Manassah, Electromagnetic decay modes in a spherical sample of two-level atoms. Phys. Lett. A **372**, 6833–6842 (2008)
14. D. Porras, J.I. Cirac, Collective generation of quantum states of light by entangled atoms. Phys. Rev. A **78**, 053816 (2008)
15. M.O. Scully, A.A. Svidzinsky, The super of superradiance. Science **325**, 1510–1511 (2009)
16. M.O. Scully, A.A. Svidzinsky, The effects of the N atom collective Lamb shift on single photon superradiance. Phys. Lett. A **373**, 1283–1286 (2009)
17. L.H. Pedersen, K. Mølmer, Few qubit atom-light interfaces with collective encoding. Phys. Rev. A **79**, 012320 (2009)
18. A.A. Svidzinsky, M.O. Scully, Evolution of collective N atom states in single photon superradiance: effect of virtual Lamb shift processes. Opt. Commun. **282**, 2894–2897 (2009)
19. A.A. Svidzinsky, M.O. Scully, On the evolution of N-atom state prepared by absorption of a single photon. Opt. Commun. **283**, 753–757 (2010)
20. R. Friedberg, Refinement of a formula for decay after weak coherent excitation of a sphere. Annals Phys. **325**, 345–358 (2010)
21. A.A. Svidzinsky, J.-T. Chang, M.O. Scully, Cooperative spontaneous emission of n atoms: many-body eigenstates, the effect of virtual Lamb shift processes, and analogy with radiation of n classical oscillators. Phys. Rev. A **81**, 053821 (2010)
22. P.R. Berman, J.-L. Le Gouët, Phase-matched emission from an optically thin medium following one-photon pulse excitation: Energy considerations. Phys. Rev. A **83**, 035804 (2011)
23. A.M. Afanas'ev, Yu. Kagan, Theory of hyperfine structure of the Mössbauer line in paramagnetic substances. Sov. Phys. JETP **18**, 1139–1149 (1964)
24. J.P. Hannon, G.T. Trammell, Mössbauer diffraction. I. Quantum theory of gamma-ray and x-ray optics. Phys. Rev. **169**, 315–329 (1968)
25. J.P. Hannon, G.T. Trammell, Mössbauer diffraction. II. Dynamical theory of Mössbauer optics. Phys. Rev. **186**, 306–325 (1969)
26. J.P. Hannon, G.T. Trammell, Coherent γ-ray optics. Hyperfine Interactions **123**, 127–274 (1999)
27. R. Röhlsberger, K. Schlage, B. Sahoo, S. Couet, R. Rüffer, Collective Lamb shift in single photon superradiance. Science **328**, 1248 (2010)
28. R. Röhlsberger, H.C. Wille, K. Schlage, B. Sahoo, Electromagnetically induced transparency with resonant nuclei in a cavity. Nature **482**, 199–203 (2012)
29. K.P. Heeg, H.-C. Wille, K. Schlage, T. Guryeva, D. Schumacher, I. Uschmann, K.S. Schulze, B. Marx, T. Kämpfer, G.G. Paulus, R. Röhlsberger, J. Evers, Vacuum-assisted generation and control of atomic coherences at x-ray energies. Phys. Rev. Lett. **111**, 073601 (2013)

30. R.H. Dicke, Coherence in spontaneous radiation processes. Phys. Rev. **93**, 99–110 (1954)
31. R. Friedberg, S.R. Hartmann, J.T. Manassah, Frequency shifts in emission and absorption by resonant systems ot two-level atoms. Phys. Reports **7**, 101–179 (1973)
32. R. Wiegner, J. von Zanthier, G.S. Agarwal, Quantum-interference-initiated superradiant and subradiant emission from entangled atoms. Phys. Rev. A **84**, 023805 (2011)
33. A.I. Lvovsky, B.C. Sanders, W. Tittel, Optical quantum memory. Nature Photon. **3**, 706 (2009)
34. M. Sarovar, A. Ishizaki, G.R. Fleming, K.B. Whaley, Quantum entanglement in photosynthetic light-harvesting complexes. Nature Phys. **6**, 462 (2010)
35. K.P. Heeg, J. Evers, X-ray quantum optics with Mössbauer nuclei embedded in thin-film cavities. Phys. Rev. A **88**, 043828 (2013)
36. K.P. Heeg, J. Haber, D. Schumacher, L. Bocklage, H.-C. Wille, K.S. Schulze, R. Loetzsch, I. Uschmann, G.G. Paulus, R. Rüffer, R. Röhlsberger, J. Evers, Tunable subluminal propagation of narrow-band x-ray pulses. Phys. Rev. Lett. **114**, 203601 (2015)
37. K.P. Heeg, J. Evers, Collective effects between multiple nuclear ensembles in an x-ray cavity-QED setup. Phys. Rev. A **91**, 063803 (2015)
38. J. Haber, K.S. Schulze, K. Schlage, R. Loetzsch, L. Bocklage, T. Gurieva, H. Bernhardt, H.C. Wille, R. Rüffer, I. Uschmann, G.G. Paulus, R. Röhlsberger, Collective strong coupling of x-rays and nuclei in a nuclear optical lattice. Nature Photon. **10**, 445 (2016)
39. J. Haber, X. Kong, C. Strohm, S. Willing, J. Gollwitzer, L. Bocklage, R. Rüffer, A. Pálffy, R. Röhlsberger, Rabi oscillations of x-ray radiation between two nuclear ensembles. Nature Photon. **11**, 720–725 (2017)
40. D. Lentrodt, K.P. Heeg, C.H. Keitel, Jörg Evers, Ab initio quantum models for thin-film x-ray cavity QED. Phys. Rev. Res. **2**, 023396 (2020)
41. M.O. Scully, Collective Lamb shift in single photon Dicke superradiance. Phys. Rev. Lett. **102**, 143601 (2009)
42. R. Friedberg, J.T. Manassah, Cooperative Lamb shift and the cooperative decay rate for an initially detuned phased state. Phys. Rev. A **81**, 043845 (2010)
43. J.T. Manassah, Cooperative radiation from atoms in different geometries: decay rate and frequency shift. Adv. Opt. Photon. **4**, 108–156 (2012)
44. J. Ruostekoski, J. Javanainen, Emergence of correlated optics in one-dimensional waveguides for classical and quantum atomic gases. Phys. Rev. Lett. **117**, 143602 (2016)
45. J. Keaveney, A. Sargsyan, U. Krohn, I.G. Hughes, D. Sarkisyan, C.S. Adams, Cooperative Lamb shift in an atomic vapor layer of nanometer thickness. Phys. Rev. Lett. **108**, 173601 (2012)
46. Z. Meir, O. Schwartz, E. Shahmoon, D. Oron, R. Ozeri, Cooperative Lamb shift in a mesoscopic atomic array. Phys. Rev. Lett. **113**, 193002 (2014)
47. S.J. Roof, K.J. Kemp, M.D. Havey, I.M. Sokolov, Observation of single-photon superradiance and the cooperative Lamb shift in an extended sample of cold atoms. Phys. Rev. Lett. **117**, 073003 (2016)
48. T. Peyrot, Y.R.P. Sortais, A. Browaeys, A. Sargsyan, D. Sarkisyan, J. Keaveney, I.G. Hughes, C.S. Adams, Collective Lamb shift of a nanoscale atomic vapor layer within a sapphire cavity. Phys. Rev. Lett. **120**, 243401 (2018)
49. N. Auerbach, V. Zelevinsky, Super-radiant dynamics, doorways and resonances in nuclei and other open mesoscopic systems. Reports Progr. Phys. **74**, 106301 (2011)
50. U. van Bürck, Coherent pulse propagation through resonant media. Hyperfine Interactions **123**(124), 483–509 (1999)
51. E. Gerdau, R. Rüffer, H. Winkler, W. Tolksdorf, C.P. Klages, J.P. Hannon, Nuclear Bragg diffraction of synchrotron radiation in yttrium iron garnet. Phys. Rev. Lett. **54**, 835–838 (1985)
52. J.B. Hastings, D.P. Siddons, U. van Bürck, R. Hollatz, U. Bergmann, Mössbauer spectroscopy using synchrotron radiation. Phys. Rev. Lett. **66**, 770–773 (1991)
53. E. Gerdau, H. de Waard (eds.), Nuclear resonant scattering of synchrotron radiation, part A, in *Hyperfine Interactions*, vol. 123/124 (Springer, 1999)

54. E. Gerdau, H. de Waard (eds.), Nuclear resonant scattering of synchrotron radiation, part B, in *Hyperfine Interactions*, vol. 125 (Springer, 2000)
55. W. Sturhahn, E. Gerdau, Evaluation of time-differential measurements of nuclear-resonance scattering of x rays. Phys. Rev. B **49**, 9285–9294 (1994)
56. W.E. Lamb, Robert C. Retherford, Fine structure of the hydrogen atom by a microwave method. Phys. Rev. **72**, 241–243 (1947)
57. R. Friedberg, J.T. Manassah, The dynamical cooperative Lamb shift in a system of two-level atoms in a slab-geometry. Phys. Lett. A **373**, 3423–3429 (2009)
58. J.T. Manassah, The dynamical cooperative Lamb shift in a system of two-level atoms in a sphere in the scalar photon theory. Laser Phys. **20**, 259–269 (2010)
59. C. Greiner, B. Boggs, T.W. Mossberg, Superradiant emission dynamics of an optically thin material sample in a short-decay-time optical cavity. Phys. Rev. Lett. **85**, 3793–3796 (2000)
60. D. Fröhlich, A. Kulik, B. Uebbing, A. Mysyrowicz, V. Langer, H. Stolz, W. von der Osten, Coherent propagation and quantum beats of quadrupole polaritons in Cu_2O. Phys. Rev. Lett. **67**, 2343–2346 (1991)
61. R. Röhlsberger, Cooperative emission from nuclei: the collective Lamb shift and electromagnetically induced transparency. Fortschritte der Physik **61**, 360–376 (2013)
62. A.A. Svidzinsky, J.-T. Chang, Marlan O. Scully, Dynamical evolution of correlated spontaneous emission of a single photon from a uniformly excited cloud of n atoms. Phys. Rev. Lett. **100**, 160504 (2008)
63. A.A. Svidzinsky, Nonlocal effects in single-photon superradiance. Phys. Rev. A **85**, 013821 (2012)
64. Yu.V. Shvyd'ko, G.V. Smirnov, Experimental study of time and frequency properties of collective nuclear excitations in a single crystal (gamma-ray resonance). J. Phys. Condensed Matter **1**, 10563 (1989)
65. J.T. Manassah, Giant cooperative Lamb shift in a density-modulated slab of two-level atoms. Phys. Lett. A **374**, 1985–1988 (2010)
66. R. Röhlsberger, Theory of x-ray grazing incidence reflection in the presence of nuclear resonance excitation. Hyperfine Interactions **123**(124), 301–325 (1999)
67. R. Röhlsberger, J. Evers, S. Shwartz. *Quantum and Nonlinear Optics with Hard X-Rays*, ed. by E. Jaeschke, S. Khan, J.R. Schneider, J.B. Hastings (Springer International Publishing, Cham, 2014), pp. 1–28
68. P. Longo, C.H. Keitel, J. Evers, Tailoring superradiance to design artificial quantum systems. Sci. Rep. **6**, 23628 (2016)
69. F. Pfeiffer, C. David, M. Burghammer, C. Riekel, T. Salditt, Two-dimensional x-ray waveguides and point sources. Science **297**, 230–234 (2002)
70. K.J. Vahala, Optical microcavities. Nature **424**, 839–846 (2003)
71. K.-J. Boller, A. Imamoğlu, S.E. Harris, Observation of electromagnetically induced transparency. Phys. Rev. Lett. **66**, 2593–2596 (1991)
72. M. Fleischhauer, A. Imamoglu, J.P. Marangos, Electromagnetically induced transparency: optics in coherent media. Rev. Mod. Phys. **77**, 633–673 (2005)
73. L.G. Parratt, Surface studies of solids by total reflection of x-rays. Phys. Rev. **95**, 359–369 (1954)
74. W. Sturhahn, CONUSS and PHOENIX: evaluation of nuclear resonant scattering data. Hyperfine Interactions **125**, 149–172 (2000)
75. A. Taflove, S.C. Hagness, *Computational Electrodynamics: The Finite-difference Time-domain Method*, 3rd edn. (Artech House, Norwood MA, 2005)
76. D. Lentrodt, J. Evers, Ab initio few-mode theory for quantum potential scattering problems. Phys. Rev. X **10**, 011008 (2020)
77. C. Gardiner, P. Zoller (eds.), *Quantum Noise: A Handbook of Markovian and Non-Markovian Quantum Stochastic Methods with Applications to Quantum Optics* (Springer Series in Synergetics, Springer, Heidelberg, 2004)
78. B. Marx, K.S. Schulze, I. Uschmann, T. Kämpfer, R. Lötzsch, O. Wehrhan, W. Wagner, C. Detlefs, T. Roth, J. Härtwig, E. Förster, T. Stöhlker, G.G. Paulus, High-precision x-ray polarimetry. Phys. Rev. Lett. **110**, 254801 (2013)

79. K.P. Heeg, C. Ott, D. Schumacher, H.-C. Wille, R. Röhlsberger, T. Pfeifer, J. Evers, Interferometric phase detection at x-ray energies via fano resonance control. Phys. Rev. Lett. **114**, 207401 (2015)
80. U. Fano, Effects of configuration interaction on intensities and phase shifts. Phys. Rev. **124**, 1866–1878 (1961)
81. A.E. Miroshnichenko, S. Flach, Yuri S. Kivshar, Fano resonances in nanoscale structures. Rev. Mod. Phys. **82**, 2257–2298 (2010)
82. M.F. Limonov, M.V. Rybin, A.N. Poddubny, Y.S. Kivshar, Fano resonances in photonics. Nat. Photon. **11**, 543 (2017)
83. C. Ott, A. Kaldun, P. Raith, K. Meyer, M. Laux, J. Evers, C.H. Keitel, C.H. Greene, T. Pfeifer, Lorentz meets Fano in spectral line shapes: a universal phase and its laser control. Science **340**, 716–720 (2013)
84. K.P. Heeg, A. Kaldun, C. Strohm, P. Reiser, C. Ott, R. Subramanian, D. Lentrodt, J. Haber, H.-C. Wille, S. Goerttler, R. Rüffer, C.H. Keitel, R. Röhlsberger, T. Pfeifer, J. Evers, Spectral narrowing of x-ray pulses for precision spectroscopy with nuclear resonances. Science **357**, 375–378 (2017)
85. P.M. Anisimov, J.P. Dowling, B.C. Sanders, Objectively discerning autler-townes splitting from electromagnetically induced transparency. Phys. Rev. Lett. **107**, 163604 (2011)
86. H. Yueyun, W. Liu, Y. Sun, X. Shi, J. Jiang, Y. Yang, S. Zhu, J. Evers, H. Chen, Electromagnetically-induced-transparency-like phenomenon with resonant meta-atoms in a cavity. Phys. Rev. A **92**, 053824 (2015)
87. G.S. Agarwal, *Quantum Statistical Theories of Spontaneous Emission and Their Relation to Other Approaches*. Springer Tracts in Modern Physics (Springer, 1974)
88. Z. Ficek, S. Swain, *Quantum Interference and Coherence: Theory and Experiment. Springer Series in Optical Sciences* (Springer, Heidelberg, 2005)
89. M. Kiffner, M. Macovei, J. Evers, C.H. Keitel, Vacuum-induced processes in multilevel atoms. Progress Opt. **55**, 85–197 (2010)
90. M. Kiffner, J. Evers, C.H. Keitel, Quantum interference enforced by time-energy complementarity. Phys. Rev. Lett. **96**, 100403 (2006)
91. G.S. Agarwal, Anisotropic vacuum-induced interference in decay channels. Phys. Rev. Lett. **84**, 5500–5503 (2000)
92. P. Zhou, S. Swain, Cavity engineering of quantum interference. Opt. Commun. **179**, 267–274 (2000)
93. G.V. Smirnov, U. van Bürck, A.I. Chumakov, A.Q.R. Baron, R. Rüffer, Synchrotron Mössbauer source. Phys. Rev. B **55**, 5811–5815 (1997)
94. T. Mitsui, M. Seto, R. Masuda, K. Kiriyama, Y. Kobayashi, Synchrotron radiation Mössbauer spectroscopy using Doppler-shifted 14.4 keV single-line ^{57}Fe-Mössbauer radiation. Japan. J. Appl. Phys. **46**(7L), 703 (2007)
95. V. Potapkin, A.I. Chumakov, G.V. Smirnov, J.-P. Celse, R. Rüffer, C. McCammon, L. Dubrovinsky, The ^{57}Fe synchrotron Mössbauer source at the ESRF. J. Sync. Radiat. **19**, 559–569 (2012)
96. T.S. Toellner, E.E. Alp, W. Sturhahn, T.M. Mooney, X. Zhang, M. Ando, Y. Yoda, S. Kikuta, Polarizer/analyzer filter for nuclear resonant scattering of synchrotron radiation. Appl. Phys. Lett. **67**, 1993–1995 (1995)
97. X. Kong, A. Pálffy, Stopping narrow-band x-ray pulses in nuclear media. Phys. Rev. Lett. **116**, 197402 (2016)
98. C. Weisbuch, M. Nishioka, A. Ishikawa, Y. Arakawa, Observation of the coupled exciton-photon mode splitting in a semiconductor quantum microcavity. Phys. Rev. Lett. **69**, 3314–3317 (1992)
99. M.G. Raizen, L.A. Orozco, M. Xiao, T.L. Boyd, H.J. Kimble, Squeezed-state generation by the normal modes of a coupled system. Phys. Rev. Lett. **59**, 198–201 (1987)
100. S. Haroche, J.-M. Raimond, *Exploring the Quantum: Atoms* (Oxford University Press, Cavities and Photons, 2006)
101. B. Dromey, Strong coupling of light goes nuclear. Nature Photon. **10**, 436 (2016)

102. D. Goldberg, L.I. Deych, A.A. Lisyansky, Z. Shi, V.M. Menon, V. Tokranov, M. Yakimov, S. Oktyabrsky, Exciton-lattice polaritons in multiple-quantum-well-based photonic crystals. Nature Photon. **3**, 662 (2009)
103. A. Askitopoulos, L. Mouchliadis, I. Iorsh, G. Christmann, J.J. Baumberg, M.A. Kaliteevski, Z. Hatzopoulos, P.G. Savvidis, Bragg polaritons: strong coupling and amplification in an unfolded microcavity. Phys. Rev. Lett. **106**, 076401 (2011)
104. A. Schilke, C. Zimmermann, P.W. Courteille, W. Guerin, Optical parametric oscillation with distributed feedback in cold atoms. Nature Photon. **6**, 101 (2012)
105. N. Kaina, M. Fink, G. Lerosey, Composite media mixing Bragg and local resonances for highly attenuating and broad bandgaps. Sci. Reports **3**, 3240 (2013)
106. I.H. Deutsch, R.J.C. Spreeuw, S.L. Rolston, W.D. Phillips, Photonic band gaps in optical lattices. Phys. Rev. A **52**, 1394–1410 (1995)
107. M. Weidemüller, A. Hemmerich, A. Görlitz, T. Esslinger, T.W. Hänsch, Bragg diffraction in an atomic lattice bound by light. Phys. Rev. Lett. **75**, 4583–4586 (1995)
108. M. Hübner, J.P. Prineas, C. Ell, P. Brick, E.S. Lee, G. Khitrova, H.M. Gibbs, S.W. Koch, Optical lattices achieved by excitons in periodic quantum well structures. Phys. Rev. Lett. **83**, 2841–2844 (1999)
109. E.L. Ivchenko, A.N. Poddubny, Resonant diffraction of electromagnetic waves from solids (a review). Phys. Solid State **55**, 905–923 (2013)
110. R. Röhlsberger, E. Witthoff, E. Gerdau, E. Lüken, Observation of nuclear diffraction from multilayers with a Fe/^{57}Fe superstructure. J. Appl. Phys. **74**, 1933–1937 (1993)
111. L. Deak, G. Bayreuther, L. Bottyan, E. Gerdau, J. Korecki, E.I. Kornilov, H.J. Lauter, O. Leupold, D.L. Nagy, A.V. Petrenko, V.V. Pasyuk-Lauter, H. Reuther, E. Richter, R. Röhloberger, E. Szilagyi, Pure nuclear Bragg reflection of a periodic ^{56}Fe/^{57}Fe multilayer. J. Appl. Phys. **85**, 1–7 (1999)
112. A.I. Chumakov, G.V. Smirnov, A.Q.R. Baron, J. Arthur, D.E. Brown, S.L. Ruby, G.S. Brown, N.N. Salashchenko, Resonant diffraction of synchrotron radiation by a nuclear multilayer. Phys. Rev. Lett. **71**, 2489–2492 (1993)
113. A.I. Chumakov, L. Niesen, D.L. Nagy, E.E. Alp, Nuclear resonant scattering of synchrotron radiation by multilayer structures. Hyperfine Interactions **123**, 427–454 (1999)
114. Y.D. Chong, D.E. Pritchard, M. Soljačić, Quantum theory of a resonant photonic crystal. *Phys. Rev. B* **75**, 235124 (2007)
115. S. Rist, P. Vignolo, G. Morigi, Photonic spectrum of bichromatic optical lattices. Phys. Rev. A **79**, 053822 (2009)
116. T. Ikawa, K. Cho, Fate of the superradiant mode in a resonant Bragg reflector. Phys. Rev. B **66**, 085338 (2002)
117. A. Salomon, R.J. Gordon, Y. Prior, T. Seideman, M. Sukharev, Strong coupling between molecular excited states and surface plasmon modes of a slit array in a thin metal film. Phys. Rev. Lett. **109**, 073002 (2012)
118. J. Haber, Hard X-ray Quantum Optics in Thin Film Nanostructures. *Ph.D. Thesis*, University of Hamburg, Germany (2017)
119. S. Jorda, U. Rössler, D. Broido, Fine structure of excitons and polariton dispersion in quantum wells. Phys. Rev. B **48**, 1669–1677 (1993)
120. E.L. Ivchenko, A.I. Nesvizhskii, S. Jorda, Resonant Bragg reflection from quantum-well structures. Superlattices Microstruct. **16**, 17–20 (1994)
121. M. Hübner, J. Kuhl, T. Stroucken, A. Knorr, S.W. Koch, R. Hey, K. Ploog, Collective effects of excitons in multiple-quantum-well Bragg and anti-Bragg structures. Phys. Rev. Lett. **76**, 4199–4202 (1996)
122. J.P. Prineas, C. Ell, E.S. Lee, G. Khitrova, H.M. Gibbs, S.W. Koch, Exciton-polariton eigenmodes in light-coupled $In_{0.04}Ga_{0.96}As/GaAs$ semiconductor multiple-quantum-well periodic structures. Phys. Rev. B **61**, 13863–13872 (2000)
123. M.V. Erementchouk, L.I. Deych, A.A. Lisyansky, Optical properties of one-dimensional photonic crystals based on multiple-quantum-well structures. Phys. Rev. B **71**, 235335 (2005)

124. J. Majer, M. Chow, J.M. Gambetta, J. Koch, B.R. Johnson, J.A. Schreier, L. Frunzio, D.I. Schuster, A.A. Houck, A. Wallraff, A. Blais, M.H. Devoret, S.M. Girvin, R.J. Schoelkopf, Coupling superconductive circuits via a cavity bus. Nature **449**, 443 (2007)
125. K. Baumann, C. Guerlin, F. Brennecke, T. Esslinger, Dicke quantum phase transition with a superfluid gas in an optical cavity. Nature **464**, 1301–1306 (2010)
126. J. Simon, W.S. Bakr, R. Ma, M. Eric Tai, P.M. Preiss, M. Greiner, Quantum simulation of antiferromagnetic spin chains in an optical lattice. Nature **472**, 307–312 (2011)
127. I. Bloch, J. Dalibard, W. Zwerger, Many-body physics with ultracold gases. Rev. Mod. Phys. **80**, 885–964 (2008)
128. I. Bloch, J. Dalibard, S. Nascimbene, Quantum simulations with ultracold quantum gases. Nat. Phys. **8**, 267–276 (2012)
129. Y.O. Dudin, L. Li, F. Bariani, A. Kuzmich, Observation of coherent many-body Rabi oscillations. Nature Phys. **8**, 790–794 (2012)
130. M. Brune, F. Schmidt-Kaler, A. Maali, J. Dreyer, E. Hagley, J.M. Raimond, S. Haroche, Quantum Rabi oscillation: a direct test of field quantization in a cavity. Phys. Rev. Lett. **76**, 1800–1803 (1996)
131. J.-M. Raimond, M. Brune, S. Haroche, Manipulating quantum entanglement with atoms and photons in a cavity. Rev. Mod. Phys. **73**, 565 (2001)
132. T. Yoshie, A. Scherer, J. Hendrickson, G. Khitrova, H.M. Gibbs, G. Rupper, C. Ell, O.B. Shchekin, D.G. Deppe, Vacuum Rabi splitting with a single quantum dot in a photonic crystal nanocavity. Nature **432**, 200–203 (2004)
133. G. Khitrova, H.M. Gibbs, M. Kira, S.W. Koch, A. Scherer, Vacuum Rabi splitting in semiconductors. Nature Phys. **2**, 81–90 (2006)
134. Yu. Kagan, A.M. Afanas'ev, V.G. Kohn, On excitation of isomeric nuclear states in a crystal by synchrotron radiation. J. Phys. C Solid State Phys. **12**, 615 (1979)
135. B. Sahoo, K. Schlage, J. Major, U. von Hörsten, W. Keune, H. Wende, R. Röhlsberger, Preparation and characterization of ultrathin stainless steel films. AIP Conf. Proc. **1347**, 57–60 (2011)
136. P. Emma et al., First lasing and operation of an Ångström-wavelength free-electron laser. Nature Photon. **4**, 641 (2010)
137. Tetsuya Ishikawa et al., A compact x-ray free-electron laser emitting in the sub-Ångström region. Nature Photon. **6**, 540 (2012)
138. M. Altarelli et al., The European x-ray free-electron laser, technical design report, DESY 2006–097, in *Technical report* (Deutsches Elektronen-Synchrotron, 2007)
139. A. Madsen, H. Sinn, Europe enters the extreme x-ray area. CERN Courier **57**, 19–23 (2017)
140. H. Weise, W. Decking, The world's longest superconducting linac. CERN Courier **57**, 25–27 (2017)
141. T. Tschentscher, C. Bressler, J. Grünert, A. Madsen, A.P. Mancuso, M. Meyer, A. Scherz, H. Sinn, U. Zastrau, Photon beam transport and scientific instruments at the European XFEL. Appl. Sci. **7**, 592 (2017)
142. A.I. Chumakov, A.Q.R. Baron, I. Sergueev, C. Strohm, O. Leupold, Y. Shvyd'ko, G.V. Smirnov, R. Rüffer, Y. Inubushi, M. Yabashi, K. Tono, T. Kudo, T. Ishikawa, Superradiance of an ensemble of nuclei excited by a free electron laser. Nature Phys. **14**, 261–264 (2018)
143. D. Einfeld, M. Plesko, J. Schaper, A pure insertion device synchrotron light source utilizing the MBA-optics. J. Phys. IV France **04**(C9), 373–376 (1994)
144. D. Einfeld, M. Plesko, J. Schaper, First multi-bend achromat lattice consideration. J. Sync. Rad. **21**, 661–856 (2014)
145. ESRF Upgrade Programme Phase II (2015–2022), Technical Design Study, The Orange Book, December 2014
146. M. Eriksson, et al., Commissioning of the MAX IV light source, in *Proc. IPAC 2016* (Busan, Korea, 2016)
147. S.M. Liuzzo, et al., Updates on lattice modeling and tuning for the ESRF-EBS lattice, in *Proceedings IPAC 2016* (Busan, Korea, 2016), pp. 2818–2821

148. M. Borland, Preliminary expected performance characteristics of an APS multi-bend achromat lattice, white paper, in *Technical Report ANL/APS/LS-337, Advanced Photon Source* (2014)
149. C.G. Schroer, I. Agapov, W. Brefeld, R. Brinkmann, Y.-Y. Chae, H.-C. Chao, M. Eriksson, J. Keil, X. Nuel Gavaldà, R. Röhlsberger, O.H. Seeck, M. Sprung, M. Tischer, R. Wanzenberg, E. Weckert, The ultralow-emittance source project at DESY PETRA IV. J. Synch. Radiat. **25**, 1277–1290 (2018)
150. K.-J. Kim, Yu.V. Shvyd'ko, Tunable optical cavity for an x-ray free-electron-laser oscillator. Phys. Rev. ST Accel. Beams **12**, 030703 (2009)
151. R.R. Lindberg, K.-J. Kim, Yu. Shvyd'ko, W.M. Fawley, Performance of the x-ray free-electron laser oscillator with crystal cavity. Phys. Rev. ST Accel. Beams **14**, 010701 (2011)
152. Y. Shvyd'ko, V. Blank, S. Terentyev, Diamond x-ray optics: transparent, resilient, high-resolution, and wavefront preserving. MRS Bulletin **42**, 437–444 (2017)
153. K.P. Heeg, C.H. Keitel, J. Evers, Inducing and detecting collective population inversions of Mössbauer nuclei. arXiv:1607.04116 [quant-ph] (2016)
154. B.L. Bax, G.E. Maciel Hawkins, N.M. Szeverenyi, Two-dimensional NMR spectroscopy, in *Magnetic Resonance*, ed. by L. Petrakis, J.P. Fraissard, vol. 124 of *NATO ASI series* (Springer, 1984)
155. M. Cho, *Two-Dimensional Optical Spectroscopy* (CRC Press, 2009)
156. S. Doniach, Fourth-generation X-ray sources: some possible applications to biology. J. Sync. Rad. **7**, 116–120 (2000)
157. L. von der Wense, B. Seiferle, M. Laatiaoui, J.B. Neumayr, H.J. Maier, H.-F. Wirth, C. Mokry, J. Runke, K. Eberhardt, C.E. Düllmann, N.G. Trautmann, Peter G. Thirolf, Direct detection of the ^{229}Th nuclear clock transition. Nature **533**, 47 (2016)
158. V.V. Flambaum, Variation of fundamental constants. AIP Conf. Proc. **869**, 29–36 (2006)
159. W.G. Rellergert, D. DeMille, R.R. Greco, M.P. Hehlen, J.R. Torgerson, E.R. Hudson, Constraining the evolution of the fundamental constants with a solid-state optical frequency reference based on the ^{229}Th nucleus. Phys. Rev. Lett. **104**, 200802 (2010)
160. W.-T. Liao, A. Pálffy, C.H. Keitel, Nuclear coherent population transfer with x-ray laser pulses. Phys. Lett. B **705**, 134–138 (2011)
161. W.-T. Liao, A. Pálffy, C.H. Keitel, Three-beam setup for coherently controlling nuclear-state population. Phys. Rev. C **87**, 054609 (2013)
162. T.J. Bürvenich, J. Evers, Christoph H. Keitel, Nuclear quantum optics with x-ray laser pulses. Phys. Rev. Lett. **96**, 142501 (2006)
163. J. Haber, J. Gollwitzer, S. Francoual, M. Tolkiehn, J. Strempfer, R. Röhlsberger, Spectral control of an x-ray L-edge transition via a thin-film cavity. Phys. Rev. Lett. **122**, 123608 (2019)

Chapter 4
From Small Molecules to Complex Systems: A Survey of Chemical and Biological Applications of the Mössbauer Effect

Volker Schünemann

Abstract Mössbauer spectroscopy and synchrotron based nuclear resonance scattering are ideal tools to investigate electronic and dynamic properties of iron centers in chemical and biological systems. These methods have reached a level of sophistication during the last decades so that it is now possible to hunt for particular functional active iron sites even in very complex systems like iron based heterogeneous catalysts or even in some cases in biological cells. This book chapter will try to give a comprehensive overview of what can be achieved by using experimental techniques using the Mössbauer effect when combining different evaluation strategies like e.g. relatively straight forward analysis using lorentzian lines or hyperfine field distributions and more sophisticated investigations of paramagnetic iron sites by means of the spin Hamiltonian formalism. In addition the possibilities of synchrotron techniques based on the Mössbauer effect like nuclear forward and nuclear inelastic scattering will be shown. Special emphasis lies also on the sample requirements and on theoretical methods like quantum chemical density functional theory which nowadays is also available coupled with molecular mechanic shells which enables the treatment of very large systems like iron proteins. In addition to laboratory-based Mössbauer spectroscopy recent progress using synchrotron based nuclear inelastic scattering (NIS) to detect iron based vibrational modes in iron proteins and chemical systems will be described. In combination with quantum mechanical calculations for example, the iron ligand modes of NO transporter proteins have been explored. Via NIS it has been possible to detect iron ligand modes in powders and single crystals, but also in thin solid films of iron(II) based spin crossover (SCO) compounds. In addition, nuclear forward scattering (NFS) has been applied to monitor the spin switch between the $S = 0$ and $S = 2$ state of SCO microstructures. Furthermore, recent work on polynuclear iron(II) SCO compounds, iron based catalysts as well as biological cells will be discussed.

V. Schünemann (✉)
Department of Physics, Technische Universität Kaiserslautern, Kaiserslautern, Germany
e-mail: schuene@physik.uni-kl.de

Y. Yoshida and G. Langouche (eds.), *Modern Mössbauer Spectroscopy*,
Topics in Applied Physics 137, https://doi.org/10.1007/978-981-15-9422-9_4

4.1 Iron Centers in Chemical Complexes and Biomolecules: Structural Overview, Biological Relevance and Physical Properties

The trace element iron is indispensable for almost all life forms on earth. The best known example is the function of iron as an oxygen carrier in the red blood pigment hemoglobin. However, by exploiting its redox properties, iron also plays an important role in biological electron transport chains, e.g. in mitochondria or in photosynthetic reaction centers of thylakoid membranes of green plants [1]. Many enzymes, in turn, use the catalytic activity of iron, such as the enzyme cytochrome P450, which is also vital for humans and occurs in liver cells of mammals [2]. Iron is present in mononuclear form either in heme centers [3, 4] or as non-heme iron [5], which is directly bound to amino acid chains of the protein. Iron ions can also be linked by oxygen or carboxyl bridges to dimeric centers such as methane monooxygenase [6] or ribonucleotide reductase [7].

Iron centers bridged by sulfur can be found in iron-sulfur proteins. The function of these centers ranges from electron transfer to catalysis [8]. Unbound iron can produce free radicals in the presence of water, which are a danger to any cell. For this reason, iron metabolism in any organism is strictly regulated. Ferritins, proteins containing amorphous iron hydroxide clusters with diameters up to 8 nm [9], are used for iron storage. Biological complexing agents, transferrins, which contain iron in monomeric form and bind so strongly that its catalytic activity cannot lead to the formation of free radicals even in the presence of water, are used for iron transport [10].

The most common coordination environment of iron ions in biomolecules and chemical complexes are the sixfold octahedral and the fourfold tetrahedral coordination (Fig. 4.1) [11]. Sixfold coordination is found in heme proteins involved in electron transfer reactions like the cytochrome c family and also in the iron center of hemo and myoglobins when O_2 is bound. Iron sulfur proteins which are involved in electron transfer, but can also have catalytic (enzymatic) functions have tetrahedral sulfur coordinated iron centers in most cases.

The type of coordination defines the order of the 3d-orbitals of the iron center due to the electronic interaction with the surrounding ligands. These are in ligand field theory regarded as negatively charged point charges (Fig. 4.2). This is quite a simplification of the "real" electronic situation, but describes the electronic properties of iron centers often quite astonishing well. However, this simple picture fails when strongly covalent bonds are formed like metalorganic bonds.

Iron in biomolecules occurs in different spin and oxidation states. Divalent iron with 4 unpaired electrons in the 3d shell often occurs in the $S = 2$ state, where one 3d orbital is double and the others are single occupied according to Hund's rule (Fig. 4.3). Such a state is also often called a ferrous high spin state. If the ligands exert a strong ligand field, this leads to the diamagnetic $S = 0$ form of the divalent iron [12]. The latter state is also often called a ferrous low spin state.

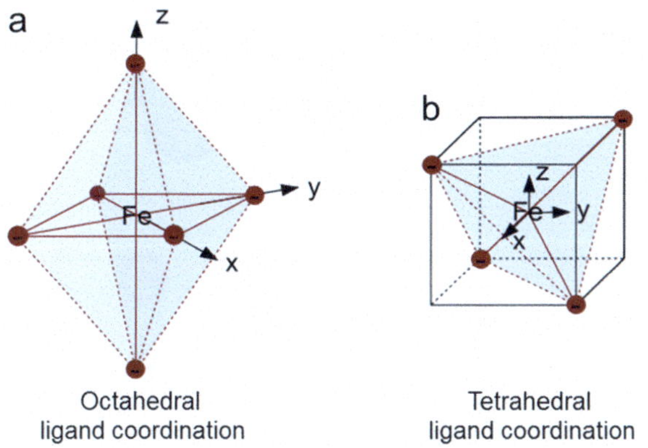

Fig. 4.1 Structures of octahedral and tetrahedral chemical complexes with iron as the central atom. **a** Schematic drawing of an octahedral complex. Six ligands lie at the same distance from the central ion on the coordinate axes. **b** Schematic drawing of a tetrahedral complex. Four ligands coordinate with the central ion so that the connection between the ligands results in a tetrahedron. The ligands are located between the coordinate axes. Note that this is a very simplified picture of the real chemical situation. The iron ligands are simplified in a way that they present negative point charges instead e.g. N and/or O atoms in (**a**) or S in (**b**). Adapted by permission of Shaker-Verlag GmbH, Düren, Germany [11]

Trivalent iron with 5 unpaired electrons is present in the S = 5/2 spin state. This also called ferric high spin state occurs when the ligand field splitting generated by the negative electronic charge of the surrounding ligands is smaller than the spin pairing energy. This is e.g. the case for fivefold coordinated iron centers in heme proteins. If a sixth ligand is added, the crystal field energy is higher than the spin pairing energy and a ferric low spin state (S = 1/2) is present [13–15]. Sometimes, however, the ligand field strength is just so high that S = 3/2 [16] or even electron configurations resulting from linear combinations of S = 5/2 and S = 3/2 spin states [17] form the ground state.

High-valent catalytic intermediates such as formally tetravalent iron have also been characterized spectroscopically and detected in enzymatic reactions. The first tetravalent iron with four 3d electrons detected in reaction intermediates of the peroxidase of horseradish has a spin state S = 1 [18, 19].

Iron centers, with the exception of diamagnetic divalent iron with S = 0, are molecular paramagnets whose structures are precisely defined. Biomolecules are produced in an absolutely reproducible way based on their genetic code. For this reason, each iron center in a protein or enzyme is in a well-defined state essential to its function. These iron centers, in turn, are ideal for biophysical studies of their structural, electronic and dynamic properties. The goal of Mössbauer spectroscopic studies of these biomolecules, is not only academic-biophysical, but understanding the function of biological processes at the molecular level using spectroscopic methods.

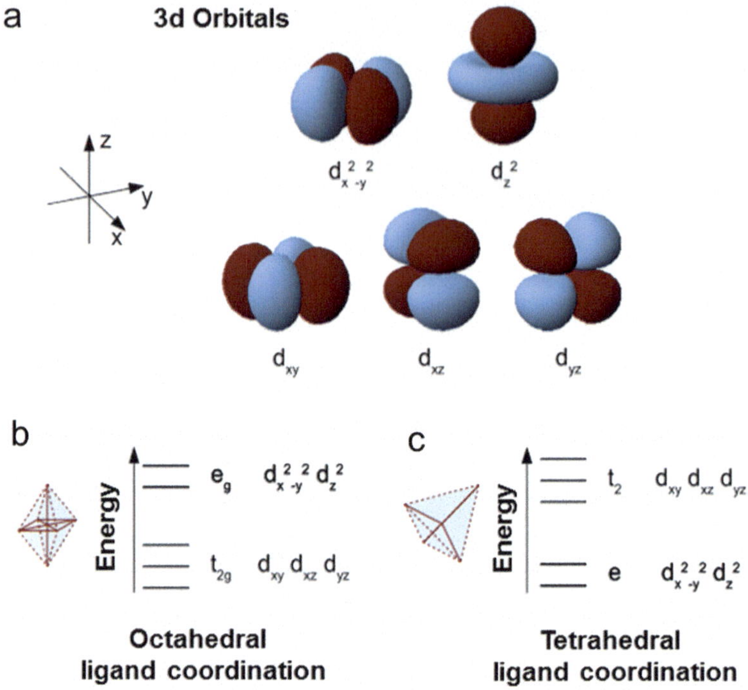

Fig. 4.2 Visualization and energetic splitting of d-orbitals. **a** Shape of the d-orbitals. **b** Energetic splitting of the d-orbitals in an octahedral complex. The d_{x2-y2} and d_{z2} orbitals are energetically elevated and are called e_g orbitals. The other three orbitals (d_{xy}, d_{xz} and d_{yz}) are energetically lower and are called t_{2g} orbitals. **c** Energetic splitting in a tetrahedral complex. The energetic splitting of the orbitals takes place inversely to that in an octahedral complex. The higher orbitals are called t_2 orbitals and the lower orbitals are called e orbitals. Adapted by permission of Shaker-Verlag GmbH, Düren, Germany [11]

In the last decades many studies have jointly being carried out by chemically, biochemically and biophysically oriented research groups with the aim of elucidating the role of metal centers in proteins [20, 21]. Thus, the field of bioinorganic chemistry was established which deals not only with the metal proteins themselves, but also structural and functional chemical models of metal centers are investigated. In many cases, this has made it possible to understand the function of metal and in particular also iron containing proteins. For the spectroscopist, who deals with iron proteins, the study of chemical models offers two tasks: (i) The characterization of the structural and electronic properties of model complexes and the comparison with corresponding metal proteins. (ii) The possibility to test new spectroscopic methods on chemical models, since model complexes in general can be produced easily in high concentrations (of some mM) often necessary for spectroscopic investigations. The aim of bioinorganic research, however, is not only to understand the structure and function of metal proteins and their models, but also to make use of the properties of these proteins.

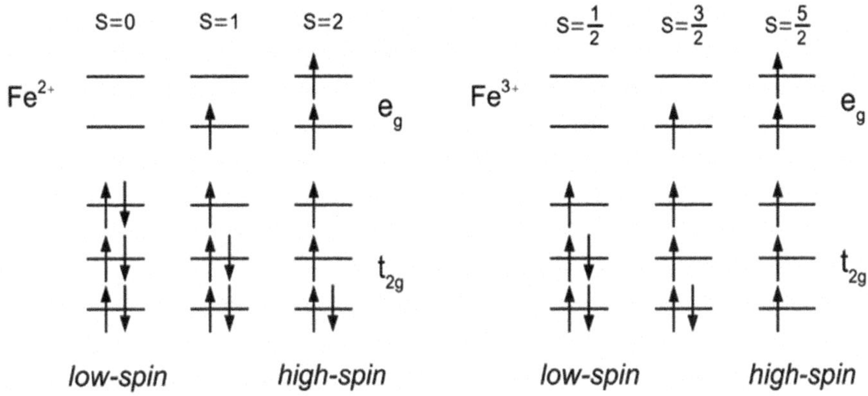

Fig. 4.3 Electronic occupation of the d-orbitals of Fe^{2+} and Fe^{3+} in a symmetrical octahedral complex. On the left is always the configuration of a low-spin complex and on the right the configuration of a high-spin complex. In between there are configurations of intermediate spin states, which rarely occur. Low spin complexes are formed in a strong, high spin complexes in a weak ligand field. Reproduced by permission of Shaker-Verlag GmbH, Düren, Germany [11]

In addition to examples dealing with Mössbauer spectroscopic techniques on biological systems this chapter will also discuss some recent work on artificial chemical iron(II) complexes the spin state of which can be switched by external parameters like change of temperature, pressure and/or light. For more general information on those spin crossover systems we refer to Chap. 6 of this book.

4.2 Spectroscopic Techniques to Investigate Iron Centers in Chemistry and Biology Based on the Mössbauer Effect: Strategy and Requirements

In this chapter we will discuss how much sample or better how many ^{57}Fe nuclei are necessary to acquire Mössbauer spectra in a reasonable amount of time. In the laboratory measuring a Mössbauer spectrum at one experimental condition may even last about a month, if highly diluted samples have to be investigated. In case of synchrotron experiments beam time is available only for some days and the question of sample amount and ^{57}Fe concentration may be even more important than for an in-house experiment. There will be also a short discussion about data evaluation procedures for conventional Mössbauer spectroscopy. As far as paramagnetic iron centers in proteins are concerned the analysis of field and temperature dependent Mössbauer spectra by means of the spin Hamiltonian formalism [22, 23] needs to be applied. When performing synchrotron based nuclear inelastic scattering (NIS) experiments on iron complexes and/or iron proteins quantum chemical density functional calculations can nowadays be run already by talented users which have no in deep knowledge of quantum chemistry and numerical methods. More sophisticated calculations have to be performed when also the protein environment should be taken into account.

4.2.1 *Sample Requirements for Conventional Mössbauer Spectroscopy and Synchrotron Based Techniques*

The Mössbauer isotope ^{57}Fe has a natural relative abundance of only 2.2%. Nevertheless, this is enough to obtain Mössbauer spectra in transmission mode with samples having a total iron content down to 1% by weight. Powder samples of inorganic iron compounds can be investigated with samples that have a sample thickness of ~0.5 mm and an area of ~1 cm. There are research groups which prefer to measure pressed pellets, but if the samples are sensitive to pressure, it is better to use small samples holders made of e.g. Delrin®. If one measures iron containing metal foils at room temperature, these can also be taped on the detector. This is a habit in some laboratories in order to perform calibration of the Mössbauer spectrometer with α-Fe foils. For any sample preparation it is important taking special care that there are no holes in the sample. In case there are holes in the sample the γ-rays go through the sample holder, but do not penetrate sample material. This can lead to a drastic decrease in absorption of the γ-quanta by the ^{57}Fe nuclei in the sample specimen and thus to a decrease in the intensity of the Mössbauer lines. Such a situation can easily occur with powder samples. A good test is checking whether light is shining through the sample holder. In case when only a tiny amount of sample is present, one can put one or several thin sheets of paper in between the sample holder and the sample. But also in case the sample holder is prepared correctly and there is sufficient ^{57}Fe in the sample, it is possible that only a very weak noisy Mössbauer spectrum is obtained at room temperature. In this case, the Lamb-Mößbauer factor f is small and therefore one should measure at cryogenic temperatures $\leq$120 K. One should also be aware of the fact that heavy atoms like e.g. Pt and Au have a high electronic absorption coefficient for 14.4 keV radiation which results in low count rates during the measurement.

Compared to optical spectroscopies taking a Mössbauer spectrum requires long acquisition times. The measurement can take in between several hours (e.g. for a calibration spectrum with an unenriched α-Fe foil of 25 μm thickness) up to weeks for samples very dilute in ^{57}Fe like brain *substancia nigra* tissue [24]. To give an example from our laboratory: Collecting a Mössbauer spectrum of a powder sample of an iron compound with an atomic mass of around 500–1000 amu at 77 K in a sample cup made of Delrin®, like those displayed in Fig. 4.4 normally takes 24 h.

There are only a few biological molecules which can be studied with their natural ^{57}Fe content. For example, the oxygen transporter protein hemoglobin in the blood can be measured within the red blood cells by Mössbauer spectroscopy after the blood has been thickened by centrifugation. A way to enhance the ^{57}Fe concentration in

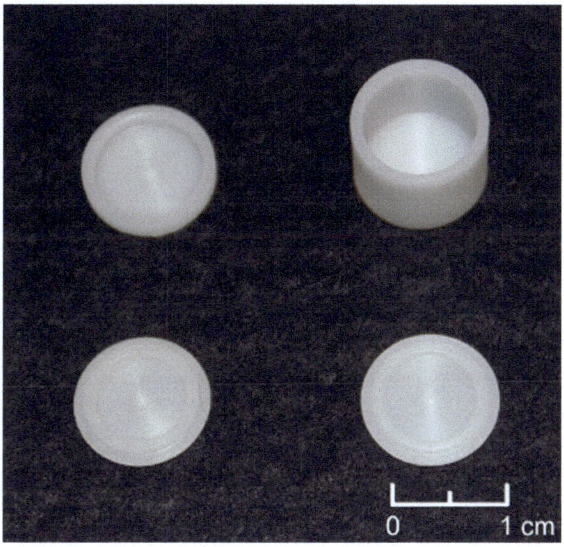

Fig. 4.4 Sample cups for Mössbauer spectroscopy of powder samples (left) and frozen solution samples (right) used in the laboratory of the author

hemoglobin of rats has been done by injecting ^{57}Fe citrate into their blood stream. In such a way one of the early Mössbauer spectroscopic studies of hemoglobin has been performed [25]. Nowadays, if iron proteins are to be investigated, an enrichment with ^{57}Fe is unavoidable. This can be done via ^{57}Fe reconstitution of e.g. iron sulfur clusters by adding iron salts to the apo protein. This method is often used but has the disadvantage that unspecific iron can be formed. Therefore, it is advisable to grow a bacterial expression system (e.g. *E. coli* cells) in nutrient medium enriched with ^{57}Fe and purify the protein one wants to study from the cell extracts. One has to keep in mind that for liquid solutions containing iron compounds or proteins, the Lamb-Mößbauer factor goes to zero, and therefore ^{57}Fe Mössbauer spectroscopy is only possible on frozen solutions. For such experiments sample volumes of 0.25–0.5 ml are needed. In some cases, frozen solution studies can be performed down to an ^{57}Fe concentration of ~0.1 mM (1 M = 1 Mol/l), but taking a Mössbauer spectrum of such a sample can take weeks depending on the shape of the spectrum. Therefore, it is strongly recommended that the concentration of ^{57}Fe to be at least 1 mM. Sometimes, it is possible to have even protein samples with ^{57}Fe concentrations of for example, 5–10 mM. Such samples can very conveniently be measured in a couple of hours.

For synchrotron experiments the sample volume can be less than for conventional Mössbauer spectroscopy in transmission mode. Both, for nuclear forward scattering (NFS) and for nuclear inelastic scattering (NIS) experiments sample holders which have sample volumes of 50–70 μl are used. The sample volume can be decreased even more if a focused beam (e.g. 10 μm × 15 μm) is used. Figure 4.5 shows a sample holder designed for liquid samples. After the sample solution is filled the holes can be closed with vacuum grease. In this way the sample can be frozen and stored e.g. at liquid nitrogen temperatures, if the sample is sensitive to air. It should be mentioned that for the sample holders discussed above top loading cryostats are advisable which

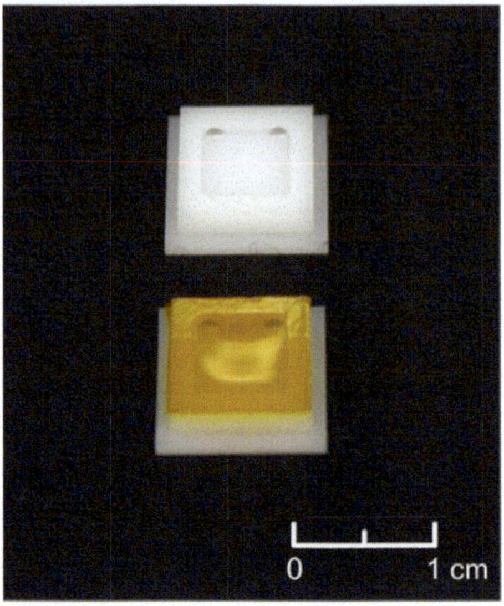

Fig. 4.5 NIS sample holder for liquid samples (e.g. ^{57}Fe enriched protein solution). The liquid can be filled via the holes seen on the upper part of the holder. A thin self adhesive Kapton® film is glued on the sample holder prior to filling and acts as window material for the NIS signal. Since in case of ^{57}Fe 6.4 keV radiation contributes to ~90% of the NIS signal the film has to be as thin as possible

allow fast sample change at cryogenic temperatures. In order to achieve reasonable acquisition times iron containing chemical complexes should be prepared with 50–100% enrichment in ^{57}Fe even when working with solid chemical compounds. In rare cases when ^{57}Fe enrichment is not possible and the scientific case is very strong one might also get beam time at a synchrotron to work with unenriched samples. However, for synchrotron experiments with iron protein solutions enrichment with ^{57}Fe is still a must to date. Also the concentration of ^{57}Fe should be as high as possible in order to minimize acquisition time. It is desirable to have ^{57}Fe concentration of 5–10 mM but even in that case a NIS experiment at one experimental condition may take several hours with nowadays synchrotron sources.

4.2.2 Spectral Analysis: Thin Absorber Approximation and Transmission Integral

Conventional Mössbauer spectra of biological systems and model complexes which are not enriched in ^{57}Fe can be conveniently analyzed by means of the thin absorber approximation using lorentzian line shape for the absorption lines observed. The same holds at least to the experience of the author for ^{57}Fe enriched iron proteins with concentrations up to the range of some mM. There is no need to create novel software for data analysis since both commercial software packages like e.g. NORMOS [26], MOSWINN [27], WMOSS [28] and public domain software like VINDA [29] are available. By means of computational fitting one obtains quite conveniently the isomer shift δ, the quadrupole splitting ΔE_Q, and the line width Γ with all parameters given in velocity units, mostly mms^{-1}. In the case of magnetically split spectra a further parameter, the magnetic hyperfine field at the nucleus B_{hf}, can be obtained. However, when it comes to samples which have higher concentration of ^{57}Fe, e.g. ^{57}Fe labelled chemical complexes in powder form, the thin absorber approximation is not valid anymore and self-absorption of resonantly absorbed γ-quanta lead to significant line broadening. In this case the transmission integral formalism should be used. In order to account for self-absorption this analysis requires the knowledge of the so called effective thickness which can be calculated from the density of ^{57}Fe nuclei and the volume of the sample [30]. Furthermore, the effective thickness is proportional to the Lamb-Mößbauer-factor f of the compound under study which is often not known and thus is treated as a fit parameter in addition to δ, ΔE_Q and B_{hf}. Note that the line width which goes into the transmission integral analysis is not a free parameter anymore like in the lorentzian line shape analysis. Based on the natural line width of a ^{57}Fe Mössbauer experiment which is 0.19 mms^{-1} the transmission integral formalism calculates the observed line width. It is obvious, but might be overseen by new users of Mössbauer spectroscopy, that it is not possible to have a line width which is lower or even equal to 0.19 mms^{-1}. This value is actually twice the natural line width of the 14.4 keV first excited nuclear state of ^{57}Fe since in a Mössbauer spectrometer one has to regard the natural line width of the source and of the absorber material. Typical experimental values of Γ are 0.24–0.36 mm^{-1}. If one observes higher linewidths some structural inhomogeneity may be present in the sample which leads to distributions of the Mössbauer parameters. Common are distributions of B_{hf} in small particles with a high surface to volume ratio, but also in amorphous metallic compounds [31]. Sometimes, also distributions of isomer shifts and quadrupole splittings are reported. However, it should be noted that also vibrations of the experimental set-up due to vacuum pumps or cryogenic coolers in the vicinity of the experiment can lead to significant line broadening effects. Therefore special care has to be taken, if one wants to investigate phenomena which might be related to line width broadening.

4.2.3 Mössbauer Spectroscopy of Iron in Molecules: The Spin Hamiltonian Concept

Iron centers in proteins—with the exception of diamagnetic iron ions or magnetically coupled diamagnetic clusters—are paramagnetic having a spin quantum number S. An ideal isolated spin system S has a spin multiplicity of $2S + 1$ and can be described by its spin functions $|S; m_s>$ with m_s being the magnetic spin quantum number which runs from $-S; -S+1;\ldots S-1; S$. The spin-orbit interaction of the paramagnetic iron ion manifests itself in the spin Hamiltonian concept in two parameters, the zero field splitting *D* and the rhombicity parameter *E/D* since the orbital momentum in 3d transition metals is nearly quenched. In ions which exert a ligand field, caused by the negative charge of the surrounding ligands like in chemical complexes (s. Figure 4.1) the orbital momentum is not totally quenched and can be treated as a second order effect. This leads to an energy splitting of the $|S; m_s>$ states even in zero magnetic fields. These energy splittings due to zero field splitting are quite small, normally in the order of some Kelvin.

The coupling of the total orbital momentum $\vec{L}$ to the total spin momentum $\vec{S}$ can be treated via perturbation theory. This concept was first introduced in order to describe electron paramagnetic resonance (EPR) signals of paramagnetic ions [32] and was later applied to paramagnetic Mössbauer spectroscopy by Wickmann et al. [33]. A paramagnetic iron with spin quantum number *S* in a ligand field is than usually described by the electronic spin Hamiltonian:

$$\hat{H}_S = D\left[\hat{S}_z^2 - S(S+1)/3\right] + \frac{E}{D}\left[\hat{S}_x^2 - \hat{S}_y^2\right] + \mu_B \vec{S} \cdot \overleftrightarrow{g} \cdot \vec{B} \tag{4.1}$$

Hereby denotes $\overleftrightarrow{g}$ the electronic g-tensor and μ_B the Bohr magneton. In order to calculate the energy eigenvalues E_S of the spin Hamiltonian above one needs to choose a quantum mechanical basis. Since we are dealing with spin states we choose as a quantum mechanical basis the pure spin functions $|S; m_s>$. If one rewrites Eq. 4.1 in terms of spin up and spin down operators it is possible to write down Eq. 4.1 in matrix representation. The exact procedure can be found e.g. in [34]. The diagonalization of the spin Hamiltonian matrix than leads to the energy eigenvalues E_{si} (Fig. 4.6) and once these are known also the corresponding eigenfunctions can be calculated. For e.g., $S = 2$ we have $2S + 1$ energy levels and thus $i = 1, 2, ...,5$. The spin orientation of every energy level E_{si} can be calculated via the determination of the spin expectation value.

$$\left\langle \vec{S} \right\rangle = \left(\langle S_x\rangle, \langle S_y\rangle, \langle S_z\rangle\right) \tag{4.2}$$

It is important to note that every sublevel E_{Si} might have a different spin expectation value $\langle \vec{S} \rangle_i$. The electronic spin of the iron ion is interacting with the ^{57}Fe nucleus by producing the magnetic hyperfine field $\vec{B}_{hf}$. Here, we introduce the hyperfine coupling tensor $\overleftrightarrow{A}$ which describes the interaction between the electron spins and the nuclear spin:

$$\vec{B}_{hf} = \frac{\left\langle \vec{S} \right\rangle \overleftrightarrow{A}}{g_n \mu_n} \tag{4.3}$$

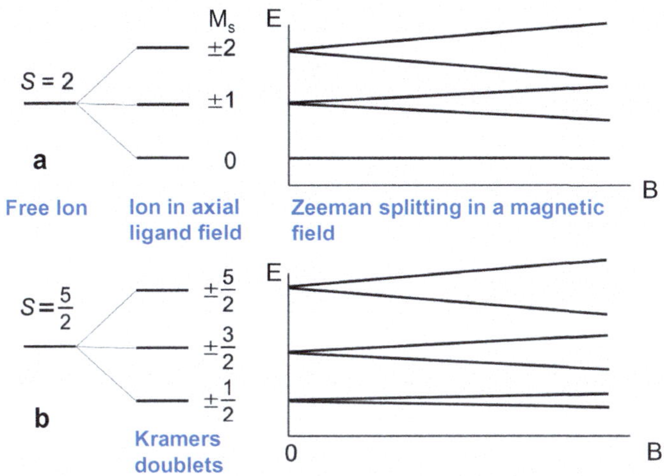

Fig. 4.6 Example of energy splittings of a spin system S caused by non totally quenched spin–orbit coupling and interaction with an external magnetic field B for (**a**) $S = 2$ and **b** $S = 5/2$ [34, 35]. In a ligand field the $\pm m_s$ states initially remain partially degenerate. For a system with half-integer spin (**b**) they always remain degenerate because of the Kramers theorem. The application of an external field B lifts the degeneracy by the Zeeman effect. Adapted by permission of Shaker-Verlag GmbH, Düren, Germany [11]

Very often $\overset{\leftrightarrow}{A}$ is given in units of the nuclear g factor g_n times the nuclear magneton μ_n which results in units of Tesla or Gauss. A high spin ferric ion with $S = 5/2$ for example in a sixfold oxygen coordination has a quite isotropic hyperfine coupling tensor of $A/g_n\ \mu_n \sim -22$ T [22].

With Eq. 4.3 the nuclear Hamiltonian can be written in the form

$$\hat{H}_N = \frac{eQV_{zz}}{4I(2I-1)}\left[3\hat{I}_z^2 - I(I+1) + \eta(\hat{I}_x^2 - \hat{I}_y^2\right] - g_n\mu_n\vec{I}\cdot\vec{B} + <\vec{S}>\cdot\overset{\leftrightarrow}{A}\cdot\vec{I} \tag{4.4}$$

Here I denotes the spin quantum number of the nuclear states, Q the nuclear quadrupole moment, V_{zz} the z-component of the electric-field gradient (efg) tensor, and $\eta = (V_{xx} - V_{yy})/V_{zz}$ the asymmetry parameter of the efg.

Figure 4.7 shows a simplified scheme for simulating field dependent Mössbauer spectra of paramagnetic iron centers [4, 23, 34, 35]. As pointed out above, first the electronic problem is solved by calculating the energy eigenvalues and eigenfunctions. The spin functions are selected as the basis. These wave functions are determined by the quantum numbers of the total spin S of the iron ion. The eigenfunctions are used to calculate the spin expectation values, which in turn determine the hyperfine field (see Eq. 4.3). If the hyperfine field is known, the nuclear Hamilton operator can be solved whose energy eigenvalues represent the energy of the nuclear

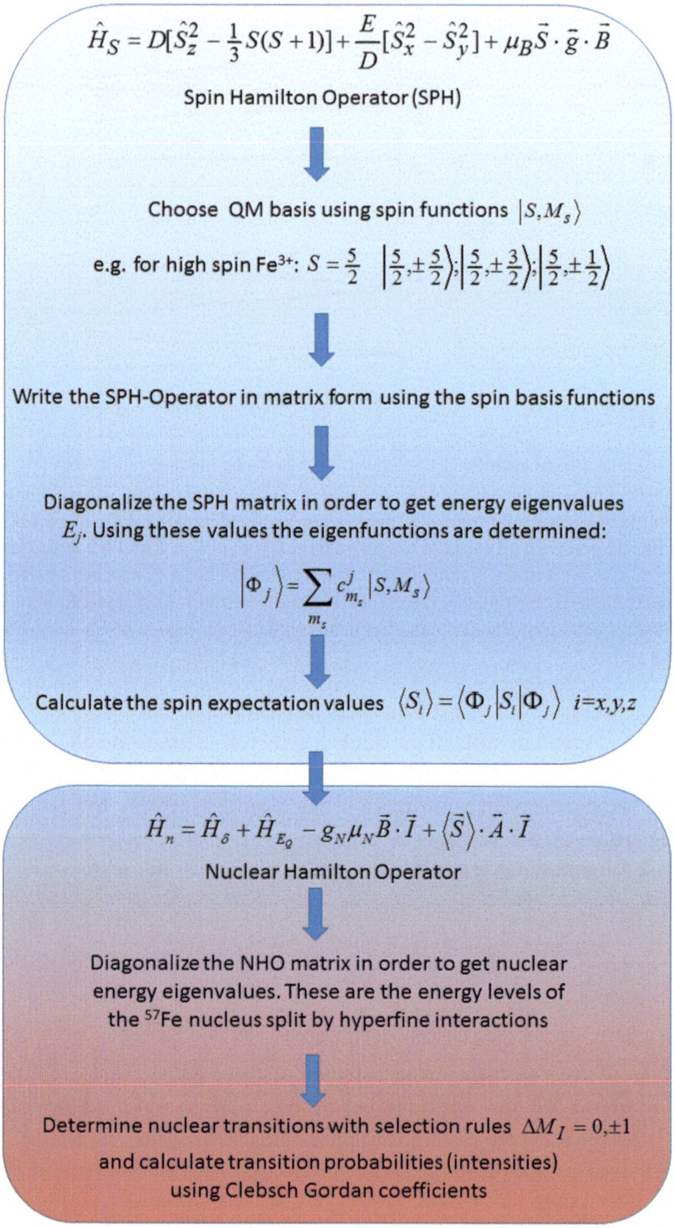

Fig. 4.7 Simplified scheme for the analysis of paramagnetic Mössbauer spectra with the help of the spin Hamiltonian formalism. The upper, blue area visualizes the calculation of the electronic problem. In the lower part, highlighted in red, the energy eigenvalues of the nuclear Hamiltonian operator are calculated which are used to obtain nuclear transition energies and probabilities. Note that for a powder sample this has to be done for every orientation of the molecule with respect to the external field direction which serves as the quantization axis

levels. Using the dipole selection rules possible transitions can then be determined. However, not only one Mössbauer spectrum is calculated, but one for each of the states split by zero field splitting [35]. The occupancy of these $2S + 1$ states depends on the temperature T, so that the spectra are still weighted with a temperature-dependent factor. Here the borderline cases of fast or slow relaxation of the electronic spin are usually taken into account. The limit of slow relaxation holds if the nuclear Larmor frequency $\frac{\langle \vec{S} \rangle \overleftrightarrow{A}}{\hbar}$ is considerably slower than the relaxation rates of the electronic spin system S. In this case the calculated Mössbauer spectra for each eigenfunction $|\varphi_j\rangle$ need to be weighted with the Boltzmann factor p_j and finally summed up:

$$p_j = \frac{\exp\left(-\frac{E_j}{k_B T}\right)}{\sum_K \exp\left(-\frac{E_K}{k_B T}\right)} \tag{4.5}$$

If, on the contrary, the spin transition rates are considerably larger than the nuclear larmor frequency the fast relaxation limit applies. In this case the spin expectation value in Eq. 4.4 is replaced by the thermal average of the $2S + 1$ spin expectation values:

$$\langle \vec{S} \rangle_{fast} = \frac{\sum_j \langle \vec{S} \rangle_j \exp\left(-\frac{E_j}{k_B T}\right)}{\sum_K \exp\left(-\frac{E_K}{k_B T}\right)} \tag{4.6}$$

Now only a single Mössbauer spectrum is being calculated with a hyperfine field which is the average of the hyperfine fields caused by population of the $2S + 1$ electronic sublevels. If the electronic relaxation rate is comparable to the Larmor frequency complicated Mössbauer spectra with relatively broad lines arise. For the spectroscopist who is not interested in the determination of electronic relaxation rates via Mössbauer spectroscopy it is advisable to push the system under study to the slow relaxation limit by performing Mössbauer experiments around liquid He temperatures. But it should also be mentioned here that there are suitable theoretical models to treat the case of intermediate relaxation. Examples on myoglobin have been treated by Winkler et al. [36]. Also for the case of iron containing magnetic nanoparticles it might be interesting to determine iron dependent spin relaxation rates which can be performed by the model of Blume and Tjon [37].

It should be noted that the procedure discussed above is sufficient for the calculation of non-interacting paramagnetic iron sites in a single crystal of a chemical complex (or a protein) when the unit cell has only one molecule. Only in this case there exists exactly one orientation of the molecular frames with respect to the external magnetic field B. However, since the large majority of Mössbauer spectroscopic studies are performed on powder samples or in the case of iron proteins on frozen solution samples it is necessary to calculate one Mössbauer spectrum for every molecular orientation. This so called powder averaging is done numerically by varying the Euler angles of the molecular axes with respect to the laboratory frame given by the direction of the external field and the direction of the γ-ray. Details of this procedure can be found in the classic work of Münck et al. [38].

4.2.4 Calculation of Mössbauer Parameters with Quantum Chemical Methods

Nowadays Mössbauer parameters for iron molecules can be calculated using theoretical models based on density functional theory (DFT). These calculations do not necessarily need to be performed by quantum chemists. Using the software packages ORCA [39], Gaussian [40] and/or Turbomole [41] it is possible also for the spectroscopist to achieve reasonable results. After the installation of these software packages the user can choose suitable models to perform calculations without further major programming skills.

DFT represents a possibility to solve the many-particle Schrödinger equation of complex molecules numerically. It is important to note that the central quantum mechanical observable of DFT is the ground state electron density ρ_0. The theoretical basis of DFT is given by two theorems formulated by Hohenberg and Kohn [42]. The first states that every system of interacting electrons in an external potential is determined by its electron density ρ_0. Thus, the ground state energy is also determined by the density ρ_0, in other words, the electronic ground state energy E is a function of ρ_0 and a functional exists of the form:

$$E[\rho] = V_{ne}[\rho] + V_{ee}[\rho] + T[\rho] \tag{4.7}$$

The first term $V_{ne}[\rho]$ represents the potential energy between the nuclei n and the electrons e, the second one $V_{ee}[\rho]$ describes the electron–electron interaction and the last term is the kinetic energy of the electrons $T[\rho]$.

The second theorem states that the ground state electron density ρ_0 among all possible electron densities for the molecule is the one at which the energy becomes minimal:

$$E[\rho] \geq E[\rho_0] \tag{4.8}$$

Kohn and Sham made the theorems usable for the calculation of concrete quantum mechanical problems with the approach that approximations must be made for the kinetic energy T and the potential energy V_{ee}. They divided the two approximated energies into explicitly calculable parts (J and T_S) and a residual contribution. The sum of these residual contributions was combined to the exchange correlation energy E_{XC}. The energy functional is represented by the following expression:

$$E[\rho] = V_{ne}[\rho] + J[\rho] + T_S[\rho] + E_{XC}[\rho] \tag{4.9}$$

And the electron density writes as

$$\rho(\vec{r}) = \sum_{i=1}^{N} |\psi_i(\vec{r})|^2 \tag{4.10}$$

The electron density is calculated from one-electron wave functions ψ_i and after the insertion in Eq. 4.9 provides an expression for the energy depending on one-electron wave functions ψ_i. After applying the second theorem of Hohenberg and Kohn, i.e. variation of ψ_i in order to minimize $E[\rho]$ one obtains a set of Schrödinger equations on the basis of one-particle wave functions. First, one-particle wave functions are advised and the energy is determined from them. Then, the Schrödinger equations are again solved and the corresponding energy is determined again on the basis of the Schrödinger equation just determined. This process is continued until self-consistency is achieved. The functional E_{XC} cannot be determined exactly. For this reason, various approximations exist which are implemented in DFT programs.

In order to perform a DFT calculation one needs a structure which is in most cases obtained from X-ray crystallography. Such structures can be obtained from the corresponding data banks. Protein structures can be found in the RCBS Protein Data Bank [43] and structures of chemical complexes in the Cambridge Structural Database (CSD) [44] or structures provided by collaborators. Since the DFT calculation is a self-consistent process, which involves finding the ground state energy minimum, the atom coordinates in the structure are varied until an energy minimum according to Eq. 4.8 is found. This process is called energy minimization and requires the choice of a functional and a basis set. Different functionals and basis sets exist, but in our experience the functionals B3LYP and/or TPSSH and the basis set TZVP lead to a reasonable compromise between computer time required for the calculation and accuracy. Once the minimized molecular structure is found one can go ahead and calculate experimental observables. If one wants to calculate Mössbauer parameters like isomer shifts and quadrupole splittings and, if paramagnetic iron centers are concerned also hyperfine coupling tensors $\overleftrightarrow{A}$, one has to make sure that the electron density near the ^{57}Fe nucleus is calculated with better accuracy than e.g. with TZVP. For that purpose one uses e.g. the CP(PPP) basis set for Fe and TZVP for all other atoms. For more details see e.g. [45].

In proteins very often the ligand sphere of the iron sites is influenced by the protein matrix via interactions with amino acid residues. It is impossible to perform DFT calculations for a whole protein, but one can use an approach which combines DFT calculations of the iron center and its ligands—the quantum mechanics (QM) part—with classical molecular mechanics (MM) calculations using empirical force fields. In Gaussian this approach is called ONIOM [46]. Using the force field uff for the MM part which is also called "low layer" and the above mentioned functionals and basis set for the iron center and its coordinating ligands ("high layer") it is possible to determine minimized structures of proteins. Recently, we have used this approach in order to calculate isomer shift and quadrupole splittings of iron in the NO carrier protein nitrophorin as will be discussed in Sect. 4.4.

4.2.5 Calculation of Iron Ligand Modes in Chemical Complexes and Proteins

Vibrational properties of molecules can also be calculated using theoretical models based on DFT. The first step is finding a structure which resembles at least in the computer the structure resembling the energy minimum of the molecule. In case the molecule is large like in case of an iron protein the above mentioned QM/MM approach can be used to find the ground state of the system. Such a calculation is performed according to the procedure described in the previous chapter. After that a normal mode analysis of the whole ground state structure is being performed which gives all normal modes of the molecule. Once the normal modes have been obtained vibrational spectra like Infrared and/or Raman spectra can be calculated. These features are nowadays implemented in DFT software packages. In our case of special interest are those modes which involve non-null movement of the iron center, because such modes are directly experimental accessible via nuclear inelastic scattering (NIS) experiments (see Chap. 1 of this book). It should be noted that this method is often denoted also as Nuclear Resonance Vibrational Spectroscopy (NRVS). These techniques deliver the partial density of phonon states (pDOS) and DFT calculations can be used for mode assignment. Examples for calculating spin marker bands in spin crossover complexes will be presented in Sect. 4.3.

4.3 Exploring Spin States in Iron(II) Containing Compounds

The spin state of iron(II) spin crossover (SCO) compounds can be switched reversibly from the low-spin state ($S = 0$) to a high-spin state ($S = 2$) by variation of temperature, pressure or by irradiation with light [47, 48]. More recently the spin dependent charge transport properties of SCO molecules have also generated interest for their use in spintronic devices [49, 50]. Since SCO compounds are also discussed in Chap. 6 of this book by N. Kojima and A. Okazawa, we restrict ourselves to some recent examples which concern the characterization of polynuclear SCO complexes, light excited spin states and the characterization of spin marker bands via synchrotron based NIS [51].

4.3.1 Thermal Spin Crossover (SCO) and Mössbauer Spectroscopy

The spin crossover between the $S = 0$ and the $S = 2$ state of ferrous ions can be ideally followed by Mössbauer spectroscopy. This has been shown in a plethora of cases by the group of Gütlich who pioneered the research of the SCO effect by means of the Mössbauer effect [52, 53]. Iron(II) containing SCO complexes are mostly in an octahedral ligand field which is caused by six nitrogen ligands. When the iron is in its ferrous low spin state the three t_{2g} orbitals are fully occupied with 6 electrons (see Fig. 4.3). This situation causes a symmetric charge distribution around the iron nucleus and thus only a small or sometimes even zero quadrupole splitting is observed. On the other hand, when the complex is switched to its $S = 2$ state the iron ligand distances increase. The partial negatively charged ligands are farer away from the iron ion and thus the crystal field splitting between the t_{2g} and the e_g orbitals decreases. This is caused by the occupation of the σ–antibonding e_g d-orbitals which leads to a reduced bond strength in comparison to the $t_{2g}{}^6$ low spin state. As a consequence the electronic population of the orbitals changes. Now the iron ion faces a weaker ligand field and the electrons populate the d-orbitals according to Hund's rule with a maximum spin multiplicity resulting in a $S = 2$ state.

In order to understand the increase in quadrupole splitting it is advantageous to consider first the symmetry of the negative charge density caused by the five spin-up electrons. This charge density is spherical symmetric around the iron ion and thus does not significantly contribute to the quadrupole splitting. This is actually the reason that octahedral coordinated iron(III) ions with $S = 5/2$ in most cases have a small quadrupole splitting <0.8 mms^{-1}. For ferrous high spin ions the main electronic contribution comes from the excess spin down electron in the double occupied t_{2g} orbital. The charge distribution caused by one electron has been estimated to app. 3–4.5 mms^{-1} [4, 54] which easily explains the high value of the quadrupole splitting in ferrous high spin ions.

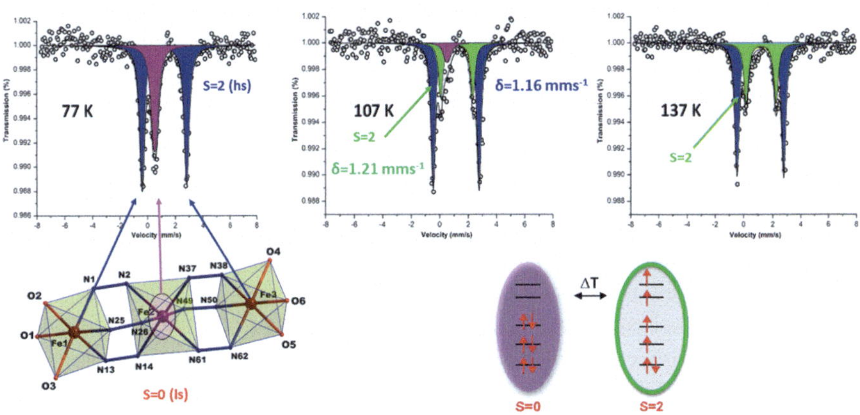

Fig. 4.8 Mössbauer spectra at indicated temperatures of a trinuclear iron(II) complex undergoing SCO analyzed with three doublets having lorentzian line shape. The purple doublet represents the S = 0 fraction of the central iron (Fe2). The blue doublet is due to the S = 2 fraction of the two terminal irons (Fe1 and Fe3) which stay in their high spin state at all temperatures. Raising the temperature leads to the green doublet representing the high spin fraction of the central iron (Fe2). Also shown is the labelled core of the complex with the iron(II) sites (spheres) being colored differently to distinguish the spin state (S = 2: red, S = 0: purple). Adapted from [55]

An advantage of using Mössbauer spectroscopy for the characterization of SCO compounds lies in the fact that this technique is able to discriminate between different iron sites and/or conformers of the corresponding complex under study. This is particular important for the investigation of polynuclear SCO complexes. Recently, we have shown that only the middle iron site of a trinuclear iron linear N1, N2-triazole bridged trinuclear Fe(II) complex undergoes a spin transition [55]. The fact that the two outer iron of the complex are in a 3N3O environment and the inner iron is in a 6 N coordination lead to slightly different isomer shifts of $\delta = 1.16$ mms^{-1} for the outer high spin iron(II) in comparison to the inner iron(II) which undergoes SCO around 100 K and displays $\delta = 1.21$ mms^{-1} in its high spin state (Fig. 4.8). More examples of this kind can be found in Chap. 6 of this book.

4.3.2 *Light Induced Excited Spin State Trapping (LIESST)*

For the possible application of SCO molecules in novel storage devices chemists have started to combine photoactive metal ions with iron ions with suitable ligand fields which allow SCO. Such a trinuclear complex containing two Pt^{II} ions and one Fe^{II} ion has been prepared by the group of M. Ruben [56].

We have studied this complex by conventional Mössbauer spectroscopy the results of which will be described in the following. Fig. 4.9 shows the structure of two conformers of the Pt_2Fe complex which have been investigated by temperature dependent Mössbauer spectroscopy. The presence of two molecular conformers has been shown by X-ray crystallography. One conformer (1H) is bent and the other is straight (1L). A Mössbauer spectrum taken at low temperature of a powder sample of this compound shows two components. One component with $\delta = 0.40$ mms^{-1} and $\Delta E_Q = 0.75$ mms^{-1} represents the ferrous low spin iron and the other component with $\delta = 1.11$ mms^{-1} and $\Delta E_Q = 3.43$ mms^{-1} is characteristic for ferrous high spin iron (Fig. 4.10a).

When the $FePt_2$-complex is irradiated with light at T = 4.2 K, almost all 1L iron sites can be transferred to the high spin state. This phenomenon is called Light Induced Excited Spin State Trapping (LIESST) . It has been extensively investigated by the group of Gütlich and coworkers [52] and is also discussed in Chap. 6 of this book. In order to investigate the electronic properties of LIESST states Mössbauer spectroscopy in large applied fields can be used. In order to realize such experiments a homebuilt sample rod for a high field cryostat in our laboratory [57] has been equipped with optical fibers as displayed in Fig. 4.11. In this way, Mössbauer spectroscopic measurements can be performed during irradiation of a sample at very low temperatures ($T \lesssim 4.2$ K) and high applied magnetic fields (in our case up to 5 T). In principle every cryostat suitable for Mössbauer spectroscopy with top loading sample capabilities can be equipped with such a modification.

Figure 4.12 shows Mössbauer spectra of Fe_2Pt_2 after 17 h irradiation with red light at 635 nm using a CW laser with the set-up displayed in Fig. 4.11 [58]. The analysis of the field dependent data by means of the spin Hamiltonian formalism (see Sect. 4.2.3)

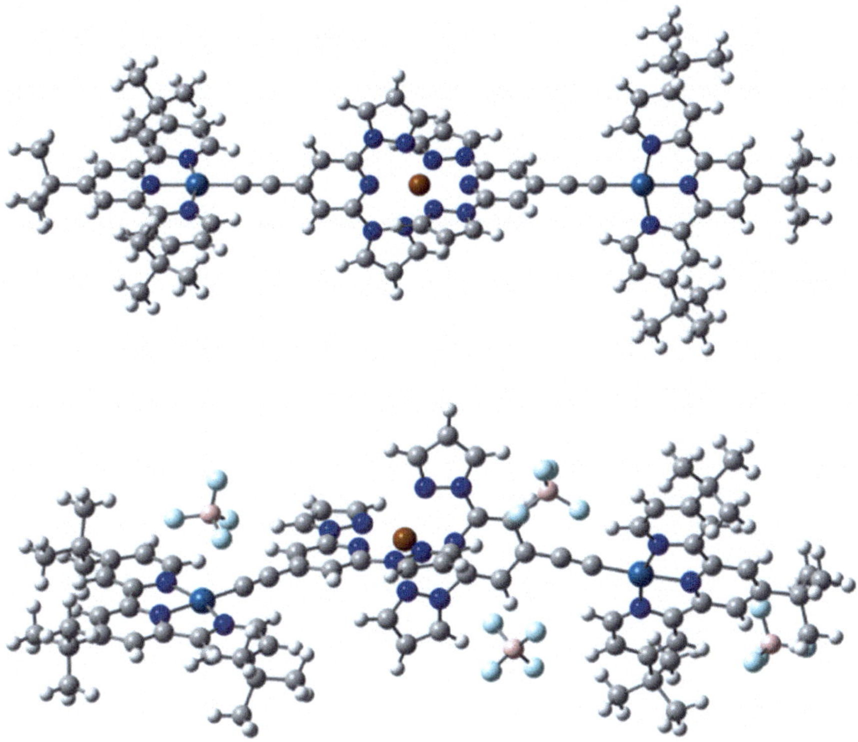

Fig. 4.9 Structural view of two conformations of a chemical complex containing two Pt^{II} ions and one Fe^{II}. The straight conformation 1L (top) is SCO active and shows LIESST after irradiation with red light. The bend conformation 1H (bottom) does not undergo SCO and exhibits a high spin state. Reprinted by permission from Springer-Nature: Hyperfine Interact. Copyright (2017) [58]

shows that the LIESST $S = 2$ state has a ligand field which leads to a zero-field splitting of $D \approx 5.3\ cm^{-1}$ and a rhombicity of $E/D = 0.14$. These values are different from those of the non-SCO active 1H conformation ($D = 2.1\ cm^{-1}$; $E/D = 0.23$) which were also determined by field dependent Mössbauer spectroscopy presented in [56]. The reason for this difference lies in the fact that there are two different conformers of the molecule and each confirmation leads to a different ligand field strength and symmetry which results in different values of D and E/D. Interestingly also the isomer shift of the LIESST state 1H′ is somewhat less than that of the 1H conformation which might be the result of the difference in conformation, since the 1H′ conformer is linear but the 1H conformer is not. Based on the latter argument also the possibility of a Ligand-Driven Light-Induced Spin Change (LD-DISC) was excluded.

It should be noted, that the analysis of the Mössbauer data displayed in Fig. 4.12 could only be done because the Mössbauer parameters of the 1H confirmation have been determined also by field dependent Mössbauer spectroscopy [56]. This is a

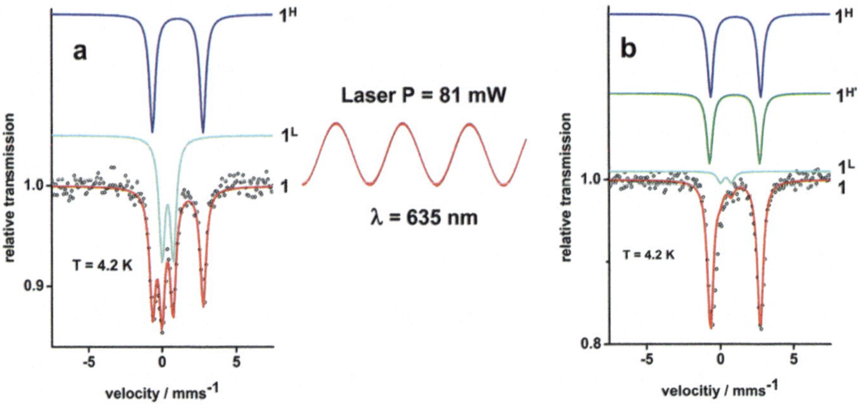

Fig. 4.10 Mössbauer spectra of the Pt_2Fe complex (Fig. 4.9) obtained at $T = 4.2$ K **a** before and **b** after irradiating with red laser light ($\lambda = 635$ nm using the set-up shown in Fig. 4.11. Comparison of (**a**) and (**b**) indicates a significant decrease of the low spin component 1L and the rise of a new component, 1H′, with a relative percentage share of 43%. Reproduced from [56] with permission from The Royal Society of Chemistry

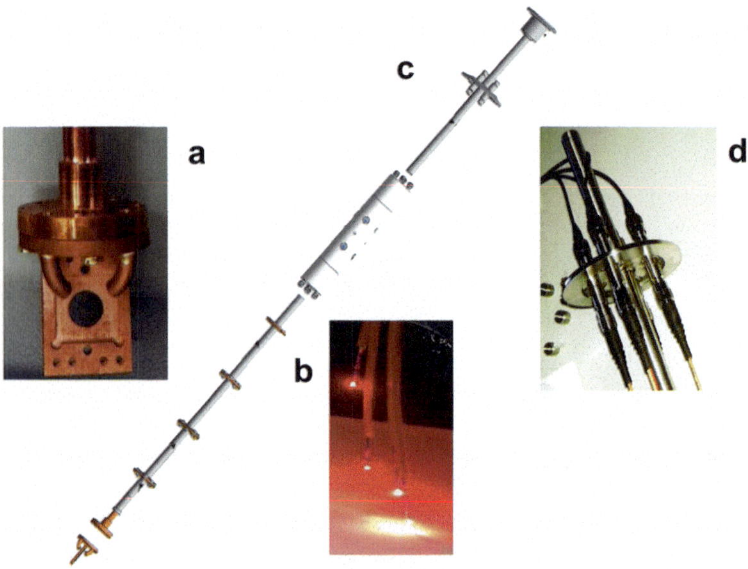

Fig. 4.11 Sample holder which shows copper tubing (**a**) which during the experiment contain scissor cut optical fibers (**b**) illuminating the sample. The sample rod (**c**) is equipped with four special optic feedthroughs (**d**). Reprinted by permission from Springer-Nature: Hyperfine Interact. Copyright (2017) [58]

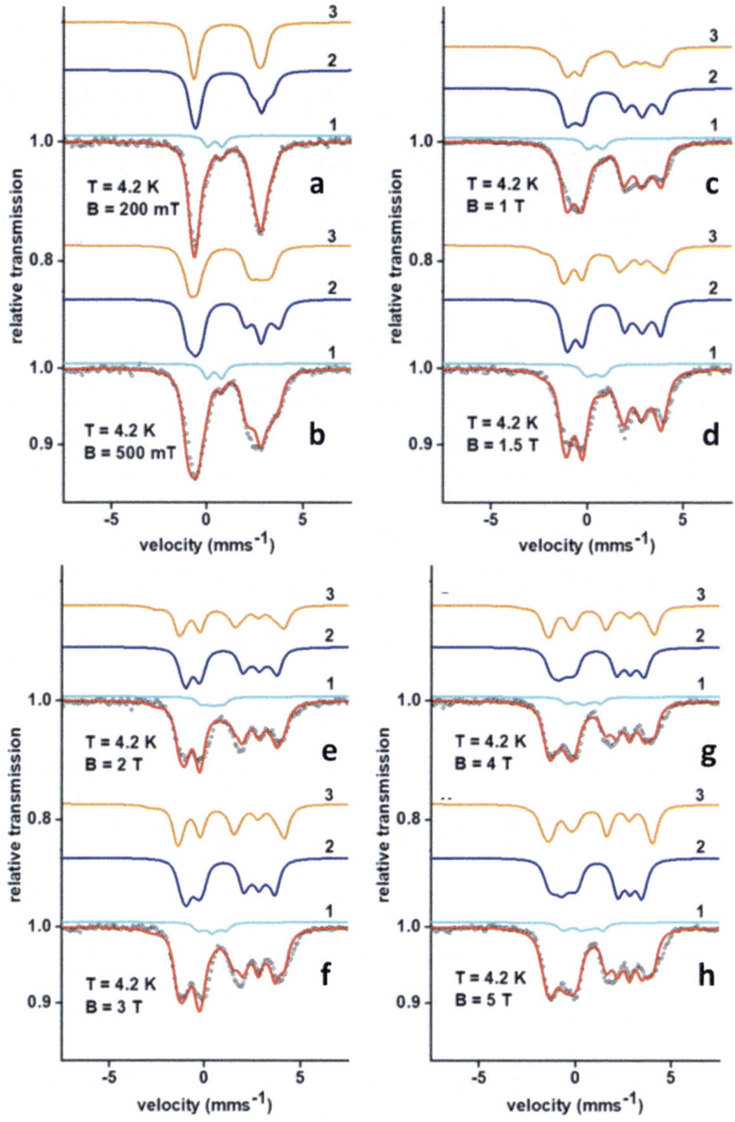

◂**Fig. 4.12** Field dependent Mössbauer spectra of the $FePt_2$ complex displayed in Fig. 4.9 after irradiation with red light obtained at $T = 4.2$ K and the indicated external fields. The simulations (red solid lines) have been obtained by a spin Hamiltonian analysis employing simultaneous fitting of all displayed data sets. For parameters see [58]; each simulated spectrum consists of three components (cyan 1 (1L conformer), blue 2 (1H conformer) and yellow 3 (1H′, the LIESST state of the 1L conformer)). Reprinted by permission from Springer-Nature: Hyperfine Interact. Copyright (2017) [58]

general strategy in our laboratory in order to analyze Mössbauer spectra of complex systems with multiple iron sites or iron phases. In order to reduce the number of unknown parameters it is advisable to fix those Mössbauer parameters originating from iron phases which are known or have been investigated in their pure state before. Pure free unsupervised parameter fitting of Mössbauer spectra from samples with more than one iron site is at least to our experience in almost all cases not successful, since the analysis is being very much over parametrized.

4.3.3 Exploration of Iron Ligand Modes by Synchrotron Based Nuclear Inelastic Scattering (NIS)

Synchrotron based Nuclear Inelastic Scattering (NIS) as well as Nuclear Forward Scattering (NFS)—the latter of which can be regarded as Mössbauer spectroscopy in the time domain—are ideal spectroscopic tools to study nanostructures and thin films of isotopically enriched ^{57}Fe containing materials since these techniques have a sensitivity down to a sub monolayer ^{57}Fe [59]. The two methods play an exceptional unique role compared to other synchrotron based techniques like X-ray diffraction and absorption, because not only the structural and dynamic but also the spin dependent magnetic properties of solids, surfaces and nanostructures can be investigated in one experimental set-up (see Chap. 1 of this book). NIS has been successfully applied to measure the density of phonon states (pDOS) in iron containing materials. From the measured DOS elastic and thermodynamic parameters including compressional and shear velocities, heat capacity, entropy, kinetic energy, zero point energy and Debye temperature can be obtained. Recently these parameters have been reported for SCO nanoparticles with diameter of 50 nm [60].

In order to investigate size effects in the SCO behaviour of nanostructures we have started to investigate ^{57}Fe enriched SCO microstructures via NFS. For this purpose we have used an iron(II) containing aminotriazole polymer. The microstructures have been prepared by pre-patterning with photolithography. A solution of the SCO compound was drop casted onto the pre-patterned substrates and the resist was lifted off using acetone. Figure 4.13 shows a micrograph of such a structure together with NFS data which show a clear beating structure characteristic for a species with a high quadrupole splitting of $\Delta E_Q = 2.73$ mms^{-1} indicative for a ferrous high spin state at 283 K. Cooling the structure to 243 K leads to an NFS pattern, with no clear beating structure showing the absence of quadrupole splitting indicating that the major species is now in the ferrous low-spin state. These results clearly indicate a change of the spin state due to the SCO effect [61].

Figure 4.14 shows NFS data of a [2 × 2] Fe(II) grid-complex with four iron ions which undergoes SCO [62]. Complementary susceptibility data show that at 80 K all 4 iron(II) ions are in their low-spin (LS) state. Iron(II) LS ions generally have a low quadrupole splitting which does not lead to quantum beats as visible in Fig. 4.13c. The minimum of the NFS-signal displayed in Fig. 4.14a occurring at ~40 ns is what

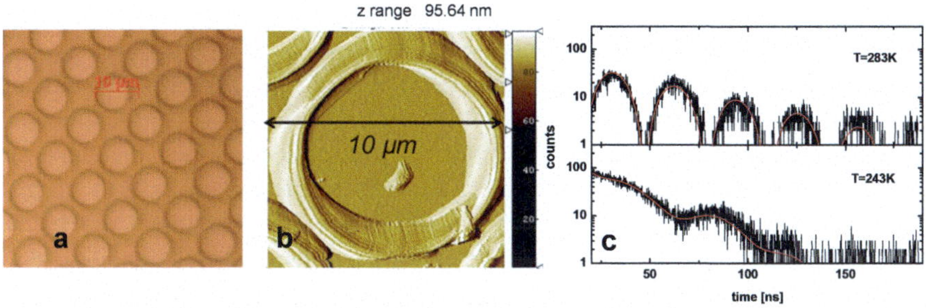

Fig. 4.13 **a** Optical view and **b** an AFM picture of micro structured $[Fe(atz)_3](CH_3SO_3)_2$ rings with 10 μm in diameter and with a height of ca. 100 nm. **c** NFS data for the $[Fe(atz)_3](CH_3SO_3)_2$ microstructures. The change of the beating indicates a spin crossover process from S = 2 at 283 K to S = 0 at 243 K. Reprinted by permission from Springer-Nature: Hyperfine Interact. Copyright (2014) [61]

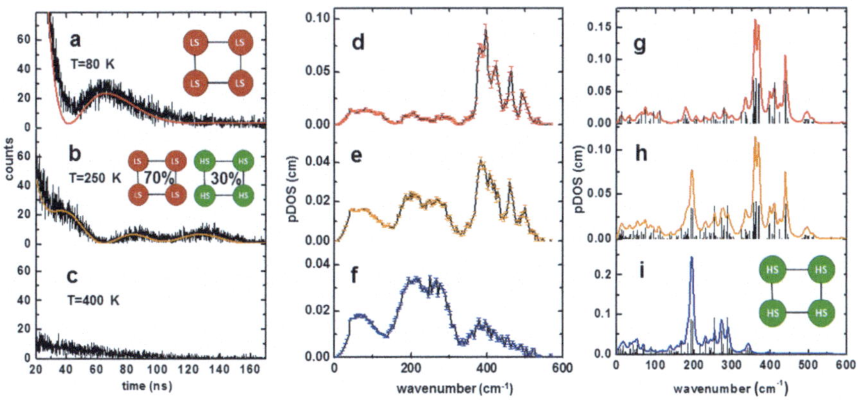

Fig. 4.14 NFS of a [2 × 2] iron(II) complex. Red line in (**a**) is a simulation with 100% LS iron(II); orange line in (**b**) represents 70% LS and 30% HS iron(II) sites. At 400 K (**c**) the NFS signal diminishes due to the low Lamb-Mössbauer-factor. **d**–**f** Experimental pDOS at corresponding temperatures and DFT simulated pDOS of an all LS state in (**g**), 70% LS and 30% HS iron(II) in (**h**) and an all HS state in (**i**). Adapted by permission of John Wiley and Sons [62]

is called a "dynamical beat" which is not due to the quadrupole splitting of 0.66 ± 0.01 mms^{-1} but caused by the large effective thickness of the sample which has been enriched with ^{57}Fe to almost 100%. However, when raising the temperature to 250 K (Fig. 4.14b) a beat pattern evolves which is indicative for a second iron species having significant quadrupole splitting characteristic of a high spin (HS) iron(II) ion. The simulation of the NFS data has been done in this case with the software MOTIF [63] yielding $\Delta E_Q = 0.56 \pm 0.03$ mms^{-1} and a relative contribution of 70% for the LS iron(II) sites and $\Delta E_Q = 1.87 \pm 0.05$ mms^{-1} and a relative contribution of 30% for the HS sites. It should be noted, that the time dependent NFS-signal is due to a coherent scattering process and therefore the shape of the signal is only sensitive to the difference of the isomer shifts of the two species present in the sample. The simulation yields $|\delta_{HS}-\delta_{LS}| = 0.7 \pm 0.15$ mms^{-1} which is in line with the isomer shifts reported for 6 N coordinated HS iron(II) centers (0.9–1.05 mms^{-1}) and for 6 N coordinated LS iron(II) centers (0.3–0.4 mms^{-1}) reported at ~300 K. In order to drive the system into the full HS state, the temperature needs to be increased even more. At 400 K the low Lamb-Mößbauer factor of chemical complexes in general leads to an almost vanishing NFS-Signal (4.14c). In terms of conventional Mössbauer spectroscopy that means that the signal disappears in the experimental noise of the background signal.

However, this is not the case for the NIS signal, which gives direct access to the iron-ligand modes. During the transition from a $t_{2g}{}^6$ electron configuration in the low spin state to a $t_{2g}{}^4e_g{}^2$ electron configuration in the excited high spin state not only the spin state is changed, but also a significant increase of the metal-ligand bond length by ~0.2 Å occurs. This is caused by the occupation of the σ-antibonding $e_g{}^*$ d-orbitals which leads to a reduced bond strength in comparison to the $t_{2g}{}^6$ LS state. This reduced bond strength can be ideally detected e.g. by vibrational spectroscopy techniques like NIS.

At $T = 80$ K the pDOS of the complex exhibits a complex band structure in the range of 400–550 cm^{-1} typical for iron(II) LS iron stretching and bending modes. These so called LS marker bands are also visible at 250 K, but now in addition features occur in the pDOS ranging from 180 to 320 cm^{-1} reflecting HS marker bands. At an even higher temperature of 400 K, the HS marker bands dominate the pDOS. This indicates that SCO has occurred to almost but not exact 100%, since residual LS marker bands are still visible. In order to learn something about the character of the iron-ligand modes, the experimental pDOS can be simulated by DFT-based normal mode analysis as mentioned in Sect. 2.5. The simulation of the pDOS of the molecule in its LS state performed using the software package Gaussian 09 is shown in Fig. 4.14 g–i. The height of the block bars is proportional to the mean square displacement of the iron in a particular mode. The position of every bar represents the energy of a particular normal mode of the molecule expressed in wavenumber units. In this way the whole vibrational features of the molecule can be addressed. Of course, this holds only for the assumption of harmonic potentials, when anharmonic effects have to be taken into account life becomes more difficult.

4.4 Electronic and Vibrational Properties of a Heme Protein: The NO Transporter Protein Nitrophorin

Proteins containing heme are among the most well-known proteins. The oxygen (O_2) carrier protein hemoglobin occurring in the blood of all higher organisms contains 4 subunits each having one heme unit. The heme unit contains an iron(II) center which serves to bind O_2 and transports this molecule to the cells of the organisms. In recent decades it has become known that also other small molecules like nitric

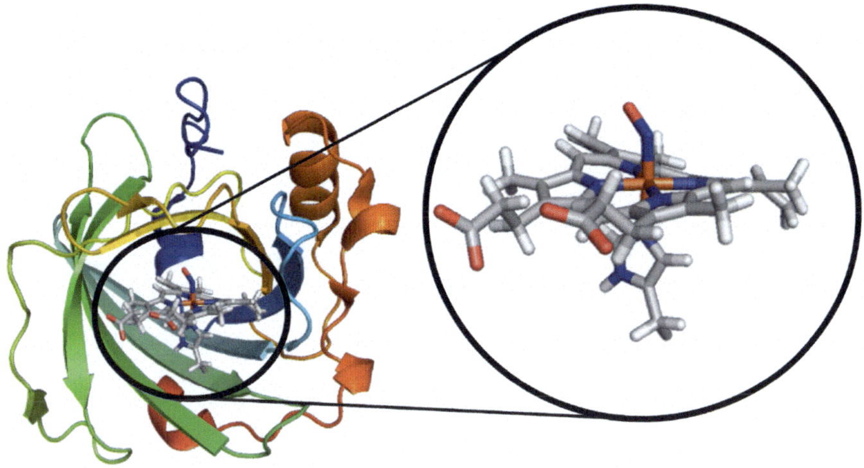

Fig. 4.15 Structural view of the heme protein nitrophorin 2 bound to nitric oxide based on the structure file 1T68.pdb. For the visualization the software package PyMOL Version 1.3 was used. The iron ion is located in the middle of the plane of the heme which is zoomed in. The structure of a free heme would be almost planar, but due to the special arrangement of the heme inside the protein the heme is strongly ruffled

oxide (NO) are important messenger molecules which serve e.g. to regulate the blood pressure by vasodilatation capabilities in the blood stream [64]. In mammals the signal molecule NO, although highly toxic in high concentrations, is not only involved in vasodilation, but also in inflammatory processes. In cooperation with F. Ann Walker we have investigated the interaction of NO with several isoforms of the NO transporter heme protein nitrophorin (NP). NPs occur e.g. in the saliva of the blood-sucking Amazon river-based kissing bug *Rhodnius prolixus*. In order to feed, the saliva containing the nitrophorin is injected into the victim's tissues, and due to the dilution and pH change (from ~5.5 in the salivary glands to 7.35 in the tissues) the iron ligand NO is released. NO then causes dilation of the victim's capillaries and inhibition of platelet aggregation so that more blood flows to the insect [65] (Fig. 4.15).

4.4.1 Probing Small Ligand Binding to Nitrophorin with Mössbauer Spectroscopy

NO is not the only small molecule which is relevant to the function of nitrophorin. After the NO has been released into the blood stream, there is a free binding site at the axial ligand position of the heme which can serve to bind histamine produced by the organism in response to the bite. Therefore, we have also studied the binding of histamine to nitrophorin and in addition, also the binding of the very toxic molecule

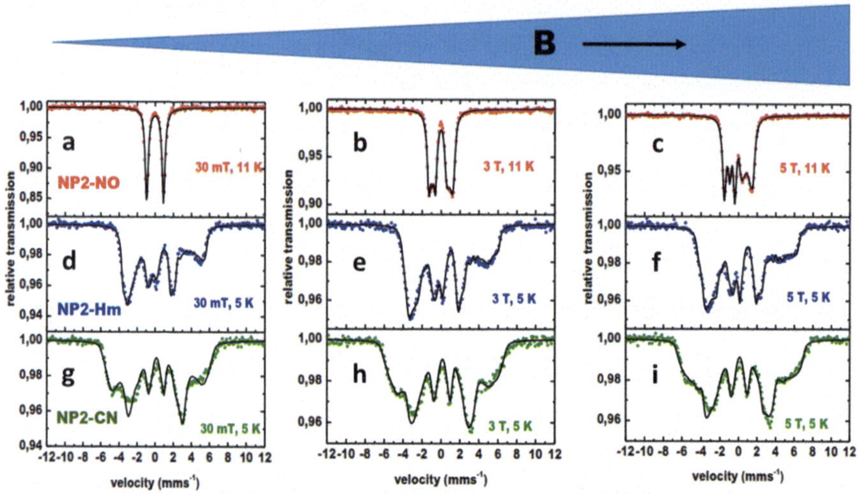

Fig. 4.16 Mössbauer spectra of NP2 coordinated with NO **a–c**, with histamine **d–f** and with cyanide **g–i**. The spectra were obtained at indicated temperatures in the presence of several different magnetic fields up to 5 T applied perpendicular to the γ-beam. The solid lines are fits performed in the limit of slow relaxation with parameters given in Table 4.1. Adapted with permission from [66]. Copyright (2012) American Chemical Society

cyanide (CN^-) which served as a further reference model [66]. Figure 4.16a–c shows the Mössbauer spectra of the isoform nitrophorin 2 (NP2) bound to NO taken at 4.2 K and increasing external fields. The spectrum in a low field of 20 mT shows a doublet with $\delta = -0.01$ mms^{-1} and $\Delta E_Q = 1.86$ mms^{-1}. Such a Mössbauer spectrum can be conveniently evaluated by an analysis using a doublet with lorentzian line shape. However, when increasing the external field a magnetic splitting becomes more and more prominent. Here, a simple lorentzian line shape analysis would not be sufficient anymore and one has to use the spin Hamiltonian formalism (see Sect. 4.2.3). The analysis shows that the observed magnetic splitting is only due to the external applied field. Therefore, the binding of NO to the heme unit of the nitrophorin creates a diamagnetic unit with a spin of the iron NO moiety of in total $S = 0$. This can be rationalized by taking into account that the spin of the NO is $S_{NO} = 1/2$ and the spin of the heme iron is also $S_{Fe} = 1/2$. An antiparallel spin alignment of the iron and the NO spins thus explains the diamagnetism of the NP2–NO system. The question may arise why it is necessary to use the rather complex spin Hamiltonian formalism in order to simulate the experimental data shown in Fig. 4.16. Obviously, when setting $S = 0$ in the spin Hamiltonian given in Eq. 4.1 the spin expectation values (Eq. 4.3) will also be zero. Therefore the magnetic field which is seen by the ^{57}Fe nucleus is only the external magnetic field. Why do we not observe a simple sextet? The reason is that all the proteins are randomly embedded in a frozen solutions. Therefore the orientation of the external field direction is not fixed with respect to the main axis system of the electric field gradient tensor of the ^{57}Fe in the sample. We say that the ^{57}Fe nucleus

Table 4.1 Mossbauer parameters of the simulations shown in Fig. 4.16. Values in square brackets given for the hyperfine coupling tensor $\overleftrightarrow{A}$, and the asymmetry parameter η have been calculated according to Oosterhuis and Lang [68] in the reference frame of Taylor [69] using $P = 62$ T and $\kappa = 0.35$. [a] The $\overleftrightarrow{A}$-tensor for NP2–CN is turned with respect to the electric field gradient tensor by the Euler angles $\alpha = 25°$ and $\beta = 56°$. For more details see [66]. Adapted with permission from [66]. Copyright (2012) American Chemical Society

	S	δ (mms^{-1})	ΔE_Q (mms^{-1})	η	g	$\overleftrightarrow{A}/g_n\mu_n$ (T)
NP2–NO	0	−0.01	1.86	0	–	–
NP2–CN	1/2	+0.17	1.03	−1.74 [−1.71]	(0.67, 1.96, 3.55)	(−41, 44, 47)[a] [−36, 27.6, 85.3]
NP2–His	1/2	+0.27	2.25	−1.57 [−2.59]	(1.52, 2.24, 2.92)	(−27, 19, 60) [−40.5, 16.7, 50]

sees magnetic fields the orientation of which are randomly distributed. This rather complex situation is taken care of by calculating powder averaged Mössbauer spectra as also explained in Sect. 4.2.3.

In addition, the reader may note that the trace of the Mössbauer spectrum obtained at 5 T is quite asymmetric (Fig. 4.16c). This is caused by the fact that the anisotropy parameter of the electric field gradient η has a value which is close to zero. In the case of $\eta = 1$ a fully symmetric trace would be observed. In the case of η being non unity also the sign of the quadrupole splitting and therefore the sign of the directly related main component of the electric field gradient V_{zz} can be determined. In the case here it turns out that $\Delta E_Q = +1.86$ mms and thus V_{zz} is also positive. In the case of a negative quadrupole splitting one would observe a mirror image of the trace shown in Fig. 4.16c with the broad double line feature being in the negative velocity range. Collins was the first who reported these effects and thus sometimes such diamagnetic high field Mössbauer spectra are also called "Collins-type" Mössbauer spectra [67].

The addition of histamine (Hm) to NP2 causes very different signatures of the Mössbauer spectra as displayed in Fig. 4.16d–f. In contrast to the well-defined doublet of NP2–NO a rather unstructured magnetic pattern is observed at a low field of 30 mT. In such a case it is of advantage to have information about the spin state of the heme histamine unit from other methods like electron paramagnetic resonance (EPR) spectroscopy. EPR spectroscopy performed at X-band frequency (~9.4–9.6 GHz) clearly shows that NP2-Hm has a spin state of $S = 1/2$ with an anisotropic g-Tensor $\overleftrightarrow{g} = (1.52, 2.24, 2.92)$. If we now look at the electronic spin Hamiltonian given in Eq. 4.1 we can deduct the following information: Since NP2-Hm has $S = 1/2$ it cannot have a zero field splitting D and a rhombicity parameter E/D. This simplifies the spin Hamiltonian considerably and what is left of Eq. 4.1 is just the electronic Zeeman-Term

$$\hat{H}_{S=1/2} = \mu_B \vec{S} \cdot \overleftrightarrow{g} \cdot \vec{B} \tag{4.11}$$

Since in the case of NP2-Hm the g-tensor is quite anisotropic it turns out that also the spin expectation value $\langle \vec{S} \rangle = (\langle S_x \rangle, \langle S_y \rangle, \langle S_z \rangle)$ is quite anisotropic. In addition, also the hyperfine coupling tensor tensor $\overleftrightarrow{A}$ of ferric low spin centers is quite anisotropic. The components of $\overleftrightarrow{A}$ can be determined by simulation of Mössbauer spectra obtained at different magnetic fields at low temperature (e.g. ~4.2 K) in order to make sure that the spectra are not blurred by electronic relaxation effects. The simulations than should fit the different spectra equally well, a task which had to be done by manual simulations in the early times of Mössbauer spectroscopy. Nowadays this can be performed with the help of a simultaneous fitting of several spectra, a feature that can be implemented in e.g. VINDA [29]. This has been done by means of the simulations displayed in Fig. 4.16d–f. It should be noted that it is allowed only to vary the parameter which has been varied in the experiment, in this case the external field, all other parameters should be the same for all simulations. Such a simulation requires quite a high number of parameters like the three components of $\overleftrightarrow{A}$, the asymmetry parameter η, the sign of ΔE_Q as well as its absolute value. The latter can be also determined by Mössbauer experiments at temperatures where the electronic spin relaxation rate is fast in the Mössbauer time window and only a quadrupole splitting is observed. This value can be taken as an input for the simulation. However, one should be aware that the quadrupole splitting can be temperature dependent when excited electronic states are populated. In some cases it might be even necessary to introduce the Euler angles between the main axis system of the g-tensor and those of the electric field gradient and the magnetic hyperfine tensor.

There is also a possibility to calculate relatively straightforward theoretically values of the hyperfine tensor components $\overleftrightarrow{A}$ from experimentally obtained g-values for ferric low spin species. Oosterhuis and Lang [68] developed analytical equations which describe at least to our experience the experimentally observed components of the $\overleftrightarrow{A}$ tensor components of ferric low spin hemes quite well.

4.4.2 *Investigation of Vibrational Properties of Nitrophorins*

In this chapter we will discuss the investigation of ligand binding to nitrophorins using NIS. The application of this technique to heme proteins started with several studies of the O_2 storage protein myoglobin [70–76]. One may ask the question what information can be gained by using a complicated synchrotron technique in contrast to laboratory based Raman- and or Infrared-Spectroscopy. The advantage of using the ^{57}Fe nucleus as a nuclear probe also for the study of the vibrational dynamics is that there are no optical selection rules other than that the iron has to move with respect to the direction of the incoming synchrotron beam. In addition, Raman and IR spectroscopy do detect a whole variety of vibrations of the protein environment which may mask the relevant modes one is hunting for especially when Fe ligand modes are concerned. Figure 4.17 shows the NIS raw data of NP2–NO, NP2–CN

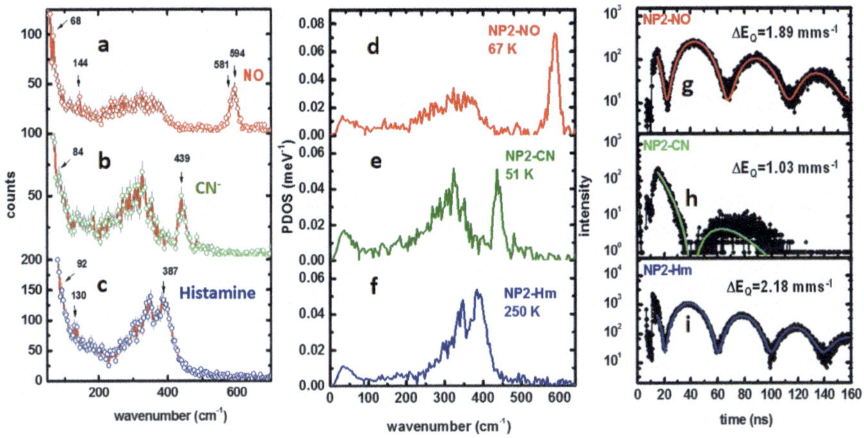

Fig. 4.17 Experimental NIS data obtained from NP2 under three different conditions, from top to bottom: **a** NO-bound (NP2–NO) obtained at 67 K, **b** CN^--bound (NP2–CN) obtained at 51 K, **c** Histamine-bound (NP2 − Hm) obtained at 250 K. **d–f** corresponding pDOS as obtained from the experimental NIS data. **g–i** NFS data obtained under the same conditions as the NIS data in (**a–c**). The protein concentration was ~10 mM in all three cases. Adapted with permission from [66]. Copyright (2012) American Chemical Society

and NP2-Hm as obtained during the experiment (Fig. 4.16a–c) and the corresponding partial density of vibrational states pDOS (Fig. 4.16d–f) as obtained from the raw data [66]. In addition to the relatively unstructured region where heme modes occur one can clearly identify strong vibrational bands due to NO, cyanide (CN^-), and histamine binding to the heme center of NP 2.

The change of the vibrational signature of the protein upon ligand binding can also be calculated via quantum–mechanical DFT methods, coupled with molecular mechanics (QM/MM) methods. The QM/MM calculations show that the heme ruffling in NP2 (see Fig. 4.15) is a consequence of the interaction with the protein matrix. Treating the protein matrix with the MM approach and the heme, including its axial ligands, with DFT methods, retains the heme ruffling of the NO, CN^-, and histamine complexes of NP2. If the protein matrix is not included, energy minimization of just the heme including its axial ligands always leads to almost flat, non-ruffled heme structures. It should be noted here that with the so obtained minimized structures QM/MM derived pDOS are presented in [66]. In this way the iron ligand vibrational modes could be identified.

4.4.3 Calculation of Mössbauer Parameters of Nitrophorins Using DFT Based Methods to Proof Structural Models of Heme Centers

The bug *Rhodnius prolixus* occurring in South America is not the only insect which uses nitrophorins in order to achieve an easier blood meal. Also the bedbug *cimex lectularius* uses a nitrophorin (cNP) as a NO carrying protein. Like the nitrophorins from the *Rhodnius prolixus*, cNP forms a stable heme iron(III)-NO complex, where the NO can be stored reversibly for a long period of time. Furthermore, the cNP can bind a second NO molecule to the proximal heme cysteine when NO is present at higher concentrations. Upon binding of the second NO molecule an S-nitrosyl conjugate with the proximal cysteine is formed [77]. In order to confirm this proposed reaction scheme Mössbauer spectroscopy was used to characterize the binding of the first NO to the ferric heme unit, a situation comparable to that for the NP from *Rhodnius prolixus* discussed in the previous chapter. In a second step the binding of the second NO was investigated. During this process a reduction of the ferric heme iron to a ferrous heme iron occurs which is well reflected in the corresponding Mössbauer spectroscopic signatures as will be discussed below. DFT calculations have been performed based on structural models of both scenarios [78].

Figure 4.18 shows the Mössbauer spectrum of the cNP bound to NO at pH 5.5 taken at $T = 12$ K with a small external field of $B = 20$ mT (Fig. 4.18a) and at $T = 15$ K with a high external field $B = 5$ T (Fig. 4.18b). Spectral analyses with two components shows that the majority component 1 has a relative intensity of 88% and exhibits $\delta_1 = 0.11$ mms^{-1} and $\Delta E_{Q1} = 1.33$ mms^{-1}. The observed symmetric doublet at low field shows in a high field of 5 T a magnetic splitting which is due only to the external magnetic field as confirmed by the simulation representing a diamagnetic iron species the efg of which is randomly distributed with respect to the magnetic field direction in the frozen protein solution. Thus it can be concluded that cNP binds one NO in the form of a diamagnetic ferric low spin heme NO just like the nitrophorin from *rhodnius prolixus* as discussed in the previous chapter. When cNP is exposed to twice the amount of NO as above and the pH is raised to pH 7.5 the Mössbauer spectrum changes drastically at low temperatures even at a low field of 20 mT (Fig. 4.18c). Now a complex magnetic splitting is observed which reflects a paramagnetic iron species. A spin Hamiltonian simulation using $S = 1/2$ gives $\delta_2 = 0.34$ mms^{-1} and $\Delta E_{Q2} = 1.42$ mms^{-1}. The increase of the isomer shift reflects the reduction of the ferric low spin heme to a ferrous low spin heme center. But ferrous low spin iron is diamagnetic, why do we observe a paramagnetic iron species? The NO itself has S = 1/2 and binds to the diamagnetic ferrous low spin heme creating a paramagnetic entity. In order to simulate both the Mössbauer spectra at low (Fig. 4.18c) and high field (Fig. 4.18d) a hyperfine coupling tensor $\overleftrightarrow{A}/g_n\mu_n =$ (24.4, −24.2, 12.0) T has been used. Such parameters are characteristic of a low-spin ferrous heme iron(II)-NO complex which has been formed after the reduction of the heme iron in the presence of high NO concentrations.

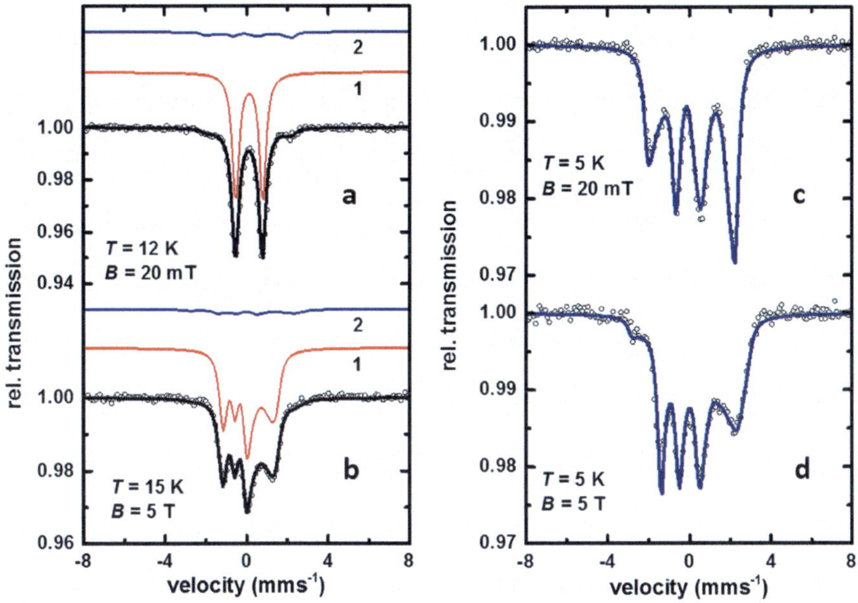

Fig. 4.18 Mössbauer spectra of cNP–NO at pH 5.5 taken at T = 12 K with B = 20 mT applied perpendicular to the γ-beam **a** and at T = 15 K with B = 5 T (**b**) in comparison to the Mössbauer spectra of cNP–NO at pH 7.5 taken at T = 5 K (**c, d**) with the same field conditions as in (**a**) and (**b**). The *solid lines* are the result of a spin Hamiltonian analysis with parameters given in the text and in [78]. Reprinted by permission from Springer-Nature: Hyperfine Interact. Copyright (2016) [78]

The example above shows how Mössbauer spectroscopy can be used to obtain spin states and oxidation numbers of iron centers. As has been pointed out in Sect. 4.2.4 it is also possible to calculate Mössbauer parameters using DFT methods. In this way it is possible to proof or disprove model structures of the iron environment since both, δ and ΔE_Q can be calculated nowadays in a reliable manner. In the case of cNP Weichsel et al. reported that in the presence of high NO concentrations a second NO molecule forms an S-nitrosyl conjugate with the proximal cysteine [77]. In order to suggest structural models and to test this proposed reaction scheme DFT calculations were performed in conjunction with molecular mechanics calculations. Figure 4.19 shows the structural models which have been generated after energy minimization of the whole protein structure using the QM/MM approach of Gaussian 09. For these QM (DFT) calculations of the heme environment the functional TPSSH in combination with the basis set TZVP was used. For the protein environment the universal force field UFF was applied. A subsequent DFT calculation using Orca [39] and the CP(PPP) basis set for Fe and TZVP basis set for all other atoms lead to $\delta_{1calc} = 0.07$ mms^{-1} and $\Delta E_{Q1calc} = 0.97$ mms^{-1} for the structural model including a diamagnetic NO-heme unit with protonated heme carboxyl groups representing component 1 in Fig. 4.18a, b. For the reduced ferrous low spin heme bound to NO at pH 7.5 the calculations yield $\delta_{2calc} = 0.25$ mms^{-1} and $\Delta E_{Q2calc} = 1.12$ mms^{-1} for

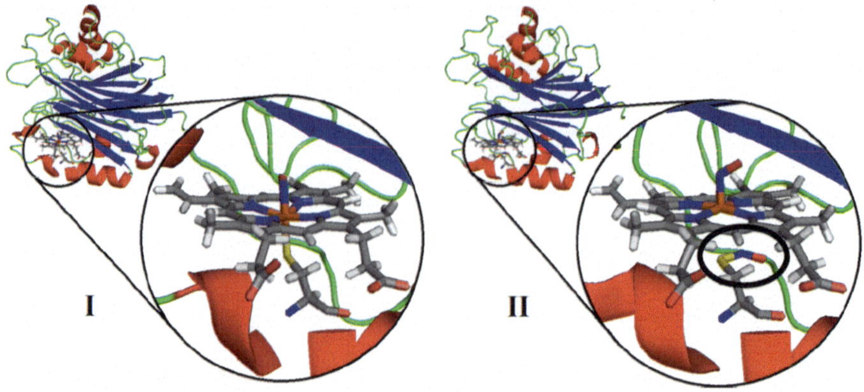

Fig. 4.19 Molecular viewgraph of cNP complexed with NO. The zoomed-in regions show the heme with the molecular bonds represented as sticks. The iron (*brown*) and the NO ligand on top of the iron with the nitrogen (*blue*) and the oxygen (red) are visible in the center of the zoomed-in regions. The structures were obtained via geometry optimization of the models of cNP–NO/pH 5.5 (Fe(III) **I**, *left*) and cNP–NO/pH 7.5 (Fe(II) **II** with Cys–SNO, *right*). In **II** the Cys–SNO is circled. Reprinted by permission from Springer-Nature: Hyperfine Interact. Copyright (2016) [78]

deprotonated heme carboxyl groups. These parameters are in reasonable agreement with the experimentally-determined ones (see above). What is even more important is that the characteristic change of the Mössbauer parameters upon reduction from Fe(III) to Fe(II) is fully reproduced by calculations using the structural models I and II which are displayed in Fig. 4.19.

4.5 Investigation of Iron-Sulfur Proteins

Mössbauer spectroscopy has contributed a lot to the understanding of structural and electronic properties of iron-sulfur centers in biological systems ranging from relatively simple $Fe–S_4$ sites in rubredoxins over 2Fe–2S and 4Fe–4S centers to far more complicated structures in hydrogenases and nitrogenases [79]. Mössbauer spectroscopic investigations of frozen protein samples at very low temperatures down to the liquid helium regime and the application of high magnetic fields up to 7 T and above lead to the identification of phenomena like electron delocalization over $Fe^{2.5+}$–$Fe^{2.5+}$ pairs e.g. in 3Fe–4S and 4Fe–4S proteins and allow the study of magnetic exchange interactions like Heisenberg and double exchange interactions [80].

4.5.1 *Identification of Fe–S-Centers by Mössbauer Spectroscopy*

In collaborations with the laboratory of Alfred X. Trautwein and colleagues from biology and biochemistry we have contributed to this field by the characterization of rubredoxin-type proteins both with the iron in the Fe^{2+}–S_4 and Fe^{3+}–S_4 state [81], the identification of a 2Fe–2S center in the FhuF protein which is part of a plant siderophore-reductase system [82], and the identification of unusual 4Fe–4S proteins with non-sulfur ligands like the radical S-adenosylmethionine enzyme coporphyrinogen III oxidase HemN [83]. High-field Mössbauer spectroscopy also helped to identify two essential diamagnetic 4Fe–4S centers in the evolutionarily highly conserved ATP-binding cassette protein ABCE1 [84].

As an example high field Mössbauer spectra obtained from rubredoxin-type single iron sites are displayed in Fig. 4.20. The Mössbauer spectra of the Fe^{3+}–S_4 site show a complex magnetic sextet pattern which arises from partial population of the three Kramers doublets of the $S = 5/2$ spin system [81]. The solid lines are spin Hamiltonian simulations calculated with $D = 2.13\ \text{cm}^{-1}$, $E/D = 0.23$, $\delta = 0.24\ \text{mms}^{-1}$, $\Delta E_Q = -0.36\ \text{mms}^{-1}$ as well as an almost axial A-Tensor of ~ −17 T. These parameters are typical for single high spin iron(III) sites in tetrahedral sulphur coordination. Although the function of rubredoxin has not been fully elucidated yet, it is assumed and highly likely that these proteins act as electron transfer proteins. In fact the iron site can accommodate a further electron and can be fully converted to a ferrous high spin Fe^{2+}–S_4 center. Fe^{2+}–S_4 centers show at liquid helium temperatures a doublet with $\delta = 0.7\ \text{mms}^{-1}$ and $\Delta E_Q = -3.25\ \text{mms}^{-1}$ (not shown here, but see [85]) which

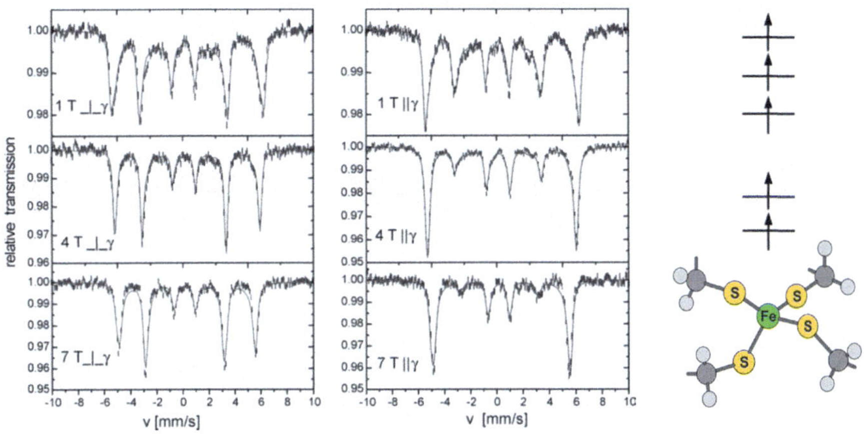

Fig. 4.20 Mössbauer spectra of a rubredoxin type protein with its iron being in the ferric high spin state. The solid lines are simulations using the spin Hamiltonian formalism for $S = 5/2$ and the following parameter set: $\overleftrightarrow{A}/g_N\mu_N = (-17, -15.7, -16)$ T, $\Delta E_Q = -0.36$ mm/s, $\eta = 0.8$, $\delta = 0.24$ mm/s, $D = 2.13\ \text{cm}^{-1}$, and $E/D = 0.23$. Reprinted by permission from Springer-Nature: Hyperfine Interact. Copyright (2004) [81]

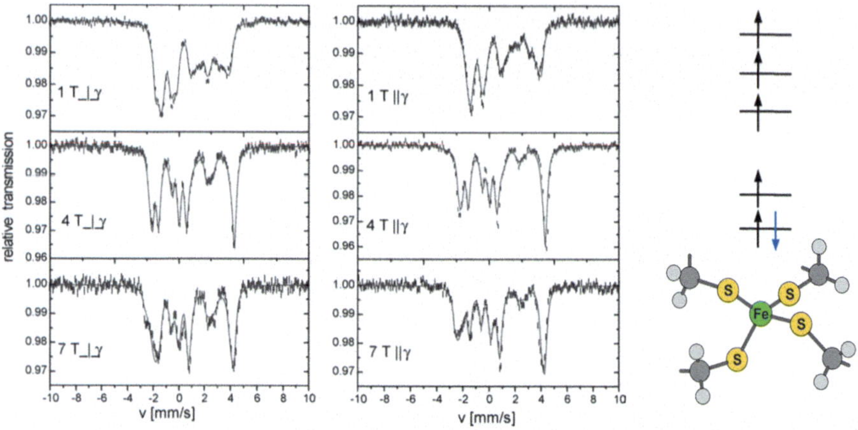

Fig. 4.21 Mössbauer spectra of a rubredoxin type protein in the ferrous high spin state. The solid lines are simulations using the spin Hamiltonian formalism for $S = 2$ and the following parameter set: $\overleftrightarrow{A}/g_N\mu_N = (-14.5, -9.2, -27.5)$ T, $\Delta E_Q = -3.25$ mms^{-1}, $\eta = 0.74$, $\delta = 0.7$ mms^{-1}, $D = 7.2$ cm^{-1}, and $E/D = 0.16$. Reprinted by permission from Springer-Nature: Hyperfine Interact. Copyright (2004) [81]

by the application of external fields can lead to a more and more magnetically split Mössbauer pattern as depicted in Fig. 4.21.

The difference in the development of magnetic splitting in the case of ferric and ferrous high spin centers can be understand under the consideration of the field dependence of the spin expectation values of the $S = 5/2$ versus the $S = 2$ spin systems. For most if not all $S = 5/2$ systems the spin expectation values do almost saturate even under very small fields. Therefore each Kramers doublet causes a magnetically split sextet for the S = 5/2 systems. Note that this holds only for the slow relaxation limit. In the fast relaxation limit only one sextet is observable. There is another situation which can influence the relaxation rate of the electronic spin *S*. When the temperature is raised spin lattice relaxation is more and more effective so that the slow relaxation limit does not apply and considerable line broadening is observed. Or, for example in a powder sample of a small synthetic ferric high spin chemical complex the spin–spin relaxation between two iron containing molecules becomes effective which can also lead to an acceleration of the spin relaxation rate. In this case also considerable line broadening can be observed.

For the $S = 2$ case the spin expectation values do—very often—not saturate at small fields and thus one can observe a more and more magnetically split pattern with increasing external field as is shown in Fig. 4.21.

To our experience it is a good experimental strategy to acquire field dependent Mössbauer spectra at very low temperature—say at 4.2 K—in order to avoid considerably influence of relaxation effects on the Mössbauer spectra. By simultaneous fitting of the experimental data using the spin Hamiltonian formalism it is than possible to determine *D*, *E/D* and the hyperfine coupling tensor. The quadrupole splitting ΔE_Q

and the isomer shift δ can—as a first input for the simulation be taken from measurements in zero field—this holds for the S = 2 case—or from measurements taken at higher T for those paramagnetic centers where magnetic splitting is observed at 4.2 K. For δ one can use e.g. measurements at T = 77 K since the second order Doppler shift between 77 and 4.2 K is often negligible, nevertheless a temperature dependence of ΔE_Q can occur due to population of excited orbital states.

Quite a lot of Fe–S proteins have 2Fe–2S iron centers. These centers occur in the $[2Fe–2S]^{2+}$ state and can be reduced to the $[2Fe–2S]^{1+}$ state. Thus, also 2Fe–2S centers support electron transfer in proteins. A typical example are ferredoxins. These proteins have 2Fe–2S centers which are coordinated to the protein via 4 cystein residues and which are connected via a disulfide bridge. There are also 2Fe–2S proteins which have one iron site coordinated to two histidine ligands, the most well-known of them being the Rieske-cluster of the bc_1 complex [86]. Very recently a bishistidinyl-coordinated non-Rieske [2Fe–2S] protein was identified also by means of Mössbauer spectroscopy [87]. Figure 4.22 shows the Mössbauer spectrum of the wild-type protein APD showing two doublets. One doublet with $\delta_1 = 0.24$ mms^{-1} is typical for a ferric FeS_4 center whereas the other doublet has a slightly higher isomer shift of $\delta_2 = 0.35$ mms^{-1}. This points to a coordination with two sulfurs and two nitrogen ligands from histidin like in the Rieske potein (FeS_2N_2). The fact that APD has also a diferric center can be interfered by the analysis of a Mössbauer spectrum obtained in a high external field (Fig. 4.22b). The so observed Mössbauer pattern can be reproduced by a simulation which assumes that the total spin of the $[2Fe–2S]^{2+}$ center is zero. This can be explained by an antiparallel spin coupling of two ferric high spin sites as depicted in Fig. 4.22.

After reduction of the $[2Fe–2S]^{2+}$ center to a $[2Fe–2S]^{1+}$ center by one electron the Mössbauer spectrum at low field changes completely (Fig. 4.22c). Now already at low fields a magnetic splitting can be observed. From EPR spectroscopy one deduces that the total spin of the $[2Fe–2S]^{1+}$ center is $S = 1/2$. This spin state together with the g-tensor components obtained from EPR can be used as input for the spin Hamiltonian simulations which are drawn as solid curves in Fig. 4.22c, d. The resulting total spin, which often also is called system spin $S = 1/2$ originates from an antiparallel spin coupling of the ferric high spin $S_1 = 5/2$ and the ferrous high spin $S_2 = 2$. For a more in deep discussion of the spin coupling between mixed valent iron sulfur sites we refer to reviews [34, 35] and the references given therein.

Figure 4.23 displays the Mössbauer spectra of an iron sulfur protein having a $[4Fe–4S]^{2+}$ center. Also these clusters are quite common in iron sulfur proteins. In the majority of cases known to date these also function as electron transfer centers, but in some 4Fe–4S proteins the cluster itself is also catalytically active as shown in Sect. 5.2. The protein the Mössbauer spectrum of which is displayed in Fig. 4.23 is called PhrB [88] and it belongs to the cryptochrome/photolyase family (CPF). The function of PhrB has been shown to be related to DNA repair.

Despite there are four iron sites in the $[4Fe–4S]^{2+}$ cluster there is only one doublet in the spectrum of the ^{57}Fe-enriched PhrB both at T = 77 K (Fig. 4.23a) and T = 4.2 K (Fig. 4.23b) with $\delta = 0.42$ mms^{-1} and $\Delta E_Q = 1.26$ mms^{-1} (Fig. 4.23a). This shows that all four iron sites are equivalent and have the same coordinated ligands. The

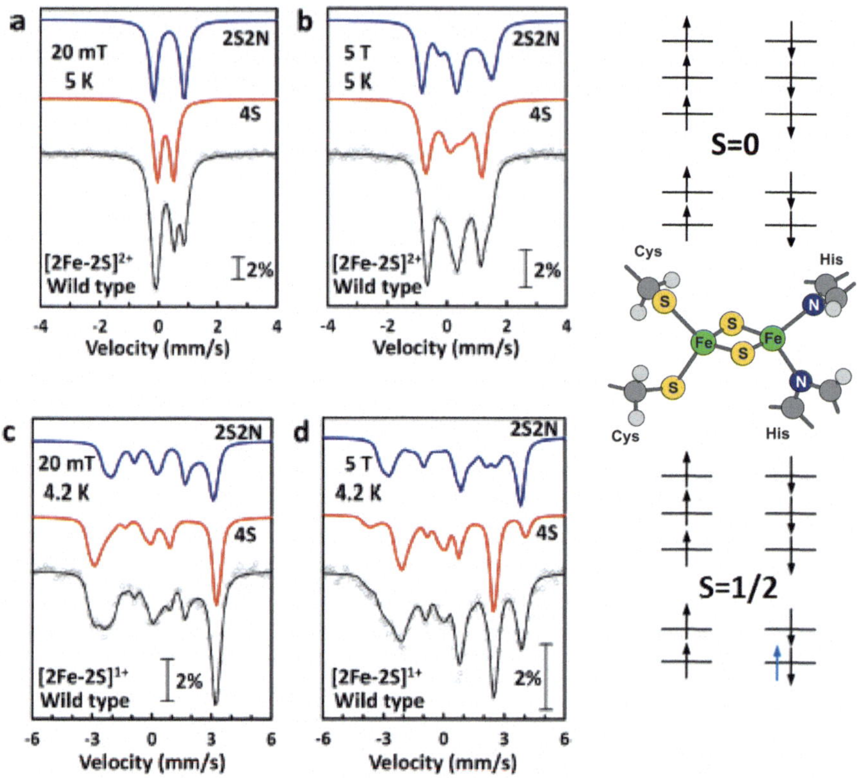

Fig. 4.22 Mössbauer spectra taken at 5 K of the native bishistidinyl-coordinated non-Rieske protein APD having a diamagnetic $[2Fe–2S]^{2+}$ cluster obtained in a field of 20 mT **a** and a field of 5 T (**b**) applied parallel to the γ-ray. The solid lines are simulations using the total spin $S = 0$, $\delta_1 = 0.24$ mms^{-1}, $\Delta E_{Q1} = -0.54$ mms^{-1}, $\eta_1 = 0.6$, (red) and $\delta_2 = 0.35$ mms^{-1}, $\Delta E_{Q2} = +1.06$ mms^{-1}, $\eta_2 = 0.6$ (blue). Reduction with dithionite yields a protein with a paramagnetic $[2Fe–2S]^{1+}$ center showing magnetic splitting both even at a low field of 20 mT **c** and at 5 T **d**. Now the simulations are calculated with the total spin $S = 1/2$, $\delta_1 = 0.32$ mms^{-1}, $\Delta E_{Q1} = 0.81$ mms^{-1}, $\eta_1 = 0$, $\overleftrightarrow{A}_1/g_N\mu_N$ = (−36.3, −41.8, −31) T, (red) and $\delta_2 = 0.75$ mms^{-1}, $\Delta E_{Q2} = -3.16$ mms^{-1}, $\eta_2 = 3$, $\overleftrightarrow{A}_2/g_N\mu_N$ = (16.5, 8.4, 24.9) T, (blue). The g-tensor $\overleftrightarrow{g}$ = (1.861, 1.906, 2.009) was from complimentary EPR measurements. Reprinted with permission from [87]. Copyright (2019) American Chemical Society

isomer shift is just in between that of the δ-values of a Fe^{3+}–S_4 and a Fe^{2+}–S_4 center. This has been interpreted as a consequence of a quantum mechanical delocalization of the excess electron in a mixed valent Fe^{2+}–Fe^{3+} pair (see scheme in Fig. 4.23). Since the Pauli principle has to be obeyed such an electron delocalization can only occur if the iron spins of the two sites are aligned in a parallel fashion resulting in a pair spin of 9/2. When PhrB with its $[4Fe–4S]^{2+}$ cluster is subjected to a high external field a magnetic splitting is observed which is only due to the external magnetic field. This is shown by the simulation given in Fig. 4.23c using the spin Hamiltonian analysis

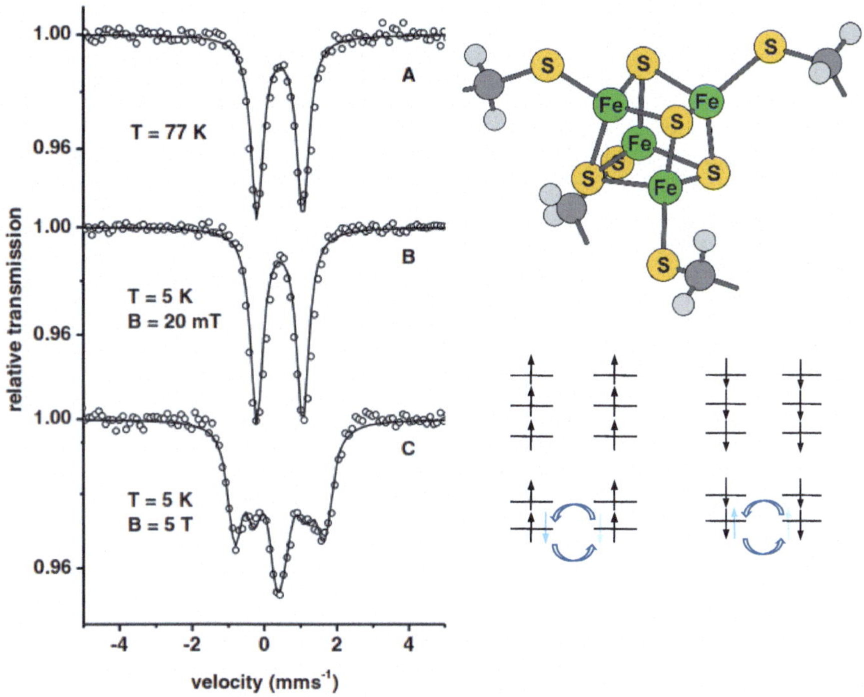

Fig. 4.23 Mössbauer spectra of PhrB taken at (**a**) T = 77 K, **b** T = 5 K with an external field of 20 mT and **c** T = 5 K with an external magnetic field of 5 T. The magnetic fields were applied perpendicular to the γ-beam. The solid lines are spin Hamiltonian simulations assuming a diamagnetic ground state and all four iron being equivalent. PhrB has a $[4Fe–4S]^{2+}$ cluster the structural view is given on the right. The diamagnetic ground state can be rationalized as follows: An electron is delocalized between two mixed valent iron ions, resulting in a mean charge of $Fe^{2.5+}$. The two $Fe^{2.5+}$–$Fe^{2.5+}$ pairs have a spin of $S = 9/2$ each, but those couple antiparallel to each other and the total spin becomes zero. Mössbauer data reprinted by permission from Springer-Nature: Hyperfine Interact. Copyright (2013) [88]

assuming a diamagnetic $S = 0$ ground state. Such behavior is characteristic for a diamagnetic $[4Fe–4S]^{2+}$ cluster. The cluster possesses two mixed-valence iron pairs with two excess electrons each one delocalized over one pair ($Fe^{2.5+}$–$Fe^{2.5+}$). The spins of the two $Fe^{2.5+}$–$Fe^{2.5+}$ pairs $S_{12} = 9/2$ and $S_{34} = 9/2$ are antiparallel coupled, yielding a total cluster spin $S = 0$.

4.5.2 Exploration of the Unusual 4Fe–4S Center of the LytB Protein

The LytB protein also called isoprenoid synthase (IspH) is an air sensitive iron sulfur enzyme which is crucial for the biosynthesis of isoprenoids in many bacteria and in the

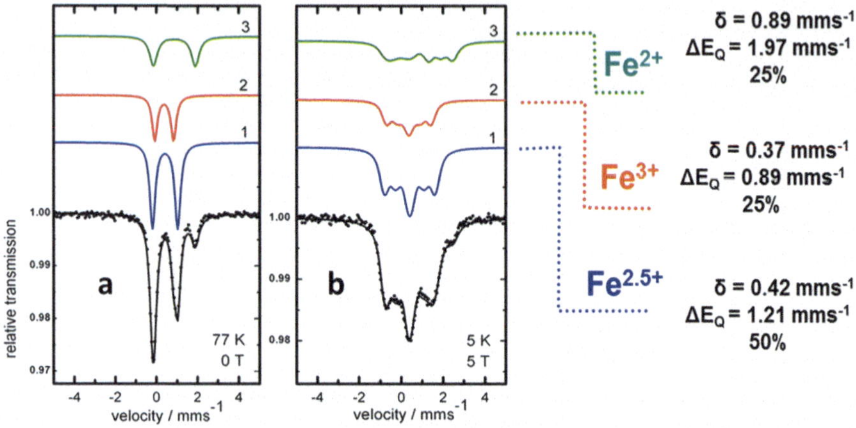

Fig. 4.24 Mössbauer spectra of LytB (IspH) taken at 77 K with no applied field **a** and at 5 K in a field of 5 T. The solid lines are spin Hamiltonian simulations assuming a diamagnetic $[4Fe–4S]^{2+}$ cluster. Reprinted with permission from [90]. Copyright (2009) American Chemical Society

malaria parasite *Plasmodium falciparum.* In these pathogens, isoprenoid synthesis occurs according to the methylerythritol phosphate (MEP) pathway, an alternative to the mevalonate pathway in higher organisms [89]. LytB catalyses the last step of the MEP pathway. After initial crystal structure data indicate that this enzyme has an 3Fe–4S cluster Mössbauer spectroscopy clearly showed that LytB indeed harbours a 4Fe–4S center as the active site [90]. Unlike other 4Fe–4S proteins the cluster of LytB has a unique site which serves has a substrate binding site for (E)-4-hydroxy-3-methylbut-2-en-1-yl diphosphate (HMBPP) in order to initiate HMBPP conversion into a mixture of isopentenyl diphosphate (IPP) and dimethylallyl diphosphate (DMAPP) [91].

The inspection of the Mössbauer spectrum of LytB shown in Fig. 4.24 indicates the presence of a component with $\delta = 0.89$ mms^{-1} and $\Delta E_Q = 1.97$ mms^{-1}. These parameters are typical for high spin iron(II) and one could easily assume the presence of non-protein bound iron(II) in the sample under investigation. In this case a Mössbauer spectrum taken at high field would induce spin expectation values and a magnetically split pattern would be expected similar to the case of reduced rubredoxin (see Fig. 4.21). However, for LytB the Mössbauer spectrum taken at high fields could be simulated with the spin Hamiltonian formalism under the assumption that all components are diamagnetic. This can only be explained by the fact that the ferrous high spin site is part of an unusual 4Fe–4S cluster with one $Fe^{2.5+}$–$Fe^{2.5+}$ pair having a spin of 9/2 and another more valence trapped "Fe^{2+}–Fe^{3+}" pair having also a spin of 9/2. An antiparallel spin arrangement than causes the diamagnetism of the $[4Fe–4S]^{2+}$ cluster of LytB.

When the chemical substrate HMBP is added to the LytB protein the Mössbauer spectrum chances significantly (Fig. 4.25). An analysis of the data displayed in Fig. 4.25b indicates that only the doublet representing the special iron(II) site is

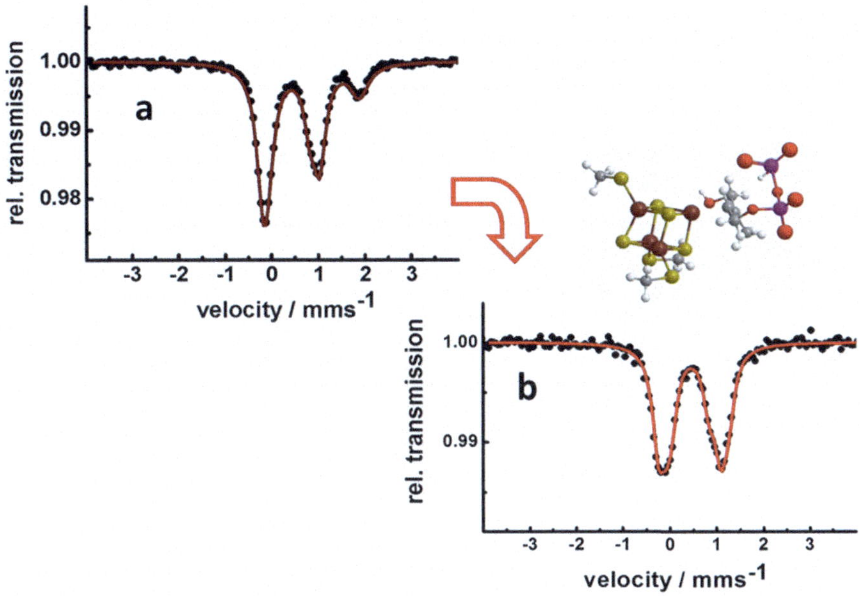

Fig. 4.25 Mössbauer spectrum of LytB (IspH) **a**. The change in the pattern of the Mössbauer spectrum obtained after substrate HMBP (see inset) addition (**b**) shows that the substrate is coordinating to the unique fourth iron site of LytB. Adapted with permission from [90]. Copyright (2009) American Chemical Society

affected after substrate binding. This clearly shows that the special iron(II) site of LytBs $[4Fe–4S]^{2+}$ cluster is the substrate binding site. Moreover, Mössbauer spectroscopy could also show complementary to protein crystallography [92] and EPR spectroscopy [91] that the special iron(II) site of the $[4Fe–4S]^{2+}$ cluster of LytB is also a binding site of enzyme inhibitors which are potential candidates for new antibiotics since this biochemical pathway does not exist in higher organisms [93].

4.5.3 In Vivo Mössbauer Spectroscopy of Iron-Sulfur Proteins Inside E. Coli *Cells*

The investigation of whole cells with Mössbauer spectroscopy has shown that the uniqueness of the substrate binding site in LytB is not a consequence of treatments during the purification process (Fig. 4.26). Via subtraction of Mössbauer spectra from LytB overexpressing *E.coli* cells and spectra which originate from cells which do not overexpress LytB the spectral signature of the LytB protein could be isolated and compared to the data obtained from the purified protein [90]. Spectral subtraction procedures for the identification of iron centers in complicated mixtures of several iron species do not necessarily require Mössbauer spectra obtained at high fields.

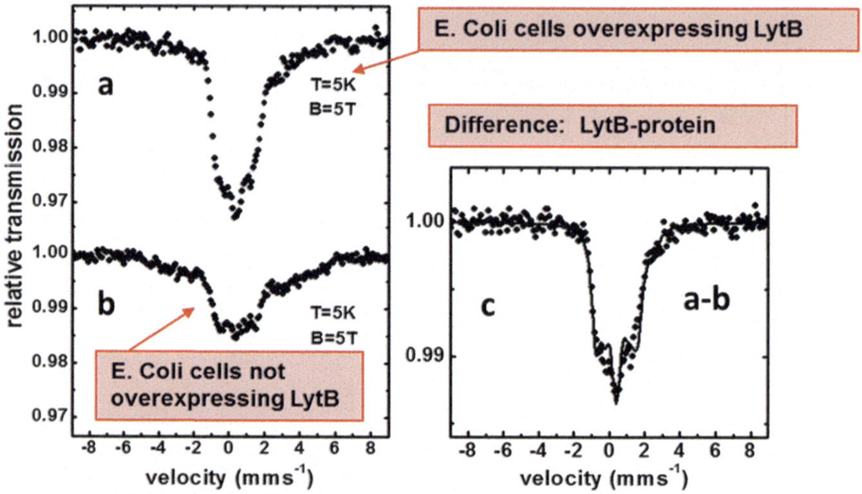

Fig. 4.26 High-field Mössbauer spectra of LytB overexpressing *E. coli* cells *E. coli* M15 [pREP4, pQE30-LytB] **a** and of cells which do not overexpress LytB (*E. coli* M15 [pREP4]) **b**. The difference spectrum is shown in (**c**). The solid line represents a simulation with the parameters of pure LytB [90]. The small deviation from the simulation might be caused by a different iron(III) content, probably ferritin, within the two samples. Adapted with permission from [90]. Copyright (2009) American Chemical Society

However, such experiments as shown in Fig. 4.26 can be used to support conclusions drawn from low field Mössbauer spectra.

4.6 Future Applications of the Mössbauer Effect in Chemistry and Biology

Since decades Mössbauer spectroscopy has been an indispensable spectroscopic tool for the determination of electronic properties like iron oxidation numbers and spin states of iron centers in proteins. Since there are a thousands of iron proteins many of them not fully understood there will be a continuing demand of conventional Mössbauer spectroscopy on these systems. Still even new discoveries with respect to iron storage and iron metabolism are made which await characterization by Mössbauer spectroscopy. One example is an iron storage protein in anaerobic microorganisms that therefore cannot use oxygen to oxidize ferrous iron. Very recently it was shown that the archaeon *Pyrococcus furiosus*, which grows optimally near 100 °C in hydrothermal marine vents stores iron in form of thioferrate protein nanoparticles near their mitochondria [94]. This protein is called IssA and forms large nanoparticles with diameters up to 300 nm and is one of the largest natural metalloprotein complex known to date.

More and more understanding will be gathered by studying complex systems which have several iron species like paramagnetic iron ions, agglomerates of a small number of iron sites or even clusters with hundreds of iron atoms or more. The evaluation of temperature and field dependent Mössbauer spectra by combining different evaluation routes like e.g. simple lorentzian line shape analysis, fitting with magnetic hyperfine field distributions and spin Hamiltonian analysis, the latter being very important as far as paramagnetic iron centers are concerned have been used to understand complex systems like e.g. heterogeneous catalysts. In these systems the identification of active iron-sites is a demanding task. Such supported catalysts contain iron in many phases, e.g. iron oxide nanoparticles, as single ions, but also agglomerates of only a few iron centers.

Iron loaded zeolites (e.g. Fe–ZSM-5) are potent candidates for the catalytic abatement of nitrogen oxides from car exhaust, e.g. from Diesel engines. This catalytic process, the reduction of nitrogen oxides (NO_x) to N_2 by NH_3 in presence of oxygen is called "selective catalytic reduction, SCR"). Recently Fe–ZSM-5 catalysts prepared via solid-state ion exchange have been investigated by using field dependent Mössbauer spectroscopy at low temperature in order to identify the different iron species present in this type of catalyst in the fresh state and after use in catalysis or more precisely before and after standard SCR (Fig. 4.27) [95]. Mössbauer spectroscopy proves to be the key technique for a full understanding of the iron species, but due to the complexity of structures, guidance by complimentary methods like EPR experiments and control by SQUID magnetometry were essential to prove reliability of derived species distributions. The combination with complimentary spectroscopic techniques like EPR spectroscopy allowed quantifying the amount of paramagnetic iron sites in Fe–ZSM-5 and the detection of EPR-silent diamagnetic iron species like μ-oxo-bridged dinuclear iron sites even in these highly complex systems. The simulations in Fig. 4.27 allow to identify and quantify the relative amount of the different iron species: K1 is assigned to iron oxide/hydroxide nanoparticles, K2 represents diamagnetic most probably μ-oxo bridged dinuclear iron(III) centers, K3 are paramagnetic iron(II) high spin sites and K4 and K5 are due to single high spin ferric sites. A similar methodical approach has been used very recently in order to elucidate the structural composition of an Fe–N–C catalyst related to fuel cell batteries [96].

Synchrotron based NFS will allow to obtain Mössbauer parameters of small species like micro- and nanocrystals or even films of biological cells which have been feeded with ^{57}Fe containing nutrients. Today the maximum foci which can be achieved for e.g. 14.4 keV with meV energy resolution are in the range of some micrometers. With the help of the now developed diffraction limited synchrotron sources it can be expected that nanofocussing can be done. However, care must be taken, because the investigation of single micro- and nano-objects can also be influenced by radiation damage which needs to be investigated expecially for biological systems.

Synchrotron based NIS has added the possibility to trace down iron centred molecular modes in molecular switches like SCO materials and proteins. Only some few examples have been discussed in this contribution. These applications will certainly

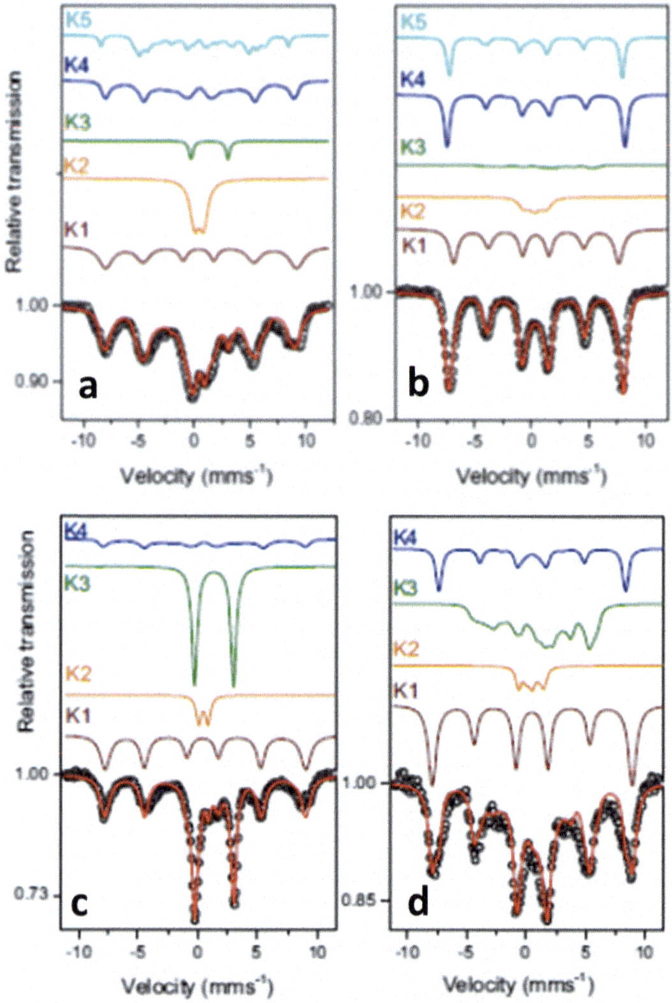

Fig. 4.27 An example of a spectral analysis using lorentzian lines and the spin Hamiltonian formalism for para- and diamagnetic iron sites in a complex system. The Mössbauer spectra of iron loaded ZSM-5 before SCR were obtained at $T = 5$ K and $B = 10$ mT **a** and $B = 5$ T **b**. The red solid line is the result of a simulation procedure using 5 different iron species K1–K5 with parameters given in Ref. [95]. K4 and K5 represent $S = 5/2$ species with $E/D = 0$ and $E/D = 0.33$ as indicated by complimentary EPR experiments. After SCR Mössbauer spectra obtained at $T = 5$ K and $B = 10$ mT **c** and $B = 5$ T **d** show 4 different iron species K1–K4 as described in the text. The relative contribution of K1–K4 including the obtained Mössbauer parameter sets are given in Table 4.1 of Ref. [95]. Reprinted by permission from Springer-Nature: Hyperfine Interact. Copyright (2017) [95]

continue. Especially very complex iron centers like in the N_2 fixing protein nitrogenases have been investigated already, but how biological nitrogen fixation really works has not been clarified yet and NIS is certainly one of the spectroscopic methods which can shine light on the N_2 binding to the active site FeMo-cofactor of nitrogenase.

But there are also novel experimental technique like time dependent optical pump NRS probe techniques. With such experiments it will be possible to monitor the response of optical excitable iron containing systems after photostimulation like in molecular switches and magneto optical nanomaterials [97] or possibly even in iron proteins. With X-ray free electron lasers delivering fs pulses of 14.4 keV radiation on the horizon optical pump NIS probe experiments would enable to trace ultrafast changes of the vibrational properties of iron ligand modes in chemical complexes and proteins even in the fs time range.

4.7 Summary

The proceeding chapter tries to give a flavor to the reader of what can be achieved with conventional and synchrotron-based Mössbauer spectroscopic techniques, both in biology and chemistry. Since this contribution is meant as an introduction to the broad field of chemical and biological applications of the Mössbauer effect, it is clear that not all aspects of the field could be covered. Instead the reader is guided after some practical aspects both in sample requirements and in evaluation techniques through examples of scientific cases, which have been worked on in the group of the author during the last two decades. These concern Mössbauer spectroscopic studies of synthetic chemical complexes like SCO compounds and studies on NO transporter heme proteins, the nitrophorins. Since Mössbauer spectroscopy is the method of choice for the characterization of Fe–S centers in proteins, especially when they occur in their EPR silent diamagnetic or integer-spin states, some examples are presented starting from simple mononuclear Fe–S centers over dinuclear Fe_2S_2 centers to Fe_4S_4 centers. More complex cases like Fe–S centers in nitrogenases are only briefly mentioned since the discussion of such complex structures like the FeMo-Cofactor are behind the scope of the text. Nevertheless, also some views on the future of Mössbauer spectroscopy in chemistry and biology are presented. The author hopes that the reader at the end is getting an impression about the intriguing possibilities of the method. Considering the fact that the first biological and chemical applications of the Mössbauer effect have been performed in the 1960s, a lot has been learned since that time. On this footing there is a bright perspective of the technique especially when it comes to the study of complex systems like iron containing reaction mixtures, heterogeneous catalysts and biological cells.

Many older Mössbauer spectroscopists consider the 1970s and 80s as the golden years of Mössbauer spectroscopy. At least to the opinion of the author the method has bright future and the really shining golden years may still be ahead of us, especially since novel concepts for the use of synchrotron radiation and XFELs in nuclear resonance are on the horizon.

Acknowledgements The experimental and theoretical work presented in this chapter has been performed by the members of the Biophysics and Medical Physics group at the Department of Physics at TU Kaiserslautern and the many collaborators of the group. Special thanks goes to Dr. H. Auerbach and Dr. A. Reinhardt for their support with the drawing of several figures. The author would like to thank Dr. J. A. Wolny for his continuing work in the field of SCO research and quantum chemical calculations. Several students contributed to the publications referenced in this chapter. In addition to Dr. H. Auerbach and Dr. A. Reinhardt mentioned already above, I want to name Dr. B. Moeser, Dr. A. Janoschka, Dr. S. Rackwitz, Dr. I. Faus and Dr. T. O. Bauer. On behalf of the many outstanding biochemists and chemists with whom the author has worked over the last two decades, special thanks are due to Prof. F. Ann Walker for her continuous support and guidance.

This work is dedicated in memoriam of two longtime companions. My academic teacher and mentor Prof. Dr. A. X. Trautwein as well as to one of his students with whom I had the pleasure to work with, Dr. P. Wegner, who passed away much too early.

References

1. G. Hauska, M. Schütz, M. Büttner, in *Oxygenic Photosynthesis: The Light Reactions*. ed. by D.R. Ort, C.F. Yocum (Kluwer Academic Publishers, Dordrecht, The Netherlands, 1996), pp. 377–398
2. D.F.V. Lewis, *Cytochromes P450—Structure, Function and Mechanism* (Taylor & Francis Ltd., London, 1996)
3. A. Trautwein, Struct. Bond. **20**, 101–167 (1975)
4. P.G. Debrunner, in *Iron Porphyrins*. ed. by A.B.P. Lever, H.B. Gray (Part III. VCH Publishing, Heidelberg, 1989), pp. 139–234
5. E.I. Solomon, T.C. Brunold, M.I. Davis, J.N. Kemsley, S.K. Lee, N. Lehnert, F. Neese, A.J. Skulan, Y.S. Yang, J. Zhou, Chem. Rev. **100**, 235–349 (2000)
6. A.C. Rosenzweig, C.A. Frederick, S.J. Lippard, P. Nordlund, Nature **366**, 537–543 (1993)
7. B.M. Sjöberg, Struct. Bond. **88**, 139–173 (1997)
8. R.H. Holm, P. Kennepohl, E.I. Solomon, Chem. Rev. **96**, 2239–2314 (1996)
9. P.M. Harrison, T.H. Lilley, in T.M. Loehr (ed.), Physical bioinorganic chemistry 5, in *Iron Carriers and Iron Proteins* (VCH-Publishers, New York, 1989), pp. 123–238
10. B.F. Matzanke, in *Transition Metals in Microbial Metabolism*. ed. by G. Winkelmann, C.J. Carrano (Harwood Academic Publishers, New York, 1997), pp. 117–157
11. A.S. Reinhardt, Mössbauer-Spektroskopie und Dichte-funktional-Theorie-Untersuchungen an Modellkomplexen der [Fe]-Hydrogenase und dem Protein LytB, Dissertation TU Kaiserslautern, zugleich Shaker Verlag GmbH Aachen, (2012)
12. H.L. Schläfer, G. Gliemann, *Einführung in Die Ligandenfeldtheorie* (Akademische Verlagsgesellschaft, Frankfurt Am Main, 1980)
13. V. Schünemann, A.X. Trautwein, J. Illerhaus, W. Haehnel, Biochemistry **38**, 8981–8991 (1999)
14. V. Schünemann, A.M. Raitsimring, R. Benda, A.X. Trautwein, T.K. Shokireva, F.A. Walker, JBIC **4**, 708–716 (1999)
15. G. Simmoneaux, V. Schünemann, C. Morice, L. Carel, L. Toupet, H. Winkler, A.X. Trautwein, F.A. Walker, J. Am. Chem. Soc. **122**, 4366–4377 (2000)

16. H. Keutel, I. Käpplinger, E.G. Jäger, M. Grodzicki, V. Schünemann, A.X. Trautwein, Inorg. Chem. **38**, 2320–2327 (1999)
17. V. Schünemann, M. Gerdan, A.X. Trautwein, N. Haoudi, D. Mandon, J. Fischer, R. Weiss, A. Tabard, R. Guillard, Angew. Chem. Int. Ed. **38**, 3181–3183 (1999)
18. C. Schulz, P.W. Devaney, H. Winker, P.G. Debrunner, N. Doan, R. Chiang, R. Rutter, L.P. Hager, FEBS Lett. **103**, 102–105 (1979)
19. C. Schulz, R. Rutter, J.T. Sage, P.G. Debrunner, L.P. Hager, Biochemistry **23**, 4743–4754 (1984)
20. W. Kaim, B. Schwederski, *Bioanorganische Chemie* (Teubner Taschenbücher Chemie, Stuttgart, 1995)
21. S.J. Lippard, J.M. Berg, in *Principles of Bioinorganic Chemistry* (University Science Books, Mill Valley CA 94941, USA, 1994)
22. V. Schünemann, H. Paulsen, Mössbauer spectroscopy, in *Application of Physical Methods to Inorganic and Bioinorganic Chemistry*, ed. By R.A. Scott, C.M. Lukehart (John Wiley and Sons, Ltd., 2007)
23. P. Gütlich, E. Bill, A.X. Trautwein, See Chapter 4 in: *Mössbauer Spectroscopy and Tansition Metal Chemistry* (Springer, Berlin, 2011)
24. L. Zecca, M. Gallorini, V. Schünemann, A.X. Trautwein, M. Gerlach, P. Riederer, P. Vezzoni, D. Tampellini, J. Neurochem. **76**, 1766 (2001)
25. G. Lang, W. Marshall, Proc. Phys. Soc. **87**, 3 (1966)
26. https://www.wissel-gmbh.de/index.php?option=com_content&task=view&id=55&Itemid=116
27. https://mosswinn.com/
28. https://www.wmoss.org/
29. H.P. Gunnlaugsson, Spreadsheet based analysis of Mössbauer spectra. Hyperfine Interact. **237**, 79 (2016)
30. P. Gütlich, E. Bill, A.X. Trautwein, See Chapter 2.6 in: *Mössbauer Spectroscopy and Tansition Metal Chemistry* (Springer, Berlin, 2011)
31. C. Le Caer, J.M. Dubois, J. Phys. E: Sci Instrum. **12**, 1083 (1979)
32. A. Abragam, M.H.L. Pryce, Proc. R. Soc. A **250**, 135 (1951)
33. H.H. Wickman, M.P. Klein, D.A. Shirley, Phys. Rev. **152**, 345–357 (1966)
34. A.X. Trautwein, E. Bill, E.-L. Bominaar, H. Winkler, Struct. Bond. **78**, 1–95 (1991)
35. V. Schünemann, H. Winkler, Rep. Prog. Phys. **63**, 263–353 (2000)
36. H. Winkler, X.Q. Ding, M. Burkhardt, A.X. Trautwein, Hyperfine Interact. **91**, 875 (1994)
37. N. Blume, J.A. Tjon, Phys. Rev. **165**, 446 (1968)
38. E. Münck, J.L. Groves, T.A. Tumolillo, P.G. Debrunner, Comput. Phys. Commun. **5**, 225–238 (1973)
39. F. Neese, WIREs Comput. Mol. Sci. **8**, e1327 (2018)
40. https://gaussian.com/
41. https://www.turbomole.org/
42. W. Koch und M. C. Holthausen, A chemist's guide to density functional theory, (Wiley, 2001)
43. https://www.rcsb.org/
44. https://www.ccdc.cam.ac.uk/
45. P. Gütlich, E. Bill, A.X. Trautwein, See Chapter 5 by F. Neese and T. Petrenko in: Mössbauer Spectroscopy and Tansition Metal Chemistry (Springer, Berlin, 2011)
46. F. Clemente, T. Vreven, M.J. Frisch, in *C.* ed. by Q. Biochemistry (Matta Wiley VCH, Weinheim, 2010), pp. 61–84
47. P. Gütlich, H.A. Goodwin, Spin crossover—an overall perspective, in *Spin Crossover in Transition Metal Complexes. Topics in Current Chemistry*, vol. 233, 1, ed. by P. Gütlich, H.A. Goodwin (Springer, Berlin, 2004)
48. O. Kahn, C.J. Martinez Sci. **279**, 44 (1998)
49. E. Ruiz, Phys. Chem. Chem. Phys. **16**, 14 (2014)
50. T.G. Gopakumar, F. Matino, H. Naggert, A. Bannwarth, F. Tuczek, R. Berndt, Angew. Chem. Int. Ed. **51**, 6262 (2012)
51. J.A. Wolny, R. Diller, V. Schünemann, Eur. J. Inorg. Chem. 2635–2648 (2012)

52. P. Gütlich, A. Hauser, H. Spiering, Angew. Chem. Int. Ed. **13**, 2024–2054 (1994)
53. P. Gütlich, H.A. Goodwin (eds.), *Spin Crossover in Transition Metal Complexes. Topics in Current Chemistry*, vol. 233, 1 (Springer , Berlin, 2004)
54. P. Gütlich, E. Bill, A.X. Trautwein, in *Mössbauer Spectroscopy and Tansition Metal Chemistry* (Springer, Berlin, 2011)
55. A.-M. Li, T. Hochdörffer, J.A. Wolny, V. S., E. Rentschler, Magnetochemistry **4**(3), 34 (2018)
56. B. Schäfer, T. Bauer, I. Faus, J. A. Wolny, F. Dahms, O. Fuhr, S. Lebedkin, H. C. Wille, K. Schlage, K. Chevalier, F. Rupp, R. Diller, V. S., M. M. Kappes, M. Ruben, Dalton Trans. **46**, 2289–2302 (2017)
57. A. Janoschka, G. Svenconis, V. Schünemann, J. Phys. Conf. Ser. **217**, 012005 (2010)
58. T. Bauer, A. Omlor, H. Auerbach, K. Jenni, B. Schäfer, R. Diller, M. Ruben, V. Schünemann, Hyperfine Interact. **238**, 98 (2017)
59. S. Stankov, R. Röhlsberger, T. Ślęzak, M. Sladecek, B. Sepiol, G. Vogl, A.I. Chumakov, R. Rüffer, N. Spiridis, J. Łażewski, K. Parliński, J. Korecki, Phys. Rev. Lett. **99**, 185501 (2007)
60. G. Félix, M. Mikolasek, H. Peng, W. Nicolazzi, G. Molnár, A.I. Chumakov, L. Salmon, A. Bousseksou, Phys. Rev. B **91**, 024422 (2015)
61. S. Rackwitz, I. Faus, B. Lägel, J. Linden, J. Marx, E. Oesterschulze, K. Schlage, H.-C. Wille, S. Wolff, J.A. Wolny, V. Schünemann, Hyperfine Interact. **226**, 667 (2014)
62. B. Schäfer, J.-F. Greisch, I. Faus, T. Bodenstein, I. Salitros, O. Fuhr, K. Fink, V. Schünemann, M.M. Kappes, M. Ruben, Angew. Chem. Int. Ed. **55**, 10881–10885 (2016)
63. Y.V. Shvyd'ko, Hyperfine Interact. **125**, 173–188 (2000)
64. The Nobel Prize in Physiology or Medicine 1998. NobelPrize.org. Nobel Media AB 2019. Sun. 1 Dec 2019. https://www.nobelprize.org/prizes/medicine/1998/summary/
65. J.M.C. Ribeiro, J.M.H. Hazzard, R. Nussenzveig, D. Champagne, F.A. Walker, Science **260**, 539 (1993)
66. B. Moeser, A. Janoschka, J.A. Wolny, H. Paulsen, I. Fillipov, R.E. Berry, H. Zhang, A.I. Chumakov, F.A. Walker, V. Schünemann, J. Am. Chem. Soc. **134**(9), 4216–4228 (2012)
67. R.L. Collins, J. Chem. Phys. **42**(3), 1072–1080 (1965)
68. W.T. Oosterhuis, G.J. Lang, Chem. Phys. **50**, 4381–4387 (1969)
69. C.P.S. Taylor, Biochim. Biophys. Acta **491**, 137–149 (1977)
70. C. Keppler, K. Achterhold, A. Ostermann, U. van Bürck, A.I. Chumakov, R. Rüffer, W. Stuhrhahn, E.E. Alp, F.G. Parak, Eur. Biophys. J. **29**, 146 (2000)
71. F.G. Parak, Rep. Prog. Phys. **66**, 103 (2003)
72. C. Herta, H. Winkler, R. Benda, M. Haas, A.X. Trautwein, Eur. Biophys. J. **31**, 478 (2002)
73. J.T. Sage, S.M. Durbin, W. Stuhrhahn, D.C. Wharton, P.M. Champion, P. Hession, J. Sutter, E.E. Alp, Phys. Rev. Lett. **86**, 4966 (2001)
74. K. Achterhold, C. Keppler, A. Ostermann, U. van Bürck, W. Stuhrhahn, E.E. Alp, F.G. Parak, Phys. Rev. E **65**, 051916 (2002)
75. K. Achterhold, W. Stuhrhahn, E.E. Alp, F.G. Parak, Hyperfine Interact. **141/142**, 3 (2002)
76. K. Achterhold, F.G. Parak, J. Phys, Condens. Matter **15**, 1683 (2003)
77. A. Weichsel, E.M. Maes, J.F. Andersen, J.G. Valenzuela, TKh. Shokhireva, F.A. Walker, W.R. Montfort, PNAS **102**(3), 594 (2005)
78. R. Christmann, H. Auerbach, R.E. Berry, F.A. Walker, V. Schünemann, Hyperfine Interact. **237**, 19 (2016)
79. H. Beinert, J. Biol. Inorg. Chem. **5**, 2–15 (2000)
80. H. Beinert, R.H. Holm, E. Münck, Science **277**, 653–659 (1997)
81. P. Wegner, M. Bever, V. Schünemann, A.X. Trautwein, C. Schmidt, H. Bönisch, M. Gnida, W. Meyer-Klaucke, Hyperf. Interact. **156**(157), 293–298 (2004)
82. B.F. Matzanke, S. Anemüller, V. Schünemann, A.X. Trautwein, K. Hantke, Biochemistry **43**(5), 1386–1392 (2004)
83. G. Layer, K. Grage, T. Teschner, V. Schünemann, D. Breckau, A. Masoumi, M. Jahn, P. Heathcote, A.X. Trautwein, D. Jahn. J. Biol. Chem. **280**(32), 29038–29046 (2005)
84. D. Barthelme, U. Scheele, S. Dinkelaker, A. Janoschka, F. MacMillan, S.V. Albers, A.J.M. Driessen, M.S. Stagni, W. Meyer-Klaucke, V. Schünemann, R. Tampe, J. Biol. Chem. **282**(19), 14598–14607 (2007)

85. C.E. Schulz, P.G. Debrunner J. Physique, Colloq. **37**, 153 (1976)
86. J.A. Fee, K.L. Findling, T. Yoshida, R. Hille, G.E. Tarr, D.O. Hearshen, W.R. Dunham, E.P. Day, T. Kent, E. Münck, J. Biol. Chem. **259**, 124–133 (1984)
87. K. Stegmaier, C.M. Blinn, D.F. Bechtel, C. Greth, H. Auerbach, C.S. Müller, V. Jakob, E.J. Rejerze, D. Netz, V. Schünemann, A.J. Pierik, J. Am. Chem. Soc. **141**, 5753 (2019)
88. T.O. Bauer, D. Graf, T. Lamparter, V. Schünemann, Hyperfine Interact. **226**, 445–449 (2014)
89. W. Eisenreich, F. Rohdich, A. Bacher, Trends Plant Sci. **6**, 78–84 (2001)
90. M. Seemann, K. Janthawornpong, J. Schweizer, L.H. Böttger, A. Janoschka, A. Ahrens-Botzong, M.N. Tambou, O. Rotthaus, A.X. Trautwein, M. Rohmer, V. Schünemann, J. Am. Chem. Soc. **131**(37), 13184 (2009)
91. W. Wang, J. Li, K. Wang, T.I. Smirnova, E. Oldfield, J. Am. Chem. Soc. **133**, 6525 (2011)
92. I. Span, K. Wang, W. Wang, J. Jauch, W. Eisenreich, A. Bacher, E. Oldfield, M. Groll, Angew. Chem. Int. Ed. **52**, 2118 (2013)
93. A. Ahrens-Botzong, K. Janthawornpong, J.A. Wolny, E.N. Tombou, M. Rohmer, S. Krasutzsky, C.D. Poulter, V. Schünemann, M. Seemann, Angew. Chem. Int. Ed. **50**, 11976 (2011)
94. B.J. Vaccaro, S.M. Clarkson, J.F. Holden, D.-W. Lee, C.-H. Wu, F.L. Poole, J.J.H. Cotelesage, M.J. Hackett, S. Mohebbi, J. Sun, H. Li, M.K. Johnson, G.N. George, M.W.W. Adams, Nat. Commun. **8**, 16110 (2017)
95. K.G. Padmalekha, H. Huang, I. Ellmers, R. Perez Velez, J. van Leusen, A. Brückner, W. Grünert, V. Schünemann, Hyperfine Interact. **238**, 80 (2017)
96. S. Wagner, H. Auerbach, Dr. C.E. Tait, I. Martinaiou, S.C.N. Kumar, C. Kübel, I. Sergeev, H.-C. Wille, J. Behrends, J.A. Wolny, V. Schünemann, U.I. Kramm, Angew. Chemie Int. Ed. **58** (31), 10486 (2019)
97. S. Sakshath, K. Jenni, L. Scherthan, P. Würtz, M. Herlitschke, I. Sergeev, C. Strohm, H.-C. Wille, R. Röhlsberger, J.A. Wolny, V. Schünemann, Hyperfine Interact. **238**, 89 (2017)

Chapter 5
Mössbauer Spectroscopy with High Spatial Resolution: Spotlight on Geoscience

Catherine McCammon

Abstract Mössbauer spectroscopy is an important tool in geoscience. Samples are often small or heterogeneous, requiring measurements with small beam size. Such measurements can be carried out in the laboratory using a radioactive source, or at a synchrotron facility in either time or energy domain. This chapter covers practical aspects of carrying out Mössbauer measurements with high spatial resolution in a tutorial style that covers radioactive source properties, how to choose between time and energy domain, practical aspects associated with preparing the sample and setting up a measurement, useful free software for data analysis and analytical data mapping, and concludes with a showcase of research questions in geoscience that have been addressed using Mössbauer spectroscopy with high spatial resolution.

List of Acronyms

APS	Advanced photon source
EFG	Electric field gradient
ESRF	European synchrotron radiation facility
FIB	Focussed ion beam
GIS	Geographic information system
GKE	Goldanskii-Karyagin effect
GUI	Graphical user interface
NFS	Nuclear forward scattering
PETRA III	Positron-electron tandem ring accelerator III
QGIS	Quantum geographic information system
RPS	Radioactive point source
RCS	Radioactive conventional source
SMS	Synchrotron Mössbauer source
SNR	Signal to noise ratio

C. McCammon (✉)
Bayerisches Geoinstitut, Universität Bayreuth, 95440 Bayreuth, Germany
e-mail: catherine.mccammon@uni-bayreuth.de

Y. Yoshida and G. Langouche (eds.), *Modern Mössbauer Spectroscopy*,
Topics in Applied Physics 137, https://doi.org/10.1007/978-981-15-9422-9_5

SPring-8	Super photon ring-8 GeV
XANES	X-ray absorption near-edge structure

5.1 Introduction

Mössbauer spectroscopy is practically unique among other methods used in geoscience. The signal is specific to the nucleus being examined, so only those phases containing the Mössbauer isotope give a signal. It is one of the few methods that is able to distinguish different valence states, and one of the even fewer that require no calibration to determine accurate values of relative abundance. Mössbauer spectroscopy provides structural information, such as the coordination and geometry of crystallographic sites. The method is a short range probe, so it allows the study of poorly crystallised materials such as glasses and provides information complementary to data from long-range probes such as X-ray diffraction. Finally, there is practically no limitation on pressure and also to a certain extent temperature, which means that Mössbauer spectroscopy can be carried out on samples at most conditions within Earth's interior.

One aspect of Mössbauer spectroscopy that has improved greatly throughout its history is spatial resolution. Technical advances have reduced the beam size dramatically, initially as radioactive point sources became commercially available and subsequently through advances in focussing capabilities at synchrotron facilities. The main driver of these developments has been research questions involving high pressure, where the quest for ever higher pressures has driven the need to measure ever smaller samples. High-pressure studies have not been the sole benefactor of higher spatial resolution, however. Research questions in geoscience are usually best addressed using analytical methods with high spatial resolution, since natural samples may be small and their history can often be deciphered from inhomogeneous variations in composition and oxidation state.

The Mössbauer effect has been observed in a large number of elements (more than 40) in about 80 different nuclides for a total of nearly 100 different nuclear transitions, e.g., [1]. Applications of Mössbauer spectroscopy are concentrated on isotopes with convenient transitions, and by far the most popular (particularly for geoscience applications) has been the 14.4 keV transition in ^{57}Fe. Iron is a popular Mössbauer isotope because the half-life of its common parent, ^{57}Co, is reasonably long (270 days), recoil-free fractions are generally high at room temperature and the absorption cross section is sufficiently large that reasonable spectra can be obtained even for absorbers with low iron concentrations. For ^{57}Fe, the lifetime of the excited state is 98 ns, which gives a natural linewidth of 4.7 neV. Compared to the energy of the γ-ray, the energy resolution is roughly one part in 10^{13}, which enables very small changes in the local atomic environment to be detected. The natural abundance of ^{57}Fe is 2.14%, which provides the possibility to enhance the Mössbauer signal through isotopic enrichment. More than 95% of Mössbauer studies of minerals have used ^{57}Fe [2], so this chapter will focus exclusively on iron.

Many important research questions in geoscience can be addressed using Mössbauer spectroscopy with high spatial resolution. Originally questions focussed on the behaviour of materials at high pressure, particularly phase transitions, since Mössbauer spectroscopy is sensitive to a wide range of transitions, including magnetic, structural, spin and insulator–metal. Studies also address the properties of materials at high pressure, such as density and magnetic, transport and elastic properties. This knowledge can be used in combination with bulk geophysical measurements to answer questions about Earth, such as its chemistry, structure, dynamics and history. Mössbauer spectroscopy is highly sensitive to iron oxidation state, hence it can answer questions about defect structure and charge balance that are important for studying water (hydration) and oxygen (redox) in both minerals and bulk Earth. Mössbauer spectroscopy is consequently in high demand for calibration of other methods to determine oxidation state such as XANES and the flank method. Natural samples such as inclusions in diamond or those with strong heterogeneity provide a window into their geological history through Mössbauer spectroscopy that is not available from any other method.

Chapter 5 is aimed at an audience that includes users experienced with conventional sources who want to learn about point sources, users experienced with lab-based measurements who want to learn about synchrotron possibilities, geoscientists experienced with other microanalytical methods who want to learn about Mössbauer spectroscopy, and researchers from other fields who are interested in exciting applications of the method. The tutorial style assumes only a basic knowledge of Mössbauer spectroscopy that can be easily acquired from short course chapters or text books, e.g., [1, 3, 4]. The chapter covers topics related to properties of radioactive sources, considerations in choosing between energy and time domain measurements, practical aspects of small beam size (geometry, sample preparation, texture and thickness), useful free software, and a survey of research problems in geoscience that high spatial resolution measurements have been applied to. In keeping with the tutorial style, the absorber in the Mössbauer experiment is referred to in this chapter as the sample.

5.2 Mössbauer Sources for High Spatial Resolution

The nature of the radiation source used in a Mössbauer experiment is one of the main factors in determining the spatial resolution of the measurement. Section 5.2 describes the most commonly used types of sources and some considerations in their use.

5.2.1 Conventional Radioactive Source

^{57}Co is the commonly used radioactive parent of ^{57}Fe 14.4 keV γ-rays. The magnetic ordering temperature of cobalt is around 1404 K [5], however, so pure ^{57}Co sources are not used since they would emit γ-rays with multiple energies. The strategy used to obtain ^{57}Co sources that emit a single narrow line is to diffuse ^{57}Co into a non-metallic matrix, e.g., [6]. Properties to consider in choosing a matrix include good chemical stability, high cobalt and iron solubility, favourable crystal and electronic structure (isotropic and non-magnetic), high recoilless fraction and negligible X-ray fluorescence near 14.4 keV [7]. Further considerations relate to the intended use of the source, for example at low temperature or in high external magnetic fields. While no matrix perfectly satisfies all requirements, rhodium is the best all-around choice [7] and indeed is currently the most commonly used matrix in commercial sources.

The strength and flux of the source is determined by its physical dimensions as well as the amount of ^{57}Co contained in the source matrix. The thickness of the source foil affects the emission efficiency of γ-rays, where thinner foils allow more γ-rays to escape the source matrix. The thickness also affects the source line width, where thinner foils have a higher degree of internal resonance (excitation of ^{57}Fe in the source by emitted γ-rays). A useful metric for internal resonance is the effective source thickness, t_s, given by:

$$t_s(\text{dimensionless}) = \sigma_0 f_s n_s \tag{5.1}$$

where σ_0 is the resonance cross section of the transition ($= 2.56 \times 10^{-18}$ cm^2 for the 14.4 keV ^{57}Fe transition), f_s is the recoil-free fraction of the source, and n_s is the number of ^{57}Fe atoms per cm^2 in the source. When a ^{57}Co source is first produced, t_s is essentially zero because there are no ^{57}Fe atoms in the source. As the source ages, however, t_s increases because ^{57}Co decays to ^{57}Fe and builds up in the source, causing progressively more internal resonance. The consequence of this internal resonance is broadening of the source line width and reduction of recoilless fraction, which can ultimately limit the working life of the source. These effects are negligible in the thin source limit ($t_s << 1$), which can be achieved (even for thin source foils) if the starting concentration of ^{57}Co in the source is low.

^{57}Co Mössbauer sources with low concentrations of ^{57}Co that remain within the thin source limit for all of their working life are referred to as conventional sources. The specific activity of conventional sources is generally below 7 GBq/cm^2 (0.2 Ci/cm^2), which means that a 1.85 GBq (50 mCi) Co(Rh) source would have an active diameter of around 6 mm. Such sources are best suited to beam sizes with comparable diameters. Figure 5.1 illustrates the trade-offs between the specific activity of a source and its performance.

Higher spatial resolution for imaging applications can be obtained using a conventional source. Smith and colleagues [8] used a position-sensitive detector to record 256 spectra simultaneously over a sample with 50 mm length, while a Mössbauer

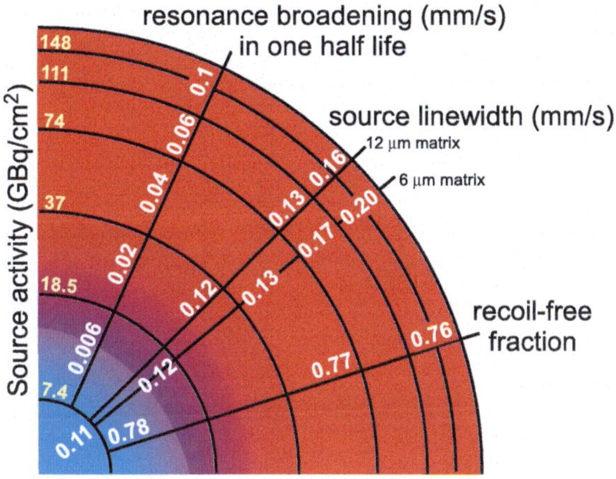

Fig. 5.1 Performance data for a freshly produced ^{57}Co Mössbauer source in a rhodium matrix. Low specific activity sources (<7 GBq/cm^2 or 0.2 Ci/cm^2), i.e., so-called conventional sources (*blue*), have line widths close to the theoretical minimum (0.097 mm/s) that do not increase appreciably with time, in contrast to point sources (*red*). Adapted from [6]

microscope [9] focuses γ-rays using a multi-capillary X-ray lens and detects conversion electrons using a scanning electron microscope over a sample up to 10 × 10 mm^2. Each method reported spatial resolutions of 50–100 μm. More exotic approaches to imaging include tomographic analysis of a sample subjected to a velocity gradient through rotation [10], achieving a resolution of 250 μm on a 10 mm long sample [11].

5.2.2 High Specific Activity (Point) Radioactive Source

The active diameter of a ^{57}Co source can be reduced without loss of flux by increasing the specific activity of the source. The theoretical maximum specific activity is determined by the properties of ^{57}Co used to fabricate Mössbauer sources, which is typically 222 TBq/g (7500 Ci/g). Such material could be used to produce a pure ^{57}Co source with 6 μm thickness of 1.5 TBq/cm^2 (40 Ci/cm^2) specific activity. The theoretical minimum diameter of a source with 0.37 GBq (10 mCi) overall activity would be ~200 μm. Such a source would be highly impractical, however, since it would emit multiple γ-ray energies due to magnetic interactions and would have an extremely short working life due to internal resonance. While 1.5 TBq/cm^2 (40 Ci/cm^2) represents the theoretical limit of maximum specific activity for a ^{57}Co Mössbauer source, smaller values of specific activity offer the capability for small beam size without substantial sacrifice of source performance.

^{57}Co Mössbauer sources with specific activities higher than conventional sources (and hence higher concentrations of ^{57}Co) are referred to as point sources. A specific activity of 185 GBq/cm^2 (5 Ci/cm^2) is generally considered the upper limit for a ^{57}Co point source, which represents a compromise between limited broadening of source line width due to magnetic interactions and maximum flux. Figure 5.2 illustrates a comparison of the time evolution of the effective source thickness (Eq. 5.1) for a

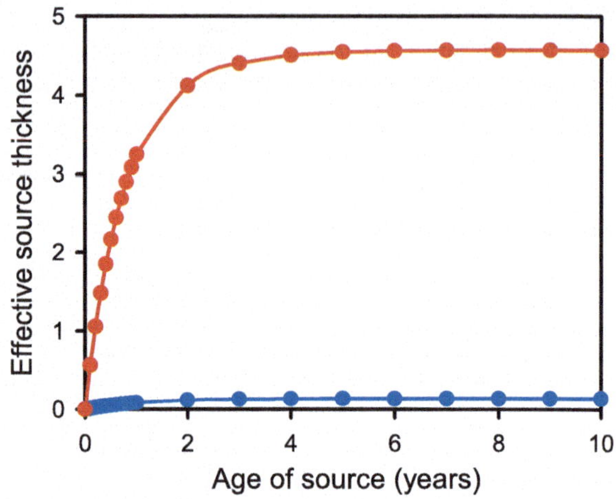

Fig. 5.2 Calculated time evolution of effective source thickness for ^{57}Co Mössbauer sources. Calculations are based on specific activities of 7 and 185 GBq/cm^2 (0.2 and 5 Ci/cm^2) for conventional (*blue*) and point (*red*) sources, respectively. Adapted from [14]

conventional source and a point source. The former is limited only by the declining count rate, since this can be compensated by longer counting times without substantial loss of signal quality. For the point source, however, the effective source thickness increases rapidly with source age, leading to greater source line width, smaller recoil-free fraction, and distortion of spectral line shape and area fraction, e.g., [12, 13]. Most spectroscopists consider a ^{57}Co source unusable after reaching an effective thickness of 3, hence a point source should be renewed roughly every year.

A point source delivers a substantially higher flux compared to a conventional source. For example, a point source with specific activity 5 Ci/cm^2 distributed over a diameter of 500 μm would have an overall activity of 0.37 GBq (10 mCi). Calculations have shown that comparable signal/noise ratios can be obtained using a conventional source with a large diameter sample (12 mm) and a point source with a small diameter sample (0.5 mm) [14]. Key parameters that influence the quality of spectra collected with a point source include the age of the source and the amount of electronic absorption in the sample.

5.2.3 Synchrotron Source

Synchrotron radiation bypasses the specific activity limitations of ^{57}Co sources and offers a number of advantages for high spatial resolution compared to radioactive sources. The most obvious is that photon flux is much greater (photons s^{-1} eV^{-1}), but more relevant is that synchrotron radiation can be focussed, leading to substantially higher photon flux densities (photons s^{-1} eV^{-1} mm^{-2}). For example the photon flux density emitted by a third-generation synchrotron focussed to 6×6 μm^2 is calculated to be 10^{11} higher compared to a 3.7 GBq (100 mCi) ^{57}Co conventional source at 10 cm distance [15]. A further advantage of synchrotron radiation applied

Table 5.1 Summary of third-generation synchrotron facilities with nuclear resonance capability

Facility	Location	Techniques[a]	Website
Advanced photon source (APS)	Argonne, USA	NFS	https://www.aps.anl.gov/
European synchrotron radiation facility (ESRF)	Grenoble, France	NFS, SMS	https://www.esrf.eu/
Positron–electron tandem ring accelerator (PETRA) III	Hamburg, Germany	NFS	https://petra3.desy.de/
Super photon ring-8 GeV (SPring-8)	Sayo, Japan	NFS, SMS	https://www.spring8.or.jp/

[a] As of 2018. Each synchrotron has further capabilities, but only those relevant to the current chapter are listed

to nuclear resonance is the absence of non-resonant background [16], which means that the signal to noise ratio is substantially higher compared to radioactive source experiments, even if all other parameters are equal.

The nuclear resonance techniques most commonly used at synchrotrons are nuclear forward scattering (NFS) and the synchrotron Mössbauer source (SMS).[1] An excellent introduction to these techniques is given by [17]. Both techniques require a third-generation synchrotron, of which there are currently four that offer nuclear resonance methods (Table 5.1). All synchrotrons listed have user access programmes, where applications for beamtime are generally evaluated on the basis of scientific merit and feasibility. Additional restrictions on principal investigators may apply at individual facilities, such as country of affiliation, and websites should be consulted for up to date details.

5.2.4 Mössbauer Source Comparison

The choice of Mössbauer source for a specific measurement depends on a number of factors. The most important are beam size and sample composition and ^{57}Fe enrichment, logistics (number of samples and measuring time available), and financial considerations. There can be trade-offs between factors that may influence the choice of source, which is illustrated by the following two cases where the same sample was measured with different types of Mössbauer source.

[1] Note that NFS is referred to in some regions as synchrotron Mössbauer spectroscopy, which is a time domain method. This should not be confused with the synchrotron Mössbauer source, which is an energy domain method. This chapter uses the acronym SMS to refer to the synchrotron Mössbauer source.

5.2.4.1 Conventional and Point Source Comparison

Samples with diameters as small as 2 mm can be measured with a conventional source if sufficient time is available for the measurement. Synthetic clinopyroxene with composition $(Fe_{0.1}Mg_{0.9})(Ca_{0.7}Li_{0.3})Si_2O_6$ was prepared from a glass starting material at 3 GPa and 950 °C in a piston-cylinder apparatus for 18 h. Iron with natural abundance of ^{57}Fe (2.14%) was used. A sample holder was constructed from 1 mm thick lead sheet with a 2 mm diameter hole into which 6 mg of powder sample was placed, giving a dimensionless effective sample thickness of 2 (Sects. 5.4.2 and 5.4.4). The same sample mount was measured using a conventional source and point source, where the source-sample distance was adjusted for each measurement to give appropriate solid angles (Sect. 5.4.1). Data were collected for approximately 24 h each.

The Mössbauer spectra of clinopyroxene from the conventional and point sources show substantially different signal to noise ratios (SNR) (Fig. 5.3). Assuming Poisson counting statistics, the SNR can be determined from:

$$\mathrm{SNR} = \frac{(N_\infty - N_0)}{\sqrt{N_\infty}} \tag{5.2}$$

where N_∞ and N_0 are the baseline counts and counts at resonance (minimum counts in spectrum), respectively, e.g., [1]. The main reason for the difference in SNR is the much higher specific activity of the point source (154 GBq/cm^2 or 4.2 Ci/cm^2) compared to the conventional source (3.5 GBq/cm^2 or 0.1 Ci/cm^2) at the time of measurement. While the overall count rate for the conventional source is lower, this can be compensated for by extending the measuring time. The SNR of the point source

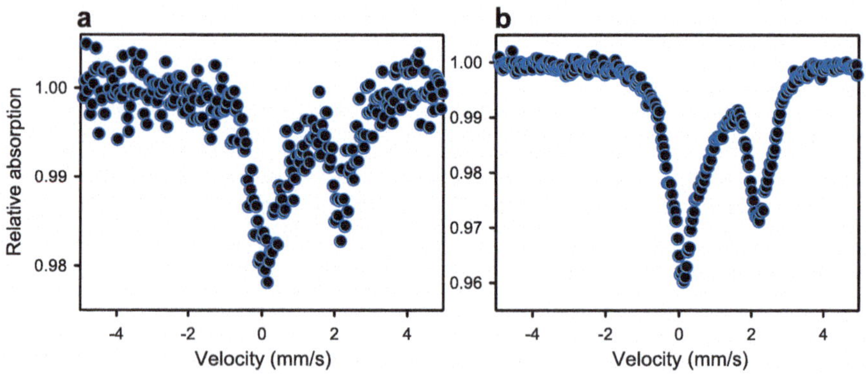

Fig. 5.3 Room temperature Mössbauer spectra of synthetic $(Fe_{0.1}Mg)(Ca_{0.7}Li_{0.3})Si_2O_6$ clinopyroxene **a** collected with a conventional source and **b** point source. The same sample mount was used for each measurement and the solid angles and measuring times were similar. The baseline counts for the conventional and point source spectra are 2×10^5 and 2×10^6 counts, respectively, while the respective SNRs are 10 and 58

spectrum (Fig. 5.3b) can be reached by measuring the sample using a conventional source for 36 days, assuming that the count rate remains constant with time. This possibility may be a feasible alternative to purchasing a point source if measuring time is not constrained, since the thickness of a conventional source does not change appreciably with time (Fig. 5.2).

5.2.4.2 Point Source and SMS Comparison

Samples less than 100 μm in diameter can be measured with a point source if effective sample thickness is high (Sect. 5.4.4), sufficient time is available for the measurement and good spectrum quality is not a priority. Synthetic bridgmanite with composition $Fe_{0.4}Mg_{0.6}Si_{0.63}Al_{0.37}O_3$ was prepared at high pressure using iron enriched with 90% ^{57}Fe [18].

The sample used was approximately 30 μm in diameter and 15 μm thick, giving a dimensionless effective sample thickness of 20 (Sect. 5.4.4). The sample was loaded into a diamond anvil cell and compressed to 93 GPa, and spectra were collected until an acceptable quality was reached, which was 7 days for the point source and 10 min for SMS.

The Mössbauer spectra of bridgmanite from the point source and SMS show substantially different SNRs (Fig. 5.4). Not only is the count rate substantially higher for SMS (~6000 counts/s) compared to the point source (~150 counts/s), but also the absence of non-resonant background gives a much higher resonant absorption for SMS (83%) compared to the point source (2%). It would be challenging to extend the point source measuring time sufficiently to match the SNR of SMS, since this would

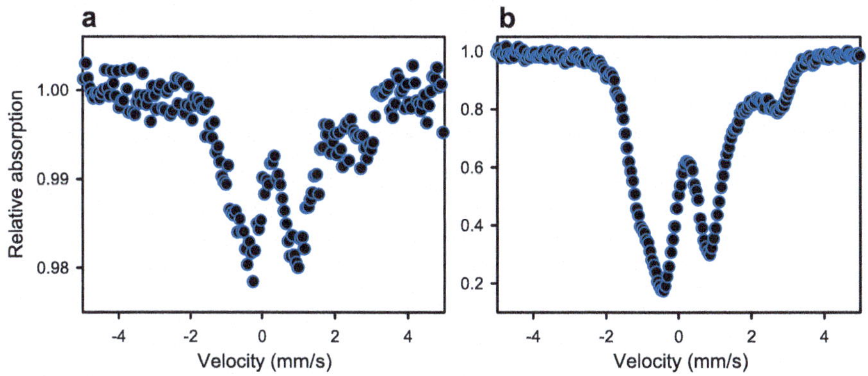

Fig. 5.4 Room temperature Mössbauer spectra of synthetic $^{57}FeMg_{0.6}Si_{0.63}Al_{0.37}O_3$ bridgmanite in a diamond anvil cell at 93 GPa **a** collected with a point source and **b** SMS. In both cases the same sample at the same pressure in the same diamond anvil cell was measured. Measuring times were 7 days for the point source and 10 min for SMS, and the respective SNRs are 13 and 98. Baseline counts are 3.6×10^5 and 6.9×10^3 for the point source and SMS spectra, respectively. SMS data are taken from [18]

require nearly 400 days if the count rate remained constant. The count rate decreases, however, and source thickness would increase substantially (Fig. 5.2), necessitating the replacement of the point source several times during the measurement. However if a lower quality spectrum were sufficient for the research question being addressed, this exercise demonstrates the possibilities for collecting Mössbauer data with beam sizes as small as 50 μm diameter using a point source.

There are a number of geoscience applications for which a point source is not appropriate and the only choice is synchrotron radiation. These cases include very small samples or when high spatial resolution is needed (<50 μm diameter for samples with high dimensionless effective thickness but much larger if samples contain less ^{57}Fe), as well as experiments where time is an important parameter. Examples include measurements during laser heating at high pressure [19], time-differentiated measurements [20], and in situ kinetic studies [21].

5.3 Measurement Approach

Mössbauer spectra collected using a radioactive source are traditionally measured by changing the velocity of the sample relative to the source (or vice versa) and measuring counts as a function of velocity (v), which can be converted to energy through the Doppler shift:

$$E = E_0 v/c \tag{5.3}$$

where E_0 is the unperturbed energy of the γ-ray (14.4 keV) and c is the velocity of the γ-ray (i.e., the speed of light $= 2.998 \times 10^{11}$ mm/s). The variable is therefore energy, and such measurements are said to be carried out in the energy domain.

Mössbauer spectra collected using a synchrotron source can be carried out in the energy domain using SMS, but can also be carried out in the time domain using NFS. In the latter method all resonant nuclear transitions are excited at time zero by a sharp pulse, after which interference between the decaying energy levels in the quiet time between pulses produces an intensity spectrum as a function of time. More details are available in the literature, for example [17], but for the purpose of the present chapter, Sect. 5.3 focuses specifically on aspects of energy and time domain measurements that are relevant to small or heterogeneous samples in geoscience investigations, namely spectral deconvolution and counting time.

5.3.1 *Energy and Time Domain Comparison: Spectral Deconvolution*

The same hyperfine interactions are probed during measurements in both energy and time domain, but they appear differently in the resulting spectrum (Fig. 5.5). In the energy domain, each transition gives one absorption line (middle row in Fig. 5.5). In the time domain, however, the spectrum shows the interference of radiation from all resonant nuclear transitions as "quantum beats" (bottom row in Fig. 5.5). Note that a single nuclear transition (i.e., a singlet in energy domain) cannot be distinguished in time domain measurements. This limitation can be overcome by inserting a reference material in the synchrotron beam to produce quantum beats between the reference and the sample (e.g., [22]).

Energy domain measurements are generally more sensitive for detecting small quantities of spectral components. For example a study of a ferropericlase [(Mg,Fe)O] inclusion in diamond was able to detect the presence of a magnetic component (likely magnesioferrite, (Mg,Fe)Fe_2O_4) [23], which yielded important constraints on the

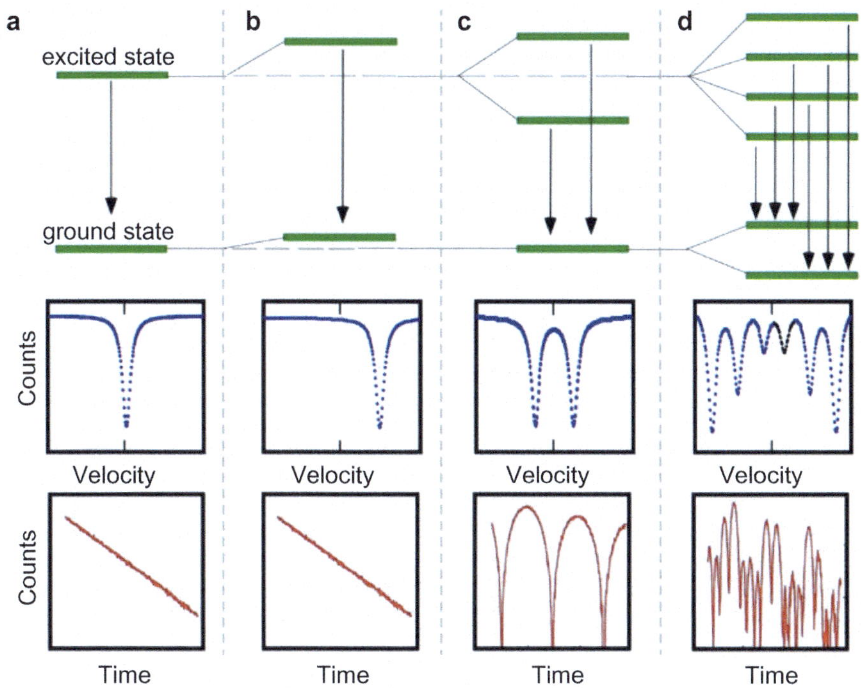

Fig. 5.5 Hyperfine interactions for ^{57}Fe nuclei showing the energy level diagram for each interaction: **a** unperturbed nucleus; **b** isomer shift; **c** quadrupole splitting; **d** hyperfine magnetic splitting. The resulting Mössbauer spectra in the energy domain and time domain are shown in the *middle* and *bottom* rows, respectively

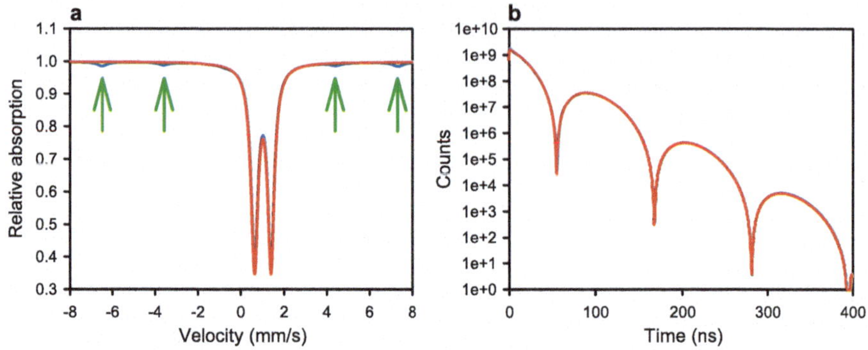

Fig. 5.6 Comparison of simulated **a** energy and **b** time domain spectra for a ferropericlase inclusion in diamond. The spectra were calculated based on hyperfine parameters reported by [23]. Each panel shows spectra for pure ferropericlase (*red*) and 95% ferropericlase with 5% magnetite (*blue*). The presence of magnetite can be detected in the energy domain spectrum (indicated by *green arrows*; also seen in Fig. 5.6 of [23]), but not in the time domain spectrum, where *red* and *blue lines* are almost identical. Spectra were simulated using MOTIF [24]

history of the diamond. The study used an energy domain approach (SMS) for the study, which enabled the small amount of magnesioferrite to be quantified. If a time domain approach (NFS) had been used instead, it would have been challenging, if not impossible, to detect the presence of magnesioferrite (Fig. 5.6).

Hyperfine parameters can generally be unambiguously determined from energy domain spectra if absorption lines do not overlap substantially or if valid constraints can be applied during spectral deconvolution. Significant line overlap, however, causes ambiguities in spectral fitting when such constraints are not available, e.g., [25]. Similar ambiguities are encountered when determining hyperfine parameters from time domain spectra if the number of unknowns is large, for example in the case of multiple iron sites and electronic states (valence and spin). Additional factors such as distribution of hyperfine parameters, texture effects, inhomogeneous thickness, pressure gradients in diamond anvil cell measurements and so on will further add to ambiguities.

Silicate perovskite, i.e., the mineral bridgmanite, presents a good case for the use of energy domain measurements instead of time domain. Iron is distributed between two sites in the crystal structure, and can occur in two different valence states (Fe^{2+} and Fe^{3+}) and three different spin states (high, intermediate and low spin). Absorption lines in energy domain spectra are generally sufficiently resolved such that spectra can be unambiguously deconvoluted and, with the help of constraints from crystal structure refinements, doublets can be correctly assigned [26]. The presence of low-spin Fe^{3+}, for example, could be recognised in energy domain (SMS) spectra after laser heating of bridgmanite at 68 GPa [19]. In this study the appearance of additional absorption that was distinct from other spectral components allowed the amount of low-spin Fe^{3+} to be quantified. Time domain (NFS) spectra of bridgmanite can generally be deconvoluted with an appropriate model using energy domain spectra

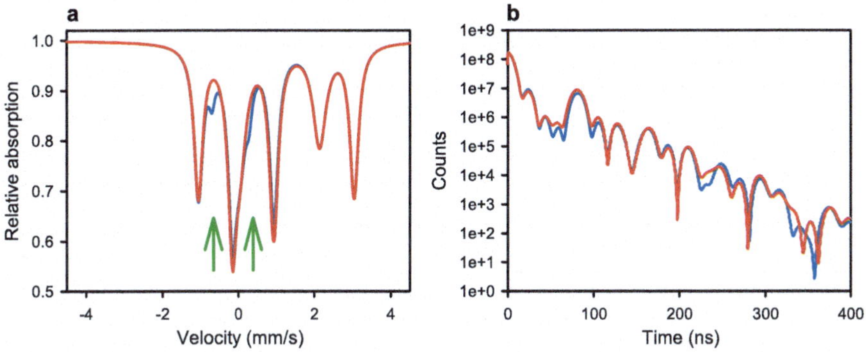

Fig. 5.7 Comparison of simulated **a** energy and **b** time domain spectra for silicate perovskite at 68 GPa. The spectra were calculated based on hyperfine parameters reported by [19]. Each panel shows spectra for 40% high-spin Fe^{3+} (*red*) and 35% high-spin Fe^{3+} and 5% low-spin Fe^{3+} (*blue*). Low-spin Fe^{3+} can be identified in the energy domain spectrum by additional absorption at −0.7 and 0.2 mm/s (indicated by *green arrows*; also seen in Fig. 5.2 of [19]); however the difference in time domain spectra cannot be unambiguously assigned as low-spin Fe^{3+}, especially below 200 ns which is the typical time window for NFS data collection. Spectra were simulated using MOTIF [24]

as a guide, e.g., [27]; however unambiguous identification of new components for which there are no constraints available (for example low-spin Fe^{3+}) are not possible (Fig. 5.7). The situation is particularly challenging when the time between bunches is short (<200 ns), for example during the commonly used 16-bunch mode at ESRF and 24-bunch mode at APS.

Time domain measurements are a good choice for materials with relatively simple spectra. For example, iron in ferropericlase occupies only one crystallographic site and occurs almost exclusively as Fe^{2+}. The transition of Fe^{2+} from high to low spin has been often studied in the time domain (NFS), for example [28]. The change from a quadrupole doublet (high-spin state) to a singlet (low-spin state) is easily recognised in time domain spectra, even below 200 ns (Fig. 5.8).

5.3.2 Energy and Time Domain Comparison: Counting Time

Time domain measurements (i.e., NFS) generally give a higher SNR than synchrotron-based energy domain measurements (i.e., SMS) for the same sample and counting time. Time domain spectra collect from zero background, so each count contributes to the signal. In contrast, energy domain spectra collect down from background (the baseline) and counts are divided over all channels, so only a portion of counts contributes to the signal.

Simulated spectra in both time and energy domain are illustrated in Fig. 5.9 for a simple quadrupole interaction where one thousand counts have been collected for each. The SNR for the time domain spectrum is 5.7 (Fig. 5.9a), while only 1.3 is

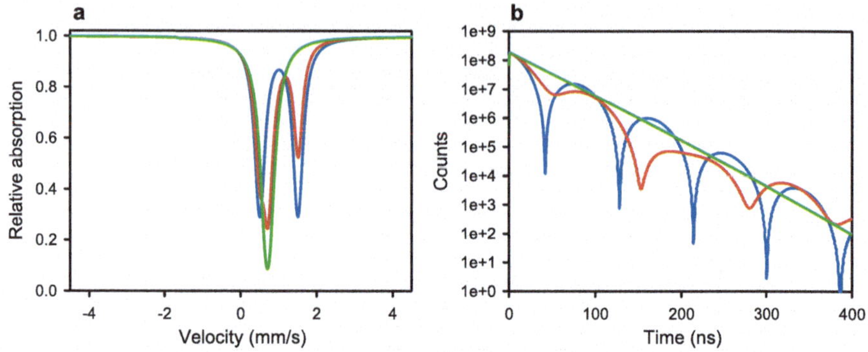

Fig. 5.8 Comparison of simulated **a** energy and **b** time domain spectra for ferropericlase as a function of pressure. The spectra were calculated based on hyperfine parameters reported by [28]. Each panel shows spectra for 100% high spin Fe^{2+} (*red*), 50% high spin & 50% low spin Fe^{2+} (*blue*) and 100% low spin Fe^{2+} (*green*). The difference between spectra is straightforward to recognise in both energy and time domain. Spectra were simulated using MOTIF [24]

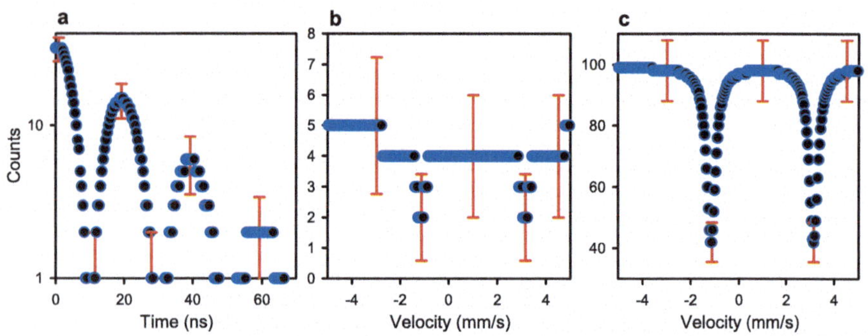

Fig. 5.9 Comparison of simulated **a** time and **b**, **c** energy domain spectra for a phase with a quadrupole splitting (ΔE_Q) of 4.3 mm/s (for example intermediate-spin Fe^{2+} in silicate perovskite [26]). The *left* and *middle* spectra each contain 1000 counts. To reach the same SNR in the energy domain as in time domain, the spectrum must be collected for 22 times as long (*right*). Error bars are shown in *red* for selected channels. Spectra were simulated using MOTIF [23]

achieved for the same number of counts in the energy domain (Fig. 5.9b). The time domain spectrum allows quadrupole splitting to be determined within a reasonable error, while the signal in the energy domain is barely detectable above the background.

The obvious way to improve SNR in the energy domain spectrum is to increase the counting time. If the spectrum in Fig. 5.9b were collected for 22 times longer, a SNR of 5.7 is reached (Fig. 5.9c), which is sufficient for an accurate determination of quadrupole splitting. The difference in counting time between time and energy domain for the same sample depends on the time and energy resolution of each spectrum (i.e., number of channels) as well as its complexity, where complex time domain spectra may need to be collected for longer to record counts at high delay times. One example is provided by two studies of the same sample, one in time domain [29] and the other in energy domain [19]. The time domain spectra were collected for 10–60 min each, while energy domain spectra were collected for 1–2 h, representing a maximum time difference factor of 12.

5.4 Practical Considerations for Small Beam Size

Many of the practical aspects associated with collecting Mössbauer spectra with high spatial resolution are common to all experiments, for which there are many excellent references, e.g., [1, 4]. In Sect. 5.4 we focus on a few aspects that are of particular importance for small beam size.

5.4.1 Spectrometer Geometry

The geometry of a transmission Mössbauer spectrometer with a radioactive source is one of the simplest among all spectroscopies. The main components—the source, sample and detector—are aligned linearly, so the only adjustable parameters are the distance between the source and the sample (D_1) and the sample and the detector (D_2) (Fig. 5.10). Selecting these distances appropriately enables high quality measure-

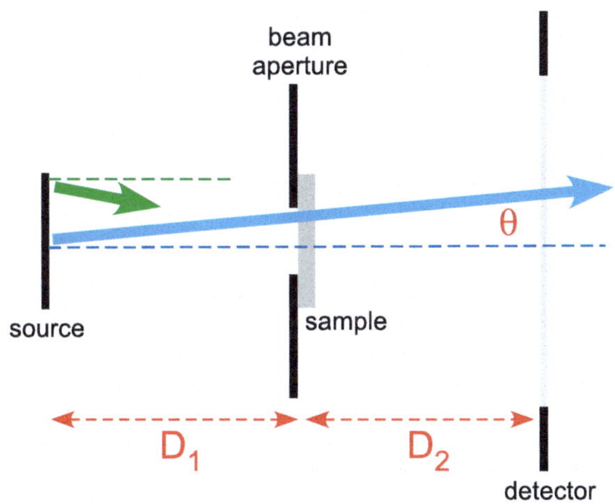

Fig. 5.10 Schematic diagram of a transmission Mössbauer spectrometer with a radioactive source. Vertical dimensions are shown to scale for a 10 mm diameter source, 5 mm diameter beam aperture, and 25 mm diameter detector window, while horizontal dimensions are not to scale. *Blue* and *green* arrows show potential ray paths for γ-rays, where the angle between the *blue* ray path and the horizontal is θ. The source to sample and sample to detector distances are D_1 and D_2, respectively

ments to be made with small beam size, even when using a conventional source, while inattention to setting these distances can significantly reduce the quality of the measurement.

There is a trade-off in choosing the optimum distance between source and sample. A short distance increases the count rate due to a larger solid angle of radiation (good), but at the same time increases distortion of the velocity (energy) scale (bad). While the former effect can be compensated for by longer counting times, there is no possibility to correct spectra for the latter effect after they have been collected (H. Spiering, personal communication).

Distortion of the velocity scale occurs because γ-rays travelling at angle θ to the horizontal have an energy shift given by:

$$E = E_0(v\cos\theta)/c \tag{5.4}$$

which reduces to Eq. 5.3 when $\theta = 0$. The $\cos\theta$ term changes the energy distribution of emitted γ-rays, which ultimately affects the shapes, widths and positions of the absorption lines in the resulting Mössbauer spectrum. These effects have been calculated in detail [30] where, for example, line shift distortions are less than 1% at a source velocity of 5 mm/s when $\theta < 10°$ (distortion increases at higher source velocities). This calculation assumes a point source (i.e., the source diameter is much less than the beam size), but since conventional sources can be considered sheet emitters when beam size is smaller (Fig. 5.10), the appropriate limit in this case is $\theta < 5°$ (see also [1]).

The source to absorber distance (D_1) can be simply calculated from θ using the following relation:

$$D_1 = d/(2\ \tan\theta) \tag{5.5}$$

where d is the beam aperture diameter (Fig. 5.10).

Equation 5.5 can be used to calculate the appropriate source to sample distances for the measurements shown in Fig. 5.3 (recall that the beam size for those measurements was 2 mm diameter). For the conventional source measurement, θ should be less than 5°, so D_1 should be greater than 11 mm, and for the point source measurement, θ should be less than 10°, so D_1 should be greater than 6 mm. Equation 5.5 is most useful when measuring with small beam size using a conventional source, since in this case the source to sample distance can be reduced to compensate for the smaller beam size.

Equation 5.5 can also be used to calculate the appropriate source to sample distance for the point source measurement shown in Fig. 5.4a. Here the beam size is 30 μm diameter, which is considerably smaller than the point source diameter (500 μm). To achieve $\theta = 5°$ would require a source to sample distance of only 100 μm, which is not possible due to the thickness of the diamonds since the sample is inside a diamond anvil cell. Indeed, Eq. 5.5 demonstrates that velocity scale distortion is unlikely to cause problems in high-pressure experiments. Similarly, measurements using synchrotron radiation do not require such considerations due to the minimal divergence of the X-ray beam compared to a radioactive source.

The other distance to be considered when setting up a Mössbauer measurement, namely the sample to detector distance (D_2), does not influence the solid angle of radiation since the detector window is usually much larger than the beam size (Fig. 5.10). Here the distance D_2 should be sufficiently large to minimise the amount of radiation scattered from the sample that reaches the detector, which contributes only to the background, not the signal.

5.4.2 Sample Preparation

Mössbauer spectroscopy is extremely flexible with regard to sample form and measurements with small beam size are no exception. Beam diameters for Mössbauer spectroscopy using a radioactive source typically start at 100 μm and go upwards for unenriched samples [14], and ^{57}Fe enrichment allows even smaller samples to be measured (for example the spectrum in Fig. 5.4a was collected on a 30 μm diameter sample). Focussing optics at synchrotron sources enable beam diameters as small as 10 μm to be used [31, 32].

Many samples can be used with minimal preparation, although there may be advantages to investing some effort into cutting and polishing, particularly in the case of inhomogeneous samples. Ultimately the choice of sample form depends on the nature and amount of sample available, as well as the desired sample thickness (Sect. 5.4.4). A final step for all sample types before collecting spectra with a radioactive source is to ensure that the detector records only γ-rays that have passed through the sample, which is usually accomplished by masking the sample with a thick absorbing foil (for example Pb or Ta).

A common sample form, especially for large samples measured using a conventional source, is powder, meaning that the grain size is much smaller than the beam size. Conventionally a standard sample holder is used, into which an appropriate weight of sample is placed (Fig. 5.11a). For smaller quantities of sample, a holder can be made by drilling a hole in lead foil and securing the contents with cellophane tape (Fig. 5.11b), or by mixing the powder with glue and affixing it to a plastic sheet (Fig. 5.11c). Sample diameters of 500 μm or less can be easily achieved using the latter two approaches, and are suitable for use both with a radioactive source and for synchrotron measurements.

Samples in the form of grains or shards can be used without alteration if their thickness is appropriate for their composition. The grain can be simply affixed onto a plastic sheet (Fig. 5.11d) or multiple grains can be glued in the form of a mosaic if a larger diameter is required (Fig. 5.11e). In this case empty space should be minimised and thickness should not vary substantially across the mosaic (Sect. 5.4.4). It is also possible to measure rare or precious samples such as polished gemstones if at least one part of the stone has a thickness appropriate for its composition.

Fig. 5.11 Sample forms and holders for Mössbauer spectroscopy. **a** plastic sample holder (diameter 12 mm). **b** hole drilled in lead foil (hole diameter 1 mm). **c** powder mixed with glue and affixed onto plastic sheet (sample diameter 4 mm). **d** single grain on plastic sheet (grain dimensions 900 × 600 μm^2) (enlarged view shown in inset). **e** multiple grains formed into mosaic (sample diameter 7 mm). **f** glass shards in epoxy (disk dimensions 9 × 6 mm^2). **g** capsule from high-pressure experiment in epoxy (capsule dimensions 1400 × 900 μm^2). **h** single crystal clinopyroxene cut into a cube (700 × 700 × 700 μm^3). **i** oriented garnet single crystal cut and polished into a slice and glued onto an epoxy disk (sample dimensions 1.5 mm diameter × 150 μm thick) (garnet data reported in [36]). **j** suboceanic lava thick section (3.5 × 2 × 0.5 mm^3) (provided by N. Hirano). **k** metasomatised eclogite thick section (2 × 2 × 0.5 mm^3). **l** inclusions within intact diamond (diamond dimensions 3 × 3 × 3 mm^3) (adapted from [23]). **m** inclusions exposed by polishing diamond slice (diamond dimensions 2.2 × 1.8 × 0.5 mm^3) (data reported in [39]). **n** single crystal (*centre*) inside diamond anvil cell with ruby chip for pressure calibration (*upper left*) (gasket hole diameter 100 μm). **o** polycrystals inside a diamond anvil cell (gasket hole diameter 120 μm). The latter two images are adapted from [40]

Samples can be prepared by mounting them in epoxy and grinding and polishing the disks to the appropriate thickness. The epoxy provides support and easy handling for small grains and the same epoxy mount can be used for chemical determination (for example using the electron microprobe) as well as Mössbauer spectroscopy. This capability is particularly valuable for studies where Mössbauer spectroscopy is used to calibrate standards for the flank method [33] and X-ray Absorption Near Edge Structure (XANES) spectroscopy [34] (Fig. 5.11f). Preparation in epoxy also allows optimal exposure of the region to be measured, such as in capsules recovered from large volume high-pressure experiments [35] (Fig. 5.11g).

Single crystals provide information not available from powder measurements, but require additional preparation to realise these benefits. Generally it is advantageous to orient crystals prior to cutting, for example using cleavage directions or X-ray diffraction. Crystals can then be cut along desired directions with appropriate thickness. For measurements in three orthogonal directions, a cube shape is ideal (Fig. 5.11h), particularly for samples with low iron concentration that require substantial thickness. A slice is useful for measurements that involve rotating the crystal at an angle inclined to the source radiation, e.g., [36] (Fig. 5.11i).

Inhomogeneous samples provide a wealth of information, but require more attention to preparation in order to facilitate documentation of the regions that are measured. Thick sections offer the possibility to use optical images, and if sections are polished, the same regions on the sections can be analysed using other methods such as electron microprobe analysis, e.g., [37] (Fig. 5.11j, k). Important to note, however, is that Mössbauer measurements in transmission geometry record iron phases throughout the entire thickness of the sample section, not only on the surface. One strategy to minimise inhomogeneity along the radiation path is to select regions that show similar characteristics on both upper and lower surfaces of the section.

Inclusions in diamond can be measured in situ since carbon is a relatively light element (Sect. 5.4.4). There are numerous advantages to keeping diamonds intact (Fig. 5.11l), including preserving features such as diamond microstructure and inclusion associations that can be studied using non-destructive methods, for example micro-computed tomography [38]. However chemical composition is difficult to determine quantitatively when inclusions are enclosed in diamond, so one compromise is to polish the diamond until inclusions are partly exposed (Fig. 5.11m). Of course inclusions can also be released entirely from their diamond hosts, usually by breaking the diamond, in which case inclusions can be mounted on plastic foil (Fig. 5.11d) or in epoxy (Fig. 5.11f).

Diamond anvil cells for Mössbauer measurements are generally prepared in the same way as for other diamond anvil cell experiments. Special considerations for Mössbauer spectroscopy include ensuring an appropriate sample thickness (Sect. 5.4.4) and using a gasket with high atomic weight (to collimate source radiation) that does not contain iron. Samples can be loaded as single crystals (Fig. 5.11n) or as powder (Fig. 5.11o).

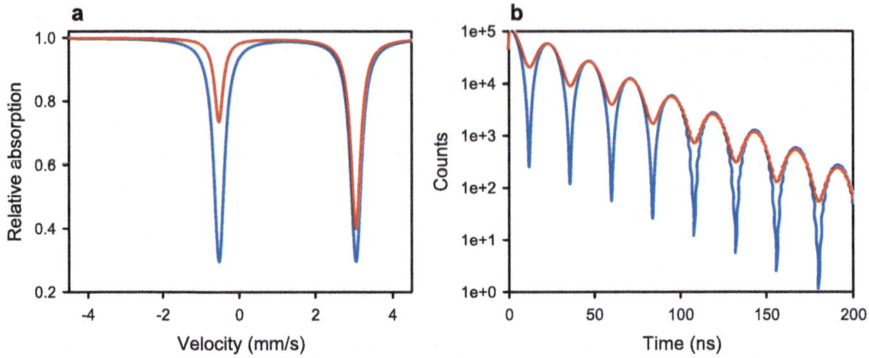

Fig. 5.12 Comparison of simulated **a** energy and **b** time domain spectra when texture effects are present. Spectra were calculated for a quadrupole splitting (ΔE_Q) of 3.6 mm/s. The *blue lines* in each panel correspond to randomly oriented crystals (i.e., powder) while the *red lines* show the case of a single crystal where the principal axis of the electric field gradient (EFG) is parallel to the direction of source radiation. Spectra were simulated using MOTIF [23]

5.4.3 Texture Effects

Texture effects can occur if samples are single crystals or if there is strong preferred orientation. These effects will be particularly pronounced if the source is polarised, for example in the case of synchrotron radiation. The manifestation of texture effects is different for energy and time domain spectra. In the energy domain, areas of doublet, sextet and octet components no longer follow ideal ratios (for example, 1:1 for quadrupole doublets) (Fig. 5.12a), while in the time domain the depth of quantum beats changes (Fig. 5.12b). Texture effects occur in all phases with crystal structure symmetry higher than cubic, but are generally not present in amorphous materials, such as glass.

Texture effects should not be ignored when analysing Mössbauer spectra. In some cases symmetry relations constrain the electric field gradient (EFG) orientation of different sites so that additional constraints may be applied, e.g., [18, 41] and in other cases it is possible to remove texture effects by measuring spectra with the crystal tilted at the magic angle, 54.7° [42, 43]. A further possibility to remove texture is to superimpose four spectra collected at different orientations of the sample in the same plane [44].

5.4.4 Sample Thickness

The effective sample thickness, t_a, is given by:

$$t_a(\text{dimensionless}) = \sigma_0 f_a n_a \tag{5.6}$$

where σ_0 is the resonance cross section of the transition (= 2.56×10^{-18} cm^2 for the 14.4 keV ^{57}Fe transition), f_a is the recoil-free fraction of the sample, and n_a is the number of ^{57}Fe atoms per cm^2 in the sample. Equation 5.6 is the equivalent of Eq. 5.1 for the sample. In the same way that the thin source limit is defined as $t_s << 1$, the thin absorber approximation is valid for $t_a << 1$. However while the ideal value for t_s is zero, the value of t_a must be non-zero in order to obtain a signal. There are analytical advantages to values of t_a within the thin absorber approximation [12], but such values are rarely the best choice for optimising the quality of the spectrum [45]. Extensive discussions of how to choose t_a are given in [1, 4].

A trade-off occurs when sample thickness is increased: a greater thickness means more iron and hence a stronger signal, but it also increases the amount of other elements in the sample which attenuates the signal due to electronic absorption and scattering. The latter is quantified through the total sample mass absorption coefficient which is given by:

$$\mu = \sum_i f_i \mu_i \tag{5.7}$$

where f_i is the mass fraction of the ith element and μ_i is the mass absorption coefficient of the ith element at 14.4 keV. The mass absorption coefficients for all elements are listed for selected energies in [46]. If sample thickness is expressed as an area density t' (for example in units of g/cm^2), the maximum signal to noise ratio for a sample as a function of t′ (optimum thickness) occurs at $t' = 2/\mu$ when non-resonant background is low (i.e., the sample contains no heavy elements) and $t' = 1/\mu$ when non-resonant background is high [1, 46].

These relations can be applied to $Fe_{0.4}Mg_{0.6}Si_{0.63}Al_{0.37}O_3$, whose spectra are shown in Fig. 5.4. The total sample mass absorption coefficient calculated from Eq. 5.7 is 17.3 cm^2/g for 14.4 keV radiation, which means the optimum thickness t' is 115.9 mg/cm^2 of sample. The physical thickness can be calculated from $10\, t'/\rho$ where ρ is the density of the sample. For the density 5 g/cm^3, the physical thickness is 232 μm, which corresponds to a dimensionless effective thickness of 9.3 according to Eq. 5.6 for iron with natural isotopic abundance (2.14% ^{57}Fe). However the sample used in Fig. 5.4 was constrained to a physical thickness of 15 μm because of diamond anvil cell geometry, which would have been a dimensionless effective thickness of 0.6 for natural ^{57}Fe abundance. To obtain a stronger signal, the sample was synthesised with 90% enriched ^{57}Fe, giving a dimensionless effective thickness of 25.4 according to Eq. 5.6. While this value is high and causes substantial thickness effects, a strong signal is required to overcome the loss of intensity due to absorption by the diamonds.

In addition to electronic absorption by elements within the sample, signal will be lost due to surrounding material that γ-rays pass through. Examples include substrates that the sample is mounted on, liquid if the sample is suspended in solution, and chambers and/or windows associated with samples in extreme environments (for example a diamond anvil cell) . The intensity of transmitted radiation can be calculated from:

Table 5.2 Transmission of 14.4 keV radiation through selected materials as a function of physical sample thickness (in mm)

Material	μ (cm^2/g)	ρ (g/cm^3)	Transmission (I/I_0)		
			99%	50%	1%
Diamond (C)	0.87	3.5	0.033	2.3	15
Glass (SiO_2)	6.97	2.2	0.007	0.45	3
Polycarbonate ($C_{16}H_{18}O_5$)	1.2	1.2	0.070	4.8	32
Epoxy resin ($C_{21}H_{25}ClO_5$)	2.84	1.2	0.028	2.0	13
Water (H_2O)	2.00	1.9	0.050	3.5	23

$$I = I_0 e^{-(l\mu/\rho)} \tag{5.8}$$

where I is the transmitted intensity, I_0 is the incident intensity, l is the length of the sample in the direction of radiation (i.e., the physical thickness) and ρ is the sample density.

Table 5.2 shows γ-ray transmission for selected materials calculated from Eq. 5.8 based on values of μ taken from [46]. More than 50% of 14.4 keV radiation is lost to diamonds in diamond anvil cell experiments where diamond thickness exceeds 2.3 mm. This thickness is usually exceeded, especially since diamonds are used in pairs. For preparation of small samples, glass is a poor choice of mounting substrate due to its high absorption (Sect. 5.4.2). Instead a plastic such as polycarbonate should be used. Epoxy resin can also be used, but the thickness of epoxy underlying the sample should be minimised. Finally, Eq. 5.8 can be used to demonstrate the benefits of ^{57}Fe enrichment. For an unenriched $Fe_{0.4}Mg_{0.6}Si_{0.63}Al_{0.37}O_3$ sample with physical thickness of 232 μm ($t_a = 9.3$), only 13% of the signal would be transmitted through the sample, while a 90% ^{57}Fe enriched sample with 15 μm physical thickness ($t_a = 25.4$) would transmit 88% of 14.4 keV radiation.

High values of t_a cause internal resonance within the sample that manifests itself differently in energy and time domain spectra. In the energy domain, linewidths become broadened, lineshapes deviate from Lorentzian, and relative areas become distorted from the relative abundance of their iron species. In the time domain, there is a speed-up effect (intensity increases at short delay times) and dynamical beats are superimposed over the quantum beats (Chap. 1, this volume). There are analytical solutions for these thickness effects in both energy and time domain, but not all cases yield comparable results to thinner samples. Such situations occur primarily in energy domain measurements, where linewidths may be too broad to resolve individual components and the overall intensity may be so saturated that information is lost at velocities where absorption is highest.

5.5 Analysis Software

Software plays an important role in Mössbauer investigations since deconvolution of Mössbauer spectra is required to extract quantitative information from the data. While this task is common to all experiments, there are particular aspects that are important for measurements with small beam size. Thickness effects should be taken into account, generally by fitting using the full transmission integral, e.g., [1, 4]. Thickness effects arise because radioactive point sources have high effective thickness (Sect. 5.2.2) and samples can also have high effective thickness, especially when enriched with ^{57}Fe (Sect. 5.4.4). SMS spectra have high absorption and negligible non-resonant background, which also requires a full transmission integral fit, e.g., [18]. Texture effects should be part of the fitting procedure (Sect. 5.4.3). Section 5.5 focuses on two different types of free software: (1) spectral deconvolution of spectra from small beam size measurements, and (2) mapping of microanalytical data obtained from heterogeneous samples.

5.5.1 Spectral Deconvolution

MossA fits and simulates energy domain data from both radioactive source (Lorentzian line shape) and synchrotron (Lorentzian-squared line shape) measurements [47], incorporates full transmission integral fitting and allows for texture effects (unequal component areas). MossA has a graphical user interface (GUI) where users define sites by mouse clicks and the fitting process is displayed in real time. The program is written in MATLAB and offers stand-alone executables for a number of different operating systems. The source code is also available and the software is distributed under the GNU General Public License. MossA can be downloaded from https://www.clemensprescher.com/.

CONUSS is highly versatile, allowing simulation and fitting of both energy and time domain spectra as well as data from other techniques (for example nuclear Bragg/Laue scattering and grazing incidence nuclear resonant scattering experiments) [48] and incorporates both thickness and texture effects. The program consists of different modules that are accessed through a command line interface, and offers a graphical display option. The package is written in Fortran90 and can be installed on many different operating systems. The source code is distributed under the GNU General Public License, and the software can be downloaded on request from https://nrixs.com/.

The simulations of energy and time domain spectra in this chapter were performed using MOTIF [24], which also performs fitting and incorporates both thickness and texture effects. The program uses a command line interface but leaves the graphical presentation of results up to the user. Binary files containing compiled executables for different operating systems are available on request from the author, Y. Shvyd'ko.

5.5.2 Mapping on a Microscopic Scale

Geographic information system (GIS) software is most commonly applied at the kilometre scale (cities and continents), but the GIS approach can also be used at the microscale [49]. Like most analytical methods, Mössbauer spectroscopy provides information that is most useful when combined with results from other techniques, for example structure determination or chemical analysis. Curating all information collected on a material is straightforward for homogeneous samples, but becomes more complicated for inhomogeneous samples because spatial relationships are important.

The free and open source software package QGIS (https://qgis.org/) provides a means to integrate all collected information and display visual maps through a user-friendly GUI. The Wisconsin Secondary Ion Mass Spectrometer Laboratory has developed a knowledge base that includes a tutorial on applying QGIS software to the integration of analytical data with spatial information (e.g., analysis locations, regions of interest or pictures) (https://sites.google.com/a/wisc.edu/wiscsims-micro-qgis/home). To illustrate its application to inhomogeneous samples, QGIS was applied to data measured on metasomatised eclogite from the same locality as the sample shown in Fig. 5.11k.

The map file, which is the integrated output of QGIS, consists of multiple layers that contain images and analyses, all keyed to a universal reference frame (Fig. 5.13a). For the metasomatised eclogite, the workflow consisted of the following: (1) a

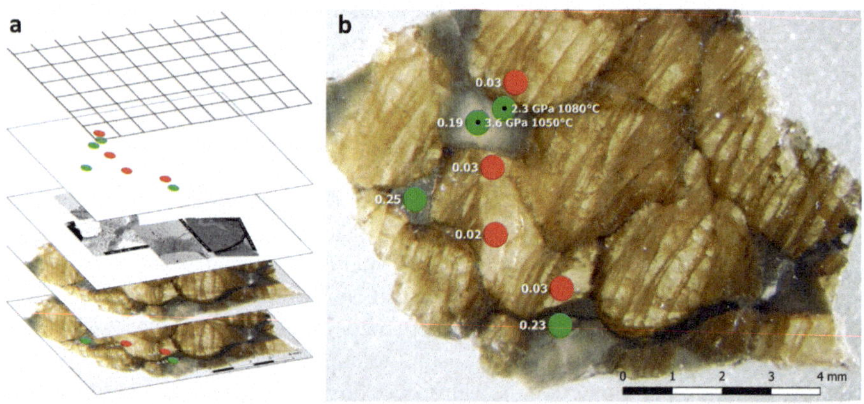

Fig. 5.13 QGIS software applied to metasomatised eclogite showing a different layers of **a** QGIS map, and **b** one of many possible exported images. The *left panel* shows (from *bottom* to *top*) the QGIS map file, the reflected light image, the backscattered electron images, the Mössbauer points, and the spatial reference frame (adapted from [49]). Not shown are the electron microprobe data analysis points. The *right panel* shows where Mössbauer spectra were collected (500 μm diameter *circles* shaded *green* for clinopyroxene and *red* for garnet) labelled with $Fe^{3+}/\Sigma Fe$ values, as well as pressures, temperatures and oxygen fugacities determined from clinopyroxene analyses at the *black dots* and adjacent garnet

reflected light image of the thick section was imported and used to define the coordinate system; (2) the x–y locations of the Mössbauer analyses were identified and entered into a data table with all fitting results and then imported; (3) scanning electron images at the micron scale were imported and keyed to their locations within the reference frame; (4) chemical analysis data from the electron microprobe were imported and aligned with the images; and (5) determinations of pressure, temperature and oxygen fugacity from adjoining garnet and clinopyroxene grains (based on chemical analyses, Mössbauer $Fe^{3+}/\Sigma Fe$ ratios and thermodynamic models) were imported and keyed to the relevant clinopyroxene grains. The resulting QGIS map file can be displayed within the QGIS environment with any combination of images and data and facilitates interrogation of all information and exploration of spatial relationships. Clicking on individual layers brings up a list of every piece of data attached to objects within that layer and allows individual analysis points to be identified.

After exploring the QGIS maps they can be exported in different formats for publication. Figure 5.13b illustrates an image from the eclogite example that shows the location and diameter of Mössbauer analyses combined with two determinations of pressure, temperature and oxygen fugacity. These results combined with analyses of other grains allowed the history of the rock to be elucidated—from subduction to residence in the mantle to final exhumation (C. McCammon, unpublished).

5.6 Applications of High Spatial Resolution Measurements

The myriad of applications using the techniques described above are the showcase of this chapter. Section 5.6 summarises studies organised according to the broad research topic that was investigated, the property of the sample targeted in the study, the specific observable that was measured, and the technique that was used. The compilation targets all studies reported in the peer-reviewed literature involving small beam size that address geoscience research questions or investigate materials of geological importance.

5.6.1 In Situ High-Pressure Studies with a Radioactive Source

The first Mössbauer spectra collected using a radioactive point source (RPS) were in situ measurements under high pressure. Indeed, the very first high-pressure Mössbauer study [50] was conducted not long after the Mössbauer effect was first discovered. The source was prepared by diffusing 74 MBq (2 mCi) of ^{57}Co into an iron disk of approximately 700 μm diameter, giving a specific activity of 20 GBq/cm^2 (0.5 Ci/cm^2), which was essentially the first ^{57}Co point source. Initial interest in high pressure came from laboratories focussed on physics and chemistry. For example, many notable contributions were made by the University of Illinois at Urbana-Champaign where H.G. Drickamer and collaborators applied high-pressure Mössbauer spectroscopy to the study of electronic structure and transitions [51]. Early studies used a press with hardened steel anvils to reach up to 30 GPa [52], but later adaptation of diamond anvil cell technology to high pressure Mössbauer studies allowed pressures in excess of 200 GPa to be attained (reviewed, for example, in [53]).

Radioactive point sources were fabricated initially by individual Mössbauer laboratories, e.g., [7], which restricted the number of installations due to the logistics of working with open radioactive material. The subsequent development of commercially available sealed radioactive point sources allowed the number of laboratories carrying out point source Mössbauer spectroscopic measurements to expand substantially, in particular to the geoscience community.

Research topics involving high-pressure phase transitions in geological materials greatly dominate the list of studies using a radioactive point source (76% of publications) [54–101] (Table 5.3). Phase transitions are generally easy to recognise in Mössbauer spectra, so in principle qualitative observation would suffice to map the location of phase transitions. It is notable, however, that all studies listed in Table 5.3 used hyperfine parameters derived from detailed fits to the Mössbauer spectra to

Table 5.3 Mössbauer studies on geological materials at high pressure using a radioactive source

Topic	Property measured	Observable	Method	References
Magnetic transition	Magnetism	Hyperfine parameters	RPS	[54–71]
Structure transition	Cation distribution, magnetism, spin state, site geometry	Hyperfine parameters	RPS	[72–84]
Spin transition	Magnetism, spin state	Hyperfine parameters	RPS	[85–92]
Insulator–metal transition	Cation distribution, magnetism	Hyperfine parameters	RPS	[93–101]
Charge transfer	Cation distribution	Hyperfine parameters	RPS	[102]
Chemical reaction	Cation distribution	Hyperfine parameters	RPS	[111]
Chemistry of Earth's core	Cation distribution	Hyperfine parameters	RPS	[112]
Crystal chemistry	Site geometry	Hyperfine parameters	RPS	[103–106]
Elastic properties	Spin state	Hyperfine parameters	RPS	[109]
Electronic structure	Site geometry	Hyperfine parameters	RPS	[107, 108]
Impact history	Cation distribution	Hyperfine parameters	RPS	[114–116]
Phase stability	Cation distribution	Hyperfine parameters	RPS	[113]
Seismic anomalies	Spin state	Hyperfine parameters	RPS	[110]

obtain additional information about the electronic environment around the nucleus [102–108]. Determination of hyperfine parameter variations are especially crucial for identifying spin and structure transitions in the non-magnetic state. Key to all studies listed in Table 5.3 is that energy domain Mössbauer spectra can usually be deconvoluted uniquely into individual components.

Studies carried out by geoscience laboratories generally address research questions beyond observing material behaviour of geologically relevant phases at high pressure. For example, spin transitions can change density and elastic properties [109] which can be used to interpret seismic anomalies [110]. Phase transitions can involve chemical transformation as well as structural and electronic changes [111]. Deciphering the chemistry of Earth's core relies on knowledge of phase relations involving iron and potential light alloying elements [112, 113]. Finally, high-pressure studies of meteorites provide information on impact processes that occurred in their history [114–116].

5.6.2 *In Situ High-Pressure Studies with a Synchrotron Source*

The advent of nuclear resonance beam lines at third-generation synchrotron sources greatly expanded possibilities for in situ high-pressure Mössbauer studies on geological materials. The smaller beam size and higher flux enabled studies at a wider range of pressures, dramatically shorter counting times made combined high-temperature high-pressure measurements possible, and generous access programmes expanded the user base to include those without in house facilities. The result is a wider range of research topics addressed, where studies focussed exclusively on high-pressure phase transitions make up only 32% of the total number of publications [19, 28, 29, 40, 117–139] (Table 5.4).

Other studies reported in Table 5.4 also involve phase transitions, but address research questions related to Earth properties. For example, a large number of studies focus on the chemistry of Earth's core, since phase transitions change properties such as density and sound velocity [140–148]. These can be compared with bulk geophysical measurements (for example from seismology) to place constraints on which elements could be alloyed with iron in Earth's core. Other studies focus on properties of Earth's mantle, in particular elastic properties [149–161] and how velocity anomalies observed by seismology can be explained [162–167]. Iron oxidation state influences mantle chemistry, both in the present day and throughout Earth's history [18, 168–172]. Determination of electronic structure and crystal chemistry of mantle minerals at high pressure can place constraints on their behaviour in Earth's mantle [173–180], and studies of magnetic properties provide insight into planetary magnetic fields [181].

Research questions addressed by high-pressure Mössbauer studies also target dynamic processes in Earth's interior. For example, spin transitions can change

Table 5.4 Mössbauer studies on geological materials at high pressure using a synchrotron source

Topic	Property measured	Observable	Method	References
Magnetic transition	Magnetism	Hyperfine parameters	NFS	[117]
Structure transition	Magnetism, site geometry	Hyperfine parameters	NFS	[118–121]
Spin transition	Cation distribution, spin state	Hyperfine parameters	NFS, SMS	[19, 28, 29, 40, 122–139]
Charge transfer	Magnetism	Hyperfine parameters	SMS	[175]
Chemical reaction	Cation distribution, oxidation state	Hyperfine parameters, signal intensity	NFS	[21, 192]
Chemistry of Earth's core	Magnetism	Hyperfine parameters	NFS, SMS	[140–148]
Temperature of Earth's core	Melting point	Signal intensity	NFS	[190, 191]
Crystal chemistry	Cation distribution, spin state	Hyperfine parameters	NFS, SMS	[176–180]
Elastic properties	Cation distribution, site geometry, spin state, magnetism	Hyperfine parameters, Fe^{3+} area	NFS, SMS	[149–161]
Electronic structure	Cation distribution, site geometry	Hyperfine parameters	NFS	[173, 174]
Mantle chemistry	Cation distribution, oxidation state, spin state	Hyperfine parameters	SMS	[18, 168–172]
Mantle magnetism	Magnetism	Hyperfine parameters	NFS	[181]
Mantle melt	Site geometry, spin state	Hyperfine parameters	NFS, SMS	[182–187]
Method development	Magnetism, cation distribution	Hyperfine parameters	SMS	[31, 193, 194]
Phase stability	Cation distribution	Hyperfine parameters	SMS	[188, 189]
Seismic anomalies	Spin state	Hyperfine parameters	NFS	[162–167]

melt densities and their relative buoyancy compared to bulk mantle material; hence analogue studies of glass can place constraints on the dynamics of molten mantle [182–187]. Also, stability studies of phases that undergo transformations involving exchange of oxygen provide information on deep Earth volatile cycles [188, 189].

Most studies reported in Table 5.4 use detailed fits of NFS and/or SMS spectra to monitor the electronic environment around the nucleus. The challenge of non-unique fitting models for NFS spectra (Sect. 5.3.1) is addressed through consistency checks with energy domain spectra collected on the same sample (usually with RPS) and using fitting software (Sect. 5.5) that incorporates realistic models for atomic environments (including hyperfine parameter distributions) as well as exploration of parameter space (for example using a Monte Carlo approach) to find other possible solutions [48]. In some cases, however, a simpler approach can be used for NFS measurements, particularly when monitoring processes in situ. For example, chemical reaction rates can be determined with high temporal resolution (30 s) by integrating counts within time windows [21] and melting points can be determined by monitoring signal intensity [190, 191].

5.6.3 Ex Situ High-Pressure and/or High-Temperature Studies

High-pressure experiments allow controlled studies of minerals at conditions relevant to Earth's interior. Mineral phases can be quenched to preserve chemistry, oxidation state, isotopic composition and often defect structure; hence *post mortem* analysis can provide important information about mineral history. Such studies have focussed primarily on samples quenched from large volume high-pressure experiments, mainly due to the high degree of control that can be maintained during experiments (for example temperature, oxygen fugacity, hydrogen fugacity, and so on) as well as straightforward procedures for sample recovery. Not surprisingly, therefore, most studies have used RPS Mössbauer spectroscopy since beam sizes can be large (>100 μm diameter) and lab-based Mössbauer measurements are more convenient (Table 5.5).

Research questions involving hydration dominate the list of studies. The incorporation of hydrogen into iron-bearing phases (both crystalline and glass) can alter charge balance; hence determination of $Fe^{3+}/\Sigma Fe$ values provides information on charge balance mechanisms that help for inferring hydrogen solubility, calculating deep water reservoirs and quantifying the deep Earth water cycle [195–210]. Sample characterisation combined with physical property measurements is also a common application. For example electrical conductivity studies of mantle minerals can be combined with Mössbauer measurements of $Fe^{3+}/\Sigma Fe$ to determine charge carrier

Table 5.5 Mössbauer studies on samples quenched from extreme conditions

Topic	Property measured	Observable	Method	References
Calibration XANES/flank method	Oxidation state	Hyperfine parameters, Fe^{3+} area	RPS	[33, 34, 257, 258]
Carbon chemistry	Oxidation state, site geometry	Hyperfine parameters	RPS	[248]
Charge transfer	Magnetism	Hyperfine parameters	RPS	[247]
Chemical reaction	Defect structure	Fe^{3+} area	RPS	[246]
Crystal chemistry	Cation distribution	Hyperfine parameters, Fe^{3+} area	RPS	[242–245]
Elastic properties	Cation distribution, defect structure	Hyperfine parameters, Fe^{3+} area	RPS	[220–227]
Hydration	Defect structure, oxidation state	Fe^{3+} area	RPS	[195–210]
Isotopic fractionation	Oxidation state	Hyperfine parameters	NFS	[249–251]
Mantle chemistry	Cation distribution, defect structure, oxidation state, spin state	Hyperfine parameters, Fe^{3+} area	RPS	[35, 232–241]
Mantle conductivity	Charge carrier density, defect structure	Fe^{3+} area	RPS, SMS	[211–219]
Phase stability	Oxidation state, phase identification	Hyperfine parameters, Fe^{3+} area	RPS	[252–254]
Seismic anomalies	Oxidation state	Fe^{3+} area	RPS	[228]
Spin transition	Cation distribution, defect structure	Hyperfine parameters, Fe^{3+} area	RPS	[229–231]

concentration and hence charge mobility, which provides important insight into transport properties of Earth's interior [211–219]. Elastic properties of minerals are influenced by their cation distribution and defect structure; hence Mössbauer measurements also play an important role in sample characterisation [220–227] and can be used to construct velocity models for comparison with geophysical data [228]. Although spin transitions cannot be directly probed in quenched samples, characterisation of cation distribution and iron oxidation state of minerals studied in situ at high pressure using other methods allows a more robust interpretation of the bulk data [229–231].

Mössbauer measurements of $Fe^{3+}/\Sigma Fe$ in lower mantle minerals have delivered many profound revelations regarding mantle chemistry [35, 232–241]. These include the discovery that ferric iron is predominant in the lower mantle instead of ferrous iron, that disproportionation of ferrous iron produces metallic iron throughout the lower mantle, and that deep blobs of material near the core-mantle boundary can be explained by the density contrast caused by different iron oxidation states. Mössbauer spectroscopy also provides information on crystal chemistry that can be used to constrain mantle properties and speciation of volatiles [242–248] as well as isotopic fractionation [249–251]. Finally, measurements of phase stability are important ingredients for determining the composition of different regions in Earth's interior [252–254].

There are a number of other micro-techniques used to determine Fe^{3+}/Fe ratios in geological samples (for example XANES [255] and the flank method [256]), but Mössbauer spectroscopy has an important advantage over these methods in that it requires no calibration to obtain accurate results. It is therefore ideally suited for calibrating synthetic XANES and flank method standards, and a small beam size enables the same grain to be used for all methods (Sect. 5.4.2) [33, 34, 257, 258].

5.6.4 Inclusions in Diamond

Diamonds are natural samples from Earth's interior. When they crystallise in Earth's mantle, they can incorporate mineral phases from their surroundings and protect them from alteration during rapid exhumation of diamonds during kimberlite eruptions. Mössbauer measurements of mineral inclusions therefore provide important information about properties and processes inside Earth. Since Mössbauer spectroscopy is non-destructive and measurements can be made directly through host diamonds, it is ideally suited for measuring these rare and precious samples (Sect. 5.4.2). Initially measurements were carried out using RPS Mössbauer spectroscopy, which limited the diameter of inclusions that could be studied to greater than around 200 μm (since there is no possibility to enrich samples in ^{57}Fe), but the use of SMS provides the capability to measure smaller inclusions.

Most studies have focussed on determination of Fe^{3+}/Fe ratios of minerals enclosed by diamond (Table 5.6). One obvious application is to decipher conditions during diamond formation [23, 41, 259, 260], but studies have also been carried out to determine oxygen fugacity [39, 261, 262] and hydration conditions [263] in regions of the mantle where diamond formation takes place. Iron oxidation state is also influenced by other volatile species such as carbon, which provides an ancient tracer of the deep carbon cycle [264].

Table 5.6 Mössbauer studies of inclusions in natural diamonds

Topic	Property measured	Observable	Method	References
Diamond formation	Magnetism, oxidation state	Hyperfine parameters, Fe^{3+} area	RPS, SMS	[23, 259, 260]
Hydration	Cation distribution	Fe^{3+} area	RPS	[263]
Mantle carbon	Oxidation state	Fe^{3+} area	SMS	[264]
Mantle chemistry	Cation distribution, oxidation state	Fe^{3+} area	RPS	[39, 261, 262]
Method development	Oxidation state	Fe^{3+} area	RPS	[41]

5.6.5 *Rare and/or Unusual Natural Samples*

Natural samples deliver important information about themselves and their history, but often are available only as small grains. Since Mössbauer spectroscopy generally requires minimal sample preparation, a wide spectrum of sample types can be measured (Sect. 5.4.2) (Table 5.7).

One common application of measuring small grains using Mössbauer spectroscopy is when reporting new minerals, where oxidation state and cation distribution can be extremely helpful for describing crystal structures [265–271]. Similarly, Mössbauer data can help decipher the crystal chemistry of iron-containing minerals with complex crystal structures [272] or their state of hydration [273]. Fe^{3+}/Fe ratios are important inputs to thermodynamic models used to infer oxygen fugacities [274–279] and pressure–temperature paths during metamorphism [280], and can be correlated with iron isotopic ratios to constrain ancient processes such as partial melting [281–286]. The cooling history of minerals is often written in their cation distributions [287, 288], and iron oxidation state and/or the presence of nanophases provide insight to the impact history of meteorites [289] and tektites [290]. Magnetic properties of minerals can constrain ancient magnetic fields, both on Earth [291] and other planets [292].

Small grains are often single crystals, which means strong texture effects are present that cause asymmetric component areas in non-cubic materials, which can complicate spectral fitting (Sect. 5.4.3). Such challenges can be overcome if the principal EFG axis direction of individual sites are known, and these can be determined through studies of oriented crystals [293]. Similarly, the effect of the Goldanskii-Karyagin effect (GKE), which also causes asymmetric component areas, can also be documented [36]. Such studies are useful not only from a theoretical perspective, but also they enable highly accurate measurement of Fe^{3+}/Fe ratios on single crystals. Natural grains can be used for calibration of XANES and flank method standards in the same way as for synthetic phases as mentioned above, since the use of standards whose compositions closely match the actual samples to be studied reduces systematic errors in those methods [294].

Table 5.7 Mössbauer studies of rare and/or unusual natural samples

Topic	Property measured	Observable	Method	References
Aqueous geochemistry	Phase identification	Hyperfine parameters	SMS	[295]
Calibration flank method	Oxidation state	Fe^{3+} area	RPS	[294]
Cooling history	Cation distribution	Fe^{3+} area	RPS	[287, 288]
Crystal chemistry	Cation distribution	Fe^{3+} area	RPS	[272]
Crystal structure	Cation distribution	Fe^{3+} area	RPS	[265–270]
Electronic structure	EFG orientation, GKE	Area asymmetry	RPS	[36, 293]
Hydration	Defect structure	Fe^{3+} area	RPS	[273]
Impact history	Cation distribution, defect structure, oxidation state	Fe^{3+} area	RPS	[289, 290]
Isotopic fractionation	Oxidation state	Fe^{3+} area	RPS	[281–286]
Mantle chemistry	Cation distribution, oxidation state	Hyperfine parameters, Fe^{3+} area	RPS	[274–279]
Metamorphism	Oxidation state	Fe^{3+} area	RPS	[280]
Method development	Cation distribution	Hyperfine parameters	RPS	[271]
Paleomagnetism	Magnetism	Hyperfine parameters, Fe^{3+} area	RPS	[291, 292]
Technology	Magnetism	Hyperfine parameters	RPS	[296]

Natural samples may have low iron abundance that can be enhanced through concentrating measures that nevertheless yield only a small quantity of material. For example, dissolved iron in aquatic environments can be concentrated through vacuum filtration and measured using Mössbauer spectroscopy, even if the sample volume is microscopic [295]. The use of SMS provides a significantly enhanced signal to noise ratio compared to RPS, which is important for dilute samples.

5.6.6 Heterogeneous Samples

Mössbauer studies of geologically relevant samples are almost always carried out on single phases, despite the rich information that can be obtained from heterogeneous samples. This focus on single phase measurements is a consequence of the beam size: when it is large compared to the length scale of heterogeneities, Mössbauer spectroscopy records only the average assemblage and information can be lost. Natural samples contain heterogeneities on many length scales, ranging

Table 5.8 Mössbauer studies of samples with sub-mm heterogeneity

Topic	Property measured	Observable	Method	References
Deformation	Oxidation state	Fe^{3+} area	RPS	[300]
Magmatic history	Oxidation state	Fe^{3+} area	RPS	[301]
Metasomatism	Oxidation state	Fe^{3+} area	RPS	[37]
Method development	Phase identification, oxidation state	Hyperfine parameters	RPS, RCS, NFS	[8, 32, 297]
Planetary magnetism	Magnetism	Hyperfine parameters	SMS	[302]

from the very large (for example rocks with centimetre-sized grains) to the very small (for example nanoscale exsolution lamellae). Methodological development has been steadily advancing down this scale. The radioactive point source enabled single measurements down to around 100 μm diameter beam size that could be repeated on different spots of a heterogeneous sample [297], while the use of a position-sensitive detector with a conventional source opened the possibility to record many spectra simultaneously with spatial resolution down to 50 μm over a heterogeneous sample [8]. Current synchrotron capabilities offer spatial resolutions down to 5 μm coupled with a rastering approach to scan heterogeneous samples [32], while upgrades currently in progress have targeted nanoscale beam sizes (for example the ESRF upgrade anticipates a beam size of about 200 nm for the nuclear resonance beamline [298, 299]).

Studies applying Mössbauer spectroscopy to heterogeneous samples are sparse but diverse (Table 5.8). Oxygen fugacity gradients can be monitored during deformation experiments through measurements of iron oxidation state in ferropericlase [300]. Magmatic history (specifically oxygen fugacity) can be deciphered in subliquidus (i.e., not fully molten) systems from measurements of iron oxidation state in assemblages where glass and crystal can be individually distinguished [301]. Zoning profiles in minerals provide information on chemical processes that occurred during the history of the assemblage. For example, progressive changes in oxygen fugacity associated with metasomatism can be determined from iron oxidation state of zoned garnets that ultimately led to dissolution of diamond [37]. Finally, the sequence of magnetic phases in meteorites provides a time-series record of ancient magnetic fields in the solar system [302].

5.7 Conclusions and Outlook

This chapter has presented topics relevant to Mössbauer measurements with high spatial resolution specifically directed to research problems in geoscience. Such measurements are challenging, mainly in obtaining data with sufficiently high signal to noise ratio that provide a meaningful answer to the research questions being investigated. This chapter demonstrates how experiments can be optimised with regard to spectrometer geometry and choice of radioactive source for lab-based measurements, choice of time or energy domain for synchrotron measurements, and how to prepare samples (form and thickness) for both types of measurements. Calculations of thickness and geometry are straightforward, and free software is available that enables simulation of potential spectra to facilitate the best decisions. High spatial resolution is therefore not a barrier to obtaining high quality spectra.

Applications of high spatial resolution Mössbauer spectroscopy in geoscience cover a wide range of research questions, but also show a number of trends. In situ high-pressure investigations using a radioactive source focus mainly on transitions and material properties, while synchrotron studies address a broader scope of research questions and focus deeper within Earth. This trend may reflect the predominance of physics and chemistry laboratories carrying out high-pressure studies in the early days of radioactive source experiments, while synchrotron facilities allowed the user base to expand to include those without in house facilities. *Post mortem* investigations involve predominantly samples quenched from large-volume apparatus, where large and homogeneous samples offer the convenience of in house (radioactive source) studies. All high-pressure studies, both in situ and ex situ, focus nearly exclusively on single phase samples due to constraints on beam size. This constraint also restricts the size of diamond inclusions that can be studied, as well as the spatial resolution that can be achieved in studies of inhomogeneous samples.

The dominant role of iron in myriads of processes that take place on and within Earth as well as the diversity of iron species means that many important research questions are still waiting to be addressed with current capabilities. Furthermore, recent developments have expanded possibilities, for example the focussed ion beam (FIB) technique combined with SMS enables the collection of Mössbauer spectra on single-phase samples quenched from diamond anvil cell experiments.

Future developments will expand possibilities further and allow entirely new areas of science to be explored. For example the upgrade at ESRF [298, 299] is expected to offer users a sub-micron beam size for SMS measurements, opening many new possibilities, including studies of multiphase samples at high pressure. Such experiments include the study of chemical reactions, the motor of most Earth processes, not to mention life itself. Further applications of sub-micron beam size are presented in Chap. 1 of this volume. In the same way that reducing the beam size of Mössbauer measurements by two orders of magnitude (radioactive point source) or three orders of magnitude (synchrotron source) brought incomparable new insights described in the more than 250 studies reported in Sect. 5.6, the further reduction in beam size by one to two orders of magnitude will surely bring new surprises and understanding of how our planet works.

References

1. P. Gütlich, E. Bill, A. Trautwein, *Mössbauer Spectroscopy and Transition Metal Chemistry* (Springer, Berlin, 2011)
2. C. A. McCammon, Hyperfine Interact. **144/145**, 289 (2003)
3. F.C. Hawthorne, in *Spectroscopic Methods in Mineralogy and Geology*. ed. by F.C. Hawthorne (Mineralogical Society of America, Washington, D.C., 1988), p. 255
4. E. Murad, J.D. Cashion, *Mössbauer Spectroscopy of Environmental Materials and Their Industrial Utilization* (Kluwer, Boston, 2004)
5. D. E. Gray, *American Institute of Physics Handbook* (McGraw-Hill, 1963)
6. Z.M. Stadnik, Mössbauer Effect Reference and Data Journal **1**, 217 (1978)
7. G. Longworth, B. Window, J. Phys. D Appl. Phys. **4**, 835 (1971)
8. P.R. Smith, J.D. Cashion, L.J. Brown, Hyperfine Interact. **71**, 1503 (1992)
9. Y. Yoshida, T. Kamimura, M. Ichino, K. Hayakawa, K. Yukihira, H. Soejima, J. Phys. Conference Series **217** (2010). https://doi.org/10.1088/1742-6596/217/1/012003
10. S.J. Norton, Nature **330**, 151 (1987)
11. U. Atzmony, S.J. Norton, L.J. Swartzendruber, L.H. Bennett, Nature **330**, 153 (1987)
12. S. Margulies, J.R. Ehrman, Nuclear Instruments & Methods **12**, 131 (1961)
13. D.G. Rancourt, Nucl. Instrum. Methods Phys. Res. **B44**, 199 (1989)
14. C.A. McCammon, Hyperfine Interact. **92**, 1235 (1994)
15. W. Sturhahn, Journal of Physics-Condensed Matter **16**, S497 (2004)
16. R. Rüffer, A. I. Chumakov, Hyperfine Interactions 97/98, 589 (1996)
17. R. Rüffer, A. I. Chumakov, in E. Jaeschke, S. Khan, J. R. Schneider, J. B. Hastings (eds.), *Synchrotron Light Sources and Free-Electron Lasers* (Springer, 2016), p. 1759
18. V. Potapkin, C. McCammon, K. Glazyrin, A. Kantor, I. Kupenko, C. Prescher, R. Sinmyo, G. V. Smirnov, A. I. Chumakov, R. Rüffer, L. Dubrovinsky, Nature Commun. **4** (2013). https://doi.org/10.1038/ncomms2436
19. I. Kupenko, C. McCammon, R. Sinmyo, V. Cerantola, V. Potapkin, A.I. Chumakov, A. Kantor, R. Rüffer, L. Dubrovinsky, Earth Planet. Sci. Lett. **423**, 78 (2015)
20. I. Kupenko, C. Strohm, C. McCammon, V. Cerantola, K. Glazyrin, S. Petitgirard, D. Vasiukov, G. Aprilis, A. Chumakov, R. Rüffer, L. Dubrovinsky, Rev. Scienti. Instru. **86** (2015). https://doi.org/10.1063/1.4935304
21. J.E. Amonette, R.K. Kukkadapu, E.E. Alp, W. Sturhahn, T.S. Toellner, Geochim. Cosmochim. Acta **67**, 2109 (2003)
22. H. Grünsteudel, M. Haas, O. Leupold, D. Mandon, B. Matzanke, W. Meyer-Klaucke, H. Paulsen, E. Realo, H. Ruter, A. Trautwein, R. Weiss, H. Winkler, Inorg. Chim. Acta **276**, 334 (1998)
23. F. Nestola, V. Cerantola, S. Milani, C. Anzolini, C. McCammon, D. Novella, I. Kupenko, A. Chumakov, R. Rüffer, J.W. Harris, Lithos **265**, 328 (2016)
24. Y. V. Shvyd'ko, Hyperfine Interact. **125**, 173 (2000)
25. W.A. Dollase, Am. Miner. **60**, 257 (1975)
26. C. McCammon, K. Glazyrin, A. Kantor, I. Kantor, I. Kupenko, O. Narygina, V. Potapkin, C. Prescher, R. Sinmyo, A. Chumakov, R. Rüffer, I. Sergueev, G. Smirnov, L. Dubrovinsky, High Pressure Research **33**, 663 (2013)
27. J.-F. Lin, E.E. Alp, Z. Mao, T. Inoue, C. McCammon, Y. Xiao, P. Chow, J. Zhao, Am. Miner. **97**, 592 (2012)
28. J. F. Lin, A. G. Gavriliuk, V. V. Struzhkin, S. D. Jacobsen, W. Sturhahn, M. Y. Hu, P. Chow, C. S. Yoo, Phys. Rev. B 73 (2006). https://doi.org/10.1103/PhysRevB.73.113107
29. I. Kupenko, C. McCammon, R. Sinmyo, C. Prescher, A.I. Chumakov, A. Kantor, R. Rüffer, L. Dubrovinsky, Lithos **189**, 167 (2014)
30. J.J. Bara, B.F. Bogacz, Mössbauer Effect Reference and Data Journal **3**, 154 (1980)
31. V. Potapkin, A.I. Chumakov, G.V. Smirnov, J.P. Celse, R. Ruffer, C. McCammon, L. Dubrovinsky, Journal of Synchrotron Radiation **19**, 559 (2012)

32. L.F. Yan, J.Y. Zhao, T.S. Toellner, R. Divan, S.L. Xu, Z.H. Cai, J.S. Boesenberg, J.M. Friedrich, S.P. Cramer, E.E. Alp, Journal of Synchrotron Radiation **19**, 814 (2012)
33. M. Longo, C.A. McCammon, S.D. Jacobsen, Contrib. Miner. Petrol. **162**, 1249 (2011)
34. F. Sorbadere, V. Laurenz, D.J. Frost, M. Wenz, A. Rosenthal, C. McCammon, C. Rivard, Geochim. Cosmochim. Acta **239**, 235 (2018)
35. D.J. Frost, C.A. McCammon, Am. Miner. **94**, 872 (2009)
36. J.N. Bull, W.C. Tennant, T.B. Ballaran, F. Nestola, C.A. McCammon, Phys. Chem. Miner. **39**, 561 (2012)
37. C.A. McCammon, W.L. Griffin, S.R. Shee, H.S.C. O'Neill, Contrib. Miner. Petrol. **141**, 287 (2001)
38. D.E. Jacob, R. Wirth, F. Enzmann, A. Kronz, A. Schreiber, Earth Planet. Sci. Lett. **308**, 307 (2011)
39. C. McCammon, M. Hutchison, J. Harris, Science **278**, 434 (1997)
40. V. Cerantola, C. McCammon, I. Kupenko, I. Kantor, C. Marini, M. Wilke, L. Ismailova, N. Solopova, A. Chumakov, S. Pascarelli, L. Dubrovinsky, Am. Miner. **100**, 2670 (2015)
41. C.A. McCammon, W.C. Tennant, R. Miletich, Hyperfine Interact. **126**, 241 (2000)
42. T. Ericsson, R. Wäppling, Journal De Physique Colloques **37**(C6), 719 (1976)
43. H.D. Pfannes, U. Gonser, Appl. Phys. **1**, 93 (1973)
44. J.M. Greneche, F. Varret, Journal De Physique Lettres **43**, 233 (1983)
45. D.G. Rancourt, A.M. Mcdonald, A.E. Lalonde, J.Y. Ping, Am. Miner. **78**, 1 (1993)
46. G.L. Long, T.E. Cranshaw, G. Longworth, Mössbauer Effect Reference and Data Journal **6**, 42 (1983)
47. C. Prescher, C. McCammon, L. Dubrovinsky, J. Appl. Crystallogr. **45**, 329 (2012)
48. W. Sturhahn, Hyperfine Interact. **125**, 149 (2000)
49. B. J. Linzmeier, K. Kitajima, A. C. Denny, J. N. Cammack, Eos **99** (2018). https://doi.org/10.1029/2018EO099269
50. R.V. Pound, G.B. Benedek, R. Drever, Phys. Rev. Lett. **7**, 405 (1961)
51. H.G. Drickamer, C.W. Frank, Annu. Rev. Phys. Chem. **23**, 39 (1972)
52. P. Debrunner, R.W. Vaughan, A.R. Champion, J. Cohen, J. Moyzis, H.G. Drickamer, Reviews of Scientific Instruments **37**, 1310 (1966)
53. M.P. Pasternak, R.D. Taylor, in *Mössbauer Spectroscopy Applied to Magnetism and Materials Science*, vol. 2, ed. by G.J. Long, F. Grandjean (Plenum Press, New York and London, 1996), p. 167
54. I. Choe, R. Ingalls, J. M. Brown, Y. Satosorensen, R. Mills, Phys. Rev. B **44** (1991). https://doi.org/10.1103/PhysRevB.44.1
55. M. Hayashi, I. Tamura, O. Shimomura, H. Sawamoto, H. Kawamura, Phys. Chem. Miner. **14**, 341 (1987)
56. A. P. Kantor, S. D. Jacobsen, I. Y. Kantor, L. S. Dubrovinsky, C. A. McCammon, H. J. Reichmann, I. N. Goncharenko, Phys. Rev. Lett. **93** (2004). https://doi.org/10.1103/PhysRevLett.93.215502
57. I. Kantor, L. Dubrovinsky, C. McCammon, Phase Transitions **80**, 1151 (2007)
58. I.Y. Kantor, C.A. McCammon, L.S. Dubrovinsky, J. Alloy. Compd. **376**, 5 (2004)
59. C.A. McCammon, Hyperfine Interact. **93**, 1511 (1994)
60. C.A. McCammon, Am. Miner. **80**, 1 (1995)
61. S. Nasu, K. Kurimoto, S. Nagatomo, S. Endo, F.E. Fujita, Hyperfine Interact. **29**, 1583 (1986)
62. S. Nasu, T. Sasaki, T. Kawakami, T. Tsutsui, S. Endo, Hyperfine Interact. **151**, 245 (2003)
63. M.P. Pasternak, R.D. Taylor, R. Jeanloz, X. Li, J.H. Nguyen, C.A. McCammon, Phys. Rev. Lett. **79**, 5046 (1997)
64. S. Takele, G.R. Hearne, Nuclear Instruments & Methods in Physics Research Section B-Beam Interactions with Materials and Atoms **183**, 413 (2001)
65. R.D. Taylor, M.P. Pasternak, R. Jeanloz, J. Appl. Phys. **69**, 6126 (1991)
66. R.W. Vaughan, H.G. Drickamer, Journal of Chemical Physics **47**, 1530 (1967)
67. Q.G. Wei, C. McCammon, S.A. Gilder, Geochem. Geophys. Geosyst. **18**, 4646 (2017)
68. D.L. Williamson, S. Bukshpan, R. Ingalls, Physical Review B **6**, 4194 (1972)

69. D.N. Pipkorn, C.K. Edge, P. Debrunner, G.D. Pasquali, H.G. Drickamer, H. Frauenfelder, Phys. Rev. **135**, 1604 (1964)
70. H. King, D. Virgo, H. Mao, Carnegie Institute of Washington Yearbook **77**, 830 (1978)
71. H.K. Mao, D. Virgo, P.M. Bell, Carnegie Institute of Washington Yearbook **76**, 522 (1977)
72. K. Glazyrin, C. McCammon, L. Dubrovinsky, M. Merlini, K. Schollenbruch, A. Woodland, M. Hanfland, Am. Miner. **97**, 128 (2012)
73. E. Greenberg, L.S. Dubrovinsky, C. McCammon, J. Rouquette, I. Kantor, V. Prakapenka, G.K. Rozenberg, M.P. Pasternak, Am. Miner. **96**, 833 (2011)
74. E. Greenberg, W. M. Xu, M. Nikolaevsky, E. Bykova, G. Garbarino, K. Glazyrin, D. G. Merkel, L. Dubrovinsky, M. P. Pasternak, G. K. Rozenberg, Phys. Rev. B **95** (2017). https://doi.org/10.1103/PhysRevB.95.195150
75. G. Hearne, V. Pischedda, J. Solid State Chem. **187**, 134 (2012)
76. I.Y. Kantor, L.S. Dubrovinsky, A.P. Kantor, V.S. Urusov, C. McCammon, W. Crichton, Dokl. Phys. **50**, 343 (2005)
77. T. Kawakami, S. Nasu, T. Tsutsui, T. Sasaki, T. Yamada, S. Endo, M. Takano, T. Katamoto, J. Phys. Soc. Jpn. **72**, 2640 (2003)
78. M.P. Pasternak, S. Nasu, K. Wada, S. Endo, Physical Review B **50**, 6446 (1994)
79. G. K. Rozenberg, Y. Amiel, W. M. Xu, M. P. Pasternak, R. Jeanloz, M. Hanfland, R. D. Taylor, Phys. Rev. B **75** (2007). https://doi.org/10.1103/PhysRevB.75.020102
80. S. Winell, H. Annersten, V. Prakapenka, Am. Miner. **91**, 560 (2006)
81. X. Wu, G. Steinle-Neumann, O. Narygina, I. Kantor, C. McCammon, S. Pascarelli, G. Aquilanti, V. Prakapenka, L. Dubrovinsky, Phys. Rev. B **79** (2009). https://doi.org/10.1103/PhysRevB.79.094106
82. X. Wu, G. Steinle-Neumann, O. Narygina, C. McCammon, L. Dubrovinsky, High Pressure Research **30**, 395 (2010)
83. L. Zhang, S.S. Hafner, Am. Miner. **77**, 462 (1992)
84. F.E. Huggins, H.-K. Mao, D. Virgo, Carnegie Institute of Washington Yearbook **75**, 756 (1976)
85. I. Kantor, L. Dubrovinsky, C. McCammon, A. Kantor, S. Pascarelli, G. Aquilanti, W. Crichton, M. Mattesini, R. Ahuja, J. Almeida, V. Urusov, Phys. Chem. Miner. **33**, 35 (2006)
86. I. Y. Kantor, L. S. Dubrovinsky, C. A. McCammon, Phys. Rev. B **73** (2006). https://doi.org/10.1103/PhysRevB.73.100101
87. J. Rouquette, I. Kantor, C.A. McCammon, V. Dmitriev, L.S. Dubrovinsky, Inorg. Chem. **47**, 2668 (2008)
88. R. Sinmyo, C. McCammon, L. Dubrovinsky, Am. Miner. **102**, 1263 (2017)
89. S. Takele, G.R. Hearne, Journal of Physics-Condensed Matter **13**, 10077 (2001)
90. W. M. Xu, G. R. Hearne, S. Layek, D. Levy, J. P. Itie, M. P. Pasternak, G. K. Rozenberg, E. Greenberg, Phys. Rev. B **96** (2017). https://doi.org/10.1103/PhysRevB.96.045108
91. W. M. Xu, G. R. Hearne, S. Layek, D. Levy, J. P. Itie, M. P. Pasternak, G. K. Rozenberg, E. Greenberg, Phys. Rev. B **95** (2017). https://doi.org/10.1103/PhysRevB.95.045110
92. W. M. Xu, G. R. Hearne, S. Layek, D. Levy, M. P. Pasternak, G. K. Rozenberg, E. Greenberg, Phys. Rev. B **97** (2018). https://doi.org/10.1103/PhysRevB.97.085120
93. L.S. Dubrovinsky, N.A. Dubrovinskaia, C. McCammon, G.K. Rozenberg, R. Ahuja, J.M. Osorio-Guillen, V. Dmitriev, H.P. Weber, T. Le Bihan, B. Johansson, Journal of Physics-Condensed Matter **15**, 7697 (2003)
94. E. Greenberg, I. Leonov, S. Layek, Z. Konopkova, M. P. Pasternak, L. Dubrovinsky, R. Jeanloz, I. A. Abrikosov, G. K. Rozenberg, Phys. Rev. X **8** (2018). https://doi.org/10.1103/PhysRevX.8.031059
95. E. Greenberg, G. K. Rozenberg, W. M. Xu, M. P. Pasternak, C. McCammon, K. Glazyrin, L. S. Dubrovinsky, Phys. Rev. B **88** (2013). https://doi.org/10.1103/PhysRevB.88.214109
96. M.P. Pasternak, G.K. Rozenberg, G.Y. Machavariani, O. Naaman, R.D. Taylor, R. Jeanloz, Phys. Rev. Lett. **82**, 4663 (1999)
97. M.P. Pasternak, W.M. Xu, G.K. Rozenberg, R.D. Taylor, R. Jeanloz, J. Magn. Magn. Mater. **265**, L107 (2003)

98. M.P. Pasternak, W.M. Xu, G.K. Rozenberg, R.D. Taylor, R. Jeanloz, J. Phys. Chem. Solids **65**, 1531 (2004)
99. G.K. Rozenberg, M.P. Pasternak, G.R. Hearne, C.A. McCammon, Phys. Chem. Miner. **24**, 569 (1997)
100. S. Takele, G.R. Hearne, Physical Review B **60**, 4401 (1999)
101. W. M. Xu, G. Y. Machavariani, G. K. Rozenberg, M. P. Pasternak, Phys. Rev. B **70** (2004). https://doi.org/10.1103/PhysRevB.70.174106
102. T. Seda, G.R. Hearne, Journal of Physics-Condensed Matter **16**, 2707 (2004)
103. I. Choe, R. Ingalls, J.M. Brown, Y. Satosorensen, Phys. Chem. Miner. **19**, 236 (1992)
104. F. E. Huggins, Dissertation, Massachusetts Institute of Technology (1975)
105. I. Kantor, L. Dubrovinsky, C. McCammon, G. Steinle-Neumann, A. Kantor, N. Skorodumova, S. Pascarelli, G. Aquilanti, Phys. Rev. B **80** (2009). https://doi.org/10.1103/PhysRevB.80.014204
106. F.E. Huggins, H.-K. Mao, D. Virgo, Carnegie Institute of Washington Yearbook **74**, 405 (1975)
107. O.V. Narygina, I.Y. Kantor, C.A. McCammon, L.S. Dubrovinsky, Phys. Chem. Miner. **37**, 407 (2010)
108. L. Zhang, S.S. Hafner, Am. Miner. **77**, 474 (1992)
109. S. Speziale, V. E. Lee, S. M. Clark, J. F. Lin, M. P. Pasternak, R. Jeanloz, J. Geophys. Res. Solid Earth **112** (2007). https://doi.org/10.1029/2006jb004730
110. S. Speziale, A. Milner, V.E. Lee, S.M. Clark, M.P. Pasternak, R. Jeanloz, Proc. Natl. Acad. Sci. U.S.A. **102**, 17918 (2005)
111. X. Wu, G. Steinle-Neumann, O. Narygina, I. Kantor, C. McCammon, V. Prakapenka, V. Swamy, L. Dubrovinsky, Phys. Rev. Lett. **103** (2009). https://doi.org/10.1103/PhysRevLett.103.065503
112. T.T. Gu, X.A. Wu, S. Qin, L. Dubrovinsky, Phys. Earth Planet. Inter. **184**, 154 (2011)
113. J. Rouquette, D. Dolejs, I. Y. Kantor, C. A. McCammon, D. J. Frost, V. B. Prakapenka, L. S. Dubrovinsky, Appl. Phys. Lett. **92** (2008). https://doi.org/10.1063/1.2892400
114. U. Chandra, K.K. Pandey, G. Parthasarathy, S.M. Sharma, Geosci. Front. **7**, 265 (2016)
115. U. Chandra, G. Parthasarathy, N.V.C. Shekar, P.C. Sahu, Chemie Der Erde-Geochemistry **73**, 197 (2013)
116. U. Chandra, P. Sharma, G. Parthasarathy, B. Sreedhar, Am. Miner. **95**, 870 (2010)
117. L. Zhang, J. Stanek, S.S. Hafner, H. Ahsbahs, H.F. Grünsteudel, J. Metge, R. Rüffer, Am. Miner. **84**, 447 (1999)
118. H. F. Grünsteudel, H. J. Heese, A. I. Chumakov, H. Grünsteudel, O. Leupold, J. Metge, R. Rüffer, G. Wortmann, in *Proceedings of the International Conference on the Applications of the Mossbauer Effect—ICAME-95*, ed. by I. Ortalli (SIF, Bologna, 1996), p. 853 (1996)
119. J. F. Lin, J. S. Tse, E. E. Alp, J. Y. Zhao, M. Lerche, W. Sturhahn, Y. M. Xiao, P. Chow, Phys. Rev. B **84** (2011). https://doi.org/10.1103/PhysRevB.84.064424
120. D.Z. Zhang, J.M. Jackson, B. Chen, W. Sturhahn, J.Y. Zhao, J.Y. Yan, R. Caracas, Journal of Geophysical Research-Solid Earth **118**, 4071 (2013)
121. D.Z. Zhang, J.M. Jackson, W. Sturhahn, Y.M. Xiao, Am. Miner. **96**, 1585 (2011)
122. S.M. Dorfman, S.E. Dutton, V. Potapkin, A.I. Chumakov, J.P. Rueff, P. Chow, Y.M. Xiao, R.J. Cava, T.S. Duffy, C.A. McCammon, P. Gillet, Am. Miner. **101**, 1659 (2016)
123. B. Grocholski, S. H. Shim, W. Sturhahn, J. Zhao, Y. Xiao, P. C. Chow, Geophys. Res. Lett. **36** (2009). https://doi.org/10.1029/2009gl041262
124. M. Hamada, S. Kamada, E. Ohtani, T. Mitsui, R. Masuda, T. Sakamaki, N. Suzuki, F. Maeda, M. Akasaka, Phys. Rev. B **93** (2016). https://doi.org/10.1103/PhysRevB.93.155165
125. L. Ismailova, M. Bykov, E. Bykova, A. Bobrov, I. Kupenko, V. Cerantola, D. Vasiukov, N. Dubrovinskaia, C. McCammon, M. Hanfland, K. Glazyrin, H.P. Liermann, A. Chumakov, L. Dubrovinsky, Am. Miner. **102**, 184 (2017)
126. J.M. Jackson, W. Sturhahn, G.Y. Shen, J.Y. Zhao, M.Y. Hu, D. Errandonea, J.D. Bass, Y.W. Fei, Am. Miner. **90**, 199 (2005)
127. J. Li, W. Sturhahn, J.M. Jackson, V.V. Struzhkin, J.F. Lin, J. Zhao, H.K. Mao, G. Shen, Phys. Chem. Miner. **33**, 575 (2006)

128. J.F. Lin, E.E. Alp, Z. Mao, T. Inoue, C. McCammon, Y.M. Xia, P. Chow, J.Y. Zhao, Am. Miner. **97**, 592 (2012)
129. J.F. Lin, A.G. Gavriliuk, W. Sturhahn, S.D. Jacobsen, J.Y. Zhao, M. Lerche, M. Hu, Am. Miner. **94**, 594 (2009)
130. J.F. Lin, Z. Mao, J. Yang, J. Liu, Y.M. Xiao, P. Chow, T. Okuchi, Geophys. Res. Lett. **43**, 6952 (2016)
131. J.F. Lin, H. Watson, G. Vanko, E.E. Alp, V.B. Prakapenka, P. Dera, V.V. Struzhkin, A. Kubo, J.Y. Zhao, C. McCammon, W.J. Evans, Nat. Geosci. **1**, 688 (2008)
132. I.S. Lyubutin, J.F. Lin, A.G. Gavriliuk, A.A. Mironovich, A.G. Ivanova, V.V. Roddatis, A.L. Vasiliev, Am. Miner. **98**, 1803 (2013)
133. C. McCammon, I. Kantor, O. Narygina, J. Rouquette, U. Ponkratz, I. Sergueev, M. Mezouar, V. Prakapenka, L. Dubrovinsky, Nat. Geosci. **1**, 684 (2008)
134. N.V. Solomatova, J.M. Jackson, W. Sturhahn, J.K. Wicks, J.Y. Zhao, T.S. Toellner, B. Kalkan, W.M. Steinhardt, Am. Miner. **101**, 1084 (2016)
135. E.C. Thompson, A.H. Davis, W. Bi, J. Zhao, E.E. Alp, D. Zhang, E. Greenberg, V.B. Prakapenka, A.J. Campbell, Geochem. Geophys. Geosyst. **19**, 305 (2018)
136. X. Wu, Y. Wu, J.F. Lin, J. Liu, Z. Mao, X.Z. Guo, T. Yoshino, C. McCammon, V.B. Prakapenka, Y.M. Xiao, Journal of Geophysical Research-Solid Earth **121**, 6411 (2016)
137. Y. Wu, F. Qin, X. Wu, H.J. Huang, C.A. McCammon, T. Yoshino, S.M. Zhai, Y.M. Xiao, V.B. Prakapenka, Journal of Geophysical Research-Solid Earth **122**, 5935 (2017)
138. Y. Wu, X. Wu, J.F. Lin, C.A. McCammon, Y.M. Xiao, P. Chow, V.B. Prakapenka, T. Yoshino, S.M. Zhai, S. Qin, Earth Planet. Sci. Lett. **434**, 91 (2016)
139. I.S. Lyubutin, A.G. Gavriliuk, K.V. Frolov, J.F. Lin, I.A. Troyan, JETP Lett. **90**, 617 (2010)
140. B. Chen, L. L. Gao, B. Lavina, P. Dera, E. E. Alp, J. Y. Zhao, J. Li, Geophys. Res. Lett. **39** (2012). https://doi.org/10.1029/2012gl052875
141. L. L. Gao, B. Chen, J. Y. Wang, E. E. Alp, J. Y. Zhao, M. Lerche, W. Sturhahn, H. P. Scott, F. Huang, Y. Ding, S. V. Sinogeikin, C. C. Lundstrom, J. D. Bass, J. Li, Geophys. Res. Lett. **35** (2008). https://doi.org/10.1029/2008gl034817
142. J.F. Lin, Y.W. Fei, W. Sturhahn, J.Y. Zhao, H.K. Mao, R.J. Hemley, Earth Planet. Sci. Lett. **226**, 33 (2004)
143. W. L. Mao, W. Sturhahn, D. L. Heinz, H. K. Mao, J. F. Shu, R. J. Hemley, Geophys. Res. Lett. **31** (2004). https://doi.org/10.1029/2004gl020541
144. M. Mookherjee, Y. Nakajima, G. Steinle-Neumann, K. Glazyrin, X. A. Wu, L. Dubrovinsky, C. McCammon, A. Chumakov, J. Geophys. Res. Solid Earth **116** (2011). https://doi.org/10.1029/2010jb007819
145. O. Narygina, L.S. Dubrovinsky, C.A. McCammon, A. Kurnosov, I.Y. Kantor, V.B. Prakapenka, N.A. Dubrovinskaia, Earth Planet. Sci. Lett. **307**, 409 (2011)
146. C. Prescher, L. Dubrovinsky, E. Bykova, I. Kupenko, K. Glazyrin, A. Kantor, C. McCammon, M. Mookherjee, Y. Nakajima, N. Miyajima, R. Sinmyo, V. Cerantola, N. Dubrovinskaia, V. Prakapenka, R. Rüffer, A. Chumakov, M. Hanfland, Nat. Geosci. **8**, 220 (2015)
147. C. Prescher, L. Dubrovinsky, C. McCammon, K. Glazyrin, Y. Nakajima, A. Kantor, M. Merlini, M. Hanfland, Phys. Rev. B **85** (2012). https://doi.org/10.1103/PhysRevB.85.140402
148. A. Shahar, E.A. Schauble, R. Caracas, A.E. Gleason, M.M. Reagan, Y. Xiao, J. Shu, W. Mao, Science **352**, 580 (2016)
149. K. Catalli, S.H. Shim, V.B. Prakapenka, J.Y. Zhao, W. Sturhahn, P. Chow, Y.M. Xiao, H.Z. Liu, H. Cynn, W.J. Evans, Earth Planet. Sci. Lett. **289**, 68 (2010)
150. G.J. Finkelstein, J.M. Jackson, W. Sturhahn, D.Z. Zhang, E.E. Alp, T.S. Toellner, Am. Miner. **102**, 1709 (2017)
151. J.M. Jackson, E.A. Hamecher, W. Sturhahn, Eur. J. Mineral. **21**, 551 (2009)
152. Z. Mao, J.F. Lin, J. Yang, T. Inoue, V.B. Prakapenka, Geophys. Res. Lett. **42**, 4335 (2015)
153. Z. Mao, F. Wang, J.F. Lin, S.Y. Fu, J. Yang, X. Wu, T. Okuchi, N. Tomioka, V.B. Prakapenka, Y.M. Xiao, P. Chow, Am. Miner. **102**, 357 (2017)
154. D.M. Vasiukov, L. Ismailova, I. Kupenko, V. Cerantola, R. Sinmyo, K. Glazyrin, C. McCammon, A.I. Chumakov, L. Dubrovinsky, N. Dubrovinskaia, Phys. Chem. Miner. **45**, 397 (2018)

155. H. Giefers, R. Lübbers, K. Rupprecht, G. Wortmann, D. Alfé, A.I. Chumakov, High Pressure Research **22**, 501 (2002)
156. K. Glazyrin, L.V. Pourovskii, L. Dubrovinsky, O. Narygina, C. McCammon, B. Hewener, V. Schünemann, J. Wolny, K. Muffler, A.I. Chumakov, W. Crichton, M. Hanfland, V. Prakapenka, F. Tasnádi, M. Ekholm, M. Aichhorn, V. Vildosola, A.V. Ruban, M.I. Katsnelson, I.A. Abrikosov, Phys. Rev. Lett. **110**, 117206 (2013). https://doi.org/10.1103/PhysRevLett.110.117206
157. R. Lübbers, H.F. Grünsteudel, A.I. Chumakov, G. Wortmann, Science **287**, 1250 (2000)
158. H.K. Mao, J. Xu, V.V. Struzhkin, J. Shu, R.J. Hemley, W. Sturhahn, M.Y. Hu, E.E. Alp, L. Vocadlo, D. Alfé, G.D. Price, M.J. Gillan, M. Schwoerer-Bohning, D. Häusermann, P. Eng, G. Shen, H. Giefers, R. Lübbers, G. Wortmann, Science **292**, 914 (2001)
159. C. McCammon, R. Caracas, K. Glazyrin, V. Potapkin, A. Kantor, R. Sinmyo, C. Prescher, I. Kupenko, A. Chumakov, L. Dubrovinsky, Progress Earth Planetary Sci. **3** (2016). https://doi.org/10.1186/s40645-016-0089-2
160. G. Shen, W. Sturhahn, E.E. Alp, J. Zhao, T.S. Toellner, V.B. Prakapenka, Y. Meng, H.R. Mao, Phys. Chem. Miner. **31**, 353 (2004)
161. V.V. Struzhkin, H.K. Mao, J.Z. Hu, M. Schwoerer-Bohning, J.F. Shu, R.J. Hemley, W. Sturhahn, M.Y. Hu, E.E. Alp, P. Eng, G.Y. Shen, Phys. Rev. Lett. **87**, 255501 (2001). https://doi.org/10.1103/PhysRevLett.87.255501
162. K. Catalli, S.H. Shim, P. Dera, V.B. Prakapenka, J.Y. Zhao, W. Sturhahn, P. Chow, Y.M. Xiao, H. Cynn, W.J. Evans, Earth Planet. Sci. Lett. **310**, 293 (2011)
163. Y.Y. Chang, S.D. Jacobsen, J.F. Lin, C.R. Bina, S.M. Thomas, J.J. Wu, G.Y. Shen, Y.M. Xiao, P. Chow, D.J. Frost, C.A. McCammon, P. Dera, Earth Planet. Sci. Lett. **382**, 1 (2013)
164. Z. Mao, J.F. Lin, H.P. Scott, H.C. Watson, V.B. Prakapenka, Y. Xiao, P. Chow, C. McCammon, Earth Planet. Sci. Lett. **309**, 179 (2011)
165. Z. Mao, J.F. Lin, J. Yang, H. Bian, J. Liu, H.C. Watson, S. Huang, J.H. Chen, V.B. Prakapenka, Y.M. Xiao, P. Chow, Earth Planet. Sci. Lett. **403**, 157 (2014)
166. C. McCammon, L. Dubrovinsky, O. Narygina, I. Kantor, X. Wu, K. Glazyrin, I. Sergueev, A.I. Chumakov, Phys. Earth Planet. Inter. **180**, 215 (2010)
167. J. K. Wicks, J. M. Jackson, W. Sturhahn, Geophys. Res. Lett. **37** (2010). https://doi.org/10.1029/2010gl043689
168. D. Andrault, M. Munoz, G. Pesce, V. Cerantola, A. Chumakov, I. Kantor, S. Pascarelli, R. Rüffer, L. Hennet, Geochemical Perspectives Letters **6**, 5 (2018)
169. I. Mashino, E. Ohtani, N. Hirao, T. Mitsui, R. Masuda, M. Seto, T. Sakai, S. Takahashi, S. Nakano, Am. Miner. **99**, 1555 (2014)
170. J. Liu, N. Dauphas, M. Roskosz, M.Y. Hu, H. Yang, W.L. Bi, J.Y. Zhao, E.E. Alp, J.Y. Hu, J.F. Lin, Nature Communications **8**, 14377 (2017). https://doi.org/10.1038/ncomms14377
171. S.H. Shim, B. Grocholski, Y. Ye, E.E. Alp, S.Z. Xu, D. Morgan, Y. Meng, V.B. Prakapenka, Proc. Natl. Acad. Sci. U.S.A. **114**, 6468 (2017)
172. H. Yang, J.F. Lin, M.Y. Hu, M. Roskosz, W.L. Bi, J.Y. Zhao, E.E. Alp, J. Liu, J.C. Liu, R.M. Wentzowitch, T. Okuchi, N. Dauphas, Earth Planet. Sci. Lett. **506**, 113 (2019)
173. Z. Mao, J.F. Lin, S. Huang, J.H. Chen, Y.M. Xia, P. Chow, Am. Miner. **98**, 1146 (2013)
174. Z. Mao, J. F. Lin, C. Jacobs, H. C. Watson, Y. Xiao, P. Chow, E. E. Alp, V. B. Prakapenka, Geophys. Res. Lett. **37** (2010). https://doi.org/10.1029/2010gl045021
175. S. V. Ovsyannikov, M. Bykov, E. Bykova, K. Glazyrin, R. S. Manna, A. A. Tsirlin, V. Cerantola, I. Kupenko, A. V. Kurnosov, I. Kantor, A. S. Pakhomova, I. Chuvashova, A. I. Chumakov, R. Rüffer, C. McCammon, L. S. Dubrovinsky, Nature Commun. **9** (2018). https://doi.org/10.1038/s41467-018-06457-x
176. K. Catalli, S.H. Shim, V.B. Prakapenka, J. Zhao, W. Sturhahn, Am. Miner. **95**, 418 (2010)
177. L. Ismailova, A. Bobrov, M. Bykov, E. Bykova, V. Cerantola, I. Kantor, I. Kupenko, C. McCammon, V. Dyadkin, D. Chernyshov, S. Pascarelli, A. Chumakov, N. Dubrovinskaia, L. Dubrovinisky, Am. Miner. **100**, 2650 (2015)
178. L. Ismailova, E. Bykova, M. Bykov, V. Cerantola, C. McCammon, T. B. Ballaran, A. Bobrov, R. Sinmyo, N. Dubrovinskaia, K. Glazyrin, H. P. Liermann, I. Kupenko, M. Hanfland, C.

Prescher, V. Prakapenka, V. Svitlyk, L. Dubrovinsky, Sci. Adv. **2** (2016). https://doi.org/10.1126/sciadv.1600427
179. R. Sinmyo, E. Bykova, S. V. Ovsyannikov, C. McCammon, I. Kupenko, L. Ismailova, L. Dubrovinsky, Scienti. Reports **6** (2016). https://doi.org/10.1038/srep32852
180. A.S. Wolf, J.M. Jackson, P. Dera, V.B. Prakapenka, Journal of Geophysical Research-Solid Earth **120**, 7460 (2015)
181. S.H. Shim, A. Bengtson, D. Morgan, W.G. Sturhahn, K. Cataiii, J.Y. Zhao, M. Lerche, V. Prakapenka, Proc. Natl. Acad. Sci. U.S.A. **106**, 5508 (2009)
182. C. Gu, K. Catalli, B. Grocholski, L. L. Gao, E. Alp, P. Chow, Y. M. Xiao, H. Cynn, W. J. Evans, S. H. Shim, Geophys. Res. Lett. **39** (2012). https://doi.org/10.1029/2012gl053950
183. F. Maeda, S. Kamada, E. Ohtani, N. Hirao, T. Mitsui, R. Masuda, M. Miyahara, C. McCammon, Am. Miner. **102**, 2106 (2017)
184. Z. Mao, J.F. Lin, J. Yang, J.J. Wu, H.C. Watson, Y.M. Xiao, P. Chow, J.Y. Zhao, Am. Miner. **99**, 415 (2014)
185. M. Murakami, A. F. Goncharov, N. Hirao, R. Masuda, T. Mitsui, S. M. Thomas, C. R. Bina, Nature Commun. **5** (2014). https://doi.org/10.1038/ncomms6428
186. C. Prescher, C. Weigel, C. McCammon, O. Narygina, V. Potapkin, I. Kupenko, R. Sinmyo, A.I. Chumakov, L. Dubrovinsky, Earth Planet. Sci. Lett. **385**, 130 (2014)
187. N.V. Solomatova, J.M. Jackson, W. Sturhahn, G.R. Rossman, M. Roskosz, Journal of Geophysical Research-Solid Earth **122**, 6306 (2017)
188. E. Bykova, L. Dubrovinsky, N. Dubrovinskaia, M. Bykov, C. McCammon, S. V. Ovsyannikov, H. P. Liermann, I. Kupenko, A. I. Chumakov, R. Rüffer, M. Hanfland, V. Prakapenka, Nature Communi. **7** (2016). https://doi.org/10.1038/ncomms10661
189. V. Cerantola, E. Bykova, I. Kupenko, M. Merlini, L. Ismailova, C. McCammon, M. Bykov, A. I. Chumakov, S. Petitgirard, I. Kantor, V. Svitlyk, J. Jacobs, M. Hanfland, M. Mezouar, C. Prescher, R. Rüffer, V. B. Prakapenka, L. Dubrovinsky, Nature Commun. **8** (2017). https://doi.org/10.1038/ncomms15960
190. J.M. Jackson, W. Sturhahn, M. Lerche, J.Y. Zhao, T.S. Toellner, E.E. Alp, S.V. Sinogeikin, J.D. Bass, C.A. Murphy, J.K. Wicks, Earth Planet. Sci. Lett. **362**, 143 (2013)
191. D.Z. Zhang, J.M. Jackson, J.Y. Zhao, W. Sturhahn, E.E. Alp, M.Y. Hu, T.S. Toellner, C.A. Murphy, V.B. Prakapenka, Earth Planet. Sci. Lett. **447**, 72 (2016)
192. J. M. Jackson, W. Sturhahn, O. Tschauner, M. Lerche, Y. W. Fei, Geophys. Res. Lett. **36** (2009). https://doi.org/10.1029/2009gl037815
193. T. Mitsui, M. Seto, N. Hirao, Y. Ohishi, Y. Kobayashi, S. Higashitaniguchi, R. Masuda, Japan. J. Appl. Phys. Part 2-Lett. Express Lett. **46**, L382 (2007)
194. T. Mitsui, N. Hirao, Y. Ohishi, R. Masuda, Y. Nakamura, H. Enoki, K. Sakaki, M. Seto, Journal of Synchrotron Radiation **16**, 723 (2009)
195. N. Bolfan-Casanova, S. Mackwell, H. Keppler, C. McCammon, D. C. Rubie, Geophys. Res. Lett. **29** (2002). https://doi.org/10.1029/2001gl014457
196. N. Bolfan-Casanova, M. Munoz, C. McCammon, E. Deloule, A. Ferot, S. Demouchy, L. France, D. Andrault, S. Pascarelli, Am. Miner. **97**, 1483 (2012)
197. G. Bromiley, N. Hilaret, C. McCammon, Geophys. Res. Lett. **31** (2004). https://doi.org/10.1029/2004gl019430
198. G.D. Bromiley, H. Keppler, C. McCammon, F.A. Bromiley, S.D. Jacobsen, Am. Miner. **89**, 941 (2004)
199. Y.Y. Chang, S.D. Jacobsen, C.R. Bina, S.M. Thomas, J.R. Smyth, D.J. Frost, T.B. Ballaran, C.A. McCammon, E.H. Hauri, T. Inoue, H. Yurimoto, Y. Meng, P. Dera, Journal of Geophysical Research-Solid Earth **120**, 8259 (2015)
200. F. Gaillard, M. Pichavant, S. Mackwell, R. Champallier, B. Scaillet, C. McCammon, Am. Miner. **88**, 308 (2003)
201. F. Gaillard, B. Schmidt, S. Mackwell, C. McCammon, Geochim. Cosmochim. Acta **67**, 2427 (2003)
202. T. Kawazoe, A. Chaudhari, J.R. Smyth, C. McCammon, Am. Miner. **101**, 1236 (2016)

203. Z. Mao, S.D. Jacobsen, D.J. Frost, C.A. McCammon, E.H. Hauri, T.S. Duffy, Am. Miner. **96**, 1606 (2011)
204. C.A. McCammon, D.J. Frost, J.R. Smyth, H.M.S. Laustsen, T. Kawamoto, N.L. Ross, P.A. van Aken, Phys. Earth Planet. Inter. **143**, 157 (2004)
205. E. Medard, C.A. McCammon, J.A. Barr, T.L. Grove, Am. Miner. **93**, 1838 (2008)
206. M. Mrosko, M. Koch-Muller, C. McCammon, D. Rhede, J. R. Smyth, R. Wirth, Contribut. Mineral. Petrol. **170** (2015). https://doi.org/10.1007/s00410-015-1163-2
207. M. Mrosko, S. Lenz, C.A. McCammon, M. Taran, R. Wirth, M. Koch-Muller, Am. Miner. **98**, 629 (2013)
208. J.R. Smyth, C.M. Holl, D.J. Frost, S.D. Jacobsen, F. Langenhorst, C.A. McCammon, Am. Miner. **88**, 1402 (2003)
209. J.R. Smyth, C.M. Holl, F. Langenhorst, H.M.S. Laustsen, G.R. Rossman, A. Kleppe, C.A. McCammon, T. Kawamoto, P.A. van Aken, Phys. Chem. Miner. **31**, 691 (2005)
210. S.M. Thomas, K. Wilson, M. Koch-Muller, E.H. Hauri, C. McCammon, S.D. Jacobsen, J. Lazarz, D. Rhede, M.H. Ren, N. Blair, S. Lenz, Am. Miner. **100**, 1084 (2015)
211. G.M. Manthilake, N. de Koker, D.J. Frost, C.A. McCammon, Proc. Natl. Acad. Sci. U.S.A. **108**, 17901 (2011)
212. C. Romano, B.T. Poe, N. Kreidie, C.A. McCammon, Am. Miner. **91**, 1371 (2006)
213. R. Sinmyo, G. Pesce, E. Greenberg, C. McCammon, L. Dubrovinsky, Earth Planet. Sci. Lett. **393**, 165 (2014)
214. Y. S. Xu, C. McCammon, J. Geophys. Res.-Solid Earth **107** (2002). https://doi.org/10.1029/2001jb000677
215. Y.S. Xu, C. McCammon, B.T. Poe, Science **282**, 922 (1998)
216. X.Z. Yang, H. Keppler, C. McCammon, H.W. Ni, Contrib. Miner. Petrol. **163**, 33 (2012)
217. X. Z. Yang, H. Keppler, C. McCammon, H. W. Ni, Q. K. Xia, Q. C. Fan, J. Geophys. Res.-Solid Earth **116** (2011). https://doi.org/10.1029/2010jb008010
218. X.Z. Yang, C. McCammon, Geology **40**, 131 (2012)
219. T. Yoshino, S. Kamada, C.C. Zhao, E. Ohtani, N. Hirao, Earth Planet. Sci. Lett. **434**, 208 (2016)
220. J. Buchen, H. Marquardt, T.B. Ballaran, T. Kawazoe, C. McCammon, Am. Miner. **102**, 2494 (2017)
221. J. Chantel, D. J. Frost, C. A. McCammon, Z. C. Jing, Y. B. Wang, Geophys. Res. Lett. **39** (2012). https://doi.org/10.1029/2012gl053075
222. S.Y. Fu, J. Yang, Y.Y. Zhang, T. Okuchi, C. McCammon, H.I. Kim, S.K. Lee, J.F. Lin, Geophys. Res. Lett. **45**, 4725 (2018)
223. G.D. Gatta, I. Kantor, T.B. Ballaran, L. Dubrovinsky, C. McCammon, Phys. Chem. Miner. **34**, 627 (2007)
224. S. D. Jacobsen, H. J. Reichmann, H. A. Spetzler, S. J. Mackwell, J. R. Smyth, R. J. Angel, C. A. McCammon, J. Geophys. Res. Solid Earth 107 (2002). https://doi.org/10.1029/2001jb000490
225. H.J. Reichmann, S.D. Jacobsen, S.J. Mackwell, C.A. McCammon, Geophys. Res. Lett. **27**, 799 (2000)
226. K. Schulze, H. Marquardt, T. Kawazoe, T.B. Ballaran, C. McCammon, M. Koch-Muller, A. Kurnosov, K. Marquardt, Earth Planet. Sci. Lett. **498**, 9 (2018)
227. I.G. Wood, L. Vocadlo, D.P. Dobson, G.D. Price, A.D. Fortes, F.J. Cooper, J.W. Neale, A.M. Walker, W.G. Marshall, M.G. Tucker, D.J. Francis, H.J. Stone, C.A. McCammon, J. Appl. Crystallogr. **41**, 886 (2008)
228. K. Glazyrin, T.B. Ballaran, D.J. Frost, C. McCammon, A. Kantor, M. Merlini, M. Hanfland, L. Dubrovinsky, Earth Planet. Sci. Lett. **393**, 182 (2014)
229. S.M. Dorfman, J. Badro, J.P. Rueff, P. Chow, Y.M. Xiao, P. Gillet, Am. Miner. **100**, 2246 (2015)
230. K. Glazyrin, R. Sinmyo, E. Bykova, M. Bykov, V. Cerantola, M. Longo, C. McCammon, V. B. Prakapenka, L. Dubrovinsky, Physical Review B 95 (2017). https://doi.org/10.1103/PhysRevB.95.214412

231. M. Nunez Valdez, I. Efthimiopoulos, M. Taran, J. Muller, E. Bykova, C. McCammon, M. Koch-Muller, M. Wilke, Phys. Rev. B 97 (2018). https://doi.org/10.1103/PhysRevB.97.184405
232. D.J. Frost, C. Liebske, F. Langenhorst, C.A. McCammon, R.G. Tronnes, D.C. Rubie, Nature **428**, 409 (2004)
233. C.K. Gessmann, D.C. Rubie, C.A. McCammon, Geochim. Cosmochim. Acta **63**, 1853 (1999)
234. T.T. Gu, M.M. Li, C. McCammon, K.K.M. Lee, Nat. Geosci. **9**, 723 (2016)
235. T. Irifune, T. Shinmei, C.A. McCammon, N. Miyajima, D.C. Rubie, D.J. Frost, Science **327**, 193 (2010)
236. C. McCammon, J. Peyronneau, J.P. Poirier, Geophys. Res. Lett. **25**, 1589 (1998)
237. C.A. McCammon, S. Lauterbach, F. Seifert, F. Langenhorst, P.A. van Aken, Earth Planet. Sci. Lett. **222**, 435 (2004)
238. O. Narygina, L.S. Dubrovinsky, H. Samuel, C.A. McCammon, I.Y. Kantor, K. Glazyrin, S. Pascarelli, G. Aquilanti, V.B. Prakapenka, Phys. Earth Planet. Inter. **185**, 107 (2011)
239. K. Otsuka, C.A. McCammon, S.I. Karato, Phys. Earth Planet. Inter. **180**, 179 (2010)
240. V. Stagno, D. J. Frost, C. A. McCammon, H. Mohseni, Y. Fei, Contributions to Mineralogy and Petrology 169 (2015). https://doi.org/10.1007/s00410-015-1111-1
241. V. Stagno, D.O. Ojwang, C.A. McCammon, D.J. Frost, Nature **493**, 84 (2013)
242. D.R. Hummer, Y.W. Fei, Am. Miner. **97**, 1915 (2012)
243. T. Kawazoe, I. Ohira, T. Ishii, T.B. Ballaran, C. McCammon, A. Suzuki, E. Ohtani, Am. Miner. **102**, 1953 (2017)
244. A.C. Palke, J.F. Stebbins, D.J. Frost, C.A. McCammon, Am. Miner. **97**, 1955 (2012)
245. R. Sinmyo, E. Bykova, C. McCammon, I. Kupenko, V. Potapkin, L. Dubrovinsky, Phys. Chem. Miner. **41**, 409 (2014)
246. L. Dubrovinsky, N. Dubrovinskaia, I. Kantor, C. McCammon, W. Crichton, V. Urusov, J. Alloy. Compd. **390**, 41 (2005)
247. S.V. Ovsyannikov, M. Bykov, E. Bykova, D.P. Kozlenko, A.A. Tsirlin, A.E. Karkin, V.V. Shchennikov, S.E. Kichanov, H.Y. Gou, A.M. Abakumov, R. Egoavil, J. Verbeeck, C. McCammon, V. Dyadkin, D. Chernyshov, S. van Smaalen, L.S. Dubrovinsky, Nature Chemistry **8**, 501 (2016)
248. T. Yoshioka, C. McCammon, S. Shcheka, H. Keppler, Am. Miner. **100**, 1641 (2015)
249. N. Dauphas, M. Roskosz, E.E. Alp, D.R. Neuville, M.Y. Hu, C.K. Sio, F.L.H. Tissot, J. Zhao, L. Tissandiere, E. Medard, C. Cordier, Earth Planet. Sci. Lett. **398**, 127 (2014)
250. K.B. Prissel, M.J. Krawczynski, N.X. Nie, N. Dauphas, H. Couvy, M.Y. Hu, E.E. Alp, M. Roskosz, Geochim. Cosmochim. Acta **238**, 580 (2018)
251. M. Roskosz, C.K.I. Sio, N. Dauphas, W.L. Bi, F.L.H. Tissot, M.Y. Hu, J.Y. Zhao, E.E. Alp, Geochim. Cosmochim. Acta **169**, 184 (2015)
252. N. Dubrovinskaia, L. Dubrovinsky, C. McCammon, Journal of Physics-Condensed Matter **16**, S1143 (2004)
253. O. Narygina, L.S. Dubrovinsky, N. Miyajima, C.A. McCammon, I.Y. Kantor, M. Mezouar, V.B. Prakapenka, N.A. Dubrovinskaia, V. Dmitriev, Phys. Chem. Miner. **38**, 203 (2011)
254. V. Stagno, Y. Tange, N. Miyajima, C. A. McCammon, T. Irifune, D. J. Frost, Geophys. Res. Lett. **38** (2011). https://doi.org/10.1029/2011gl049560
255. M. Wilke, F. Farges, P. Petit, G.E. Brown, F. Martin, Am. Miner. **86**, 714 (2001)
256. H.E. Höfer, G.P. Brey, B. Schulz-Dobrick, R. Oberhänsli, Eur. J. Mineral. **6**, 407 (1994)
257. N.M. McGowan, W.L. Griffin, J.M. Gonzalez-Jimenez, E. Belousova, J.C. Afonso, R.D. Shi, C.A. McCammon, N.J. Pearson, S.Y. O'Reilly, Geology **43**, 179 (2015)
258. K. Otsuka, M. Longo, C.A. McCammon, S.I. Karato, Earth Planet. Sci. Lett. **365**, 7 (2013)
259. C.A. McCammon, I.L. Chinn, J.J. Gurney, M.E. McCallum, Contrib. Miner. Petrol. **133**, 30 (1998)
260. A.A. Shiryaev, Y.V. Zubavichus, A.A. Veligzhanin, C. McCammon, Russ. Geol. Geophys. **51**, 1262 (2010)
261. D. Lenaz, H. Skogby, A.M. Logvinova, N.V. Sobolev, F. Princivalle, Phys. Chem. Miner. **40**, 671 (2013)

262. C.A. McCammon, T. Stachel, J.W. Harris, Earth Planet. Sci. Lett. **222**, 423 (2004)
263. M. Palot, S.D. Jacobsen, J.P. Townsend, F. Nestola, K. Marquardt, N. Miyajima, J.W. Harris, T. Stachel, C.A. McCammon, D.G. Pearson, Lithos **265**, 237 (2016)
264. E.S. Kiseeva, D.M. Vasiukov, B.J. Wood, C. McCammon, T. Stachel, M. Bykov, E. Bykova, A. Chumakov, V. Cerantola, J.W. Harris, L. Dubrovinsky, Nat. Geosci. **11**, 144 (2018)
265. P. Elliott, U. Kolitsch, G. Giester, E. Libowitzky, C. McCammon, A. Pring, W.D. Birch, J. Brugger, Mineral. Mag. **73**, 131 (2009)
266. S. Gentile, C. Biagioni, P. Comodi, M. Pasero, C. McCammon, C. Bonadiman, Am. Miner. **101**, 461 (2016)
267. R.P. Liferovich, E.V. Sokolova, F.C. Hawthorne, K.V.O. Laajoki, S. Gehor, Y.A. Pakhomovsky, N.V. Sorokhtina, Can. Mineral. **38**, 1477 (2000)
268. J. Schilling, F.Y. Wu, C. McCammon, T. Wenzel, M.A.W. Marks, K. Pfaff, D.E. Jacob, G. Markl, Mineral. Mag. **75**, 87 (2011)
269. E. Sokolova, F.C. Hawthorne, C. McCammon, R.P. Liferovich, Can. Mineral. **39**, 1121 (2001)
270. Y. Uvarova, E. Sokolova, F.C. Hawthorne, C.A. McCammon, V.I. Kazansky, K.V. Lobanov, Mineral. Mag. **71**, 651 (2007)
271. T. Ruskov, I. Spirov, H.W. Green, D. Kovacheva, P. Tzvetkov, M. Georgieva, L. Dobrzhinetskaya, Phys. Chem. Miner. **35**, 485 (2008)
272. F. Bosi, F. Camara, M.E. Ciriotti, U. Halenius, L. Reznitskii, V. Stagno, Eur. J. Mineral. **29**, 445 (2017)
273. H. Skogby, M. Janak, I. Broska, Eur. J. Mineral. **28**, 631 (2016)
274. A.P. Martin, R.C. Price, A.F. Cooper, C.A. McCammon, J. Petrol. **56**, 193 (2015)
275. T. Ruskov, I. Spirov, M. Georgieva, S. Yamamoto, H.W. Green, C.A. McCammon, L.F. Dobrzhinetskaya, J. Metamorph. Geol. **28**, 551 (2010)
276. V.N. Sobolev, C.A. McCammon, L.A. Taylor, G.A. Snyder, N.V. Sobolev, Am. Miner. **84**, 78 (1999)
277. V. Stahle, M. Koch, C.A. McCammon, U. Mann, G. Markl, Can. Mineral. **40**, 1609 (2002)
278. C. McCammon, M.G. Kopylova, Contrib. Miner. Petrol. **148**, 55 (2004)
279. H.M. Williams, C.A. McCammon, A.H. Peslier, A.N. Halliday, N. Teutsch, S. Levasseur, J.P. Burg, Science **304**, 1656 (2004)
280. A. Proyer, E. Dachs, C. McCammon, Contrib. Miner. Petrol. **147**, 305 (2004)
281. H.M. Williams, S.G. Nielsen, C. Renac, W.L. Griffin, S.Y. O'Reilly, C.A. McCammon, N. Pearson, F. Viljoen, J.C. Alt, A.N. Halliday, Earth Planet. Sci. Lett. **283**, 156 (2009)
282. H.M. Williams, A.H. Peslier, C. McCammon, A.N. Halliday, S. Levasseur, N. Teutsch, J.P. Burg, Earth Planet. Sci. Lett. **235**, 435 (2005)
283. M. Blanchard, N. Dauphas, M.Y. Hu, M. Roskosz, E.E. Alp, D.C. Golden, C.K. Sio, F.L.H. Tissot, J. Zhao, L. Gao, R.V. Morris, M. Fornace, A. Floris, M. Lazzeri, E. Balan, Geochim. Cosmochim. Acta **151**, 19 (2015)
284. N. Dauphas, M. Roskosz, E.E. Alp, D.C. Golden, C.K. Sio, F.L.H. Tissot, M.Y. Hu, J. Zhao, L. Gao, R.V. Morris, Geochim. Cosmochim. Acta **94**, 254 (2012)
285. V.B. Polyakov, R.N. Clayton, J. Horita, S.D. Mineev, Geochim. Cosmochim. Acta **71**, 3833 (2007)
286. V.B. Polyakov, D.M. Soultanov, Geochim. Cosmochim. Acta **75**, 1957 (2011)
287. M.C. Domeneghetti, A.M. Fioretti, F. Camara, C. McCammon, M. Alvaro, Geochim. Cosmochim. Acta **121**, 571 (2013)
288. I.S. McCallum, M.C. Domeneghetti, J.M. Schwartz, E.K. Mullen, M. Zema, F. Camara, C. McCammon, J. Ganguly, Geochim. Cosmochim. Acta **70**, 6068 (2006)
289. U.W. Bläß, F. Langenhorst, C. McCammon, Earth Planet. Sci. Lett. **300**, 255 (2010)
290. R. Skala, L. Strnad, C. McCammon, M. Cada, Geochim. Cosmochim. Acta **73**, 1145 (2009)
291. C.A. McCammon, S.A. McEnroe, P. Robinson, K. Fabian, B.P. Burton, Earth Planet. Sci. Lett. **288**, 268 (2009)
292. B. van de Moortele, B. Reynard, P. Rochette, M. Jackson, P. Beck, P. Gillet, P.F. McMillan, C.A. McCammon, Earth Planet. Sci. Lett. **262**, 37 (2007)
293. W.C. Tennant, C.A. McCammon, R. Miletich, Phys. Chem. Miner. **27**, 156 (2000)

294. S. Creighton, T. Stachel, S. Matveev, H. Hofer, C. McCammon, R.W. Luth, Contrib. Miner. Petrol. **157**, 491 (2009)
295. C. Schröder, I. Köhler, F. L. L. Muller, A. I. Chumakov, I. Kupenko, R. Rüffer, A. Kappler, Hyperfine Interact. **237** (2016). https://doi.org/10.1007/s10751-016-1289-2
296. S.A. McEnroe, B. Carter-Stiglitz, R.J. Harrison, P. Robinson, K. Fabian, C. McCammon, Nat. Nanotechnol. **2**, 631 (2007)
297. C.A. McCammon, V. Chaskar, G.G. Richards, Meas. Sci. Technol. **2**, 657 (1991)
298. ESRF Upgrade Programme Phase II (2015–2019) White Paper (2013)
299. ESRF Upgrade Programme Phase II (2015–2022) Technical Design Study (2014)
300. F. Heidelbach, M.P. Terry, M. Bystricky, C. Holzapfel, C. McCammon, Earth Planet. Sci. Lett. **278**, 386 (2009)
301. G.M. Partzsch, D. Lattard, C. McCammon, Contrib. Miner. Petrol. **147**, 565 (2004)
302. R. Blukis, R. Rüffer, A.I. Chumakov, R.J. Harrison, Meteorit. Planet. Sci. **52**, 925 (2017)

Chapter 6
Molecular Magnetism of Metal Complexes and Light-Induced Phase Transitions

Norimichi Kojima and Atsushi Okazawa

Abstract One of the most attractive research in the field of molecular solids is multifunctional molecular magnets coupled with spin, charge and photon, and single-molecular magnet/single-chain magnet toward spintronics. In this chapter, we focus on the molecular magnetism and its related light-induced phase transitions from the viewpoint of Mössbauer spectroscopy. In the Sect. 6.1, the outline of this chapter is described. In the Sect. 6.2, static and dynamic spin crossover phenomena between the high-spin and low-spin states, and the spin frustration system induced by dynamic spin crossover phenomena for $[Mn^{II}Fe^{III}(C_2O_3S)_3]$ complex are described. In the Sect. 6.3, metal complexes showing charge transfer phase transitions such as Prussian blue analogous complexes and $[Fe^{II}Fe^{III}(C_2O_2S_2)_3]$ complexes and their photo-induced phase transitions are described. In the Sect. 6.4, various kinds of molecular magnets including single molecular chain magnets are described.

List of Abbreviations and Symbols

CTPT	Charge transfer phase transition
CTIST	Charge transfer induced spin transition
IVCT	Inter-valence charge transfer
HS	High spin
LS	Low spin
HTP	High temperature phase
LTP	Low temperature phase
IS	Isomer shift
QS	Quadrupole splitting

N. Kojima (✉)
Toyota Physical and Chemical Research Institute, Yokomichi 41-1, Nagakute, Aichi 480-1192, Japan
e-mail: cnori-kojima@g.ecc.u-tokyo.ac.jp

A. Okazawa
Division of Chemistry, Institute of Liberal Education, Nihon University School of Medicine, Oyaguchi Kamimachi 30-1, Itabashi-ku, Tokyo 173-8610, Japan

Y. Yoshida and G. Langouche (eds.), *Modern Mössbauer Spectroscopy*, Topics in Applied Physics 137, https://doi.org/10.1007/978-981-15-9422-9_6

LIESST	Light induced excited spin state trapping
T_{CT}	Critical temperature of charge transfer phase transition
SP	Spiropyran
SMMs	Single-molecule magnets
SCMs	Single-chain magnets
SIMs	Single-ion magnets
QTM	Quantum tunneling of magnetization.

6.1 Introduction

Transition-metal complexes consist of transition-metal elements and various kinds of ligands with the ability to control the structural dimensionality, which have produced wide variety of magnetic materials such as low-dimensional magnets, photo-induced magnets, single-molecule magnets (SMMs) and so forth, which have stimulated the theory of molecular magnetism. Thus, the development of advanced magnetic materials and the theory of molecular magnetism have mutually stimulated each other. Especially, the discoveries of light induced excited state spin trapping (LIESST) for spin crossover system (1984), SMM of $[Mn_{12}]$ cluster complex (1993), and photo-induced ferrimagnet of Prussian blue analogous complex (1996) have significantly expanded the field of molecular magnetism. These recent topics of molecular magnetism are reviewed in the following excellent books [1].

In this chapter, we focus on the molecular magnetism and its related light-induced phase transitions from the viewpoint of Mössbauer spectroscopy. Among various kinds of molecular compounds, in the case of transition-metal complexes whose spin states are situated in the spin crossover region, new types of synergetic phenomena coupled with spin and charge are expected. Based on this viewpoint, we have synthesized organic–inorganic hybrid systems, $A[Fe^{II}Fe^{III}X_3]$ [$A = (C_nH_{2n+1})_4N$, spiropyran, etc.; X = dto ($C_2O_2S_2$), mto (C_2O_3S), tto (C_2OS_3)] and have found novel multifunctional phenomena coupled with spin, charge and photon [2].

In general, the Fe^{III} site coordinated by six S atoms is in the low spin (LS) state, while that coordinated by six O atoms is in the high spin (HS) state. Therefore, the spin state of Fe^{III} coordinated by three S atoms and three O atoms is expected to be situated in the spin crossover region. Based on this viewpoint, we have synthesized $(C_6H_5)_4P[Zn^{II}Fe^{III}(mto)_3]$ consisting of $Fe^{III}O_3S_3$ and $Zn^{II}O_6$ octahedra, and the ^{57}Fe Mössbauer spectroscopy and the electron spin resonance (ESR) revealed the dynamic spin equilibrium phenomenon occurs at the $Fe^{III}O_3S_3$ site in which the HS and LS states exchange alternately in the time scale of $10^{-10} < \tau < 10^{-7}$ s [3]. On the other hand, in the case of $(C_6H_5)_4P[Mn^{II}Fe^{III}(mto)_3]$ consisting of $Fe^{III}O_3S_3$ and $Mn^{II}O_6$, there exists a dynamic spin equilibrium ($\tau < 10^{-7}$ s) between the HS and LS states at the $Fe^{III}O_3S_3$ site, which induces the frustration of internal magnetic field between the ferromagnetic and antiferromagnetic interactions at the $Mn^{II}O_6$ site. Owing to the frustration of internal magnetic field at the Mn^{II} site caused by the rapid spin equilibrium at the Fe^{III} site, $(C_6H_5)_4P[Mn^{II}Fe^{III}(mto)_3]$ undergoes the successive magnetic phase transitions at 30 and 23 K.

In the case of $(n\text{-}C_nH_{2n+1})_4N[Fe^{II}Fe^{III}(dto)_3]$, in addition to the ferromagnetic phase transition, a spin-entropy driven charge transfer phase transition (CTPT) takes place, in which the electrons of Avogadro's constant transfer between the Fe^{II} and Fe^{III} sites [4]. At the CTPT, the Fe valence state is dynamically fluctuated, which was revealed by means of muon spin relaxation (μSR) [5]. The CTPT and the ferromagnetic phase transition strongly depend on the honeycomb ring size in $[Fe^{II}Fe^{III}(dto)_3]_\infty$ [6]. The increase of counter-cation size expands the honeycomb ring, which stabilizes and destabilizes the high temperature phase (HTP) and the low temperature phase (LTP), respectively. In order to control the ferromagnetic properties and the CTPT by means of light irradiation, we have synthesized a photo-responsive organic–inorganic hybrid system, $(SP\text{-}Me)[Fe^{II}Fe^{III}(dto)_3]$ (SP = spiropyran), and discovered that the photo-isomerization of intercalated SP-Me by UV light irradiation induces the CTPT from the LTP to the HTP in the two-dimensional $[Fe^{II}Fe^{III}(dto)_3]$ layer and the change of T_C from 5 to 22 K, by means of magnetization measurement and ^{57}Fe Mössbauer spectroscopy [7].

In the Sect. 6.4, we show various kinds of molecular magnets such as SMMs including transition-metal clusters, low-coordinated Fe complexes, and single-chain magnets (SCMs) with easy-plane anisotropy.

6.2 Spin Crossover Phenomena

6.2.1 *Static and Dynamic Spin Crossover Phenomena*

Octahedral transition-metal complexes with d^4–d^7 configurations have a possibility of spin crossover transition between the LS and HS states in the ground state. If the ligand field splitting energy is smaller than the spin-pairing energy in the d orbitals, the d electrons occupy the t_{2g} (d_{xy}, d_{yz}, d_{zx}) and the e_g ($d_{x^2-y^2}$, d_{z^2}) orbitals being followed by Hund's rule, in which the spin configuration shows the maximum spin multiplicity (HS state). On the other hand, if the ligand field splitting energy is larger than the spin-pairing energy, Hund's rule is broken down, in which the spin configuration shows the minimum spin multiplicity (LS state). Therefore, the ground state of the transition metal ion with d^n (n = 4–7) is ruled by the competition between ligand field splitting energy and spin-pairing energy in the d electrons, which is schematically shown in Fig. 6.1.

The energy diagram for d^n system called Tanabe–Sugano diagram is the most effective tool to analyze the competition between the HS and LS states as the ground state [8]. The Tanabe–Sugano diagram exhibits the energies of multiplets for $3d^n$ (n = 2–8) system as a function of the ratio of ligand field (Dq) to the Racah parameter (B) representing the strength of Coulomb interaction between 3d electrons. In the case of $3d^n$ (n = 4–7), if the ground energies of HS and LS states are close to each

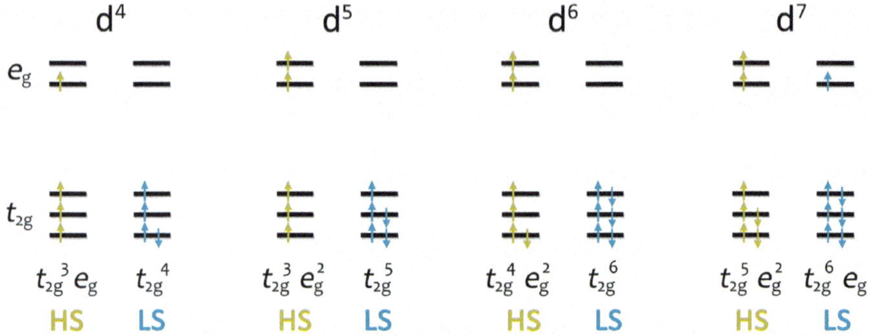

Fig. 6.1 Spin arrangements in the ground states for d^n (n = 4–7) configuration with octahedral symmetry

other, the system has a possibility to convert the spin states between the HS and LS states by external perturbations such as heat, applied pressure or light irradiation. Such the HS–LS conversion is called spin crossover phenomenon.

The spin crossover phenomenon was observed for the first time by L. Cambi et al. in 1930s for tris(dithiocarbamato)iron(III) complexes, $Fe^{III}(S_2CNR_2)_3$ (R = *n*-C_4H_9, *iso*-C_4H_9, etc.) [9]. In this system, as shown in Fig. 6.2, the Fe^{III} ion is coordinated by six S atoms. According to L. Cambi et al., the magnetic susceptibilities of $Fe^{III}(S_2CNR_2)_3$ significantly change depending on temperature caused by the gradual LS ($S = 1/2$)–HS ($S = 5/2$) transition, which is shown in Fig. 6.3. If the spin conversion between the LS ($S = 1/2$) and HS ($S = 5/2$) states is slower than the time scale ($\tau = 10^{-7}$ s) of ^{57}Fe Mössbauer spectroscopy, two kinds of quadrupole doublets corresponding to the HS and LS states should be clearly distinguished. On the other hand, if the spin conversion is faster than 10^{-7} s, an averaged quadrupole doublet

Fig. 6.2 Molecular structure of $Fe^{III}(S_2CNR_2)_3$ (R = *n*-C_nH_{2n+1}, etc.)

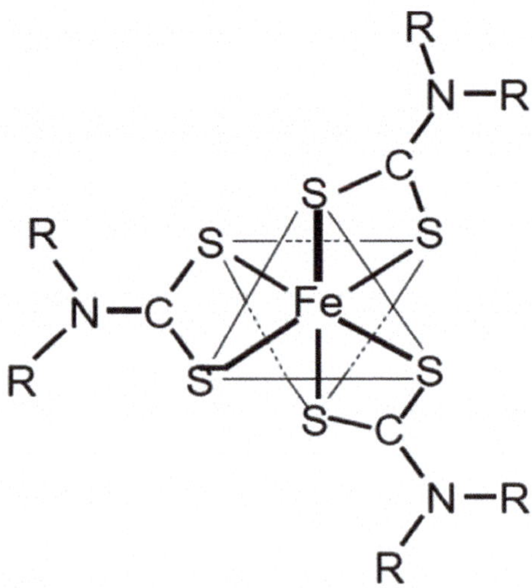

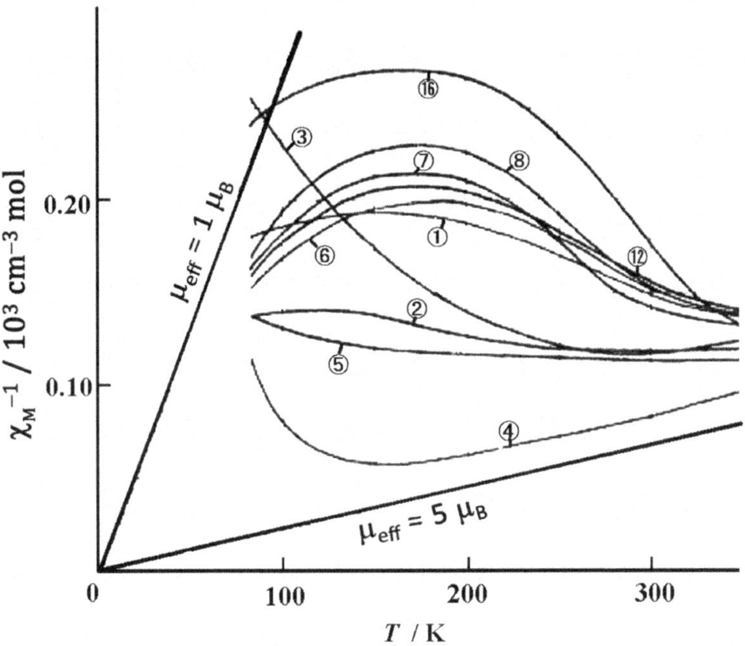

Fig. 6.3 Inverse magnetic susceptibilities as a function of temperature for $Fe^{III}(S_2CNR_2)_3$ (R = n-C_nH_{2n+1}, etc.). (Reprinted from [9(c)]. Copyright Wiley-VCH Verlag GmbH & Co. KGaA)

should be observed. In the case of $Fe^{III}(S_2CNR_2)_3$ (R = C_2H_5), as shown in Fig. 6.4, the ^{57}Fe Mössbauer spectrum exhibits a broad quadrupole doublet in the temperature range of the gradual LS ($S = 1/2$) ↔ HS ($S = 5/2$) transition, which shows that the rapid spin equilibrium between the HS and LS states is realized in the time scale of $\tau < 10^{-7}$ s [10]. Until now, various kinds of Fe^{III} complexes consisting of $Fe^{III}S_6$, $Fe^{III}O_3S_3$, or $Fe^{III}N_4O_2$ octahedra exhibiting rapid spin equilibrium phenomena have been reported [11].

Thirty years after the discovery of spin-crossover phenomenon for $Fe^{III}(S_2CNR_2)_3$, W.A. Baker et al. reported the spin-crossover phase transition for $[Fe^{II}(phen)_2X_2]$ (phen = 1,10-phenanthroline, X = NCS or NCSe) [12]. In $[Fe^{II}(phen)_2(NCS)_2]$, the Fe^{II} ion is coordinated by six N atoms of phen and NCS molecules, and this complex undergoes the first-order LS ($^1A_{1g}$, $S = 0$)–HS ($^5T_{2g}$, $S = 2$) transition at 176 K with a small thermal hysteresis. Since the discovery of the spin-crossover phase transition for $[Fe^{II}(phen)_2X_2]$ (X = NCS, NCSe), various kinds of spin crossover complexes have been reported for Cr^{II} ($3d^4$), Mn^{III} ($3d^4$), Mn^{II} ($3d^5$), Fe^{III} ($3d^5$), Fe^{II} ($3d^6$), Co^{III} ($3d^6$), and Co^{II} ($3d^7$) complexes [13–15], in which most of them are Fe^{II} ($3d^6$) and Fe^{III} ($3d^5$) complexes.

The spin-crossover phenomena have attracted much attention since the discovery of the light-induced spin transition for $[Fe^{II}(ptz)_6](BF_4)_2$ (ptz = 1-propyltetrazole) in 1984 [16(b)]. This complex exhibits a sharp spin transition at 135 K with a remarkable

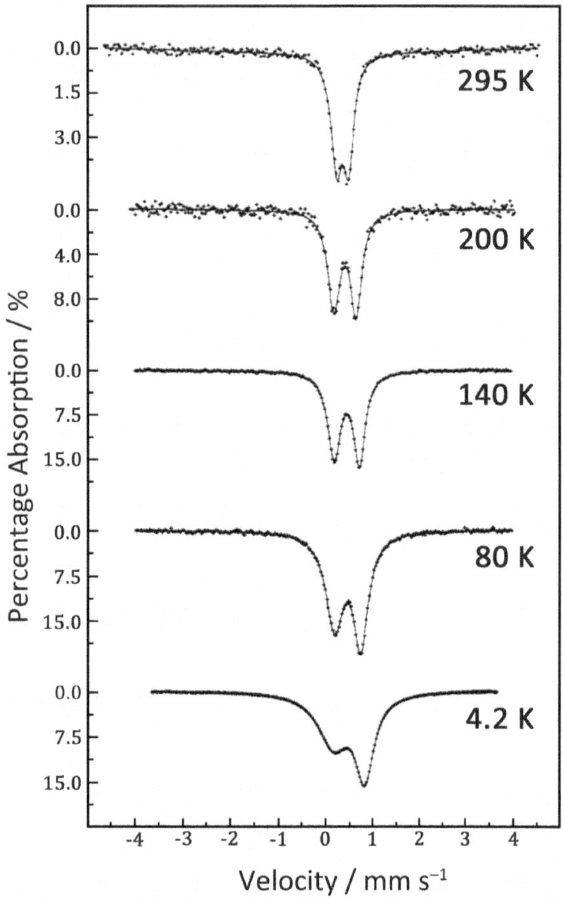

Fig. 6.4 ^{57}Fe Mössbauer spectra of $Fe^{III}(S_2CNR_2)_3$ ($R = C_2H_5$) at various temperatures. Reprinted with permission from [10]. Copyright 1987 American Chemical Society

change of color [16(a)]. The color in the LS state is purple, while that in the HS state is colorless. S. Decurtins, P. Gütlich, et al. discovered a persistent light-induced HS state for $[Fe^{II}(ptz)_6](BF_4)_2$ when they measured the optical absorption spectra at 8 K with white light including the spin-allowed absorption region ($^1A_{1g}(t_{2g}^6) \rightarrow {}^1T_{1g}, {}^1T_{2g}(t_{2g}^5e_g)$ transitions), in which the purple color was completely "bleached" and the colorless remained unchanged as long as the temperature was kept below 50 K [16(c)]. This light-induced spin-crossover phenomenon was named LIESST (Light Induced Excited Spin State Trapping) by them. The trapping $^5T_{2g}$ state of the lowest HS state could also be optically pumped back to the $^1A_{1g}$ ground state via intermediate triplet states $^3T_{1g}, {}^3T_{2g}(t_{2g}^5e_g)$ by the light irradiation corresponding to the $^5T_{2g} \rightarrow {}^5E_g$ transition, which was called reverse-LIESST. The discovery of LIESST and reverse-LIESST have attracted much attention from the viewpoint of photonic molecular devices, and various kinds of spin crossover complexes showing LIESST have been developed mainly for Fe^{II} and Fe^{III} complexes [17]. Figure 6.5

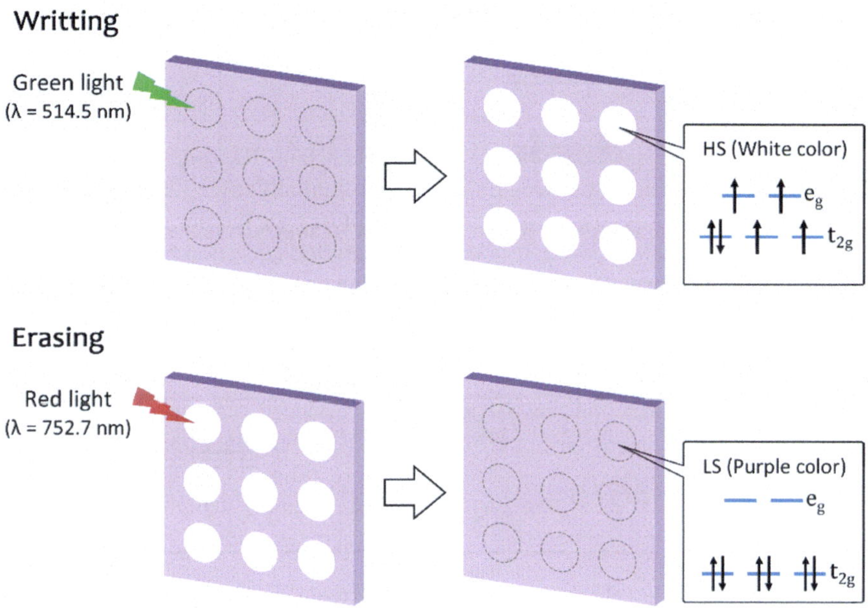

Fig. 6.5 Schematic demonstration of the erasable optical recording based on the LIESST and reverse-LIESST for $[Fe^{II}(ptz)_6](BF_4)_2$

shows the schematic demonstration of the erasable optical recording based on the LIESST and reverse-LIESST for $[Fe^{II}(ptz)_6](BF_4)_2$.

6.2.2 *Dynamic Spin Crossover Phenomena of $A[M^{II}Fe^{III}(mto)_3]$ (A = Counter Cation; M = Zn, Cd; mto = C_2O_3S)*

In this section, we focus on the spin crossover phenomenon of Fe^{III} coordinated by three S atoms and three O atoms, $Fe^{III}O_3S_3$. In general, the Fe^{III} site coordinated by six S atoms tends to be in the LS state (t_{2g}^5, $S = 1/2$), while the Fe^{III} site coordinated by six O atoms is in the HS state ($t_{2g}^3e_g^2$, $S = 5/2$). Therefore, the spin state of $Fe^{III}O_3S_3$ has a possibility to exhibit the spin-crossover phenomenon between the LS and HS states, which is schematically shown in Fig. 6.6.

As a matter of fact, in the case of tris(monothio-β-diketonato)iron(III) complexes, the LS ($S = 1/2$) and HS ($S = 5/2$) states coexist between 300 and 80 K, and two kinds of quadrupole doublets corresponding to the LS and HS states are clearly distinguished in the ^{57}Fe Mössbauer spectra, in which the component of LS state increases with decreasing temperature from 300 to 80 K [18]. In the case of

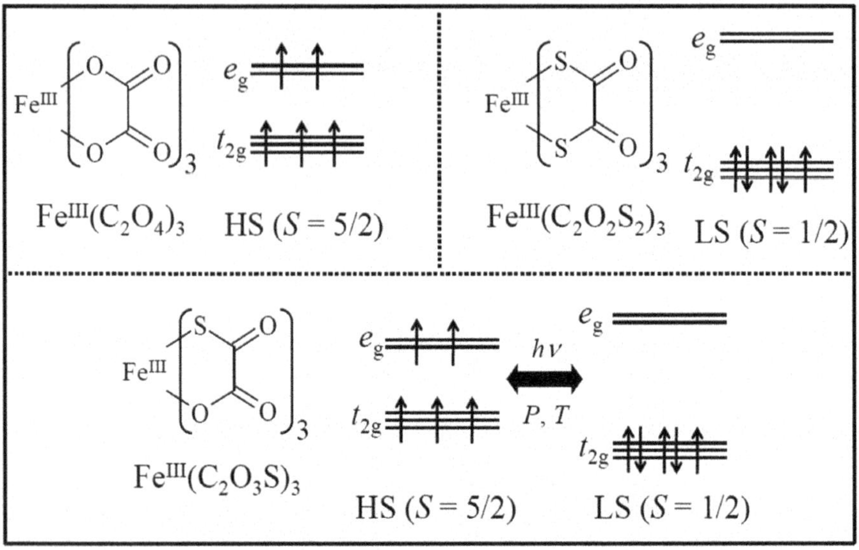

Fig. 6.6 Spin states of the Fe^{III} sites for $[Fe^{III}(ox)_3]$, $[Fe^{III}(dto)_3]$, and $[Fe^{III}(mto)_3]$

tris(monothiocarbamato)iron(III) complexes, on the other hand, the rapid spin equilibrium occurs in which the HS and LS states exchange in the time scale of less than 10^{-7} s [19]. In this case, an averaged quadrupole doublet between the HS and LS states is observed. However, the rapid spin equilibrium phenomenon has not yet been found for assembled metal complex systems. For this viewpoint, in order to search the rapid spin equilibrium phenomenon for assembled metal complex system including $Fe^{III}O_3S_3$ site, we have synthesized $(C_6H_5)_4P[M^{II}Fe^{III}(mto)_3]$ (M = Mn, Zn) consisting of the $Fe^{III}O_3S_3$ and $M^{II}O_6$ sites and investigated the spin state of the $Fe^{III}O_3S_3$ site [3].

Based on the crystal structure of $(n\text{-}C_3H_7)_4N[Fe^{II}Fe^{III}(dto)_3]$ analyzed by the single-crystal X-ray diffraction [20], the Rietveld analysis was performed for the powder X-ray diffraction pattern of $(C_6H_5)_4P[Mn^{II}Fe^{III}(mto)_3]$ at room temperature. In this complex, the Mn^{II} and Fe^{III} ions are alternately arrayed by the bridging ligand, mto, which forms the honeycomb network structure of $[Mn^{II}Fe^{III}(mto)_3]$. The $(C_6H_5)_4P^+$ layer is intercalated between two adjacent $[Mn^{II}Fe^{III}(mto)_3]$ layers.

In the case of $(n\text{-}C_4H_9)_4N[Zn^{II}Fe^{III}(ox)_3]$ with a similar structure of $(C_6H_5)_4P[Zn^{II}Fe^{III}(mto)_3]$, the χT value (= 4.22 cm^3 K mol^{-1}) is almost constant between 300 and 4.2 K, whose value is regarded as the spin-only χT value (4.375 cm^3 K mol^{-1}) of the HS state (S = 5/2) for Fe^{III} [21]. On the other hand, the χT value for $(C_6H_5)_4P[Zn^{II}Fe^{III}(mto)_3]$ decreases from 3.60 (at 300 K) to 2.66 cm^3 K mol^{-1} (at 4.2 K) [3]. Therefore, the χT value of $(C_6H_5)_4P[Zn^{II}Fe^{III}(mto)_3]$ is considered to be situated in the middle value between those of the HS and LS states for Fe^{III}.

Figure 6.7a shows the X-band ESR spectra for $(C_6H_5)_4P[Zn^{II}Fe^{III}(mto)_3]$ at 300, 77 and 10 K, in which the ESR signals corresponding to the HS and LS states

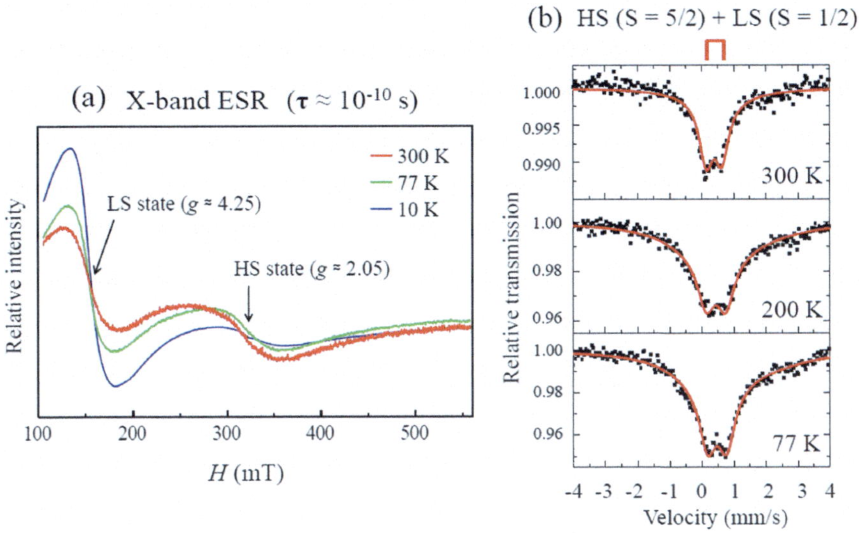

Fig. 6.7 **a** X-band ESR spectra, and **b** ^{57}Fe Mössbauer spectra for $(C_6H_5)_4P[Zn^{II}Fe^{III}(mto)_3]$ at various temperatures. Reprinted with permission from [3]. Copyright 2010 the Chemical Society of Japan

of Fe^{III} site are clearly distinguished at about 300 mT ($g \approx 2.05$) and 150 mT ($g \approx 4.25$), respectively. The ESR intensity ratio of the LS state to the HS state increases with decreasing temperature, which is consistent with the temperature dependence of $\chi_M T$. Therefore, these results imply that the spin state of $Fe^{III}O_3S_3$ in $(C_6H_5)_4P[Zn^{II}Fe^{III}(mto)_3]$ is the spin equilibrium between the HS and LS states, and the HS and LS states are clearly distinguished under the time scale (10^{-10} s) of X-band ESR spectroscopy. Figure 6.7b shows the ^{57}Fe Mössbauer spectra of $(C_6H_5)_4P[Zn^{II}Fe^{III}(mto)_3]$ at 300, 77 and 10 K, where only one quadrupole doublet of Fe^{III} appears in spite of the coexistence of HS and LS states in the ESR signal between 300 and 10 K. Therefore, the ^{57}Fe Mössbauer spectra of $(C_6H_5)_4P[Zn^{II}Fe^{III}(mto)_3]$ indicates the rapid spin equilibrium at the Fe^{III} site whose time scale is faster than that (10^{-7} s) of the ^{57}Fe Mössbauer spectroscopy. The relaxation process between the HS and LS states at the $Fe^{III}O_3S_3$ site is considered to be a tunneling process. The time scale of spin equilibrium at the Fe^{III} site in $(C_6H_5)_4P[Zn^{II}Fe^{III}(mto)_3]$ is estimated at $10^{-10} < \tau < 10^{-7}$ s. The tunneling probability for a non-radiative multi-phonon process from a given vibrational levels, m of the HS state and m' of the LS state is given by the following equation [22],

$$W_{mm'} = \frac{2\pi}{\hbar^2\omega}\beta_{\mathrm{HL}}^2 |< \chi_{m'}|\chi_m > |^2 \delta(E_{m'}, E_m), \tag{6.1}$$

where the electronic coupling matrix element $\beta_{\mathrm{HL}} = < \psi_{\mathrm{LS}}|H_{\mathrm{SO}}|\psi_{\mathrm{HS}} >$ is the second order spin-orbit coupling, $\hbar\omega$ is the energy of metal-ligand vibration, $\delta(E_{m'}, E_m)$ is the delta function ensuring energy conservation, $|<\chi_{m'}|\chi_m>|^2$ is the Franck–Condon factor of the overlap of the vibrational functions between the HS state (χ_m) and LS state ($\chi_{m'}$). At $T \approx 0$ K, since the vibrational ground state of the HS state is populated, the relaxation rate constant between the HS and LS states is expressed as follows [22],

$$k_{\mathrm{HL}}(T \approx 0) = \frac{2\pi}{\hbar^2\omega}\beta_{\mathrm{HL}}^2 |< \chi_n|\chi_0 > |^2, \qquad n = \frac{\Delta E_{\mathrm{HL}}^0}{\hbar\omega}, \tag{6.2}$$

where $|<\chi_n|\chi_0>|^2$ is expressed as follows,

$$|< \chi_n|\chi_0 > |^2 = \frac{S^n e^{-S}}{n!}, \qquad S = \frac{\frac{1}{2}f\Delta Q_{\mathrm{HL}}^2}{\hbar\omega}, \tag{6.3}$$

where S is the Huang-Rhys factor [23], ΔQ_{HL} is the difference of horizontal displacement between the HS and LS potential wells in the metal-ligand coordinate geometry, and f is the force constant. According to Eqs. (6.2) and (6.3), with decreasing the Huang–Rhys factor, the Frank–Condon factor increases, which increases the relaxation rate constant between the HS and LS states. In general, the ΔQ_{HL} of spin crossover Fe^{III} complex is shorter than that of Fe^{II} complex. Indeed, the change in the Fe^{II}–ligand distance due to the HS–LS transition is 0.16–0.21 Å, while that in the Fe^{III}–ligand distance is about 0.12 Å [24], which is the reason why the rapid spin equilibrium phenomenon has been found for Fe^{III} complexes consisting of $Fe^{III}S_6$, $Fe^{III}O_3S_3$, or $Fe^{III}N_4O_2$ octahedra.

6.2.3 Spin Frustration Induced by Dynamic Spin Crossover Phenomena for A[$Mn^{II}Fe^{III}(mto)_3$]

In the case of mto bridged hetero-metal complex system, [$Mn^{II}Fe^{III}(mto)_3$] consisting of $Mn^{II}O_6$ and $Fe^{III}O_3S_3$ octahedra, the spin states of the Mn^{II} and Fe^{III} sites are considered to be HS state ($S = 5/2$) and the spin equilibrium state of HS ($S = 5/2$) $\leftrightarrow$ LS ($S = 1/2$), respectively. If the spin state of Fe^{III} site is LS state ($S = 1/2$), there exist four potential exchange interactions (J_{P}) and one kinetic exchange interaction (J_{K}) between the Fe^{III} and Mn^{II} sites. The sum of the potential exchange interaction is considered to be stronger than the kinetic exchange interaction, which is responsible for the ferromagnetic ordering. In connection with this, the following should be mentioned. The ferromagnetic ordering of (n-$C_3H_7)_4N$[$Mn^{II}Fe^{III}(dto)_3$] with Mn^{II} ($S = 5/2$) and Fe^{III} ($S = 1/2$) has been reported [25], in which the Curie temperature (T_{C}) and the Weiss temperature (θ) were estimated at 4 and 10 K, respectively, from the analysis of magnetization and magnetic susceptibility. On the other hand, if the spin state of Fe^{III} site is HS state ($S = 5/2$), there exist five J_{K} between the Fe^{III} and Mn^{II} sites, which are considered to be stronger than the potential exchange interactions between the Fe^{III} and Mn^{II} sites. Indeed, the ferrimagnetic ordering of

$(n\text{-}C_nH_{2n+1})_4N[Mn^{II}Fe^{III}(ox)_3]$ with Mn^{II} ($S = 5/2$) and Fe^{III} ($S = 5/2$) has been reported [26], in which T_N was estimated at 27–28 K.

Therefore, if the spin state of the Fe^{III} site in the $[Mn^{II}Fe^{III}(mto)_3]$ system behaves as the dynamic spin equilibrium phenomenon, the internal magnetic field at the Mn^{II} site should be frustrated between the ferromagnetic and antiferromagnetic interactions, which is schematically shown in Fig. 6.8.

Based on this viewpoint, we have synthesized $(C_6H_5)_4P[Mn^{II}Fe^{III}(mto)_3]$ and investigated the magnetic properties [27]. Figure 6.9 shows the temperature dependence of the magnetization and ac magnetic susceptibility for $(C_6H_5)_4P[Mn^{II}Fe^{III}(mto)_3]$. As shown in Fig. 6.9a, with decreasing temperature, the field cooled magnetization (FCM) remarkably increases just below 30 K, then slightly increases below 23 K. In the heating process, the remnant magnetization (RM) disappears at 30 K. On the other hand, both of the real (χ') and imaginary (χ'') parts of ac magnetic susceptibility exhibit a steep peak at 23 K indicating a magnetic phase transition, which is shown in Fig. 6.9b, c. Therefore, it is obvious that $(C_6H_5)_4P[Mn^{II}Fe^{III}(mto)_3]$ undergoes two successive magnetic phase transitions at 30 and 23 K. As shown in Fig. 6.10, the ^{57}Fe Mössbauer spectra of $(C_6H_5)_4P[Mn^{II}Fe^{III}(mto)_3]$ imply that the spin state at the Fe^{III} site is still paramagnetic between 30 and 24 K. Then, both of the Mn^{II} and Fe^{III} spins are eventually ordered at 23 K. In this manner, it is concluded that the successive magnetic phase

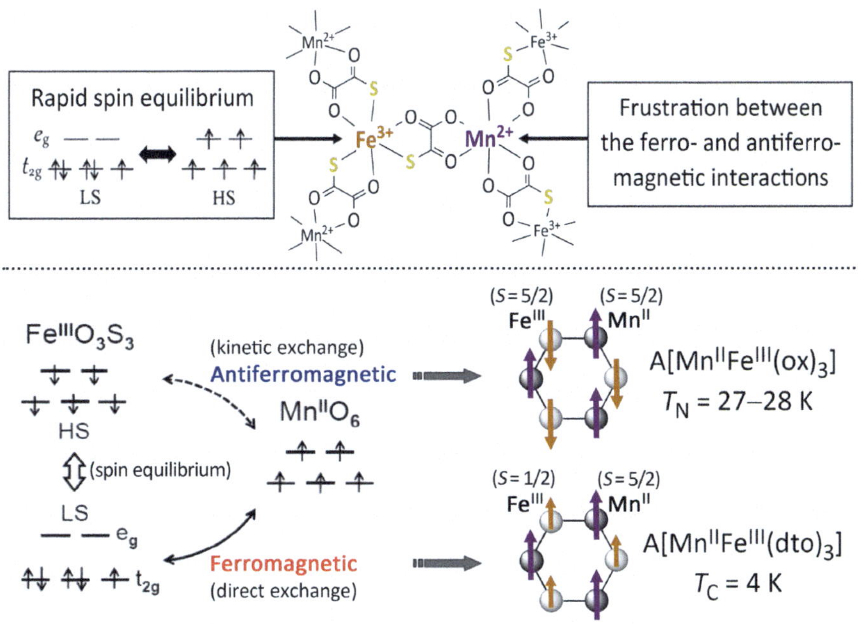

Fig. 6.8 Schematic representation of the relationship between the rapid spin equilibrium of the Fe^{III} site and the dynamic frustration of internal magnetic field at the Mn^{II} site in $[Mn^{II}Fe^{III}(mto)_3]$

Fig. 6.9 **a** Magnetization, **b** in-phase ac magnetic susceptibility (χ'), and **c** out-phase ac magnetic susceptibility (χ'') of $(C_6H_5)_4P[Mn^{II}Fe^{III}(mto)_3]$ as a function of temperature. FCM: field-cooled magnetization, ZFCM: zero-field-cooled magnetization, RM: remnant magnetization. FCM and ZFCM were measured at 3.0 mT. The ac magnetic susceptibility measurements were performed under the ac magnetic field of 0.3 mT and the frequency between 20 and 1000 Hz

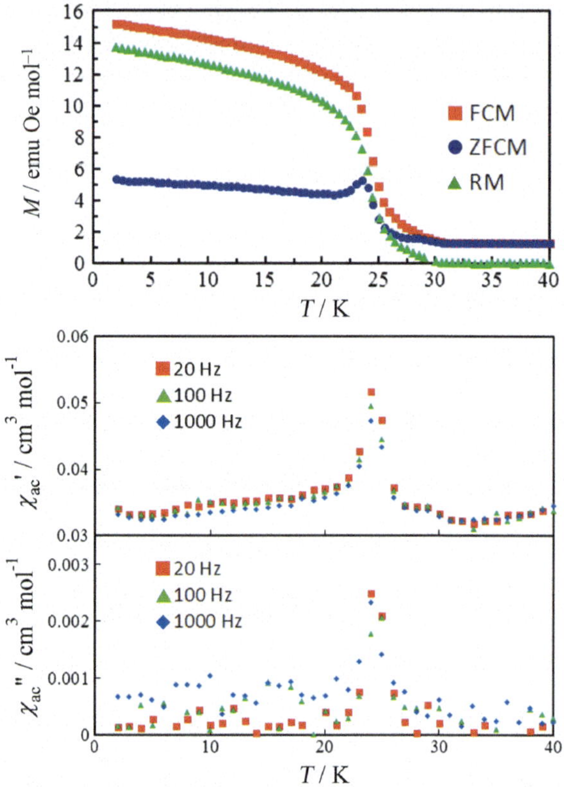

transitions at 30 and 23 K for $(C_6H_5)_4P[Mn^{II}Fe^{III}(mto)_3]$ are induced by the rapid spin equilibrium at the Fe^{III} site.

Taking account of the temperature dependences of the magnetization, ac magnetic susceptibility, and ^{57}Fe Mössbauer spectra for $(C_6H_5)_4P[Mn^{II}Fe^{III}(mto)_3]$, the Gibbs energies of the A, B and C phases as a function of temperature for $(C_6H_5)_4P[Mn^{II}Fe^{III}(mto)_3]$ are schematically represented in Fig. 6.11, in which the A, B and C phases are defined as follows,

Phase A The spins at the Mn^{II} and Fe^{III} sites are paramagnetic.
Phase B The spin at the Mn^{II} site is ordered, while that at the Fe^{III} site is still paramagnetic.
Phase C Both of the spins at the Mn^{II} and Fe^{III} sites are ordered.

As shown in Fig. 6.11, with decreasing temperature, the first magnetic phase transition from the phase A to the phase B takes place at 30 K, then the second magnetic phase transition from the phase B to the phase C takes place at 23 K.

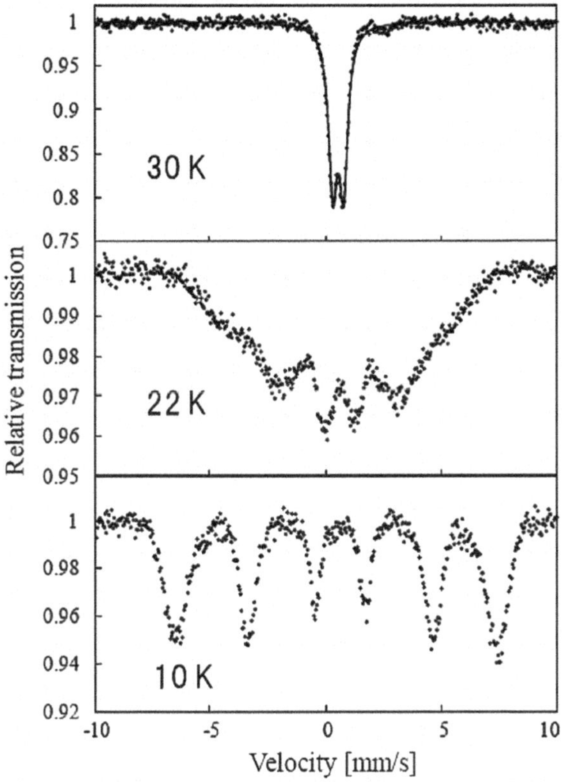

Fig. 6.10 ^{57}Fe Mössbauer spectra of $(C_6H_5)_4P[Mn^{II}Fe^{III}(mto)_3]$ at 30, 22 and 10 K

6.3 Mixed-Valence System and Charger Transfer Phase Transition

6.3.1 Classification of Mixed-Valence System

Mixed-valence compounds, defined as compounds containing an element with more than one oxidation state, are classified into three groups according to the Robin–Day classification [28],

Class I — Compounds whose valence states occupy different crystallographic sites and are too distant from each other to allow the electronic interaction between the mixed-valence states.

Class II — Compounds in which an electron exchange between the different valence states takes place as a thermally activated process. The inter-valence charge transfer (IVCT) between different valence states is observable in optical absorption spectra.

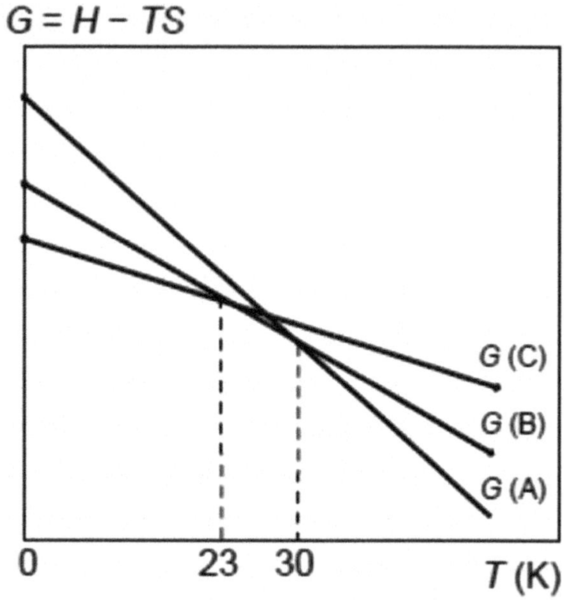

Fig. 6.11 Schematic Gibbs energies of the A, B and C phases as a function of temperature for $(C_6H_5)_4P[Mn^{II}Fe^{III}(mto)_3]$. *G*(A), *G*(B) and *G*(C) indicate the Gibbs energies for the A, B and C phases, respectively. Phase A: The spins at the Mn^{II} and Fe^{III} sites are paramagnetic, Phase B: The spin at the Mn^{II} site is ordered, while that at the Fe^{III} site is still paramagnetic, Phase C: Both of the spins at the Mn^{II} and Fe^{III} sites are ordered

Class III(A) Compounds in which valence electrons are delocalized between atoms within discrete polynuclear ions.

Class III(B) Compounds in which valence electrons are delocalized between identically crystallographic sites.

6.3.2 Prussian Blue and Its Analogues Salts Showing Photo-Induced Magnetism

Among these mixed-valence compounds, the boundary compounds between the class II and III are the leading candidates exhibiting field-responsive multifunctional properties coupled with transport, optical or magnetic properties. In this section, we describe the characteristic properties of Prussian blue and its analogous compounds as an example.

Prussian blue has a three dimensional network structure with an alternating array of Fe^{II} and Fe^{III} ions through CN bridges [29], which is schematically shown in Fig. 6.12a. The spin state of the Fe^{II} site coordinated by six C atoms is LS ($S = 0$) state, and that of the Fe^{III} site coordinated by six N atoms is HS ($S = 5/2$) state. Prussian blue has a broad strong IVCT band at around 730 nm, which is responsible for the deep blue pigment. Figure 6.12b shows "The Great Wave off Kanagawa" painted by Hokusai, in which Prussian blue was used as a deep blue pigment.

The soluble Prussian blue, $KFe^{II}Fe^{III}(CN)_6$, is obtained by adding Fe^{3+} to the aqueous solution of $K_4[Fe^{II}(CN)_6]$, as following scheme, $Fe^{3+}(aq) + K^+(aq) + [Fe^{II}(CN)_6]^{4-}(aq) \rightarrow KFe^{II}Fe^{III}(CN)_6(s)$. On the other hand, by adding Fe^{2+} to the

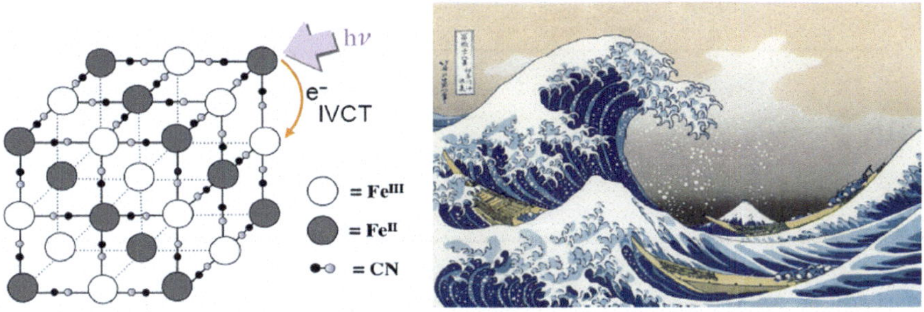

Crystal structure of Prussian blue, $Fe^{III}{}_4[Fe^{II}(CN)_6]_3 \cdot 15H_2O$

The Great Wave off Kanagawa, Hokusai Katsushika (1831)

Fig. 6.12 **a** Schematic representation of the crystal structure of $Fe_4^{III}[Fe^{II}(CN)_6]_3$. **b** "The Great Wave off Kanagawa" painted by Hokusai, in which Prussian blue was used as deep blue pigment

aqueous solution of $K_3[Fe^{III}(CN)_6]$, $KFe^{II}Fe^{III}(CN)_6$ called Turnbull's blue is also obtained as follows, $Fe^{2+}(aq) + K^+(aq) + [Fe^{III}(CN)_6]^{3-}(aq) \rightarrow KFe^{II}Fe^{III}(CN)_6(s)$. A long historical debate had continued in relation to Prussian blue and Turnbull's blue, whether Prussian blue obtained by mixing the solutions of ferric compound and ferrocyanide, and Turnbull's blue obtained by mixing the solutions of ferrous compound and ferricyanide, are the same compound or not? This problem was solved by means of ^{57}Fe Mössbauer spectroscopy in 1968 [30]. According to the ^{57}Fe Mössbauer spectroscopy, Prussian blue and Turnbull's blue are essentially the same compound, in spite of the different synthesis processes. Figure 6.13 shows the ^{57}Fe Mössbauer spectra for the soluble Prussian blue, insoluble Prussian blue, and Turnbull's blue in the ferromagnetic ordered phase ($T_C = 5.5$ K) at 1.6 K [30]. These spectra show the superposition of a central peak corresponding to the LS state ($S = 0$) of Fe^{II} coordinated by six C atoms and six magnetically split branches corresponding to the HS state ($S = 5/2$) of Fe^{III} coordinated by six N atoms. The estimated internal magnetic fields of Fe^{III} ($S = 5/2$) at 1.6 K were estimated at 53.6 ± 2, 54.1 ± 2 and 54.3 ± 2 T for the soluble Prussian blue, insoluble Prussian blue and Turnbull's blue, respectively.

In connection with Turnbull's blue, the following should be noted. As shown in Fig. 6.14, in the precipitation process of Turnbull's blue by adding $^{57}Fe^{2+}$ to the aqueous solution of $K_3[Fe^{III}(CN)_6]$, an electron of $^{57}Fe^{2+}$ transfers to the Fe^{III} site in $[Fe^{III}(CN)_6]^{3-}$ and Turnbull's blue, $KFe^{II}\,{}^{57}Fe^{III}(CN)_6$, is precipitated [31].

In connection with Prussian blue, one of the most fascinating phenomena is the photo-induced magnetism of Prussian blue analogues by means of IVCT. In 1996, O. Sato, K. Hashimoto, et al. first reported a photo-induced ferrimagnetic phase for $K_{0.2}Co_{1.4}[Fe(CN)_6]\cdot 6.9H_2O$ [32] and $K_{0.4}Co_{1.3}[Fe^{II}(CN)_6]\cdot 5H_2O$ [33]. In the case of $K_{0.4}Co^{II}_{0.3}Co^{III}[Fe^{II}(CN)_6]\cdot 5H_2O$, the electronic state of the Fe–CN–Co framework before light irradiation is $Fe^{II}(t_{2g}^6e_g^0,\ S = 0)-CN-Co^{III}(t_{2g}^6e_g^0,\ S = 0)$ at 5 K. Visible light ($\lambda = 500$–750 nm) irradiation at 5 K induces the IVCT, forming the excited state of $Fe^{III}(t_{2g}^5e_g^0,\ S = 1/2)-CN-Co^{II}(t_{2g}^6e_g^0,\ S = 1/2)$. Subsequently, this excited state

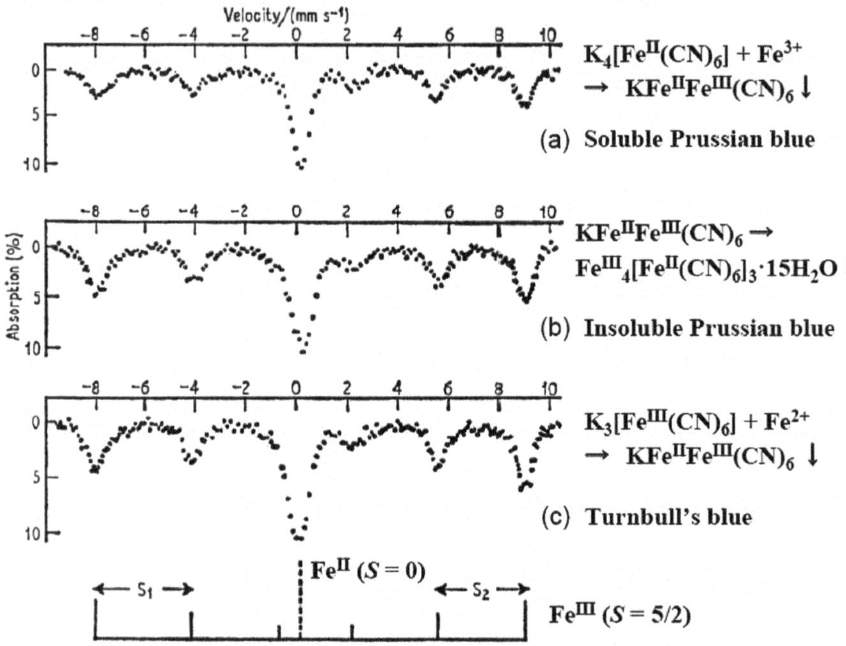

Fig. 6.13 ^{57}Fe Mössbauer spectra of **a** soluble Prussian blue, **b** insoluble Prussian blue and **c** Turnbull's blue at 1.6 K. Reprinted from [30], with the permission of AIP Publishing

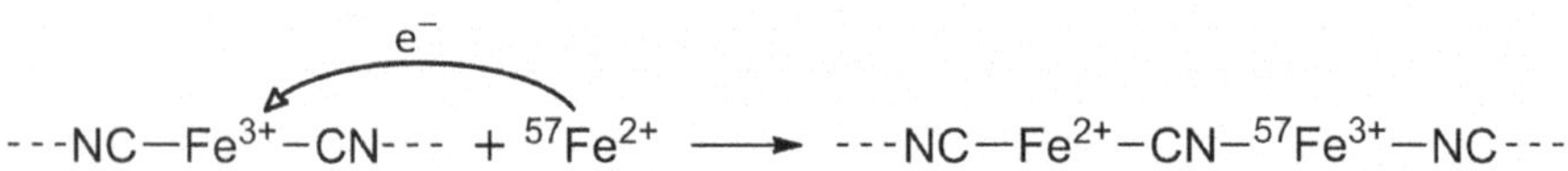

Fig. 6.14 Formation process of Turnbull's blue

relaxes to the meta-stable state of $Fe^{III}(t_{2g}^5e_g^0, S = 1/2)$–CN–$Co^{II}(t_{2g}^5e_g^0, S = 3/2)$. While the compound before light irradiation is paramagnetic, that after the visible light irradiation exhibits a meta-stable ferrimagnetic phase, which can be reconverted to the initial phase by near-IR light ($\lambda = 1319$ nm) irradiation or a thermal treatment [33]. Since then, not only Prussian blue analogues but also various CN-bridged compounds exhibiting photo-induced magnetism have been reported [34]. In addition to the photo-induced magnetism, since the discovery of thermally induced charge-transfer induced spin transition (CTIST) for $Na_{0.4}Co_{1.3}[Fe(CN)_6]\cdot 5H_2O$ in 1997 [35], several Prussian blue analogues and CN-bridged compounds exhibiting thermally induced CTIST or CTPT have been reported [36].

Figure 6.15 shows the photo-induced ferrimagnetic transition for $Rb_{0.66}Co_{1.25}[Fe(CN)_6]\cdot 4.3H_2O$ by using IVCT [37]. While $Rb_{0.66}Co_{1.25}[Fe(CN)_6]\cdot 4.3H_2O$ before the light irradiation is paramagnetic in

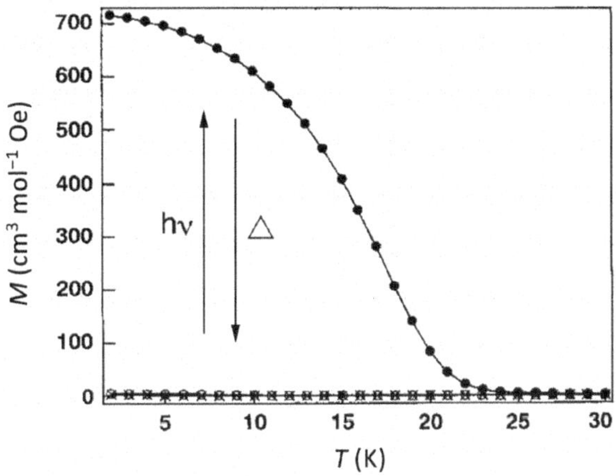

Fig. 6.15 Field-cooled magnetization (FCM) curves for $Rb_{0.66}Co_{1.25}[Fe(CN)_6]·4.3H_2O$ before and after light irradiation ($h\nu$) at $H = 0.5$ T. ○: before light irradiation, •: after light irradiation, ×: after thermal treatment (Δ) at 150 K. Reprinted with permission from [37]. Copyright 1999 American Chemical Society

the whole temperature region, that after the light irradiation corresponding to the IVCT (500–750 nm) exhibits a meta-stable ferrimagnetic phase with $T_N = 22$ K.

Figure 6.16 shows the ^{57}Fe Mössbauer spectra before and after the light irradiation corresponding to the IVCT (500–750 nm) at 25 K [37]. Before the light irradiation, the single peak with $IS = 0.02 \pm 0.01$ mm/s and $QS = 0.15 \pm 0.01$ mm/s is assigned to the LS state ($S = 0$) of Fe^{II}. After the light irradiation at 25 K, the single peak corresponding to the LS state of Fe^{II} was significantly reduced and a new doublet with $IS = 0.09 \pm 0.01$ mm/s and $QS = 1.13 \pm 0.01$ mm/s assigned to the LS state ($S = 1/2$) of Fe^{III}. As a result of the light-induced CT from the Fe^{II} site to the Co^{III} site, $Rb_{0.66}Co_{1.25}[Fe(CN)_6]·4.3H_2O$ was converted from the paramagnet to the ferrimagnet with $T_N = 22$ K, which is schematically shown in Fig. 6.17.

6.3.3 *Mixed-Valence System, $A[Fe^{II}Fe^{III}(dto)_3]$ (A = Counter Cation; dto = $C_2O_2S_2$), and the Charge Transfer Phase Transition*

Among various kinds of multifunctional compounds, in the case of mixed-valence system whose spin states are situated in the spin crossover region, it is expected that new types of synergetic phenomena induced by both of spin and charge appear between different metal ions in order to minimize the Gibbs energy. Based on this viewpoint, we have developed a ferromagnetic organic–inorganic hybrid system, $A[Fe^{II}Fe^{III}(dto)_3]$ ($A = (n\text{-}C_nH_{2n+1})_4N$, spiropyran, etc.), and investigated their synergetic properties coupled with spin, charge and photon [4]. The single-crystal X-ray structural analysis of $(n\text{-}C_3H_7)_4N[Fe^{II}Fe^{III}(dto)_3]$ reveals the existence of the honeycomb network structure of $[Fe^{II}Fe^{III}(dto)_3]$, and the $(n\text{-}C_3H_7)_4N^+$ layer is intercalated between two adjacent $[Fe^{II}Fe^{III}(dto)_3]$ layers [20], whose structure is common for $A[M^{II}M'^{III}X_3]$ (A = cation, X = ox, dto, mto).

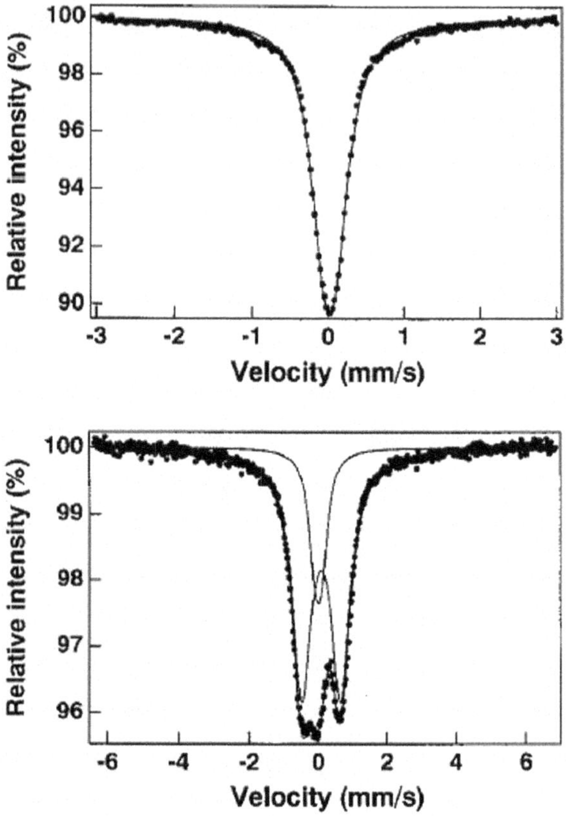

Fig. 6.16 ^{57}Fe Mössbauer spectra of $Rb_{0.66}Co_{1.25}$ $[Fe(CN)_6]\cdot 4.3H_2O$ before and after the light irradiation of 500–750 nm at 25 K [37]

In order to investigate the electronic properties and the valence states of Fe^{II} and Fe^{III} of $(n\text{-}C_3H_7)_4N[Fe^{II}Fe^{III}(dto)_3]$ by means of ^{57}Fe Mössbauer spectroscopy, we employed ^{57}Fe enriched complexes, $(n\text{-}C_3H_7)_4N[^{57}Fe^{II}Fe^{III}(dto)_3]$ and $(n\text{-}C_3H_7)_4N[Fe^{II\,57}Fe^{III}(dto)_3]$, in which Fe^{II} or Fe^{III} was substituted with ^{57}Fe (96%) [38]. In the case of $(n\text{-}C_3H_7)_4N[^{57}Fe^{II}Fe^{III}(dto)_3]$, a solution of $^{57}FeCl_2\cdot 4H_2O$ and $(n\text{-}C_3H_7)_4NBr$ in a methanol/water mixture was added to a solution of $KBa[Fe^{III}(dto)_3]$, then black powdered crystals of $(n\text{-}C_3H_7)_4N[^{57}Fe^{II}Fe^{III}(dto)_3]$ were precipitated. $(n\text{-}C_3H_7)_4N[Fe^{II\,57}Fe^{III}(dto)_3]$ was also obtained by using $FeCl_2\cdot 4H_2O$, $(n\text{-}C_3H_7)_4NBr$ and $KBa[^{57}Fe^{III}(dto)_3]$. Figure 6.18 shows the ^{57}Fe Mössbauer spectra of $(n\text{-}C_3H_7)_4N[^{57}Fe^{II}Fe^{III}(dto)_3]$ and $(n\text{-}C_3H_7)_4N[Fe^{II\,57}Fe^{III}(dto)_3]$ at 200 and 60 K. In the case of the ^{57}Fe Mössbauer spectra of $(n\text{-}C_3H_7)_4N[^{57}Fe^{II}Fe^{III}(dto)_3]$ at 200 K, the dominant doublet (74.5% area) with $IS = 1.18$ mm s^{-1} and $QS = 1.36$ mm s^{-1} is assigned to the $Fe^{II}(t_{2g}^4e_g^2, S = 2)$ site coordinated by six O atoms. These values of IS and QS are quite similar to those ($IS = 1.18$ mm s^{-1}, $QS = 1.26$ mm s^{-1}) of the ^{57}Fe Mössbauer spectrum for the $Fe^{II}(t_{2g}^4e_g^2, S = 2)$ site in $(n\text{-}C_4H_9)_4N[Fe^{II}Cr^{III}(ox)_3]$

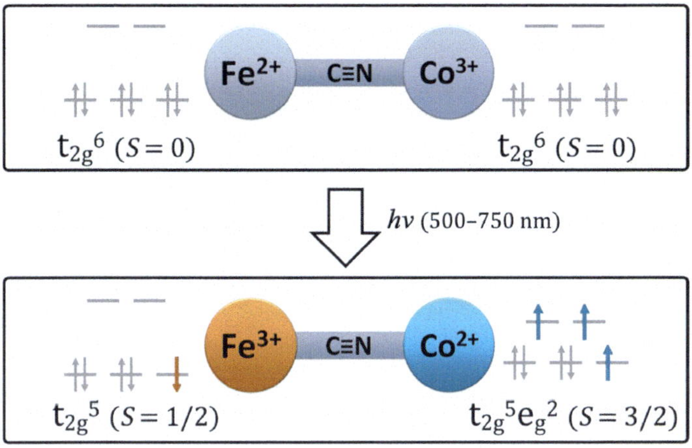

Fig. 6.17 Schematic representation of the photo-induced magnet for $Rb_{0.66}Co_{1.25}[Fe(CN)_6]\cdot 4.3H_2O$

[39]. As shown in Fig. 6.18, the line profile of the ^{57}Fe Mössbauer spectra of (n-$C_3H_7)_4N[^{57}Fe^{II}Fe^{III}(dto)_3]$ dramatically changes between 200 and 60 K. The intensity of the dominant doublet corresponding to the Fe^{II} ($S = 2$) site decreases and a new doublet with $IS = 0.55$ mm s^{-1} and $QS = 0.91$ mm s^{-1} appears. The IS and QS values of the new doublet are quite similar to those ($IS = 0.49$ mm s^{-1}, $QS = 0.68$ mm s^{-1}) of the ^{57}Fe Mössbauer spectrum for the $Fe^{III}(t_{2g}^3e_g^2, S = 5/2)$ site in (n-$C_4H_9)_4N[Ni^{II}Fe^{III}(ox)_3]$ at 78 K [40]. On the other hand, in the case of (n-$C_3H_7)_4N[Fe^{II}\,^{57}Fe^{III}(dto)_3]$ at 200 K, the dominant doublet with $IS = 0.33$ mm s^{-1} and $QS = 0.36$ mm s^{-1} is assigned to the $Fe^{III}(t_{2g}^5, S = 1/2)$ site coordinated by six S atoms. These values of IS and QS are quite similar to those ($IS = 0.33$ mm s^{-1}, $QS = 0.35$ mm s^{-1} at 196 K) for the $Fe^{III}(t_{2g}^5, S = 1/2)$ site coordinated by six S atoms in $KBa[Fe^{III}(dto)_3]$ [41]. On the other hand, the dominant doublet with $IS = 0.41$ mm s^{-1} and $QS = 0.51$ mm s^{-1} at 60 K is assigned to the $Fe^{II}(t_{2g}^6, S = 0)$ site coordinated by six S atoms. These IS and QS values are quite similar to those ($IS = 0.41$ mm s^{-1}, $QS = 0.64$ mms^{-1} at 77 K) for the $Fe^{II}(t_{2g}^6, S = 0)$ site coordinated by six S atoms in pyrite (FeS_2) [42]. In this manner, we have elucidated the CTPT for (n-$C_3H_7)_4N[Fe^{II}Fe^{III}(dto)_3]$ by means of ^{57}Fe Mössbauer spectroscopy.

As mentioned above, (n-$C_3H_7)_4N[Fe^{II}Fe^{III}(dto)_3]$ undergoes the CTPT between 200 and 60 K. In order to determine the critical temperature (T_{CT}) of CTPT, ^{57}Fe Mössbauer spectroscopy was carried out in the temperature range between 130 and 90 K. As shown in Fig. 6.19, the HT phase with $Fe^{II}O_6$ ($S = 2$)–$Fe^{III}S_6$ ($S = 1/2$) and the LT phase with $Fe^{II}S_6$ ($S = 0$)–$Fe^{III}O_6$ ($S = 5/2$) are clearly distinguished in the vicinity of CTPT, which implies that the frequency of electron transfer between the Fe^{II} and Fe^{III} sites at the CTPT is at least slower than the time scale (10^{-7} s) of ^{57}Fe Mössbauer spectroscopy. When the temperature decreases from 130 to 90 K, the fractions of HT and LT phases decreases and increases, respectively. The crossing

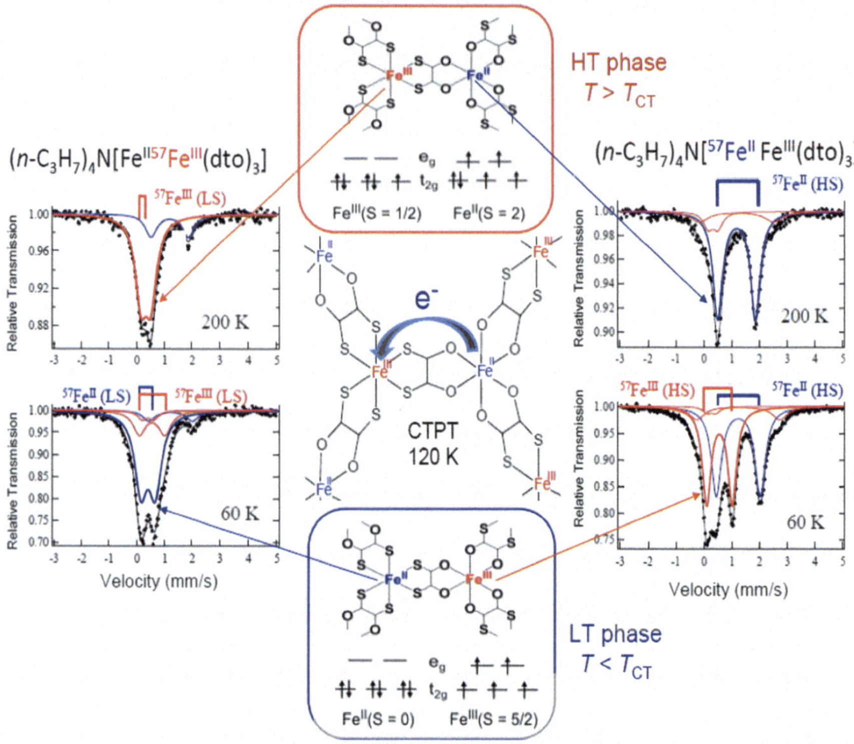

Fig. 6.18 ^{57}Fe Mössbauer spectra below and above the CTPT ($T_{CT} \approx$ 120 K) for (n-$C_3H_7)_4N[^{57}Fe^{II}Fe^{III}(dto)_3]$ and (n-$C_3H_7)_4N[Fe^{II}\,^{57}Fe^{III}(dto)_3]$, and the schematic representation of CTPT. The HT phase ($T > T_{CT}$) and LT phase ($T < T_{CT}$) indicate the high temperature phase with $Fe^{III}S_6(t_{2g}^5) - Fe^{II}O_6(t_{2g}^4e_g^2)$ and the low temperature phase with $Fe^{II}S_6(t_{2g}^6) - Fe^{III}O_6(t_{2g}^3e_g^2)$, respectively

point between the HT phase and LT phase components is about 122 K, which is consistent with the critical temperature (122.4 K) of CTPT determined by the heat capacity [43].

In order to confirm the direct evidence of charge transfer between the Fe^{II} and Fe^{III} sites in (n-$C_3H_7)_4N[Fe^{II}Fe^{III}(dto)_3]$, the electrical resistivity was measured along the parallel and perpendicular directions to the $[Fe^{II}\ Fe^{III}(dto)_3]$ layer under several applied pressures [44]. At 0.5 GPa, (n-$C_3H_7)_4N[Fe^{II}Fe^{III}(dto)_3]$ behaves as a semiconductor as shown in Fig. 6.20. Above 0.9 GPa, on the other hand, both of the intra- and inter-layer resistivities show an anomalous drop with thermal hysteresis due to the CTPT. The intra-layer resistivity as a function of temperature at 0.9 GPa exhibits a plateau in the middle range of hysteresis loop, while an abrupt drop of resistivity appears in the inter-layer resistivity. Similar behavior appears at 1.5 GPa except for the temperature shift of the hysteresis loop. The temperature of hysteresis loop increases with increasing applied pressure, which is consistent with that of the magnetic susceptibility under applied pressures [45]. The activation energies at 0,

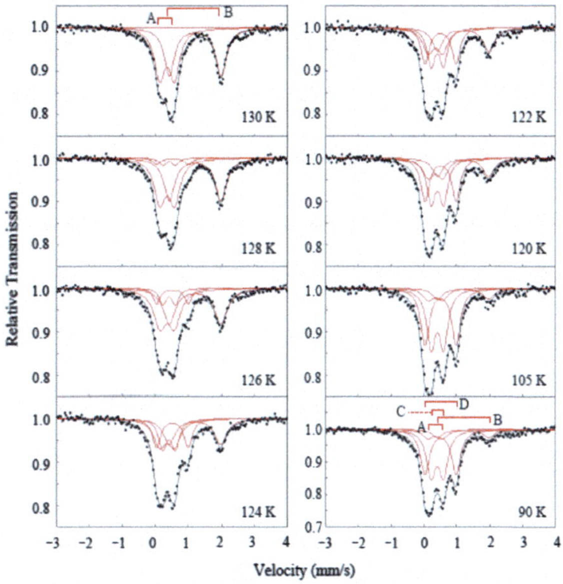

Fig. 6.19 ^{57}Fe Mössbauer spectra of $(n\text{-}C_3H_7)_4N[Fe^{II}Fe^{III}(dto)_3]$ in the vicinity of CTPT (T_{CT} = 122 K). A: Fe^{III} ($S = 1/2$), B: Fe^{II} ($S = 2$), C: Fe^{II} ($S = 0$), D: Fe^{III} ($S = 5/2$)

0.9 and 1.5 GPa are 1700, 1400 and 1200 K, respectively. The anomalous drop in the resistivity as a function of temperature is attributed to the electron transfer between the t_{2g} orbitals of the Fe^{II} and Fe^{III} sites in the temperature region of CTPT.

6.3.4 Dynamics of Charge Transfer Phase Transition in A[$Fe^{II}Fe^{III}$ (dto)$_3$] by Means of Muon Spectroscopy

Muon spin relaxation (μSR) is one of the most powerful methods to investigate the magnitude, distribution or fluctuation of internal magnetic field. Therefore, the μSR technique has been applied to the study of magnetic phase transitions, various kinds of spin frustrations, superconducting phenomena and so forth [46]. The asymmetry parameter of μSR is defined as $A(t) = [N_B(t) - N_F(t)]/[N_B(t) + N_F(t)]$, in which $N_F(t)$ and $N_B(t)$ are the total muon events of the forward and backward counters in the muon beam line, respectively. The initial asymmetry, $A(0)$, is defined as the asymmetry at $t = 0$. In order to analyze the time spectra of $A(t)$, the following function with two components was used [46],

$$A_0 G_Z(\Delta, H_{LF}, t) \exp(-\lambda_0 t) + A_1 \exp(-\lambda_1 t) \tag{6.4}$$

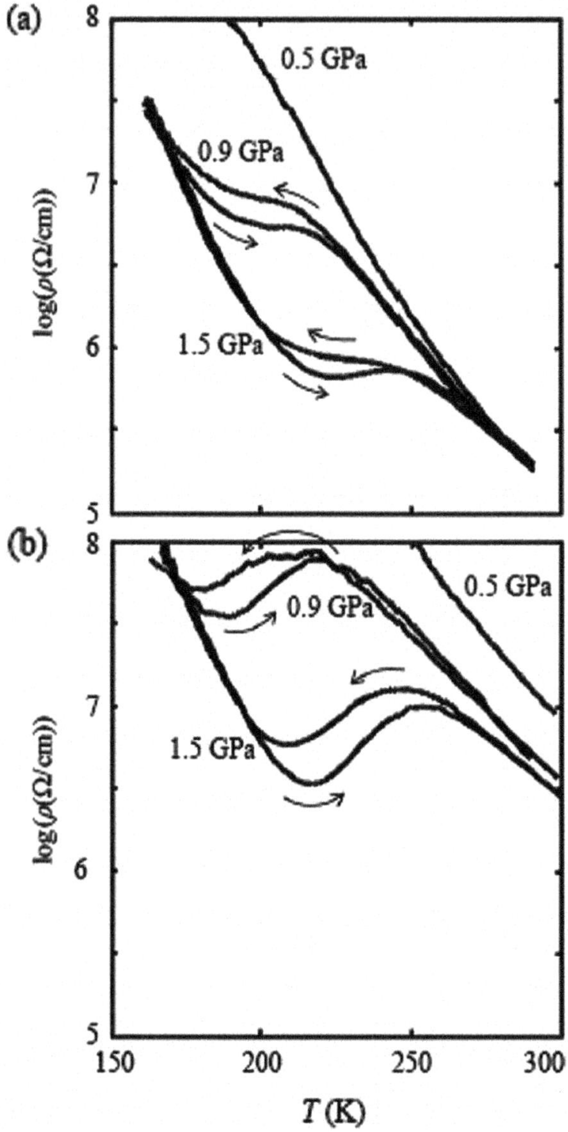

Fig. 6.20 Temperature dependence of **a** intra-layer and **b** inter-layer electrical resistivities of (n-$C_3H_7)_4N[Fe^{II}Fe^{III}(dto)_3]$ under several applied pressures. Arrows denote the direction of heating or cooling process. Reprinted from [44], Copyright 2003, with permission from Elsevier

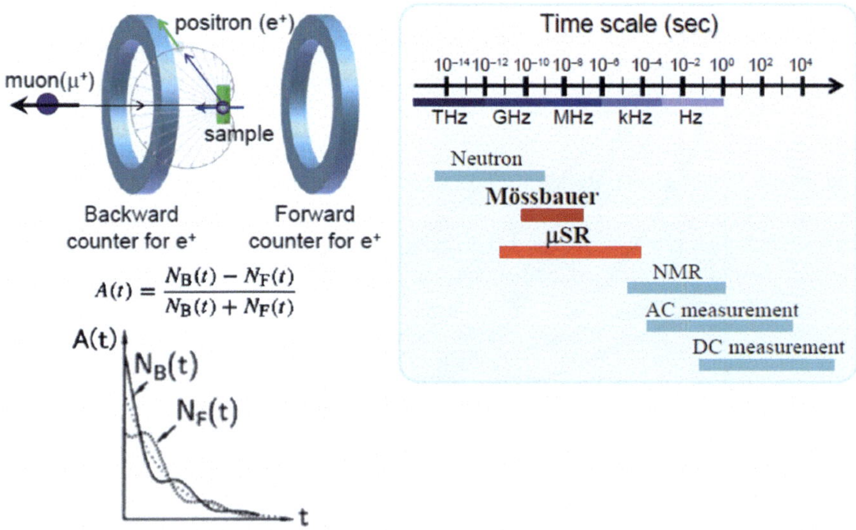

Fig. 6.21 Schematic representation of muon spectroscopy and its time scale

where A_0 and A_1 are the initial asymmetries of the slow and fast relaxation components, λ_0 and λ_1 are the respective muon spin depolarization rates. $G_Z(\Delta, H_{LF}, t)$ is the static Kubo–Toyabe function [47]. Δ/γ_μ is the distribution width of the nuclear-dipole fields at the muon sites, and γ_μ is the gyromagnetic ratio of muon spin. H_{LF} is the longitudinal magnetic field. Figure 6.21 shows the time scale and the schematic representation of muon spectroscopy.

Figure 6.22 shows the time dependence of the asymmetry parameter, $A(t) = [N_B(t) - N_F(t)]/[N_B(t) + N_F(t)]$ of μSR for $(n\text{-}C_3H_7)_4N[Fe^{II}Fe^{III}(dto)_3]$ at various temperatures under zero-field [5]. From the analysis of time spectra of μSR by using Eq. (6.4), we obtained the depolarization rate (λ_0) as a function of temperature for $(n\text{-}C_nH_{2n+1})_4N[Fe^{II}Fe^{III}(dto)_3]$.

Figure 6.23 shows the depolarization rate (λ_0) of muon spin as a function of temperature under various longitudinal magnetic fields for $(n\text{-}C_nH_{2n+1})_4N[Fe^{II}Fe^{III}(dto)_3]$ (n = 3, 5). In the case of n = 3, an anomalous enhancement of depolarization rate appears at around 80 K. This anomalous peak decreases with increasing the longitudinal field, and eventually disappears at about 10 mT (= 100 Oe). In the case of n = 5, on the other hand, any anomalous peak does not appear between 200 and 40 K. Therefore, the anomalous enhancement of depolarization rate at around 80 K for n = 3 is attributed to the frequency of electron transfer between the Fe^{II} and Fe^{III} sites at the CTPT, which induces the fluctuating internal magnetic field at the muon site.

As shown in Fig. 6.23, the depolarization rates for n = 3 and n = 5 decrease with increasing longitudinal field up to 10 mT and become almost constant above 10 mT. The constant value increases with decreasing temperature from 200 to 40 K, which

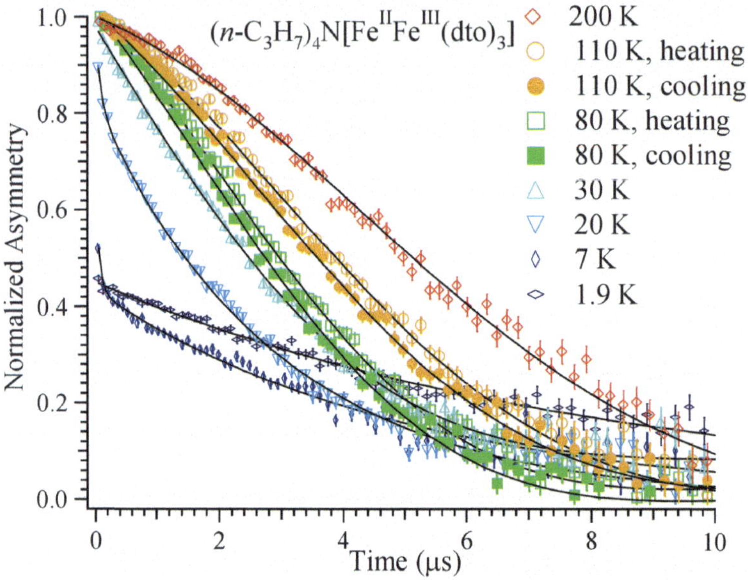

Fig. 6.22 Time dependence of the asymmetry parameter, $A(t) = [N_B(t) - N_F(t)]/[N_B(t) + N_F(t)]$ of μSR for $(n\text{-}C_3H_7)_4N[Fe^{II}Fe^{III}(dto)_3]$ at several temperatures under zero-field [5]. Reprinted figure with permission from [5]. Copyright 2008 by the American Physical Society

suggests the existence of two components. One component is easily suppressed by 10 mT, while the other component remains even at 400 mT (= 4000 Oe). The former component implies that there is a weak and slowly fluctuating internal magnetic field at the muon site, which is easily masked by 10 mT, which is attributed to the fluctuating component of nuclear dipoles such as MnSi [48]. On the other hand, considering that the latter component increases with decreasing temperature and similar values of the depolarization rate are observed for n = 3 and 5 in the same temperature region, which implies that the latter component is attributed to the dipole field due to the fluctuation of paramagnetic Fe spins.

As for the anomalous enhancement of the depolarization rate of muon spin induced by the CTPT, we extract the difference of depolarization rate between those of n = 3 and 5 to analyze the frequency of electron transfer between the Fe^{II} and Fe^{III} sites at the CTPT. The longitudinal field dependence of this subtracted depolarization rate at 80 K is plotted in Fig. 6.24. By using the Redfield equation [49] for the longitudinal field (H_{LF}) dependence of the subtracted depolarization rate (λ_{CT}) between n = 3 and 5, we determined the correlation time of muon spins (τ_c) and the amplitude of fluctuating internal magnetic field (H_{loc}) at the muon site, respectively. The Redfield equation is expressed as follows,

$$\lambda_{CT} = \frac{2\gamma_\mu^2 H_{loc}^2 \tau_c}{1 + \gamma_\mu^2 H_{LF}^2 \tau_c^2} \tag{6.5}$$

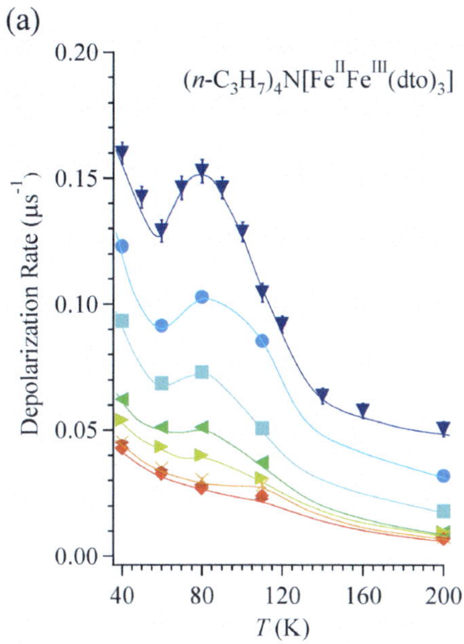

Fig. 6.23 Temperature and longitudinal-field dependence of the muon spin depolarization rate, λ_0, for $(n\text{-}C_3H_7)_4N[Fe^{II}Fe^{III}(dto)_3]$ and $(n\text{-}C_5H_{11})_4N[Fe^{II}Fe^{III}(dto)_3]$. 1 Oe = 0.1 mT. The solid lines are guides for eye. Reprinted figure with permission from [5]. Copyright 2008 by the American Physical Society

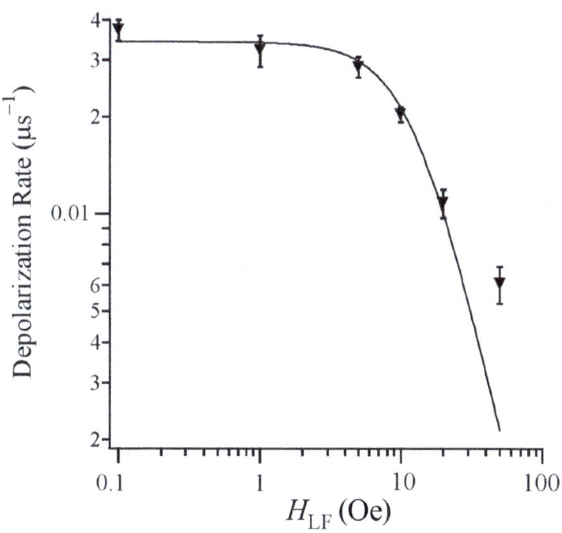

Fig. 6.24 Subtracted depolarization rate (λ_{CT}) between n = 3 and 5 as a function of applied longitudinal magnetic field for $(n\text{-}C_nH_{2n+1})_4N[Fe^{II}Fe^{III}(dto)_3]$ at 80 K in the cooling process. The solid line shows the best fit of the Redfield equation. 1 Oe = 0.1 mT. Reprinted figure with permission from [5]. Copyright 2008 by the American Physical Society

Table 6.1 Parameters obtained by fitting the Redfield equation to the subtracted depolarization rate (λ_{CT}) between n = 3 and 5 for $(n\text{-}C_nH_{2n+1})_4N[Fe^{II}Fe^{III}(dto)_3]$ [5]

T (K)	τ_c (μs)	H_{loc} (mT)
60	–	0
80	5.7	0.40
110	10.6	0.19

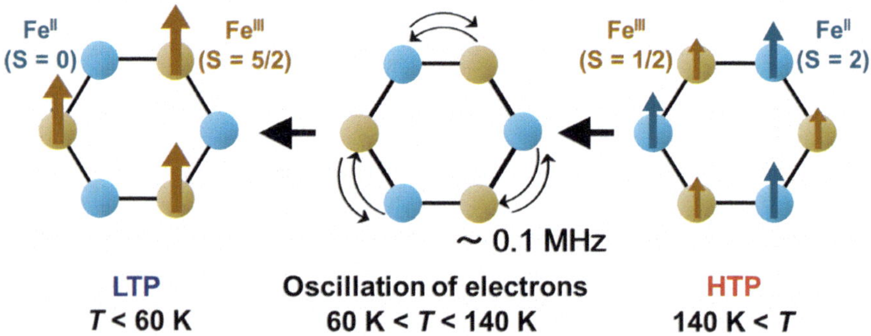

Fig. 6.25 Schematic representation of the valence fluctuation between the Fe^{II} and Fe^{III} sites at the CTPT for $(n\text{-}C_3H_7)_4N[Fe^{II}Fe^{III}(dto)_3]$

where γ_μ is the gyromagnetic ratio of muon spin. These parameters at several temperatures are summarized in Table 6.1 [5].

The frequency of the additional internal field at the muon site induced by the CTPT, which is given as $\nu = 1/\tau_c$, is the order of 0.1 MHz as the maximum value at 80 K and zero at 60 K [5]. The magnitude of H_{loc} is between 0.19 and 0.40 mT, which is larger than that of nuclear dipole field (~0.1 mT). These results imply the fluctuation of electron transfer between the Fe^{II} and Fe^{III} sites at the CTPT, which is schematically shown in Fig. 6.25. The time scale of τ_c is consistent with the result of the ^{57}Fe Mössbauer measurement. The fluctuation of electron transfer between the Fe^{II} and Fe^{III} sites induced by the CTPT for $(n\text{-}C_3H_7)_4N[Fe^{II}Fe^{III}(dto)_3]$ is responsible for the anomalous drop with thermal hysteresis in the electrical resistivity at the CTPT.

6.3.5 Size Effect of Intercalated Cation on the Charge Transfer Phase Transition and Ferromagnetism for A[$Fe^{II}Fe^{III}$ (dto)$_3$]

Figure 6.26 shows the ^{57}Fe Mössbauer spectra at 200, 77 and 4 K for $(n\text{-}C_nH_{2n+1})_4N[Fe^{II}Fe^{III}(dto)_3]$ (n = 3–6) [50]. At 200 K, the line profiles of all the complexes are quite similar to each other, where two quadrupole doublets (A and

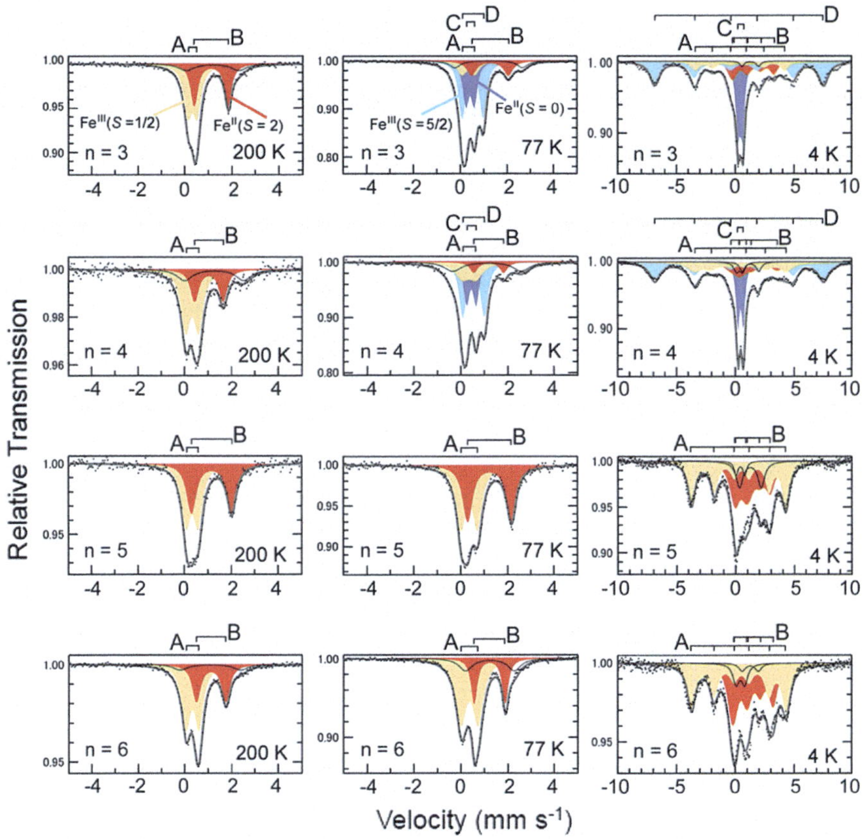

Fig. 6.26 ^{57}Fe Mössbauer spectra of $(n\text{-}C_nH_{2n+1})_4N[Fe^{II}Fe^{III}(dto)_3]$ (n = 3–6) at 200 K, 77 K and 4 K. A: Fe^{III} (S = 1/2), B: Fe^{II} (S = 2), C: Fe^{II} (S = 0), D: Fe^{III} (S = 5/2) [50]. Reprinted with permission from [50]. Copyright 2012 Mössbauer Effect Data Center

B) mainly exist. The wider doublet (B) is assigned to the Fe^{II} ($S = 2$) site coordinated by six O atoms. The *IS* and *QS* of (B) fall in the range of 1.05–1.17 mm s^{-1} and 1.20–1.69 mm s^{-1}, respectively. On the other hand, the narrower doublet (A) with IS = 0.29–0.35 mm s^{-1} and QS = 0.38–0.56 mm s^{-1} is assigned to the Fe$^{III S = 1/2)}$ site coordinated by six S atoms.

In the cases of n = 3 and 4, with decreasing temperature from 200 to 77 K, the ^{57}Fe Mössbauer spectra corresponding to the HTP decrease by about 80%. Instead of these spectra, two doublets (C and D) appear. The doublet (C) with IS = 0.43–0.46 mm s^{-1} and QS = 0.35–0.41 mm s^{-1} is assigned to the Fe^{II} ($S = 0$) site coordinated by six S atoms. On the other hand, the doublet (D) with IS = 0.54–0.55 mm $s^{-1}$and QS = 0.89–0.92 mm s^{-1} is assigned to the Fe^{III} ($S = 5/2$) site coordinated by six O atoms. At 4 K, a central strong doublet (C) with IS = 0.43–0.45 mm $s^{-1}$and QS = 0.39–0.46 mm s^{-1}, and a sextet (D) divided from −8 to +8 mm s^{-1} with a large hyperfine field of H_{int} = ~45.0 T are observed in the spectra of n = 3 and 4. The former can be assigned to the diamagnetic state of Fe^{II} (S = 0) and the latter is typical of the magnetically ordered state of Fe^{III} (S = 5/2). From the analysis of ^{57}Fe Mössbauer spectra for $(n\text{-}C_nH_{2n+1})_4N[Fe^{II}Fe^{III}(dto)_3]$, it is obvious that the charge transfer between the Fe^{II} and Fe^{III} sites takes place between 200 and 77 K for n = 3 and 4. At 4 K, in addition to the ^{57}Fe Mössbauer spectra corresponding to the Fe^{II} (S = 0) and Fe^{III} (S = 5/2) sites, the sextets of the Fe^{II} (S = 2) and Fe^{III} (S = 1/2) sites corresponding to the HTP are also required as the relative area of *ca.* 15% (for n = 3) or 30% (for n = 4). In the case of n = 4, two ferromagnetic phases with T_C = 7 and 13 K are clearly observed in the magnetization measurement, being consistent with the large area of the HTP at 4 K in the ^{57}Fe Mössbauer spectra.

The ^{57}Fe Mössbauer spectra in the ferromagnetic phases for n = 5 and 6 are completely different from those for n = 3 and 4. In the cases of n = 5 and 6, the CTPT does not take place, so that the spectra of n = 5 and 6 at 4 K can be assigned to the Fe^{II} (S = 2) and Fe^{III} (S = 1/2) sites in the magnetically ordered phase. The sextets of the Fe^{III} (S = 1/2) sites have relatively large H_{int} of 24.6 T (for n = 5) and 24.8 T (for n = 6) at 4 K, while the Fe^{II} (S = 2) sites have relatively small H_{int} of 7.3 T (for n = 5) and 9.9 T (for n = 6).

Figure 6.27 shows the ^{57}Fe Mössbauer spectra for $(n\text{-}C_nH_{2n+1})_4N[Fe^{II}Fe^{III}(dto)_3]$ (n = 3 and 5) in the magnetically ordered phase. In the case of n = 3, the LTP with $Fe^{II}S_6$ (S = 0)–$Fe^{III}O_6$ (S = 5/2) undergoes the ferromagnetic transition with T_C = 7.1 K. In the case of n = 5, on the other hand, the HTP with $Fe^{II}O_6$ (S = 2)–$Fe^{III}S_6$ (S = 1/2) undergoes the ferromagnetic transition with T_C = 19.5 K. As shown in Fig. 6.13, the ^{57}Fe Mössbauer spectrum of the ferromagnetically ordered state in

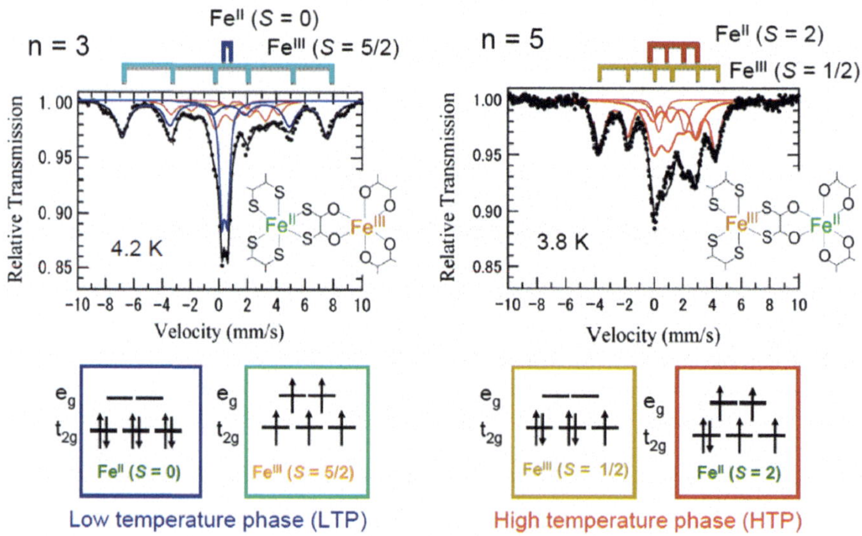

Fig. 6.27 ^{57}Fe Mössbauer spectra of $(n\text{-}C_nH_{2n+1})_4N[Fe^{II}Fe^{III}(dto)_3]$ (n = 3 and 5) in the magnetically ordered phase

Prussian blue is a very informative guide to the assignment of the ^{57}Fe Mössbauer spectra in magnetically ordered state for n = 3 and 4. In the case of Prussian blue, the spin state of the Fe^{II} site coordinated by six C atoms of cyanide ligands is the LS state (t_{2g}^6, $S = 0$), and that of the Fe^{III} site coordinated by six N atoms is the HS state ($t_{2g}^3e_g^2$, $S = 0$). The estimated H_{int} for $Fe^{II}(t_{2g}^6, S = 0)$ and $Fe^{III}(t_{2g}^3e_g^2, S = 5/2)$ in $Fe_4^{III}[Fe^{II}(CN)_6]_3$ at 1.6 K are 0 and 54.0 T, respectively [30]. ^{57}Fe Mössbauer paramers of $(n\text{-}C_nH_{2n+1})_4N[Fe^{II}Fe^{III}(dto)_3]$ (n = 3–6) are listed in Table 6.2.

In order to confirm the ferromagnetic phase transition, we investigated the magnetization as a function of temperature for n = 3–6, which is shown in Fig. 6.28. The Curie temperature were estimated at 7, (7 and 13), 19.5 and 23 K for n = 3, 4, 5, and

Table 6.2 ^{57}Fe Mössbauer parameters of $(n\text{-}C_nH_{2n+1})_4N[Fe^{II}Fe^{III}(dto)_3]$ (n = 3–6)

Doublet	*T* (K)	*IS* (mm s^{-1}) n = 3	4	5	6	*QS* (mm s^{-1}) n = 3	4	5	6	H_{int} (T) n = 3	4	5	6
(A)	200	0.35	0.29	0.35	0.31	0.38	0.55	0.51	0.56				
$Fe^{III}S_6$	77	0.38	0.36	0.39	0.38	0.55	0.60	0.68	0.74				
(*S* = 1/2)	4	0.33	0.33	0.40	0.42	0.55	0.51	0.73	0.71	23.4	24.1	24.6	24.8
(B)	200	1.14	1.05	1.17	1.14	1.43	1.20	1.69	1.24				
$Fe^{II}O_6$	77	1.28	1.20	1.24	1.23	1.57	1.27	1.84	1.31				
(*S* = 2)	4	1.26	1.26	1.24	1.26	1.41	1.41	1.84	1.37	10.2	10.0	7.3	9.9
(C)	77	0.43	0.46	–	–	0.35	0.41	–	–				
$Fe^{II}S_6$	4	0.43	0.46	–	–	0.39	0.46	–	–	0	0	–	–
(*S* = 0)													
(D)	77	0.54	0.55	–	–	0.89	0.92	–	–				
$Fe^{III}O_6$	4	0.55	0.55	–	–	0.90	0.95	–	–	44.8	44.8	–	–
(*S* = 5/2)													

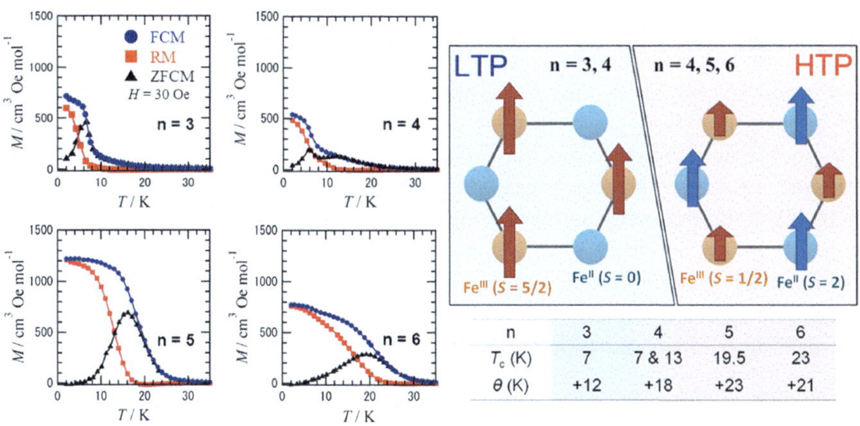

Fig. 6.28 Magnetization of $(n\text{-}C_nH_{2n+1})_4N[Fe^{II}Fe^{III}(dto)_3]$ (n = 3–6) as a function of temperature. FCM: field-cooled magnetization, ZFCM: zero-field-cooled magnetization, RM: remnant magnetization. FCM and ZFCM were measured at 3.0 mT. (= 30 Oe). T_C and θ denote the Curie temperature and Weiss temperature, respectively

6, respectively, from the heat capacity measurements for n = 3 and 4, [2(b)] and from the analysis of Arrott plot for n = 5 and 6 [6]. In the case of n = 4, the coexisting LTP and HTP undergo the ferromagnetic phase transitions at 7 K and 13 K, respectively. The critical temperature of CTPT, the Curie temperature and Weiss temperature of $(n\text{-}C_nH_{2n+1})_4N[Fe^{II}Fe^{III}(dto)_3]$ (n = 3–6) are summarized in Fig. 6.28.

6.3.6 New Type of Photo-Induced Magnetism Induced by the Photo-Isomerization of Intercalated Cation

As mentioned in 6.3.4, the CTPT and the ferromagnetic phase transition of $(n\text{-}C_nH_{2n+1})_4N[Fe^{II}Fe^{III}(dto)_3]$ remarkably depend on the size of intercalated cation, which implies a possibility to control the CTPT and the ferromagnetic phase transition in the $\left[Fe^{II}Fe^{III}(dto)_3\right]_\infty^-$ layer by means of a photo-isomerization of intercalated cation as well as a photo-induced IVCT. In organic–inorganic hybrid systems, it is effective to use an organic photochromic molecule for producing photo-switchable materials. On the basis of this strategy, we have synthesized $(SP\text{-}R)[Fe^{II}Fe^{III}(dto)_3]$ (SP-R = cationic spiropyran, R = alkyl group) [7]. In general, the cationic spiropyran can be converted from the yellow-colored closed form (CF) to the red-colored open form (OF) under UV light irradiation (330–370 nm). The OF is usually less stable and returns to the CF thermally or photo-chemically (500–600 nm). This photo-isomerization is associated with a volume change of molecule, which is schematically shown in Fig. 6.29.

Figure 6.30 shows the change of absorption spectra for $(SP\text{-}Me)[Fe^{II}Fe^{III}(dto)_3]$ in a KBr pellet under UV light irradiation (350 nm with 40 mW cm^{-2}) at 70 K [7]. At 70 K, the absorption band between 550 and 600 nm corresponding to the $\pi-\pi^*$ transition of the OF is continuously enhanced with the increase of UV light irradiation time, meanwhile the initial black pellet changes to a deep purple pellet. After the UV light irradiation for 3 h, its intensity is almost saturated. The UV light induced OF in $(SP\text{-}Me)[Fe^{II}Fe^{III}(dto)_3]$ is stable even at room temperature on dark condition. This absorption band almost disappears under visible light irradiation (white light, 600 mW cm^{-2}) for 2 h. The photo-isomerization of cationic SP-Me from the CF to the OF by UV light irradiation and that from the OF to the CF by visible light irradiation reversibly occurs at 300 K as well as at 70 K [7].

Figure 6.31 shows the temperature dependence of the magnetization for $(SP\text{-}Me)[Fe^{II}Fe^{III}(dto)_3]$ before and after the UV light irradiation at 300 and 77 K. As shown in Fig. 6.31, on the heating process, the FCM and RM curves decrease stepwise

Fig. 6.29 Photochromic reaction of cationic spiropyran (SP-R)

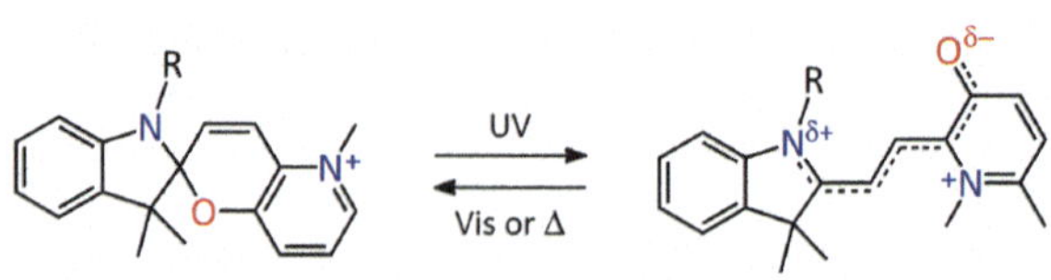

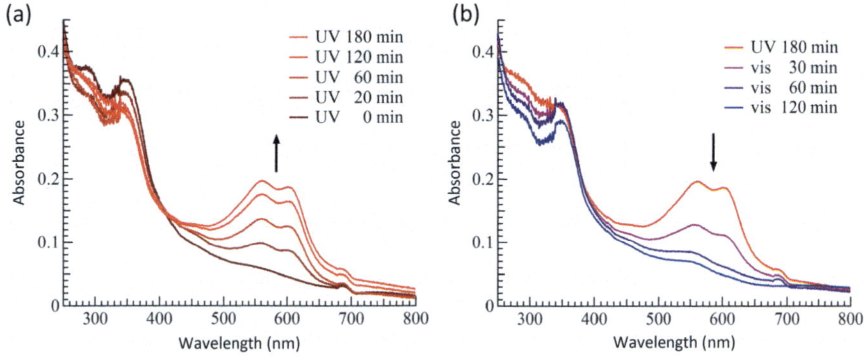

Fig. 6.30 Optical absorption of (SP-Me)[$Fe^{II}Fe^{III}(dto)_3$] in a KBr pellet under **a** UV light irradiation and **b** visible-light irradiation at 70 K. UV light irradiation (350 nm with 40 mW cm^{-2}) on the KBr pellet was performed at first. After the saturation of light-induced absorption spectra at around 550–600 nm, visible-light irradiation (white light with 600 mW cm^{-2}) on the pellet was performed. Reprinted with permission from [7(a)]. Copyright 2009 American Chemical Society

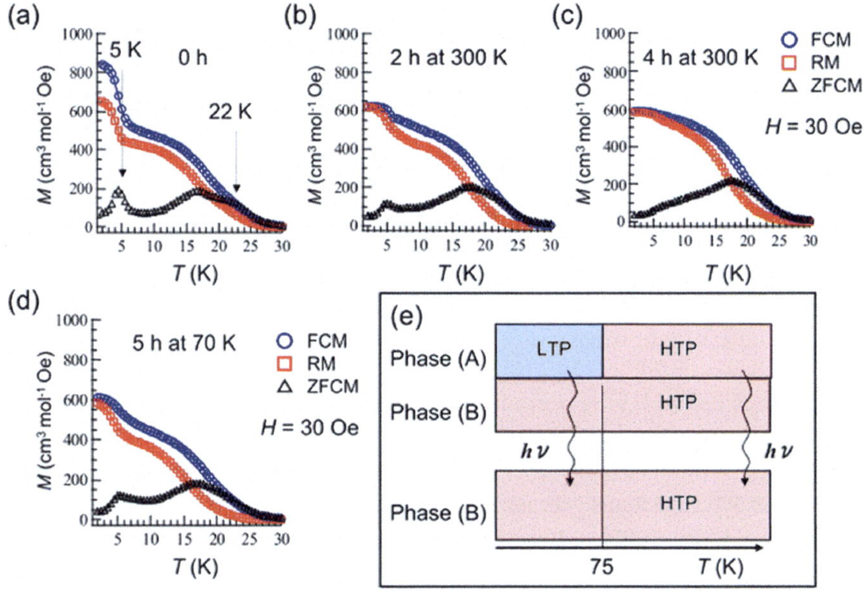

Fig. 6.31 **a–c** Temperature dependence of the magnetization for (SP-Me)[$Fe^{II}Fe^{III}(dto)_3$] before and after the UV light irradiation (350 nm with 40 mW cm^{-2}) for 2 and 4 h at 300 K. **d** Temperature dependence of the magnetization after the UV light irradiation for 5 h at 70 K. **e** Schematic representation of the photo-induced phase conversion from the phase (A) to the phase (B). $T = 75$ K denotes the CTPT. Reprinted with permission from [7(a)]. Copyright 2009 American Chemical Society

at about 7 K and disappears at about 25 K. On the other hand, the ZFCM curve has two maxima (5 K and 17 K) and a hump at 22 K. The LTP and HTP of (SP-Me)[$Fe^{II}Fe^{III}(dto)_3$] individually undergo the ferromagnetic phase transitions at 5 K and 22 K, respectively.

When the UV light (350 nm with 40 mW cm^{-2}) is irradiated for 4 h at 300 K, the FCM and RM curves change from the stepwise magnetization curves to normal magnetization curves, and the peak of ZFCM at around 5 K disappears, being reflected by the disappearance of LTP. On the other hand, being reflected by the growth of HTP, the FCM and RM values between 5 and 22 K slightly increase after the UV light irradiation. These photo-induced effects can be explained as follows. The photo-isomerization of cationic SP-Me from CF to OF by UV irradiation leads to the expansion of its own volume, which expands the unit cell volume and destabilizes the LTP. The photo-induced phase conversion of (SP-Me)[$Fe^{II}Fe^{III}(dto)_3$] is schematically shown in Fig. 6.31e, where we name two phases, (A) and (B). Only in the phase (A), the CTPT takes place. Suppose that the UV light irradiation induces the transformation from the phase (A) to the phase (B), the LTP in the phase (A) is forced to convert into the HTP in the phase (B) by UV light irradiation below 95 K. In order to confirm the CTPT in [$Fe^{II}Fe^{III}(dto)_3$] induced by the photo-isomerization of spiropyran, we performed UV light irradiation at 70 K, which corresponds to the arrow of the left side in Fig. 6.31e. The FCM, RM, and ZFMC curves after UV light irradiation for 5 h at 70 K are shown in Fig. 6.31d. Indeed, the photo-induced change in the ferromagnetic property shows the same tendency as in the case of UV light irradiation for 2 h at 300 K. Therefore, It is concluded that the destabilization of LTP and the stabilization of HTP by the photo-isomerization of intercalated $SP\text{-}Me^+$ below 70 K induces the CTPT in the [$Fe^{II}Fe^{III}(dto)_3$] layer.

Moreover, in order to prove the disappearance of LTP by UV light irradiation, we measured the ^{57}Fe Mössbauer spectra of (SP-Me)[$Fe^{II}Fe^{III}(dto)_3$] at 6 K before and after the UV light irradiation (350 nm with 40 mW cm^{-2}) for 3 h at 300 K, which is shown in Fig. 6.32 [7(b)]. As shown in Fig. 6.32a, the ^{57}Fe Mössbauer spectra at 6 K before UV light irradiation shows two kinds of magnetically ordered states. The sextet with a large hyperfine field from −8 to +8 mm s^{-1} is typical of the magnetically ordered spectrum of $Fe^{III}(t_{2g}^3e_g^2, S = 5/2)$ and the narrow quadrupole doublet at around 0.3 mm s^{-1} is assigned to the diamagnetic $Fe^{II}(t_{2g}^6, S = 0)$, which indicates the magnetically ordered LTP. The magnetic sextets of $Fe^{II}(t_{2g}^4e_g^2, S = 2)$ and $Fe^{III}(t_{2g}^5, S = 1/2)$ corresponding to the HTP are also observed. However, as shown in Fig. 6.32b, after the UV light irradiation for 3 h at 300 K, the ^{57}Fe Mössbauer spectra corresponding to the LTP vanishes entirely at 6 K. In this way, the remarkable change of ^{57}Fe Mössbauer spectra derived from the UV light irradiation proves that the photo-isomerization of $SP\text{-}Me^+$ in (SP-Me)[$Fe^{II}Fe^{III}(dto)_3$] induces the transformation from the phase (A) to the phase (B).

The photo-isomerization induced CTPT for (SP-Me)[$Fe^{II}Fe^{III}(dto)_3$] is illustrated in Fig. 6.33. This new type of synergetic phenomenon coupled with spin, charge and photon is triggered by the photo-isomerization of intercalated spiropyran from the CF to the OF, which seems to have a significant similarity with the first event for

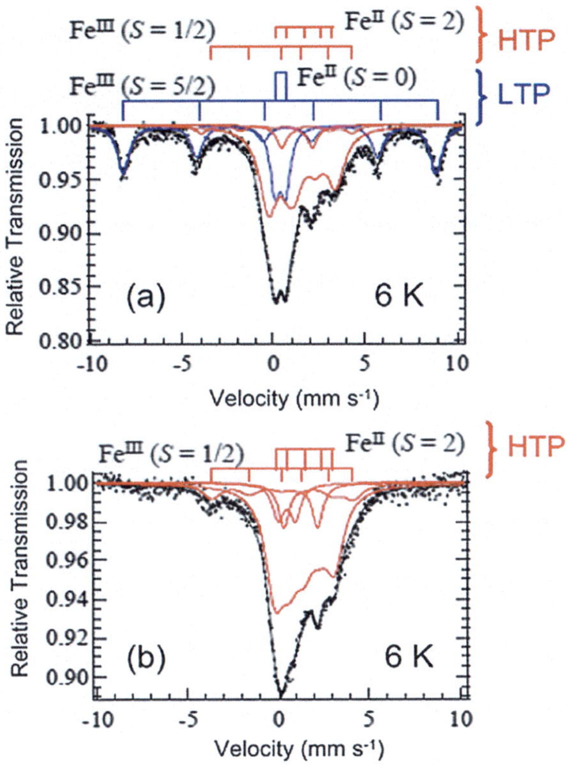

Fig. 6.32 ^{57}Fe Mössbauer spectra of (SP-Me)[$Fe^{II}Fe^{III}(dto)_3$] at 6 K before and after UV light irradiation (350 nm with 40 mW cm^{-2}) for 3 h at 300 K. Reprinted from [7(b)], Copyright 2009, with permission from Elsevier

the perception of light in rhodopsin where the photo-isomerization of 11-*cis*-retinal to the all-*trans*-retinal induces a conformational change in opsin and activates the associated G protein and triggers a second messenger cascade [51].

6.4 Single-Molecule Magnets and Single-Chain Magnets

6.4.1 Single-Molecule Magnets of Transition-Metal Clusters

Mössbauer spectroscopy have been applied to various kinds of molecular magnets as well as classical inorganic bulk magnets. Molecular magnets include not only classical magnets showing a long-range magnetic order but also quantum molecular magnets such as SMMs and SCMs the latter of which resemble superparamagnetic compounds in magnetism. These molecular nanomagnets have a possibility for

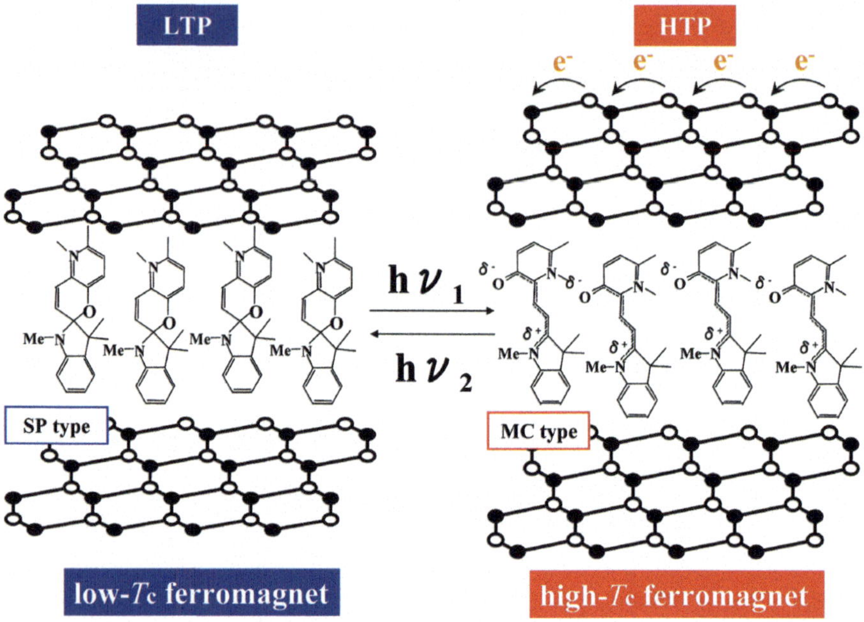

Fig. 6.33 Schematic representation of the synergetic CTPT induced by the photo-isomerization of spiropyran in (SP-Me)[$Fe^{II}Fe^{III}(dto)_3$]. White and black colored circles represent the Fe^{II} and Fe^{III} sites, respectively. Reprinted with permission from [7(a)]. Copyright 2009 American Chemical Society

ultra-high density memory, spintronics, and quantum computing devices. However, thermal and quantum relaxation processes limit them to afford such an application. Therefore, it is much important to elucidate slow relaxation dynamics of magnetization in quantum magnets. Varied measurements with individual time windows (Fig. 6.21) have been applied to SMMs and SCMs in order to reveal such spin dynamics. Especially, Mössbauer spectroscopy is a powerful tool to investigate spin dynamics for target atoms even in the condition of zero field as well as applied dc magnetic fields.

The dodecanuclear mixed-valence manganese cluster, $[Mn_{12}O_{12}(AcO)_{16}(H_2O)_4]\cdot 2AcOH\cdot 4H_2O$ (AcOH = acetic acid) abbreviated as $[Mn_{12}]_{ac}$ (Fig. 6.34), is well known as the first SMM reported in 1993 [52], while the synthesis of $[Mn_{12}]_{ac}$ had already been described in 1980 [53]. In the precedent papers, slow magnetization relaxation like superparamagnetic bahavior was revealed in ac susceptibility measurements [54, 55], accompanied by that the nature of the ground spin state is shown to be $S_{\text{total}} = 10$ by high-field electron paramagnetic resonance [54]. The relaxation time (τ) of magnetization was found to follow the Arrhenius law:

$$(\omega =)\frac{1}{\tau} = \frac{1}{\tau_0}\exp\left(-\frac{\Delta}{k_B T}\right), \tag{6.6}$$

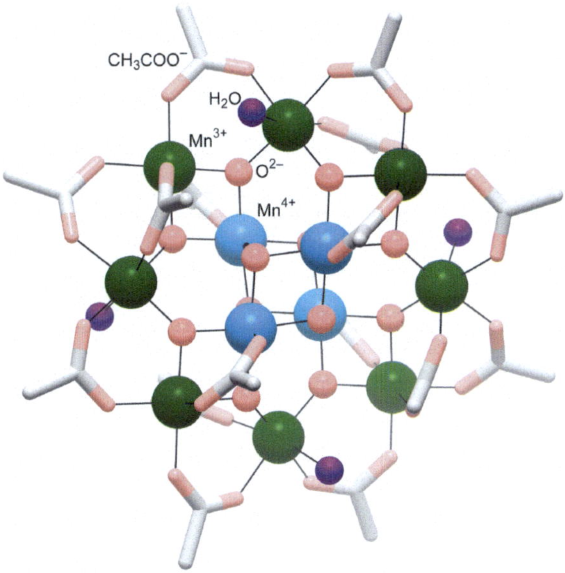

Fig. 6.34 Molecular structure of $[Mn_{12}O_{12}(AcO)_{16}(H_2O)_4]$ ($[Mn_{12}]_{ac}$)

where ω is the relaxation rate at the temperature of T, $1/\tau_0$ is the pre-exponential factor, Δ is the effective activation energy, and k_B is the Boltzmann constant. This temperature dependence is the characteristics of the thermally-activated relaxation process by Orbach mechanism. The Δ is corresponding to $|D|S_z^2$ for an integer spin or $|D|(S_z^2 - 1/4)$ for a half integer spin, where D represents the axial zero-field splitting parameter. For $[Mn_{12}]_{ac}$, the Arrhenius parameters were estimated as $\tau_0 = 2.1 \times 10^{-7}$ s and $\Delta/k_B = 61$ K [52], where the blocking temperature (T_B) was estimated as about 3 K for $\tau \sim 10^2$ s. Actually, magnetic hysteresis in *MH* curves was found below 4 K. Subsequently, some succeeding reports demonstrated clear magnetization steps (Fig. 6.35) in the hysteresis loops [56–58]. They are the first observations of quantum tunnelling of magnetization (QTM). A step occurs at every 0.44–0.46 T in an equally -spaced manner. The QTM originates in the level crossing between negative and positive m_S states as shown in Fig. 6.36. For example, at zero field the pair of $+m_S$ and $-m_S$ levels degenerates, affording a fast relaxation of magnetization because of the QTM. By applying external magnetic field, a non-zero field mismatching the energies of the $+m_S$ and $-m_S$ levels breaks down the tunnel relaxation. Some applied fields again lead to the degeneration of the energy levels when the $m_S = +S$ level corresponds to the $m_S = -S + n$ level (n is an integer), resulting in a fast relaxation. The interval field of the magnetization steps approximates the value of $-nD/g\mu_B$, giving the D value of ~0.60 K for $[Mn_{12}]_{ac}$. From the Arrhenius analysis, the D value is given as 0.61 K using $\Delta = |D|S_{total}^2$ ($S_{total} = 10$), indicating an excellent agreement of those estimated values.

The $[Mn_{12}]_{ac}$ has diversified into various derivatives, and wide variety of spectroscopic measurements have been performed to probe their SMM characteristics

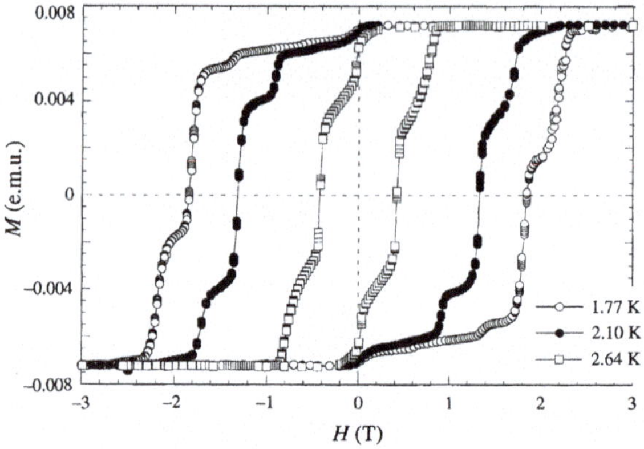

Fig. 6.35 Magnetization versus magnetic field (*MH*) curves at 1.77, 2.10, and 2.64 K for a single crystal (volume: 0.066 ± 0.008 mm^3) of $[Mn_{12}]_{ac}$. The magnetic field was applied along the magnetization axis, which aligns with the longest crystal dimension. Reprinted by permission from Springer Nature: [58], Copyright 1996

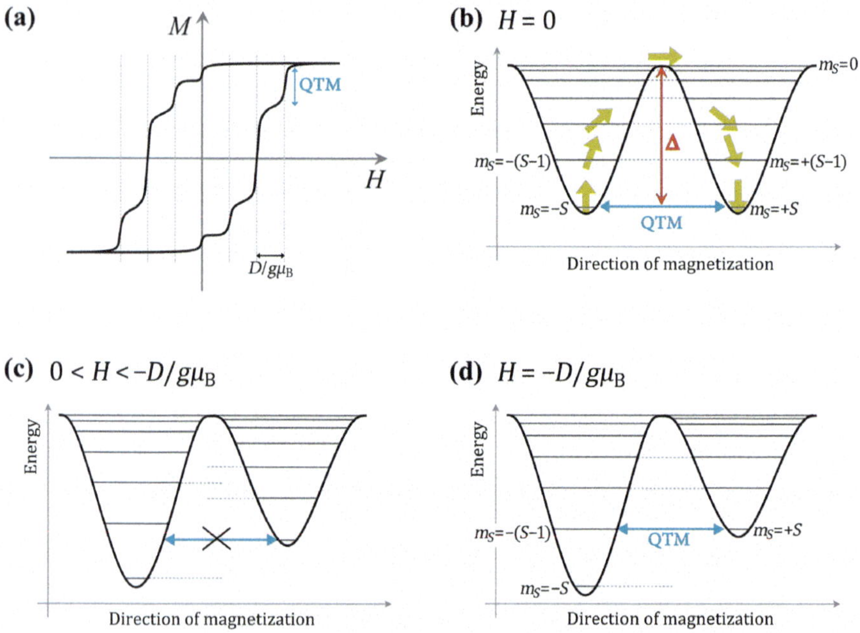

Fig. 6.36 **a** Magnetic hysteresis curve and **b**–**d** double-well potentials with energy levels for a spin state of S for a single-molecule magnet having easy-axis magnetic anisotropy ($D < 0$). The levels of positive m_S states are localized in the right well and the negative ones in the left well. Applied dc fields (H) are **b** zero, **c** $0 < H < -D/g\mu_B$, and **d** $-D/g\mu_B$. The QTM occurs when energy levels between the two wells come to the same level

such as ground and excited spin states, magnetic anisotropy, and QTM. No Mössbauer study has been reported for $[Mn_{12}]_{ac}$ because of less applicability for Mn atoms. A one-electron reduced analog, $[Fe(C_5H_5)_2][Mn_{12}O_{12}(O_2CC_6F_5)_{16}(H_2O)_4]$ ($Fe(C_5H_5)_2$ = ferrocenium cation), was reported by Kuroda et al. [59] to determine the relaxation of magnetization derived from the magnetic field of the $[Mn_{12}]^-$ core via ^{57}Fe Mössbuaer spectra of the ferrocenium site. The complex showed the effective energy barrier (U_{eff}) of magnetization reversal estimated as 50 K. In ^{57}Fe Mössbauer spectra (Fig. 6.37) below 13.5 K, a broad magnetic hyperfine splitting was found, which is a proof of slow paramagnetic relaxation originating in the SMM behavior of the $[Mn_{12}]^-$ unit.

The $[Fe_4]$ [60, 61] and $[Fe_8]$ [60, 62, 63] coordination clusters are representative SMMs, the chemical formulae of which represent as $[Fe_4(OCH_3)_6(dpm)_6]$ (Hdpm = dipivaloylmethane) and $[Fe_8O_2(OH)_{12}(tacn)_6]Br_8$ (Fig. 6.38; tacn = 1,4,7-

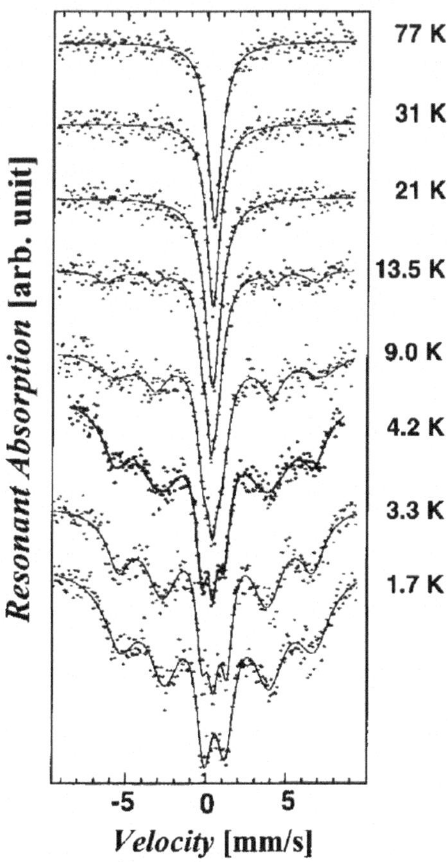

Fig. 6.37 Temperature-varied ^{57}Fe Mössbauer spectra of 11% ^{57}Fe-enriched $[Fe(C_5H_5)_2][Mn_{12}O_{12}(O_2CC_6F_5)_{16}(H_2O)_4]$ sample [59]. Reprinted with permission from [59]. Copyright 2001 American Chemical Society

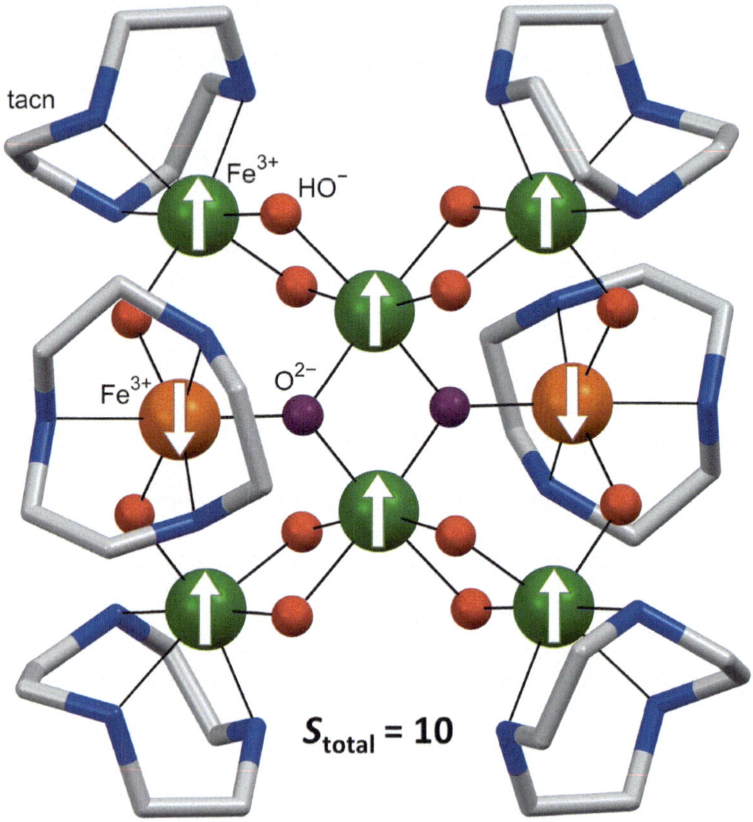

Fig. 6.38 Molecular structure of $[Fe_8O_2(OH)_{12}(tacn)_6]Br_8$. The arrows represent the spin structure

triazacyclononane), respectively. These typical SMMs have been examined by means of ^{57}Fe Mössbauer spectroscopy, which were well summarized in a preceding book [64].

The $[Fe_8]$ complex consists of eight Fe^{III} ions to form the ground $S_{total} = 10$ state [65] and shows SMM behavior with frequency dependence of ac magnetic susceptibilities giving $\tau_0 = 1.9 \times 10^{-7}$ s and $\Delta/k_B = 22.2$ K [63]. The relaxation time of magnetization was first investigated by means of ^{57}Fe Mössbauer spectroscopy for a SMM. As shown in Fig. 6.39a, a clear magnetic sextet superimposed three six-line patterns was observed in the ^{57}Fe Mössbauer spectrum at 2 K also under the absence of an external field. On heating, the splitting collapses gradually and results in partially-resolved quadrupole doublets at 300 K. A simplified line shape model was applied to the Mössbauer spectra in order to compare the estimated relaxation time with the values deduced from the ac magnetic measurements. The τ values were reasonably evaluated as they are compatible with the Arrhenius law in the $\ln(\tau)$ versus T^{-1} plot of ac measurements.

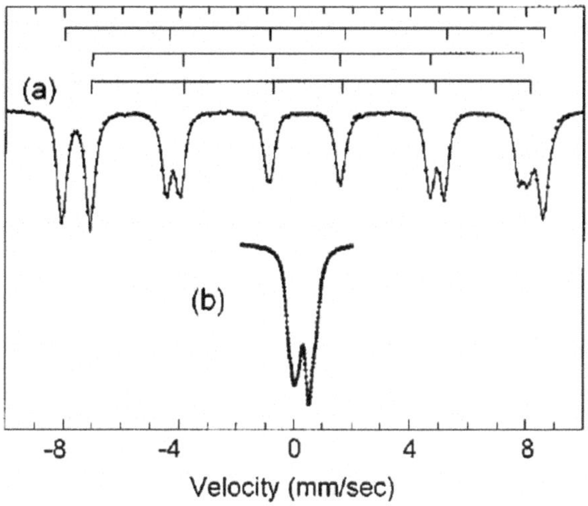

Fig. 6.39 ^{57}Fe Mossbauer spectra of $[Fe_8]$ **a** at 2 K in an applied field of 0.22 T and **b** at 300 K in zero field. The rake-like bar graphs represent three sextet patterns attributable to the three iron sites. Reprinted with permission from [63]. Copyright 1987 EDP Sciences

Very recently, the development of synchrotron techniques enables us to employ ^{57}Fe Mössbauer spectroscopy of a monolayer SMM grafted on a gold substrate. This technique was first applied to a monolayered sample containing a ^{57}Fe-enriched $[Fe_4]$ moiety [66], revealing slow magnetization relaxation of the iron cluster at zero magnetic field. In the monolayer, individual Fe^{III} sites lay in different chemical environments due to structural deformations induced by the interaction with the substrate, although such a modification tenuously affected SMM behavior. The zero-field splitting parameter, D, was estimated as $|D| = -0.49(6)$ K by fitting ^{57}Fe Mössbauer spectrum at the lowest temperature of 2.2 K, which is reduced compared to the typical values of ~-0.6 K for the $[Fe_4]$ family.

6.4.2 *Linear Two-Coordinate Fe^{II} and Fe^{I} Complexes*

Linear two-coordinate Fe^{II} complexes have been recently developed in the viewpoint of remarkably slow magnetization relaxation to afford SMMs due to a large magnetic anisotropy. In particular, a species of SMM originated in magnetic anisotropy of a single metal ion are called single-ion magnets (SIMs). A representative two-coordinate Fe^{II} compound, $[Fe(C(SiMe_3)_3)_2]$, has impressed with a huge internal hyperfine field of 152 T in ^{57}Fe Mössbauer spectra at 4.2 K on both chemists and physicists [66]. Such a giant magnetic anisotropy is attributable to the unquenched orbital angular momentum corresponding to the degenerate $d_{xy} - d_{x^2-y^2}$ orbital pair. The $[Fe(C(SiMe_3)_3)_2]$ compound possesses the strictly linear C–Fe^{II}–C geometry in the crystal [67, 68], being free from the Renner–Teller effect leading to a reduction of the orbital contribution due to a break of the local $D_{\infty h}$ symmetry.

The ^{57}Fe Mossbauer spectrum at room temperature shows an asymmetric doublet with much broadening in order to a magnetic relaxation, where the *IS* of ~0.4 mm/s and the *QS* of 1.3 mm/s. The temperature-independent characteristics of quadrupole splitting is observed, indicating a large separation between the ground orbital states and the excited ones. Below about 50 K, the broad doublet becomes a well-resolved sextet, and then ends up with the sharp spectrum having overall hyperfine splitting of 48.8 mm s^{-1} at 4.2 K. The result indicates slower magnetization relaxation than the Mössbuaer time scale ($\tau \sim 10^{-8}$ s). Magnetic hyperfine field, H_n, is estimated as 152 T. According to $H_n = H_F + H_D + H_L$, where H_F is the Fermi contact roughly determined by $-12.7 \times (2S)$ T [69] and H_D is the dipolar contribution, the value of the orbital contribution term, H_L, is estimated as 203 T when H_D is assumed to be zero.

In the precedent paper [70], saturation magnetization of this compound was reported to reach ~32,500 cm^3 Oe mol^{-1} (or 5.82 μ_B) at 1.85 K (for field-oriented sample), which is larger than the spin-only value (4 μ_B) that would be expected for the Fe^{II} ion with $S = 2$. After this preliminary study, the details of dynamic magnetization in the SIM compound have been examined by means of ac and dc magnetic property measurements and *ab initio* calculations [71], followed by the analysis of ^{57}Fe Mössbauer relaxation spectra [72]. Low-temperature saturation magnetization for a fixed sample of $[Fe(C(SiMe_3)_3)_2]$ is of $M_{sat} = 3.24\ \mu_B$. This value is lower than the spin-only value of the HS Fe^{II} ion because of random orientation of strong anisotropic species. Actually, the $\chi_M T$ value at room temperature is 4.79 cm^3 K mol^{-1} (the effective magnetic moment: $\mu_{eff} = 6.19\ \mu_B$), indicating strong anisotropy with a large g value of 2.53. At zero applied dc field, $[Fe(C(SiMe_3)_3)_2]$ shows no slow relaxation of magnetization in ac magnetic susceptibility, indicating no SIM behavior. However, such a relaxation can be detectable by means of ^{57}Fe Mössbauer spectroscopy as shown in Fig. 6.40, and analysed using a relaxation model formalized by Dattagupta and Blume [73]. Relaxation time constant (τ) ranges from 2.5×10^{-7} to 1.2×10^{-10} s within the measuring temperature range of 5–295 K. From the analysis of Arrhenius plots, the physical parameters on quantum tunnelling and Orbach relaxation processes were extracted as $\tau_{QTM}^{-1} = 4.2(3) \times 10^6\ s^{-1}$ and $U_{eff} = 178(9)$ and 553(68) cm^{-1}, respectively. Two obtained U_{eff} values correspond to the energy splitting from the ground $M_J = \pm 4$ to the first excited $M_J = \pm 3$ or the second excited $M_J = \pm 2$ doublets among the Stark sublevels of $^5D_4$. It should be mentioned that ac susceptibility measurements ($\tau <$ ~10^4 s) cannot detect the magnetic relaxation for this system due to the fast QTM process ($\tau_{QTM} = 2.4 \times 10^{-6}$ s) at the level crossing between the ground $M_J = \pm 4$ states under zero field. Applying a small dc field induced suppression of the level crossing between the M_J doublets, and thus magnetization dynamics was observable as a nonzero signal in the out-of-phase ac susceptibility (χ_M''). For Cole–Cole plots, the relaxation dynamics can be evaluated by using a generalized Debye model [74, 75]. Their results reveal clearly the field-induced SIM nature of $[Fe(C(SiMe_3)_3)_2]$.

Following the pioneering compound, some of *ca.* 30 linear two-coordinate Fe^{II} complexes [76] have currently been investigated from the viewpoint of SIM

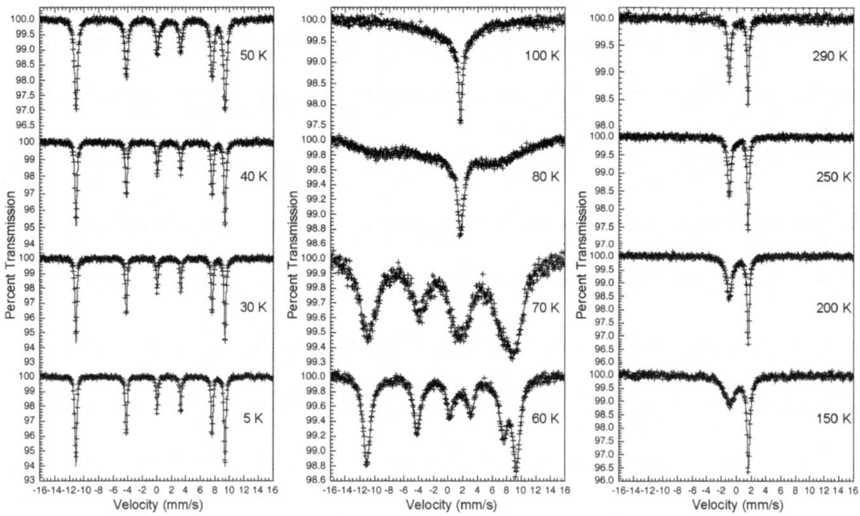

Fig. 6.40 Temperature-varied ^{57}Fe Mössbauer spectra of [$Fe(C(SiMe_3)_3)_2$] at zero dc magnetic field. Solid lines are the best fit profiles of a relaxation model. Reprinted with permission from [72]. Copyright 2013 American Chemical Society

behavior [71, 77, 78] as well as magnetization dynamics by using Mössbauer spectroscopy [78–81]. The development of linear two-coordinate metal complexes frequently needs the introduction of bulky ligands. A fused-ring 1,1,3,3,5,5,7,7-octa-R-substituted *s*-hydrindacene skeleton (Rind) is useful as a sterically-hindered ligand to isolate stably low-coordinated complexes [82]. Linear two-coordinate iron(II) complexes with various Rinds have been developed and investigated for an extremely large internal hyperfine field in the Mössbauer spectrum [78]. Owing to bulkiness of ligands, the crystal structure of $Fe(Eind)_2$ (Fig. 6.41; Eind denotes Rind of R = ethyl) adopts a highly linear two-coordination geometry with the C–Fe–C bond angles of 174.24(15)–177.73(13)°. The effective magnetic moment of 5.82 μ_B at 300 K exceeds the spin-only value of 4.90 μ_B estimated for a HS ($S = 2$) Fe^{II} species, indicating virtually unquenched orbital angular momentum on the 3d electrons of Fe^{II}.

Such a large contribution of angular momentum leads to slow magnetization reversal, which is detectable by means of ac magnetic measurements (about 10–1000 Hz) as well as ^{57}Fe Mössbauer spectroscopy. Without a dc magnetic field, no frequency dependence was observed in the ac magnetic susceptibilities, which is consistent with a broad signal derived from paramagnetic relaxation in the ^{57}Fe Mössbauer spectrum at 4.2 K. This Mössbauer result indicates that the relaxation time approaches the time scale of 10^{-9} s. The rapid relaxation is due to QTM. An applied dc field of 0.1 T suppressed the quantum tunnelling effectively, and consequently the spectrum split into a remarkably wide sextet with an internal hyperfine field of H_n = 140.3 T. The H_n value reached up to 143.1 T at an applied field of 3.0 T. The *IS* is 0.368(9) mm s^{-1} and *QS* −1.706(13) mm s^{-1}. The *QS* value is rather

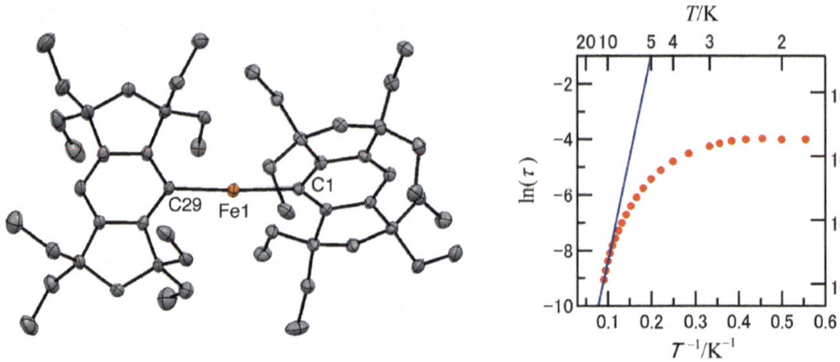

Fig. 6.41 (left) Crystal structure of $Fe(Eind)_2$ with 50% probability ellipsoids. Hydrogen atoms are omitted for clarity. (right) Arrhenius plot showing the temperature dependence of the relaxation time (τ). The solid line represents a data fit to the Arrhenius law. Reprinted with permission from [78]. Copyright 2016 the Chemical Society of Japan

negative, reflecting a larger electric field gradient at the iron(II) ion compared with that of $[Fe(C(SiMe_3)_3)_2]$ showing $QS = -1.08(1)$ to -1.27 mm s^{-1}. As with ^{57}Fe Mössbauer results, slow relaxation was observed in the χ_M'' versus frequency plots below 12 K in the ac frequency range of 5–1512 Hz with the applied dc field of 1200 Oe. The analysis of Arrhenius plots (Fig. 6.41) gave $U_{eff} = 51.2(7)$ cm^{-1}. At lower temperatures ($T < 10$ K), a dominant relaxation process gradually changed from thermally-activated Orbach mechanism to direct and/or Raman ones before quantum tunnelling regime in the temperature range less than 3 K.

Very recently, the first two-coordinate Fe^I complex, $[Fe^I(C(SiMe_3)_3)_2]^-$, was developed via the chemical reduction of $[Fe^{II}(C(SiMe_3)_3)_2]^0$ using KC_8, and its excellent SIM behavior was reported [83]. This complex has a highly anisotropic $S = 3/2$ spin state that is regarded as a Kramers ion with the ground $M_J = \pm 7/2$ doublets. Single-crystal X-ray diffraction analysis revealed the almost perfectly linear structure with the C–Fe–C angle of 179.2(2)° [83]. Extremely slow magnetization relaxation to behave as SIM was observed in the absence of an applied dc field, resulting in a large magnetic hysteresis loop due to magnetic blocking below 4.5 K. The effective activation barrier of spin reversal is estimated as $U_{eff} = 226(4)$ cm^{-1} from ac susceptibility measurements, which is consistent with the value of $U_{eff} =$ 246(3) cm^{-1} evaluated by the analyses of ^{57}Fe Mössbauer relaxation spectra [72].

6.4.3 Single-Chain Magnets: Unique Chain Magnet with Easy-Plane Anisotropy

Although pure one-dimensional (1D) spin systems never show magnetic phase transition based on any long-range ordering at a finite temperature, they can afford to behave like a magnet below a blocking temperature in the case of the presence of strong uniaxial magnetic anisotropy. Such compounds are categorized as SCMs, where slow relaxation dynamics of magnetization follows the Arrhenius law [Eq. (6.6)]. Dynamics of spin reversal for Ising chain systems is formulated by Glauber [84], whose model manages to explain quantitatively experimental results for some tangible SCMs. There are currently some practical reviews for SCMs [85, 86], interpreting anisotropic Heisenberg model applicable to the real materials, based on Glauber dynamics of 1D Ising chain. The spin system can be described by the following Hamiltonian:

$$H = -2J\sum_{i=1}^{L} S_i \cdot S_{i+1} + D\sum_{i=1}^{L} S_i^2 - g\mu_B H\sum_{i=1}^{L} S_i^2, \tag{6.7}$$

where J is the exchange coupling constant between adjacent spins and L the effective chain length. In the case of $|D/J| > 4/3$ (Ising limit), the spin system can be considered to be pseudo-Ising chain as shown in Fig. 6.42. At finite temperatures, the thin domain walls are created with the energy barrier $2\Delta_\xi = 8|J|S^2$ if the system is regarded as the infinite chain with the small domain size (2ξ) compared to L (that is $2\xi < L$, Fig. 6.42a). Additionally, each spin has an anisotropic energy of $\Delta_A = |D|S^2$, and therefore the relaxation time can be described as follows,

$$\tau_1 = \tau_0 \exp\left(\frac{2\Delta_\xi + \Delta_A}{k_B T}\right) = \tau_0 \exp\left(\frac{(8J + |D|)S^2}{k_B T}\right). \tag{6.8}$$

On cooling, ξ increases exponentially and gets comparable to L, which results in the occurrence of the finite size effect (Fig. 6.42b). In such a case, the formation

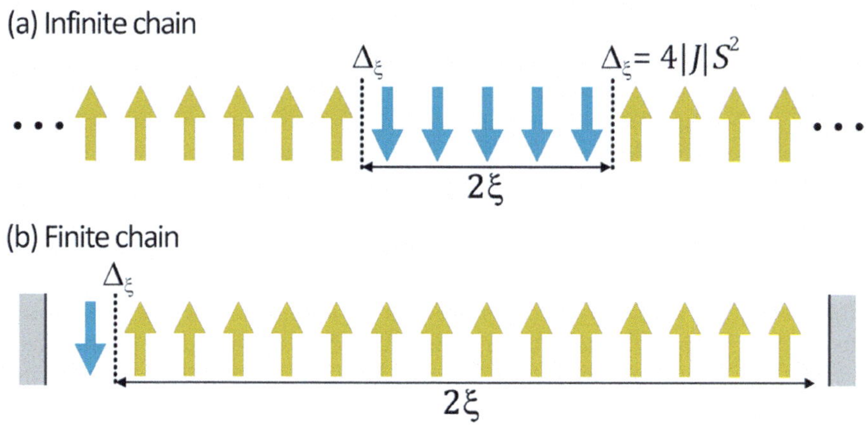

Fig. 6.42 Schematic represents for spin reversal in **a** infinite and **b** finite Ising chain systems. The parameters of Δ_ξ and ξ denote the energy barrier of spin reversal and the correlation length (2ξ: the width of domain), respectively

energy of the domain is $\Delta_\xi = 4|J|S^2$ due to the creation of the domain wall in only one side. Thus, the thermal activation energy can be estimated as $(4J + |D|)S^2$, and the dynamics of magnetization reversal obeys Eq. (6.9).

$$\tau_2 = \tau_0 \exp\left(\frac{\Delta_\xi + \Delta_A}{k_B T}\right) = \tau_0 \exp\left(\frac{(4J + |D|)S^2}{k_B T}\right) \tag{6.9}$$

In 2001, a slow relaxation of magnetization like SMMs was observed in 1D spin systems. The first SCM is the metal-radical compound [Co(hfac)$_2$(NITPhOMe)] (Hhfac = hexafluoroacetylacetone, NITPhOMe = *p*-methoxyphenyl nitronyl nitroxide) [87]. This compound consists of the effective $S = 1/2$ spin of Co^{2+} ($g_{Co} = 7.4$) and the $S = 1/2$ spin of radical ($g_R = 2$) connected via an antiferromagnetic coupling ($2J/k_B = -220$ K for $-2J\Sigma S_{Co} \cdot S_R$). The ac measurements revealed frequency dependences of χ_M' and χ_M'' below 17 K, indicating apparently a slow relaxation of magnetization. The relaxation time was perfectly accordance with the Arrhenius law, and the analysis gave the parameters of $\tau_0 = 3.0(2) \times 10^{-11}$ s and $\Delta/k_B = 154(2)$ K.

Almost coincidentally, a heterometallic chain [Mn$_2$(saltmen)$_2$Ni(pao)$_2$(py)$_2$](ClO$_4$)$_2$ (H$_2$saltmen = *N,N'*-bis(salicylidene)tetramethylethylenediamine, Hpao = pyridine-2-aldoxime) was developed in 2002 [87], which is comprised of ferromagnetically-coupled $S_T = 3$ spin units to form an Ising-like chain. Varied magnetic analyses were performed to scrutinize Glauber dynamics for the compound [88, 89]. Below 6.5 K, the frequency dependence of χ_M' and χ_M'' were clearly observed, and a semicircle in the Cole–Cole plot indicates a practical single relaxation process according to the Debye model. From a detailed Arrhenius plot, such a thermally-activated relaxation process switched in the vicinity of ~2.7 K (Fig. 6.43). The Arrhenius parameters of two relaxation manners were given as $\tau_0 = 5.5(1) \times 10^{-11}$ s, $\Delta_1/k_B = 72(1)$ K in a high-temperature region and τ_0 ~10^{-8} s, Δ_2/k_B ~55 K in a low-temperature one. The former is considered as the infinite chain regime according to Eq. (6.8), and the latter seems to obey Eq. (6.9) due to the finite size effect. The exchange coupling between the spin units was estimated as $J/k_B = 0.77$ K by means of dc magnetic measurements, while the anisotropy parameter of each unit was $D/k_B = -2.5$ K. The Δ_1 and Δ_2 are excellent consistent with the expected values of $(8J + |D|)S_T^2$ (= 78 K) and $(4J + |D|)S_T^2$ (= 50 K), respectively.

Among various SCMs developed, there are few examples on Mössbauer relaxation study compared with an overwhelming number of ac susceptibility measurements. An alternating Fe^{II}–Fe^{III} chain compound, *catena*-[Fe^{II}(ClO$_4$)$_2${Fe^{III}(bpca)$_2$}]ClO$_4$ (Hbpca = bis(2-pyridylcarbonyl)amine) as shown in Fig. 6.44, is one of representative SCMs, where ^{57}Fe Mössbauer and μSR spectroscopic measurements were performed to investigate the spin relaxation based on Glauber dynamics in the chain structure [90, 91]. The material can be described as a unique 1D Ising system composed of metal ions in the nature of easy-plane magnetic anisotropy ($D > 0$). In general, SMMs and SCMs prefer consisting of an easy-axis anisotropic ($D < 0$) metal ions in order to overcome a thermal relaxation of magnetization. However, the overall easy-axis anisotropy of the molecule can stem from the

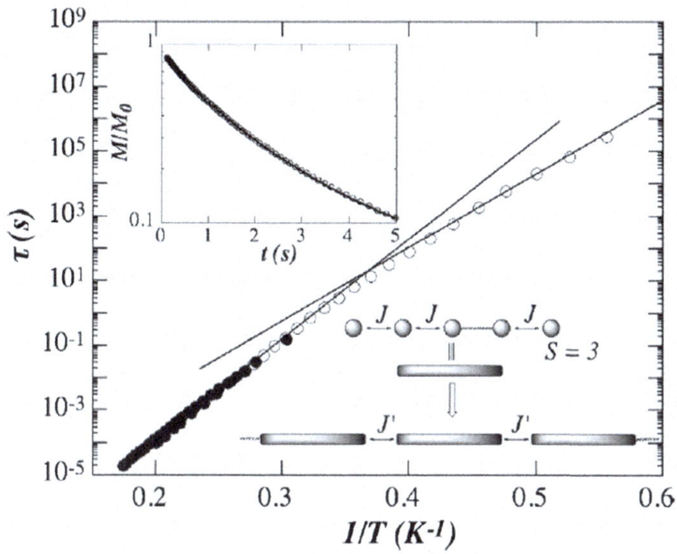

Fig. 6.43 Arrhenius plot of the relaxation time τ (logarithmic scale) versus $1/T$. Inset graph is time-dependent magnetization relaxation at 3 K. Solid lines represent the best fits using the model for Glauber dynamics. Schematic view indicates the spin model consisting of $S = 3$ units connected by an inter-unit ferromagnetic coupling of J'. Reprinted figure with permission from [89]. Copyright 2004 by the American Physical Society

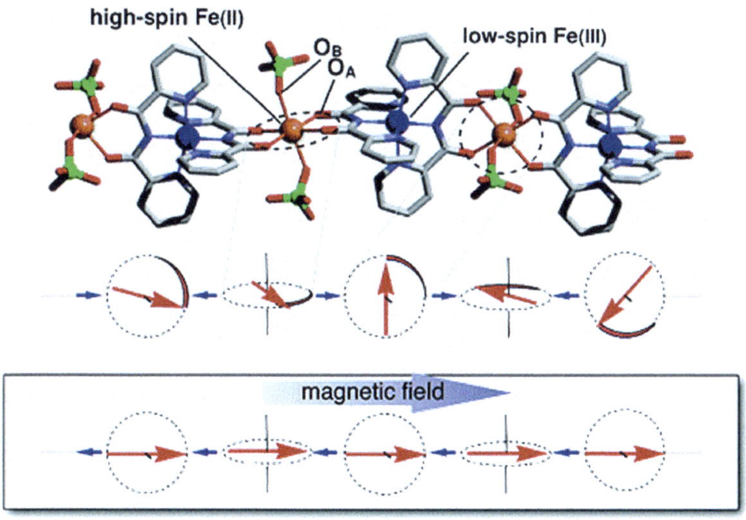

Fig. 6.44 Crystal structure of the *catena*-$[Fe^{II}(ClO_4)_2\{Fe^{III}(bpca)_2\}]ClO_4$ (Hbpca = bis(2-pyridylcarbonyl)amine) and the spin alignment of Fe^{II} ($S = 2$)–Fe^{III} ($S = 1/2$) in the chain. Reprinted with permission from [90]. Copyright 2005 American Chemical Society

orthogonal arrangement of each metal ion with easy-plane anisotropy. In *catena*-[$Fe^{II}(ClO_4)_2\{Fe^{III}(bpca)_2\}]ClO_4$, the HS Fe^{II} ions have a octahedral coordination geometry with the axial bonds elongated. This situation can provides an easy-plane anisotropy (XY spin) in the equatorial plane. The XY spin of each Fe^{II} site is coupled antiferromagnetically with Fe^{III} ($S = 1/2$) spins in an alternately-twisted array of the easy-plane, which affords a net easy-axis anisotropy along the chain direction. This compound showed a slow magnetization relaxation in ac susceptibility, and activation energy U_{eff}/k_B of 27 K was estimated from the Arrhenius plot [90]. The U_{eff} value is smaller than the $(8J + |D|)S^2$ expected for pure Glauber dynamics theory, suggesting that the transverse magnetization of the Fe^{II} spins in the easy plane is responsible for the magnetic relaxation.

A short-range ordering was observed in ^{57}Fe Mössbauer spectroscopy as well as μSR spectroscopy [91]. The μSR technique is one of the methods to approach a ground magnetic state at zero field (ZF). The depolarization of spin-polarized muons was observed in ZF-μSR time spectra at 20 K, which are reproducible by using two Gaussian-type relaxations. On cooling down to 6 K, a gradual loss of the initial asymmetry was found, which was simultaneously followed by the enhancement of the relaxation rate for the fast depolarizing component. Below 6 K, the initial asymmetry became almost constant, and the relaxation rate declined rapidly after the divergence around 6 K. This behavior indicates a critical slowing down of magnetic fluctuation. No oscillation signal was observed in the decay spectra even at 0.3 K, indicating that the muon is located in a randomly distributed magnetic field.

The ^{57}Fe Mössbauer spectroscopy also accepts a short-range ordering nature of magnetic moment on the SCM compound (Fig. 6.45). Above 20 K, two doublets are

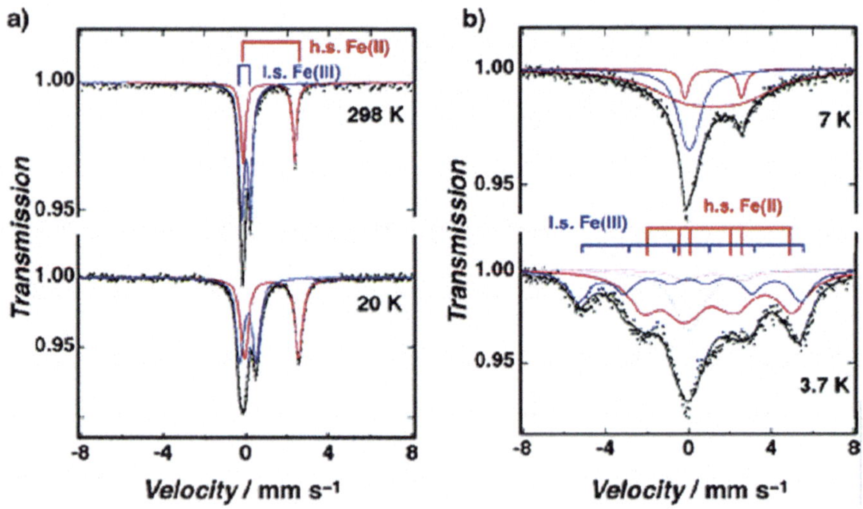

Fig. 6.45 Temperature-varied ^{57}Fe Mössbauer spectra of *catena*-[$Fe^{II}(ClO_4)_2\{Fe^{III}(bpca)_2\}]ClO_4$ **a** at 298 and 20 K, **b** at 7 and 3.7 K. Reprinted with permission from [90]. Copyright 2005 American Chemical Society

observed, which are assignable as the HS Fe^{II} and LS Fe^{III} species. These paramagnetic signals turn into broadening before splitting into two magnetic sextet with a partially residual paramagnetic components below 7 K. The hyperfine fields were estimated at 19.2 T for Fe^{II} and 33.5 T for Fe^{III}.

6.5 Conclusion

In this chapter, we have focused on the molecular magnetism and its related light-induced phase transitions from the viewpoint of Mössbauer spectroscopy. In the Sect. 6.2, we have described mainly the dynamic spin equilibrium phenomena for $[M^{II}Fe^{III}(mto)_3]$ (M = Zn, Mn) complexes. In the case of $(C_6H_5)_4P[Mn^{II}Fe^{III}(mto)_3]$ consisting of $Fe^{III}O_3S_3$ and $Mn^{II}O_6$ octahedra, there exists a rapid spin equilibrium ($\tau < 10^{-7}$ s) between the HS and LS states at the $Fe^{III}O_3S_3$ site, which induces the fluctuation of internal magnetic field on the Mn^{II} site. Owing to the fluctuation of internal magnetic field caused by the rapid spin equilibrium, $(C_6H_5)_4P[Mn^{II}Fe^{III}(mto)_3]$ undergoes the successive magnetic phase transitions at 30 and 23 K. In the Sect. 6.3, we have described the CTPT and the ferromagnetism of $(n\text{-}C_nH_{2n+1})_4N[Fe^{II}Fe^{III}(dto)_3]$. At the CTPT, the electrons of Avogadro's constant transfer between the Fe^{II} and Fe^{III} sites, and the Fe valence state is dynamically fluctuated with a frequency of about 0.1 MHz, which was confirmed by means of μSR. In order to control the ferromagnetism and the CTPT by means of photo-irradiation, we have synthesized a photo-responsive organic-inorganic hybrid system, (SP-Me)$[Fe^{II}Fe^{III}(dto)_3]$, and discovered that the photo-isomerization of intercalated SP-Me by UV light irradiation induces the CTPT and the enhancement of the Curie temperature from 7 to 22 K. In the Sect. 6.4, we have described the spin dynamics of various kinds of molecular magnets such as SMMs with transition-metal clusters, SIMs with linear coordination, and SCMs with easy-plane anisotropy.

Acknowledgements In this chapter, our contributed original research has been created in collaboration with Profs. M. Enomoto (Tokyo University of Science), M. Itoi (Nihon University), M. Okubo (The University of Tokyo), K. Kagesawa (Tohoku University), T. Matsuo (Kinki University), Y. Kobayashi (University of Electro-Communications), T. Kajiwara (Nara Women's University), K. Tamao (Toyota Riken), M. Yamashita (Tohoku University), M. Seto (Kyoto University), H. Sawa (Nagoya University), E. Nishibori (Tsukuba University), Drs. N. Kida (Mitsubishi Chemical Co.), Y. Ono (Mitsubishi Chemical Co.), I. Watanabe (RIKEN). We wish to thank all the collaborators. This work has partly been supported by Toyota Physical and Chemical Research Institute, and a Grant-in-Aid for Science Research from the Ministry of Education, Science, Sports and culture.

References

1. (a) P. Gütlich, H.A. Goodwin (Ed.), *Spin Crossover in Transition Metal Compounds I, II, III* (Springer, 2004). (b) M.A. Halcrow (Ed.), *Spin-Crossover Materials: Properties and Applications* (Wiley, 2013). (c) J.-P. Launay, M. Verdaguer, *Electronics in Molecules: From Basic Principles to Molecular Electronics* (Oxford University Press, 2013). (d) D. Gatteschi, R. Sessoli, J. Villain, *Molecular Nanomagnets* (Oxford University Press, 2006). (e) R. Winpenny, G. Aromi (Ed.), *Single-Molecule Magnets and Related Phenomena* (Springer, 2006). (f) M. Holynska (Ed.), *Single-Molecule Magnets: Molecular Architectures and Building Blocks for Spintronics* (Wiley, 2018)
2. (a) N. Kojima, M. Enomoto, N. Kida, K. Kagesawa, *Materials,* **3**, 3141 (2010). (b) N. Kojima, M. Itoi, Y. Miyazaki, *Curr. Inorg. Chem.* **4**, 85 (2014). (c) N. Kojima, A. Sugahara, Spin-crossover and related phenomena coupled with spin, photon and charge, in *Mössbauer Spectroscopy: Applications in Chemistry, Biology and Nanotechnology*. ed. by V.K. Sharma, C. Klingelhöfer, T. Nishida (Wiley Inc., 2013), pp. 152–176
3. K. Kagesawa, A. Okazawa, M. Enomoto, N. Kojima, Chem. Lett. **39**, 872 (2010)
4. N. Kojima, W. Aoki, M. Itoi, M. Seto, Y. Kobayashi, Yu. Maeda, Solid State Commun. **120**, 165 (2001)
5. N. Kida, M. Enomoto, I. Watanabe, T. Suzuki, N. Kojima, Phys. Rev. B **77**, 144427 (2008)
6. M. Itoi, Y. Ono, N. Kojima, K. Kato, K. Osaka, M. Takata, *Eur. J. Inorg. Chem.* 1198 (2006)
7. (a) N. Kida, M. Hikita, I. Kashima, M. Okubo, M. Itoi, M. Enomoto, K. Kato, M. Takata, N. Kojima, *J. Am. Chem. Soc.* **131** 212 (2009). (b) N. Kida, M. Hikita, I. Kashima, M. Enomoto, M. Itoi, N. Kojima, *Polyhedron* **28**, 1694 (2009)
8. Y. Tanabe, S. Sugano, J. Phys. Soc. Jpn. **9**, 766 (1954)
9. (a) L. Cambi, A. Cagnasso, *Atti Accad. Naz. Lincei.* **13** 809 (1931). (b) L. Cambi, L. Szegö, A. Cagnasso, *Atti Accad. Naz. Lincei.* **15**, 266 (1932). (c) L. Cambi, L. Szegö, *Ber. Deutsch. Chem. Ges.* **66**, 656 (1933)
10. J.M. Fiddy, I. Hall, F. Gradjean, U. Russo, G.J. Long, Inorg. Chem. **26**, 4138 (1987)
11. (a) K.R. Kunze, D.L. Perry, L.J. Wilson, *Inorg. Chem.* **16**, 594 (1977). (b) H. Oshio, Y. Maeda, Y. Takahashi, *Inorg. Chem.* **22**, 2684 (1983). (c) Y. Maeda, N. Tsutsumi, Y. Takahashi, *Inorg. Chem.* **23**, 2440 (1984). (d) M.D. Timken, A.M. Abdel-Mawgoud, D.N. Hendrickson, *Inorg. Chem.* **25**, 160 (1985). (e) M. Nihei, T. Shiga, Y. Maeda, H. Oshio, *Coord. Chem. Rev.* **251**, 2606 (2007)
12. W.A. Baker, H. Bobonich, Inorg. Chem. **3**, 1184 (1964)
13. P. Gütlich, H.A. Goodwin (Ed.), *Spin Crossover in Transition Metal Compounds I* (Springer-Verlag, Berlin Heidelberg, Germany, 2004), and related references therein
14. (a) I. Krivokapic, M. Zerara, M.L. Daku, A. Vargas, C. Enachescu, C. Ambrus, P. Tregenna-Piggott, N. Amstutz, E. Krausz, A. Hauser, *Coord. Chem. Rev.* **251**, 364 (2007). (b) H.A. Goodwin, in *Spin Crossover in Transition Metal Compounds II*, ed. by P. Gütlich, H.A. Goodwin (Springer-Verlag, Berlin Heidelberg, Germany, 2004), pp. 23–48
15. Y. Garcia, P. Gütlich, in *Spin Crossover in Transition Metal Compounds II*, ed. by P. Gütlich, H.A. Goodwin (Springer-Verlag, Berlin Heidelberg, Germany, 2004), pp. 49–62
16. (a) P.L. Franke, J.G. Haasnoot, A.P. Zuur, *Inorg. Chim. Acta* **59**, 5 (1982). (b) S. Decurtins, P. Gütlich, C.P. Köhler, H. Spiering, A. Hauser, *Chem. Phys. Lett.* **105**, 1 (1984). (c) S. Decurtins, P. Gütlich, K.M. Hasselbach, A. Hauser, H. Spiering, *Inorg. Chem.* **24**, 2174 (1985)
17. (a) P. Gütlich, H.A. Goodwin (Ed.) *Spin Crossover in Transition Metal Compounds I, II, IIII*, (Springer-Verlag, Berlin Heidelberg, Germany, 2004). (b) O. Sato, J. Tao, Y.-Z. Zhang, *Angew. Chem. Int. Ed.* **46**, 2152 (2007), and related references therein
18. M. Cox, J. Darken, B.W. Fitzsimmons, A.W. Smith, L.F. Larkworthy, K.A. Rogers, *J. Chem. Soc. Dalton Trans.* 1192 (1972)
19. K.R. Kunze, D.L. Perry, L.J. Wilson, Inorg. Chem. **16**, 594 (1977)
20. M. Itoi, A. Taira, M. Enomoto, N. Matsushita, N. Kojima, Y. Kobayashi, K. Asai, K. Koyama, T. Nakamoto, Y. Uwatoko, J. Yamamura, Solid State Commun. **130**, 415 (2004)

21. H. Ōkawa, N. Matsumoto, H. Tamaki, M. Ohba, Mol. Cryst. Liq. Cryst. **233**, 257 (1993)
22. A. Hauser, in *Spin Crossover in Transition Metal Compounds II*, ed. by P. Gütlich, H.A. Goodwin (Springer-Verlag, Berlin Heidelberg, Germany, 2004), pp. 155–198
23. T.C. Brunold, H.U. Güdel, in *Inorganic Electronic Structure and Spectroscopy I*, ed. by E.I. Solomon, A.B.P. Lever (Wiley, 1999), pp. 259–306
24. (a) J.-F. Létard, G. Chastanet, P. Guionneau, C. Desplanches, in *Spin-Crossover Materials: Properties and Applications*, ed. by M.A. Halcrow (Wiley, 2013), pp. 475–506. (b) Y. Moritomo, K. Kato, A. Nakamoto, N. Kojima, E. Nishibori, M. Takata, M. Sakata, *J. Phys. Soc. Jpn.* **71**, 1015 (2002)
25. (a) S.G. Carling, J.M. Bradley, D. Visser, P. Day, *Polyhedron* **22**, 2317 (2003). (b) M. Enomoto, H. Ida, A. Okazawa, N. Kojima, *Crystals* **8**, 446 (2018)
26. C. Mathonière, C.J. Nuttall, S.G. Carling, P. Day, Inorg. Chem. **35**, 1201 (1996)
27. K. Kagesawa, Doctoral Thesis, (The University of Tokyo, 2011)
28. M.B. Robin, P. Day, Adv. Inorg. Chem. Radiochem. **10**, 247 (1967)
29. F. Herren, P. Fischer, A. Ludi, W. Hālg, Inorg. Chem. **19**, 956 (1980)
30. A. Ito, M. Suenaga, K. Ôno, J. Chem. Phys. **48**, 3597 (1968)
31. K. Maer Jr., M.L. Beasley, R.L. Collins, W.O. Milligan, J. Am. Chem. Soc. **90**, 3201 (1968)
32. O. Sato, T. Iyoda, A. Fujishima, K. Hashimoto, Science **272**, 704 (1996)
33. O. Sato, Y. Einaga, T. Iyoda, A. Fujishima, K. Hashimoto, J. Electrochem. Soc. **144**, L11 (1967)
34. O. Sato, J. Tao, Y.-Z. Zhang, *Angew. Chem. Int. Ed.* **46**, 2152 (2007), and related references therein
35. O. Sato, Y. Einaga, T. Iyoda, A. Fujishima, K. Hashimoto, J. Phys. Chem. B **101**, 3903 (1997)
36. (a) N. Shimamoto, S. Ohkoshi, O. Sato, K. Hashimoto, *Inorg. Chem.* **41**, 678 (2002). (b) V. Escax, A. Bleuzen, C. Cartier dit Moulin, F. Villain, A. Goujon, F. Varret, M. Verdaguer, *J. Am. Chem. Soc.* **123**, 12536 (2001). (c) A. Carolina, G.H. Matthew, P. Andrey, S. Codi, S. Darryl, R.D. Kim, *J. Am. Chem. Soc.* **132**, 13123 (2010). (d) H. Tokoro, S. Ohkoshi, T. Matsuda, K. Hashimoto, *Inorg. Chem.* **43**, 5231 (2004). (e) A. Bleuzen, V. Marvaud, C. Mathoniere, B. Sieklucka, M. Verdaguer, *Inorg. Chem.* **48**, 3453 (2009). (f) S. Ohkoshi, Y. Hamada, T. Matsuda, Y. Tsunobuchi, H. Tokoro, *Chem. Mater.* **20**, 3048 (2008). (g) R. Podgajny, S. Chorazy, W. Nitek, M. Rams, A.M. Majcher, B. Marszalek, J. Zukrowski, C. Kapusta, B. Sieklucka, *Angew. Chem., Int. Ed.* **52**, 896 (2013), and related references therein
37. O. Sato, Y. Einaga, A. Fujishima, K. Hashimoto, Inorg. Chem. **38**, 4405 (1999)
38. Y. Ono, Doctoral Thesis, (The University of Tokyo, 2005)
39. S. Iijima, T. Katsura, H. Tamaki, M. Mitsumi, N. Matsumoto, H. Ôkawa, Mol. Cryst. Liq. Cryst. **233**, 263 (1993)
40. S. Iijima, F. Mizutani, M. Mitsumi, N. Matsumoto, H. Ôkawa, Inorg. Chim. Acta **253**, 47 (1996)
41. T. Birchall, K.M. Tun, Inorg. Chem. **15**, 376 (1976)
42. B.J. Evans, R.G. Johnson, F.E. Senftle, C.B. Cecil, F. Dulong, Geochim. Cosmochim. Acta **46**, 761 (1982)
43. T. Nakamoto, Y. Miyazaki, M. Itoi, Y. Ono, N. Kojima, M. Sorai, *Angew. Chem., Int. Ed.* **40**, 4716 (2001)
44. M. Enomoto, M. Itoi, Y. Ono, M. Okubo, N. Kojima, Synth. Metals **137**, 1231 (2003)
45. Y. Kobayashi, M. Itoi, N. Kojima, K. Asai, J. Phys. Soc. Jpn. **71**, 3016 (2002)
46. K. Nagamine, *Introductory Muon Science* (Cambridge University Press, 2003)
47. R.S. Hayano, Y.J. Uemura, J. Imazato, N. Nishida, T. Yamazaki, R. Kubo, Phys. Rev. B **20**, 850 (1979)
48. R. Kadono, T. Matsuzaki, T. Yamazaki, S.R. Kreitzman, J.H. Brewer, Phys. Rev. B **42**, 6515 (1990)
49. C.P. Slichter, *Principles of magnetic resonance* (Harper International, New York, 1965)
50. N. Kojima, N. Kida, A. Okazawa, M. Enomoto, Mössbauer Eff. Ref. Data J. **35**, 154 (2012)
51. H. Nakamichi, T. Okada, *Angew. Chem., Int. Ed.* **45**, 4270 (2006), and related references therein
52. R. Sessoli, D. Gatteschi, A. Caneschi, M.A. Novak, Nature **365**, 141 (1993)
53. T. Lis, Acta Crystallogr. B **36**, 2042 (1980)

54. A. Caneschi, D. Gatteschi, R. Sessoli, A.L. Barra, L.C. Brunel, M. Guillot, J. Am. Chem. Soc. **113**, 5873 (1991)
55. R. Sessoli, H.L. Tsai, A.R. Schake, S. Wang, J.B. Vincent, K. Folting, D. Gatteschi, G. Christou, D.N. Hendrickson, J. Am. Chem. Soc. **115**, 1804 (1993)
56. J.R. Friedman, M.P. Sarachik, J. Tejada, J. Maciejewski, R. Ziolo, J. Appl. Phys. **79**, 6031 (1996)
57. J.R. Friedman, M.P. Sarachik, J. Tejada, R. Ziolo, Phys. Rev. Lett. **76**, 3830 (1996)
58. L. Thomas, F. Lionti, R. Ballou, D. Gatteschi, R. Sessoli, B. Barbara, Nature **383**, 145 (1996)
59. T. Kuroda-Sowa, M. Lam, A.L. Rheingold, C. Frommen, W.M. Reiff, M. Nakano, J. Yoo, A.L. Maniero, L.-C. Brunel, G. Christou, D.N. Hendrickson, Inorg. Chem. **40**, 6469 (2001)
60. D. Gatteschi, R. Sessoli, A. Cornia, *Chem. Commun.* 725 (2000)
61. A.L. Barra, A. Caneschi, A. Cornia, F. Fabrizi de Biani, D. Gatteschi, C. Sangregorio, R. Sessoli, L. Sorace, J. Am. Chem. Soc. **121**, 5302 (1999)
62. C. Sangregorio, T. Ohm, C. Paulsen, R. Sessoli, D. Gatteschi, Phys. Rev. Lett. **78**, 4645 (1997)
63. A.L. Barra, P. Debrunner, D. Gatteschi, C.E. Schulz, R. Sessoli, Europhys. Lett. **35**, 133 (1996)
64. L. Cianchi, G. Spina, *Studies of spin fluctuations in single molecular magnets by using Mössbauer spectroscopy*, ed. by P. Carrette, A. Lascialfari, NMR-MRI, μSR and Mössbauer Spectroscopies in Molecular Magnets (Springer-Velrag, Milano, 2007)
65. C. Delfs, D. Gatteschi, L. Pardi, R. Sessoli, K. Wieghardt, D. Hanke, Inorg. Chem. **32**, 3099 (1993)
66. A. Cini, M. Mannini, F. Totti, M. Fittipaldi, G. Spina, A. Chumakov, R. Rüffer, A. Cornia, R. Sessoli, Nat. Commun. **9**, 480 (2018)
67. A.M. LaPointe, Inorg. Chim. Acta **345**, 359 (2003)
68. T. Viefhaus, W. Schwarz, K. Hübler, K. Locke, J. Weidlein, Z. Anorg, Allg. Chem. **627**, 715 (2001)
69. N.N. Greenwood, T.C. Gibb, *Mössbuaer Spectroscopy* (Chapman and Hall, London, 1971)
70. W.M. Reiff, A.M. LaPointe, E.H. Witten, J. Am. Chem. Soc. **126**, 10206 (2004)
71. J.M. Zadrozny, M. Atanasov, A.M. Bryan, C.-Y. Lin, B.D. Rekken, P.P. Power, F. Neese, J.R. Long, Chem. Sci. **4**, 125 (2013)
72. J.M. Zadrozny, D.J. Xiao, J.R. Long, M. Atanasov, F. Neese, F. Grandjean, G.J. Long, Inorg. Chem. **52**, 13123 (2013)
73. S. Dattagupta, M. Blume, Phys. Rev. B **10**, 4540 (1974)
74. K.S. Cole, R.H. Cole, J. Chem. Phys. **9**, 341 (1941)
75. S.M.J. Aubin, Z. Sun, L. Pardi, J. Krzystek, K. Folting, L.-C. Brunel, A.L. Rheingold, G. Christou, D.N. Hendrickson, Inorg. Chem. **38**, 5329 (1999)
76. P.P. Power, Chem. Rev. **112**, 3482 (2012)
77. N.F. Chilton, H. Lei, A.M. Bryan, F. Grandjean, G.J. Long, P.P. Power, Dalton Trans. **44**, 11202 (2015)
78. S. Goda, M. Nikai, M. Ito, D. Hashizume, K. Tamao, A. Okazawa, N. Kojima, H. Fueno, K. Tanaka, Y. Kobayashi, T. Matsuo, Chem. Lett. **45**, 634 (2016)
79. A.M. Bryan, C.-Y. Lin, M. Sorai, Y. Miyazaki, H.M. Hoyt, A. Hablutzel, A. LaPointe, W.M. Reiff, P.P. Power, C.E. Schulz, Inorg. Chem. **53**, 12100 (2014)
80. W.M. Reiff, C.E. Schulz, M.-H. Whangbo, J.I. Seo, Y.S. Lee, G.R. Potratz, C.W. Spicer, G.S. Girolami, J. Am. Chem. Soc. **131**, 404 (2009)
81. W.A. Merrill, T.A. Stich, M. Brynda, G.J. Yeagle, J.C. Fettinger, R. De Hont, W.M. Reiff, C.E. Schulz, R.D. Britt, P.P. Power, J. Am. Chem. Soc. **131**, 12693 (2009)
82. T. Matsuo, K. Suzuki, T. Fukawa, B. Li, M. Ito, Y. Shoji, T. Otani, L. Li, M. Kobayashi, M. Hachiya, Y. Tahara, D. Hashizume, T. Fukunaga, A. Fukazawa, Y. Li, H. Tsuji, K. Tamao, Bull. Chem. Soc. Jpn **84**, 1178 (2011)
83. J.M. Zadrozny, D.J. Xiao, M. Atanasov, G.J. Long, F. Grandjean, F. Neese, J.R. Long, Nat. Chem. **5**, 577 (2013)
84. R.J. Glauber, J. Math. Phys. **4**, 294 (1963)
85. L. Bogani, A. Vindigni, R. Sessoli, D. Gatteschi, J. Mater. Chem. **18**, 4750 (2008)
86. W.-X. Zhang, R. Ishikawa, B. Breedlove, M. Yamashita, RSC Adv. **3**, 3772 (2013)

87. A. Caneschi, D. Gatteschi, N. Lalioti, C. Sangregorio, R. Sessoli, G. Venturi, A. Vindigni, A. Rettori, M.G. Pini, M.A. Novak, *Angew. Chem., Int. Ed.* **40**, 1760 (2001)
88. R. Clérac, H. Miyasaka, M. Yamashita, C. Coulon, J. Am. Chem. Soc. **124**, 12837 (2002)
89. C. Coulon, R. Clérac, L. Lecren, W. Wernsdorfer, H. Miyasaka, Phys. Rev. B **69**, 132408 (2004)
90. T. Kajiwara, M. Nakano, Y. Kaneko, S. Takaishi, T. Ito, M. Yamashita, A. Igashira-Kamiyama, H. Nojiri, Y. Ono, N. Kojima, J. Am. Chem. Soc. **127**, 10150 (2005)
91. T. Kajiwara, I. Watanabe, Y. Kaneko, S. Takaishi, M. Enomoto, N. Kojima, M. Yamashita, J. Am. Chem. Soc. **129**, 12360 (2007)

Chapter 7
Application of Mössbauer Spectroscopy to Li-Ion and Na-Ion Batteries

Pierre-Emmanuel Lippens

Abstract Electrochemical energy storage is essential in everyday life and is crucial for energy transition requiring sustainable energy sources and clean transportation. Li-ion batteries are widely used due to their high energy density while other technologies are now emerging, as Na-ion batteries, to offer a variety of energy storage systems. The electrode materials are one of the key components that should be improved for better battery performance. This requires the investigation of the electrochemical reactions that take place within the batteries. Mössbauer spectroscopy is a powerful tool to probe the local environment at the atomic scale in electrode materials. This technique can be used for ex situ analyses or to follow in situ the electrochemical reactions. In this chapter, some general aspects of currently commercialized Li-ion batteries are briefly described to introduce electrode materials, electrochemical mechanisms and their characterizations. The basic concepts of the Mössbauer effect and hyperfine interactions are presented to define the Mössbauer parameters that provide information about structural, electronic and magnetic properties, with a special attention to in situ measurements. Then, some selected examples of insertion, alloying and conversion materials are considered to illustrate the application of ^{57}Fe and ^{119}Sn Mössbauer spectroscopies to elucidate reaction mechanisms in Li-ion and Na-ion batteries.

7.1 Introduction

In 2019, the Nobel Prize in Chemistry was awarded to John B. Goodenough, M. Stanley Whittingham and Akira Yoshino for their contributions to the development of Li-ion batteries. Li-ion batteries are now used in a wide range of applications from small electronic devices to large storage systems for sustainable and renewable energy sources, including electric transportation [1–4]. This technology has already changed the life of many people worldwide and is certainly a key element for energy

P.-E. Lippens (✉)
Institut Charles Gerhardt, UMR 5253 CNRS Université de Montpellier, Place Eugène Bataillon, 34095 Montpellier Cedex 5, France
e-mail: pierre-emmanuel.lippens@umontpellier.fr

Y. Yoshida and G. Langouche (eds.), *Modern Mössbauer Spectroscopy*,
Topics in Applied Physics 137, https://doi.org/10.1007/978-981-15-9422-9_7

transition. The success of Li-ion batteries is mainly due to their high energy density, arising from the low atomic mass of lithium and its high electropositivity that give high electrode capacity and high cell voltage, respectively. But such lightweight, rechargeable and powerful batteries are also the result of a long series of fundamental research and technology advances [5].

The early use of lithium metal as negative electrode for lithium batteries encountered severe safety issues due to Li dendritic growth, leading to internal short-circuits [6]. In a Li-ion battery, the negative electrode is usually made from an intercalation material like graphite, while Li^+ ions are provided by the positive electrode mostly comprising a layered lithium transition metal oxide. The Li^+ ions move between the two electrodes through an electrolyte that usually consists of a lithium salt and a liquid organic solvent, although solid electrolytes should be a promising alternative to enhance safety and energy density [7–9].

Depending on their applications, Li-ion batteries should have high power and/or high energy density, but in all cases they should be safe, with long calendar and cycle life, environmentally friendly, and as cheap as possible. Everyone would like to use small and light batteries that can be charged fast and do not explode! The performance of conventional Li-ion batteries based on the "negative electrode-liquid electrolyte-positive electrode" configuration depends on the properties of each component and on their combination. It is a real challenge to improve significantly such complex systems, which not only requires to optimize the electrode materials and electrolyte, but also the interfaces, the design of electrodes and cells, the current collectors, etc. [10–16]. Such a task needs a multidisciplinary approach combining different fields of chemistry and physics, but above all, the application of different techniques of characterization [17–23].

Mössbauer spectroscopy can be used in different ways to improve the performance of Li-ion batteries [24–27]. The most obvious application concerns the characterization of the structural and electronic properties of pristine electrode materials, following the methodologies developed in solid-state chemistry. The technique provides a lot of information about the local environment of atoms through electric and magnetic interactions. Mössbauer spectroscopy is also of high interest for the characterization of poorly crystalline materials, amorphous phases and nanoparticles since X-ray diffraction cannot be used due to the loss of long-range order. This aspect is not only important for pristine electrode materials but also for the products of electrochemical reactions that can be, under certain experimental conditions, studied ex situ, i.e. by extracting the electrode material out of the electrochemical cell at a certain stage of lithiation or sodiation.

Most importantly, Mössbauer spectroscopy is a powerful technique to investigate reaction mechanisms under operating conditions as shown, for example, in catalysis [28]. By considering specific electrochemical cells that make γ-ray transmission possible, the technique can be used for in situ or operando studies of electrochemical reactions. Although these two terms are often used indistinctly in the literature, one can consider that operando measurements refer to batteries in charge or discharge operations while in situ characterizations concern all studies where batteries are not disassembled. In situ Mössbauer spectroscopy is of high interest for the characterization of lithiated and sodiated materials that are often air and moisture sensitive, but also and above all, this technique provides information about the reactions taking place in electrochemical cells. It should be noted that such reactions depend on the experimental conditions such as the current imposed to the cell in galvanostatic regime. While a high current usually enhances kinetic effects, a low current should be used to approach thermodynamic equilibrium. This means that the experimental protocol used for in situ measurements should be carefully defined. Finally, it is worth noting that in situ experiments are not always needed to investigate reaction mechanisms as shown in Sect. 7.6.4.

There are some restrictions in the application of Mössbauer spectroscopy to Li-ion and Na-ion batteries. The number of isotopes that can be used with a conventional in-lab Mössbauer technique (*vs.* synchrotron radiation-based Mössbauer spectroscopy) is limited. Fortunately, many positive and negative electrode materials contain iron and tin atoms, respectively, that are the most studied elements by Mössbauer spectroscopy. Both ^{57}Fe and ^{119}Sn isotopes can be considered for room temperature investigations of Li-ion or Na-ion batteries. Other isotopes have been used such as ^{121}Sb [29] or ^{99}Ru [30] but these studies mainly focused on the characterization of pristine electrode materials or ex situ measurements due to the required specific experimental conditions. For electrode materials that do not contain any Mössbauer isotopes, doping with a Mössbauer probe was also used as reported for Sn doped TiO_2 [31] or Fe doped $Li_2Ti_3O_7$ [32].

This chapter primarily deals with the application of Mössbauer spectroscopy to investigate electrochemical mechanisms in Li-ion and Na-ion batteries. This means that the characterization of pristine electrode materials is not really considered here although it is an important task for the optimization of these materials. However, this aspect is similar to the well-documented applications of Mössbauer spectroscopy in materials science and the methodology is not different for electrode materials. Since this book is addressed to a large audience, some aspects of Li-ion batteries and Mössbauer spectroscopy are first introduced to give the basic knowledge and the key parameters for elucidating the electrochemical mechanisms in batteries.

This chapter covers general and specific aspects of Li-ion batteries and Mössbauer spectroscopy. The main features of Li-ion batteries are given in Sect. 7.2, including a presentation of the most commonly used electrode materials, the main types of electrochemical reactions and some methods of characterization. Section 7.3 deals with basic considerations on Mössbauer spectroscopy and is focused on electric and magnetic hyperfine interactions that are at the origin of the Mössbauer parameters. Some specific aspects of in situ Mössbauer measurements are also given. Finally, the application of Mössbauer spectroscopy to the study of electrochemical reactions in Li-ion and Na-ion batteries are illustrated by selected examples of insertion reactions in Sect. 7.4, alloying reactions in Sect. 7.5 and conversion reactions in Sect. 7.6.

7.2 Electrochemical Energy Storage

7.2.1 Li-Ion Batteries

A Li-ion battery is composed of one or several electrochemical cells that are connected in parallel and/or in series to increase the current and/or the voltage, respectively. For instance, the Li-ion battery pack of the Toyota Prius Plug-in hybrid cars consists of 5 battery stacks, each containing 19 cells for a total energy of 8.8 kW h and a voltage of 352 V. There are different types of commercialized electrochemical cells, including cylindrical, prismatic, pouch and coin cells [33]. A typical Li-ion electrochemical cell is formed by a positive electrode, a negative electrode, a separator to avoid electrical contacts between the electrodes and an electrolyte for internal ionic charge transfer. Each electrode consists of a metallic film, copper for the negative electrode and aluminum for the positive electrode, connected to an external cell terminal and coated (single or double side) by a polymer film containing the electrochemically active particles and electronic conductive additives. In cylindrical and prismatic cells, the two electrodes and the separator are coiled, providing higher surface areas and storage capacities (Fig. 7.1).

During the charge of a Li-ion cell, the energy provided by a power source connected to the cell generates a current of electrons that flows from the positive to the negative electrode in the external circuit. This produces a movement of Li^+ ions through the electrolyte within the cell in the same direction (Fig. 7.2). Thus, the first charge of the cell requires that the positive electrode material contains and provides to the negative electrode the Li^+ ions. During the discharge, the electrons and Li^+ ions move from the negative to the positive electrode in the external circuit and in the cell, respectively.

The first lithium batteries were affected by serious problems due to the combined use of pure lithium metal as negative electrode and solvent-based organic electrolytes. The reaction of metallic lithium with liquid electrolyte leads to the progressive growth of Li dendrites at the surface of the negative electrode towards the positive electrode at each charge-discharge cycle, ending with internal electrical short-circuits [6, 9].

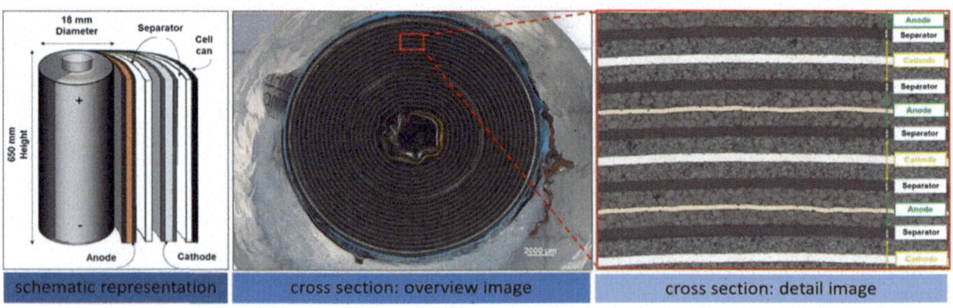

Fig. 7.1 Schematic view of a cylindrical Li-ion cell and cross section images showing coiled negative (anode) and positive (cathode) electrodes with separator. Copyright: ©ZEISS Microscopy

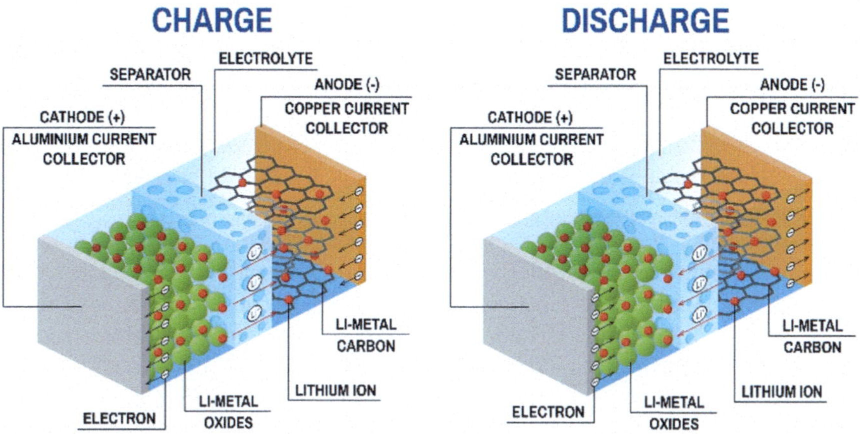

Fig. 7.2 Schematic view of a Li-ion cell showing that Li^+ ions and electrons move from the positive (cathode) to the negative (anode) electrode during the charge and back when discharging. The Li^+ ions move in the cell through the electrolyte and the electrons in the external circuit. The separator prevents electrical contacts between electrodes

To avoid lithium electrodeposition and resolve safety issues, metallic lithium was replaced by an intercalation material and combined with a positive electrode material as a source of lithium. The first Li-ion batteries, commercialized by Sony in 1991, were made up of carbon and $LiCoO_2$ as negative and positive electrode materials, respectively. Since then, many works have been devoted to electrode materials and electrolytes to improve the performance of Li-ion batteries including energy density, rate capability, lifespan and safety. The energy density of an electrochemical cell can be enhanced by increasing the electrode capacity and the cell voltage. In order to increase the cell voltage, much effort has been expended to increase the potential of the positive electrode since the potential of the commonly used carbonaceous negative electrodes is close to that of metallic lithium and cannot be decreased further. For both positive and negative electrode materials, the capacity, but also the rate capability and the cycle life should be improved for better battery performance.

Most of the commercialized positive electrode materials are currently lithium transition metal oxides and phosphates. Layered compounds $LiMO_2$, where $M = Co$, Mn, Ni or mixtures thereof, are often used due to their high potential versus Li/Li^+ and their high specific capacity. However, there are some issues with the stability of layered compounds and the use of Co in the composition of electrode materials. This metal is expensive and cobalt mining has bad environmental and social consequences. A lot of works have been devoted to decrease the cobalt content or replace layered metal oxides by other types of metal oxides such as spinel LiM_2O_4. Transition metal phosphates, such as olivine $LFePO_4$, are also an interesting alternative to $LiMO_2$. $LFePO_4$ based electrodes offer good cycle life, high specific power and safety but have lower operating potential and specific capacity than layered metal oxides.

The negative electrodes of Li-ion batteries are mainly based on carbon. Graphite can intercalate Li^+ ions, has a low cost and can be easily found or produced. Graphite can intercalate between the carbon layers up to 1 Li per 6 C, leading to specific and volumetric capacities of 372 mA h g^{-1} and ~800 mA h cm^{-3}, respectively. The voltage profile of a graphite based electrode, i.e., the electrode potential vs. Li/Li^+ as a function of the number of inserted Li, shows well-defined plateaus and a good reversibility. The mechanism is more complex for disordered forms of carbon such as hard carbons, soft unorganized carbons or natural graphite that contain a lot of defects, leading to different voltage profiles and electrode performances [34–36]. One issue with carbon is the possible formation of lithium dendrites under certain conditions due to its operating potential close to that of metallic lithium. In addition, specific and volumetric capacities should be increased. This is obtained by the addition of a small amount of silicon since this element has a specific capacity about ten times higher than carbon, but the increase of capacity is still too low for future applications.

Titanates (TiO_2, $Li_4Ti_5O_{12}$) have been commercialized as negative electrode materials for high power batteries. The high operating potential of titanate based electrodes provides a high level of safety but a lower energy density than carbon. Finally, other families of materials with higher operating potentials and capacities than carbon have been proposed but encountered different critical issues. For example, tin based materials were previously regarded as negative electrode materials for high energy density batteries with the trade names STALION by Fuji in 1997 [37] and NEXELION by Sony in 2005 [38]. Tin is of particular interest within the framework of this chapter since it is involved in different electrochemical mechanisms that can be studied by ^{119}Sn Mössbauer spectroscopy as shown in Sects. 7.5 and 7.6.

7.2.2 Electrochemical Mechanisms

Three main types of electrochemical reactions that take place in electrode materials are usually distinguished: insertion, alloying and conversion reactions.

Insertion or intercalation reactions: Li^+ ions are intercalated between layers (graphite, $LiMO_2$) or inserted into diffusion channels (Nasicon, $LiFePO_4$, $Li_4Ti_5O_{12}$) of a host material to form a solid solution without strong structural changes or a new phase (two-phase reaction). In the first case, the voltage profile has a typical S-shape whereas a two-phase reaction often shows a voltage plateau reflecting first-order phase transition. These reactions are often highly reversible.

Alloying reactions: Li atoms form chemical bonds with an element A (Si, Sn, etc.) leading to Li_xA alloys or compounds. Such reactions come along with strong volume variations, causing mechanical and chemical instabilities of the electrode. The formed species are often nanoparticles but they can coalesce during cycling. The generally observed staircase voltage profile reflects the succession of two-phase alloying reactions but kinetic limitations or the formation of metastable phases often smooth the voltage curve. These reactions are reversible but reducing the effect of volume variations is still a challenge for long-term cycling.

Conversion reactions: Li atoms react with AB (intermetallics, oxides, sulfides, etc.) to give A and Li_xB nanoparticles that usually form a composite. For example, the lithiation of SnO produces Sn and Li_2O nanoparticles. Conversion reactions can be reversible, irreversible and partially reversible as observed for transition metal oxides, tin intermetallics and tin oxides, respectively. The voltage profile is formed by two plateaus for lithiation and delithiation, respectively, showing a large voltage polarization. Some conversion reactions are followed by alloying reactions.

It should be noted that the number of Li (or Na) per electrochemically active element involved in alloying and conversion reactions is generally higher than for insertion reactions, leading to high capacity electrode materials.

7.2.3 Characterization of Electrochemical Reactions

In Li-ion batteries using a lithium salt and organic solvent as electrolyte, the electrodes consist of metallic foils as current collectors that are coated with a film of several tens to hundreds micrometers obtained from a slurry containing powdered electrochemically active materials, electronic conductive additives and binder. All these components should be optimized for better performance. For instance, the electrode formulation is essential to form a mechanically and chemically stable composite film containing the active particles with electron and lithium conductive pathways. Concerning the electrochemically active materials, not only their intrinsic properties should be improved, but also their interfaces with the electrolyte and the other components of the film. Thus, the characterization of pristine electrode materials, as obtained after synthesis, should be combined with the analysis of the changes in these materials during charge-discharge cycles.

A lot of techniques used in solid-state chemistry and electrochemistry can be considered for the characterization of electrode materials and reaction mechanisms at different length and time scales. Electrode performance, such as volumetric and specific capacities, operating voltage, cycling and rate capabilities, and cycle life, can be evaluated from the measurements of the cell voltage or the electric current with different experimental protocols based on galvanostatic (constant current) or potentiostatic (constant voltage) regimes. Electrochemical experiments also provide useful information about the reactions taking place in the cell from the voltage profile, current or voltage relaxations, impedance, etc. For such experiments, but also for in situ characterizations, Li half-cells are used instead of Li-ion full-cells. In a Li half-cell, the negative electrode is lithium metal that should be considered as the reservoir of lithium while the positive electrode contains the active material under study. The latter material can be the positive or negative electrode material in a Li-ion full-cell. Thus, the discharge and charge of a Li half-cell correspond to the lithiation and delithiation of the electrode material under study, respectively, while the situation is somewhat more ambiguous for a Li-ion full-cell that contains the two types of electrodes. To avoid confusion, the terms lithiation and delithiation (sodiation and desodiation) are used along this chapter instead of charge and discharge, respectively.

One of the most commonly used techniques for studying electrode materials is X-ray diffraction (XRD). XRD has often been considered for the characterization of crystalline pristine materials and their evolution during the lithiation-delithiation cycles. This is of particular interest to follow changes in the lattice parameters and atomic positions of crystalline insertion materials, resulting from the insertion of ions, or to identify new formed crystalline phases. However, this technique fails for poorly crystalline, nanosized and amorphous phases as often encountered in pristine electrode materials or detected as products of alloying and conversion reactions. Such characterizations require local probe techniques, including X-ray photoelectron spectroscopy (XPS), which is a surface sensitive technique, X-ray absorption spectroscopy (XAS), nuclear magnetic resonance (NMR), electron paramagnetic resonance (EPR) or Mössbauer spectroscopy. These techniques not only give information about the local environment of the probe elements, but also on the electronic, magnetic and dynamic properties of the electrode materials. For instance, they are used to determine the oxidation and spin states, charge anisotropy and bond properties of atoms in the electrode particles, but they are also used to identify the products formed during insertion-deinsertion processes within the electrode and at the interface with the electrolyte. All these techniques go along with imaging techniques, such as transmission electron microscopy (TEM), scanning electron microscopy (SEM) or atomic force microscopy (AFM), that provide direct visualized information about changes in the size, morphology and aggregation of the electrode particles caused by electrochemical reactions.

7.3 Basic Aspects of Mössbauer Spectroscopy

7.3.1 The Mössbauer Effect

The Mössbauer effect is briefly described here, while more details can be found in many textbooks [39–42]. This effect is based on transitions between excited and ground nuclear states at energy E_γ (~keV). An excited free nucleus can reach equilibrium by emitting a γ-photon (γ decay) of energy $E_\gamma - E_R$, where E_R (~meV) is the nuclear recoil energy, while a nucleus of the same isotope can be excited by absorbing a γ-photon of energy $E_\gamma + E_R$. The energy distribution of γ-rays is a Lorentzian curve whose the linewidth Γ (~neV) is related to the mean lifetime of the nuclear excited state by the Heisenberg uncertainty principle. The perfect overlap of the emission and absorption Lorentzian curves (resonance) is not possible for free or weakly bound atoms because the nuclear recoil energy is several orders of magnitude higher than Γ. However, R.L. Mössbauer observed such a resonance in solids that can be explained as follows. In a solid, the recoil energy of nuclei during the emission of γ-rays should be transferred to the lattice, but the quantization of atomic vibrations requires that such transfer is only possible for energy values corresponding to the creation of phonons. Thus, a fraction of the excited nuclei can emit γ-photons without energy transfer to the lattice (zero-phonon transition), i.e. with a Lorentzian energy distribution centered at E_γ. Conversely, a fraction of the nuclei at ground state in a solid can absorb γ-photons without recoil, i.e., at the same energies as the source of γ-rays and, therefore, resonance is possible. This fraction of nuclei, f, also known as the recoil-free fraction, corresponds to the probability of recoil-free emission or absorption of γ-rays. The recoil-free fraction is a fundamental quantity, not only to explain the Mössbauer effect, but also for the applications of Mössbauer spectroscopy since f depends on the γ-ray energy and the lattice dynamical properties. To summarize, the Mössbauer effect can be defined as the resonant and recoil-free emission or absorption of γ-rays by a fraction of the nuclei of atoms bound in a solid. In this chapter, these atoms are called the "Mössbauer atoms".

Although changes in the energy of nuclear states due to interactions with the surrounding electrons and atoms are weak (~neV), the extremely small linewidth Γ prevents resonance except if both source and absorber of γ-rays are identical materials. However, R.L. Mössbauer proposed to use the Doppler effect to modulate the energy of γ-rays by small movements of the source relative to the absorber, making possible the detection of the Mössbauer effect in any absorbing samples containing the same Mössbauer isotope as the source. Thus, a conventional in-lab Mössbauer spectroscopy in transmission geometry consists basically of a radioactive source with a drive system, an absorber which is the sample under investigation and a γ-ray detector. The relation between the velocity of the source relative to the absorber, v, and the energy shift, E, is given by the relation

$$v = \frac{c}{E_\gamma} E \tag{7.1}$$

where c is the light velocity. For the 1/2−3/2 nuclear transition in ^{57}Fe (E_γ = 14.4 keV) and ^{119}Sn (E_γ = 23.9 keV), v = 1 mm s^{-1} corresponds to energy shifts of 4.8 10^{-8} eV and 8.0 10^{-8} eV, respectively. This velocity can be compared to the natural linewidths of ^{57}Fe (Γ = 0.1 mm s^{-1}) and ^{119}Sn (Γ = 0.32 mm s^{-1}) and is typical of nuclear energy changes for these two isotopes, leading to velocity ranges for ^{57}Fe and ^{119}Sn Mössbauer spectra usually less than ±10 mm s^{-1}. The high intrinsic energy resolution (Γ/E_γ ~ 10^{-12}) is really the most remarkable feature of Mössbauer spectroscopy, making possible the detection of very weak changes in nuclear transition energy arising from electric and magnetic interactions with the surrounding charges. These interactions are localized within the nucleus of the Mössbauer atoms that should be regarded as atomic probes. The use of such local probes is of high interest for electrode materials. For instance, Li can randomly occupy the vacant sites of an insertion material and modify the electronic configuration of the Mössbauer atoms, leading to changes in oxidation state or causing structural relaxation. For alloying and conversion reactions, the electrochemically formed species are usually nanosized and disordered, so they can be more easily identified by Mössbauer spectroscopy than XRD.

The standard application of in-lab Mössbauer spectroscopy is restricted to a small number of isotopes. ^{57}Fe and ^{119}Sn are by far the most commonly used isotopes, but others like ^{121}Sb [29] or ^{99}Ru [30] have also been considered to study electrode materials. For other chemical elements, synchrotron radiation based Mössbauer spectroscopy [43] or atomic substitution with a Mössbauer isotope should be envisaged. In some cases, the low concentration of Mössbauer atoms requires long-term experiments that are not always compatible with operando measurements.

Information about samples under investigation are mainly obtained from the analysis of the shape of the Mössbauer spectra that are usually not a single Lorentzian line (see Sect. 7.3.6). Electric and magnetic interactions between the Mössbauer nuclei and their local environment lead to the shift and splitting of the nuclear states, which affects the number and energy of the recoil-free nuclear transitions. These interactions are at the origin of three main Mössbauer parameters that can be, in principle, obtained from the Mössbauer spectra: isomer shift, quadrupole splitting and hyperfine magnetic field. These parameters and the recoil-free fraction are described in the following subsections.

7.3.2 Isomer Shift

The hyperfine interactions between the nuclear charge density and the Coulomb potential due to the electron density are localized within the nucleus. The atomic nucleus has a small and finite size and can be non-spherical, leading to monopole and quadrupole electric interactions that are related to the average size of the nucleus considered as a sphere and the deviation from the non-spherical shape of the nucleus, respectively. The monopole and quadrupole interactions are at the origin of the isomer shift and quadrupole splitting, respectively. For the determination of the isomer shift, it is first necessary to evaluate the change in the nuclear energy level, E, due to the Coulomb interaction between the charge Ze of the nucleus, where Z is the atomic number and e is the elementary charge, and the electron density at the nucleus, $\rho(0)$:

$$E = \frac{Ze^2\langle r_n^2\rangle}{6\varepsilon_0}\rho(0) \tag{7.2}$$

where ε_0 is the vacuum permittivity and $\langle r_n^2\rangle$ is the average value of the squared nuclear radius that depends on the nuclear spin number I. Thus, the monopole interaction leads to different values of E for the two nuclear spin states involved in a nuclear transition and the energy of this transition is modified by ΔE. For the nuclear transition 1/2–3/2 (^{57}Fe, ^{119}Sn), $\Delta E = E(I = 3/2) - E(I = 1/2)$ depends on the difference between the average values of the squared nuclear radii for the excited ($I = 3/2$) and ground ($I = 1/2$) nuclear states: $\Delta\langle r_n^2\rangle = \langle r_n^2(I = 3/2)\rangle - \langle r_n^2(I = 1/2)\rangle$. This difference is positive for ^{119}Sn and negative for ^{57}Fe. The Mössbauer atoms in the source and absorber have generally different local environments, leading to different electronic densities at the nucleus, $\rho_s(0)$ (source) and $\rho_a(0)$ (absorber), and therefore, different energy transitions: ΔE_a and ΔE_s. The isomer shift is given by $\delta = \Delta E_a - \Delta E_s$, which corresponds to the energy shift of the resonance line relative to the source. The isomer shift is a relative quantity and the origin can be chosen arbitrary. A reference material is often considered instead of the source and $\rho_s(0)$ is replaced by $\rho_{ref}(0)$. This defines the zero-velocity of the isomer shift scale. In this chapter, the commonly used references α-Fe for ^{57}Fe and $BaSnO_3$ for ^{119}Sn are considered, but others can be found in the literature, which should be taken into account to compare the isomer shifts originating from different works. The expression of the isomer shift obtained from Eq. (7.2) is

$$\delta = \frac{Ze^2}{6\varepsilon_0}\Delta\langle r_n^2\rangle\big(\rho(0) - \rho_{ref}(0)\big) \tag{7.3}$$

Since $\Delta\langle r_n^2\rangle$ is constant for a given isotope, the following simplified expression is often used for chemical applications:

$$\delta = \alpha\big(\rho(0) - \rho_{ref}(0)\big) \tag{7.4}$$

where the constant α is positive for ^{119}Sn and negative for ^{57}Fe, leading to opposite variations of δ versus $\rho(0)$ for these two isotopes. The value of α can be evaluated from the correlation between the experimental values of δ and the theoretical values of $\rho(0)$ for a series of reference materials [44–49]. For example, the values of $\rho(0)$ were evaluated for a series of tin compounds with density functional theory (DFT), using the linearized augmented plane wave (LAPW) method [50]. A good $\delta-\rho(0)$ linear correlation is observed (Fig. 7.3) as expected from Eq. (7.4). The calibration constant α obtained by linear regression for this series of compounds can then be used with $\rho(0)$ calculated by the LAPW method and Eq. (7.4) to evaluate the theoretical values of δ for other tin based phases as illustrated in Sects. 7.5 and 7.6.

The isomer shift provides information about the local environment of a Mössbauer atom through the electron density at the nucleus. Many studies have been devoted to the interpretation of the isomer shift from the evaluation of $\rho(0)$ or from some more qualitative but very fruitful approaches [51–53]. For instance, it is possible to relate the variations of the isomer shift to the valence electron populations of the Mössbauer atoms. First, it is important to note that only the s electrons and some relativistic p electrons are inside the nucleus, and the main contributions to $\rho(0)$ come, in descending order, from 1s, 2s, 3s, etc. electrons. However, changes in p-type and d-type wavefunctions influence the s-type wavefunctions (shielding effect) even at the nucleus. Since most of the modifications in the local environment of the Mössbauer atoms affect the chemical bonds and the valence electron density, one can consider, as a first approximation, that the contribution of core electrons $\rho_c(0)$ is constant and $\rho(0)$ mainly changes with the valence electron contribution $\rho_v(0)$. As shown by atomic calculations, the latter term strongly increases with the number of s-type valence electrons and decreases, but to a lesser extent, with the number of

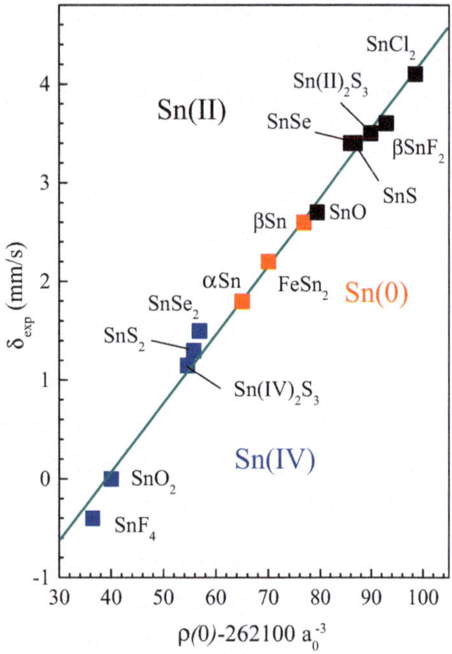

Fig. 7.3 Correlation between the experimental values of the [119]Sn isomer shift, δ_{exp}, relative to $BaSnO_3$ and the theoretical DFT-LAPW values of the electronic density at the nucleus $\rho(0)$. The values corresponding to Sn(IV), Sn(0) and Sn(II) compounds are shown in blue, red and black, respectively

p-type or d-type valence electrons. It is possible to derive an approximate analytical expression of $\rho_v(0)$ from the values obtained for different electron configurations of a free atom [51, 52]. A simple expression was derived for Sn and used for the tight-binding interpretation of the isomer shifts of tin chalcogenides [54]. This expression can be further simplified to qualitatively show the relative contributions of the Sn valence electrons

$$\rho_v(0) \sim N_{5s} - 0.1N_{5p} \tag{7.5}$$

where N_{5s} and N_{5p} are the Sn 5s and Sn 5p electron populations, respectively, obtained by integration of the local densities of occupied states within the tight-binding approximation. The oversimplified expression (7.5) should only be used to analyze some trends in the variations of the isomer shift and is just considered here to show that $\rho_v(0)$, and therefore δ, strongly increases for the series $N_{5s} = 0$–1–2 and, to a lesser extent, decreases with N_{5p}. This explains that the isomer shifts of Sn(IV), Sn(0) and Sn(II) compounds are found in three distinct ranges of values as shown in Fig. 7.3.

Interpretations based on Eq. (7.5) are rather qualitative but it is possible to relate some trends in the variations of δ in terms of oxidation state or chemical bond. A similar distinction can be made for [121]Sb but the isomer shift decreases for the sequence Sb(V)-Sb(0)-Sb(III) because of the negative sign of α. The situation is more complex for [57]Fe because there are overlaps between some ranges of isomer shifts corresponding to different iron oxidation and spin states. Other Mössbauer parameters, such as the quadrupole splitting or the hyperfine magnetic field, should be considered to resolve any ambiguity.

The isomer shift depends on the valence electron configuration of the Mössbauer atom and can be used to investigate changes in the type of ligand, oxidation state, chemical bonding, coordination number and bond lengths arising from electrochemical reactions in Li-ion or Na-ion batteries. A common situation is the reversible change of Fe oxidation state in Fe^{2+} cathode materials such as $LiFePO_4$ following the reaction

$$\mathrm{LiFe^{2+}PO_4 = Fe^{3+}PO_4 + Li^+ + e^-} \tag{7.6}$$

The values of the ^{57}Fe Mössbauer isomer shift relative to α-Fe at room temperature are very different for $LiFe^{2+}PO_4$ ($\delta = 1.2$ mm s^{-1}) and $Fe^{3+}PO_4$: ($\delta = 0.6$ mm s^{-1}). Equation (7.6) indicates that the delithiation of $LiFePO_4$ leads to the decrease of the Fe 3d electron population and, through the shielding effect, to the increase of *ρ(0)*. The negative sign of α in Eq. (7.4) for ^{57}Fe explains the observed decrease of the isomer shift from $LiFePO_4$ to $FePO_4$. In this example, the two Fe oxidation states can be easily distinguished by only considering the values of the isomer shift because ferrous and ferric ions in $LiFePO_4$ and $FePO_4$, respectively, have both high spin states. Another example is the reduction of Sn(IV) into Sn(0) during the first lithiation of SnO_2 following the conversion reaction

$$\mathrm{Sn(IV)O_2 + 4Li^+ + 4e^- \rightarrow \beta Sn(0) + 2Li_2O} \tag{7.7}$$

The ^{119}Sn Mössbauer isomer shifts relative to $BaSnO_3$ of SnO_2 and βSn at room temperature are 0 mm s^{-1} and 2.55 mm s^{-1}, respectively. They can be easily distinguished by ^{119}Sn Mössbauer spectroscopy.

7.3.3 Quadrupole Splitting

The quadrupole splitting originates from the interaction between the nuclear quadrupole moment and the electric field gradient (EFG) at the nucleus created by the anisotropic charge distribution surrounding the nucleus. The nuclear quadrupole moment exists for a nuclear spin state $I > 1/2$ and measures the deviation of the nuclear charge distribution from spherical symmetry. Both the EFG and the nuclear quadrupole moment operator $\hat{Q}$ are second rank tensors and the Hamiltonian H_{QI} resulting from quadrupole interactions is expressed as the product of these two tensors. By considering the EFG principal axis system: *X*, *Y*, *Z*, the Hamiltonian can be written [55]

$$H_{QI} = \frac{eQV_{ZZ}}{4I(2I-1)}\left[3\hat{I}_Z^2 - I(I+1) + \eta\left(\hat{I}_X^2 - \hat{I}_Y^2\right)\right] \tag{7.8}$$

where Q is an average value obtained from $\hat{Q}$ and usually defined as the nuclear quadrupole moment, which depends on the nuclear spin value I, V_{ZZ} is the EFG component of highest absolute value in the principal axis system, $\hat{I}_X$, $\hat{I}_Y$ and $\hat{I}_Z$ are the nuclear spin operators and η is the asymmetry parameter defined by

$$\eta = \frac{V_{XX} - V_{YY}}{V_{ZZ}} \tag{7.9}$$

where V_{XX} and V_{YY} are the two other components of the EFG tensor in the principal axis system such that $|V_{XX}| \leq |V_{YY}| \leq |V_{ZZ}|$. It should be noted that $0 \leq \eta \leq 1$. The nuclear energy levels are obtained by the diagonalization of the H_{QI} matrix in the basis of the nuclear spin states $|I, m_I>$, where m_I is the nuclear magnetic quantum number.

For the first excited nuclear state of ^{57}Fe and ^{119}Sn ($I = 3/2$), H_{QI} can be diagonalized analytically, giving two twofold degenerated energy levels of opposite signs for the states |3/2, ± 1/2> and |3/2, ± 3/2>, respectively, that are separated by the energy

$$\Delta = \frac{1}{2} eQV_{ZZ}\left(1 + \frac{\eta^2}{3}\right)^{1/2} \tag{7.10}$$

called the quadrupole splitting. Thus, there are two possible nuclear transitions between the ground state ($I = 1/2$) and the first excited state ($I = 3/2$). The corresponding Mössbauer spectrum is formed by a doublet of two resonant lines whose separation is given by Δ. Equation (7.10) indicates that the quadrupole splitting is proportional to the nuclear quadrupole moment of the excited state ($I = 3/2$). Its value can be determined experimentally or from the linear correlation between the experimental values of Δ and the theoretical values of EFG for a series of crystalline phases [49, 56–58]. As an example, the experimental values of the quadrupole splitting are reported against the DFT-LAPW values of the electronic term $| V_{ZZ} | \left(1 + \frac{\eta^2}{3}\right)^{1/2}$ for a series of tin compounds (Fig. 7.4). The slope of the regression line gives $Q = 10.5$ fm^2 in agreement with the experimental value $Q = 10.9(8)$ fm^2 [59].

Both V_{ZZ} and η provide information about the anisotropic charge distribution around the nucleus. For the 1/2–3/2 nuclear transition, these two terms cannot be obtained independently from the experimental value of the quadrupole splitting as shown by Eq. (7.10). This is nevertheless possible with a conventional experimental setup and the application of an external magnetic field.

For other nuclear transitions, the quadrupole interactions are more complex, requiring the numerical diagonalization of H_{QI} given by Eq. (7.8) in the basis of the nuclear spin states. This is, for example, the case of the 5/2–7/2 nuclear transition of ^{121}Sb. The 5/2 nuclear ground state and the 7/2 nuclear excited state are split into three and four substates, respectively, leading to eight allowed transitions. In that case, the values of V_{ZZ} and η can be determined from the Mössbauer spectra, but there are often large uncertainties on the values of η [60].

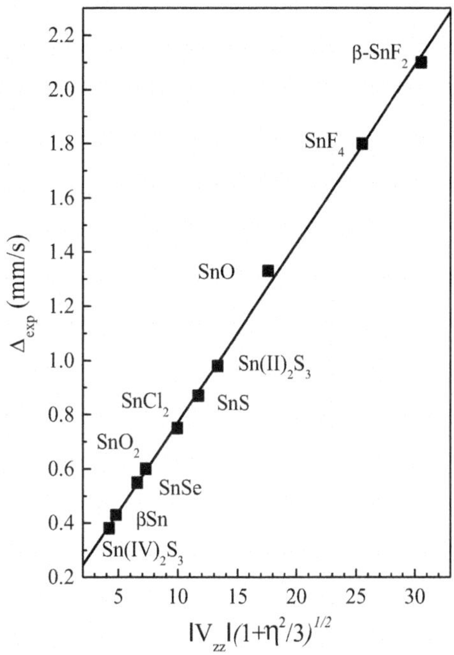

Fig. 7.4 Correlation between the experimental values of the ^{119}Sn quadrupole splitting, Δ_{exp}, and the theoretical DFT-LAPW values of the electronic term in Eq. (7.10)

Since $0 \leq \eta \leq 1$, η has often a small influence on Δ compared to V_{ZZ}, as shown by Eq. (7.10), and one can consider that the value of the quadrupole splitting is mainly given by V_{ZZ}. Different approaches have been used to interpret V_{ZZ}, including the point charge model that distinguishes between valence and lattice contributions [61]. However, the calculation of the EFG by quantum mechanical methods provides more accurate values and is often a better approach for a quantitative analysis [56].

Along the same line as the interpretation of the isomer shift described in Sect. 7.3.2, simple models were proposed to evaluate V_{ZZ} from the electron populations especially when the EFG originates from the valence electrons of the Mössbauer atom. For instance, the EFG at the Sn nucleus often arises from the Sn 5p electron anisotropy and, under certain assumptions, V_{ZZ} can be related to the Sn 5p electron asymmetry count $\Delta N_{5\mathrm{p}}$:

$$V_{ZZ} \sim \langle r^{-3} \rangle \Delta N_{5p} \tag{7.11}$$

with

$$\Delta N_{5p} = \frac{1}{2}\left(N_{5p_X} + N_{5p_Y}\right) - N_{5p_Z} \tag{7.12}$$

where $\langle r^{-3} \rangle$ is the expectation value of r^{-3} for the Sn 5p states, N_{5p_i} is the Sn 5p$_i$ orbital occupancy (or partial number of Sn 5p$_i$ electrons) along the principal axis

$i = X, Y, Z$. Equations (7.11) and (7.12) show that the sign of V_{ZZ} reflects the accumulation ($V_{ZZ} < 0$) or the depletion ($V_{ZZ} > 0$) of the Sn 5p electrons along the Z axis relative to the average number of the remaining Sn 5p electrons in the XY plane perpendicular to Z.

For the 1/2–3/2 nuclear transition, the sign of V_{ZZ} cannot be determined experimentally except for combined electric quadrupole and magnetic interactions. However, the magnitude of Δ is of particular interest since it provides information about the anisotropy of the charge density around the Mössbauer atom, which depends on the Sn local structure and chemical bonds. For battery applications, this parameter can be used for the analysis of structural relaxations induced by Li or Na insertion or for the characterization of electrochemically formed small and poorly crystallized particles in two-phase, alloying and conversion reactions. The quadrupole splitting can also help to understand differences between metastable and stable crystalline phases as described in Sect. 7.6.1 for Li_7Sn_2.

7.3.4 Magnetic Splitting

The interaction between the magnetic moment of the nucleus $\boldsymbol{\mu}$ and the magnetic field at the nucleus $\boldsymbol{B}$ is described by the Hamiltonian

$$H = -\boldsymbol{\mu}.\boldsymbol{B} \tag{7.13}$$

with

$$\boldsymbol{\mu} = g_I \mu_n \hat{\boldsymbol{I}} \tag{7.14}$$

where g_I is the nuclear g-factor that depends on the nuclear spin number I, μ_n is the nuclear magneton and $\hat{\boldsymbol{I}}$ is the nuclear spin operator. The nuclear energy levels are

$$E = -g_I \mu_n B m_I \tag{7.15}$$

where the nuclear magnetic quantum number m_I takes the $2I + 1$ values $-I, -I + 1, \ldots, I-1, I$, leading to the splitting of the nuclear states into $2I + 1$ substates (nuclear Zeeman splitting). For ^{57}Fe and ^{119}Sn isotopes, the ground (I = 1/2) and excited (I = 3/2) states are split into two ($m_{I=1/2} = \pm 1/2$) and four ($m_{I=3/2} = \pm 1/2, \pm 3/2$) substates, respectively, leading to six allowed dipole transitions by considering the selection rules $\Delta m_I = 0, \pm 1$. In that case, the Mössbauer spectrum is a sextet formed by six resonant absorption lines with intensities depending on the magnetization and γ-ray directions. For polycrystalline samples, the relative intensities are in the ratio 3:2:1:1:2:3.

The effective magnetic field at the nucleus comes from external and internal (in the Mössbauer atom) magnetic fields. In the latter case, the magnetic field at the nucleus originates from the spin imbalance at the nucleus (Fermi contact), the interaction between nuclear and electronic orbital moments, and the dipolar interaction between nuclear and electron spin moments. For example, the hyperfine magnetic field of ^{57}Fe in α-Fe is mainly due to the Fermi contact term arising from the difference between spin-up and spin-down electron densities at the nucleus $\rho_{\uparrow}(0) - \rho_{\downarrow}(0)$. The spin densities at the nucleus $\rho_{\uparrow}(0)$ and $\rho_{\downarrow}(0)$ are due to s-type electrons that are affected by the shielding effects of Fe $3d_{\uparrow}$ and Fe $3d_{\downarrow}$ valence electrons, respectively. The intensity of the hyperfine magnetic field can be obtained from the energy difference between the two external lines of the Mössbauer spectrum corresponding to the two nuclear transitions between states $|I, m_I>$: $|1/2,-1/2> - |3/2,-3/2>$ and $|1/2,1/2> - |3/2,3/2>$. According to Eq. (7.15), the observed value of 10.6 mm s^{-1} gives a hyperfine magnetic field of 33 T.

Although, the magnetic properties of electrode materials are not of direct interest for applications in batteries, they provide additional information that can help in the characterization of materials (crystallinity, particle size) or reaction mechanisms. Finally, it should be noted that the electric quadrupole and hyperfine magnetic interactions are not only described by a number, as the isomer shift, but they both depend on direction (EFG axes, magnetic field direction). If these two interactions occur simultaneously, the Hamiltonian resulting from Eqs. (7.8) and (7.13) must be generally solved numerically. In that case, the shape of the Mössbauer spectrum is more complex than line splitting. An example of such Mössbauer spectrum is given for antiferromagnetic $FeSn_2$ in Sect. 7.6.1.

7.3.5 Recoil-Free Fraction

The recoil-free fraction, f, is the probability of recoilless emission (source) or absorption (absorber) of γ-rays and is given by

$$f = exp\left(-k^2\langle x^2\rangle\right) \tag{7.16}$$

where k is the γ-ray wavenumber and $\langle x^2\rangle$ is the mean-square amplitude of the nucleus vibrations in the direction of γ-rays. A significant Mössbauer effect requires a large recoil-free fraction, which is obtained with a low value of k, i.e. of the γ-ray energy, typically lower than 150 keV, and a small value of $\langle x^2\rangle$. The nucleus displacements depend on the phonon spectrum of the solid that usually shows complex dispersion and must be evaluated by computational methods. However, analytical expressions can be derived by considering the Debye model. In the Debye model, the density of phonon states is assumed to increase as ω^2 up to the Debye frequency $\omega_D = K_B\theta_D/\hbar$, where θ_D is the Debye temperature, K_B is the Boltzmann constant and $\hbar$ is the reduced Planck constant. Simple analytical expressions can be obtained for $T \ll \theta_D$

$$f = exp\left(-\frac{3E_R}{2K_B\theta_D}\right) \quad (7.17)$$

and for $T \gg \theta_D$

$$f = exp\left(-\frac{6E_RT}{K_B\theta_D^2}\right) \quad (7.18)$$

where E_R is the nuclear recoil energy. At low temperature, *f(T)* does not vary with temperature while it strongly decreases as T increases for high temperatures. The temperature dependence of f is of high interest to determine the Debye temperature but can also be used for the analysis of Mössbauer spectra since the intensity of the resonant absorption line depends on f. For instance, the recoil-free fraction of βSn is small at room temperature, $f(300\text{ K}) \approx 0.04$, and strongly increases with decreasing temperature, $f(77\text{ K}) \approx 0.45$ and $f(4\text{ K}) \approx 0.72$, leading to a strong increase of the intensity of the corresponding resonant absorption line [62]. For tin based electrode materials, it is often difficult to detect at room temperature a small amount of βSn that coexists with other tin phases such as Li_xSn, Na_xSn or Sn intermetallics. The isomer shifts are all in the Sn(0) range and the recoil-free fractions of the tin compounds or alloys are higher than that of βSn. By decreasing the temperature, the contribution of βSn to the Mössbauer spectrum increases, making the quantitative analysis more accurate.

The recoil-free fraction decreases with decreasing Debye temperature as shown for low and high temperatures by Eqs. (7.17) and (7.18), respectively. The physical meaning of the Debye temperature is not discussed here and θ_D is just regarded as a characteristic quantity of a solid that can be related to lattice stiffness. At a given temperature, solids with very different θ_D have different recoil-free fractions, which can help in the analysis of Mössbauer spectra when different phases coexist in the same absorbing sample. For example, the recoil-free fraction of SnO_2 at room temperature, $f(300\text{ K}) = 0.55$, is more than ten times higher than that of βSn, leading to strongly different resonance line intensities. These two phases can be easily distinguished by Mössbauer spectroscopy since the difference between their Mössbauer isomer shifts is of about 2.6 mm s^{-1}. Thus, a small amount of SnO_2, as encountered in the oxide layer of βSn for instance, can be detected. Another situation concerns the reactants and products of electrochemical reactions with very different recoil-free fractions that must be taken into account to quantitatively characterize the electrochemical mechanisms as shown for conversion reactions in Sect. 7.6.

Finally, it should be noted that in a solid the Mössbauer atoms have generally different local environments, i.e., different types of neighbors, bond distances and angles. The values of f depend on the bonding forces on the Mössbauer atoms and can be significantly different for strongly different environments. In addition, f can be angular dependent for anisotropic environments. The situation is even more complex for multiphase systems as often encountered in electrode materials. In most cases, the values of f cannot be evaluated for each Mössbauer site and some simplifying approximations are made as discussed in the next subsection.

7.3.6 The Mössbauer Spectrum

In Mössbauer spectroscopy, the energy of the γ-rays emitted by the source is modified by using the Doppler effect. It is then possible to detect the Mössbauer effect in any absorbing samples containing the same Mössbauer isotope as the source. A low activity source produces γ-rays with energy distribution modulated by an electromechanical drive system. In transmission geometry, the γ-rays transmitted through the absorber are transformed into an electric signal by a detector. From this signal, the number of γ-photons is recorded as a function of the source velocity by an electronic system synchronized with the source motion, which is periodic and has usually a constant acceleration. One Mössbauer spectrum is recorded every half-period of the source motion and the spectra are accumulated during many cycles to improve the signal-to-noise ratio. The resulting spectrum not only reflects the Mössbauer effect in the absorber but also depends on the γ-ray energy distribution of the source, which is assumed to have a single Lorentzian line shape in the present chapter.

For an absorber containing Mössbauer atoms with a single environment, e.g., a single crystallographic site occupied by the Mössbauer atoms in a crystal, it is convenient to define the dimensionless effective thickness by $t_a = \sigma_0 f_a n_a$, where σ_0 is the absorption cross-section at resonance, f_a is the recoil-free fraction of the Mössbauer isotope in the absorber and n_a is the number of atoms of the Mössbauer isotope per unit area. Thus, $f_a n_a$ represents the concentration of Mössbauer atoms that resonantly absorb γ-rays. The concentration n_a is obtained from the concentration of atoms of the same element by considering the natural abundance, e.g. 2.2% for ^{57}Fe and 8.6% for ^{119}Sn. These small values indicate that some samples with a low concentration of Mössbauer atoms should be isotopically enriched to enhance the Mössbauer effect. In the thin absorber approximation ($t_a < 1$), the Mössbauer spectrum can be described by a sum of Lorentzian curves, making easier the theoretical derivations and the fitting procedures of the experimental data. If both source and absorber materials are identical, the Mössbauer spectrum is a single Lorentzian curve centered at zero-velocity relative to the source with a natural linewidth arising from emission and absorption of $2\,\Gamma \approx 0.19$ mm s^{-1} for ^{57}Fe and $2\,\Gamma \approx 0.63$ mm s^{-1} for ^{119}Sn. However, it should be noted that the linewidth is often broadened due to additional effects such as the absorber thickness, self-absorption, imperfections of the spectrometer, etc.

If the source and absorber materials are different, the shape of the Mössbauer spectrum depends on the hyperfine interactions described in Sects. 7.3.2–7.3.4. For the 1/2–3/2 nuclear transition, the observed main changes from the single Lorentzian line obtained for identical source and absorber materials are the following:

- Isomer shift: the single Lorentzian line is shifted by a value that gives the isomer shift relative to the source (or a reference material).
- Quadrupole splitting: the single Lorentzian line is split into a doublet and the line separation gives the quadrupole splitting.
- Magnetic interaction: the single Lorentzian line is split into six lines, the hyperfine magnetic field can be determined from line separation.

The intensity of the spectral lines depends on the nuclear transition probabilities, which gives for example, 3:2:1:1:2:3 intensity ratio for the magnetic sextet of powdered samples. This also depends on the directional dependence of the recoil-free fraction arising from vibrational anisotropy, leading to an asymmetric doublet for the quadrupole effect (Goldanskii-Karyagin effect) [63]. The sample thickness and its texture can also affect the line shape. Finally, it is important to note that combined quadrupole and magnetic interactions or magnetic relaxations may lead to complex spectra.

SnO_2 and SnO have been regarded as negative electrode materials for Li-ion batteries and the electrochemical mechanisms are described in Sect. 7.6.4. The Mössbauer spectra of these two compounds are discussed here to illustrate the effect of hyperfine interactions (Fig. 7.5). Stannic oxide (SnO_2) and stannous oxide (SnO) have $P4_2/mnm$ and $P4/nmm$ tetragonal structures, respectively. They both contain one Sn crystallographic site, which means that the Mössbauer spectra of these two compounds reflect a single Sn environment. In SnO_2, Sn is six-coordinated by O atoms and the Mössbauer spectrum at room temperature is formed by a single peak that can be fitted to a doublet with $\delta = 0$ mm s^{-1} and $\Delta = 0.5$ mm s^{-1}. The value of the isomer shift is typical of Sn(IV) oxidation state while the quadrupole splitting is due to the slightly distorted SnO_6 octahedral environment.

SnO is composed of Sn-O-Sn layers. The Sn local environment can be described by a SnO_4 square based pyramid. The Mössbauer spectrum at room temperature can be fitted to an asymmetric doublet with parameters $\delta = 2.62$ mm s^{-1} and

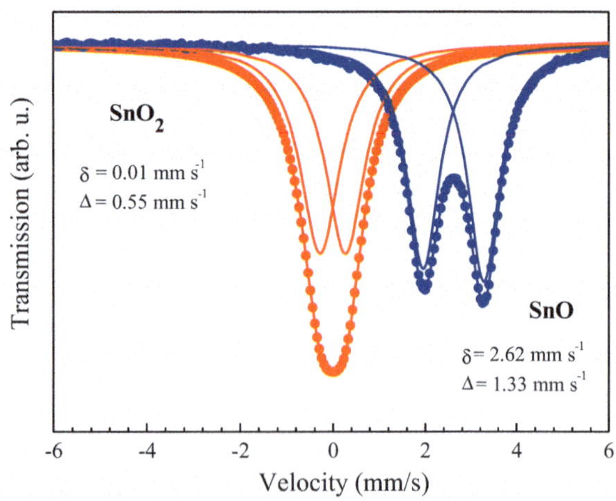

Fig. 7.5 ^{119}Sn Mössbauer spectra of SnO_2 and SnO at room temperature. The experimental data are fitted to two Lorentzian curves, showing a symmetric and unresolved doublet for SnO_2 and an asymmetric doublet for SnO due to the Goldanskii-Karyagin effect

$\Delta = 1.33$ mm s^{-1}. The asymmetry of the doublet is due to the Goldanskii-Karyagin effect arising from the vibrational anisotropy of Sn. The isomer shift of SnO is close to that of βSn (2.56 mm s^{-1}) but must be assigned to Sn(II) oxidation state. Its low value, compared to other Sn(II) species, reflects the hybridization of the Sn valence orbitals that reduces the Sn 5s electron population. The rather high quadrupole splitting is due to the asymmetry of the Sn 5p electrons arising from the SnO_4 local structure that can be related to the existence of a Sn 5p lone pair along the fourfold symmetry axis of the SnO_4 square based pyramid. This axis corresponds to the Z axis of the EFG principal axis system, which leads to $\eta = 0$. According to Eq. (7.10), V_{ZZ} is therefore the only electronic contribution to the quadrupole splitting and Δ reflects the activity of the Sn 5p lone pair.

The area of the resonance absorption line is of primary interest for the application of Mössbauer spectroscopy to batteries. For thin samples containing Mössbauer atoms with the same local environment, the area is proportional to t_a, and therefore, to f_a and n_a. However, the Mössbauer atoms have generally different environments, i.e., different Mössbauer sites as encountered, for example, in crystals with different crystallographic sites, in amorphous phases, or in the different phases of a composite. The different Mössbauer sites, labelled i, give different subspectra that contribute to the overall Mössbauer spectrum. The total area is the sum of the subspectrum areas that are each proportional to $f_i\, n_i$ and a quantitative analysis requires the knowledge or the determination of the different f_i, which cannot be usually achieved.

Many electrode materials are transformed into a multiphase system during the electrochemical reactions and a particular attention must be paid to the evaluation of the recoil-free fractions for a quantitative analysis. For instance, the relative amounts of reactants and products in a Fe^{2+}/Fe^{3+} two-phase reaction can often be obtained by considering the same value of f for the different phases, as in the case of $LiFePO_4/FePO_4$ discussed in Sect. 7.4.2. But this simple assumption fails for some reactions such as the first lithiation of $FeSn_2$ ending with Li_7Sn_2. These two phases contain Sn(0) in different chemical environments, leading to significantly different recoil-free fractions at room temperature $f(FeSn_2)$ and $f(Li_7Sn_2)$. In that case, the relative amounts of $FeSn_2$ and Li_7Sn_2 can only be evaluated at each stage of lithiation from the Mössbauer spectra by considering the ratio $f(FeSn_2)/f(Li_7Sn_2)$ as shown in Sect. 7.6.1.

The existence of a large number of overlapping subspectra can make the distinction between reactants and products difficult, especially for ^{119}Sn Mössbauer spectroscopy. Thus, it is often impossible to determine the Mössbauer parameters for each Mössbauer site in each phase. In that case, more advanced fitting procedures of the Mössbauer data that include hyperfine parameter distributions should be considered, but simplified fitting strategies combined with additional information provided by complementary techniques are often more interesting for the analysis of the electrochemical mechanisms. Such a situation is often encountered with electrode materials in batteries. For instance, Li or Na insertion reactions in positive electrode materials (solid solution) come with the occupation of different vacant sites that can affect selectively the Mössbauer atoms. Another complex situation concerns alloying reactions in tin based negative electrodes that often produce poorly crystalline and small particles with different compositions. In such cases, simplified fitting procedures are often used in order to highlight some trends in the variations of the Mössbauer parameters as described in Sect. 7.5.3.

7.3.7 *In Situ Experiments*

The application of ex situ Mössbauer spectroscopy to study electrode materials is similar to the application of this technique in solid-state chemistry as widely described in textbooks [39, 40–42]. However, the products of electrochemical reactions extracted from the electrochemical cells are in powder or film form and can be chemically unstable. Samples must be prepared and handled with care to avoid oxidation or other parasitic reactions. They must be characterized quickly after the electrochemical experiments to reduce a possible evolution due to chemical reactions that could still operate after extraction from the cell even after washing. The low concentration or the low recoil-free fractions of the Mössbauer atoms in the electrode material can be responsible for long-term experiments. In that case, it is possible to use Swagelok-type cells with higher amounts of powdered electrode material, but the measurements are often limited to few cycles. Finally, it should be noted that in many cases, such *post mortem* characterizations of the electrochemical mechanisms provide rather reliable results that can be favorably compared to in situ measurements as shown in Sects. 7.5.3 and 7.6.4.

The in situ Mössbauer experiments are conducted in the same way as the electrochemical measurements except that the commonly used coin cells or Swagelok-type cells are replaced by an electrochemical "in situ cell" for γ-ray transmission. The in situ cells are also based on the "negative electrode-electrolyte-positive electrode" configuration including a separator, but all the components in the path of the γ-rays are optimized to enhance the signal-to-noise ratio and avoid parasitic resonance absorption. This is of particular importance for electrochemical reactions since the time required to record a spectrum, typically several hours, should be consistent with the evolution of the system due to insertion/extraction of typically 0.1 Li or Na per active element and per hour. The commonly encountered difficulties are due to the electrode thickness and the existence of metals or heavy elements in the composition of the cell components.

The in situ Mössbauer measurements are usually performed at room temperature and no special cell designs are required for low or high temperatures. Dunlap et al. used a modified 2325-type coin-cell with thick (1 mm) and thin (250 μm) Be windows for γ-ray transmission (Fig. 7.6) [27, 64]. These windows are electronically conductive and act as current collectors. The thin Be window is coated with the electrode material under study while a lithium foil is placed against the thick window (half-cell configuration). The ^{57}Fe Mössbauer spectra were corrected from the presence of iron in Be windows [27]. A similar cell was proposed by Wattiaux et al. but with Mylar instead of Be windows [65].

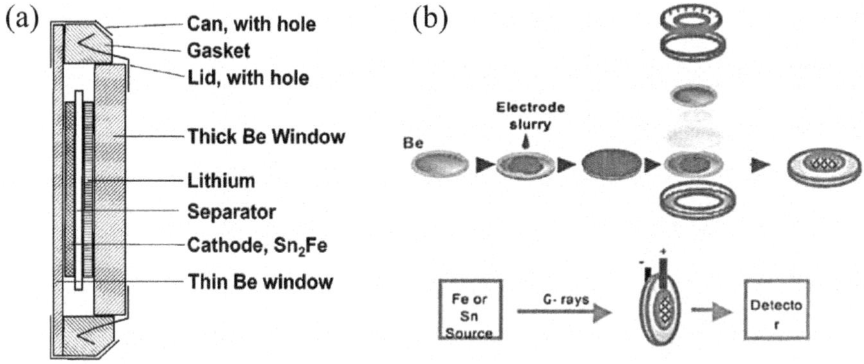

Fig. 7.6 **a** Modified 2325-type coin cell and **b** experimental set up for in situ Mössbauer measurements. **a** Reprinted with permission from Ref. [27]. Copyright 1999 American Physical Society. **b** Image courtesy Mössbauer Effect Data Centre, Dalian Institute of Chemical Physics, Chinese Academy of Sciences [64]

Ariyoshi et al. proposed an aluminum-laminated polyethylene bag containing Li metal with stainless steel sheet as current collector placed outside the γ-ray beam and working electrode on aluminum foil [66]. A pouch cell was also used by Ionica et al. consisting of an airtight aluminized plastic bag with electrodes made according to the Bellcore technology [67].

Chamas et al. used a modified Swagelok tube union with PMMA windows transparent to γ-rays, called Cell 1 in this chapter, and less fragile than Be windows to maintain the cohesion of the internal cell configuration [68]. The electrodes are lithium foil and powdered active material spread on a Be window for electronic conduction. They are both in electrical contact with hollow steel tubes as cell terminals (Fig. 7.7a). Such a cell is interesting for the characterization of powdered alloying or conversion materials that suffer from large volume variations. Another type of Swagelok cell, called Cell 2 in this chapter, as first proposed for in situ XRD measurements in reflection mode [69], was modified for γ-ray transmission (Fig. 7.7b). A similar cell but with a large Be window in contact with the active material was used in both reflection and transmission geometries, making combined XRD and Mössbauer measurements possible [70].

A potentiostat-galvanostat system is used to apply and control the voltage or current to the in situ electrochemical half-cell. Most of the operando Mössbauer experiments are carried out in galvanostatic regime, i.e. by imposing a constant current, in order to insert a constant number of Li^+ or Na^+ ions per time unit in the working electrode. The acquisition time of an operando Mössbauer spectrum should be rather short to have a snapshot of the electrochemical reaction but should also be rather long for a good signal-to-noise ratio. The acquisition time depends on the current injected to the cell that should be as low as possible to approach thermodynamic equilibrium conditions. However, a too low current leads to long-term experiments, typically more than several weeks per electrochemical cycle, which

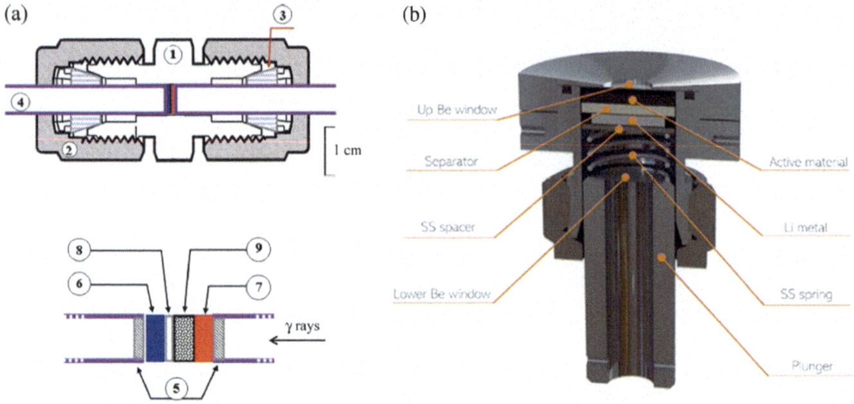

Fig. 7.7 **a** Cell 1. Modified Swagelok cell composed of PFA cell body (1), nuts (2), PFA sealing ferrules (3) and stainless steel plungers (4). The junction around the electrodes is formed by PMMA windows (5), lithium disc (6), Be based connector (7), Whatman separator (8) and active material (9). Reprinted with permission from Ref. [68]. Copyright 2013 American Chemical Society. **b** Cell 2. Modified Swagelok cell for in situ Mössbauer spectroscopy used in transmission mode designed by S4R France

can be a severe issue for the airtightness of the cell. Although it could be interesting to use high currents for the analysis of kinetic effects, such experiments are made difficult by the too long period of time required to record reliable spectra compared to the fast evolution of the insertion and extraction processes. In most cases, the mass of electrode materials is around 1–10 mg. Typical currents in the range of 10^{-2}–10^{-1} Li^+ (or Na^+) per active element and per hour and measurement times in the range of 1–10 h per spectrum generally provide a good compromise between the total duration of operando experiments and a correct signal-to-noise ratio. But this obviously depends on the experimental set-up and the concentration of Mössbauer atoms in the electrode material.

To increase the acquisition time of each operando spectrum and/or perform measurements in thermodynamic equilibrium conditions, while maintaining a reasonable time for the overall electrochemical experiment, specific current profiles should be considered. For instance, the current can be stopped at regular periods of time for voltage relaxation during which the Mössbauer spectra are recorded. Compared to the galvanostatic protocol, the acquisition time of each spectrum is longer and a better signal-to-noise ratio is obtained. However, the number of spectra recorded during the charge-discharge cycles is usually smaller, which can be a limitation for monitoring electrochemical reactions. These two approaches are compared in Sect. 7.4.1.

7.4 Insertion Reactions

7.4.1 Solid-Solution Reactions

The solid-solution reactions refer to the insertion of Li^+ or Na^+ ions on the vacant sites of a solid without phase transition. To illustrate the application of Mössbauer spectroscopy, the example considered here is the Nasicon-type (sodium super ionic conductor) compound $Na_{1.5}Fe_{0.5}Ti_{1.5}(PO_4)_3$ proposed as electrode material for Na-ion batteries and studied with two different operando protocols [71, 72]. $Na_{1.5}Fe_{0.5}Ti_{1.5}(PO_4)_3$ sub-micrometer particles were obtained by solid-state reaction and coated with carbon to improve the electronic conductivity. The crystallinity was controlled by XRD while DFT calculations confirmed the preferential occupation of some vacant sites by Na in the pristine material. The crystal has a rhombohedral $R\bar{3}c$ structure with a random occupation of the metal sites (12c) by 25% Fe and 75% Ti. The ^{57}Fe Mössbauer spectrum of $Na_{1.5}Fe_{0.5}Ti_{1.5}(PO_4)_3$ consists of a doublet due to high spin Fe^{3+} while the spectrum of the C-coated sample $Na_{1.5}Fe_{0.5}Ti_{1.5}(PO_4)_3/C$ was fitted to two doublets with Mössbauer parameters: $\delta = 0.42$ mm s^{-1}, $\Delta = 0.28$ mm s^{-1} assigned to high spin Fe^{3+} (91%) and $\delta = 1.18$ mm s^{-1}, $\Delta = 2.37$ mm s^{-1} assigned to high spin Fe^{2+} in impurities arising from pyrolysis (9%).

The voltage curve of the first cycle of $Na_{1.5}Fe_{0.5}Ti_{1.5}(PO_4)_3/C$ based electrode, with carbon black as electronic additive and PVDF binder, in a Na half-cell with galvanostatic current of 0.1 Na per fu (formula unit) and per hour shows a voltage hysteresis at about 2.2 V with a weak polarization (Fig. 7.8).

During the sodiation, the voltage decreases from 2.5 to 2.2 V for the first 0.5 Na (Region R1) and then forms a plateau until the end of sodiation at 2 Na (Region R2). The desodiation curve is similar except for a small irreversibility of 0.1 Na. The shape of the sodiation curve suggests a two-step mechanism consisting in the formation

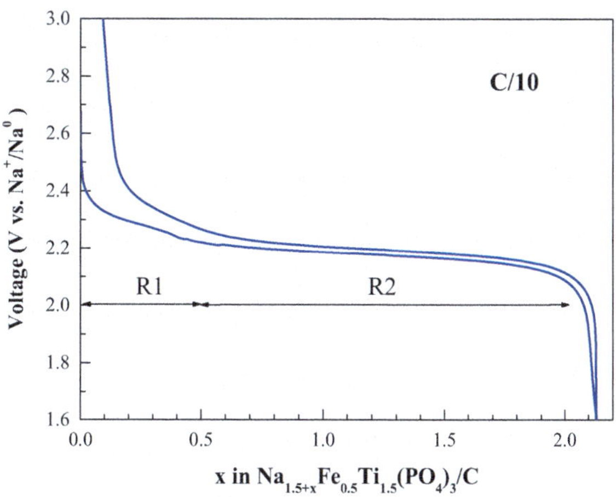

Fig. 7.8 Voltage curve of the first cycle of $Na_{1.5}Fe_{0.5}Ti_{1.5}(PO_4)_3/C$ in a Na half-cell at C/10 rate (0.1 Na per fu and per hour) showing two different mechanisms in regions R1 and R2. Reprinted with permission from Ref. [71]. Copyright 2015 American Chemical Society

of a solid solution (R1) followed by a two-phase reaction (R2). This mechanism is reversible for desodiation. The voltage curve differs from that of $NaTi_2(PO_4)_3$ formed by a plateau at 2.2 V, which was assigned to a two-phase reaction corresponding to the redox couple Ti^{4+}/Ti^{3+} [73]. The operando XRD patterns in region R1 are almost identical to the XRD pattern of the pristine material with $R\bar{3}c$ structure while the sodiation in R2 shows two phases with $R\bar{3}c$ and $P\bar{1}c$ structures. These results are consistent with the mechanism suggested by the voltage profile but do not reveal the redox mechanism.

Operando ^{57}Fe Mössbauer measurements were carried out with the Cell 2 described in Sect. 7.3.7 and two different protocols. In the first protocol, the spectra were recorded continuously during the first cycle in galvanostatic regime with a current of 0.04 Na per fu and per hour and an acquisition time of 2 h per spectrum. In the second protocol, a periodic current was imposed. Each 16-h period consisted of a constant current of 0.1 Na per fu and per hour during 4 h and a zero current (open circuit condition) during 12 h for recording the spectrum in order to improve the signal-to-noise ratio. The spectra show that in R1 the Fe^{3+} doublet of $Na_{1.5}Fe_{0.5}Ti_{1.5}(PO_4)_3$ ($\delta = 0.41$ mm s^{-1}, $\Delta = 0.34$ mm s^{-1}) is transformed into a Fe^{2+} doublet ($\delta \approx 1.2$ mm s^{-1}, $\Delta \approx 2.0$ mm s^{-1}) (Fig. 7.9).

The total area of the spectra does not vary noticeably during the sodiation-desodiation process, indicating that the recoil-free fractions of Fe^{2+} and Fe^{3+} have close values. Thus, the relative contributions of the Fe^{2+} and Fe^{3+} subspectra to the Mössbauer spectra can be regarded as the relative amounts of Fe^{2+} and Fe^{3+} ions in the electrode material. The amount of Fe^{3+} linearly decreases in R1 while the amount of Fe^{2+} increases (Fig. 7.10). In R2, the variations are not significant and the amounts of Fe^{3+} and Fe^{2+} are almost constant, showing that sodiation does not change the oxidation state of iron anymore. The variations are reversible for desodiation with strong changes at the end of the process. It should be noted that the values of the relative contributions of Fe^{2+} and Fe^{3+} are more dispersed for the galvanostatic protocol, which confirms the better quality of the data obtained with the other protocol. These results show that the mechanism of sodiation consists in the formation of a solid solution coming with the reduction of Fe^{3+} into Fe^{2+}, followed by a two-phase reaction leading to the reduction of Ti^{4+} into Ti^{3+}. The mechanism is reversible for desodiation. The cyclability and rate capability are better for $Na_{1.5}Fe_{0.5}Ti_{1.5}(PO_4)_3$ compared to $NaTi_2(PO_4)_3$, which can be attributed to enhanced Na diffusion due to the single-phase reaction in R1.

The operando Mössbauer spectroscopy provides a quantitative characterization of the redox reactions but cannot distinguish between solid-solution and two-phase reactions. This is obtained by operando XRD. Thus, the combined use of these two techniques successfully explains the two-step mechanism by the reactions

$$Na_{1.5}Fe^{3+}_{0.5}Ti^{4+}_{1.5}(PO_4)_3\left[R\bar{3}c\right] + 0.5Na = Na_2Fe^{2+}_{0.5}Ti^{4+}_{1.5}(PO_4)_3\left[R\bar{3}c\right] \quad (7.19)$$

$$Na_2Fe^{2+}_{0.5}Ti^{4+}_{1.5}(PO_4)_3\left[R\bar{3}c\right] + 1.5Na = Na_{3.5}Fe^{2+}_{0.5}Ti^{3+}_{1.5}(PO_4)_3\left[P\bar{1}c\right] \quad (7.20)$$

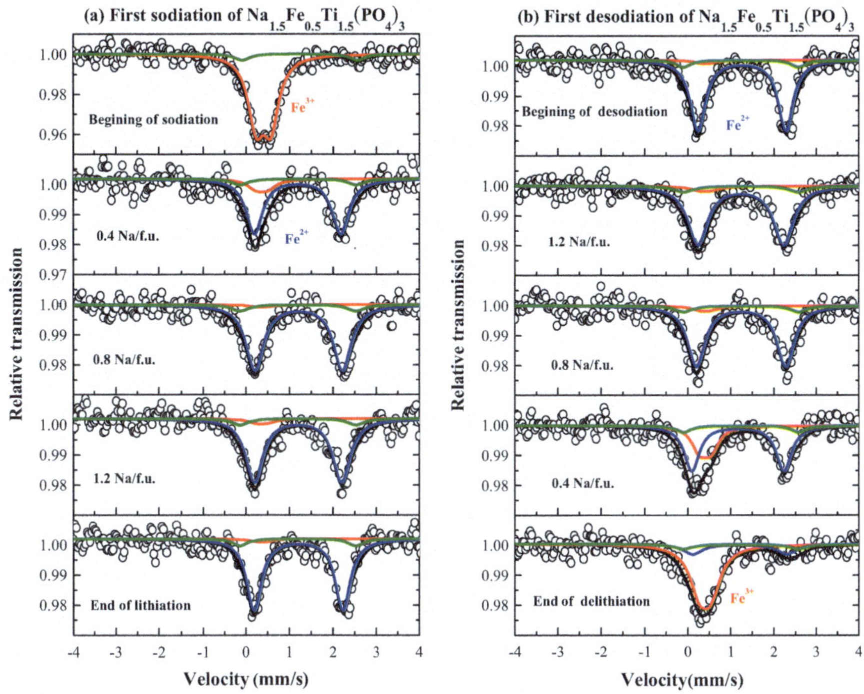

Fig. 7.9 Operando ^{57}Fe Mössbauer spectra obtained for the first sodiation and desodiation of $Na_{1.5}Fe_{0.5}Ti_{1.5}(PO_4)_3$/C in Cell 2. Galvanostatic regime (0.1 Na per fu and per hour), stopped every 0.4 Na for measurement during 12 h (open circuit). Reprinted with permission from Ref. [71]. Copyright 2015 American Chemical Society

Other applications of Mössbauer spectroscopy to solid-solution reactions in batteries can be found in Ref. [74–79].

7.4.2 Two-Phase Reactions

Lithium transition metal phosphates with olivine structure have been regarded as positive electrode materials of Li-ion batteries for a long time and $LiFePO_4$ based batteries are now widely commercialized. Although this material has lower specific capacity and operating voltage than commonly used layered metal oxides, it shows better specific power, safety and cycle life. In addition, $LiFePO_4$ is less expensive than Co-based materials and environmentally friendly.

The electrochemical mechanism can be easily studied by ^{57}Fe Mössbauer spectroscopy. The voltage profile of a $LiFePO_4$ based electrode in a Li half-cell obtained in galvanostatic regime shows two similar plateaus at about 3.4 V for delithiation and lithiation, suggesting the existence of a reversible two-phase reaction.

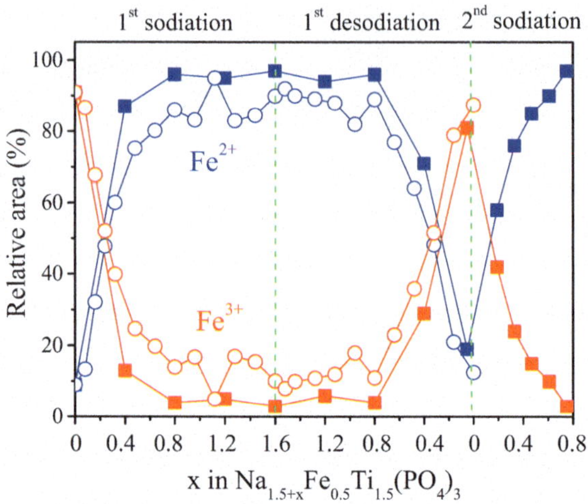

Fig. 7.10 Variations of the relative contributions of Fe^{2+} and Fe^{3+} to the ^{57}Fe Mössbauer spectra obtained for the 1st sodiation (x = 0–1.6), the 1st desodiation (x = 1.6–0) and the beginning of the 2nd sodiation (x = 0–0.8) of $Na_{1.5+x}Fe_{0.5}Ti_{1.5}(PO_4)_3$ in Cell 2. Galvanostatic regime (open circles) and mixed galvanostatic-open circuit regime (solid squares). Regions R1 and R2 shown in Fig. 7.8 correspond to x < 0.5 and x > 0.5, respectively. Reprinted with permission from Ref. [71]. Copyright 2015 American Chemical Society

The operando ^{57}Fe Mössbauer spectra recorded at room temperature at the beginning and at the end of the first delithiation are formed by a doublet with Mössbauer parameters: $\delta = 1.22$ mm s^{-1}, $\Delta = 2.96$ mm s^{-1} and $\delta = 0.42$ mm s^{-1}, $\Delta = 1.52$ mm s^{-1}, respectively, that are typical of high spin Fe^{2+} and Fe^{3+} in $LiFePO_4$ and $FePO_4$, respectively (Fig. 7.11). $LiFePO_4$ and $FePO_4$ are antiferromagnetic below 52 K and 125 K [80], respectively, and the observed doublets are consistent with the paramagnetic state for both compounds at room temperature.

The origin of the Mössbauer parameters was analyzed by DFT-LAPW calculations [46]. The calculated electron density at the nucleus increases from $LiFePO_4$ to $FePO_4$, which is consistent with the decrease of the number of Fe 3d electrons from Fe^{2+} to Fe^{3+}, as expected from the shielding effect described in Sect. 7.3.2. The observed decrease of the isomer shift from $LiFePO_4$ to $FePO_4$ is due to the negative sign of $\Delta\langle r_n^2\rangle$ for ^{57}Fe.

The EFG strongly depends on charge anisotropy, and therefore, on the occupation of the Fe $3d_\uparrow$ and Fe $3d_\downarrow$ states by the valence electrons. For instance, the calculated main EFG component of Fe in $LiFePO_4$ is lower for low spin Fe^{2+} (formally $3d_\uparrow{}^3 3d_\downarrow{}^3$): $|V_{ZZ}| \approx 3\ 10^{21}$ V m^{-2} than for high spin Fe^{2+} (formally $3d_\uparrow{}^5 3d_\downarrow{}^1$): $V_{ZZ} \approx 14\ 10^{21}$ V m^{-2}. As shown by Eq. (7.10), the value of the quadrupole splitting mainly depends on V_{ZZ}. For $LiFePO_4$, all the Fe $3d_\uparrow$ states are occupied and the anisotropy is due to the Fe $3d_\downarrow$ charge distribution. For $FePO_4$, only the Fe $3d_\uparrow$ states are occupied, leading to a small Fe 3d electron anisotropy and the quadrupole splitting mainly arises from the asymmetry of the FeO_6 octahedra distorted by the

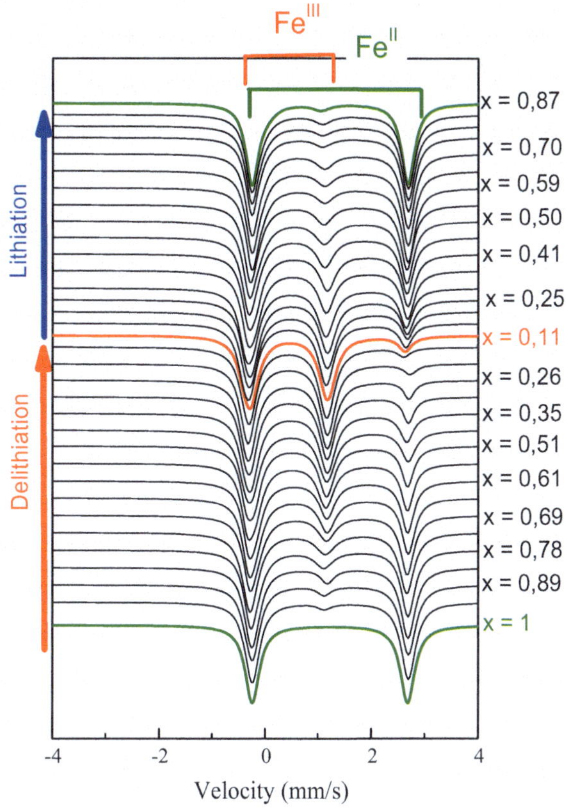

Fig. 7.11 Operando ^{57}Fe Mössbauer spectra (fitted curves) obtained for the first delithiation-lithiation cycle of Li_xFePO_4 in Cell 2

P-O bonds. This structural distortion in $FePO_4$ has a smaller effect on the value of V_{ZZ} than the Fe $3d_{\downarrow}$ valence electron anisotropy in $LiFePO_4$, which explains that the quadrupole splitting of $FePO_4$ is lower than that of $LiFePO_4$, as often encountered for high spin Fe^{3+} and Fe^{2+} in iron phosphates and oxides.

All the Mössbauer spectra recorded during the delithiation and lithiation of $LiFePO_4$ were succesfully fitted to a linear combination of the doublets of $LiFePO_4$ and $FePO_4$ [81]. Since the recoil-free fractions are similar for these two compounds [82], the relative amounts of $LiFePO_4$ and $FePO_4$ are given by the relative contributions of the two subspectra. The observed variations are linear (Fig. 7.12), in line with the reversible two-phase reaction

$$FePO_4 + Li^+ + e^- = LiFePO_4 \tag{7.21}$$

This shows that Mössbauer spectroscopy can be used as a quantitative tool to evaluate the relative amounts of reactants and products of a two-phase electrochemical reaction.

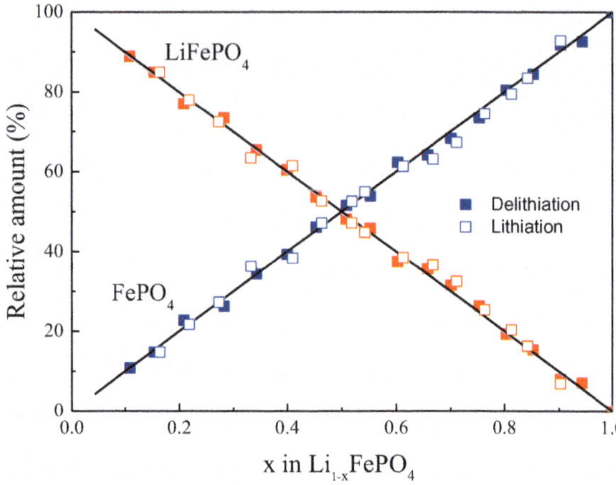

Fig. 7.12 Variations of the relative amounts of $LiFePO_4$ and $FePO_4$ obtained from operando ^{57}Fe Mössbauer spectra for the first delithiation and lithiation of $LiFePO_4$

The substitution of Mn or Co for Fe was proposed to increase the electrode potential and, consequently, the energy density of the battery. The mechanism is more complex than that of $LiFePO_4$ since it involves not only the Fe^{3+}/Fe^{2+} redox couple but also Mn^{3+}/Mn^{2+} or Co^{3+}/Co^{2+}. For example, the voltage curves of $LiFe_{0.75}Mn_{0.25}PO_4$ and $LiFe_{0.25}Mn_{0.75}PO_4$ are both formed by two successive plateaus at about 3.4 and 4 V. The analysis of the ^{57}Fe Mössbauer spectra shows that these two plateaus can be attributed to two-phase reactions corresponding to the redox couples Fe^{3+}/Fe^{2+} and Mn^{3+}/Mn^{2+}, respectively. The observed transition region between the two plateaus was attributed to a solid-solution reaction [83, 84]. Other studies of two-phase reactions in batteries by Mössbauer spectroscopy can be found in Ref. [85–88].

7.5 Alloying Reactions

7.5.1 Negative Electrode Materials for Li-Ion Batteries

Negative electrode materials containing elements that form alloys with lithium have been proposed to increase the electrode capacity and avoid the formation of lithium dendrites. Many chemical elements of the groups 13–15 of the periodic table can reversibly react with lithium at room temperature in liquid organic electrolyte based cells [89]. The two most interesting and studied elements are Si and Sn that can be used as high capacity and low voltage electrode materials. They are rather abundant and environmentally benign. The lithiation-delithiation reactions of Si and βSn lead to the reversible formation of Li_xSi and Li_ySn alloys or compounds, respectively, with the usually accepted maximum values $x = 3.75$ and $y = 4.4$. By considering these values, the volumetric capacities evaluated at the state of full lithiation are of about 2200 mA h cm^{-3} (Si) and 2100 mA h cm^{-3} (βSn) and the specific capacities are 3580 mA h g^{-1} (Si) and 990 mA h g^{-1} (βSn) [89]. These values are significantly higher than the theoretical values of ~800 mA h cm^{-3} and 372 mA h g^{-1} for graphite that accommodates only 1 Li per 6 C.

However, there is a crucial issue with alloying reactions. In contrast to Li intercalation between graphite layers that leads to about 10% volume changes, alloying reactions suffer from strong volume variations of about 300% for Si/$Li_{3.75}Si$ and Sn/$Li_{4.4}Sn$ transformations. This causes mechanical (electrode film cracks), electrical (loss of electrical contacts) and chemical (SEI instabilities) degradations, reducing significantly the cycling performance of electrodes. Different approaches have been proposed to overcome this problem by improving the electrode formulation (binder, conductive additives), the electrolyte, or the electrochemically active materials. In the latter case, this includes particle size reduction (nanoparticles), particle coating, dispersion of the active particles within the pristine material (composite), and in situ dispersion of the active species during the first electrochemical cycles. The electrochemical reactions taking place in these systems are rather complex and some examples of the application of Mössbauer spectroscopy to elucidate such mechanisms are given in Sect. 7.6.

The present section deals with Li-Sn and Na-Sn alloying reactions. Although Si is not a Mössbauer element, previous investigations of Si_xSn amorphous phases successfully provided information on Li-Si alloying reactions as shown in Sect. 7.5.4.

7.5.2 *β*Sn as Negative Electrode Material for Li-Ion Batteries

The experimental voltage curves of βSn in a Li half-cell obtained in galvanostatic regime at low current density show different plateaus that can be attributed to two-phase reactions (Fig. 7.13) [90]. These plateaus are observed for both lithiation and delithiation processes, showing the reversibility of the mechanism.

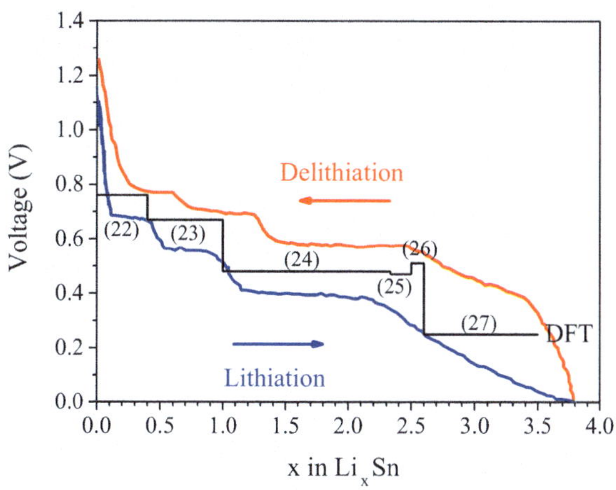

Fig. 7.13 Experimental voltage curves for the first cycle of βSn in a Li half-cell [90] and DFT-GGA voltage profile for the reactions (7.22)–(7.27)

The theoretical voltage profile of βSn was determined by DFT calculations [24, 90] by considering the different crystalline phases of the Li-Sn phase diagram [91] and the following reactions

$$5\ \beta\mathrm{Sn} + 2\ \mathrm{Li} \rightarrow \mathrm{Li_2Sn_5} \tag{7.22}$$

$$\mathrm{Li_2Sn_5} + 3\ \mathrm{Li} \rightarrow 5\ \mathrm{LiSn} \tag{7.23}$$

$$3\ \mathrm{LiSn} + 4\ \mathrm{Li} \rightarrow \mathrm{Li_7Sn_3} \tag{7.24}$$

$$2\ \mathrm{Li_7Sn_3} + \mathrm{Li} \rightarrow 3\ \mathrm{Li_5Sn_2} \tag{7.25}$$

$$5\ \mathrm{Li_5Sn_2} + \mathrm{Li} \rightarrow 2\ \mathrm{Li_{13}Sn_5} \tag{7.26}$$

$$2\ \mathrm{Li_{13}Sn_5} + 9\ \mathrm{Li} \rightarrow 5\ \mathrm{Li_7Sn_2} \tag{7.27}$$

The comparison between the experimental and theoretical voltage profiles suggests that the two first plateaus at average experimental voltages of 0.75 V and 0.65 V can be explained by Eqs. (7.22) and (7.23) corresponding to the successive formations of Li_2Sn_5 and LiSn, respectively. These results were confirmed by in situ XRD. The following plateau at about 0.5 V can be attributed to Eqs. (7.24)–(7.26) and the possible formation of Li_7Sn_3, Li_5Sn_2 and/or $Li_{13}Sn_5$. These three phases cannot be distinguished in the experimental voltage profile because they have closed compositions and formation energies. The theoretical voltage plateau arising from the formation of Li_7Sn_2, as given by Eq. (7.27), is not observed experimentally since the voltage curve shows a continuous decrease in the range 2.2–3.8 Li per Sn. Based on the simulation of the XRD patterns, this decrease was interpreted by the formation of a metastable phase with a BCC disordered structure that shows the same short-range order as $Li_{22}Sn_5$ [90]. These results show that the Sn-rich Li_xSn crystalline equilibrium phases are reversibly formed during the lithiation of βSn while the Li-rich Li_xSn phases could be metastable. It should be noted that the electrochemical mechanisms in Li-ion batteries are also affected by other processes such as low atomic diffusion within the electrode material, side reactions with the electrolyte, kinetic effects, etc. In addition, the Li_xSn phases resulting from the lithiation of Sn based electrodes are poorly crystallized and of small size, which makes the analysis by XRD difficult. Thus, the ^{119}Sn Mössbauer spectroscopy, which is sensitive to the Sn local environment, is a great alternative tool for the characterization of the Li_xSn phases.

7.5.3 Li_xSn Reference Materials

To identify the lithiated products resulting from the Li-Sn alloying reactions by ^{119}Sn Mössbauer spectroscopy, it is convenient to determine the Mössbauer parameters of the equilibrium crystalline phases considered as references. Different experimental Li-Sn phase diagrams were reported and the currently accepted one shows the existence of seven crystalline phases: Li_2Sn_5, LiSn, Li_7Sn_3, Li_5Sn_2, $Li_{13}Sn_5$, Li_7Sn_2 and $Li_{22}Sn_5$, [91], although the latter phase should be better described by $Li_{17}Sn_4$ [92]. Li_8Sn_3 was also observed and should be included in the Li-Sn phase diagram but is not considered here [93]. Finally, other compositions or structures were predicted by DFT calculations but have not been experimentally observed yet [94–96]. In line with previously reported works [97, 98], only the seven crystalline phases given above are considered here as references. The crystal structures of all the phases were experimentally determined by XRD and show the existence of one (Li_5Sn_2), two (Li_2Sn_5, LiSn, Li_7Sn_2), three (Li_7Sn_3, $Li_{13}Sn_5$) and four ($Li_{22}Sn_5$) Sn crystallographic sites. The Mössbauer spectra were reported by two groups using different synthesis methods: high temperature solid-state reactions [97] and mechanosynthesis followed by annealing [98]. The contributions of the different Sn crystallographic sites cannot be easily distinguished in the spectra that are unresolved and contain the contributions of tin based impurities (Fig. 7.14).

The comparison between the values of the Mössbauer parameters obtained by the two groups shows some small variations that can be attributed to differences in the purity and crystallinity of the materials (Table 7.1).

The relative contributions of the subspectra, when not fixed, are consistent with the crystallographic site multiplicities. The values of the quadrupole splitting reflect the different Sn local environments due to the existence of different nearest neighbors (Li, Sn), polyhedral geometries and bond lengths. Finally, the values of the isomer shift averaged over the different Sn crystallographic sites, δ_{av}, are of about 2.4 mm s^{-1} for the two Sn-rich Li_xSn phases and decrease from 2.1 to 1.8 mm s^{-1} with increasing relative amount of Li for the Li-rich Li_xSn phases (Fig. 7.15). These values are all in the range of the Sn(0) oxidation state and can be correlated to the number of Sn-Sn bonds. The two ranges of isomer shifts reflect the existence of Sn based sub-lattices for the Sn-rich Li_xSn phases and Sn single atoms or clusters of two or three Sn atoms bonded to Li atoms for the Li-rich Li_xSn phases.

It is also interesting to plot the values of the average isomer shift of the Li_xSn references as a function of the number of Li per Sn (Fig. 7.15). The observed linear correlation suggests that $\delta_{av}(x)$ strongly depends on the average composition of Li_xSn but not on crystal structure. This means that such a correlation can be used to evaluate x from the measurement of δ_{av} for Li_xSn amorphous phases or small particles as often encountered during the lithiation-delithiation processes of Sn based electrodes.

The linear variations of $\delta_{av}(x)$ can be related to changes in the numbers of Sn 5s (N_{5s}) and Sn 5p (N_{5p}) electrons. The values of $\rho(0)$, N_{5s} and N_{5p} were evaluated with the DFT-LAPW method. Following the procedure described in Sect. 7.3.2, the theoretical values of δ were obtained from $\rho(0)$ and the calibration constant

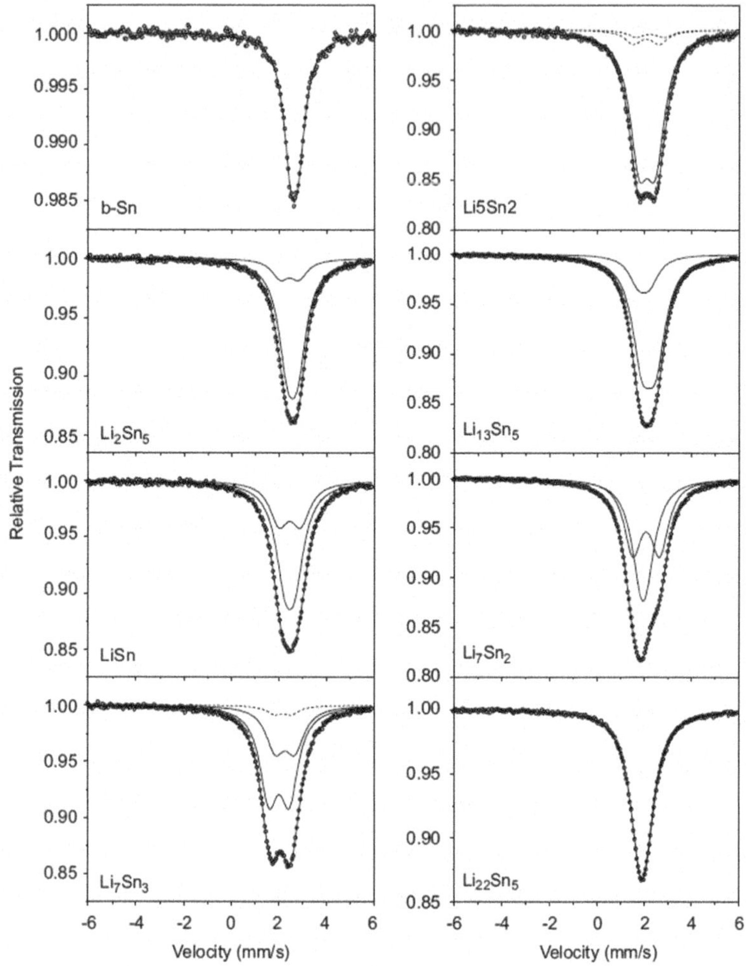

Fig. 7.14 ^{119}Sn Mössbauer spectra of Li_xSn crystalline references measured at room temperature. Reprinted from Ref. [98]. Copyright 2007, with permission from Elsevier

α (Eq. 7.4). The experimental and theoretical averaged values of the isomer shift, δ_{av}, are in good agreement and the theoretical values of $\delta_{av}(x)$ linearly increase with N_{5s} (Fig. 7.16a) and decrease with N_{5p} (Fig. 7.16b) in line with Eq. (7.5).

The variations of N_{5p} (~0.25) are larger than those of N_{5s} (~0.1) and they both linearly contribute to the variations of δ_{av}. The increase of N_{5p} with x reflects the increase of the electron transfer from Li to Sn due to the increase number of Li atoms around each Sn.

Empirical correlation rules were previously proposed for Mössbauer parameters, such as the Δ-δ correlation diagram used to predict the electrochemical activity of tin based electrode materials for Li-ion batteries [99]. In the present case, an empirical

Table 7.1 Values of the ^{119}Sn Mössbauer parameters of Li_xSn crystalline references obtained at room temperature by Dunlap et al. [97] and by Robert et al. [98]. The values of the isomer shift, δ, relative to $BaSnO_3$ and the quadrupole splitting, Δ, are given for the different Sn crystallographic sites

	βSn	Li_2Sn_5	LiSn	Li_7Sn_3	Li_5Sn_2	$Li_{13}Sn_5$	Li_7Sn_2	$Li_{22}Sn_5$
δ/Δ(mm/s) (crystallographic sites) [97]	2.56/0 (4a)	2.53/0.49 (8i) 2.23/0.67 (2d)	2.42/0.41 (2 m) 2.41/1.09 (1a)	2.17/0.46 (2e) 2.09/1.19 (2e,2e)	2.02/0.67 (6c)	1.93/0.33 (1a) 2.04/0.39 (2d) 2.05/0.97 (2d)	1.90/0.40 (4i) 1.95/1.33 (4 h)	1.84/0.27 (16e, 16e, 24f, 24g)
δ/Δ(mm/s) (crystallographic sites) [98]	2.56/0.29 (4a)	2.49/042 (8i) 2.36/0.78 (2d)	2.38/0.43 (2 m) 2.38/0.91 (1a)	2.19/0.82 (2e) 1.94/0.86 (2e,2e)	2.01/0.69 (6c)	1.86/0.48 (1a) 2.07/0.58 (2d,2d)	1.84/0.28 (4i) 1.96/1.13 (4 h)	1.83/0.31 (16e, 16e, 24f, 24g)

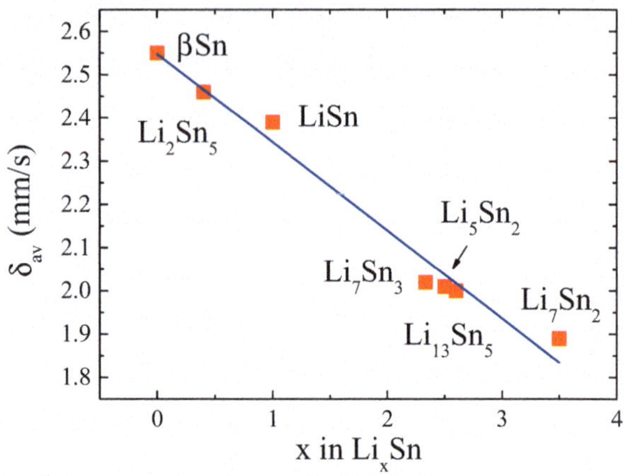

Fig. 7.15 Average experimental values of the isomer shift of Li_xSn crystalline references, δ_{av}, as a function of the number of Li per Sn, x, and regression line (blue)

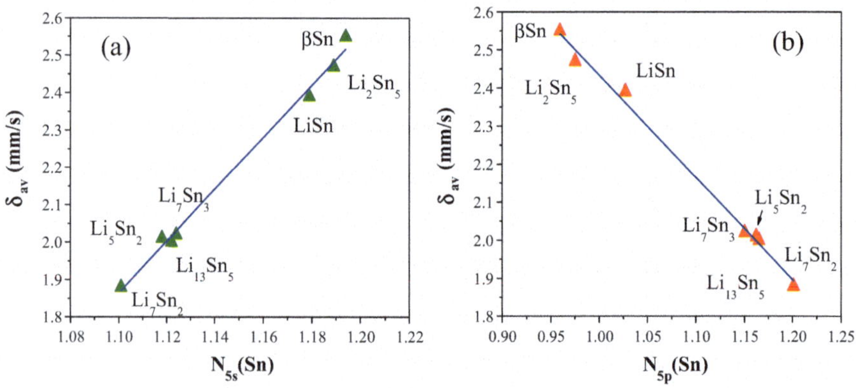

Fig. 7.16 Correlations between the experimental values of the isomer shift and the DFT numbers of Sn 5s electrons **a** and Sn 5p electrons **b** for Li_xSn crystalline references, and linear regression lines (blue)

relation between the average isomer shift of Li_xSn and x can be derived from the values obtained for the reference crystalline phases by considering the regression line shown in Fig. 7.15:

$$\delta_{av}(x) = 2.55 - 0.20x \tag{7.28}$$

From this equation, it is possible to determine x from δ_{av}, which gives the average composition of the Li_xSn species in the electrode materials when they cannot be identified as shown for the delithiation of $FeSn_2$ based electrodes in Sect. 7.6.1.

The ^{119}Sn Mössbauer spectroscopy was used for ex situ measurements at different stages of the first lithiation of βSn electrode in a Li half-cell [100]. The voltage profile is similar to that of Fig. 7.13, except at the very beginning of lithiation due to the existence of tin oxides as impurities. The spectra recorded for the insertion of 0.5, 1 and 3.4 Li per Sn were fitted to two doublets with Mössbauer parameters similar to those of Li_2Sn_5, LiSn and Li_7Sn_2, respectively, although some small differences could be tentatively attributed to variations in the composition of the electrochemically formed Li_xSn phases.

7.5.4 Si as Negative Electrode Material for Li-Ion Batteries

The electrochemical alloying reactions of Li with Si were studied by in situ ^{119}Sn Mössbauer spectroscopy for amorphous a-Si_ySn_{1-y} (y = 0.87, 0.93) powdered samples obtained by magnetron sputtering [101]. The voltage profiles of the two a-Si_ySn_{1-y} phases differ from that of βSn but are similar to that of amorphous Si (a-Si), showing two sloping plateaus with a transition at ~2.3 Li (Fig. 7.17a). The same electrochemical behaviors observed for a-Si and a-Si_ySn_{1-y} suggest that similar reactions take place in the two electrode materials, justifying the use of Sn as Mössbauer probe in a-Si [102, 103].

The operando Mössbauer spectra of $Li_xSi_{0.87}Sn_{0.13}$ obtained during the first lithiation-delithiation cycle consist of a single line or a doublet around 2 mm s^{-1}

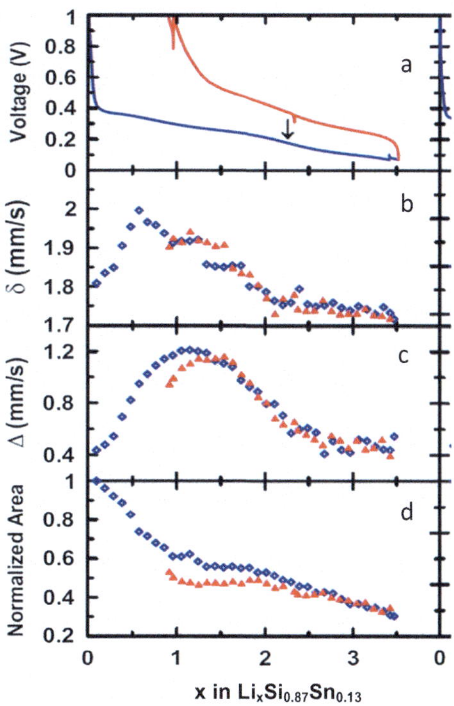

Fig. 7.17 **a** Voltage curves of a-$Si_{0.87}Sn_{0.13}$ in a Li half-cell. **b** Variations of the isomer shift δ, **c** quadrupole splitting Δ and **d** total area of the ^{119}Sn Mössbauer spectra recorded during the first lithiation (blue) and delithiation (red). Image courtesy Mössbauer Effect Data Centre, Dalian Institute of Chemical Physics, Chinese Academy of Sciences [64]

[101]. All the spectra were successfully fitted to a doublet and the Mössbauer parameters strongly change during lithiation (Fig. 7.17). Similar variations are obtained for delithiation, showing the reversibility of the mechanism.

From $x = 0$ to $x = 1$, the isomer shift increases from 1.8 to 2 mm s^{-1} and the quadrupole splitting increases from 0.4 to 1.2 mm s^{-1}. The value $\delta = 1.8$ mm s^{-1} is close to that of αSn and characteristic of Sn(0) in tetrahedral environment, as expected for sp^3 hybridization in a-Si. This confirms the substitution of Sn for Si in a-Si and the absence of βSn. The values of isomer shift obtained for $x < 1$ are lower than 2 mm s^{-1}, which excludes the formation of Sn-rich Li_xSn alloys. The increase of δ from 1.8 to 2 mm s^{-1} is mainly due to the increase of N_{5s} (see Eq. 7.5) arising from the electronic transfer from Li to Sn, which indicates the substitution of Li for Si as first nearest neighbors of Sn. The progressive increase in the number of Li atoms around Sn makes the Sn tetrahedral environments formed by Li and Si atoms more asymmetrical, in agreement with the observed increase of the quadrupole splitting.

From $x = 1$ to $x = 2.3$, the isomer shift and quadrupole splitting progressively decrease until $\delta = 1.8$ mm s^{-1} and $\Delta = 0.6$ mm s^{-1}, respectively. According to Eq. (7.28), the value $\delta = 1.8$ mm s^{-1} is characteristic of Li-rich Li_xSn alloys and shows that Sn atoms are mainly bonded to Li atoms. This is consistent with the small values of the quadrupole splitting that reflect rather symmetrical Sn environments.

From $x = 2.3$ to $x = 3.5$, both isomer shift and quadrupole splitting are almost constant, $\delta \approx 1.8$ mm s^{-1} and $\Delta \approx 0.6$ mm s^{-1}, showing there is no change in the local environment of the Sn atoms that are only surrounded by Li. The additional inserted Li atoms are expected to be located further from Sn atoms. This differs from the lithiation of βSn whose average isomer shift decreases continuously in the range $x = 0$–3.5 (Fig. 7.15). These different behaviors of βSn and a-$Si_{0.87}Sn_{0.13}$ can be related to the strong differences between the Sn local environments in the lithiated species. In lithiated $Li_xSi_{0.87}Sn_{0.13}$, the Sn atoms are tetrahedrally bonded to four first-nearest neighbors, whereas the Sn atoms in the Li_xSn alloys resulting from the lithiation of βSn are surrounded by more Li and Sn atoms. The variations of the Mössbauer parameters during the delithiation from $x = 3.5$ to $x = 1$ confirm the reversibility of the mechanism.

To conclude, the two successive sloping plateaus observed in the voltage curve of the lithiation of a-Si_ySn_{1-y} reflect the insertion of Li, first, close to Sn/Si atoms and then, close to Li atoms. This mechanism is reversible for delithiation.

7.5.5 *βSn as Negative Electrode Material for Na-Ion Batteries*

In contrast to Li-ion batteries, graphite cannot be used as negative electrode material for Na-ion batteries [104] and should be replaced by amorphous carbon. Tin could be an interesting alternative with a theoretical capacity of 850 mA h g^{-1} corresponding to the formation of $Na_{15}Sn_4$. However, the sodiation and desodiation of tin based

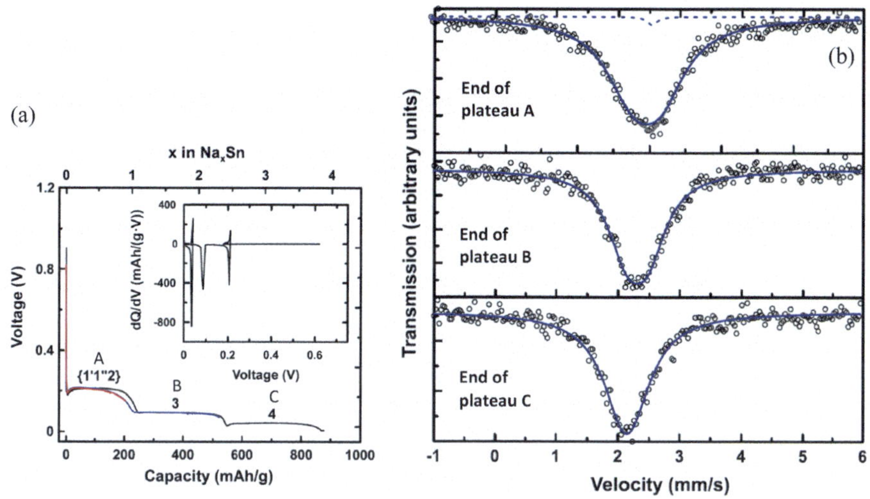

Fig. 7.18 **a** Voltage curves for the first sodiation of βSn in Na half-cells, showing three plateaus A, B and C at about 0.2 V, 0.1 V and 0.05 V. **b** Ex situ ^{119}Sn Mössbauer spectra recorded at the end of the plateaus A, B and C. Reprinted from Ref. [106]. Copyright 2014, with permission from Elsevier

electrodes also suffer from large volume variations as observed during the lithiation-delithiation cycles.

The voltage curve obtained for the first galvanostatic sodiation of βSn foil in a Na half-cell at low current rate shows three plateaus at about 0.2 V, 0.1 V and 0.05 V, respectively (Fig. 7.18a). These values are lower than those of the lithiation of βSn in a Li half-cell, in line with the difference between the standard electrode potentials of Li^+/Li (−3.04 V) and Na^+/Na (−2.71 V). The observed number of plateaus is lower than that expected from the formation of the eight equilibrium crystalline phases reported in the Na-Sn phase diagram [105].

The electrochemical reactions were characterized by ex situ ^{119}Sn Mössbauer spectroscopy, following the same approach as described for Li-ion batteries in Sect. 7.5.3 [106]. The room temperature spectra reported for crystalline NaSn, Na_9Sn_4 and $Na_{15}Sn_4$, and considered as reference materials, are all formed by a single peak. According to their number of Sn crystalline sites, the spectra of NaSn, Na_9Sn_4 and $Na_{15}Sn_4$ were successfully fitted to one doublet, two doublets and a single line, respectively. As observed for Li_xSn, the variations of the average isomer shift of Na_xSn, $\delta_{av}(x)$, can be used to distinguish Sn-rich from Na-rich Na_xSn phases. The isomer shift of NaSn (2.44 mm s^{-1}) is close to that of βSn, which corresponds to a Sn-rich Na_xSn phase. The isomer shifts of Na_9Sn_4 (2.3 mm s^{-1}) and $Na_{15}Sn_4$ (2.17 mm s^{-1}) are significantly lower than that of βSn and correspond to Na-rich Na_xSn phases. This behavior can be correlated to the number of Sn-Sn bonds as for Li_xSn. The isomer shifts of the Na_xSn phases are typical of Sn(0) oxidation state.

For a given value of x, the isomer shift of Na_xSn is higher than that of Li_xSn, which can be explained by the lower electronegativity of Na compared to Li.

The ex situ Mössbauer spectra obtained at the end of the three different voltage plateaus A, B and C are similar to those of NaSn, Na_9Sn_4 and $Na_{15}Sn_4$ reference materials, respectively (Fig. 7.18b). This indicates that the electrochemical sodiation of βSn at a very low Na rate leads to the formation of crystalline phases following the alloying reactions

$$\beta Sn + Na \rightarrow NaSn \tag{7.29}$$

$$2\,NaSn + 3\,Na \rightarrow Na_5Sn_2 \tag{7.30}$$

$$2\,Na_5Sn_2 + 5\,Na \rightarrow Na_{15}Sn_4 \tag{7.31}$$

Thus, the lithiation and sodiation of βSn in half-cells produce Li_xSn and Na_xSn crystalline phases, respectively, for $x < 2.5$. For $x > 2.5$, the formation of the $Na_{15}Sn_4$ equilibrium phase is observed instead of disordered Li-rich Li_xSn phases for lithiation. This could be due to differences in the diffusion of Li and Na atoms. The diffusion mechanisms could also explain the observed differences between the voltage profiles obtained for Sn foil and sputtered films in Na half-cells [107]. In the latter case, the mechanism is more complex, with the formation of Na_xSn amorphous and metastable phases. However, the ^{119}Sn Mössbauer spectrum of the fully sodiated electrode material is a single peak with the same Mössbauer isomer shift, $\delta = 2.14$ mm s^{-1}, as the $Na_{15}Sn_4$ reference crystalline phase. Further investigations on the phase transformations observed during the sodiation of βSn with other characterization tools led to somewhat different mechanisms but the fully sodiated phase is always $Na_{15}Sn_4$ [108–111]. This shows that, irrespective of the structure of the pristine material and the composition of the reaction intermediates, the same phase $Na_{15}Sn_4$ is obtained at the end of the sodiation of βSn, confirming the high capacity of this negative electrode material for Na-ion batteries. The observed different mechanisms suggest that, depending on the microstructure of the pristine material and the experimental conditions for electrochemical sodiation, the phase transformations can produce metastable and amorphous intermediate phases that differ from the crystalline references. For the lithiation of βSn, such a mechanism is mainly observed for Li-rich Li_xSn phases. This analysis shows that the electrochemical reactions taking place in βSn based negative electrodes of Li-ion or Na-ion batteries are rather complex due to both thermodynamic and kinetic effects. There are still some unclear aspects in these mechanisms that should be investigated by combining ^{119}Sn Mössbauer spectroscopy and other characterization tools.

7.6 Conversion Reactions

7.6.1 $FeSn_2$ as Negative Electrode Material for Li-Ion Batteries

Intermetallics composed of an electrochemically active element such as Si, Sn or Sb and an electrochemically inactive metallic element were investigated as negative electrode materials for Li-ion batteries in order to reduce the effects of the volume variations coming with alloying reactions [112–117]. Most of the tin based intermetallics, MSn_x, combine Sn with a transition metal element M that does not react with Li. The first lithiation is expected to extrude M from MSn_x to form metallic nanoparticles that maintain the dispersion of Li_xSn particles and improve the electronic conductivity. It is important to recall that the MSn_x active particles are mixed with carbon additives and a binder to form a porous film coated onto the current collector. The conductive additives play a key role in the performance of the electrodes during cycling. Carbon can also be introduced during the synthesis process to form MSn_x/C composites that were widely studied in the past in relation with the commercialization by Sony of the Nexelion Li-ion batteries [118–124]. $FeSn_2$ is regarded here as a typical example of the application of Mössbauer spectroscopy to MSn_x intermetallic based electrodes while other transition metals for M are briefly discussed in Sect. 7.6.2. $FeSn_2$ contains the highest amount of Sn among the crystalline phases reported in the Fe-Sn binary system, leading to the highest specific capacity as Fe-Sn based electrode material. Finally, $FeSn_2$ contains the two Mössbauer isotopes ^{57}Fe and ^{119}Sn to probe the electrochemical reactions at the atomic scale.

$FeSn_2$ has a tetragonal structure (*I4/mcm*) and each chemical element occupies one crystallographic site. Each Fe atom is at the center of a Sn square-based antiprism while Sn is bonded to four Fe atoms forming a $SnFe_4$ square-based pyramid. $FeSn_2$ is antiferromagnetic below 378 K [125]. Although the magnetic properties of $FeSn_2$ are not of real interest for electrochemical applications, they strongly affect the hyperfine structures, leading to rather complex and distinctive ^{57}Fe and ^{119}Sn Mössbauer spectra.

Crystalline microparticles of $FeSn_2$ were synthesized by solid-state reaction [126]. The ^{57}Fe Mössbauer spectrum of $FeSn_2$ at room temperature is formed by a sextet (Fig. 7.19a) while broad structures in the range 0–5 mm s^{-1} are observed for the ^{119}Sn Mössbauer spectrum (Fig. 7.19b), reflecting combined quadrupole effect and transferred hyperfine magnetic field [127]. The two sets of Mössbauer parameters: $\delta =$ 0.5 mm s^{-1}, $\Delta = 0$ mm s^{-1} and $B = 11$ T for ^{57}Fe and $\delta = 2.18$ mm s^{-1}, $\Delta =$ 0.83 mm s^{-1}, $B = 2.4$ T for ^{119}Sn are typical of Fe(0) and Sn(0) oxidation states, respectively, with a Sn asymmetrical environment.

The voltage profile obtained in galvanostatic regime at low current density consists of a low voltage plateau at 0.2 V for the first lithiation and voltage hystereses at the average value of 0.5 V for the subsequent delithiation-lithiation cycles [126]. This indicates that the first lithiation mechanism differs from the following reversible

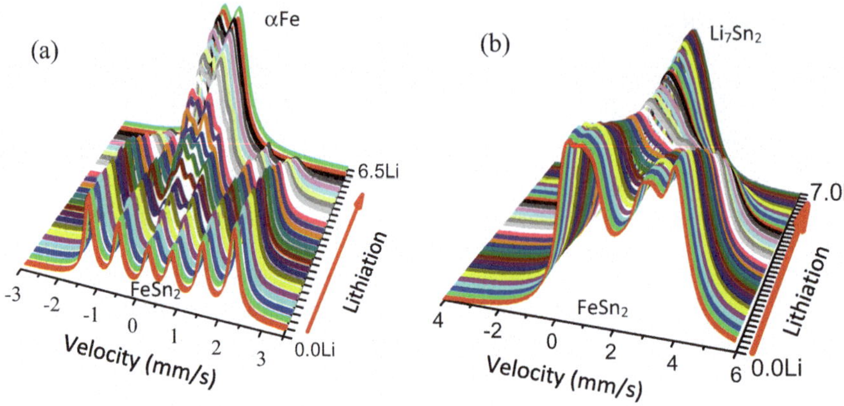

Fig. 7.19 Operando ^{57}Fe **a** and ^{119}Sn **b** Mössbauer spectra (fitted curves) obtained during the first lithiation of $FeSn_2$ in Cell 1

cycles and reflects a two-phase reaction. The operando XRD patterns show the progressive decrease of the intensity of the $FeSn_2$ Bragg peaks after the insertion of about 2 Li per $FeSn_2$ and the growth of broad peaks that can be assigned to Li-rich Li_xSn phases although the value of x cannot be determined. There are no Bragg peaks corresponding to Fe based phases.

The operando ^{119}Sn Mössbauer spectra collected during the first lithiation of $FeSn_2$ in Cell 1 show strong changes from a broadened magnetic spectrum to an asymmetrical peak at the end of the process (Fig. 7.19b). The latter spectrum was fitted to two doublets. The obtained isomer shifts of 1.88 and 1.98 mm s^{-1} are close to those of the Li_7Sn_2 crystalline reference but one quadrupole splitting, $\Delta =$ 0.72 mm s^{-1}, is smaller than the reference ($\Delta = 1.13$ mm s^{-1}). The DFT-LAPW evaluation of the Mössbauer parameters of the Li_7Sn_2 crystalline reference shows that the large quadrupole splitting originates from the Sn-Sn bond along the EFG principal axis Z for one of the two Sn crystallographic sites. For the other site, the quadrupole splitting is small due to the existence of only Li atoms as nearest neighbors. The observed experimental value, $\Delta = 0.72$ mm s^{-1}, for the fully lithiated electrode suggests that the number of Sn-Sn bonds in the electrochemically formed Li_7Sn_2 particles is lower than the reference. This could be due to the small size or the poor crystallinity of the particles. The contribution of the $FeSn_2$ sextet to the ^{57}Fe Mössbauer spectra decreases during the first lithiation while a doublet grows until the end of the process (Fig. 7.19a). The Mössbauer parameters of this doublet can be attributed to α-Fe nanoparticles in the paramagnetic state.

All the ^{57}Fe and ^{119}Sn Mössbauer spectra obtained during the first lithiation were successfully fitted to the spectra of $FeSn_2$/nano-αFe and $FeSn_2$/nano-Li_7Sn_2 phases, respectively. Since the number of atoms is constant in the in situ cell, the observed changes in the total area of the spectra correspond to the variations of the average recoil-free fraction. The total area of the ^{57}Fe Mössbauer spectra does not vary noticeably during lithiation, indicating that the ^{57}Fe recoil-free fractions of $FeSn_2$

and α-Fe have close values (Fig. 7.20a, inset). Thus, the relative amounts of $FeSn_2$ and α-Fe are directly given by their relative contributions to the ^{57}Fe Mössbauer spectra (Fig. 7.20a). The total area of the ^{119}Sn Mössbauer spectra is constant until 1 Li per $FeSn_2$ and then linearly decreases from 1 to 7 Li per $FeSn_2$ (Fig. 7.20b, inset). This can be related to the decrease of the average ^{119}Sn recoil-free fraction and shows that $f(FeSn_2) > f(Li_7Sn_2)$. By considering the full transformation of $FeSn_2$ into Li_7Sn_2 in the range 1–7 Li, one can determine the ratio $f(FeSn_2)/f(Li_7Sn_2) = 2.3$ from the slope of the regression line. The area of each ^{119}Sn Mössbauer subspectrum is proportional to the number of Sn atoms in the corresponding tin phase and to the ^{119}Sn recoil-free fraction (see Sect. 7.3.6). For $FeSn_2$ and Li_7Sn_2 that coexist during the first lithiation, the ratio between the ^{119}Sn recoil-free fractions of these two phases must be considered for an accurate evaluation of their relative amounts, transforming the non-linear variations of the relative areas of $FeSn_2$ and Li_7Sn_2 subspectra into linear variations for the relative amounts of $FeSn_2$ and Li_7Sn_2 (Fig. 7.20b).

The linear variations of the relative amounts of $FeSn_2$/α-Fe and $FeSn_2/Li_7Sn_2$ quantitatively show that the first lithiation of $FeSn_2$ can be assigned to the conversion reaction

$$FeSn_2 + 7\,Li \rightarrow Li_7Sn_2 + Fe \tag{7.32}$$

Additional information were obtained from magnetic measurements of the electrode material at different stages of lithiation. The saturation magnetization at low temperature of the fully lithiated electrode material is close to that of α-Fe. The ZFC/FC curves show the growth of a peak at 20 K with increasing number of Li that

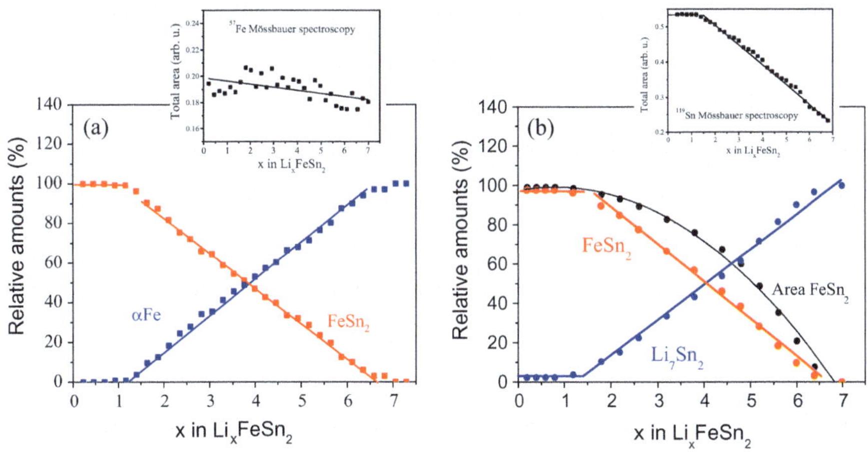

Fig. 7.20 First lithiation of $FeSn_2$ in Cell 1: **a** variations of the relative amounts of $FeSn_2$ (red) and α-Fe (blue) obtained from operando ^{57}Fe Mössbauer spectra and **b** variations of the relative amounts of $FeSn_2$ (red) and Li_7Sn_2 (blue) obtained from operando ^{119}Sn Mössbauer spectra. Total spectrum area vs. x in Li_xFeSn_2 (insets)

can be attributed to the progressive formation of superparamagnetic α-Fe nanoparticles with a constant average diameter of about 3 nm. This result is consistent with the extrusion of Fe atoms from $FeSn_2$ during the conversion reaction and the formation of α-Fe nanoparticles. The small size of the formed α-Fe particles explains the absence of XRD peaks. These results indicate that the first lithiation of $FeSn_2$ based electrodes should be considered as a restructuring step ending with the formation of α-Fe/Li_7Sn_2 nanocomposite, which is the real starting electrode material for the reversible lithiation-delithiation cycles.

Decreasing the particle size or increasing the surface area of electrochemically active materials are expected to improve the performance by enhancing the electrolyte impregnation, Li diffusion and electronic percolation. Nanostructured $FeSn_2$ particles were obtained by ball milling of $FeSn_2$ microparticles as confirmed by XRD showing broad Bragg peaks. The ^{57}Fe and ^{119}Sn Mössbauer spectra are formed by a single peak and a doublet, respectively, indicating the absence of hyperfine magnetic field as expected with the small size and poor crystallinity of the $FeSn_2$ ground particles [126]. However, the values of the isomer shift and quadrupole splitting are close to those of the antiferromagnetic $FeSn_2$ crystalline phase, showing that Fe and Sn local environments are similar in both nanostructured and crystalline $FeSn_2$. During the first lithiation, the voltage curve of nanostructured $FeSn_2$ decreases smoothly in contrast to the plateau observed for $FeSn_2$ microparticles, while ^{57}Fe and ^{119}Sn Mössbauer spectra both reflect the same conversion reaction as given by Eq. (7.32) [68]. Thus, nanostructured and crystalline $FeSn_2$ electrode materials are both transformed into an α-Fe/Li_7Sn_2 nanocomposite during the first lithiation.

The operando ^{119}Sn Mössbauer spectra obtained during the first delithiation of the $FeSn_2$ microparticle based electrode show the progressive transformation from a single peak into a broad doublet and the increase of the average isomer shift δ_{av} (Fig. 7.21a). The evolutions of the spectra and average isomer shift during the

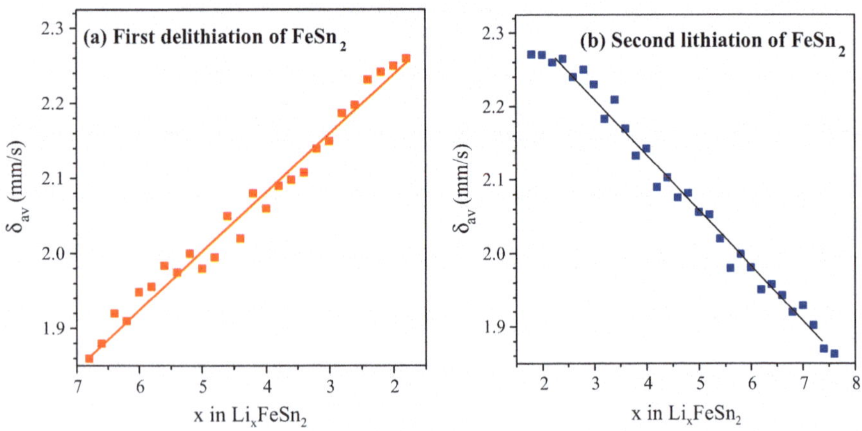

Fig. 7.21 Average experimental values of the ^{119}Sn Mössbauer isomer shift, δ_{av}, obtained during the first delithiation **a** and the second lithiation **b** of $FeSn_2$ in Cell 1

second lithiation are consistent with the reversibility of the mechanism observed during the first delithiation and strongly differs from the first lithiation (Fig. 7.21b). These spectra obtained for the first delithiation-lithiation reversible cycle cannot be unambiguously fitted by considering the Li_xSn references. However, the observed variations of δ_{av} in the range 1.8–2.3 mm s^{-1} can be interpreted from Eq. (7.28) as the variations of the Li_xSn average composition between Li_7Sn_2 and LiSn, which is consistent with the number of Li extracted during the first delithiation and inserted during the second lithiation.

A deeper delithiation can be obtained by decreasing the current density at the end of the process, leading to the back reaction of Fe with Sn extruded from Sn-rich Li_xSn and βSn to give $FeSn_2$ small particles [27]. However, for current densities commonly used in electrochemical tests, such a back reaction was not observed as confirmed by the ^{57}Fe Mössbauer spectra that do not change significantly during this first reversible cycle formed by the first delithiation and second lithiation. In that case, the reversible mechanism is based on alloying-dealloying Li-Sn reactions while α-Fe nanoparticles remain spectators and help to buffer the volume variations resulting from these reactions. The absence of back reaction between Fe and Sn during cycling was confirmed by electron paramagnetic resonance [128]. However, the existence of such reactions clearly depends on the experimental conditions used for the electrochemical tests.

7.6.2 Other Tin Based Intermetallic Compounds

The performance and electrochemical mechanisms of different tin based transition metal intermetallics were previously investigated, including $MnSn_2$ [129, 130], $CoSn_2$ [67, 131, 132], Ni_3Sn_4 [133–135] and Cu_6Sn_5 [136–139]. The voltage profiles obtained for the first three compounds are similar to that of $FeSn_2$. They are formed by a plateau for the first lithiation process and reversible voltage hystereses for the following delithiation-lithiation cycles. The main differences concern the voltage of the first-lithiation plateau and the existence of an additional plateau observed for some of these compounds at the end of delithiation. The ^{119}Sn Mössbauer spectra obtained during the first lithiation of $MnSn_2$, $CoSn_2$, and Ni_3Sn_4 based electrodes with similar galvanostatic regimes show the formation of Li_7Sn_2 according to the conversion reactions

$$MnSn_2 + 7\,Li \rightarrow Li_7Sn_2 + Mn \tag{7.33}$$

$$CoSn_2 + 7\,Li \rightarrow Li_7Sn_2 + Co \tag{7.34}$$

$$Ni_3Sn_4 + 14\,Li \rightarrow 2\,Li_7Sn_2 + 3\,Ni \tag{7.35}$$

The observed differences in the voltage values of the first lithiation plateau of $MnSn_2$, $CoSn_2$, and Ni_3Sn_4 can be mainly related to the enthalpies of the reactions (7.33)-(7.35) although significant kinetic effects can affect these values. In the three cases, the mechanisms of the first lithiation are the same as $FeSn_2$ and consist in the transformation of the pristine material into M/Li_7Sn_2 nanocomposite (M = Mn, Co, Ni). The Mössbauer spectra obtained at the end of the lithiation of $MnSn_2$, $CoSn_2$ and $FeSn_2$ are similar, reflecting the formation of poorly cristallized Li_7Sn_2 nanoparticles. For the fully lithiated Ni_3Sn_4 electrode, the Mössbauer spectrum is similar to that of the Li_7Sn_2 crystalline reference, suggesting the formation of a more stable phase.

Both $CoSn_2$ and $MnSn_2$ have the same *I4/mcm* structure as $FeSn_2$ while Ni_3Sn_4 has a monoclinic *C2/m* structure with two different Sn crystallographic sites in the primitive cell. $MnSn_2$ is antiferromagnetic below 325 K and the ^{119}Sn Mössbauer spectrum at room temperature was fitted to two sextuplets reflecting two Sn magnetic sites due to transferred hyperfine magnetic fields with parameters: $\delta = 2.34$ and 2.32 mm s^{-1}, $\Delta = 1.07$ and 1.10 mm s^{-1}, $B = 4.5$ and 2.7 T [140]. Nanostructured $MnSn_2$ was obtained by ball milling from crystalline $MnSn_2$. The Mössbauer spectrum is formed by a doublet showing the absence of transferred hyperfine magnetic field. The values of the Mössbauer parameters $\delta = 2.3$ mm s^{-1} and $\Delta = 1.1$ mm s^{-1} are close to those of crystalline $MnSn_2$ [129]. At room temperature, the Mössbauer spectrum of $CoSn_2$ can also be fitted to a doublet with $\delta = 2.14$ mm s^{-1}, $\Delta =$ 0.77 mm s^{-1} [141].

These results show that the isomer shift slightly decreases for the series $MnSn_2$-$FeSn_2$-$CoSn_2$. According to Eq. (7.5), this reflects the decrease of the number of Sn 5s electrons and the increase of the number of Sn 5p electrons. In the latter case, this is due to electron transfer from the transition metal, M, to Sn through the Sn 5p–M 3d bonds. The decrease of the quadrupole splitting for this series reflects the decrease of the Sn 5p charge anisotropy. The Mössbauer spectrum of Ni_3Sn_4 is also formed by a doublet but was fitted by considering two Sn sites, in agreement with the crystal structure of this compound. The Mössbauer parameters are $\delta = 2.01$ mm s^{-1}, $\Delta =$ 0.7 mm s^{-1} for one site and $\delta = 2.02$ mm s^{-1}, $\Delta = 1.18$ mm s^{-1} for the other site. The quadrupole splittings reflect the two different asymmetric environments of the Sn crystallographic sites [135]. It should be noted that different values of the average isomer shift of Ni_3Sn_4 can be found in the range ~1.9–2.1 mm s^{-1}, depending on the synthesis conditions. This reflects variations of the composition from Ni_3Sn_4 to $Ni_{3.5}Sn_4$ due to the existence of a solid solution [142]. In that case, Mössbauer spectroscopy can be used to evaluate the composition of $Ni_{3+x}Sn_4$ [143].

The values of the isomer shift decrease for the series $MnSn_2$-$FeSn_2$-$CoSn_2$-Ni_3Sn_4. This is due to changes in chemical bonds and electronic transfer, leading to the increase of the number of M 3d electrons. However, all the values are in the range 2.0–2.3 mm s^{-1}, which is typical of Sn(0) formal oxidation state (see Fig. 7.3). All these results explain why the first lithiation of nanostructured (paramagnetic) $MnSn_2$, nanostructured $FeSn_2$, $CoSn_2$ and Ni_3Sn_4, leads to similar changes in the ^{119}Sn Mössbauer spectra from a well-defined doublet for the pristine material to a slightly asymmetric doublet for Li_7Sn_2.

The mechanism is more complex for delithiation. The operando ^{119}Sn Mössbauer spectra obtained for the first delithiation of $CoSn_2$ [67] and nanostructured $MnSn_2$ based electrodes [130] show the same trends as $FeSn_2$. During the delithiation, the single Mössbauer peak of Li_7Sn_2 is broadened and transformed into a broadened doublet with increasing average isomer shift. This trend is typical of Li_7Sn_2 dealloying process. This suggests the formation of poorly crystallized Li_xSn nanoparticles with intermediate compositions as described in Sect. 7.6.1 for $FeSn_2$, or of more complex Li-Co-Sn and Li-Mn-Sn metastable ternary phases [130]. It is difficult to have a more reliable interpretation of the spectra since they do not show any resolved structures and the isomer shifts of the Sn based intermetallics and Li_xSn phases are all in the same range from 1.8 to 2.5 mm s^{-1}. The variations of the average isomer shift can be related to changes in the composition of the Li-Sn based species but it is impossible to identify well-defined phases as the Li_xSn crystalline references, which suggests the coexistence of metastable phases with possible different compositions, sizes and crystallinities, depending on the stage of delithiation.

For nanostructured $MnSn_2$, the voltage curve of the first delithiation shows an additional plateau at 0.75 V between about 2 and 3.5 Li per $MnSn_2$ (Fig. 7.22). Such an additional voltage plateau was also observed at the end of the delithiation of nanostructured $FeSn_2$ at very low current density and was interpreted as the back reaction of Fe with Sn to form small $FeSn_2$ particles [27]. However, the ^{119}Sn Mössbauer spectra obtained for $MnSn_2$ show a progressive line broadening with a shape typical of magnetic relaxations, which is not observed for $FeSn_2$ (Fig. 7.23). The spectra obtained for this additional plateau were successfully fitted to three doublets corresponding to remaining paramagnetic $MnSn_2$, the "Li-Mn-Sn" phase observed before this process and a magnetic subspectrum (Fig. 7.24). The latter doublet can be favourably compared to the spectrum obtained for crystalline $MnSn_2$ between room and Néel temperatures. This indicates that Sn reacts with Mn at the end of delithiation to form $MnSn_2$ with magnetic ordering although the $MnSn_2$ pristine material is paramagnetic. Such a magnetic ordering suggests that the reformation of $MnSn_2$ after the first cycle improves the crystallinity.

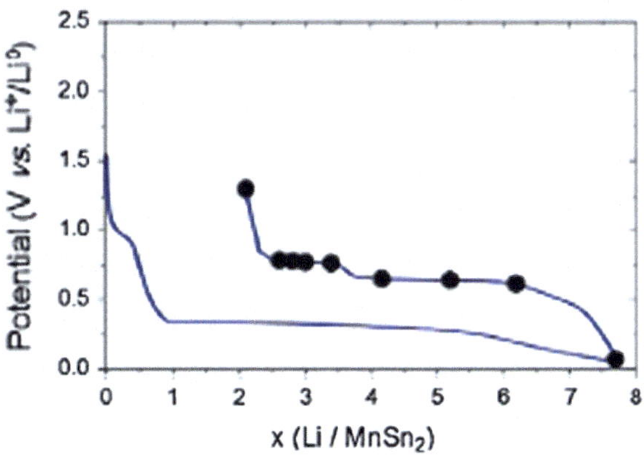

Fig. 7.22 Voltage curve for the first lithiation and delithiation of $MnSn_2$ in a Li half-cell. Reprinted from Ref. [130]. Copyright (2014), with permission from Elsevier

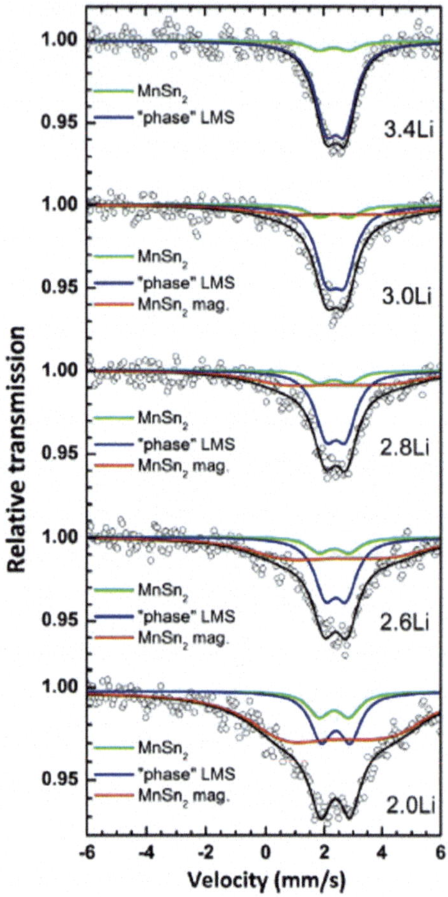

Fig. 7.23 Operando ^{119}Sn Mössbauer spectra obtained at the end of the delithiation of $MnSn_2$. Corresponding points are reported in Fig. 7.22. Reprinted from Ref. [130]. Copyright 2014, with permission from Elsevier

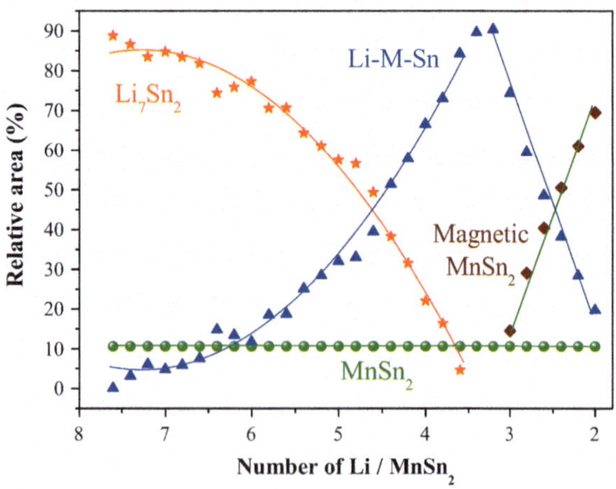

Fig. 7.24 Variations of the relative contributions to the ^{119}Sn Mössbauer spectra of the different Sn based phases formed during the first delithiation of $MnSn_2$ in a Li half-cell. Reprinted from Ref. [130]. Copyright 2014, with permission from Elsevier

For Ni_3Sn_4, the Mössbauer spectrum obtained at the end of delithiation is close to that of the pristine material and can be attributed to the reformation of Ni_3Sn_4. Such a back reaction was also observed for Ni_3Sn_4/Si/C composite where Si was introduced to increase the specific capacity [144]. In that case, both Ni_3Sn_4 and Si are electrochemically active and operando ^{119}Sn Mössbauer spectroscopy was used to follow reactions involving Sn. The spectra obtained in galvanostatic regime clearly show the back reaction of Ni with Sn to form Ni_3Sn_4 at the end of delithiation [145].

The present analysis of the Mössbauer results obtained for the first delithiation of tin intermetallics indicates that, depending on the transition metal M (Mn, Fe, Co, Ni), different mechanisms take place with different contributions of the back reactions between Sn and M. This could be due to differences in the bonding energies of M-Sn and M-M bonds, but also to interfacial and atomic diffusion properties. This aspect has not been elucidated yet and it is still not clear whether such back reactions that trap M and Sn atoms improve the electrode performance such as cycle life, which is a crucial issue for the application of intermetallic electrode materials in Li-ion batteries.

Finally, the Mössbauer spectra obtained during the second lithiation of $MnSn_2$, $CoSn_2$ and Ni_3Sn_4 differ from the spectra obtained during the first lithiation, in line with the observed different voltage profiles. This is consistent with the mechanism described for $FeSn_2$ and is obviously due to differences between the electrode materials at the beginning of the first (pristine) and second (nanocomposite) lithiations. However, both lithiation processes end with the formation of M/Li_7Sn_2 composite. The cycle formed by the first delithiation and the second lithiation is reversible and reflects the electrochemical mechanism (Li-Sn alloying reactions) of the following reversible cycles. Unfortunately, the poor mechanical stability of the electrodes during cycling limits the use of such materials and requires the optimization of the electrode formulation. In addition, all the reaction mechanisms described above can also be affected by parasitic reactions on cycling or for the cell at rest as discussed in the next subsection.

7.6.3 Aging Phenomena

Li-ion batteries can suffer from energy and power losses, voltage lowering or more severe issues with time. The performance degradation occurs for batteries at rest or during cycling [146]. The in situ Mössbauer spectroscopy can be used to study aging mechanisms if they affect the electronic properties of Mössbauer atoms. Two mechanisms are described here for tin based intermetallics as negative electrode materials for Li-ion batteries at rest and during cycling, respectively.

The first aging mechanism concerns the instability of the species formed during the lithiation or delithiation of electrode materials [147]. The example of a fully lithiated $FeSn_2$ based electrode is considered here. Such an aging process is complex since it depends not only on the macroscopic and microscopic features of the electrode as the morphology of the $FeSn_2$ particles or the electrode formulation, but also on

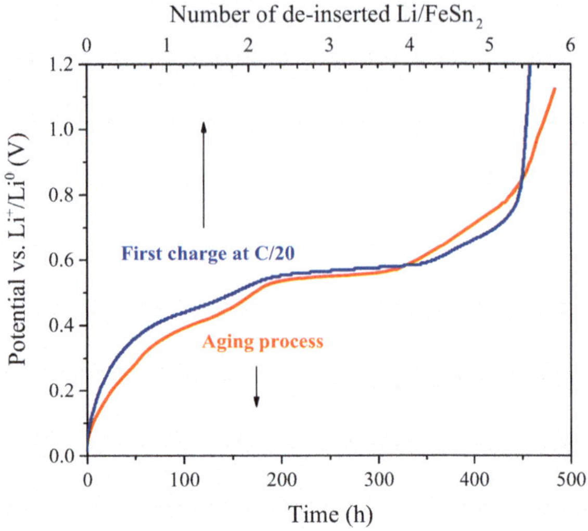

Fig. 7.25 Voltage curve of a fully lithiated $FeSn_2$ based electrode in a Li half-cell at rest as a function of time (red) compared to the galvanostatic voltage curve obtained for the first delithiation of $FeSn_2$ in a Li half-cell at the current rate of C/20 (0.05 Li per $FeSn_2$ and per hour). Reprinted with permission from Ref. [147]. Copyright 2017 American Chemical Society

the electrode-electrolyte reactivity. The present results highlight some trends in the instability of the nanocomposite α-Fe/Li_7Sn_2 formed at the end of the lithiation of $FeSn_2$ in a Li half-cell at rest, i.e., under open circuit conditions. In that case, the cell voltage increases with time from 0 to 1 V during a period of several months although there is no external current (Fig. 7.25). The observed voltage profile is similar to the profile of the galvanostatic delithiation of $FeSn_2$ in a Li half-cell, suggesting the progressive delithiation of Li_7Sn_2 with time.

The in situ ^{119}Sn Mössbauer spectra obtained during aging show that the rather symmetric peak of nano-Li_7Sn_2 centered at 2 mm s^{-1} is transformed into an asymmetric and broad line after 900 h (Fig. 7.26). The Mössbauer spectrum obtained after 170 h can be fitted to Li_7Sn_2 (2 doublets) and another doublet centered at 2.05 mm s^{-1}. The spectrum obtained after 900 h was fitted to two doublets centered at 2.05 mm s^{-1} and 2.45 mm s^{-1}, respectively. The average compositions of the formed Li_xSn phases were evaluated from Eq. (7.28). The values of the isomer shift $\delta = 2.05$ mm s^{-1} and $\delta = 2.45$ mm s^{-1} correspond to $Li_{2.5}Sn$ and $Li_{0.5}Sn$, respectively. Thus, the aging process consists in the delithiation of Li_7Sn_2 with the formation of intermediate Li_xSn phases as in the case of the galvanostatic delithiation process. The ^{57}Fe Mössbauer spectra do not show any noticeable changes even after three months, which means that iron nanoparticles were not affected by the delithiation of Li_7Sn_2. This was also confirmed by ex situ magnetic measurements [147].

The electrochemical impedance spectroscopy was also used for the characterization of the electrode-electrolyte interface [147]. Nyquist plots of the impedance spectra show two semicircles in medium and high frequency ranges attributed to charge transfer and SEI formation, respectively. During aging, the semicircles widened, revealing the increase of the resistances coming with these two phenomena. This was interpreted as the growth of the SEI layer and indicates that

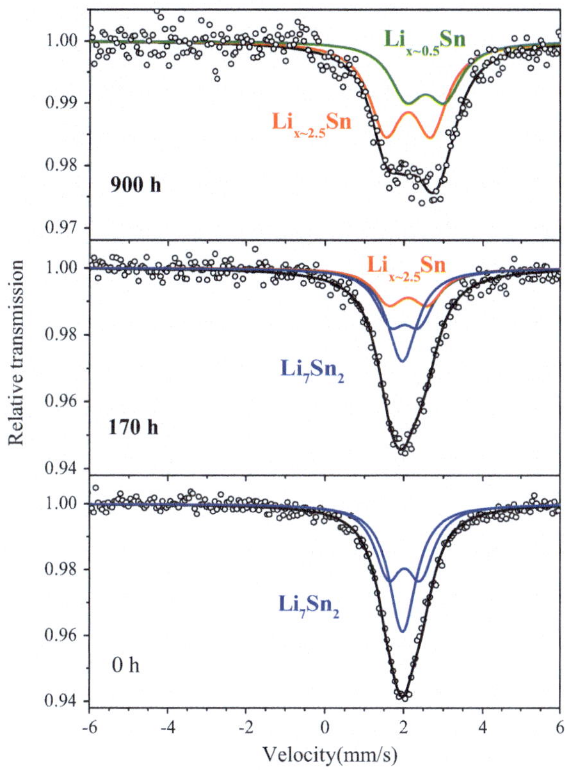

Fig. 7.26 In situ ^{119}Sn Mössbauer spectra at different stages of aging of a fully lithiated $FeSn_2$ electrode in a Li half-cell at rest. Reprinted with permission from Ref. [147]. Copyright 2017 American Chemical Society

the Li_7Sn_2 delithiation originates from reactions with the electrolyte. This is due to the metastability of the Li_xSn nanoparticles and their reactivity vs. Li.

Another aging process due to the metastability of the electrochemically formed species was revealed for delithiated $FeSn_2$ based electrodes in Li half-cells with voltage maintained during two weeks [27]. The delithiation of a fully lithiated $FeSn_2$ based electrode at very low current rate produces βSn, Sn-rich Li_xSn species and $FeSn_2$. When no more Li ions can be extracted from the electrode, and by imposing a constant voltage, the area of the Mössbauer spectrum increases with time. This can be interpreted by the increase of the amount of $FeSn_2$ in the electrode due to the back reaction of Sn with Fe.

The second aging phenomenon considered here occurs during cycling, leading to strong capacity fading [126]. The ^{57}Fe and ^{119}Sn Mössbauer spectra of a fully lithiated nanostructured $FeSn_2$ electrode obtained after 30–50 cycles show significant differences from the spectrum obtained at the end of the first lithiation.

The ^{57}Fe Mossbauer spectrum obtained after 30 cycles is formed by a magnetic sextet with the hyperfine magnetic field $B = 32.4$ T close to that of α-Fe and a single peak that can be attributed to superparamagnetic iron nanoparticles (Fig. 7.27a). The existence of magnetic particles can be explained by the growth of Fe nanoparticles during cycling. This is due to changes in the composite microstructure resulting from

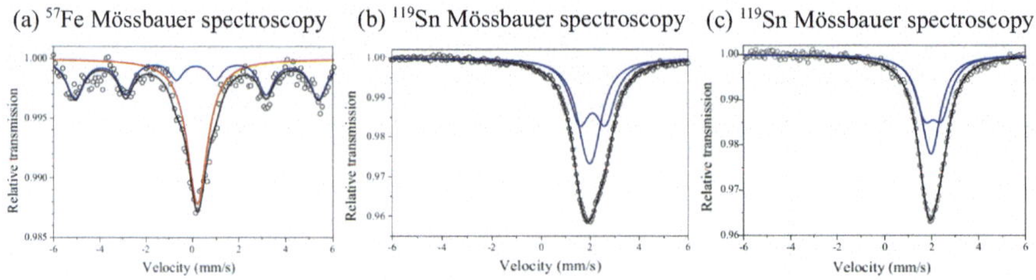

Fig. 7.27 ^{57}Fe **a** and ^{119}Sn **b** Mössbauer spectra of a fully lithiated $FeSn_2$ based electrode in a Li half-cell after 30 cycles and ^{119}Sn Mössbauer spectrum at the end of the first lithiation for comparison **c**. Reprinted with permission from Ref. [126]. Copyright 2011 from Elsevier

volume variations. The ^{119}Sn Mössbauer spectrum obtained after 30 cycles is more asymmetric than at the end of the first lithiation, but can also be fitted to two doublets (Fig. 7.27b). Compared to Li_7Sn_2 obtained at the end of the first lithiation (Fig. 7.27c), the highest of the two quadrupole splittings has increased by about 0.4 mm s^{-1} and the Mössbauer parameters are similar to those of the Li_7Sn_2 crystalline reference. Such a difference is explained in Sect. 7.6.1 and reflects here the growth of the Li_7Sn_2 particles during cycling.

The Mössbauer measurements show that the average particle size of the lithiation products α-Fe and Li_7Sn_2 increases on cycling, which reduces the electrochemical activity by worsening the effects of volume variations. The resulting mechanical, electrical and chemical instabilities are responsible for capacity fading. However, a longer cycle life can be achieved by improving the electrode microstructure and formulation.

7.6.4 Tin Oxides

Tin oxides and tin composite oxides (TCO) were proposed by Fuji in the late 1990s as high capacity negative electrode materials for Li-ion batteries [37]. The electrochemical mechanisms were studied by different techniques including operando XRD and ^{119}Sn Mössbauer spectroscopy.

The voltage curves of the first lithiation of SnO (Fig. 7.28) and SnO_2 in Li half-cells at low current density show two different regions. The region R1 corresponds to the insertion of about 2 Li per SnO (4 Li per SnO_2) and consists of a plateau at ≈ 1 V ($SnO_2 \approx 1.5$ V). The region R2 shows a continuous voltage decrease with different sloping plateaus until the end of lithiation. The voltage curve in R2 depends on the imposed current and can change from smooth to staircase profile with decreasing current, reflecting the effect of reaction kinetics as observed for βSn.

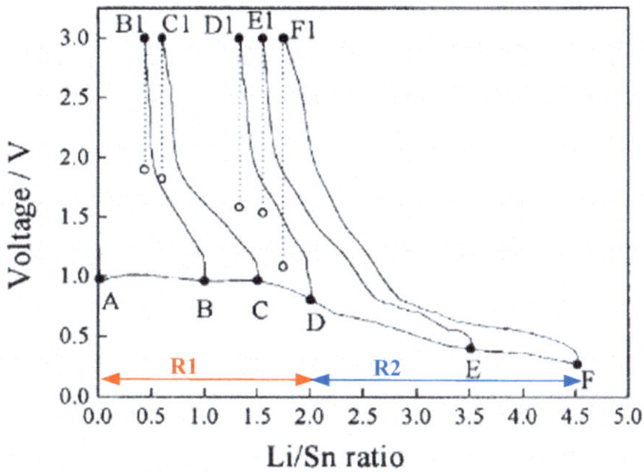

Fig. 7.28 Voltage curves of SnO in Li half-cells with the points of measurements for the first lithiation (A-F) and for different stages of delithiation at 3 V (B1-F1). Two regions R1 and R2 are distinguished for lithiation. Reprinted from Ref. [148]. Copyright 2000, with permission from Elsevier

The electrochemical reactions of SnO in Li half-cells were studied by operando [149] and ex situ [148] ^{119}Sn Mössbauer spectroscopy. The experimental conditions are different in the two works but the observed trends in the evolution of the Mössbauer spectra are similar.

For ex situ experiments, the Mössbauer spectrum of SnO formed by an asymmetric doublet centered at 2.6 mm s^{-1} is progressively broadened at the low velocity side and the doublet disappears to form a large and unresolved structure at the end of R1 (A-D in Fig. 7.29). The spectra were fitted to three components corresponding to SnO, βSn and Sn(IV) oxides, respectively, and an additional component centered at about 1.2 mm s^{-1}. The contribution of SnO to the Mössbauer spectra decreases while the contributions of βSn and the component centered at 1.2 mm s^{-1} increase during lithiation. For in situ experiments, only the Mössbauer spectra at the beginning and at the end of R1 were reported and are similar to the corresponding ex situ spectra [149]. These two studies are consistent with the transformation of SnO into βSn in R1, which corresponds to the conversion reaction

$$\mathrm{SnO} + 2\,\mathrm{Li} \rightarrow \beta\mathrm{Sn} + \mathrm{Li_2O} \tag{7.36}$$

However, the existence of the other two Mössbauer subspectra indicates that the mechanism is not a simple conversion reaction leading to isolated βSn and Li_2O particles. The observed Mössbauer isomer shift of the Sn(IV) oxide is slightly higher than that of SnO_2 and could originate from surface oxidation or impurities. The other subspectrum at 1.2 mm s^{-1} is rather intriguing since this value cannot be assigned to either Sn(IV) oxides nor Sn(0) phases (see Fig. 7.3). It can be tentatively attributed to Sn bonded to both Sn and O atoms resulting from interactions between βSn and Li_2O particles or to metastable Li-Sn-O amorphous phases. The reported variations of the total area of the operando Mössbauer spectra as a function of the number of Li per SnO, x, can be related to the variations of the average recoil-free fraction since the total amount of Sn is constant in the cell [149]. As a main result, the recoil-free fraction decreases almost linearly with increasing x and the ratio $f(x = 0)/f(x =$

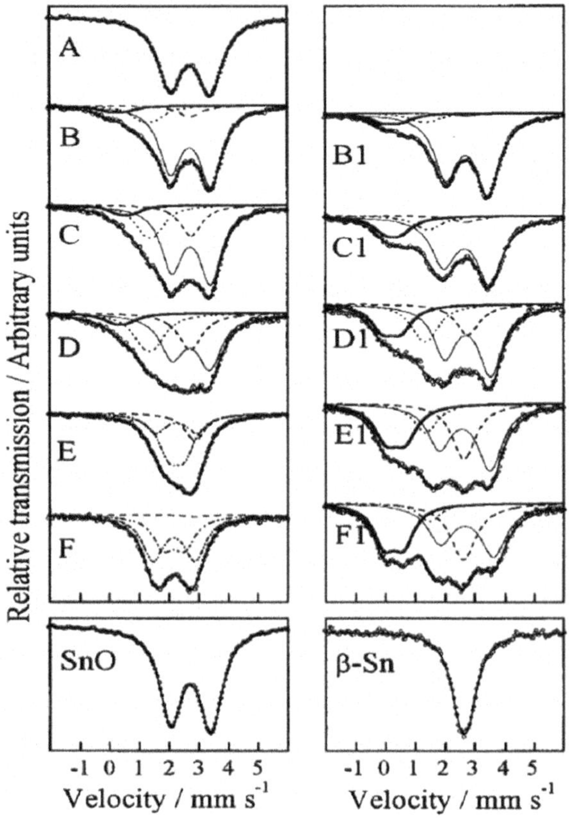

Fig. 7.29 Ex situ ^{119}Sn Mössbauer spectra at different stages of the first lithiation (A-F) and delithiation (B1-F1) of SnO in Li half-cells. The different subspectra correspond to Sn(IV) oxide (bold solid line), SnO (solid line), Li-Sn-O (dotted line), βSn (dashed line) and Li_xSn (dash-dotted line). Corresponding points are reported in Fig. 7.28. The spectra of SnO and βSn are shown for comparison. Reprinted from Ref. [148]. Copyright 2000, with permission from Elsevier

2) can be compared to *f* (SnO)/*f* (βSn), which confirms the conversion reaction given by Eq. (7.36). Thus, the lithiation of SnO in R1 brings about the dispersion of small Sn(0) electrochemically active particles within an inactive Li_2O based matrix, which should be regarded as an in situ restructuring step of the electrode material.

The operando Mössbauer spectra obtained during the lithiation in R2 are similar to those of the Li_xSn reference materials, which suggests the formation of Li_xSn equilibrium phases [149]. For Li_2Sn_5 and LiSn, there is still a low velocity component due to remaining Sn-O bonds. The agreement between the experimental data and Li_xSn references is less clear for compositions around Li_5Sn_2, probably due to the existence of different and metastable phases, while Li_7Sn_2 and $Li_{22}Sn_5$ are clearly observed at the end of R2. The latter result differs from the lithiation of βSn based electrodes that shows more disordered Li-rich Li_xSn phases as discussed in Sect. 7.5.2. This could be due to the effect of the Li_2O matrix that improves the dispersion of the Li_xSn small particles and the formation of more ordered phases.

The operando Mössbauer spectra obtained during the delithiation are similar to the Li_xSn references until LiSn [149]. Then, additional extraction of Li leads to broadened spectra at the low velocity side as voltage increases above 1 V. This can be explained by the formation of Sn-O bonds. Since all the Li atoms are extracted from Li_xSn at this stage, this is due to the partial delithiation of Li_2O and back reaction of Sn with O. The in situ spectrum obtained for the delithiated electrode at 2.5 V shows the existence of SnO_2, SnO and βSn, confirming such a back reaction. The same results are obtained for ex situ measurements at 3 V (Fig. 7.29) [148]. These reactions operate at a rather high voltage and are not expected to occur for SnO based negative electrodes in Li-ion full-cells.

To summarize, the electrochemical mechanism of SnO in a Li half-cell consists in the irreversible transformation of Sn(II)O into βSn(0) and Li_2O small particles, followed by reversible Li_xSn alloying and dealloying reactions for the first delithiation and the second lithiation, respectively, if the upper voltage cutoff is limited to about 1 V. In that case, the Li_2O particles form an electrochemically inactive matrix that maintains the dispersion of the Li_xSn particles, improving the cycling properties of the electrode compared to βSn. This is the same basic concept as described in Sect. 7.6.1 for tin intermetallic based electrodes whose metallic nanoparticles formed during the first lithiation buffer the volume variations, as the Li_2O particles in the present case. However, there is a large capacity loss at first cycle due to the transformation of SnO into βSn/Li_2O. In Li-ion full-cells, such a reaction requires the prelithiation of the negative electrode or the increase of the positive/negative electrode ratio.

The situation is even more critical for SnO_2 since 4 Li per SnO_2 are required for the conversion reaction ending with βSn/Li_2O composite. The in situ ^{119}Sn Mössbauer spectroscopy was used for SnO_2 in a plastic bag half-cell showing a three-step mechanism for the first lithiation including intercalation, conversion and alloying reactions [150]. However, the analysis of the conversion reaction is more ambiguous due to overlaps between the Sn(IV)O_2 subspectrum and broad lines due to the Sn-O bonds between βSn and Li_2O. The additional initial step reflecting Li intercalation into SnO_2 only involves 0.4 Li per SnO_2. The voltage curve of SnO_2 corresponding to Li_xSn alloying reactions (R2 for SnO) does not show well-defined plateaus and the Mössbauer spectra cannot be fitted to Li_xSn references. However, the variations of the average isomer shift are linear in this region, as expected from Eq. (7.28), reflecting reversible Li_xSn alloying reactions.

As an example of TCO, a tin borophosphate glass with the composition Sn_2BPO_6 is considered here since the mechanism was studied by operando ^{119}Sn Mössbauer spectroscopy [149]. The voltage profile of the first cycle of Sn_2BPO_6 in a Li half-cell is similar to that of SnO, except that the first plateau (R1) has a higher voltage. The Mössbauer spectrum of the pristine material consists of an asymmetric doublet that can be attributed to Sn(II) while a broad spectrum of weak intensity was obtained at the end of the first plateau. This reflects the strong decrease of the average recoil-free fraction in R1, which was attributed to the transformation of Sn_2BPO_6 into βSn small clusters embedded in a mixed borophosphate and Li_2O matrix. Such a decrease can be related to the variations of the recoil-free fraction of βSn as particle size decreases below ~10 nm when the interface with the matrix is not too constrained [149]. The proposed conversion reaction is

$$2\,Sn_2BPO_6 + 8Li \rightarrow 4\beta Sn + B_2O_3 + P_2O_5 + 4\,Li_2O \tag{7.37}$$

The Mössbauer absorption increases during the lithiation in R2 and the spectra are typical of the Li_xSn references except around Li_5Sn_2 [149]. The Mössbauer spectra obtained during the delithiation until about 1 V are similar to those obtained during the lithiation in R2 but in reverse order, ending with βSn. This shows that Li_xSn alloying reactions are reversible in R2. The spectrum of the delithiated electrode at the higher voltage of 2.5 V was successfully fitted to βSn and Sn_2BPO_6, suggesting back reactions of Sn with the oxides to form species with similar Sn local environments as the pristine glass. Thus, the electrochemical mechanism of Sn_2BPO_6 glass is similar to that of SnO except that the Sn(0) particles obtained during the initial conversion reaction given by Eq. (7.37) are significantly smaller and embedded in a more complex matrix. The small size of the βSn clusters reduces the impact of volume variations and the borophosphate matrix is more efficient to maintain the particle dispersion, which improves the cyclability of Sn_2BPO_6 glass compared to βSn. Finally, it should be noted that the upper voltage cutoff (≈ 0.8 V) is an important parameter to avoid the aggregation of mobile βSn clusters and the instability of the oxide based matrix due to back reactions between Sn and O atoms.

7.7 Conclusions

Mössbauer spectroscopy is a powerful tool to characterize pristine electrode materials and investigate reaction mechanisms in Li-ion and Na-ion batteries. This is due to the ability of this technique to probe the local environment of nuclei due its high selectivity to detect weak variations of the nuclear energy levels arising from electric and magnetic interactions. The resulting Mössbauer parameters: isomer shift, quadrupole splitting and hyperfine magnetic field, combined with the recoil-free fraction and the spectral linewidth, provide information about the structural, electronic and magnetic properties of electrode materials.

Changes in the Mössbauer parameters due to insertion and deinsertion processes can be used to follow electrochemical reactions. Although the application of Mössbauer spectroscopy could be limited due to the small number of available isotopes, the two most commonly used ones, ^{57}Fe and ^{119}Sn, are fortunately often found in the composition of electrode materials or they can be introduced as local probes. In situ and operando measurements are of high interest for the analysis of reaction mechanisms on a real-time basis and different types of in situ electrochemical cells have been described. Such cells avoid extracting the rather unstable lithiated or sodiated electrode materials although ex situ experiments can also be achieved if sample handling and measurements are made with care and shortly after the electrochemical experiments.

Some selected examples have been considered to illustrate the three main mechanisms encountered in electrode materials: insertion, alloying and conversion. These reactions can be characterized from changes in the oxidation state of the Mössbauer atoms due to redox reactions, but also from modifications in chemical bonding and local structure. Mössbauer spectroscopy is also of high interest for the identification and the quantitative evaluation of the lithiated or sodiated products. However, it should be emphasized that Mössbauer spectroscopy should be combined with other characterization tools to explore both short-range and long-range interactions, providing more accurate and quantitative analyses.

The interpretation of Mössbauer spectra is often complex due to the existence of a large variety of local environments in the electrochemically formed species. A particular attention should be devoted to the Mössbauer data analysis that can take advantages of first-principles calculations. Such calculations should also be used to investigate the structures of the lithiated or sodiated phases that are often metastable. This could be of high interest for the analysis of kinetic effects in Li-ion and Na-ion batteries, which are usually not taken into account in the interpretation of the Mössbauer experiments. The use of synchrotron radiation could also be considered for such phenomena in order to reduce the acquisition time of spectra.

However, in-lab Mössbauer spectroscopy will undoubtedly make novel contributions in the future to the development of new materials for electrochemical energy storage, which will be crucial for the energy transition and the reduction of global warming.

Acknowledgements The author is grateful to all collaborators and students who have contributed to the studies of Li-ion and Na-ion batteries, and in particular, J. C. Jumas, J. Olivier-Fourcade, M. T. Sougrati, M. Womes, L. Aldon, C. Bousquet, M. Chamas, J. Chouvin, R. Dedryvère, S. Difi, Y. He, M. El Khalifi, A. Mahmoud, M. Mouyane, S. Naille, A. Perea and F. Robert.

References

1. M. Armand, J.M. Tarascon, Nature **451**, 652 (2008)
2. B. Dunn, H. Kamath, J.-M. Tarascon, Science **334**, 928 (2011)
3. J. Jaguemont, L. Boulon, Y. Dubé, Appl. Energy **164**, 99 (2016)
4. A. Opitz, P. Badami, L. Shen, K. Vignarooban, A.M. Kannan, Renew. Sustain. Energy Rev. **68**, 685 (2017)
5. M. Winter, B. Barnett, K. Xu, Chem. Rev. **118**, 11433 (2018)
6. D. Aurbach, E. Zinigrad, Y. Cohen, H. Teller, Solid State Ionics **148**, 405 (2002)
7. C. Sun, J. Liu, Y. Gong, D.P. Wilkinson, J. Zhang, Nano Energy **33**, 363 (2017)
8. F. Zheng, M. Kotobuki, S. Song, M.O. Lai, L. Lu, J. Power Sources **389**, 198 (2018)
9. W. Xu, J. Wang, F. Ding, X. Chen, E. Nasybulin, Y. Zhang, J.-G. Zhang, Energy Environ. Sci. **7**, 513 (2014)
10. J.B. Goodenough, K.-S. Park, J. Am. Chem. Soc. **135**, 1167 (2013)
11. V. Etacheri, R. Marom, R. Elazari, G. Salitra, D. Aurbach, Energy Environ. Sci. **4**, 3243 (2011)
12. N. Nitta, F. Wu, J.T. Lee, G. Yushin, Mater. Today **18**, 252 (2015)
13. P. Rozier, J.M. Tarascon, J. Electrochem. Soc. **162**, A2490 (2015)
14. M. Yoshido, R.J. Brodd, A. Kozawa, *Li-Ion Batteries* (Springer, New York, 2009)
15. J.B. Goodenough, Y. Kim, Chem. Mater. **22**, 587 (2010)
16. G.E. Blomgren, J. Electrochem. Soc. **164**, A5019 (2017)
17. A. Barai, K. Uddin, M. Dubarry, L. Somerville, A. McGordon, P. Jennings, I. Bloom, Progr. Energy Combust. Sci. **72**, 1 (2019)

18. Z. Shadike, E. Zhao, Y.-N. Zhou, X. Yu, Y. Yang, E. Hu, S. Bak, L. Gu, X.Q. Yang, Advan. Energy Mater. **8**, 1702588 (2018)
19. R. Dedryvère, S. Leroy, H. Martinez, F. Blanchard, D. Lemordant, D. Gonbeau, J. Phys. Chem. B **110**, 12986 (2006)
20. G. Kermani, E. Sahraei, Energies **10**, 1730 (2017)
21. P. Novák, J.C. Panitz, F. Joho, M. Lanz, R. Imhof, M. Coluccia, J. Power Sources **90**, 52 (2000)
22. X. Ma, W. Luo, M. Yan, L. He, L. Mai, Nano Energy **24**, 165 (2016)
23. P.P.R.M.L. Harks, F.M. Mulder, P.H.L. Notten, J. Power Sources **288**, 92 (2015)
24. P.E. Lippens, L. Aldon, C.M. Ionica, F. Robert, J. Olivier-Fourcade, J.C. Jumas, in *Solid State Ionics-2004*, ed. by P. Knauth et al. (2005), p. 249
25. P.E. Lippens, M.E. Khalifi, M. Chamas, A. Perea, M.-T. Sougrati, C. Ionica-Bousquet, L. Aldon, J. Olivier-Fourcade, J.-C. Jumas, in *ICAME 2011: Proceedings of the 31st International Conference on the Applications of the Mössbauer Effect (ICAME 2011) held in Tokyo, Japan, 25–30 September 2011*, ed. by Y. Yoshida (Springer Netherlands, Dordrecht, 2013), p. 331.
26. P.E. Lippens, J.C. Jumas in *Nanocomposites: Ionic Conducting Materials and Structural Spectroscopies*, ed. by P. Knauth, J. Schoonman (Springer New York, 2008), p. 247
27. R.A. Dunlap, O. Mao, J.R. Dahn, Phys. Rev. B **59**, 3494 (1999)
28. F. Berry, in *Mössbauer Spectroscopy Applied to Inorganic Chemistry*, vol.1, ed. by G.J. Long (Plenum Press, New York and London, 1984), p. 391
29. C.M. Ionica, P.E. Lippens, J.O. Fourcade, J.C. Jumas, J. Power Sources **146**, 478 (2005)
30. B. Mortemard de Boisse et al., Nat. Commun. **7**, 11397 (2016)
31. L. Aldon, P. Kubiak, A. Picard, J.C. Jumas, J. Olivier-Fourcade, Chem. Mater. **18**, 1401 (2006)
32. M. Van Thournout, L. Aldon, M. Womes, B. Ducourant, J. Olivier-Fourcade, C. Tessier, S. Levasseur, J. Power Sources **174**, 1270 (2007)
33. J.M. Tarascon, M. Armand, Nature **414**, 359 (2001)
34. M. Endo, C. Kim, K. Nishimura, T. Fujino, K. Miyashita, Carbon **38**, 183 (2000)
35. Y.P. Wu, E. Rahm, R. Holze, J. Power Sources **114**, 228 (2003)
36. E. Buiel, J.R. Dahn, Electrochim. Acta **45**, 121 (1999)
37. Y. Idota, T. Kubota, A. Matsufuji, Y. Maekawa, T. Miyasaka, Science **276**, 1395 (1997)
38. Sony's new Nexelion hybrid lithium ion batteries to have thirty-percent more capacity than conventionl offering: www.sony.net/SonyInfo/News/Press/200502/05-006E/
39. T.C. Gibb, *Priciples of Mössbauer Spectroscopy* (Chapman and Hall Ltd., London, 1976)
40. G.J. Long, *Mössbauer spectroscopy applied to inorganic chemistry*, vol. 1, (Plenum Press, New York, 1984), Modern Inorganic Chemistry
41. D.P.E. Dickson, F. Berry, *Mössabauer Spectroscopy* (Cambridge University Press, Cambridge, 1986)
42. Y. Yoshida, G. Langouche, *Mössbauer Spectroscopy* (Springer-Verlag, Berlin Heidelberg, 2013)
43. T. Segi, R. Masuda, Y. Kobayashi, T. Tsubota, Y. Yoda, M. Seto, Hyperfine Interact. **237**, 7 (2016)
44. F. Neese, Inorg. Chim. Acta **337**, 181 (2002)
45. A.M. Mudarra Navarro, C.E. R. Torres, L.A. Errico, Hyperfine Interact. **202**, 117 (2011)
46. M. El Khalifi, P.E. Lippens, J. Phys. Chem. C **120**, 28375 (2016)
47. P.E. Lippens, M. El Khalifi, M. Womes, Physica Status Solidi (B) **254**, 1600194 (2017)
48. A. Svane, N.E. Christensen, C.O. Rodriguez, M. Methfessel, Phys. Rev. B **55**, 12572 (1997)
49. U.D. Wdowik, K. Ruebenbauer, Phys. Rev. B **76**, 155118 (2007)
50. K.S.P. Blaha, G.K.H. Madsen, D. Kvasnicka, J. Luitz, (2001. ISBN 3-9501031-1-2)
51. G.K. Shenoy, W.F.E., *Mössbauer isomer shift* (North-Holland Publishing Company, Amsterdam, Oxford, New York, 1978)
52. G.K. Shenoy, in *Mössbauer Spectroscopy applied to Inorganic Chemistry* ed. by G.J. Long (Plenum Press, New York and London, 1984), p. 57

53. R.V. Parish, in *Mössbauer spectroscopy* ed. by D.P.E. Dickson, F.J. Berry (Cambridge University Press, Cambridge, 1986), p. 17
54. P.E. Lippens, Phys. Rev. B **60**, 4576 (1999)
55. H. Spiering, in *Mössbauer Spectroscopy Applied to Inorganic Chemistry*. ed. by G.J. Long (Plenum Press, New York and London, 1984), p. 79
56. P. Dufek, P. Blaha, K. Schwarz, Phys. Rev. Lett. **75**, 3545 (1995)
57. P.E. Lippens, J. Olivier-Fourcade, J.C. Jumas, Hyperfine Interact. **126**, 137 (2000)
58. A. Svane, Phys. Rev. B **68**, 064422 (2003)
59. H. Haas, M. Menningen, H. Andreasen, S. Damgaard, H. Grann, F.T. Pedersen, J.W. Petersen, G. Weyer, Hyperfine Interact. **15**, 215 (1983)
60. P.E. Lippens, J.C. Jumas, J. Olivier-Fourcade, Hyperfine Interact. **141**, 303 (2002)
61. R. Ingalls, Phys. Rev. **128**, 1155 (1962)
62. C. Hohenemser, Phys. Rev. **139**, A185 (1965)
63. H.D. Pfannes, U. Gonser, Appl. Phys. **1**, 93 (1973)
64. R.A. Dunlap, M.N. Obrovac, J. Li, A. Smith, T.D. Hatchard, R.J. Sanderson, J.R. Dahn, Mössbauer Eff. Ref. Data J. **33**, 37 (2010)
65. A. Wattiaux, L. Fournès, C. Delmas, Mössbauer Eff. Ref. Data J. **33**, 32 (2010)
66. K. Ariyoshi, T. Ohzuku, Mössbauer Eff. Ref. Data J. **33**, 43 (2010)
67. C.M. Ionica-Bousquet, P.E. Lippens, L. Aldon, J. Olivier-Fourcade, J.C. Jumas, Chem. Mater. **18**, 6442 (2006)
68. M. Chamas, M.T. Sougrati, C. Reibel, P.E. Lippens, Chem. Mater. **25**, 2410 (2013)
69. J.B. Leriche et al., J. Electrochem. Soc. **157**, A606 (2010)
70. J.C. Jumas et al., Patent FR2995081A1, WO2014033402A1 (2014)
71. S. Difi, I. Saadoune, M.T. Sougrati, R. Hakkou, K. Edstrom, P.E. Lippens, J. Phys. Chem. C (2015)
72. S. Difi, I. Saadoune, M.T. Sougrati, R. Hakkou, K. Edstrom, P.E. Lippens, Hyperfine Interact. **237**, 61 (2016)
73. C. Delmas, F. Cherkaoui, A. Nadiri, P. Hagenmuller, Mater. Res. Bull. **22**, 631 (1987)
74. A. Mahmoud, C. Karegeya, M.T. Sougrati, J. Bodart, B. Vertruyen, R. Cloots, P.E. Lippens, F. Boschini, A.C.S. Appl, Mater. Interfaces **10**, 34202 (2018)
75. A. Perea, M.T. Sougrati, C.M. Ionica-Bousquet, B. Fraisse, C. Tessier, L. Aldon, J.C. Jumas, RSC Advan. **2**, 2080 (2012)
76. M. Ati, M.T. Sougrati, G. Rousse, N. Recham, M.L. Doublet, J.C. Jumas, J.M. Tarascon, Chem. Mater. **24**, 1472 (2012)
77. A. Yamada, N. Iwane, S.-I. Nishimura, Y. Koyama, I. Tanaka, J. Mater. Chem. **21**, 10690 (2011)
78. B. Mortemard de Boisse et al., J. Mater. Chem. A **3**, 10976 (2015)
79. A. Blidberg, L. Häggström, T. Ericsson, C. Tengstedt, T. Gustafsson, F. Björefors, Chem. Mater. **27**, 3801 (2015)
80. G. Rousse, J. Rodriguez-Carvajal, S. Patoux, C. Masquelier, Chem. Mater. **15**, 4082 (2003)
81. A.S. Andersson, B. Kalska, L. Häggström, J.O. Thomas, Solid State Ionics **130**, 41 (2000)
82. L. Aldon, A. Perea, M. Womes, C.M. Ionica-Bousquet, J.C. Jumas, J. Solid State Chem. **183**, 218 (2010)
83. P.E. Lippens, M. El Khalifi, M. Chamas, A. Perea, M.T. Sougrati, C. Ionica-Bousquet, L. Aldon, J. Olivier-Fourcade, J.C. Jumas, Hyperfine Interact. **206**, 35 (2012)
84. A. Perea, M.T. Sougrati, C.M. Ionica-Bousquet, B. Fraisse, C. Tessier, L. Aldon, J.C. Jumas, RSC Advances **2**, 9517 (2012)
85. N. Recham, G. Rousse, M.T. Sougrati, J.-N. Chotard, C. Frayret, S. Mariyappan, B.C. Melot, J.-C. Jumas, J.M. Tarascon, Chem. Mater. **24**, 4363 (2012)
86. R. Amisse, M.T. Sougrati, L. Stievano, C. Davoisne, G. Dražič, B. Budič, R. Dominko, C. Masquelier, Chem. Mater. **27**, 4261 (2015)
87. H. Ahouari, G. Rousse, J. Rodríguez-Carvajal, M.-T. Sougrati, M. Saubanère, M. Courty, N. Recham, J.M. Tarascon, Chem. Mater. **27**, 1631 (2015)

88. A. Nytén, S. Kamali, L. Häggström, T. Gustafsson, J.O. Thomas, J. Mater. Chem. **16**, 2266 (2006)
89. M.N. Obrovac, V.L. Chevrier, Chem. Rev. **114**, 11444 (2014)
90. I.A. Courtney, J.S. Tse, O. Mao, J. Hafner, J.R. Dahn, Phys. Rev. B **58**, 15583 (1998)
91. J. Sangster, C.W. Bale, J. Phase Equilib. **19**, 70 (1998)
92. C. Lupu, J.G. Mao, J.W. Rabalais, A.M. Guloy, J.W. Richardson, Inorg. Chem. **42**, 3765 (2003)
93. W. Gasior, Z. Moser, W. Zakulski, J. Non-Cryst, Solids **205–207**, 379 (1996)
94. R. Sen, P. Johari, A.C.S. Appl, Mater. Interfaces **9**, 40197 (2017)
95. M. Mayo, A.J. Morris, Chem. Mater. **29**, 5787 (2017)
96. M. Mayo, J.P. Darby, M.L. Evans, J.R. Nelson, A.J. Morris, Chem. Mater. (2018)
97. R.A. Dunlap, D.A. Small, D.D. MacNeil, M.N. Obrovac, J.R. Dahn, J. Alloys Compd. **289**, 135 (1999)
98. F. Robert, P.E. Lippens, J. Olivier-Foureade, J.C. Jumas, F. Gillot, M. Morcrette, J.M. Tarascon, J. Solid State Chem. **180**, 339 (2007)
99. S. Naille, J.C. Jumas, P.E. Lippens, J. Olivier-Fourcade, J. Power Sources **189**, 814 (2009)
100. J. Chouvin, J. Olivier-Fourcade, J.C. Jumas, B. Simon, O. Godiveau, Chem. Phys. Lett. **308**, 413 (1999)
101. J. Li, A. Smith, R.J. Sanderson, T.D. Hatchard, R.A. Dunlap, J.R. Dahn, J. Electrochem. Soc. **156**, A283 (2009)
102. T.D. Hatchard, J.R. Dahn, J. Electrochem. Soc. **151**, A838 (2004)
103. T.D. Hatchard, M.N. Obrovac, J.R. Dahn, J. Electrochem. Soc. **153**, A282 (2006)
104. P. Ge, M. Fouletier, Solid State Ionics **28–30**, 1172 (1988)
105. J. Sangster, C.W. Bale, J. Phase Equilib. Diffus. **19**, 76 (1998)
106. Z. Du, R.A. Dunlap, M.N. Obrovac, J. Alloys Compd. **617**, 271 (2014)
107. L. Baggetto, P. Ganesh, R.P. Meisner, R.R. Unocic, J.C. Jumas, C.A. Bridges, G.M. Veith, J. Power Sources **234**, 48 (2013)
108. S. Komaba, Y. Matsuura, T. Ishikawa, N. Yabuuchi, W. Murata, S. Kuze, Electrochem. Commun. **21**, 65 (2012)
109. L.D. Ellis, T.D. Hatchard, M.N. Obrovac, J. Electrochem. Soc. **159**, A1801 (2012)
110. J.W. Wang, X.H. Liu, S.X. Mao, J.Y. Huang, Nano Lett. **12**, 5897 (2012)
111. J.M. Stratford et al., J. Am. Chem. Soc. **139**, 7273 (2017)
112. R. Benedek, M.M. Thackeray, J. Power Sources **110**, 406 (2002)
113. M. Winter, J.O. Besenhard, Electrochim. Acta **45**, 31 (1999)
114. M. Wachtler, J.O. Besenhard, M. Winter, J. Power Sources **94**, 189 (2001)
115. C.M. Ionica, L. Aldon, P.E. Lippens, F. Morato, J. Olivier-Fourcade, J.C. Jumas, Hyperfine Interact. **156**, 555 (2004)
116. Z. Du, R.A. Dunlap, M.N. Obrovac, J. Electrochem. Soc. **163**, A2011 (2016)
117. Z. Du, S.N. Ellis, R.A. Dunlap, M.N. Obrovac, J. Electrochem. Soc. **163**, A13 (2016)
118. J. Hassoun, S. Panero, G. Mulas, B. Scrosati, J. Power Sources **171**, 928 (2007)
119. O. Mao, R.A. Dunlap, J.R. Dahn, J. Electrochem. Soc. **146**, 405 (1999)
120. P.P. Ferguson, A.D.W. Todd, J.R. Dahn, Electrochem. Commun. **10**, 25 (2008)
121. P.P. Ferguson, M.L. Martine, A.E. George, J.R. Dahn, J. Power Sources **194**, 794 (2009)
122. P.P. Ferguson, M. Rajora, R.A. Dunlap, J.R. Dahn, J. Electrochem. Soc. **156**, A204 (2009)
123. F.J. Nacimiento, P. Lavela, J.L. Tirado, J.M. Jiménez, D. Barreda, R. Santamaría, Mater. Chem. Phys. **138**, 747 (2013)
124. P.P. Ferguson, R.A. Dunlap, J.R. Dahn, J. Electrochem. Soc. **157**, A326 (2010)
125. G. Venturini, B. Malaman, G. Le Caër, D. Fruchart, Phys. Rev. B **35**, 7038 (1987)
126. M. Chamas, P.E. Lippens, J.C. Jumas, K. Boukerma, R. Dedryvere, D. Gonbeau, J. Hassoun, S. Panero, B. Scrosati, J. Power Sources **196**, 7011 (2011)
127. G.L. Caer, B. Malaman, G. Venturini, D. Fruchart, B. Roques, J. Phys. F: Met. Phys. **15**, 1813 (1985)
128. U.G. Nwokeke, R. Alcántara, J.L. Tirado, R. Stoyanova, M. Yoncheva, E. Zhecheva, Chem. Mater. **22**, 2268 (2010)

129. A. Mahmoud, M. Chamas, J.-C. Jumas, B. Philippe, R. Dedryvere, D. Gonbeau, I. Saadoune, P.E. Lippens, J. Power Sources **244**, 246 (2013)
130. B. Philippe, A. Mahmoud, J.B. Ledeuil, M.T. Sougrati, K. Edstrom, R. Dedryvere, D. Gonbeau, P.E. Lippens, Electrochim. Acta **123**, 72 (2014)
131. F. Nacimiento, R. Alcántara, J.L. Tirado, J. Electrochem. Soc. **157**, A666 (2010)
132. S. Wang, M. He, M. Walter, F. Krumeich, K.V. Kravchyk, M.V. Kovalenko, Nanoscale **10**, 6827 (2018)
133. K.K.D. Ehinon, S. Naille, R. Dedryvere, P.E. Lippens, J.C. Jumas, D. Gonbeau, Chem. Mater. **20**, 5388 (2008)
134. S. Naille, R. Dedryvere, D. Zitoun, P.E. Lippens, J. Power Sources **189**, 806 (2009)
135. S. Naille, P.E. Lippens, F. Morato, J. Olivier-Fourcade, Hyperfine Interact. **167**, 785 (2006)
136. S. Naille, R. Dedryvere, H. Martinez, S. Leroy, P.E. Lippens, J.C. Jumas, D. Gonbeau, J. Power Sources **174**, 1086 (2007)
137. L. Fransson, E. Nordström, K. Edström, L. Häggström, J.T. Vaughey, M.M. Thackeray, J. Electrochem. Soc. **149**, A736 (2002)
138. J. Wolfenstine, S. Campos, D. Foster, J. Read, W.K. Behl, J. Power Sources **109**, 230 (2002)
139. X.F. Tan, S.D. McDonald, Q. Gu, Y. Hu, L. Wang, S. Matsumura, T. Nishimura, K. Nogita, J. Power Sources **415**, 50 (2019)
140. G. Le Caer, B. Malaman, G. Venturini, I. B. Kim, Phys. Rev. B **26**, 5085 (1982)
141. S. Naille, C.M. Ionica-Bousquet, F. Robert, F. Morato, P.E. Lippens, J. Olivier-Fourcade, J. Power Sources **174**, 1091 (2007)
142. S. Furuseth, H. Fjellvag, Acta Chem. Scand. A **40** (1986)
143. T. Azib, N. Bibent, M. Latroche, F. Fischer, J.-C. Jumas, J. Olivier-Fourcade, C. Jordy, P. E. Lippens, F. Cuevas, J. Mater. Chem. A **8**, 18132 (2020)
144. Z. Edfouf, F. Cuevas, M. Latroche, C. Georges, C. Jordy, T. Hézèque, G. Caillon, J.C. Jumas, M.T. Sougrati, J. Power Sources **196**, 4762 (2011)
145. Z. Edfouf, C. Fariaut-Georges, F. Cuevas, M. Latroche, T. Hézèque, G. Caillon, C. Jordy, M.T. Sougrati, J.C. Jumas, Electrochim. Acta **89**, 365 (2013)
146. M. Broussely, P. Biensan, F. Bonhomme, P. Blanchard, S. Herreyre, K. Nechev, R.J. Staniewicz, J. Power Sources **146**, 90 (2005)
147. M. Chamas, A. Mahmoud, J. Tang, M.T. Sougrati, S. Panero, P.E. Lippens, J. Phys. Chem. C **121**, 217 (2017)
148. J. Chouvin, J. Olivier-Fourcade, J.C. Jumas, B. Simon, P. Biensan, F.J. Fernández Madrigal, J.L. Tirado, C. Pérez Vicente, J. Electroanal. Chem. **494**, 136 (2000)
149. I.A. Courtney, R.A. Dunlap, J.R. Dahn, Electrochim. Acta **45**, 51 (1999)
150. I. Sandu, T. Brousse, J. Santos-Peña, M. Danot, D.M. Schleich, Ionics **8**, 27 (2002)

Chapter 8
Mössbauer Spectroscopy in External Magnetic Fields

Michael Reissner

Abstract The application of external magnetic fields allows to broaden the knowledge, which can be gained by Mössbauer spectroscopy enormously. In combination with magnetic measurements, detailed information about local magnetic and electronic structure can be obtained. After an introduction to the influence of an external magnetic field on hyperfine interactions, examples are given. Starting from simple magnetic structures like para-, ferro-, and antiferro-magnetic ones, investigations on complex magnetic structures with spin canting are discussed. Also results on materials with inhomogeneous magnetic structures like spin glasses will be presented. Special focus will be on the influence of high external fields on the spin dynamics.

8.1 Introduction

In contrast to magnetization measurements or NMR investigations the application of an external field is not a prerequisite to get information about the magnetic behaviour of a compound in case of Mössbauer spectroscopy. Nevertheless measurements in external magnetic fields have been performed since the very early days of Mössbauer spectroscopy, because a lot of new and more in-depth information can be gained. Especially in the case of the investigation of magnetic ground state properties in-field measurements are extremely valuable. The determination of the precise hyperfine structure in magnetically ordered samples is often a non-trivial task. The simultaneous presence of magnetic dipole and electric quadrupole interaction makes modelling of the hyperfine structure quite complicate. But if external field is high the influence of electric field gradient can be reduced and interpretation of measured spectra might become easier. One of the first applications of external fields in Mössbauer spectroscopy was the measurements of the sign of the hyperfine field in α-Fe [1, 2]. For alloys with traces of iron the local susceptibility of the iron moments could be determined [3–7]. Knowledge about crystal field parameters and the atomic arrangement in complex structures, as well as the investigation of magnetic order can

M. Reissner (✉)
Institute of Solid State Physics, TU Wien, Vienna, Austria
e-mail: reissner@tuwien.ac.at

Y. Yoshida and G. Langouche (eds.), *Modern Mössbauer Spectroscopy*,
Topics in Applied Physics 137, https://doi.org/10.1007/978-981-15-9422-9_8

be obtained by high field Mössbauer spectroscopy [8, 9]. Tuning the field strength in such way that the magnetic and electrostatic hyperfine interaction is of similar strength, the sign of the electric field gradient can be obtained for 3/2 → 1/2 transitions, which is not possible without field [10, 11]. This helps to get information about the charge density distribution and to compare with theoretical calculations. Especially for intermetallic compounds high external fields are necessary, because with fields which can be produced by electromagnets (max 2 T) the resolution of the spectra is too less, to allow clear conclusions.

In this tutorial the possible great extension of knowledge about magnetic behaviour in solids by application of an external magnetic field is discussed. It is not a review about the field but should give an insight into what is possible on different levels of complexity of magnetic structures. Therefore the presented examples are all from work in which the author was involved. The focus is not only on the results of such investigations but also on how to come to these results, by also discussing possible misinterpretations. In the first part the influence of an external field on the hyperfine interactions and spectral shape is discussed. In the Applications examples from very simple to extremely complex magnetic structures are presented. The last part is devoted to the investigation of dynamic effects by high-field Mössbauer spectroscopy. This tutorial should convince that the investment in a high-field Mössbauer apparatus makes sense and should help to enter the field.

8.2 Hyperfine Field

From the three main hyperfine interactions which are detectable by Mössbauer spectroscopy the most important one for investigating the magnetic ground state properties is the magnetic hyperfine interaction. It is present if the Mössbauer nucleus has a nuclear spin $I > 1$, because in that case it has a dipole moment μ capable to interact with a magnetic field, which might be present at the site of the nucleus. This field can be an internal one, due to the surrounding of the nucleus, or an external one. The magnetic hyperfine field H_{hf}, which the nucleus senses, and which causes the nuclear Zeemann splitting, has two main components

$$H_{hf} = H_{eff} + H_{loc} \tag{8.1}$$

with H_{eff} the effective hyperfine field, and H_{loc} the local hyperfine field. Both can be decomposed into more fields, according to their origin

$$H_{eff} = H_c + H_O + H_d \tag{8.2}$$

with H_c the Fermi contact field, H_O the orbital field, and H_d the dipole field and

$$H_{loc} = H_{ext} + H_{DM} + H_L \tag{8.3}$$

with H_{ext} an external field, H_{DM} the demagnetizing field, and H_L the Lorentz field. The Fermi contact field H_c is due to the Fermi contact interaction of the nuclear moment with the spin density at the nucleus [12]. This density comes partly from an unbalanced spin density of s-electrons at the nucleus and partly from conduction electrons [13, 14]. The latter one comes from polarization produced by exchange interactions with the 3d electrons and by admixture with the 3d band. For ionic iron compounds H_c is large and negative. With increasing covalency H_c decreases in ionic compounds [15]. The orbital field H_O arises from unquenched orbital angular momentum of the parent atom for high spin ferric compounds. It is zero, because ferric iron is an S-state ion (^{6}S). In ferrous iron this term can be large and of opposite sign to the Fermi contact field H_c. The dipole field H_d is caused by the arrangement of the atomic moments in the vicinity of the Mössbauer nucleus. For most iron compounds this term is smaller than the Fermi contact and the orbital term. Whereas the Fermi field can be assumed to be independent of the crystal symmetry, because of the polarization of the inner s-shells, the orbital field and the dipole field are strongly symmetry dependent. The demagnetizing field $H_{DM} = -DM$ reduces the hyperfine field. It depends on the demagnetization factor D, which depends on the shape of the sample and the magnetization M. In contrast to measurements on bulk microcrystalline materials, where H_{DM} is negligible small, because of the multidomain structure, it can become dominant in case of monodomain nanoparticles, especially if the hyperfine field is small [16]. The usual Lorentz field $H_L = 4\pi/3$ for cubic symmetry has to be modified by small residue $H_{L'}$ for noncubic symmetry. Because of the quenched orbital moment in high-spin ferric compounds $H_O = 0$, the hyperfine field at low temperatures is always negative and rather large. In ferrous compounds where H_O can be large, the sign of the observed hyperfine field can be positive or negative. Best way to determine the sign is to apply a large external magnetic field (2 T − 5 T) and to observe the field dependence of the measured hyperfine field. In that way Hanna et al. [1] have shown for the first time that in α-Fe the hyperfine field is negative rather than positive, although at that time theory predicted a positive sign. Detailed description of the nature of the hyperfine field can be found in [8, 12, 17, 18].

By the nuclear Zeeman effect the degeneration of the ground and exited states are lifted and the levels split into $2I + 1$ energetically different levels. In case of ^{57}Fe (Fig. 8.1) the exited state ($I = 3/2$) splits into four and the ground state ($I = 1/2$) into two levels. This gives eight different transition energies, from which two are forbidden because of the selection rules for magnetic dipole transitions: $\Delta I = \pm 1$ and $\Delta m = 0$ or ± 1. As a result the Mössbauer line splits into a sextet (Fig. 8.2). The spacing of the outermost lines is the so-called hyperfine field B_{hf}. From the intensity ratio of the six lines information about the orientation of the magnetic field at the nucleus can be obtained, because the intensity of the six lines depend on the angle θ_m between direction of the magnetic field at the nucleus and the γ-ray direction. For the $\pm 3/2 \rightarrow \pm 1/2$ transitions intensity is given by $\frac{3}{4}(1 + \cos^2\theta_m)$, for $\pm\frac{1}{2} \rightarrow \pm\frac{1}{2}$ it is given by $\sin^2\theta_m$, and for $\mp\frac{1}{2} \rightarrow \pm\frac{1}{2}$ it is given by $\frac{1}{4}(1 + \cos^2\theta_m)$. According to these equations the second and fifth line vanish, if the field at the nucleus is parallel to the

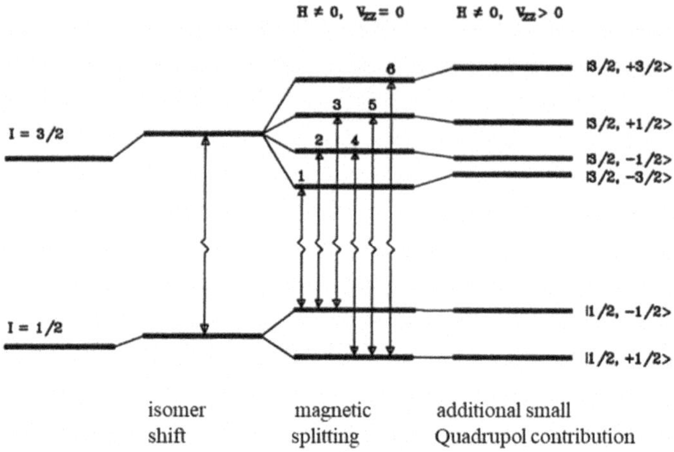

Fig. 8.1 ^{57}Fe level scheme for magnetic dipole interaction (middle) and additional small electric quadrupole interaction (right)

direction of the external field. Then the intensity ratio becomes 3:0:1:1:0:3 (Fig. 8.2b). This situation can be found, if the sample is a single crystal, which is oriented accordingly, or by a high external field, which rotates all hyperfine fields parallel to the applied field direction. In a case where the field is oriented perpendicular to the external field direction the intensity ratio becomes 3:4:1:1:4:3 (Fig. 8.2c). If the sample is polycrystalline, one has to integrate over all possible θ_m values. In this case the obtained intensity ratio is 3:2:1:1:2:3 (Fig. 8.2a). If in addition to the magnetic splitting also electric quadrupole interaction is present, spectra can become very complicate, even with appearance of the forbidden lines. In case the electrostatic interaction is much smaller than the magnetic one, the quadrupole splitting causes a shift in the position of the inner four lines against the outer ones (Fig. 8.1). To get the internal field, the applied field B_a has to be subtracted from the measured hyperfine field B_{hf}. If spectra are not fully polarized the angle θ between B_a and B_{hf} has to be taken into account. The internal field B_{int} is then given by $B_{hf}^2 = B_a^2 + B_{int}^2 - 2B_a B_{int} \cos\theta$. If the applied field is strong enough to polarize the spectra, angle θ becomes zero and the situation is easier. The value of B_{int} is than simply given by the difference $|B_{hf} - B_a|$ (Fig. 8.3). If measured B_{hf} is larger than the applied field B_a, the internal field B_{int} has to be parallel to the applied field. If B_{hf} is smaller than B_a, internal field B_{int} is antiparallel to the applied field.[1]

[1]The internal field B_{int} is sometimes also called induced field B_{ind}, or transferred field B_{trans}, because it is caused by neighbouring ligands. In this tutorial both B_{int} and B_{ind} are used synonymously.

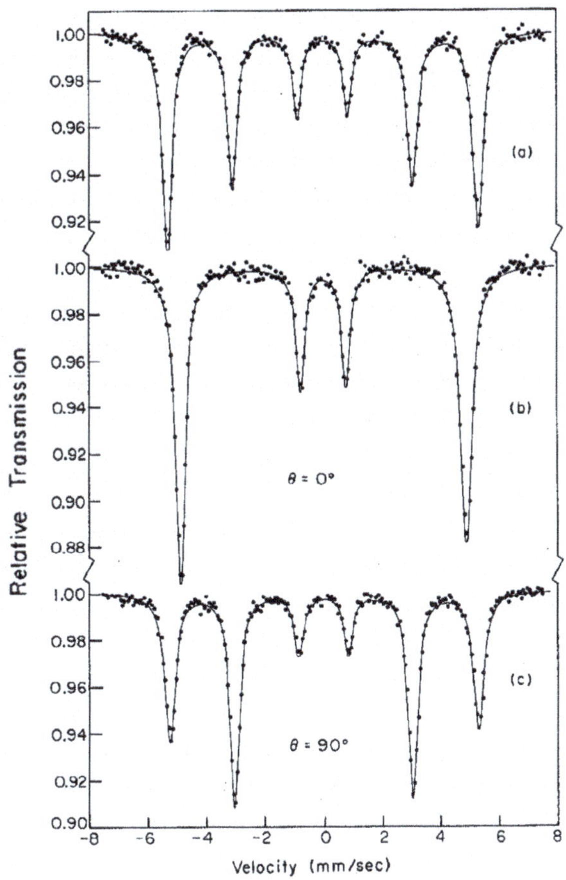

Fig. 8.2 ^{57}Fe Mössbauer spectrum for polycrystalline α-Fe in zero field (**a**), in 5 T parallel to the external field direction (**b**), and in 0.35 T perpendicular to the external field direction (**c**)

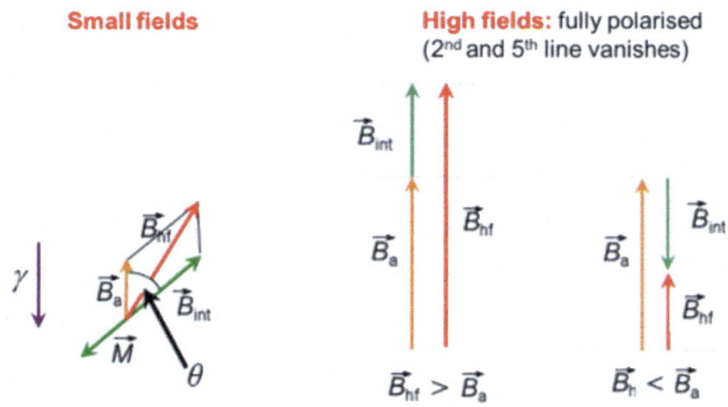

Fig. 8.3 Determination of the internal field B_{int} in case of small (left) B_a and large (right) external fields

8.3 Simple Magnetic Structures

In case of a diamagnet all electronic shells are completely filled. Thus they do not carry a magnetic moment, therefore they do not react on an applied field B_a. In that case the measured hyperfine field B_{hf} at the nucleus is identical to B_a.

In case of an isotropic paramagnet the situation is the following: without external field the spectrum is a single line, or in case of a small electric quadrupole interaction a doublet. Applying an external field induces an internal field B_{ind}. Figure 8.4 shows as a typical example spectra of $Y(Fe_{0.40}Al_{0.60})_2$ at room temperature and different external fields. For all fields the spectra are fully polarized, visible by the vanishing of the 2nd and 5th line of the sextets. The measured hyperfine field $B_{hf} = B_a - B_{ind}$ is practically identical to the applied field B_a. The line shows the calculated influence of B_a. Broadening and small shift of centre of gravity of the lines are due to differences in B_{ind} for different Fe surroundings. The change of measured B_{hf} and calculated small B_{ind} values with applied field B_a is shown in Fig. 8.5.

In ferromagnetic materials magnetization and therefore magnetic hyperfine splitting is present already at zero applied field. Figure 8.6 shows the field dependence of polycrystalline α-Fe 4.2 K for different external fields. Again the line indicates spectra, which would be obtained if only the applied field is present. With increasing

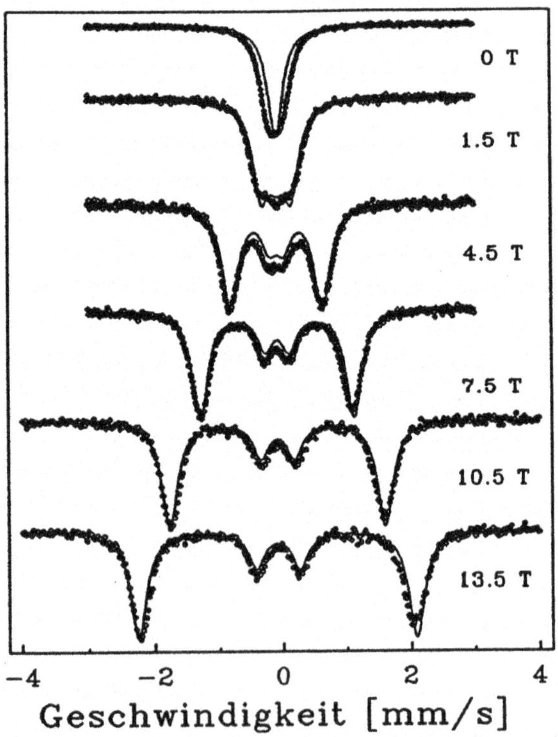

Fig. 8.4 Room temperature spectra at different applied fields for $Y(Fe_{0.4}Al_{0.6})_2$ as an example of a paramagnet. Reprinted from [19]

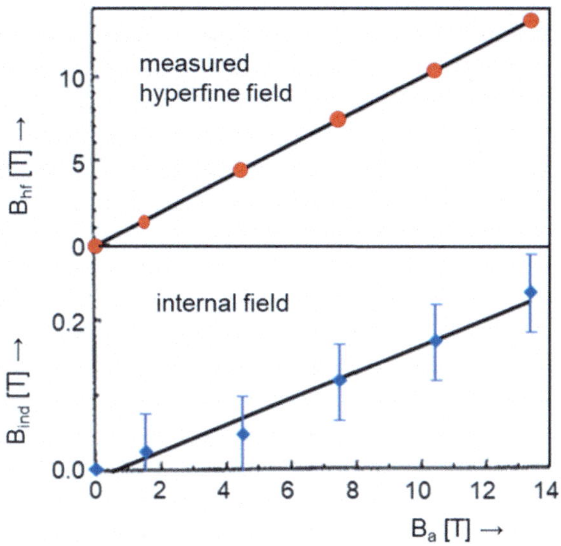

Fig. 8.5 Measured hyperfine field B_{hf} and calculated induced field B_{ind}=|B_a-B_{hf}| versus applied field B_a. Modified from [19]

field domains are rotated in direction of the applied field. Full alignment is reached at approximately 4 T, again visible by the vanishing of the 2nd and 5th line. The measured hyperfine field B_{hf} decreases with increasing B_a, but is always larger than B_a. Therefore the internal hyperfine field B_{int} is antiparallel to the external one. As in an external field the magnetic moments are always rotated into the external field direction, the hyperfine field is antiparallel to the moment in α-Fe. Figure 8.7 shows the change of B_{hf}, B_{int}, and θ with applied field B_a. At zero applied field the angle θ between B_{int} and B_a is 54.735°, which corresponds to random distribution of the fields, and reaches zero at approximately 4 T. The internal field is as expected independent of B_a. The small increase in low external fields is due to the demagnetization field, which is not taken into account in the analysis.

In antiferromagnetic materials the magnetic moments of the two magnetic sublattices compensate on a large scale, but locally magnetic fields are present. These fields lead to magnetic hyperfine split spectra. If there is only one internal field B_{int} several different cases can be found in an external field (Fig. 8.8). If internal fields are parallel to the applied field, two subspectra for the two hyperfine fields $B_{hf} = B_a \pm B_{int}$ are present (Fig. 8.8b). This case can be realized if a single crystal is oriented appropriately in field. In higher external fields it is energetically more favourable if the internal fields are perpendicular to the applied field. In this case again only one subspectrum is present (Fig. 8.8c). This case is visible also for polycrystalline powders if the external field is high enough. Is B_a too low, the spectrum of antiferromagnetic powder samples can become very complex [20].

Many magnetic systems show relaxation behaviour, which leads to time varying hyperfine fields. These relaxation behaviour can have many reasons like energy exchange between spins—so-called spin-spin-relaxation—but also energy exchange

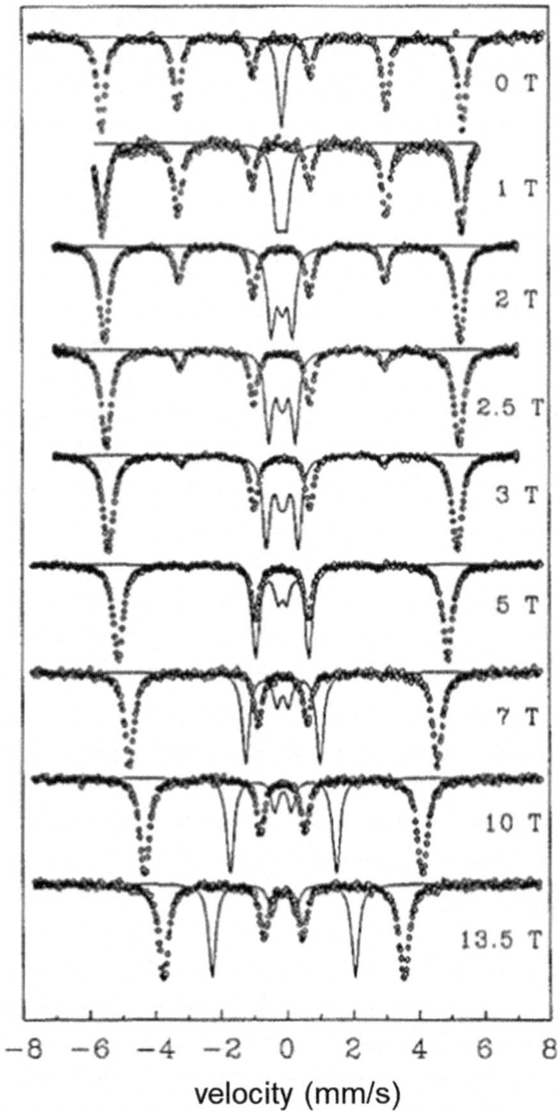

Fig. 8.6 Mössbauer spectra of an α-Fe foil 4.2 K for different applied fields, as example of a ferromagnet. The line is a calculation of the pure field effect. Reprinted from [19]

between the spins and the lattice—so-called spin-lattice-relaxations—to name only a few. These time dependence of hyperfine interactions can strongly influence the shape of the measured spectra, if the corresponding characteristic relaxation time τ_R is in the range of the characteristic time of the experiment. Each measuring method has its own characteristic time window. This is the time during which measurement is performed. In case of a bulk magnetization measurement this is in the range of 1 up to 100 s. Ac-susceptibility measurements are in the range of 1– 10^{-4}s, and

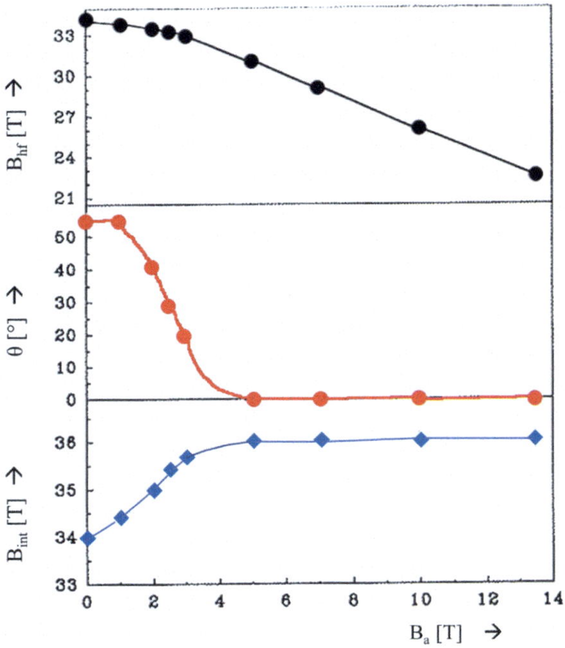

Fig. 8.7 Measured hyperfine field B_{hf}, angle between B_{hf} and B_a, as well as calculated internal field B_{int} over applied field B_a. Adapted from [19]

neutron measurements have a characteristic measuring time of $10^{-8} - 10^{-13}$s. The time window of Mössbauer spectroscopy is determined by the life time of the excited state of the nucleus τ_N. Depending on the Mössbauer isotope τ_N is between 10^{-7} and 10^{-9} s. To measure a magnetic split spectrum, there must be sufficient time for the nucleus to sense the effect of the magnetic field acting on it [21]. This means that at least one Larmor precession must take place before the nucleus decays. Therefore the Larmor precession time τ_L must be smaller than the nucleus life time τ_N. According to the relaxation time τ_R two cases can be distinguished. (i) $\tau_R \gg \tau_L$; this corresponds to slow relaxation. Here the hyperfine fields change so slow during one Larmor precession, that the nucleus senses the full hyperfine interaction. Therefore a static six line spectrum is measured. (ii) $\tau_R \ll \tau_L$; this corresponds to the fast relaxation. Here hyperfine fields change several times their direction during one Larmor precession. Therefore the nucleus senses only an averaged hyperfine interaction. In this time regime the spectra collapse and so-called motional narrowing of the lines take place. In the extreme case the spectrum collapses to a single line. Because the Larmor frequency depends on the magnetic energy, it is different for the different lines of the sextet. The broadening and collapsing of the lines appears therefore first for the inner lines of the sextet, and last for the outermost ones. A detailed discussion of the influence of time windows can be found in [21] and references therein. Figure 8.9 shows the influence of different relaxation times on the shape of the spectrum. The calculation is performed under the assumption that the hyperfine field jumps between +20 T and −20 T. Depending on the type of sample and the relaxation mechanism it

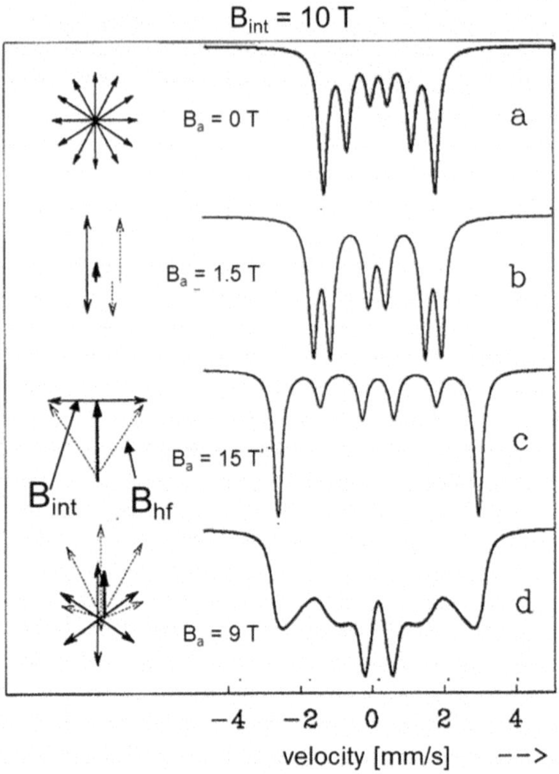

Fig. 8.8 Simulated Mössbauer spectra for different antiferromagnetic moment arrangements. For explanation see text. Reprinted from [19]

is possible with the application of a strong external field to influence the relaxation time τ_R in such a way that it comes in the range of the Larmor precession time τ_L and relaxation spectra are obtained. Information about the characteristic relaxation times and the time dependence of the hyperfine interactions can be gained. Simulation of relaxations spectra are possible by using stochastic methods [22–29], but also by perturbation theory or ab initio calculations [30–34]. In the simplest case the hyperfine field jumps between two states where the fields are antiparallel to each other. Whereas for the left figure equal occupation time of both states is assumed, an occupation of 1–2 of both states is assumed in the case shown in the right figure. No difference is found for the lower relaxation times, but large differences are present in the fast relaxation regime. Figure 8.10 shows results for field flip between +20 T and +8 T on the left, and between +20 T and −8 T on the right side. In both cases the occupation of the states was chosen to be equal. Due to the different field values in the two states the spectra are much more complicate. The influence of different orientations of the hyperfine fields relative to the γ-ray direction is shown in Fig. 8.11. As can be seen, relaxation effects can make spectra rather complicate and are not easy to be correctly analysed. Often dynamical effects in Mössbauer spectra are not

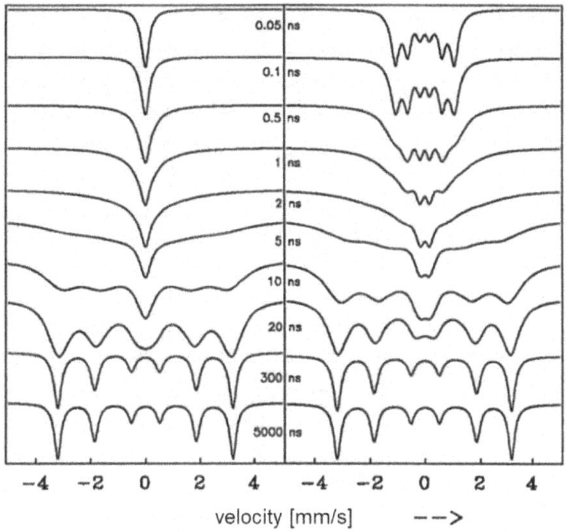

Fig. 8.9 Simulated relaxation spectra for field flip between ±20 T with an occupation probability of 1:1 left and 2:1 right for different relaxation times. Reprinted from [19]

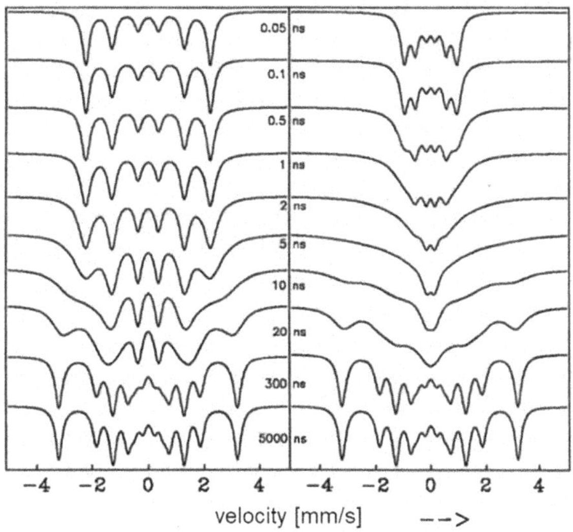

Fig. 8.10 Simulated relaxation spectra for field flip between 20 T and 8 T (left) and 20 T and −8 T (right) for different relaxation times. Reprinted from [19]

so clearly seen [35]. Alternative analyses with static hyperfine field distributions (e.g. [36]) should be taken with care [20].

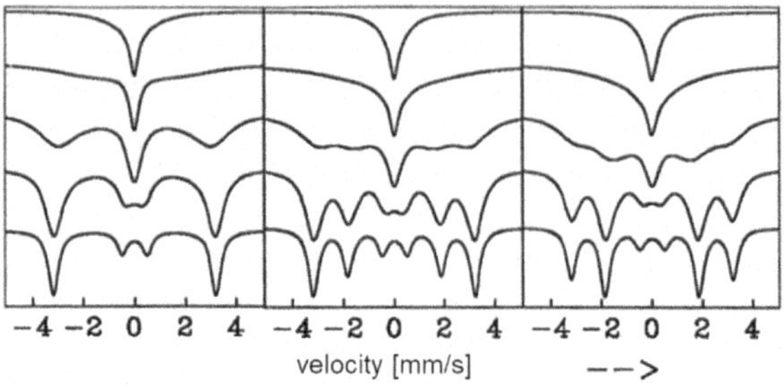

Fig. 8.11 Simulated relaxation spectra for field flip between ±20 T and different angles between the fields, assuming same occupation probabilities. Angle between fields and γ-direction left 0°(left), 54.7°(middle), and 90°(right). Relaxation times from top to bottom: 1, 3, 9, 27, 81 ns. Reprinted from [19]

8.4 Experimental

Different methods are used to apply external fields. Cheapest one, but also in many cases sufficient, is to put a permanent magnet near to the sample. By shaping the magnet as a ring around the sample, a rather homogeneous field at the absorber can be reached. Disadvantage is the rather low value of the reachable field. In general, homogeneity of the field is not critical for the Mössbauer experiment, but it should be uniform in the region where the sample is located. A uniformity of at least 1% over the measuring time, which can last more than one week, should be guaranteed. With electromagnets fields up 2 T are reachable. For higher fields Bitter magnets or superconducting coils are necessary. Commercially available superconducting solenoids for Mössbauer spectrometry have a maximum field around 15 T. For higher fields resistive solenoid magnets made by the Bitter design [37] are used. This magnets are build up by a pile of copper plates with radial slits, separated by isolating plates. The resulting distribution of the current in such coils is inversely proportional to the radius of the plates. Nearly all input energy is transformed into heat. Therefore the plates have holes and channels for transport of cooling water. A typical Bitter magnet has in the 5 cm axial bore a maximum field of 15 T consuming 5 MW power. Fields of up to 37.5 T could be reached with a Bitter magnet at the High Field Magnet Laboratory in Nijmegen, Netherlands. Thus rather high static fields can be reached with Bitter magnets, but they are available only at few places in the world. One main problem with Bitter magnets is their high level of mechanical vibrations resulting from the huge flow of cooling water. Great care has therefore to be taken in order to avoid line broadening or smearing of the spectra. Nowadays high static fields can be produced by superconducting coils made of NbTi (up to 9 T) or Nb_3Sn (up to 21 T). They are expensive and need to be cooled to liquid helium temperature, but are commercially available. Record high static fields of 45 T have been reached in a hybrid Bitter-superconducting magnet at the National High Magnetic Field Laboratory in Tallahassee, Florida. For Mössbauer spectroscopy fields 5 T are often enough to fully polarize the spectra, which makes interpretation of the measurements easier. In Figs. 8.12 and 8.13 a sketch of the 15 T high field Mössbauer equipment at the Institute of Solid State Physics, TU Wien is shown, which was installed in 1984 by Oxford Instruments. The field is generated by a system of several concentric superconducting coils. The inner ones made of Nb_3Sn and the outer ones by NbTi. The coils are in the liquid helium reservoir of a bath cryostat, which is isolated from the outside by vacuum and a liquid nitrogen shield. The field value is measured by the voltage drop over a shunt resistance placed in series with the coil set and monitored by a Hall sensor. By reducing the temperature of the liquid helium to 2.2 K by the installed lamda fridge, fields up to 15 T can be produced. The maximum field is reduced to 13.6 T if the bath temperature is 4.2 K. To hold fields constant over long time, the magnet can be switched to persistent mode, where a superconducting shortcut over the coil decouples the coil from the power supply and the current is confined in the coil. In this *persistent mode* no field reduction over a period of several days is obtained. Accuracy of field is ± 0.01 T. Field homogeneity of 1% is reached in a cylinder volume 2 mm height 15 mm diameter, where the sample is positioned. To avoid splitting of the source spectrum the ^{57}Co**Rh** source is positioned in a field compensated area, which is produced by a small compensation coil. The driving unit, based on a loudspeaker system, is situated on top of the cryostat. Because of stiffness of the rather long ($\sim$150 cm) rod which connects the source with the driving unit, only sinusoidal movement is possible. Both sample and source are mounted in the variable temperature insert (VTI), which is inserted in the bore of the magnet (Fig. 8.14). The VTI is separated from the He bath by vacuum. Via two valves liquid He can be inserted into a pot which is in thermal contact with the inner tube where the sample is located, allowing to produce temperatures at the sample between 1.5 K (by pumping above the liquid He) up to room temperature (by heating the He gas). Source and sample are in two rooms, which are separated by a window. With its own heater, the source temperature can be hold constant, independent of the temperature of the sample. Temperature of the absorber is measured by a carbon glass and a $SrTiO_3$ sensor. The first one allows to determine very precisely the temperature, whereas the second one is necessary to correct for the field dependence of the carbon glass. $SrTiO_3$ is a capacitive sensor which is not as sensitive as the carbon glass, but is practically field independent (± 1 mK at 15 T). The combination of both sensors allows to stabilize temperature during a field sweep. As detector a proportional counter is placed on bottom outside of the cryostat. Due to the large distance between source and detector of 50 cm and the seven windows, which the γ-ray has to pass on its way out of the cryostat to the detector, sources with higher activity are necessary. On the other hand the active area of the source is rather small due to the given geometry. Therefore sources with activities of approx. 35 mCi are used. Depending on the type of sample, measuring times of up to two weeks for one spectrum are not seldom, if samples are not enriched with ^{57}Fe. To change samples the whole VTI has to be removed. To avoid air entering the system, a bellow is installed at the top, which can be flushed with helium gas during the removal of the insert. For calibration of the velocity a second source is mounted on the upper side of the driving head with an α-Fe foil, a second proportional counter, and a second

Fig. 8.12 15 T high field Mössbauer equipment at the Institute of Solid State Physics, TU Wien, Austria

Fig. 8.13 Sketch of high field Mössbauer spectrometer at TU Wien

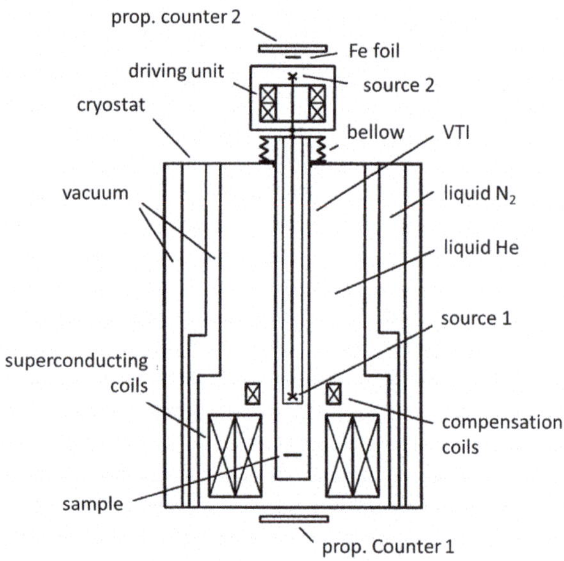

electronic system. Thus calibration spectra are taken simultaneously to the sample measurements. The stability of temperature and field are permanently checked by software, so that a temperature stability of ± 0.2 K and a field stability ± 0.01 T during the measurements is guaranteed.

Fig. 8.14 Sketch of the variable temperature insert

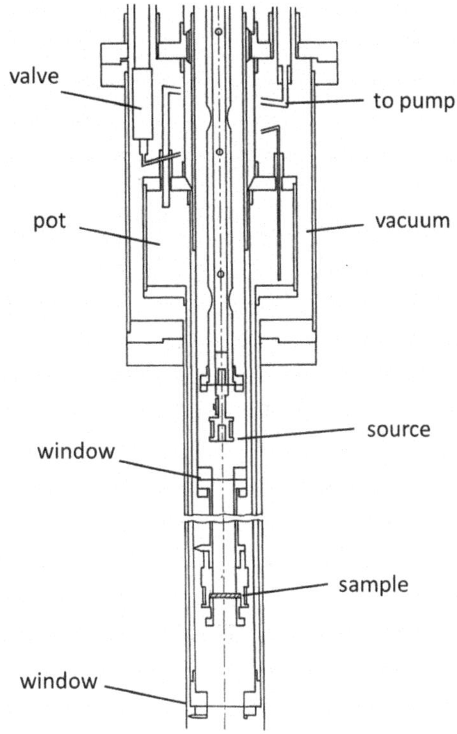

8.5 Applications

8.5.1 Ga Substituted Co Ferrite

In spinels, general structure formula AB_2O_4, magnetic ions are usually distributed over the octahedral B and the tetrahedral A sites (Fig. 8.15). In case of a two atom spinel, most prominent representative is magnetite, Fe_3O_4. For spinels, where more than one atom can occupy the A and B sites, the distribution of magnetic atoms over these sites is one of the important points for understanding the magnetic behaviour. If all A-atoms are occupying the B-sites and 50% of the B-atoms are on the A-sites than one has a so-called inverse spinel. In case of not full occupation of A-atoms on the B-sites one speaks of partial inverse spinel. For example, in case of Ga substituted Co ferrite no clear picture concerning the site distribution of the three elements Co, Ga, and Fe was obtained from different measurements. X-ray diffraction investigations indicated that Co mainly occupies only B-sites, thus suggesting an inverse spinel [38]. In contrast from neutron diffraction measurements on pure $CoGa_2O_4$ only 60% of the Ga atoms were found on the cobalt A-sites, thus pointing to partial inverse spinel [39]. ^{57}Fe Mössbauer spectroscopy on $CoGa_{2-x}Fe_xO_4$ at room temperature show two clearly separated components with different intensities, pointing to a partial inverse

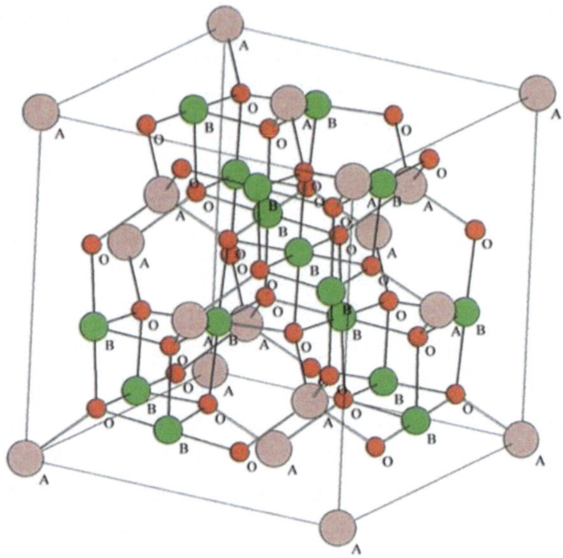

Fig. 8.15 Spinel structure AB_2O_4, with A the tetrahedral and B the octahedral sites

spinel. It was concluded that samples with higher Fe content show ferromagnetism, whereas a spin glass behaviour was proposed for the low Fe regime [40]. A final clarification was possible by ^{57}Fe Mössbauer measurements in external fields. In the following results of measurements on $CoGa_{2-x}Fe_xO_4$ samples with $x = 0.2, 0.3, 0.8$, and 1.0 in temperature 5 K to room temperature and at external fields 0, 4, 9, 13.5 T are discussed. In contrast to the zero field spectra, in-field spectra are much more complex and cannot be interpreted anymore by superposition of only two subspectra (Fig. 8.16).

Good results are obtained for the Fe-rich samples with $\Gamma/2 = 0.18$ mm/s, whereas $\Gamma/2 = 0.20$ and 0.24 mm/s are found for samples $x = 0.3$ and 0.2 on the Fe poor side. $\Gamma/2$ further increases on the Fe poor side with increasing applied field. This increase with B_a is an indication of increasing influence of relaxation effects. Especially at higher temperature typical relaxation spectra are measured (e.g. 25 K in (Fig. 8.16)). This corresponds well with findings of dc magnetic measurements, where magnetization curves typical for spin glasses are obtained. Transition temperatures of 26 and 32 K for $x = 0.2$ and 0.5 were obtained. On the Fe rich side relaxation spectra appear only above approximately 200 K (Fig. 8.16).

For analysis of the spectra below the relaxation regime (Fig. 8.17) a complex model, based on possible nearest neighbour surroundings is necessary to explain the measured Mössbauer spectra. Assuming that all three elements (Co, Ga, Fe) can occupy both A and B sites, the general formula of the compound is given by $(Co_{\lambda}Ga_{1-\lambda-y}Fe_y)_A(Co_{1-\lambda}Ga_{1+\lambda-z}Fe_z)_BO_4$ with $y + z = x$, according to a partial spinel. The final fit was performed with a superposition of several subspectra with intensity ratios determined by means of binomial distributions according to the different possible nearest neighbor (*nn*) surroundings. Taking into account only

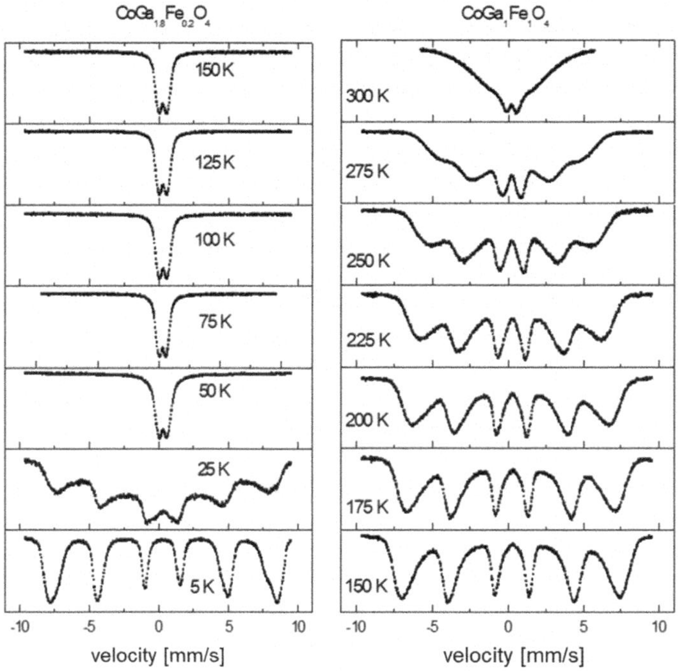

Fig. 8.16 ^{57}Fe Mössbauer spectra for $x = 0.2$ (left) and $x = 1.0$ (right) for selected temperatures

spectra with relative intensities larger than 1%, for A-sites, which have 12 *nn* (nearest neighbour) B atoms, 7–8 subspectra and for the B-sites, which have 6 *nn* B-atoms, between 5 and 6 subspectra were used in the fits. Electric quadrupole splitting, isomer shift and angle θ between external and hyperfine field were equal for the subspectra which correspond to A and B-sites respectively. $\Gamma/2$ was kept constant for all subspectra. Only B_{hf} varied between the different subspectra for the two sites. Further it was assumed, that hyperfine field increases with increasing *nn* Fe number. Under these assumptions the change of magnetic hyperfine field with applied field B_a gives detailed information about magnetic behaviour. Whereas B_{hf} increases linearly with B_a for the subspectra according to site A, B_{hf} decreases linearly with B_a for all subspectra which correspond to site B (Fig. 8.18 and lines in Fig. 8.17). With the obtained absolute value of the measured hyperfine field B_{hf} and the obtained angle θ between hyperfine field and applied field B_a the internal field B_{int} can be determined. In this way in Table 8.1 obtained B_{int} values are given together with B_{hf}, θ and B_a for both sites A and B for two samples on the Fe rich (x = 1.0 and 0.8) as well as on the Fe poor (x = 0.3 and 0.2) side. As expected, calculated B_{int} values are independent of applied field, but increase slightly with increasing Fe content x. Due to the demagnetizing field the hyperfine fields are 1 T lower at the A-site and higher at the B-site compared to the zero field values.

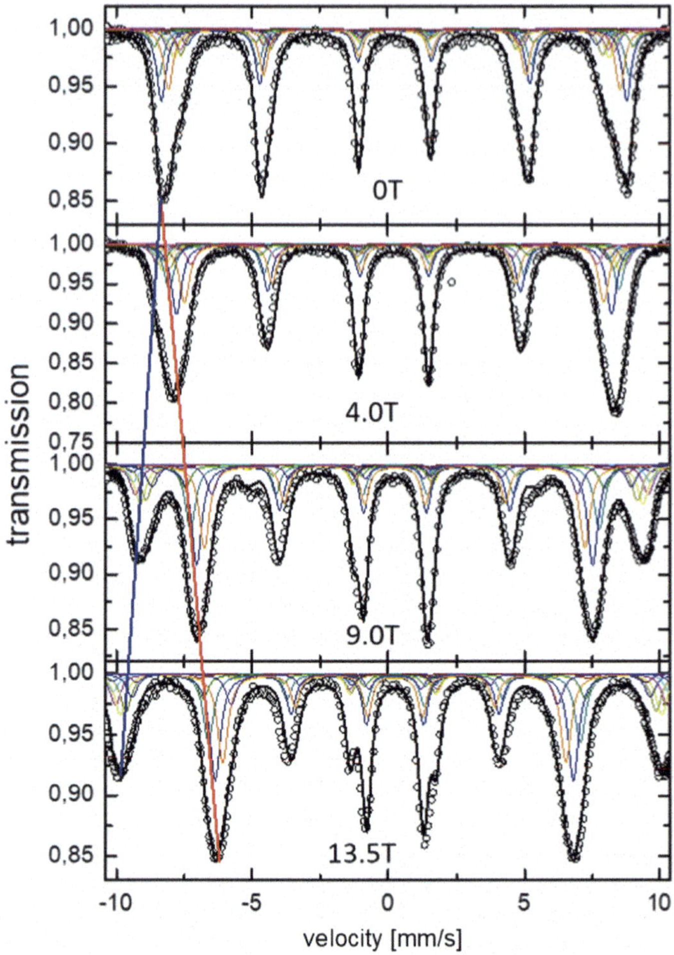

Fig. 8.17 ^{57}Fe Mössbauer spectra for $CoGa_{1.2}Fe_{0.8}O_4$ for selected fields at 4.2 K

From the obtained θ values it is seen that B_{int} is parallel to B_a, with deviations of up to 18° for A-sites. This deviations decrease with increasing Fe content x. For the B-sites B_{int} points in the direction antiparallel to B_a with deviations up to 57°. This indicates that Fe moment on A-sites are antiparallel to the Fe-moment on the B-sites. The fact that B_{int} increases with x and θ decreases with x points to an increase in coupling strength with increasing Fe content. Assuming that there is a direct proportionality between internal hyperfine field B_{int} and the magnetic moment, the fact that B_{int} is larger on B-sites than on A-sites indicates that the resultant B_{int} is antiparallel to B_a and therefore the overall moment, which is antiparallel to the internal field, is parallel to the applied field. This is in good agreement with the fact that in dc magnetic measurements a ferromagnetic like behaviour is obtained [40].

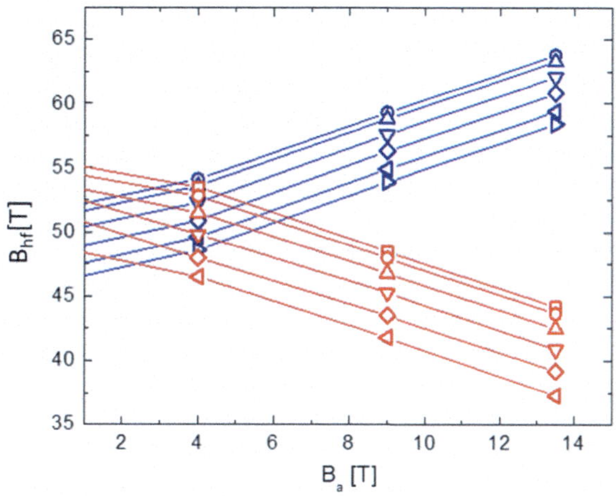

Fig. 8.18 Measured hyperfine field B_{hf} over applied field B_a for $CoGa_{1.2}Fe_{0.8}O_4$

Table 8.1 Mean values of measured hyperfine field B_{hf}, angle Θ between B_{hf} and applied field B_a and calculated internal field B_{int} for both A- and B-sites for different B_a and Fe concentration x [41]. Reprinted by permission from Springer: Hyperfine Interactions, High field Mössbauer investigations of $CoGa_{2-x}Fe_xO_4$ spinels, M. Reissner, W. Steiner, Z. Seidov, G. Sultanov, copyright (2002)

x		1,0			0,8			0,3			0,2		
	B_a	B_{hf}	θ	B_{int}	B_{hf}	θ	B_{int}	B_{hf}	θ	B_{int}	B_{hf}	θ	B_{int}
	[T]	[T]	[°]	[T]	[T]	[°]	[T]	[T]	[°]	[T]	[T]	[°]	[T]
A	0	50.1			49.0			46.8			46.2		
	4	53.2	12	49.1	51.7	18	47.9	49.2	18	45.4	49.2	18	48.0
	9	58.4	12	49.2	56.9	18	48.4	53.7	18	45.2	54.1	18	45.6
	13.5	62.9	5	49.6	61.4	18	48.7	58.4	18	45.8			
B	0	53.3			52.0			51.6			50.5		
	4	49.8	210	53.5	49.5	230	52.2	49.0	232	51.6	48.1	237	50.4
	9	45.0	205	53.3	44.8	218	52.2	44.7	230	51.0	43.0	230	49.3
	13.5	40.4	202	53.2	40.4	214	52.1	40.2	225	50.7			

With increasing x the number of iron atoms on A-sites decrease, whereas it increases on B-sites. The ratio for iron on A-sites (y) and Fe on B-sites (z) increases with x. It is 0.25, 0.33, 0.54, and 0.64 for $x = 0.2$, 0.3, 0.8, and 1.0, respectively. The same tendency was found by the X-ray investigations [38]. The application of external fields allowed to show that the different field dependence of the two components found in the spectra proves that both A- and B-sites are occupied by iron atoms. The different intensities of these two components prove that the samples are partial inverse spinels. A antiferromagnetic coupling is found between Fe on A- and B-sites. The values of the internal fields are around 10% higher for the iron atoms at the B-sites and are only mildly dependent on x.

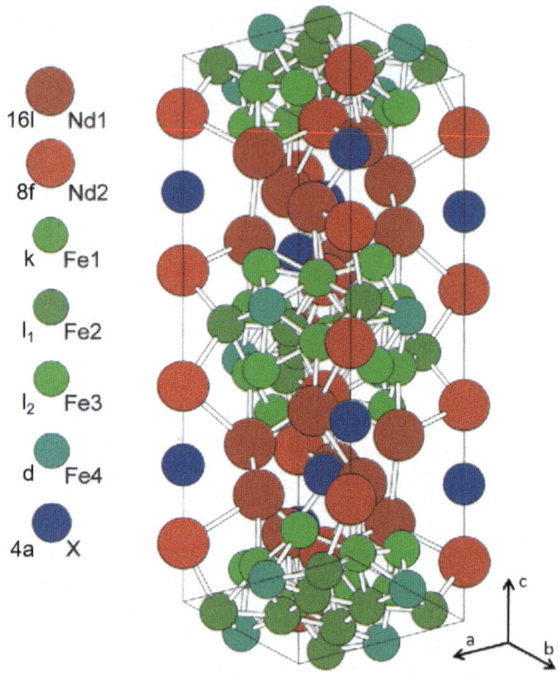

Fig. 8.19 Structure of $RE_6Fe_{13}X$ compounds

8.5.2 $RE_6Fe_{13}X$ compounds

$RE_6Fe_{13}X$ compounds where RE are light rare earth atoms and X are main group atoms from the 3rd to the 5th row of the periodic table are intensively investigated. These compounds are of interest, because they are by-products in the preparation of $Nd_2Fe_{14}B$ permanent magnets, where small additions of X metals improve wettability and corrosion resistance [42–49]. Acting as pinning centers for domain wall movement, they also have a large influence on the coercivity of the Nd-Fe-B permanent magnets [50, 51]. The crystallographic structure of $RE_6Fe_{13}X$ compounds is $Nd_2Fe_{14}B$, space group I4/mcm [52]. There are four different iron sites, namely 16k, $16l_1$, $16l_2$, and 4d (Fig. 8.19). The RE-atoms occupy the two crystallographic sites 16l and 8f. A strong influence of the RE atoms on the magnetic behaviour was found [54, 55]. From different measurements (X-ray, neutron, magnetic and Mössbauer) a ferrimagnetic or antiferromagnetic arrangement of the moments was concluded. Because of rather low magnetization values of $\mu_{sat} \leq 8\mu_B$ Weitzer et al. [52, 56] suggested a ferrimagnetic coupling within the Fe sublattice, although a canting of the RE moments could not be ruled out. On the other hand ^{57}Fe Mössbauer investigations [54] have shown, that hyperfine fields obtained from measurements take at 4 K are large and very similar for the different compounds. According to the existence of four different Fe-sites four sub-spectra with intensity ratio of 4:4:4:1 according to the k, l_1, l_2, and the d sites are expected. This holds for many of the compounds e.g.

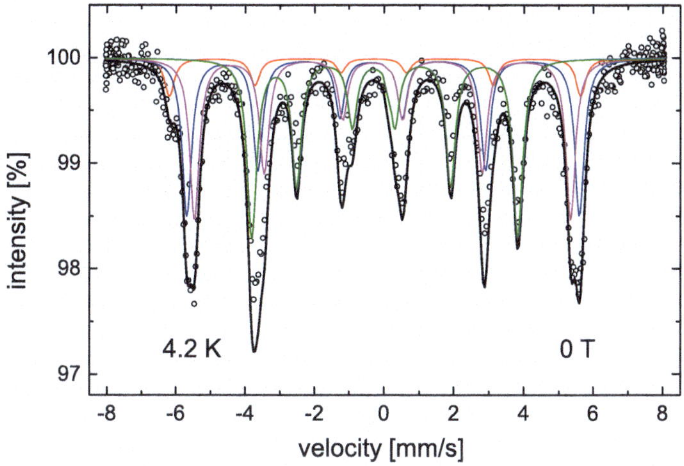

Fig. 8.20 Zero-field ^{57}Fe Mössbauer spectra of $Nd_6Fe_{13}Sn$ at 4.2 K [53]. Reprinted from J. Mag. Mag. Matter, 226–230, R. Ruzitschka, M. Reissner, W. Steiner, P. Rogl, Magnetic and high field Mössbauer investigations of $RE_6Fe_{13}X$ compounds, 1443–5. Copyright (2001), with permission from Elsevier

$Nd_6Fe_{13}X$ with X = In, Sn, Tl, Pb [54] (Fig. 8.20). Such interpretation holds, if the contribution of anisotropic dipole fields to the hyperfine fields are so small that they can be neglected. If they are not negligible small, the possibility to fit the spectra with the intensity ratio given by the occupation ratio of the lattice sites indicates, that the easy axis of the magnetization is the *c*-axis. In that case a uniform distribution of dipole fields is obtained, which only changes the magnitude of the hyperfine fields. In case of a deviation of the easy axis from the *c*-direction, due to the various dipole field contributions, the degeneracy of magnetically equivalent lattice sites may be lifted leading to several subspectra, the number of which are determined by the local symmetry [57].

In that case more subspectra are needed for one lattice site. For the given structure no change is expected for the $16l_1$, $16l_2$, and 4d sites, but a splitting in two components is expected for the 16k sites [54]. In case of the easy axis within the basal plane along one of the edges an intensity ratio of 1:1 for these sites is expected (e.g. $Pr_6Fe_{13}Pd$ Fig. 8.21). Indeed, for many compounds ($Nd_6Fe_{13}X$, with X = Cu, Ag, Au and for $Pr_6Fe_{13}X$, with X = Cu, Ag, Au, In, Sn, Tl, Pb) in contrast to four subspectra five subspectra with intensity ratio 2:2:4:4:1 are necessary to fit the measured spectra (Figs. 8.20 and 8.21). Interestingly the values of the obtained hyperfine fields for the different sites are rather independent of the RE and X elements. On the other hand there is a large spread in the value of the hyperfine fields for the different sites, varying by more than 30% [52]. With the obtained hyperfine fields and assuming a direct relation between hyperfine field and magnetic moment, various ferrimagnetic arrangements of moments on the different lattice sites were proposed to explain the rather small Fe-moments obtained in lower fields [52]. Yan et al. [58] and Wang

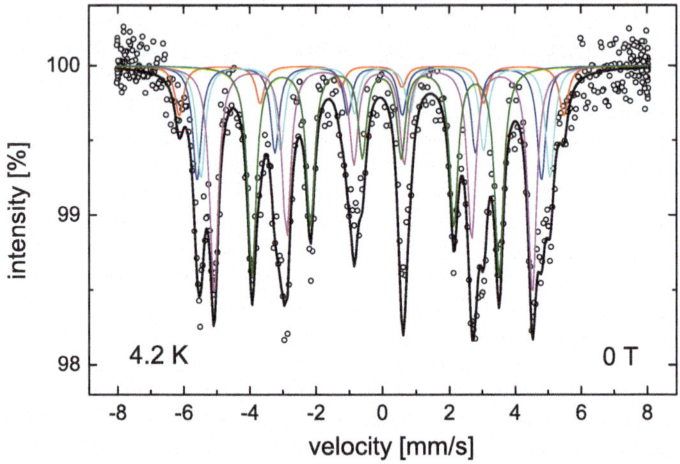

Fig. 8.21 Zero-field ^{57}Fe Mössbauer spectra of $Pr_6Fe_{13}Pd$ at 4.2 K [53]. Reprinted from J. Mag. Mag. Matter, 226–230, R. Ruzitschka, M. Reissner, W. Steiner, P. Rogl, Magnetic and high field Mössbauer investigations of $RE_6Fe_{13}X$ compounds, 1443–5. Copyright (2001), with permission from Elsevier

et al. [59] proposed from X-ray and neutron measurements that the RE moments and the Fe 4d moments are parallel and antiparallel to the other Fe atoms with an easy axis in c-direction for $Nd_6Fe_{13}Si$ and $Pr_6Fe_{13}Si$. In contrast, Schobinger-Papamantellos et al. [60, 61] proposed also from neutron investigations for $Pr_6Fe_{13}Si$ collinear antiferromagnetic ordering of the four Fe and of the two RE sublattices, with the easy axis perpendicular to the c-direction. For $Pr_6Fe_{13}Sn$ and $Nd_6Fe_{13}Sn$ best interpretation of the neutron data were obtained by assuming that all moments in the blocks defined by all atoms inbetween the X-atom layers are ferromagnetically ordered [62]. The blocks itself are antiparallel to each other. The difference between the two samples is, that in case of Pr the easy axis is perpendicular to the c-direction and parallel in case of the Nd compound. The picture becomes even more complex with the result of dc magnetic measurements in higher applied fields, where at low temperatures jumps in the magnetization curves are found. Figure 8.22 shows for example magnetization curves for $Pr_6Fe_{13}Pd$, with a strong jump around 7 T at 2.5 K and obviously saturation at 15 T. With the free ion value of 3.58 μ_B for Pr and the obtained value of 37.5 μ_B/fu at 14.5 T and 2.5 K a Fe-moment of only 1.23 μ_B is obtained. This is much lower than values of compounds with similar RE/Fe ratio like $REFe_2$ with moments of about 1.77 μ_B/Fe [63]. In contrast, for $Nd_6Fe_{13}Sn$ no saturation is found at lowest temperatures and highest fields (4.2 K, 15 T). Here at 8 T a jump in magnetization is present, followed by a second one at 13 T for $T <$ 80 K (Fig. 8.23). The field of the jumps is nearly temperature independent. Such jumps are typical for metamagnetic materials where an antiferromagnetic moment arrangement at low temperatures change with increasing field to a ferromagnetic one by reversal of the moments of one of the two sublattices into the direction of applied field. In case of the $RE_6Fe_{13}X$ compounds the jumps are too small to be explained by reversal of one of the ferromagnetically coupled sublattices present in the above mentioned proposed antiferromagnetic ordered moment arrangements, indicating that the till then proposed models are too simple. This is confirmed by ^{57}Fe Mössbauer high field measurements. Figures 8.24 and 8.25 show two typical spectra measured 4.2 K at 13.5 T. Comparison with the measurements on the same samples in zero field (Figs. 8.20 and 8.21) show quite different shape. The sharp structure of the spectra in zero field is washed out by very flat side wings in the in-field spectra. Therefore the in-field spectra can by no means be fitted by only 4 ($Nd_6Fe_{13}Sn$) or 5 ($Pr_6Fe_{13}Pd$) subspectra according to the different Fe-sites. A further subdivision of the 4 (5) subspectra was necessary. By restricting the overall number of subspectra to nine, 4, 2, 2, and 1 spectra are used for the k, l_1, l_2, and d site, respectively, keeping the intensity ratio of 4:4:4:1 for the 4 sites. The fit gives values for the magnitude of the hyperfine field and the angle θ between measured hyperfine field at the Fe nucleus and the γ-ray direction. From the difference of hyperfine field and applied field B_a, by taking into account θ, the internal field B_{int} could be determined. The obtained mean values for each crystallographic iron site are in good agreement with the values obtained from the zero-field measurements in case of $Nd_6Fe_{13}Sn$. The projection of B_{int} on the B_a direction gives 55% of the value of B_{int}. This 55% are also obtained by comparing the magnetic moment at 13.5 T and 4.2 K to the moment calculated assuming full alignment of RE^{3+} moments and Fe moments of 1.77 μ_B. For the $Pr_6Fe_{13}Pd$ compound at 13.5 T, B_{int} values which are higher than those at zero field are necessary to get agreement with the degree of saturation found in magnetization. On the other hand, taking both the RE and Fe moment of the neutron refinement [62] leads to a saturation moment which is much higher than the one observed. This together with the found deviations of hyperfine field from complete alignment with B_a indicates a further jump of magnetization at even higher applied fields. The high number of subspectra for the different Fe-sites needed to get reasonable in-field fits points to a magnetic structure which is not simply antiferromagnetic but indicates strong tilting of the spins around the antiferromagnetic alignment of the individual iron layers.

8.5.2.1 Skutterudites

The mineral skutterudite $CoAs_3$ has given its name to a large class of substances. Its structure was first solved by Oftedal in 1928 [64]. Binary skutterudites MPn_3 are formed by many atoms, with M = Co, Rh, Ir and Pn stands for pnictides (P, As, and Sb). The structure consists of a three-dimensional array of slightly distorted octahedra formed by the pnictide atoms, with the M atom in the center. The octahedra are tilted in such a way, that a rectangular arrangement of Pn atoms form, which connect the adjacent octahedra. Due to this tilting large cage-like voids are created in the structure, which can be filled by electropositive atoms A forming the large class of ternary skutterudites $A_xM_4Pn_{12}$ (Fig. 8.26).

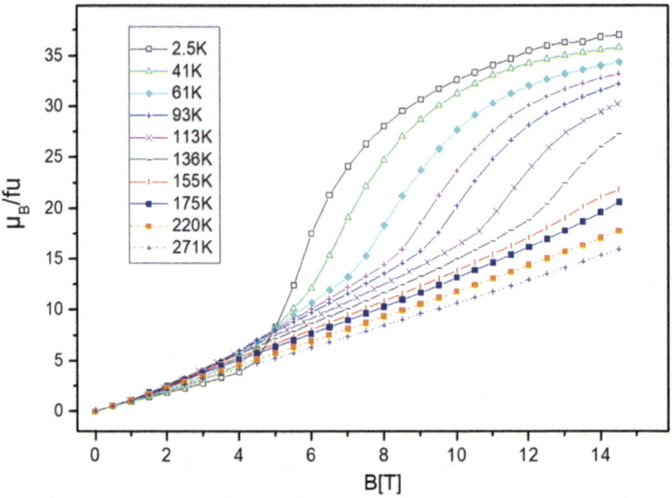

Fig. 8.22 Field dependence of magnetic moment of $Pr_6Fe_{13}Pd$ for different temperatures [55]. Reprinted from J. Mag. Mag. Matter, 242, R. Ruzitschka, M. Reissner, W. Steiner, P. Rogl, Investigation of magnetic order in RE6Fe13X (RE = Nd, Pr; X = Pd, Sn, Si), 806–8. Copyright (2002), with permission from Elsevier

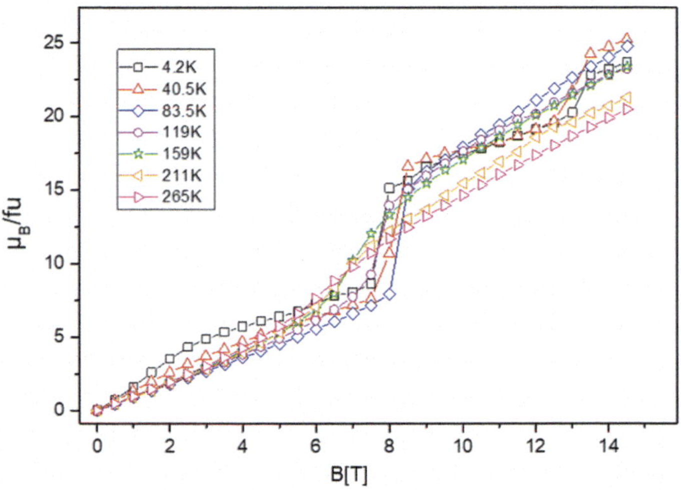

Fig. 8.23 Field dependence of magnetic moment of $Nd_6Fe_{13}Sn$ for different temperatures [55]. Reprinted from J. Mag. Mag. Matter, 242, R. Ruzitschka, M. Reissner, W. Steiner, P. Rogl, Investigation of magnetic order in RE6Fe13X (RE = Nd, Pr; X = Pd, Sn, Si), 806–8. Copyright (2002), with permission from Elsevier

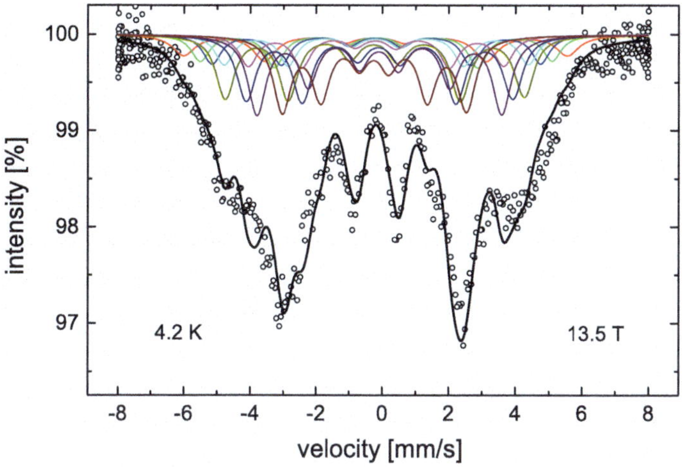

Fig. 8.24 [57]Fe Mössbauer spectrum of $Nd_6Fe_{13}Sn$ taken at 13.5 T and 4.2 K [53]. Reprinted from J. Mag. Mag. Matter, 242, R. Ruzitschka, M. Reissner, W. Steiner, P. Rogl, Magnetic and high field Mössbauer investigations of $RE_6Fe_{13}X$ compounds, 1443–5. Copyright (2001), with permission from Elsevier

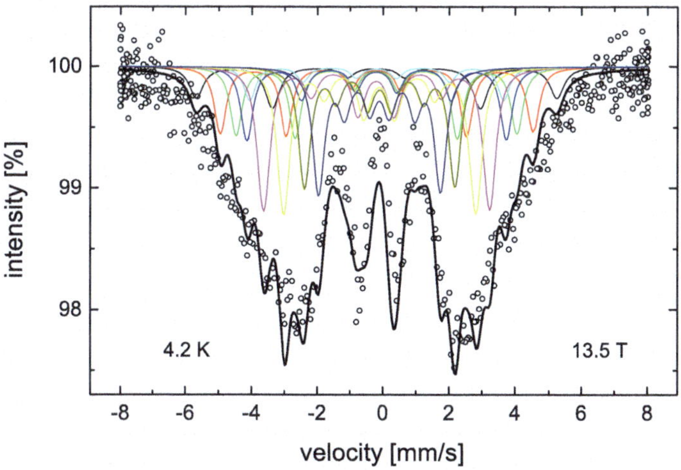

Fig. 8.25 [57]Fe Mössbauer spectrum of $Pr_6Fe_{13}Pd$ taken at 13.5 T and 4.2 K [53]. Reprinted from J. Mag. Mag. Matter, 242, R. Ruzitschka, M. Reissner, W. Steiner, P. Rogl, Magnetic and high field Mössbauer investigations of $RE_6Fe_{13}X$ compounds, 1443–5. Copyright (2001), with permission from Elsevier

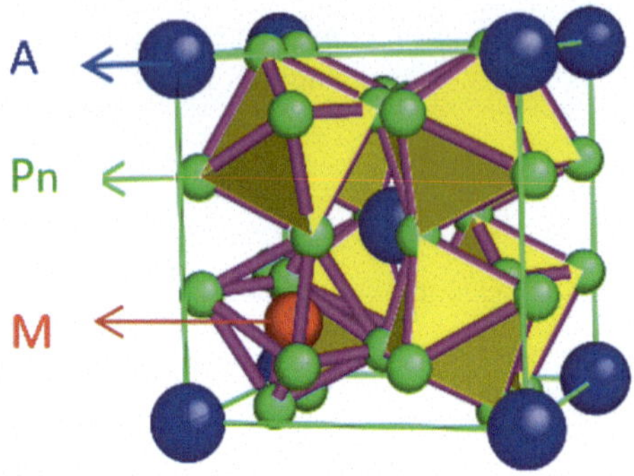

Fig. 8.26 Crystal structure of filled skutterudites

Table 8.2 Intensities of the two subspectra necessary to fit the Mössbauer spectra of $RE_xFe_4Sb_{12}$ compounds and effective magnetic moments from susceptibility measurements

RE-atom	x	Int1/%	Int2/%	μ_{eff}/μ_B
La	0.80	78	22	3.0
Pr	0.73	64	36	4.2
Nd	0.72	69	31	4.5
Eu	0.88	84	16	8.4
Yb	1.00	100	0	4.5

These ternary skutterudites are one example of so-called cage compounds. The first ternary skutterudite $LaFe_4P_{12}$ was synthesized by Jeitschko et al. in 1977 [65]. Since then a large amount of such compounds were found. As filler atoms (A) trivalent light RE (La, Ce, Pr, Nd, Sm, Eu) elements and Yb, as well as Th and U were successfully build into the structure. But also divalent ions as earth alkali (Ca, Sr, Ba) and also monovalent alkali ions (Na, K) and also Tl could be incorporated into the structure. Heavy RE skutterudites (Ho, Er, Tm) could be synthezised under high pressure. In contrast to binary skutterudites, in the ternary compounds also Fe, Os, and Ru are possible on the M site. Skutterudites are of special interest, because of their potential for thermoelectric applications [66–70]. Thermoelectric materials can convert heat in electricity. They are used as power generators in satellites, to cool computer chips and to convert heat of exhaust fumes of trucks.

To be a good thermoelectric material the figure of merit $ZT = S^2(\lambda/\sigma)$ with S the Seebeck coefficient, λ the thermal conductivity and σ the electrical conductivity should be at least 1. Such high value is possible, if good electrical conductivity is, against Wiedemann-Franz law, accompanied by low thermal conductivity, which is possible, if there are low lying Einstein modes in the phonon spectrum, which can be realized by the filler ions which are only weakly bonded in the oversized cages and therefore tend to rattle, thus hindering phonon propagation. In that sense filled skutterudites are archetypes of so-called phonon glass electron crystals, a concept introduced by Slack in 1995 [71].

Beside the high importance for applications, these materials are of interest because of the large variety of possible ground states. Depending on the RE atom features like superconductivity in $LaRu_4As_{12}$ (T_c = 10.3 K) [72], $LaOs_4As_{12}$ (T_c = 3.2 K) [72], $PrRu_4Sb_{12}$ (T_c = 1 K) [73], $PrRu_4P_{12}$ (T_c = 2.4 K) [74], where for the last compound a metal insulator transition is present at 60 K [75], long range magnetic order in $EuFe_4Sb_{12}$ (T_{mag} = 84 K) [76, 77], heavy fermion behaviour in $YbFe_4Sb_{12}$ [78], non-Fermi liquid behaviour in $CeRu_4Sb_{12}$ [79], mixed valence behaviour in $RE(Co,Fe)_4Sb_{12}$ with RE = Yb and Eu [80, 81], are found. In case of the Fe containing skutterudites also a large variety of magnetic ground states is reported. E.g. $REFe_4P_{12}$ compounds with RE = Nd, Eu, Ho are ferromagnetic with ordering temperatures 1.9 K, 80 K, and 5 K, respectively [76]. Also $SmFe_4Sb_{12}$ is ferromagnetic below 45 K. $NaFe_4Sb_{12}$, and KFe_4Sb_{12} skutterudites are itinerant ferromagnets with ordering temperatures 85 K for both [82]. $TlFe_4Sb_{12}$ was found to be a weak itinerant ferromagnet [83]. $PrFe_4Sb_{12}$ is antiferromagnetic with Neél temperature of 4.6 K [84]. For $PrFe_4P_{12}$ antiferroquadrupolar interactions play an important role below 6.2 K [85–87]. AFe_4Sb_{12} compounds with A = Ca, Sr, Ba, Tm, and Yb are paramagnetic respectively nearly ferromagnetic [83]. As mentioned above $LaFe_4P_{12}$ is superconducting below 4 K [88], whereas $LaFe_4Sb_{12}$ is an enhanced paramagnet [76, 82]. One important question is, how the Fe-atoms contribute to the magnetic behaviour. In spite of large amount of theoretical and experimental investigations a precise knowledge about Fe-moments and their interplay with the filler atoms are still missing [70]. E.g. the $LaFe_4P_{12}$ compound is as mentioned above superconducting, indicating that Fe has no moment, although from susceptibility measurements a room temperature effective moment of 1.46 μ_B/fu is obtained. Because La^{3+} has no magnetic moment, the measured one has to be attributed to the (Fe_4Sb_{12}) building blocks. Band structure calculations on $La(Co,Fe)_4P_{12}$ indicate hybridization of the La sites with Sb and Fe states resulting in an enhanced effective mass for the two highest occupied bands [89]. Furthermore a double peak structure of the 3d-DOS in the proximity of the Fermi energy was obtained, from which the presence of a non-zero moment on the Fe site was concluded [90]. Newer studies pointed out that spin fluctuations are important and that this compound seems to be near to a ferromagnetic quantum critical point [91]. The question about the contribution of Fe to the magnetization is further puzzling, if the Sb compounds are considered which show effective moments of several μ_B depending on the type of filler atom (Table 8.2). It should be mentioned that in contrast to the Fe skutterudites based on P, in the Sb based Fe skutterudites the RE sublattice is not always fully occupied. In case of the $Pr_{0.73}Fe_4Sb_{12}$ compound an effective moment of 4.19 μ_B/fu is found from magnetic measurements. Figure 8.27 shows magnetization curves at different temperatures. The compound orders around 5 K [84]. Above 52 K the bending of the magnetization curves disappears. The susceptibility determined from the slope of the $M(B_a)$ curves measured at various temperatures is shown on right side of Fig. 8.27. From this an effective moment of 4.19 μ_B and a paramagnetic Curie temperature $\theta_p = 0.5$ K is obtained in good agreement with findings of [76]. Assuming that the Pr moment is the one of the 3+ ion, the moment of the (Fe_4Sb_{12}) block can be calculated. With the assumption that the contributions are simply additive according to

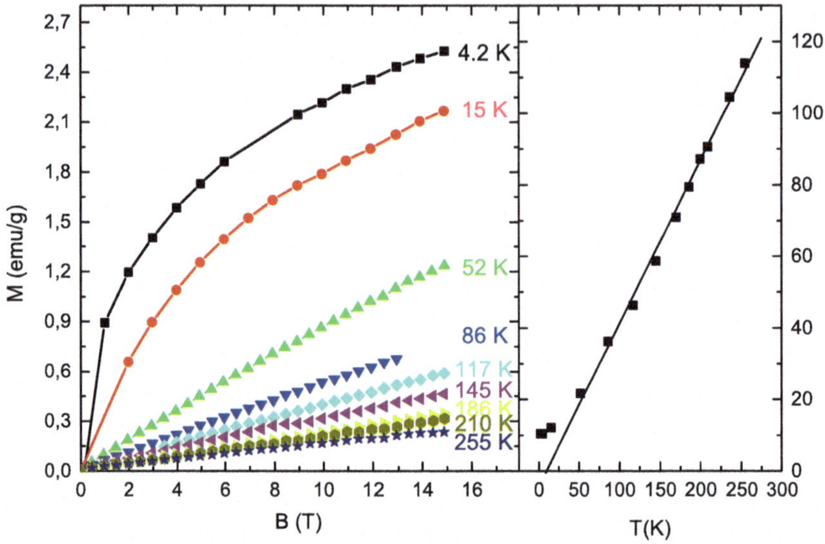

Fig. 8.27 Left: field dependence of magnetization at different temperatures for $Pr_{0.73}Fe_4Sb_{12}$. Right: temperature dependence reciprocal susceptibility [92]. Reprinted from J. Mag. Mag. Matter, 272–276, M. Reissner, E. Bauer, W. Steiner, P. Rogl, High field Mössbauer and magnetic investigations of $Pr_{0.73}Fe_4Sb_{12}$, 813. Copyright (2004), with permission from Elsevier

$$\mu_{eff}^{meas} = \sqrt{x \cdot (\mu_{eff}^{Pr})^2 + (\mu_{eff}^{Fe_4Sb_{12}})^2}$$

with x the filling factor of the RE sublattice, a rather high effective moment for the (Fe_4Sb_{12}) building block of 2.7 μ_B is obtained. Similar high effective moments of 3.0 μ_B and 3.7 μ_B are obtained for $LaFe_4Sb_{12}$ and $CaFe_4Sb_{12}$ [76], which have to be primarily attributed to the magnetic behaviour of Fe. The result that Fe carries a moment in the $Pr_xFe_4Sb_{12}$ skutterudite is in full contrast to $PrFe_4P_{12}$, where the obtained effective moment matches perfectly the Pr^{3+} value. It should be mentioned that band structure calculations of $LaFe_4Sb_{12}$ support the possibility that Fe has a moment in this compound [90]. Assuming that the DOS of $PrFe_4Sb_{12}$ resembles that of $LaFe_4Sb_{12}$ the magnetic moment ascribed to (Fe_4Sb_{12}) comes from a double peak structure of the Fe-d partial DOS below the Fermi energy. On the other hand Tanaka et al. have shown that in a full filled $Pr_1Fe_4Sb_{12}$ sample a singlet ground state and no magnetic order should be present [91]. The appearance of Fe-moments may therefore be connected to vacancies in the RE-sublattice. To check this, in field Mössbauer measurements are a good method to contribute to this debate. Shenoy et al. [93] were the first who investigated a $LaFe_4P_{12}$ compound with Mössbauer spectroscopy in field. They concluded that a possible Fe moment has to be smaller than 0.01 μ_B. Therefore a larger survey of different Fe bearing skutterudites $A_xFe_4Pn_{12}$, with A equal to trivalent La, Pr, Nd, Eu, Yb, divalent Ca, Sr, Ba, and monovalent

Na, K, Tl most with Pn = Sb, but also some phosphorous based compounds was performed. Figure 8.28 shows the low temperature spectrum of two phosphorous based skutterutides $PrFe_4P_{12}$ and $NdFe_4P_{12}$ in zero field. The spectra can be fitted by only one doublet without any sign of line broadening. This is as expected, because in the structure only one crystallographic Fe site is present and X-ray diffraction have confirmed full occupation of the RE sublattice. In contrast for the antimony based RE compounds low temperature zero-field spectra are slightly asymmetric. This asymmetry is more pronounced in the spectra taken in external fields (Figs. 8.29 and 8.30). Because of the structure type, texture as reason for the asymmetry can be ruled out. Two subspectra are necessary to fit the spectra. Only the Yb sample needs only one subspectrum to fit the data satisfactorily well within the measuring accuracy, in agreement with the fact that Yb sublattice is fully occupied [95]. The intensities of the two subspectra is given in Table 8.2. Intensity of the majority subspectrum is within measuring accuracy in good agreement with the occupation number of the RE-atoms. Looking on the local surrounding of the Fe atoms there are 6 pnictogenic atoms forming the octahedron in the first shell and 12 Sb atoms in the next nearest shell together with two electropositive filler atoms (Fig. 8.31). The fact that especially in the RE sublattice the filling factor is smaller than 1, the Fe atom may have 0, 1 or 2 filler atoms in the second shell. The other Fe atoms are in a larger distance and may thus influence the central Fe atom only by a small amount. According to the filling factor x, probabilities concerning the frequency of the respective surrounding can be calculated by binomial distribution. E.g. for $x = 0.9$ probabilities of 0.81, 0.18, and 0.01 are obtained for the case to have 2, 1, or 0 RE atoms in the next nearest neighbour shell. Since the probability to have no filler atom in the next nearest neighbour shell is rather small, it can be added to the case to have 1 filler atom in this shell, thus giving an expected intensity ratio for the two subspectra of 81 to 19. The thus obtained values are in good agreement to the ones obtained from the Mössbauer fits (Table 8.2). A similar scenario was suggested for the Co based skutterudites $Tl_{0.8}Co_3FeSb_{12}$ and $Tl_{0.5}Co_{0.35}Fe_{0.5}Sb_{12}$ by Long et al. [96], but called into question, because of deviations of the observed area ratio of the subspectra from the ones calculated by statistical distribution. Above 4 T the spectra are fully polarized—visible by the vanishing of the $\Delta m = 0$ transitions (Figs. 8.28, 8.29 and 8.30). For the La and Yb compounds which show no magnetic order [76, 95] the values of the measured hyperfine fields B_{hf} for the subspectrum allocated to the component with the high intensity either coincide with the one of the applied field B_a, or were slightly larger. Significant deviations from the value of B_a were only obtained for Fe atoms allocated to the spectra with the small area. Similar behaviour is obtained for the Pr and Nd compounds, although according to bulk magnetic measurements they are magnetically ordered (ordering 5 K and 13 K for Pr [84] and Nd [76] (Fig. 8.27)). For the Eu compound an ordering temperature of 84 K is present [76, 77]. Calculated induced hyperfine fields $B_{ind} = B_{hf} - B_a$ are shown in Fig. 8.32. For the La, Pr, and Nd compounds B_{ind} exhibits some tendency towards saturation at high applied fields for the Fe site with the low intensity subspectrum, whereas for the Fe atoms allocated to the high intensity subspectrum B_{ind} scatters around zero (Fig. 8.32). For Yb the induced hyperfine field is within measuring accuracy also zero. In case of Eu

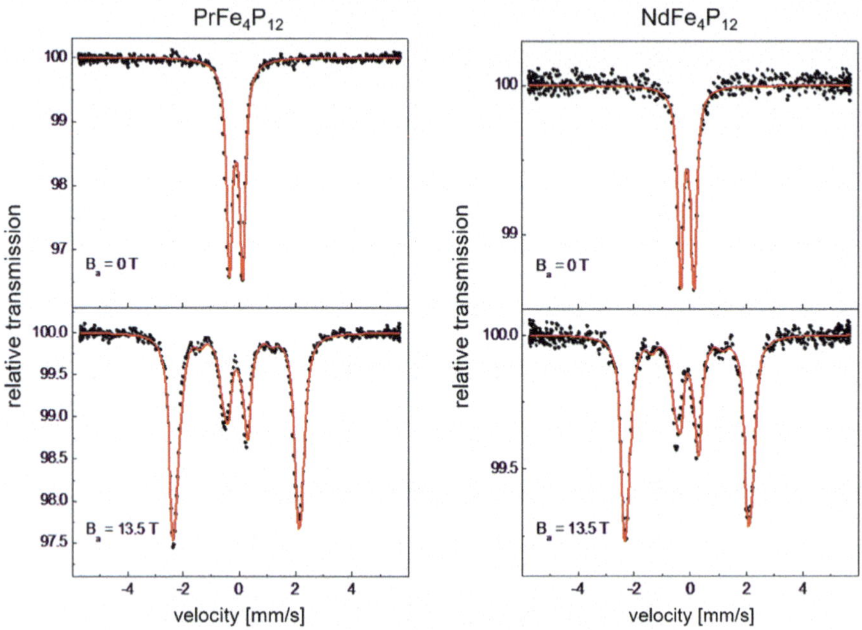

Fig. 8.28 Zero and high field spectra of $PrFe_4P_{12}$ (left) and $NdFe_4P_{12}$ (right) taken at 4.2 K

the induced hyperfine fields for both sites are negative. This indicates, that valence and core contribution to the hyperfine field are of comparable magnitude. If these contributions are of opposite sign and of similar magnitude, the measured hyperfine field can be rather small. As the magnetic moment is proportional only to the core contribution, such scenario can explain the small values of B_{ind} in comparison to the large effective moments obtained from magnetic measurements. This interpretation is strongly supported by ASW-FSM calculations [97, 98].

Changing now the trivalent RE filler atoms by monovalent Na, K, Tl and divalent Ca, Sr, Ba atoms, obtained spectra are very similar (Figs. 8.33 and 8.34). Again the spectra are asymmetric, demanding the use of at least two subspectra for interpretation of the measured spectra. The only one where one sub-spectrum is enough is the Ba compound (Fig. 8.33). An explanation for the second subspectrum in terms of voids in the filler subspectrum is in case of the di- and monovalent filler atoms not possible as the filling factor from chemical and X-ray analyses is in all cases higher than 98%, whereas the intensity of the second subspectrum is around 20%. However, with the exception of $TlCo_3FeSb_{12}$ [99] at present no experimental clue of theoretical hints exist for other interpretation of the difference in charge density at the Fe site in metallic skutterudites. For sub-stoichiometric Co-based skutterudites the existence of $CoSb_3$ gives the possiblility of another approach. A solid solution of a completely filled Fe compound in an unfilled Co compound was assumed to be realized in $Ce_xFe_{4-y}Co_ySb_{12}$ [100, 101].

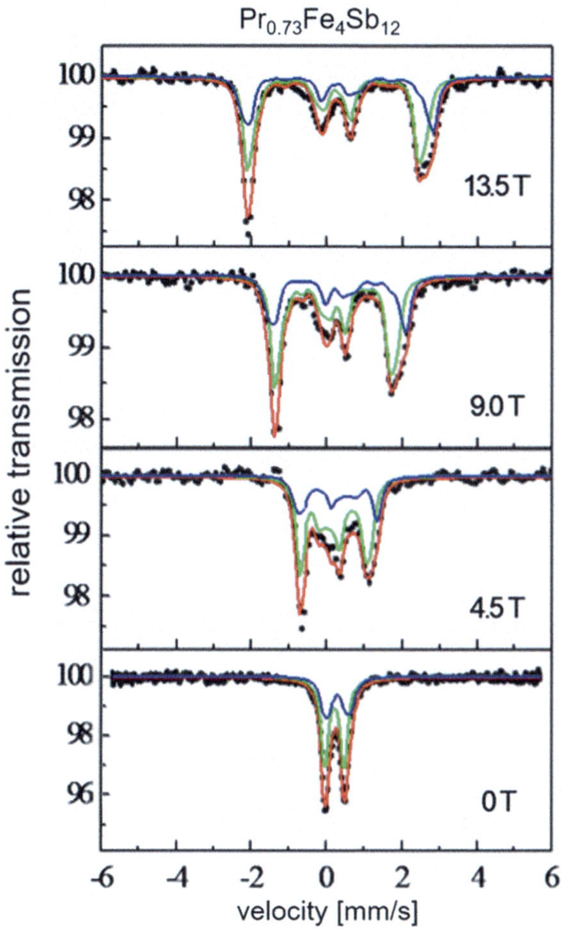

Fig. 8.29 Mössbauer spectra for $Pr_{0.73}Fe_4Sb_{12}$ at 4.2 K and different applied fields [92]. Reprinted from J. Mag. Mag. Matter, 272–276, M. Reissner, E. Bauer, W. Steiner, P. Rogl, High field Mössbauer and magnetic investigations of $Pr_{0.73}Fe_4Sb_{12}$, 813. Copyright (2004), with permission from Elsevier

The obtained induced hyperfine fields are similar to the ones for the RE skutterudites (Fig. 8.35). They all are less than 2 T. These values are too low to explain directly the effective moment assigned to the (Fe_4Sb_{12}) blocks from bulk magnetic measurements at high temperatures. Charge counting arguments that the moments on the (Fe_4Sb_{12}) blocks are due to the unpaired spins of Fe^{3+} in low spin configuration [76] are therefore wrong. The results support the picture of itinerant ferromagnetism with small ordered moments for the investigated di- and monovalent skutterudites. In summary, with the help of the in-field Mössbauer measurements it could be shown, that the magnetic structure is much more complex than expected from the simple crystallographic structure.

Fig. 8.30 Mössbauer spectra for $Eu_{0.88}Fe_4Sb_{12}$ 4.2 K and different applied fields [94]. Reprinted by permission from Springer: Hyperfine Interactions, Skutterudites, a thermoelectric material investigated by high field Mössbauer spectroscopy, M. Reissner, E. Bauer, W. Steiner, P. Rogl, A. Leithe-Jasper, Y. Grin, copyright (2008)

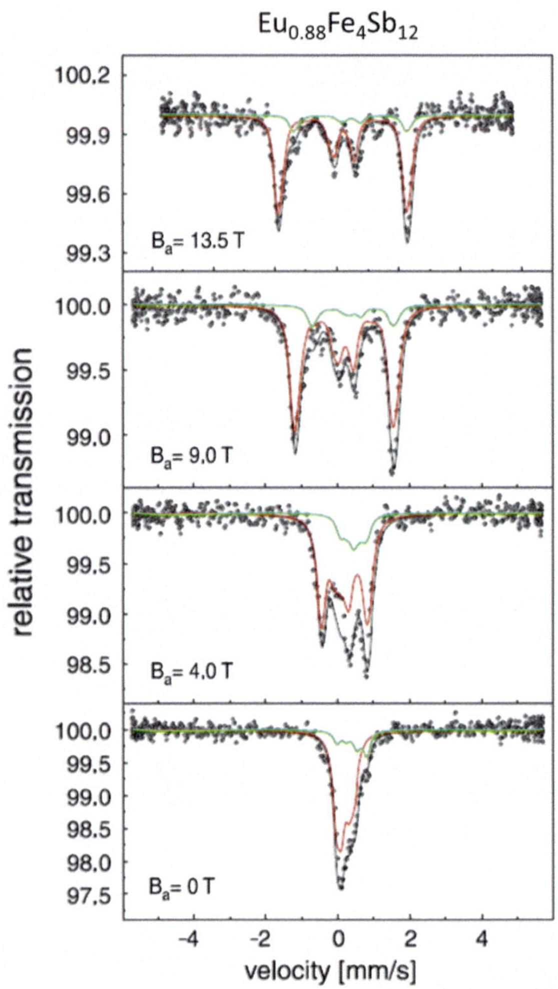

Fig. 8.31 Local surrounding of Fe atom. Large blue, medium green, and small red spheres denote A, Pn, and M atoms of AM_4Pn_{12}

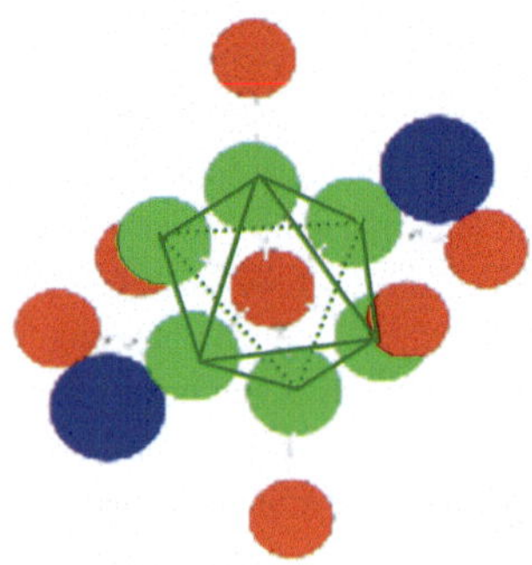

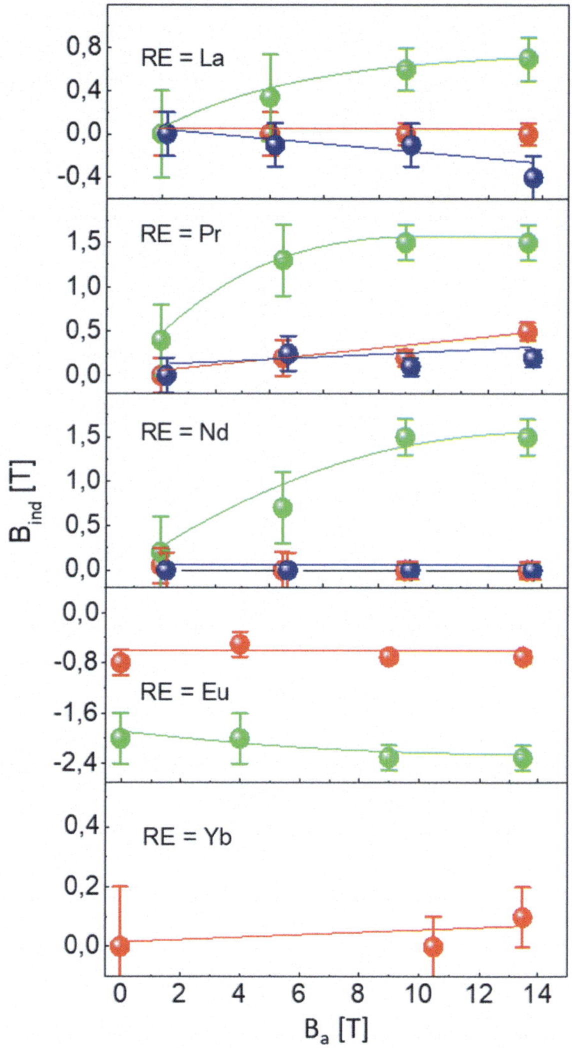

Fig. 8.32 Induced hyperfine field for $REFe_4X_{12}$ for X = Sb (red and green) and P (blue)

8.5.3 *Spin Glasses*

Spin glasses are mostly metallic alloys, showing magnetic, electronic and thermal anomalies below a characteristic temperature. Since the discovery of spin glasses in 1972 by Cannella and Mydosh [102] many good reviews and books have been published [103–107]. Spin glass behaviour was first found in noble metals with small amounts of impurities of transition metals, e.g. **Au**Fe, **Cu**Mn, and **Ag**Mn [102], but a lot of other materials turned out to show similar properties. Spin glasses are the paragon of disordered systems. There exists magnetic short range order which can

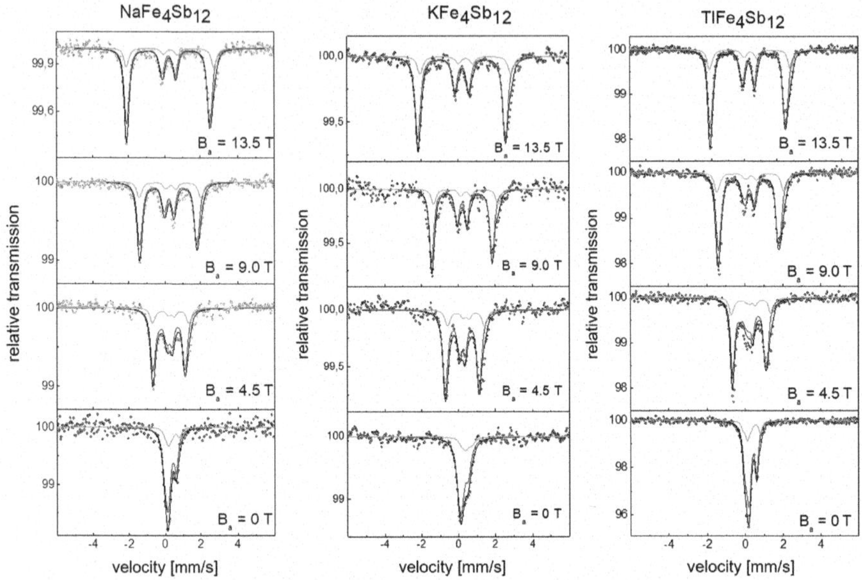

Fig. 8.33 Mössbauer spectra for monovalent filler atoms 4.2 K in different applied fields

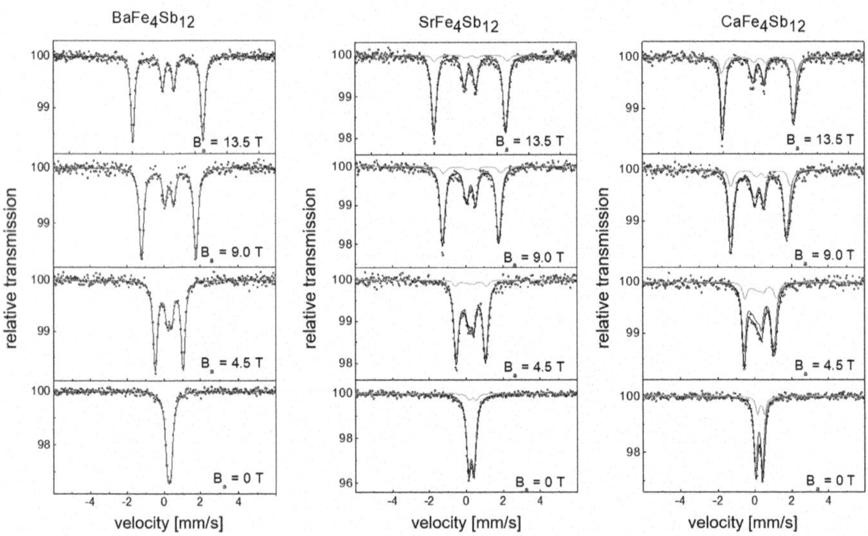

Fig. 8.34 Mössbauer spectra for divalent AFe_4Sb_{12} skutterudites for different fields at 4.2 K

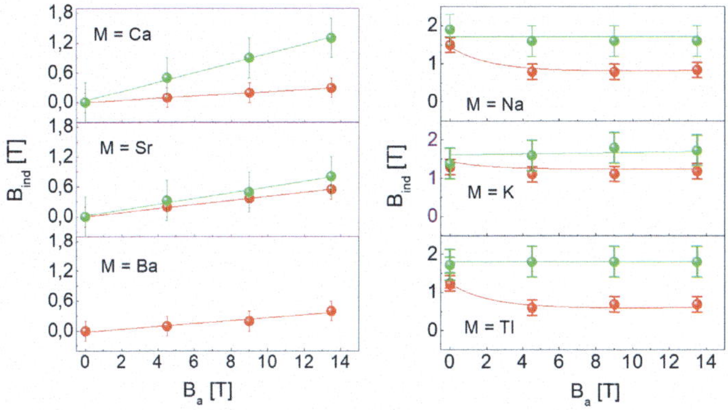

Fig. 8.35 Induced hyperfine fields for divalent (left) and monovalent (right) Fe-Sb skutterutides

Fig. 8.36 RRKY interaction J(r) between impurities at 0, A and B

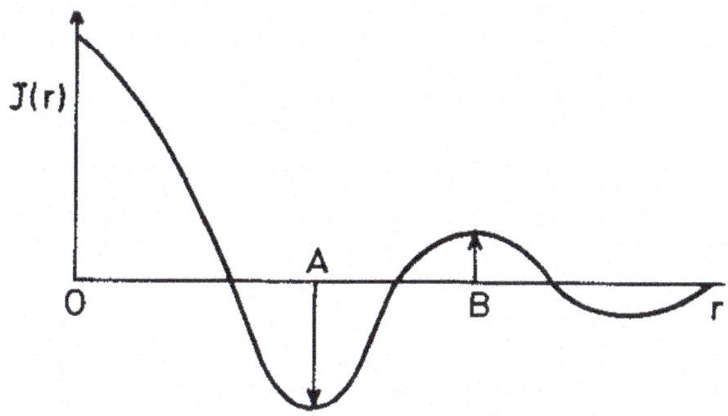

be ferro- or antiferromagnetic. On the long range the order is inhomogeneous. In the above mentioned canonical or archetypical spin glasses [108] the reason for spin glass behaviour is the so-called RKKY [109–112] interaction, where the exchange integral J depends on distance r (Fig. 8.36). E.g. a spin at the origin 0 couples antiferromagnetic with spin A and less strongly ferromagnetic with spin B (Fig. 8.36). It is caused by an indirect exchange interaction of the local moments mediated by conduction electrons. Due to a random distribution of distances between moments, it is not possible to find a spin arrangement which fulfils all exchange interactions at the same time [113]. This is called frustration [103, 114]. A further class of spin glass materials are topological spin glasses, which can appear, if an antiferromagnetic arrangement of spins should be realized on a hexagonal lattice, where the basic geometric element is a triangle. One can arrange two spins antiparallel to one another on two of the corners of the triangle, but it is impossible to put a spin on the third corner which is antiparallel to both the others at the same time. This is also one kind of frustration. Because of the random distribution of the spins in the nonmagnetic matrix, which resembles the statistical distribution of atom positions in real glasses, Coles et al. [115] introduced the notation "spin glass" for this type of magnetism. Alternatively Paul Beck [116, 117] and Tustion and Beck [118] introduced the name mictomagnet—from the greek syllabel *micto* for *mixed*—because of the simultaneous existence of ferro- and antiferro-magnetic correlations. At very low concentrations of magnetic impurity atoms, concentration independent scaling laws are present. When concentration increase, the possibility that two impurity atoms are nearest neighbours increases, and due to direct exchange interaction between d-orbitals magnetic clusters are formed. Such materials are called cluster glasses [115]. Cluster glasses are also possible due to chemical clusters, which may form during thermal treatment [116–118]. Moments of such clusters can have several thousand Bohr magnetons μ_B. If the density of the impurities is such, that the possibility that each impurity has at least one impurity in the nearest neighbour shell, the percolation limit is reached, where a path of neighbouring magnetic atoms goes from one side of the sample to the other. The probability that a moment is part of an infinite cluster is then larger than zero. The sample becomes long range ordered, but the order is strongly inhomogeneous. The main characteristic feature of spin glasses is the freezing of the moments in random orientations below a well defined freezing temperature T_f, without appearance of long range order. This freezing temperature was first discovered by a sharp peak in ac-susceptibility measurements [102] and later on confirmed also by dc-magnetization measurements [119–122]. For some spin glasses T_f is frequency dependent like for **Au**Fe and **Cu**Mn, whereas for others it is not. Below the freezing temperature strong irreversibilities are present in magnetization measurements, visible in large differences between field-cooled (FC) and zero-field cooled (ZFC) curves of temperature dependence of magnetization (Fig. 8.37). In this temperature regime also time dependence of magnetic moments is observed. Maxima are also present in specific heat and resistivity measurements, but not always at the same temperature as found in susceptibility measurements. Field dependence curves of magnetization $M(H)$ are strongly curved above T_f. They cannot be fitted by simple Brillouin function, but a fit assuming the existence of magnetic clusters of different size overlapped by a linear term from the single moments can explain the curvature $M(H, T) = \chi_0 H + \bar{\mu} c B(\bar{\mu}, (H + \lambda(M - \chi H))/T)$, with χ_0 a field independent susceptibility, $\bar{\mu}$ the mean moment with concentration c, λ the molecular field constant and B the Brillouin function. The fit shows that the mean moments of the clusters decrease with increasing temperature, whereas the number of clusters increase. This is a clear indication of dynamic magnetic behaviour. The existence of magnetic correlations above the ordering temperature is confirmed by specific heat measurements, which show that entropy at T_f is only 20 to 30% in case of Cu**Mn** [124] of the value expected in case of fully spin disorder in paramagnetic state. Under the first experiments proving a magnetic phase transition are Mössbauer experiments [125, 126], which have shown that below a temperature T_0 hyperfine splitting appears. The obtained hyperfine fields could be set in relation to local magnetic moments, with temperature dependence reminiscent to ferromagnetism. To determine, if the orientation of the spins is ferro- or antiferromagnetic, in-field Mössbauer spectra were performed on $Fe_{0.5}Au_{0.5}$ [127]. The result pointed to a weak canted antiferromagnet. Detailed analyses showed that hyperfine fields are statistical distributed in magnitude and orientation [128]. Whereas for **Au**Fe T_0 matches T_f obtained from susceptibility measurements, for other spin glasses like

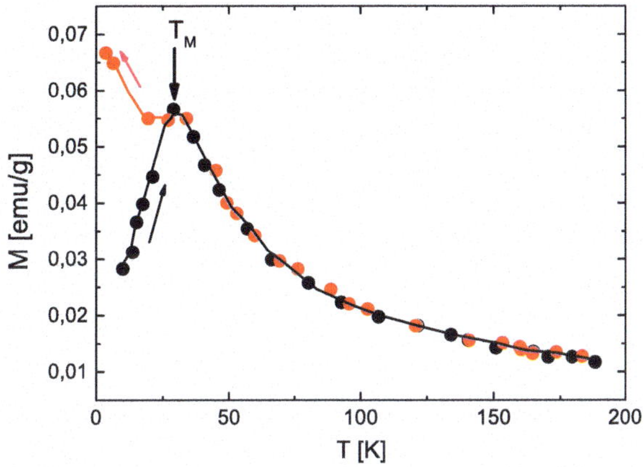

Fig. 8.37 Temperature dependence of the magnetization for $Y(Fe_{0.70}Al_{0.30})_2$ cooled without field to 4.2 K (black symbols) and with field (red symbols) [123]. © IOP Publishing. Reproduced with permission. All rights reserved

CuMn T_0 is higher than T_f. Further, small hyperfine fields are also present above T_0, proving that magnetic correlations are present above the ordering temperature [129, 130]. From small angle neutron measurements on **Au**Fe a maximum in temperature dependence of the neutron scattering cross section was found which is strongly q-dependent. Compared to susceptibility measurements T_0 is sometimes more than 10 K higher than T_f (e.g. [131]).

Very soon it was clear that the very sharp cusp in susceptibility measurements could not be explained by simple mixing of ferro- and antiferro-magnetic phases. Thus a lot of efforts were undertaken to theoretically explain these new type of magnetic behaviour. One of the first were Edwards and Anderson [132], who proposed a new ground state in which all spins are frozen in random directions below a well defined temperature. The order parameter is the autocorrelation function $q(t) = [< S_i(0)S_i(t) >_T]_{conf}$ which measure the probability that a spin has, after some time, still the same orientation. The outer bracket represents the configurational and the inner bracket the thermal average. Therefore q equals 1 in the ordered state at $T = 0$ and q becomes 0 above the ordering temperature. To calculate the free energy the Hamiltonian is build up in such a way that the spins are arranged on a regular lattice and the exchange interactions are randomly distributed. Within this model neither the cusp in susceptibility can be well reproduced, nor does the obtained specific heat agree with experiment. Subsequently many new theories based on the Edwards-Anderson model have been developed. Sherrington and Kirkpatrick [133] put this mean-field theory on quantum mechanical ground. Soukolis and Levin [134] introduced clusters and took into account both intra- and intercluster interactions. Intracluster interactions are strong and calculated exactly, whereas the weak intercluster interactions are treated in a mean field approximation. Results fit well to experimental findings. Intercluster interactions lead to the sharp peak in susceptibility and intracluster interactions lead to the smooth maximum observed in specific heat. Very interesting is the outcome of the Replica Symmetry Breaking model of Parisi [135] which proposes a multi-valley free energy landscape in the configuration space. The system can jump between different spin configurations by overcoming the barriers between the valleys. Due to the different heights of the separating barriers, different relaxation times are present. This leads to dynamic behaviour. A hierarchical distribution of time scales should be present, which could explain the experimentally observed relaxation behaviour and irreversibilities. All the theories based on the Edwards-Anderson model assume a thermodynamic phase transition into a ground state. In such case the transition temperature should be independent from experiment always the same. This is in contradiction to many experiments and also to the fact that above the ordering temperature the expected pure paramagnetism is not present, but lots of magnetic correlations are verified to exist up to temperatures several times higher than T_f. In 1974 Tholence and Tournier [136] and later Wohlfarth [137] proposed that the spin glass transition is not a true thermodynamic phase transition, but is very similar to blocking of single domain particles in rock materials. For this Néel has developed the theory of superparamagnetism. A short range order couples the spins into clusters. Because temperature counteracts the formation of such clusters, the size of the clusters increases with decreasing temperature. At high temperatures the clusters are free to rotate. They jump between easy axis directions, which are separated by energy barriers caused by anisotropy effects. This can be described in the Néel theory by τ_0, an intrinsic relaxation time in the range of $\sim 10^{-9}s$ and $E_a = KV$ the anisotropy energy with K the anisotropy constant and V the particle volume. For a particular measurement, which is characterized by a typical measuring time τ_m, clusters appear frozen, if their relaxation time τ is longer than τ_m. With decreasing temperature, volume of clusters increase and therefore the rotation frequencies decrease, and at a distinct temperature the clusters become blocked. In case of a cluster size distribution, clusters of different size will block at different temperatures. Coming from high temperatures the largest cluster will be blocked first. Within this picture the measured zero-field cooled and field-cooled curves (Fig. 8.37) are well understood. At high temperatures all clusters are free to rotate. All τ values are shorter than τ_m and the measured magnetization is zero. In lowering the temperature the cluster gradually freeze in random directions. Thus at low temperature the measured magnetization is still zero. In applying a small measuring field and by increasing the temperature (ZFC curve) the magnetization increases, because—starting with the smallest ones—more and more clusters are freed, due to the thermal energy kT and rotate in direction of the applied field. The increase of magnetization stops when the largest clusters are rotated in direction of the applied field. With further increase of temperature the thermal energy destroys successively the alignment and magnetic signal decreases like in a paramagnet thus forming the observed cusp. If temperature now decreases (FC curve), due to reduction of thermal energy, clusters begin again to rotate in direction of applied field, starting with the smallest ones and ending with the largest ones. Magnetization increases until it reaches maximum again. With further decrease of temperature the magnetization stays constant, because all clusters are now frozen in direction of the applied field. Therefore the temperature where the cusp appears is called the freezing temperature T_f. In this model all the irreversibilities and time dependences obtained in experiment can be explained. It also explains, why the value T_f is different for different measuring methods. With decreasing characteristic measuring time freezing temperature should increase. This illustrates why T_f for magnetization measurements ($\tau_m \sim 1 - 10^{-4}$ s) is smaller than for Mössbauer measurements ($\tau_m \sim 10^{-9}$ s) and much smaller for neutron measurements ($\tau_m \sim 10^{-9} - 10^{-12}$ s). Further, in this model it is clear that magnetic correlations are present far above the freezing temperature. Only at much higher temperatures, when the short range order of the clusters is destroyed, pure paramagnetic behaviour is found. Mössbauer spectroscopy in external magnetic field is an excellent method to investigate the magnetic dynamics above the freezing temperature, because the applied field changes the relaxation frequency of the clusters. If fields are large enough, they can shift the characteristic time of the dynamics into the Mössbauer time window, so that the approximation of fast relaxation limit is no longer fulfilled and the spectra show typical shape of so-called relaxation spectra. This will be shown exemplarily in the following for the spin-glass-system $Y(Fe, Al)_2$.

8.5.4 $Y(Fe, Al)_2$

$Y(Fe_xAl_{1-x})_2$ is a Laves phase, which crystallizes, with the exception of a small region near $x = 0.5$, in the cubic $MgCu_2$ structure type. Whereas YAl_2 is a Pauli paramagnet with a nearly temperature independent susceptibility of $0.8 \cdot 10^{-6}$ emu/g at room temperature, YFe_2 is a ferrimagnet with an Fe moment of $1.77 \pm 0.08\mu_B$ and a strongly delocalized yttrium moment of $-0.67 \pm 0.04\mu_B$ [63, 138]. For $x \geq 0.78$ reciprocal susceptibility χ^{-1} data are linear in temperature following a Curie-Weiss law (Fig. 8.38 right). For lower iron concentrations x the $\chi^{-1}(T)$ data are strongly curved (Fig. 8.38 left). They could be fitted by the relation $\chi = \chi_0 + C/(T - \Theta)$ where χ_0 describes the Pauli paramagnetic matrix contribution. The Weiss temperature Θ scatters around zero on the Al-rich side and increase strongly on the Fe rich side (Inset Fig. 8.39). From the Curie constant C the mean effective moment $\mu_{eff} = g\mu_B\sqrt{S(S+1)}$ is obtained, which is nearly concentration independent. Above $x \geq 0.78$ it is nearly constant equal to the value obtained for YFe_2 (3.02 μ_B) [139, 140] in contrast to the spontaneous moment which strongly decreases (Fig. 8.39). On the Al-rich side it decreases slightly to below 2 μ_B/Fe for $x = 0.1$ [123]. Although no long range magnetic order could be found down to 30 mK even at 70% Fe, the magnetization curves on the Al-rich side are still strongly curved. They could be fitted to $M = N\bar{\mu}L(\bar{\mu}H/kT) + \chi H$, with L the Langevin function. The first term describes the magnetization of iron atoms gathered in clusters with the mean cluster moment $\bar{\mu}$ and N the number of mean clusters. The second term is a susceptibility term coming from the moments which are not part of a cluster. Calculation gives mean cluster sizes of 3–4 magnetic atoms. Their concentration is a few percent. This finding is confirmed by diffuse neutron scattering on two compounds with $x = 0.25$ and 0.65, where the magnetic scattering was interpreted as arising from small clusters of less than 10 Å at temperatures far above the freezing temperature [142]. These clusters are formed by short range ferromagnetic correlations. There is

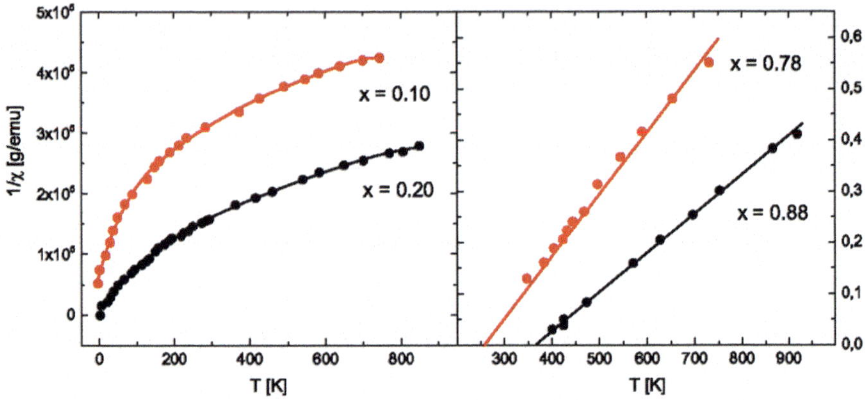

Fig. 8.38 Temperature dependence of the reciprocal susceptibility for several typical samples of $Y(Fe_xAl_{1-x})_2$ [123].

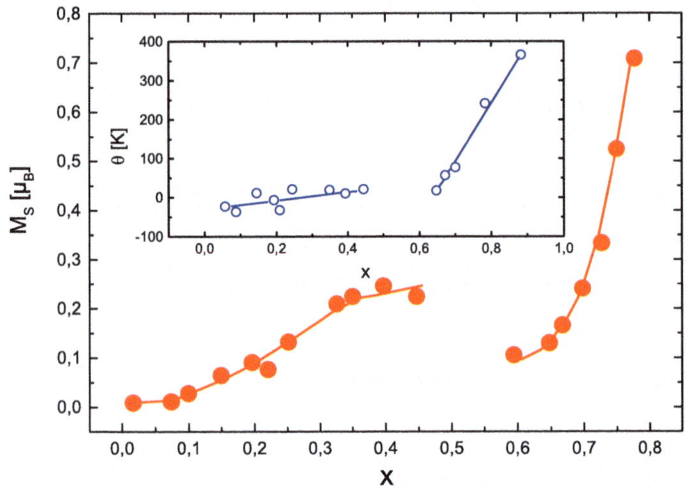

Fig. 8.39 Concentration dependence of the saturation moment $M_s = N\mu$ for $Y(Fe_xAl_{1-x})_2$ at 4.2 K. Inset: concentration dependence of Weiss temperature θ [123].

also some evidence of weak antiferromagnetic correlations. From the nuclear diffuse scattering indication of a weak anticlustering of Fe on Al sites is found [142]. The low-field magnetization against temperature curves show maxima (T_M) for $x \leq 0.8$, which broadens with increasing x.

Zero-field ^{57}Fe Mössbauer spectra at room temperature are strongly asymmetric (Fig. 8.40). They can be fitted by a superposition of quadrupole split spectra. The change of asymmetry with x points to a strong influence of the local surrounding on the electrostatic hyperfine interactions. At low temperature magnetic hyperfine split-

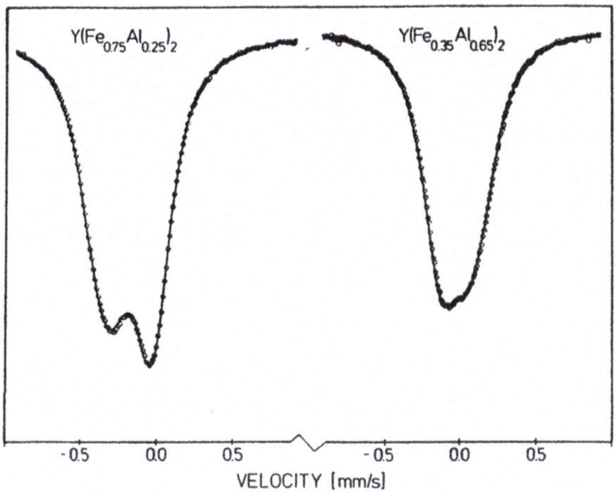

Fig. 8.40 Room temperature ^{57}Fe Mössbauer spectra of typical $Y(Fe,Al)_2$ samples [141]. Reprinted by permission from Springer: Hyperfine Interactions, Electrostatic hyperfine interactions $Y(fe, Al)_2$, M. Reissner, W. Steiner, copyright (1986)

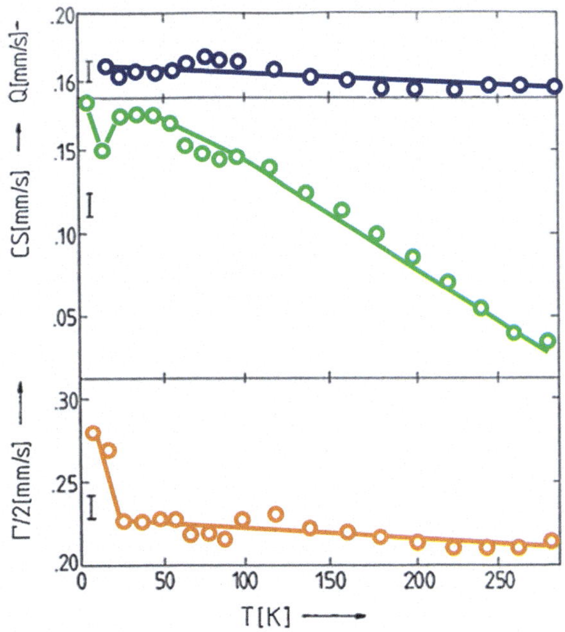

Fig. 8.41 Temperature dependence of the mean quadrupole splitting Q, mean centre shift CS (rel. ^{57}CoRh) and half width $\Gamma/2$ of $Y(Fe_{0.4}Al_{0.6})_2$. Adapted from [143]

ting is present. The transition temperature T_A, where magnetic hyperfine splitting appears, is well defined by a strong increase of half width, if spectra are still fitted by superposition of quadrupole split spectra (Fig. 8.41). T_A is in good agreement with the ordering temperature T_c obtained from the magnetic measurements on the Fe-rich side, whereas it is slightly higher than T_M on the Al-rich side, which is as mentioned above typical for spin glass behaviour and caused by the different time scales of the experiments (Fig. 8.42). The range above the ordering temperature where non-linear

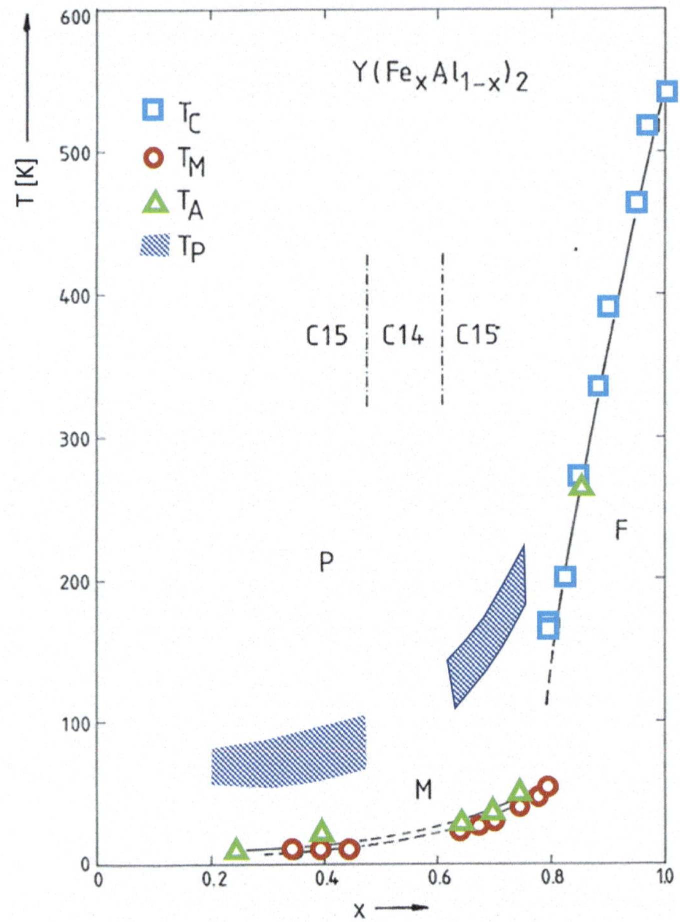

Fig. 8.42 Magnetic phase diagram of $Y(Fe_xAl_{1-x})_2$, T_c Curie temperature, T_M maximum temperature, T_A temperature where magnetic hyperfine splitting vanishes, T_P temperature above which magnetization curves are linear. Adapted from [143]

magnetization curves are still present, extends to 7 times T_f. Below T_M irreversibilities, time dependence and history effects are present. This all indicates cluster glass behaviour for $x < 0.78$. One point left open by these investigations is the question concerning the stability of the Fe moment. The only mildly concentration dependent mean effective moment points to a non-vanishing Fe moment that stays relatively constant over the entire concentration range, whereas the Mössbauer investigations indicate a strong influence of local environment on electrostatic and magnetic hyperfine interactions. Therefore the question arise, whether or not the iron moment also depends on the local coordination. There were some contradicting results in literature. Non spin-polarized band-structure calculations on $Y(Fe,Al)_2$ have shown a

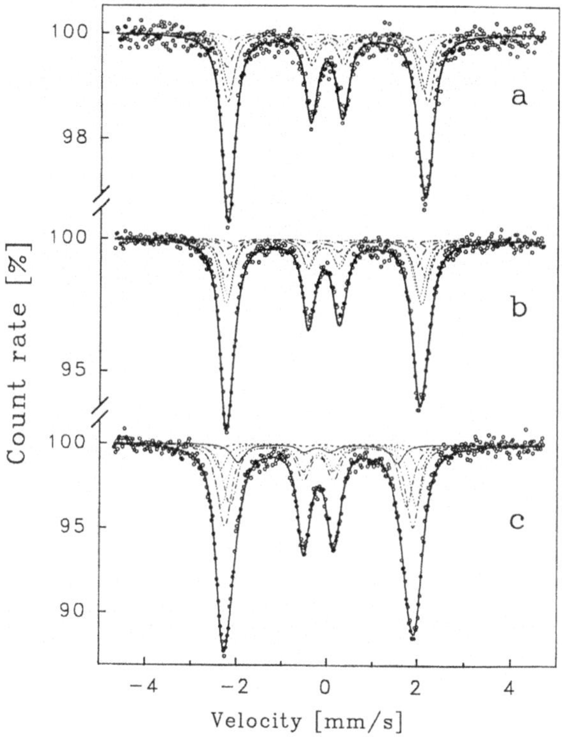

Fig. 8.43 ^{57}Fe transmission Mössbauer spectra of $Y(Fe_xAl_{1-x})_2$ at room temperature and $B_a = 13.5$ T for **a** $x = 0.25$, **b** $x = 0.40$, and **c** $x = 0.65$ [150].

reduction in the density of states at low iron concentrations [144, 145]. For Fe in Al non-magnetic Fe atoms were found from experiment [146, 147], whereas first-principle calculations [148, 149] predict an iron moment of 1.7 μ_B. Further it is shown, that in these compounds the Fe moment strongly depend on lattice constant.

To get more information about the stability and environment dependence of the iron moment in $Y(Fe_xAl_{1-x})_2$ in the paramagnetic regime, high field Mössbauer measurements were performed and compared with results of spin-polarized band-structure calculations [150]. Four representative samples with $x = 0.25$, 0.40, 0.65, and 0.75 were chosen. The Mössbauer measurements were performed at external fields up 13.5 T in temperature range 50–300 K, 100–300 K, 160–300 K, and 280–300 K for the four samples. Within these temperature ranges the respective magnetization curves are straight lines, indicating that short-range order effects are vanished. The Mössbauer spectra taken in applied fields B_a are all magnetically split (Fig. 8.43) with hyperfine fields aligned parallel to the applied field, indicated by the vanishing of the second and fifth line ($\Delta m = 0$). The spectra are asymmetric with deeper and narrower lines on the low velocity side. Therefore several subspectra are necessary to fit them. Crystallographically only one Fe/Al site exists, where Fe is surrounded by 6 Fe/Al nearest and 12 Y next nearest neighbours. Therefore a maximum of 7 subspectra, representing 0, 1, 2, 3, 4, 5, or 6 Fe atoms in the first neighbour shell, are

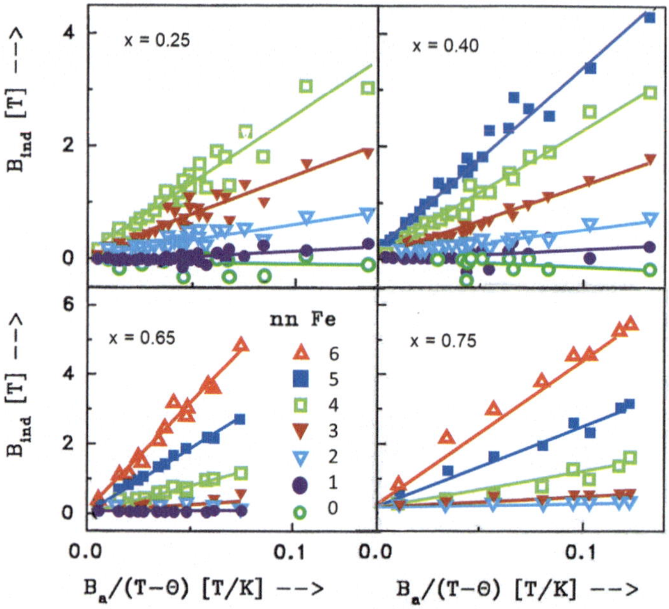

Fig. 8.44 Dependence of induced hyperfine fields in $Y(Fe_xAl_{1-x})_2$ on $B_a/(T-\theta)$ for $x = 0.25$, $x = 0.40$, $x = 0.65$, and $x = 0.75$ [150].

expected to be adequate to fit the measured spectra. From the thus obtain hyperfine fields B_{hf} the induced hyperfine fields B_{ind} can be deduced from $B_{ind} = B_a\text{-}B_{hf}$. Obtained B_{ind} values increase with increasing number of iron nearest neighbours (Fig. 8.44). From these data a Mössbauer Curie constant C_{ind}^{MB} can be calculated according to a Curie-Weiss law from $B_{ind}/B_a = C_{ind}^{MB}/(T\text{-}\Theta)$. For Θ, values obtained from the analyses of the magnetic measurements (Inset Fig. 8.39) are taken. C_{ind}^{MB} is strongly different for different Fe-surroundings. It slightly decrease with increasing iron concentration x, whereas the mean value increases with x. For zero or one Fe atom in nearest neighbour shell C_{ind}^{MB} is near to zero. Assuming, as often done and working well in many systems e.g. Fe-B alloys [151], a direct proportionality between induced hyperfine field and magnetic moment, this small C_{ind}^{MB} values would indicate vanishing Fe moments at least for zero and one Fe nearest neightbour shells, which is in full contradiction to the high, only slightly concentration dependent effective moments obtained from the magnetic measurements. LAPW and ASW band-structure calculations gave iron moments of 1.98 μ_B, 1.57 μ_B, and 1.93 μ_B for YFe_2, Y_2Fe_3Al, and Y_2FeAl_3. They also confirmed that Y carry a small moment, which is antiparallel to the Fe moments and decreases with Al concentration from $-0.42\mu_B$ for YFe_2 to $-0.31\mu_B$ for Y_2Fe_3Al and $-0.08\mu_B$ for Y_2FeAl_3. This shows that a calculation of the effective Fe moment from magnetic measurements by ignoring the Y moment leads to underestimated μ_{eff} values on the Fe side, but to more precise μ_{eff} values on the Al-rich side. Theoretically calculated hyperfine fields B_{tot}^{th} consist of two contributions: a nearly concentration independent contribution from the 3d electrons B_{core}^{th} and one from the valence electrons (4s and 4p) B_{val}^{th} which increases linearly with decreasing Fe content and has opposite sign (Fig. 8.45). For YFe_2 B_{val}^{th} is small and total hyperfine field is direct proportional to the Fe moment as usually assumed [151]. Surprisingly, with increasing Al concentration the valence electrons are strongly positive polarized. By varying the used lattice constant for the calculations allows to estimate the volume dependence of the theoretical hyperfine field. E.g. calculation of Y_2Fe_3Al with the measured lattice constant of YFe_2 (7.36 Å) instead of 7.48 Å corresponds to a concentration change from $x = 0.75$ to $x = 1.0$. This calculated concentration dependence of theoretical hyperfine fields is shown as short solid lines in Fig. 8.45. Their slopes are nearly independent of x and cannot be the reason for the concentration dependence of the hyperfine fields. The negative slope of the volume dependence of B_{tot}^{th} explains the decrease of experimentally obtained Curie constant B_{ind}^{MB} with x for fixed number of nearest neighbour Fe atoms. Because the theoretical calculations are done on ordered compounds with experimentally obtained lattice constants, the results can only describe one of the different Fe surroundings used in the Mössbauer analysis. Nevertheless B_{tot}^{th} shows the same trend on number of Fe nearest neighbours as the experimentally obtained Curie constant C_{ind}^{MB}. To get now local susceptibilities, the experimentally obtained induced hyperfine fields have to be decomposed into the core contribution B_{core} and the valence contribution B_{val}. The local susceptibility is only related to B_{core}, which can be compared with the local Fe moment, which originates from the 3d electrons. Because both B_{ind}/B_a and the susceptibility χ^{mag} determined from the magnetic measurements follow a Curie-Weiss law, one can assume that B_{core} also follows such behaviour namely $B_{core}/B_a = C_{core}^{MB}/(T\text{-}\theta)$. As a consequence B_{val}/B_a must also follow a Curie-Weiss law. With help of the band-structure calculations a separation of B_{core} and B_{val} is possible. Since the theoretical valence contribution decreases linearly with x to nearly zero for YFe_2 (Fig. 8.45), similar linear behaviour is assumed for the valence contribution of the experimentally obtained mean Curie constant $\bar{C}_{val}^{MB}$. To determine the slope of the concentration dependence of $\bar{C}_{val}^{MB}$, one assumes that the core part of the mean Curie constant obtained from Mössbauer measurements $\bar{C}_{core}^{MB}$ is directly proportional to the magnetically determined Curie constant C_{mag}. Relating C_{mag} to one Fe atom shows that both mean $\bar{C}_{core}^{MB}$ and mean $\bar{C}_{mag}$/Fe are linearly increasing with x (Fig. 8.46), indicating that the direct proportionality between mean $\bar{C}_{core}^{MB}$ and mean C_{mag} is consistent with the data. There is only one hyperfine coupling constant relating B_{ind} to μ_{Fe} for all x. The concentration dependence of the core contribution of the induced hyperfine field, determined from the Mössbauer investigations, is in perfect agreement with the concentration dependence of the effective moment, determined from the bulk magnetic measurements. If one now assumes that the valence contribution of the Curie constant for each Fe surrounding is linear, the values C_{val}^{MB} and C_{core}^{MB} for the individual Fe environments can be calculated from the respective measured induced hyperfine fields B_{ind} in the same way as was just done for the mean $\bar{C}_{val}^{MB}$ values. The local susceptibilities are now given by $\chi_{loc} = B_{core}/B_a$ with $B_{core} = C_{core}^{MB} B_a/(T - \theta)$ (Inset Fig. 8.46). For

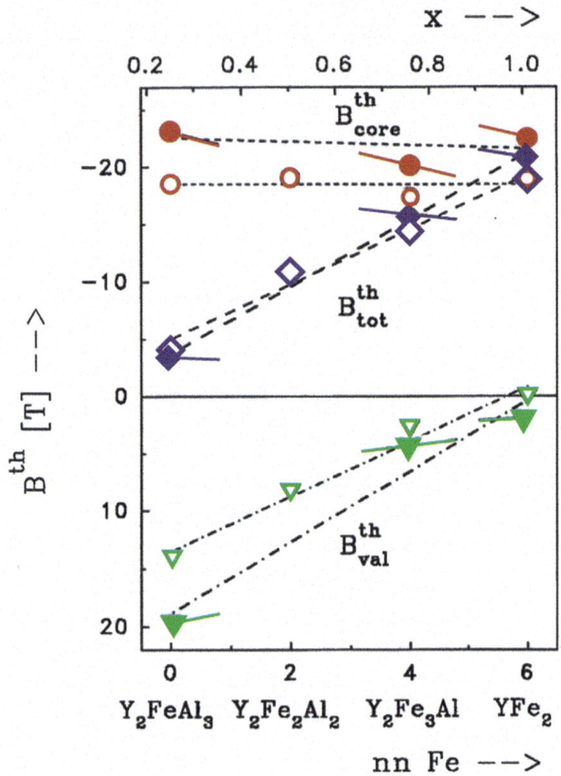

Fig. 8.45 Calculated (open symbols ASW, full symbols LAPW) hyperfine fields B^{th}_{core}(○), B^{th}_{tot}(◇) and B^{th}_{val}(▽) for $Y(Fe_xAl_{1-x})_2$ [150]. © IOP Publishing. Reproduced with permission. All rights reserved

up to three nearest Fe neighbours C^{MB}_{core} values are constant. For higher iron concentration in the nearest neighbour shell an increase is present. From this it can be concluded that Fe carries a moment even if surrounded only by Al atoms, in agreement with the band-structure calculations (Fig. 8.45). The increase of χ_{loc} for the Fe rich surroundings reflects the increase of the mean Curie constant $\bar{C}^{MB}_{core}$ and of μ_{eff}. In summary, from magnetization and high field Mössbauer measurements and with the help of band-structure calculations it could be proved that all iron atoms carry a magnetic moment in the paramagnetic regime of the $Y(Fe_xAl_{1-x})_2$ cluster glass.

Entering now the temperature regime $T_f < T <\sim 7 \cdot T_f$, where magnetic correlations are present, visible in the bended magnetization curves, the Mössbauer spectra in applied magnetic field change drastically their shape. Figure 8.47 shows spectra for $x = 0.75$ at $1.8 \cdot T_f$ (left) and $5 \cdot T_f$ (right). The shape of the spectra at 13.5 T are quite different from those obtained at room temperature in the pure paramagnetic region (Fig. 8.43). The spectra can no longer be fitted by a superposition of subspectra with intensities according to the different iron environments, calculated by binomial distribution. The nearer the temperature to T_f, the more is the deviation in shape from the one for the paramagnetic case. Often spectra with similar shape are analysed by distribution of static hyperfine fields [152, 153]. Figure 8.48 gives

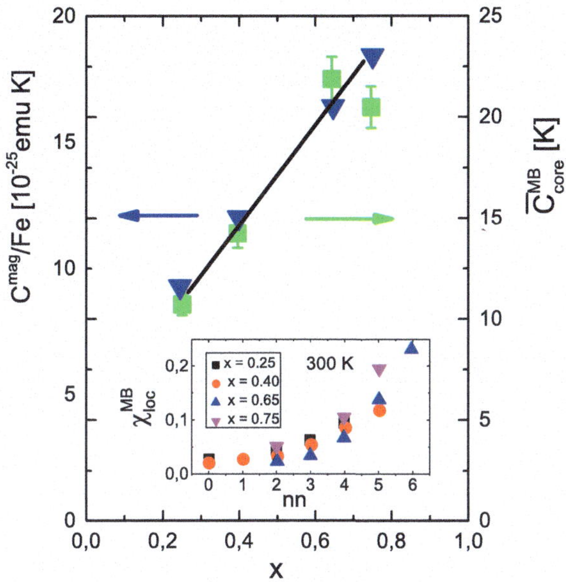

Fig. 8.46 Comparison of the x dependence of C^{mag} per Fe atom from magnetization measurements (full triangle down, left-hand scale) with that of $\bar{C}^{MB}_{core}$ from Mössbauer measurements (full squares). Inset: Dependence of local susceptibility χ^{MB}_{loc} on number of nearest Fe neighbours in $Y(Fe_xAl_{1-x})_2$ at 300 K; $x = 0.25$ (full square), 0.40 (full circle), 0.65 (full triangle up) and 0.75 (full triangle down) [150]. © IOP Publishing. Reproduced with permission. All rights reserved

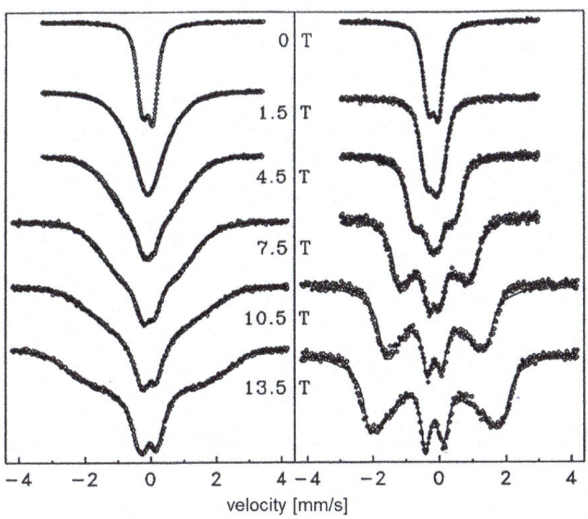

Fig. 8.47 Spectra of $Y(Fe_{0.75}Al_{25})_2$ in different fields at 70 K (left) and 200 K (right). Fits according to described model. Reprinted from [154]

an example of a fit in the fast relaxation limit with 9 subspectra according to a hyperfine field distribution. The structure in the calculated spectrum is not visible in the measured one, and could be avoided by increasing the number of subspectra. The fit is reasonable, but there is no meaningful interpretation, because the subspectra with the largest magnetic hyperfine splitting have hyperfine fields larger than the applied field. Because all spectra are fully polarized, it has to be concluded, that there are contributions to the hyperfine field which are parallel to the applied field B_a. This would imply magnetic moments which are antiparallel to B_a, a scenario not very

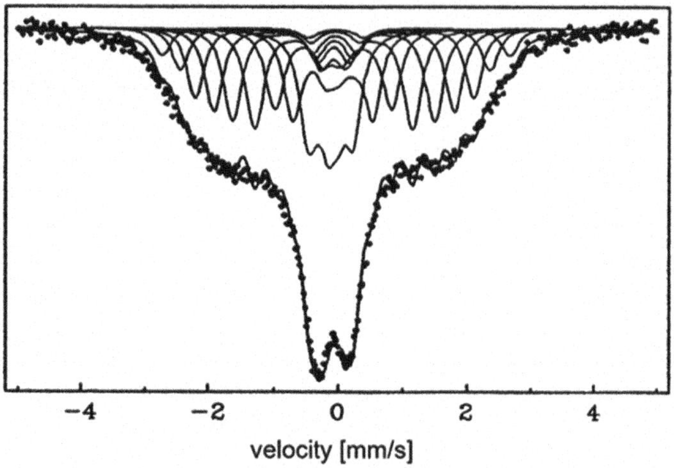

Fig. 8.48 Spectrum of $Y(Fe_{0.75}Al_{0.25})_2$ at 100 K and 13.5 T analysed in approximation of fast relaxation by hyperfine field distribution. Reprinted from [155]

plausible. On the other hand the usual case, that iron moments are parallel to the applied field would give internal hyperfine fields higher than 27 T, which is larger than the ones for undiluted YFe_2 (two subspectra 21.2 T and 20.8 T [156, 157]) and taking into account that, as mentioned before, for the shown high Fe concentration ($x = 0.75$) only core contributions to the hyperfine field are important, also too high magnetic moments would be the consequence. From the shape of the spectra, smooth inner slope of the outer lines and strongly increased intensity in the inner part of the spectra like for relaxation spectra, the dynamics of the magnetic correlations seems to be visible. Figure 8.49 shows spectra at 100 K and 13.5 T for different Fe concentration x. Here 100 K corresponds to 10, 4, and 2.5 times T_f for $x = 0.40$, 0.65, and 0.75, respectively.

The simplest model to describe dynamics is a field flip model where two magnetic hyperfine fields H_1 and H_2, which are antiparallel to one another, fluctuate between two states $|1>$ and $|2>$. If W_i is the equilibrium occupation of state $|i>$, and transition probability between the states is γ_{12} and γ_{21}, the equilibrium condition gives $W_1\gamma_{12} = W_2\gamma_{21}$. The iron nucleus senses a field jumping between B_1 and B_2. The relaxation rate is then given by $\tau = 1/(\gamma_{12} + \gamma_{21})$. The asymmetry of the spectra found in the zero field spectra (Fig. 8.40), which is due to different quadrupole splitting and centre shift according to the different iron surroundings, is also visible in the in-field spectra, and should be taken into account also in the dynamical fits. To avoid this necessity, spectra at higher fields should be preferably analysed, because at the high fields the electrostatic hyperfine interaction is much smaller than the magnetic one. The result of such simple fit, which takes into account only two subspectra, is for many temperatures reasonable (Fig. 8.50). For the occupation probability of the two states, a Boltzmann statistics was chosen with $\tau_{on}/\tau_{off} = exp(\mu_B g(B_o - B_u)/kT)$, with g set to 1. In principle the results seem to favour such field flip model. But

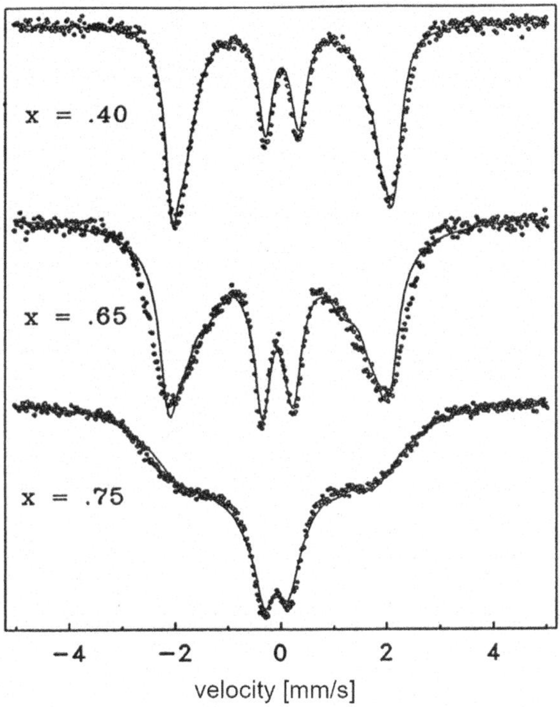

Fig. 8.49 Spectra of $Y(Fe_xAl_{1-x})_2$ at 13.5 T and 100 K for different x. Reprinted from [19]

the values of the obtained fields $B_o = B_a + B_{cl}$ and $B_u = B_a - B_{cl}$, where B_{cl} is the hyperfine field of iron moments in the correlated state, that means in a magnetic cluster, between which the hyperfine field flips, give no constant B_{cl} value (Fig. 8.51). Further the value of B_u needed to fit the inner part of the spectra—with $B_{cl} \sim 8$ T—leads to contributions resulting for B_o that are not present in the measured spectra. Therefore a different more complex model for the dynamics in these compounds is necessary. Usually the short range order, found in spin glasses above T_f is explained by a growth of size of clusters with decreasing temperature and their changing rotation frequency. What is neglected in this picture is the possibility of real dynamics, which is suggested by the shape of the Mössbauer spectra. Due to the high applied magnetic field B_a the dynamics of the clusters above T_f, which is too fast to be seen in the zero-field measurements, become visible. The mean life time of a correlated state, which results in effective magnetic moments oriented in direction of B_a comes into the time window of the Mössbauer measurement and produce relaxation spectra.

The following model, to take into account the dynamics of magnetic exchange interactions, was developed (Fig. 8.52) [19, 157, 158]. For an iron atom in a cluster (=correlated state), the measured effective hyperfine field is the vectorial sum of the applied field B_a and the cluster field B_{cl}. B_{cl} is due to the magnetization of the correlated spins and is caused by the polarization of the electrons of the individual Fe atom, which also depends on the number of nearest neighbours. The clusters are not stable,

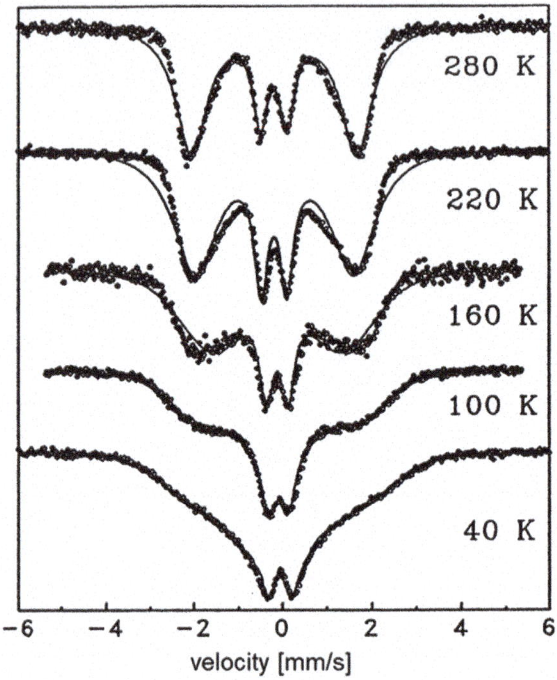

Fig. 8.50 Spectra of $Y(Fe_{0.75}Al_{0.25})_2$ for different temperatures at 13.5 T. Reprinted from [19]

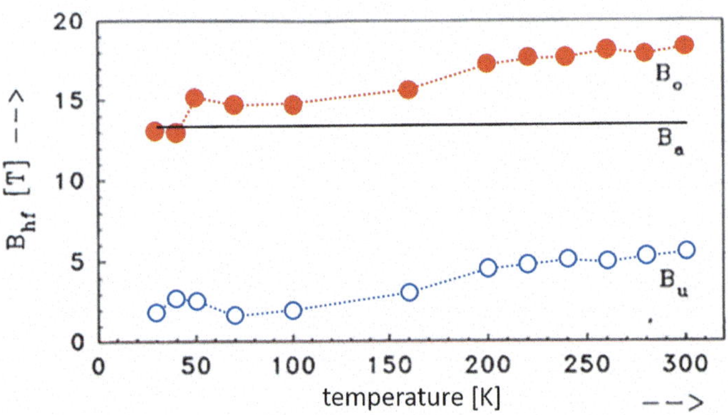

Fig. 8.51 Temperature dependence of flipping fields B_u and B_o at 13.5 T. Reprinted from [19]

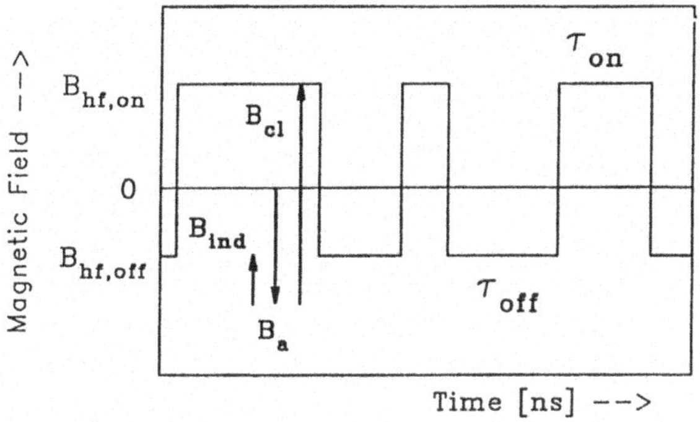

Fig. 8.52 Scheme of the used relaxation model [158].

therefore the correlation persists only for a time interval τ_{on}, afterwards the correlation vanishes. The Fe atom, which is no longer member of a cluster, now senses a hyperfine field $B_a - B_{ind}$. The induced hyperfine field B_{ind} is antiparallel to B_a and is caused by the paramagnetic iron moment. After a time τ_{off} the Fe probe atom changes from uncorrelated state back to a correlated one (Fig. 8.52). This change from correlated to uncorrelated state is a stochastic process, therefore τ_{on} and τ_{off} are average values. The applied field stabilizes the correlation in a way that a stable magnetization appears which produces B_{cl}. The probabilities for the changes are then given by $W_{on \to off} = 1/\tau_{on}$ and $W_{off \to on} = 1/\tau_{off}$. The relaxation time τ which has to be compared with the characteristic time of the Mössbauer effect—the Larmor period of the absorbing Fe nucleus—is given by $1/\tau = 1/\tau_{on} + 1/\tau_{off}$. Beside τ the ratio $a_\tau = \tau_{on}/\tau_{off}$ enters as the second dynamical parameter in the fitting procedure. Because at applied fields higher 4 T the spectra are fully polarized, with vanishing of the 2nd and 5th line, a uniaxial relaxation of the hyperfine fields could be assumed, which can be described by the relaxation equations of Wickman [159, 160]. According to the fact, that due to the crystal structure an electric quadrupole interaction is present, and because measurements are performed on polycrystalline samples, the main axis of the field gradient tensor has an arbitrary orientation in relation to the direction of the applied field. Therefore more sophisticated formula are necessary to describe the relaxation behaviour, as for example developed by Blume [22] and Clauser and Blume [23]. The Hamiltonian in terms of which the nuclear Zeeman and the electrostatic hyperfine interactions can be described is for fields acting at the nucleus which are collinear with the γ-direction, of the form.

$\mathcal{H} = -g_N \mu_N [B_a - B_{ind}\{1 - f(t)\} - B_{cl} f(t)] I_z + \frac{eQV_{\hat{z}\hat{z}}}{4I(2I-1)} \{\frac{1}{2}(3\cos^2\beta - 1)(3I_Z^2 - I^2) + \frac{3}{2}\sin 2\beta[(I_z I_x + I_x I_z)\cos\phi + (I_z I_y + I_y I_z) sin\phi] + \frac{3}{2}\sin^2\beta[(I_x^2 - I_y^2)\cos 2\phi + (I_x I_y + I_y I_x)\sin 2\phi]\}$ where β is the polar and ϕ the azimuth angle between applied field and the largest contribution of the electric field gradient tensor $V_{\hat{z}\hat{z}}$ with respect to the laboratory frame and $f(t)$ denotes a random variable taking the values 0 or 1. In case of only 2 states—correlated or paramagnetic—the size of the relaxation matrix is 16×16, which guarantees short relaxation times. The sign of $eQV_{\hat{z}\hat{z}}$ was determined from measurements in external field 1.5 T at room temperature. Different values for quadrupole splitting, center shift, B_{ind}, and B_{cl} are used for the different Fe environments, whereby the hyperfine fields were assumed to increase with increasing number of Fe atoms in the surrounding. According to the zero-field measurements quadrupole splitting is independent of temperature and centre shift changes only according to second order Doppler shift. These changes were assumed to be the same for all configurations, which means that the Debye temperature does not depend on the local environment. This assumption seems to be reasonable for a metal. Because the quadrupole interaction is much smaller than the magnetic one, in the magnetic split spectra the change in quadrupole splitting has no influence on the shape of the spectra within the given measuring accuracy. Therefore the quadrupole splitting was assumed to be field independent. Line widths are fixed to the values of the calibration α-Fe foil. Asymmetry parameter was set to zero. The number of subspectra and their intensity was fixed by the number of nearest neighbour environments and their probability calculated by binomial distribution. To take into account that an iron atom in a specific surrounding may be part of clusters of different size, two or more subspectra for each surrounding were used. Figure 8.53 gives an example. The spectrum is fitted by five subspectra, each corresponding to a specific surrounding. Each of these subspectra is the sum of two subspectra with different dynamical parameters. The sum intensity of these two subspectra are still fixed to the value obtained from the binomial distribution and the only allowed parameters to change are the two dynamical parameters τ and α_τ. Close to T_f it is necessary to admit an angle θ between B_{cl} and B_a, because in the spin glass regime below T_f the clusters are frozen and applied field is not able to align the inner fields. For simplicity θ was assumed not to change during reestablishment of the correlation. Field and temperature dependence of θ for two concentrations is shown in Fig. 8.54. All measured spectra could be fitted reasonably well within this model. Typical examples are shown in Fig. 8.47. It has to be stressed once more, that the number of subspectra remains the same for zero- and in-field measurements. Thus the distinction between individual iron environments remains. During τ_{off} the iron atom is not part of a correlated region. Therefore it must show paramagnetic behaviour. The field acting on the nucleus during this time B_{ind} is only caused by the moment of the probe atom. In that case paramagnetic behaviour is expected to show a Brillouin-like behaviour, which is really found for all Fe environments (Fig. 8.55). B_{ind} increases with the number of iron neighbours. For temperature $T <\sim T_f$ B_{cl} is antiparallel to the external field B_a. Near T_f the cluster field is not fully aligned, even at the highest field of 13.5 T. The obtained B_{cl} fields decrease with temperature and are only mildy dependent on the Fe concentration x for a given environment (Fig. 8.56). With these two fields a field B_n can be defined by $B_n = (B_{cl}\tau_{on}\cos\theta + B_{ind}\tau_{off})/(\tau_{on} + \tau_{off})$. Because B_n acts on the Fe nucleus, it should be proportional to the macroscopic measured magnetization σ. Since for $Y(Fe_{0.75}Al_{0.25})_2$, as mentioned above, the core contribution is the dominant part of the hyperfine field, B_n should be directly proportional to the bulk magnetization. In the comparison (Fig. 8.57) there is only the hyperfine field constant a free parameter, which turns out to be 11 T/μ_B for all temperatures, in good agreement with values often obtained from standard Mössbauer experiments. This strongly supports the used model. The deviations 4.5 T are due to the fact that because

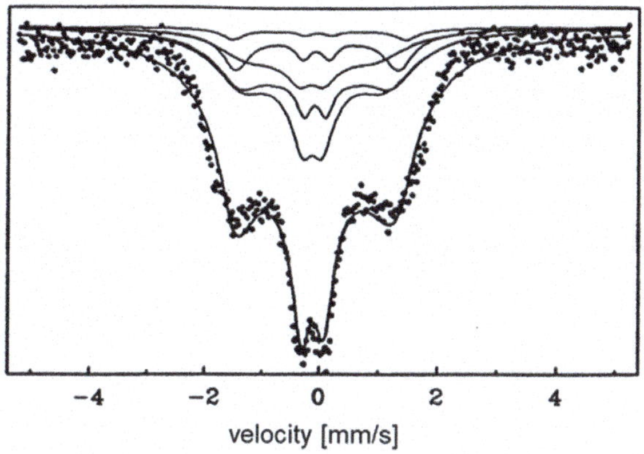

Fig. 8.53 Spectrum of $Y(Fe_{0.75}Al_{0.25})_2$ 100 K and 13.5 T. Subspectra calculated in approximation of fast relaxation. Reprinted from [19]

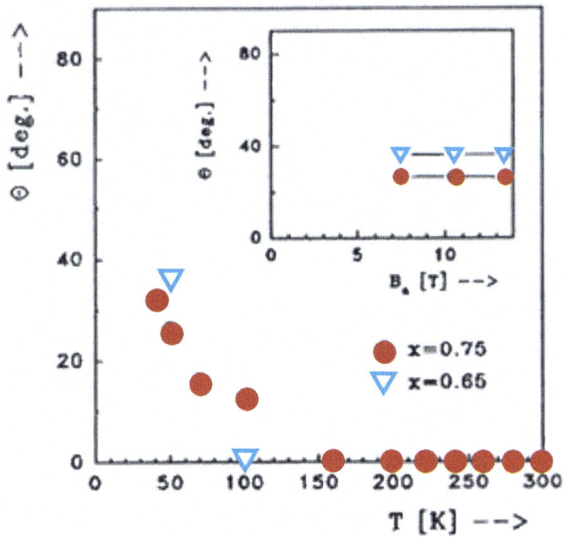

Fig. 8.54 Angle θ in $Y(Fe_xAl_{1-x})_2$ at 13.5 T. Inset: field dependence of θ at 50 K. Reprinted from [19]

of the small splitting of the spectra at low fields the obtained parameters are rather unsure. From the obtained values for τ_{on} and τ_{off} a parameter $p = \tau_{on}/(\tau_{on} + \tau_{off})$ can be calculated, which gives the fraction of time, during which the Fe atom is on average in a correlated state. Hence p also gives the fraction of Fe atoms which are on average at an arbitrary time bound in correlated regions. p increases with field and decreasing temperature for higher field, whereas it starts at lower temperatures to increase (Fig. 8.58). Although nothing can be said from these data about sizes of the correlated regions, the fact that p is about 80% 13.5 T near T_f for $x = 0.75$ shows that the correlated region has to be rather small. p is smaller for lower iron concentrations x for the same field and temperature. Looking now on the relaxation time τ, defined by $1/\tau = 1/\tau_{on} + 1/\tau_{off}$, a common, approximately linear variation with $B_a^{1/2}/t^{3/2}$, with $t = T/T_f$, is found for all investigated samples, at all measuring temperatures

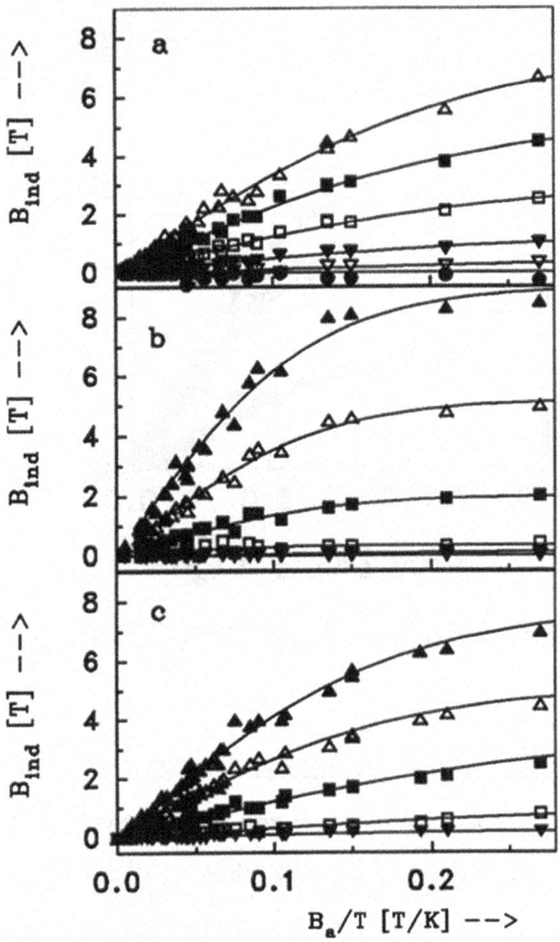

Fig. 8.55 $B_{ind}(B_a/T)$ for $Y(Fe_xAl_{1-x})_2$ $x = 0.40$ (**a**), 0.65 (**b**), and 0.75 (**c**); nearest neighbours Fe: 0 (full circle), 1 (open triangle down), 2 (full triangle down), 3 (open square), 4 (full square), 5 (open triangle up), and 6 (full triangle up). Reprinted from [19]

and all applied fields (Fig. 8.59). In this representation the slope for the higher Fe concentrations is the same, whereas for $x = 0.25$ a less steep increase is found. The analysis of the spectra above approximately $4 \cdot T_f$ was possible with only one set of τ_{on} and τ_{off} values for all different environments. This indicates that at these higher temperatures the expected distribution of correlation times [107, 161–163] are rather narrow. At lower temperatures different times have to be used for different environments, where both τ_{on} and τ_{off} increase with increasing number of iron atoms in the nearest neighbour shell.

The temperature dependence of these times averaged over all surroundings, is very similar for all field values (Fig. 8.60). The increase of $\bar{\tau}_{on}$ with B_a shows that the external field stabilises already existing correlations.

$\bar{\tau}_{off}$, which describes the probability of transition from paramagnetic to correlated state, is practical independent of B_a. That indicates that the applied field stabilizes

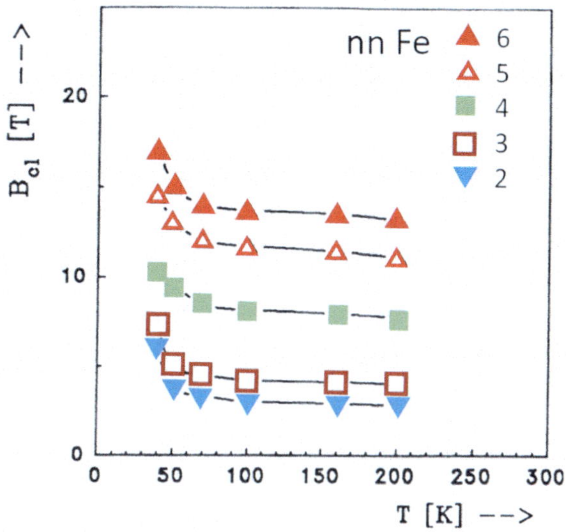

Fig. 8.56 Temperature dependence of cluster field B_{cl} in $Y(Fe_{0.75}Al_{0.25})_2$ at 13.5 T for different Fe surroundings. Reprinted from [19]

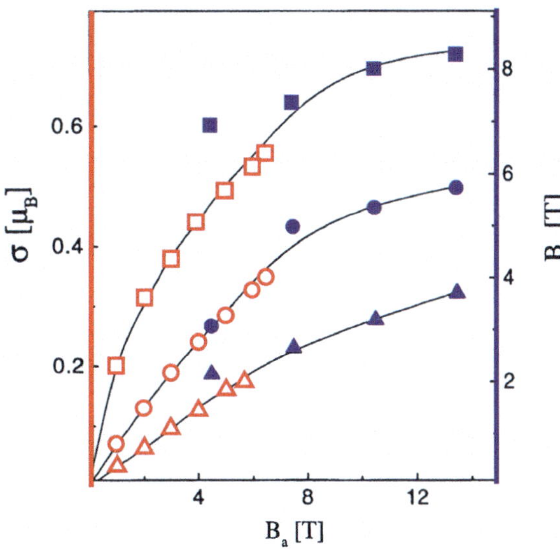

Fig. 8.57 Field dependence of magnetization σ (open symbols) and B_n (full symbols) for $Y(Fe_{0.75}Al_{0.25})_2$ at 50 K (squares), 160 K (circles) and 240 K (triangles). Adapted from [155]

the correlated state but does not support the formation of the clusters. Since the reciprocal values of τ_{on} and τ_{off} are correlated with the probability of formation and/or decay of a correlated region, this implies, following general scaling arguments (e.g. [164, 165]), that larger regions are integrated into the dynamics. Looking on the autocorrelation function, which is proportional to $\bar{\tau}_{on}$, a power law dependence on reduced temperature $t = T/T_f$ is found with an exponent of -2 for the high Fe concentration and -1 for the low Fe concentration (Fig. 8.61). The question why the magnetic correlated regions are not stable after they have formed, but are decaying

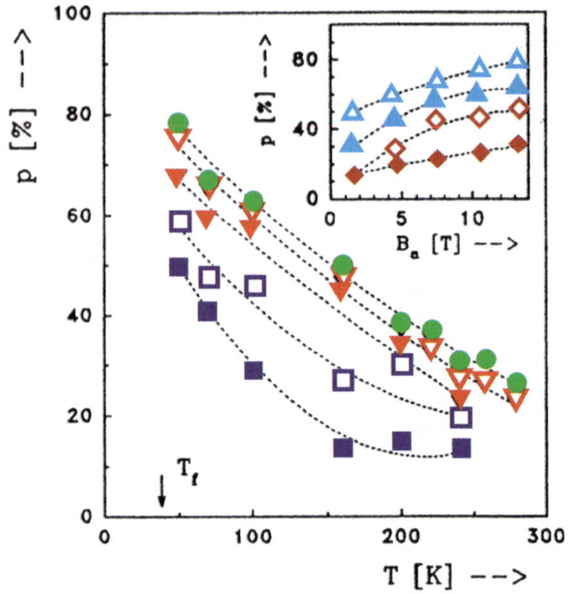

Fig. 8.58 Temperature dependence of the amount of Fe atoms in $Y(Fe_{0.75}Al_{0.25})_2$ in correlated regions at $B_a = 13.5$ T(full circle), 10.5 T (open triangle down), 7.5 T (full triangle down), 4.5 T (open square), and 1.5 T (full square). Inset: field dependence of p for $T =$ 50 K (open triangle up), 100 K (full triangle up), 160 K (open diamond), and 240 K (full diamond). Adapted from [19]

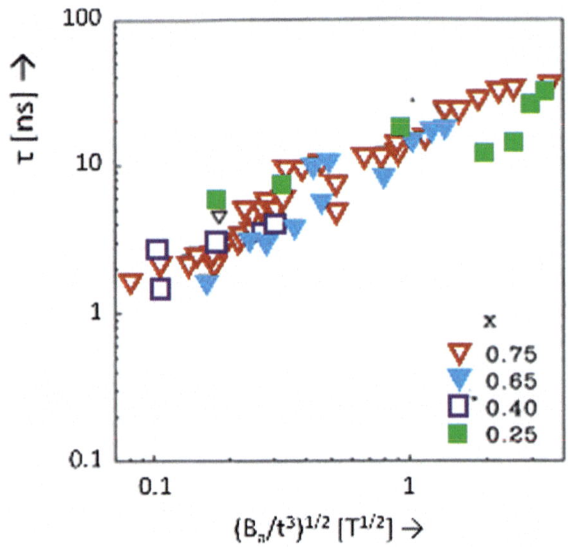

Fig. 8.59 Relaxation time τ in $Y(Fe_xAl_{1-x})_2$ [158].

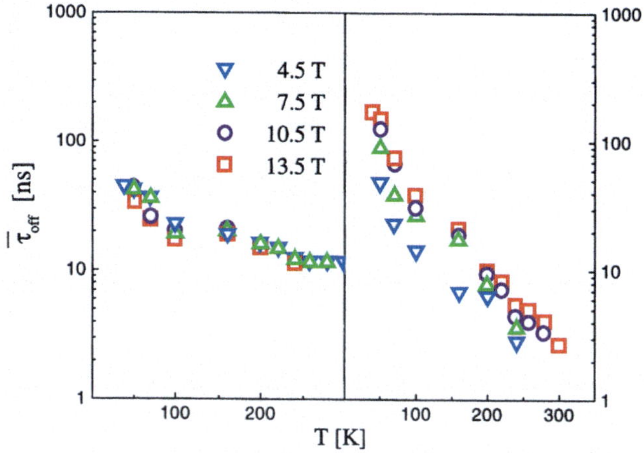

Fig. 8.60 Temperature dependence of $\bar{\tau}_{off}$ (left) and $\bar{\tau}_{on}$ (right) averaged over the different Fe environments for $Y(Fe_{0.75}Al_{0.25})_2$ at different applied fields. Adapted from [154]

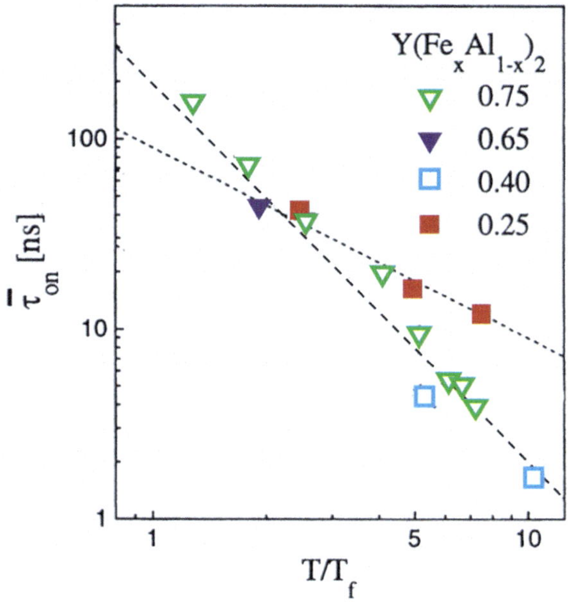

Fig. 8.61 Mean lifetime $\bar{\tau}_{on}$ of the correlated states at 13.5 T as a function of the reduced temperature for the series $Y(Fe_xAl_{1-x})_2$. Adapted from [154]

after a time $\bar{\tau}_{on}$ might be explained by taking into account the influence of entropy (Fig. 8.62). In zero field the energy of a correlated iron atom is lower than the one of an atom in the paramagnetic state due to the exchange interaction of the atom with neighbouring atoms. On the other side the entropy is much lower for an ensemble of correlated iron atoms than for the same number of atoms in the paramagnetic state. If the energy difference between the two states is not large, the entropy cannot be neglected. In an applied field the probability of states being occupied such that the Fe moment is parallel or almost parallel to B_a increases, but all magnetic substates are allowed in paramagnetic atoms and transitions between different spin directions take place so fast that only a resulting B_{ind} is visible in the Mössbauer spectra. The energy of the resulting state is shifted to lower values by $B_a < \mu_{Fe} >$, where $< \mu_{Fe} >$ is the thermal average of the Fe magnetic moment of an uncorrelated Fe atom. For correlated Fe atoms there will be an axial anisotropy inherent in the exchange interaction which permits only parallel or antiparallel orientation of μ_{cl}, like in superparamagnetic systems [166]. The state with direction of the resulting moment μ_{cl} parallel to B_a is shifted by $-\mu_{cl} B_a$ to lower values, and for antiparallel alignment by $+\mu_{cl} B_a$ to higher values. μ_{cl} is the effective magnetic moment of a correlated region. If the magnetic moment of a correlated region changes its direction from close to parallel to close to antiparallel with respect to B_a, the high entropy of the paramagnetic state would favour its occupation, even if the energy of the correlated state was comparable to that of the paramagnetic one (Fig. 8.62). In this picture a decay of correlations is possible. In this model the ratio of the occupation probabilities of the two states can be written as $p_{on}/p_{off} = exp(-\Delta S/k_B + \Delta E/k_B T + \mu_{cl} B_a/k_B T)$, where it is assumed that $\mu_{Fe} << \mu_{cl}$, ΔS is the entropy difference and ΔE is the energy difference of the two states for an Fe atom. The equilibrium condition $p_{on} W_{on \to off} = p_{off} W_{off \to on}$ and the definitions of $\tau_{on} = 1/W_{on \to off}$ and $\tau_{off} = 1/W_{off \to on}$ give $p_{on}/p_{off} = \tau_{on}/\tau_{off}$. This allows to get information about ΔS, ΔE, and μ_{cl} from the obtained τ_{on}/τ_{off} values. Fits of field and temperature dependence of τ_{on}/τ_{off} are only possible if ΔS, ΔE, and μ_{cl} are assumed to be field and temperature independent (Fig. 8.63). Although fits are not perfect, results of values of ΔS, ΔE, and μ_{cl} should be good enough to get estimation of the order of correct values. The values of ΔS are between 1.3 and 3.2 k_B, the ones of ΔE are between 1×10^{-21} and 1.7×10^{-21} J (corresponding to 72 and 123 K) and those of μ_{cl} are between 6.2 and 8.5 μ_B for all samples. Assuming an Fe moment of 1.77 μ_B as obtained for YFe_2 [63] and parallel alignment of the Fe moments in the correlated regions, gives only three to five Fe atoms per cluster, in good agreement with results of magnetic measurements.

8.6 Conclusion

The presented examples cover a broad range of different magnetic phenomena, investigated in much detail by application of external magnetic fields. In case of Ga substituted Co ferrites it was shown that the magnetically determined weak ferro-

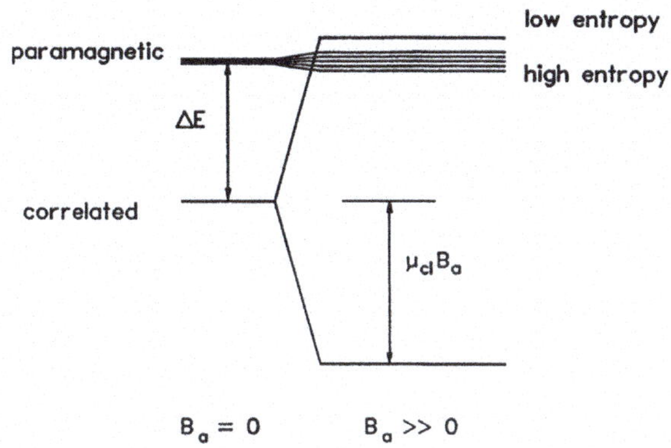

Fig. 8.62 Energy level scheme of the two states which are allowed in the model with and without B_a [158]. © IOP Publishing. Reproduced with permission.

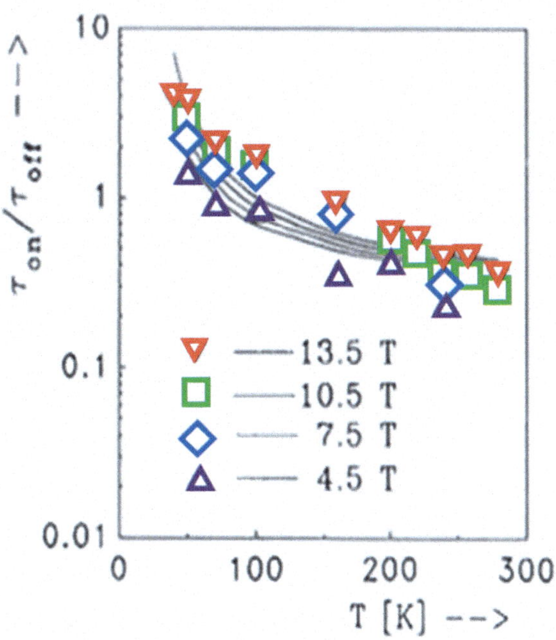

Fig. 8.63 Dependence of τ_{on}/τ_{off} on T for different B_a in $Y(Fe_{0.75}Al_{0.25})_2$ [158]. © IOP Publishing. Reproduced with permission.

magnetism is in fact a ferrimagnet. For $RE_6Fe_{13}X$ compounds the comparison of magnetic investigations with high field Mössbauer results it is shown that even in the highest applied fields the spin moments are canted. In Fe pnictides crystallographically only one Fe site is present, which should lead to only one simple Mössbauer spectrum, which is indeed found in zero field measurements. With the application of an external field, spectra become so complex, that they cannot be explained by only one subspectrum. This is only one example that the external field opens a new window for elucidation of real magnetic ground states, which are not visible without field. In the last example the magnetic behaviour of spin glasses is discussed. After introduction of the term spin glass, it is shown how very detailed information about local moments, local susceptibilities and magnetic dynamics above the ordering temperature is obtained with help of high field Mössbauer spectroscopy. In all cases it is clear that results of accompanying investigations, as for example detailed structure analysis, are necessary to interpret the results of the Mössbauer measurements correctly. In the ideal case the measurements are accompanied also by theoretical band structure calculations. In that case most information can be gained.

Acknowledgements This chapter reviews some studies accomplished in last decades in the Mössbauer group of the Institute of Solid State Physics, TU Wien, Austria, conducted by its founder Walter Steiner, who initiated these investigations. Some of the work is part of PhD theses of Adrian Pösinger, Joachim Bogner and Robert Ruzitschka. Special thanks goes to Heiner Winkler for his help in extending our program to be able to calculate relaxation spectra. Thanks also to Peter Mohn, Peter Blaha and Karlheinz Schwarz for performing theoretical band structures calculations. Peter Rogl, Andreas Leithe-Jaspers, and Zakir Seidov have prepared some of the discussed samples. Numerous other colleagues and students, who cannot all be mentioned here, have contributed to the presented results.

References

1. S.S. Hanna, J. Heberle, G.J. Perlow, R.S. Preston, D.H. Vincent, Phys. Rev. Lett. **4**(10), 513–515 (1960)
2. J. Chappert, J. de Physique **35**, 12 C6 71–88 (1974)
3. G.L. Whittle, S.J. Campbell, B.D. Maquire, Hyperfine Interactions **15–16**, 661–664 (1983)
4. P. Steiner, D. Gumprecht, W.V. Zdrojewski, S. Hüfner, J. de Physique **35**, C6 523 (1974)
5. W. Gierisch, W. Koch, W. Potzel, F.J. Litterst, P. Steiner, G.M. Kalvius, Hyperfine Interactions **4**, 407–10 (1978)
6. M. Scherg, E.R. Seidel, F.J. Litterst, W. Gierisch, G.M. Kalvius, J. de Physique **35**, C6 527 (1974)
7. S.J. Campbell, P.E. Clark, T.J. Hicks, Sol. Stat. Comm. **24**, 791–792 (1977)
8. J. Chappert, J. Teillet, F. Varret, J. Mag. Mag. Mater. **11**, 200–227 (1979)
9. J. Deportes, B. Kebe, R. Lemaire, J. Mag. Mag. Mater. **54–57**, 1089–1090 (1986)
10. S.M. Fries, C.L. Chien, J. Crummenauer, H.G. Wagner, U. Gonser, Hyperfine Interactions **23**, 273 (1985)
11. R. Coussement, P. Put, G. Scheveneels, F. Hardman, Hyperfine Interactions **23**, 273 (1985)
12. R.E. Watson, A.J. Freeman, Phys. Rev. **123**(6), 2027–2047 (1961)
13. P.W. Anderson, A.M. Clogston, Bull. Amer. Phys. Soc. **6**, 60 (1961)
14. P.W. Anderson, Phys. Rev. **124**, 41 (1961)
15. A. Okiji, J. Kanamori, J. Phys. Soc. Japan **19**(6), 908–915 (1964)
16. J.-M. Greneche, in Mössbauer Spectroscopy eds. Y. Yutaka, G. Lagouche, (Springer, Berlin, Heidelberg, 2013), pp. 187–242
17. C.E. Johnson, M.S. Ridout, T.E. Crawnshaw, Proc. Phys. Soc. **81**, 1079–1090 (1963)
18. W. Marshall, Phys. Rev. **119**(6), pp. 1280–1285
19. A. Pösinger, Dissertation TU Wien (1992)
20. Q.A. Pankhurst, R.J. Pollard, J. Phys. Cond. Matt. **2**, 7329–7337 (1990)
21. S. Dattagupta, Hyperfine Interactions **49**, 253–266 (1989)
22. M. Blume, Phys. Rev. **174**(2), 351–374 (1968)
23. M.J. Clauser, M. Blume, Phys. Rev. B **3**(3), 583–591 (1971)
24. M. Blume, Phys. Rev. B **14**(4), 96–98 (1965)

25. S. Dattagupta, M. Blume, Phys. Rev. B **10**(11), 4540–4550 (1974)
26. M.J. Clauser, Phys. Rev. B **3**(11), 7348–7353 (1971)
27. G.K. Shenoy, B.D. Dunlap, Phys. Rev. B **13**(3), 1353–1356 (1976)
28. M. Blume, J.A. Tjon, Phys. Rev. **165**(2), 446–456 (1968)
29. J.A. Tjon, M. Blume, Phys. Rev. **165**(2), 456–461 (1968)
30. F. Hartmann-Boutron, J. de Physique **37**, 537–556 (1976)
31. C. Chopin, F. Hartmann-Boutron, D. Spanjaard, J. de Physique **37**(12), C6 433–436 (1976)
32. C. Chopin, D. Spanjaard, F. Hartmann-Boutron, J. de Physique **37**(12), C6 73–78 (1976)
33. F. Hartmann-Boutron, Revue Phys. Appl. **18**, 413–430 (1983)
34. F. Hartmann-Boutron, Revue Phys. Appl. **18**, 431–455 (1983)
35. Q.A. Pankhurst, S. Suharan, M.F. Thomas, J. Phys. Cond. Matt. **4**(13), 3551–3558 (1992)
36. B. Window, J. Phys. E **4**, 401–402 (1971)
37. F. Bitter, Rev. Sci. Instr. **7**, 482–488 (1936)
38. C. Otero Arean, J.L. Rodriguez Blanco, J.M. Rubio Gonzales, F. Trobajo, J. Mat. Sci. Letts. **9**, 229 (1990)
39. J.L. Soubeyroux, D. Fiorani, E. Agostinelli, J. Mag. Mag. Mater. **54–57**, 83 (1986)
40. Z. Seidov, G. Sultanov, S. Ibragimov, A. Kaplienko, J. Phys. IV France **7**, C1–265 (1997)
41. M. Reissner, W. Steiner, Z. Seidov, G. Sultanov, Hyperfine Interactions (C) **6**, 193–196 (2002)
42. F. Weitzer, P. Rogl, J. Less-Common Metals **167**, 135 (1990)
43. T. Kajitani, K. Nagayama, T. Umeda, *Proceedings on 12th International Workshop on RE Magnets and their Applications* (Ed. R. Street, Canberra, 1992), pp. 574–583
44. K.G. Knoch, I.R. Harris, Z. Metallk. **83**, 338 (1992)
45. C. Müller, B. Reinsch, G. Petrow, Z. Metallk. **83**, 845–852 12 (1992)
46. H. Klesnar, P. Rogl, J. Mater. Res. **6**, 53 (1991)
47. H. Klesnar, K. Hiebl, P. Rogl, H. Noël, Less-Common Metals **154**, 217 (1989)
48. F. Weitzer, K. Hiebl, Yu. Grin, P. Rogl, H. Noël, J. Appl. Phys. **68**, 3504 (1990)
49. H.S. Li, B.-P. Hu, J.M. Cadogan, J.M.D. Coey, J.P. Gavigan, J. Appl. Phys. **67**, 4841 (1990)
50. P. Schrey, M. Velicescu, J. Mag. Mag. Mater. **101**, 417 (1991)
51. T. Zhao, Ph.D. Thesis, Institute of Metal Research, Shenyang, P.R. China (1995)
52. F. Weitzer, A. Leithe-Jasper, P. Rogl, K. Hiebl, A. Rainbacher, G. Wiesinger, W. Steiner, J. Friedl, F.E. Wagner, J. Appl. Phys. **75**, 7745 (1994)
53. R. Ruzitschka, M. Reissner, W. Steiner, P. Rogl, J. Mag. Mag. Mater. **226–223**, 1443–1445 (2001)
54. G. Wiesinger, A. Rainbacher, W. Steiner, A. Leithe-Jasper, P. Rogl, F. Weitzer, Hyperfine Interactions **94**, 1915–1920 (1994)
55. R. Ruzitschka, M. Reissner, W. Steiner, P. Rogl, J. Mag. Mag. Mater. **242–245**, 806–808 (2002)
56. F. Weitzer, A. Leithe-Jasper, P. Rogl, K. Hiebl, H. Noël, G. Wiesinger, W. Steiner, J. Solid State Chem. **104**, 368 (1993)
57. W. Steiner, R. Haferl, Phys. Stat. Sol. (A) **42**, 739–750 (1977)
58. Q.W. Yan, P.L. Zhang, X.D. Sun, B.P. Hu, Y.Z. Wang, X.L. Rao, G.C. Liu, C. Gou, D.F. Chen, Y.F. Cheng, J. Phys, Cond. Matt. **6**, 3101 (1994)
59. F. Wang, B.-G. Shen, H. Gong, X.D. Sun, P. Zhang, Q. Yan, J. Mag. Mag. Mater. **177–181**, 1056 (1998)
60. P. Schobinger-Papamantellos, K.H.J. Buschow, C.H. de Groot, F.R. de Boer, C. Ritter, F. Fauth, G. Boettger, J. Alloys Comp. **280**, 44 (1998)
61. P. Schobinger-Papamantellos, K.H.J. Buschow, C.H. de Groot, F.R. de Boer, G. Boettger, C. Ritter, J. Phys, Cond. Matt. **11**, 4459 (1999)
62. P. Schobinger-Papamantellos, K.H.J. Buschow, C.H. de Groot, F.R. de Boer, C. Ritter, J. Mag. Mag. Mater. **218**, 31 (2000)
63. C. Ritter, J. Phys, Cond. Matt. **1**, 2765 (1989)
64. I. Oftedal, Z. Kristllogr, A **66**, 517 (1928)
65. W. Jeitschko W, D.J. Braun, Acta Crystallogr. B **33**, 3401 (1977)
66. T.M. Tritt, *Recent Trends in Thermoelectric Materials I* (Academia Press, New York, 2001)

67. B.C. Sales, Filled skutterudites, in *Handbook on the Physics and Chemistry of Rare Earths*, vol. 33, ed. by K.A. Gschneidner Jr., J.C.G. Buenzli, V.K. Pecharsky (Elsevier Science, Amsterdam, 2003), pp. 1–34
68. T. Cichorek, A.C. Mota, F. Steglich, N.A. Frederick, W.M. Yuhasz, M.B. Maple, Phys. Rev. Lett. **94**, 107002 (2005)
69. M.B. Maple, E.D. Bauer, N.A. Frederick, P.C. Ho, W.M. Yuhasz, V.S. Zapf, Physica B **338**, 29–33 (2003)
70. C. Uher, Skutterudites prospective novel thermoelectrics, in *Recent Trends in Thermoelectric Materials Reserach I, Semiconductors and Semimetals 69, 139–253*, ed. by T.M. Tritt (Academic, New York, 2001)
71. G.A. Slack, *CRC Handbook of Thermoelectrics 1994*, ed. by D.M. Rowe (CRC Press, Boca Rator, FL, 1995), pp. 407–440
72. I. Shirotani, K. Ohno, C. Sekine, T. Yagi, T. Kawakami, T. Nakanishi, H. Takahashi, J. Tang, A. Matsushita, T. Matsumoto, Physica B **281–282**, 1021 (2000)
73. N. Takeda, M. Ishikawa, J. Phys. Soc. Jpn. **69**, 868 (2000)
74. I. Shirotani, T. Uchiumi, K. Ohno, C. Sekine, Y. Nakazawa, K. Kanoda, S. Todo, T. Yagi, Phys. Rev. B **56**, 7866 (1997)
75. C. Sekine, T. Uchiumi, I. Shirotani, T. Yagi, Phys. Rev. Lett. **79**, 3218 (1997)
76. M.E. Danebrock, C.B.H. Evers, W. Jeitschko, J. Phys. Chem. Solids **57**, 381 (1996)
77. E. Bauer, St. Berger, A. Galatanu, M. Galli, H. Michor, G. Hilscher, Ch. Paul, B. Ni, M.M. Abd-Elmeguid, V.H. Tran, A. Grytsiv, P. Rogl, Phys. Rev. B **63**, 224414 (2001)
78. N.R. Dilley, E.J. Freeman, E.D. Bauer, M.B. Maple, Phys. Rev. B **58**, 6287 (1998)
79. N. Takeda, M. Ishikawa, J. Phys. Cond. Matt. **13**, 5971 (2001)
80. E. Bauer, A. Galatanu, H. Michor, G. Hilscher, P. Rogl, P. Boulet, H. Noël, Eur. Phys. J. B **14**, 483 (2000)
81. A. Grytsiv, P. Rogl, St. Berger, Ch. Paul, E. Bauer, C. Godart, B. Ni, M.M. Abd-Elmegiud, A. Saccone, R. Ferro, D. Kaczorowski, Phys. Rev. B **66**, 094411 (2002)
82. A. Leithe-Jasper, W. Schnelle, H. Rosner, M. Baenitz, A. Rabis, A.A. Gippius, E.N. Morozova, H. Borrmann, U. Burkhardt, R. Ramlau, U. Schwarz, J.A. Mydosh, Yu. Grin, V. Ksenofontov, S. Reiman, Phys. Rev. B **70**, 214418 (2004)
83. A. Leithe-Jasper, W. Schnelle, H. Rosner, R. Cardoso-Gil, M. Baenitz, J.A. Mydosh, Yu. Grin, M. Reissner, W. Steiner, Phys. Rev. B **77**, 064412 (2008)
84. E. Bauer, St. Berger, Ch. Paul, M. Della Mea, G. Hilscher G, H. Michor, M. Reissner, W. Steiner, A. Grytsiv, P. Rogl, E.W. Scheidt, Phys. Rev. B **66**, 214421 (2002)
85. M.S. Torikachvili, J.W. Chen, Y. Dalichaouch, R.P. Guertin, M.W. McElfresh, C. Rossel, M.B. Maple, G.P. Meisner, Phys. Rev. B **36**, 8660 (1987)
86. Y. Aoki, T. Namiki, T.D. Matsuda, K. Abe, H. Sugawara, H. Sato, Phys. Rev. B **65**, 064446 (2002)
87. Y. Nakamishi, T. Simizu, M. Yoshizawa, T. Matsuda, H. Sugawara, H. Sato, Phys. Rev. B **63**, 184429 (2001)
88. P. Meisner, Physica **108B**, 763 (1981)
89. D.J. Singh, I.I. Mazin, Phys. Rev. B **56**, R1650–3 (1997)
90. K. Takegahara, H. Harima, J. Phys. Soc. Jpn. **71**(Suppl.) 240–242 (2002)
91. K. Tanaka, Y. Kawahito, Y. Yonezawa, Y. Kikuchi, H. Aoki, K. Kuwahara, M. Ichihara, H. Sugawara, Y. Aoki, H. Sato, J. Phys. Soc. Jpn. **76**, 103704 (2007)
92. M. Reissner, E. Bauer, W. Steiner, P. Rogl, J. Mag. Mag. Mater. **272–276**, 813 (2004)
93. G.K. Shenoy, D.R. Noakes, G.P. Meisner, J. Appl. Phys. **53**, 2628–2630 (1982)
94. M. Reissner, E. Bauer, W. Steiner, P. Rogl, A. Leithe-Jasper, Y. Grin, Hyperfine Interactions **182**, 15–22 (2008)
95. A. Leithe-Jasper, D. Kaczorowski, P. Rogl, J. Bogner, M. Reissner, W. Steiner, G. Wiesinger, C. Godart, Sol. Stat. Comm. **109**, 395–400 (1999)
96. G.J. Long, B. Mathieu, B.C. Sales, R.P. Hermann, F. Grandjean, J. Appl. Phys. **92**, 7236–7241 (2002)
97. P. Blaha, Private communication

98. P. Mohn, Private communication
99. P. Ghosez, M. Veithen, J. Phys. Cond. Matt. **19**, 096002 (2007)
100. G.J. Long, D. Hautot, F. Grandjean, D.T. Morelli, G.P. Meisner, Phys. Rev. B **60**, 7410–7418 (1999)
101. G.P. Meisner, D.T. Morelli, S. Hu, J. Yang, C. Uher, Phys. Rev. Lett. **80**, 3551–3554 (1998)
102. V. Cannella, J.A. Mydosh, Phys. Rev. B **6**(11), 4220 (1972)
103. P.J. Ford, Contemp. Phys. **23**(2), 141–168 (1982)
104. K. Binder, A.P. Young, Rev. Mod. Phys. **58**, 801–976 (1986)
105. D. Chowdhury, *Spin Glasses and Other Frustrated Systems* (World Scientific, Singapore, 1986)
106. J.A. Mydosh, G.J. Nieuwenhuys, Dilute transition metal alloys; spin glasses in *Ferromagnetic Materials*, vol. 1 ed. by E.P. Wohlfarth (1980)
107. J.A. Mydosh, *Spin Glasses: An Experimental Introduction* (Taylor and Francis, 1993)
108. N. Rivier, Wiss. Z., Techn. Univ. Dresden (1974)
109. M.A. Rudermann, C. Kittel, Phys. Rev. **96**, 99 (1954)
110. T. Kasuya, Prog. Theor. Phys. **16**, 45 (1956)
111. K. Yosida, Phys. Rev. **106**, 893 (1957)
112. K. Yosida, Phys. Rev. **107**, 396 (1957)
113. G. Toulouse, Comm. Phys. **2**, 115–119 (1977)
114. G. Toulouse, Lect. Notes Phys. **192**, 2–17 (1983)
115. B.R. Coles, B.V. Sarkissian, R.H. Taylor, Phil. Mag. B **37**, 489 (1987)
116. P.A. Beck, Met. Trans. **2**, 2015 (1971)
117. P.A. Beck, J. Less-Common Met. **28**, 193 (1972)
118. R.W. Tustion, P.A. Beck, Sol. Stat. Comm. **21**, 517 (1977)
119. C.N. Guy, J. Phys. F **5**, L242 (1975)
120. C.N. Guy, Physica 86-88B, 877 (1977)
121. C.N. Guy, J. Phys. F **7**, 1505 (1977)
122. C.N. Guy, J. Phys. F **8**(6), 1305–19 (1978)
123. M. Reissner, W. Steiner, J.P. Kappler, P. Bauer, M.J. Besnus, J. Phys. F Met. Phys. **14**, 1249–1259 (1984)
124. L.E. Wenger, P.H. Keesorn, Phys. Rev. B **11**, 3497 (1975)
125. R.J. Borg, R. Booth, C.E. Violet, Phys. Rev. Lett. **11**, 464 (1963)
126. V. Gonser, V.W. Grant, C.J. Meecham, A.M. Muir, H. Wiedersich, J. Appl. Phys. **36**, 2124 (1965)
127. P.P. Craig, W.A. Steyert, Phys. Rev. Lett. **13**, 802 (1964)
128. C.E. Violet, R.J. Borg, Phys. Rev. **149**, 540 (1966)
129. A.P. Murani, J. Mag. Mag. Mat. **5**, 95 (1977)
130. A.P. Murani, J. Appl. Phys. **49**, 1604 (1978)
131. A.P. Murani, J.L. Tholence, Sol. Stat. Comm. **22**, 25 (1977)
132. S.F. Edwards, P.W. Anderson, J. Phys. F **5**, 965 (1975)
133. D. Sherrington, S. Kirkpatrick, Phys. Rev. Lett. **35**(26), 1792–6 (1975)
134. C.M. Soukoulis, G.S. Grest, K. Levin, Phys. Rev. Lett. **41**, 568 (1978)
135. G. Parisi, G. Toulouse, J. Physique Lett. **41**, L361 (1980)
136. J.L. Tholence, R. Tournier, J. Physique 35, C4 229 (1974)
137. E.P. Wohlfarth, Physica 86-88B, 852–853 (1977)
138. H. Yamada, J. Inoue, K. Terao, S. Kanada, M. Shimizu, J. Phys. F Met. Phys. **14**, 1943 (1984)
139. E. Burzo, Z. Angew, Phys. **32**, 127–132 (1971)
140. E. Burzo, Phys. Rev. B **17**, 1414–1423 (1978)
141. M. Reissner, W. Steiner, Hyperfine Interactions **28**, 1017–1020 (1986)
142. J.M. Preston, J.R. Stewart, M. Reissner, W. Steiner, R. Cywinski, Appl. Phys. A **74**, S689–S691 (2002)
143. M. Reissner, Thesis TU-Wien (1990)
144. M. Aoki, H. Yamada, J. Magn. Magn. Mater. **78**, 377 (1989)
145. M. Aoki, H. Yamada, Sol. Stat. Comm. **72**, 21 (1981)

146. C. Rizzuto, Rep. Prog. Phys. **37**, 147 (1974)
147. G. Grüner, Adv. Phys. **23**, 941 (1974)
148. J. Deuz, P.H. Dederichs, R. Zeller, J. Phys. F Met. Phys. **11**, 1787 (1981)
149. O. Eriksson, A. Svane, J. Phys. Cond. Matt. **1**, 1589 (1989)
150. A. Pösinger, M. Reissner, W. Steiner, P. Blaha, P. Mohn, K. Schwarz, J. Phys. Cond. Matt. **5**, 7277–7288 (1993)
151. I. Vincze, J. Balogh, Z. Foris, D. Kaptas, T. Kemeny, Sol. Stat. Comm. **77**, 757 (1991)
152. C. Meyer, F. Hartmann-Boutron, Hyperfine Interactions **59**, 219 (1990)
153. Y. Furukawa, Y. Nakai, N. Kunitoni, J. Phys. Soc. Jpn. **59**, 2237 (1990)
154. J. Bogner, H. Schachner, M. Reissner, W. Steiner, Aust. J. Phys. **51**(2), 349–362 (1998)
155. W. Steiner, M. Reissner, A. Pösinger, H. Winkler, A.X. Trautwein, Proc. XXX Zakopane School Physica **185–195** (1995)
156. A. Pösinger, M. Reissner, W. Steiner, Physica B **155**, 211–214 (1989)
157. A. Pösinger, H. Winkler, W. Steiner, A.X. Trautwein, M. Reissner, J. Phys. Cond. Matt. **3**, 2713 (1991)
158. A. Pösinger, H. Winkler, W. Steiner, A.X. Trautwein, M. Reissner, J. Phys. C **5**, 3653 (1993)
159. H.H. Wickman, *Mössbauer Paramagnetic Hyperfine Structure in Mössbauer effect Methodology*, vol. 2, ed. by I.J. Gruverman (Plenum Press, New York, 1966), pp. S39–S66
160. H.H. Wickman, M.P. Klein, D.A. Shirley, Phys. Rev. **152**, 345 (1966)
161. A.P. Murani, J. Mag. Mag. Mater. **22**, 271 (1981)
162. Y.J. Uemura, T. Yamazaki, D.R. Harshman, M. Senba, E.J. Ansaldo, Phys. Rev. B **31**, 546 (1985)
163. I.A. Campbell, A. Amato, F.N. Gygax, D. Herlach, A. Schenk, R. Cywinski, S.H. Kilcoyne, Phys. Rev. Lett. **72**, 12 (1994)
164. A.T. Ogielsky, Phys. Rev. B **32**, 7384 (1985)
165. H. Rieger, *Annual Review of Computational Physics*, vol. II, ed. by D. Stauffer (World Scientific, Singapore, 1995), p. 295
166. L. Néel, Phil. Mag. Suppl. **4**, 191 (1955)

Chapter 9
Mössbauer Spectroscopic Studies on Atomic Diffusion in Materials

Yutaka Yoshida

Abstract Atomic jumps within the lifetime of a Mössbauer nuclear probe are known to cause a line broadening, which is directly connected to the diffusivity, as predicted theoretically by Singwi and Sjölander in 1960. Many research groups applied and further extended their theory for the studies of elementary jump processes of Fe atoms in single crystals: They analyzed Mössbauer spectrum in terms of line broadening and motional averaging as functions of temperature as well as crystal orientation. To observe such dynamical behaviours, we developed specially designed experimental set-ups such as a high-temperature Mössbauer furnace for laboratory, in-beam and on-line Mössbauer set-ups for accelerator facilities. First of all, we will introduce such measuring techniques suitable for homogeneous materials, explaining their unique features in this chapter. Secondary, we present an imaging technique in Mössbauer spectroscopy which is opening a new possibility to study diffusion processes with a diffusion-length from μm to mm after diffusion annealing at high temperature. This original method enables us to measure the diffusion profiles separately for the different spectral components in the material containing a complex microstructure. As examples, we will explain the studies on Fe diffusion in single and multi-crystalline Si materials using "Mössbauer spectroscopic microscope (MSM)," which enables us to measure the diffusion profiles separately for different chemical states with a spatial resolution of several micrometres. This new method provides a possibility to investigate a diffusion process by considering of the interactions and the correlations between Fe impurities and lattice defects such as dislocations, grain boundaries, and residual stresses in different grains of materials. Finally, we apply this technique to investigate carbon diffusion and segregation processes in Fe-steel, and the mapping images at different temperatures are obtained separately for the spectral components that originated from carbon impurities.

Y. Yoshida (✉)
Shizuoka Institute of Science and Technology, Fukuroi, Japan
e-mail: yoshida.yutaka@sist.ac.jp

Y. Yoshida and G. Langouche (eds.), *Modern Mössbauer Spectroscopy*,
Topics in Applied Physics 137, https://doi.org/10.1007/978-981-15-9422-9_9

9.1 Introduction

Atomic diffusion can be found in materials whenever a system containing many atoms tends to change toward thermal equilibrium. The atoms flow even macroscopically until the system reaches the thermal equilibrium, although they further continue to perform random jumps at a finite temperature even on the equilibrium. There appear, however, to remain many fundamental problems related to the diffusion mechanism, which must be essential to understand and to control the microstructures as well as their physical properties in many systems, including industrial materials, such as iron steel and semiconductors. Accordingly, highly desirable must be ***atomistic evaluation techniques*** accessible to such underlying problems *without causing any perturbations on the systems.*

As far as ^{57}Fe Mössbauer spectroscopy concerns, the spectroscopy provides two different possibilities to investigate the atomic diffusion corresponding to different characteristic time and space regions of (**A**) MS and (**B**) MSM presented in Fig. 9.1, respectively. We can study the first region of (**A**) through the line-broadening and the hyperfine relaxations of Mössbauer spectrum as functions of temperature and also of crystal orientation. Figure 9.2 shows a schematic drawing to give the reader an idea for such a situation, i.e., an atom is jumping on the crystal lattice within a diffusion length of a few atomic distances as well as within a time scale of 140 ns, i.e., the

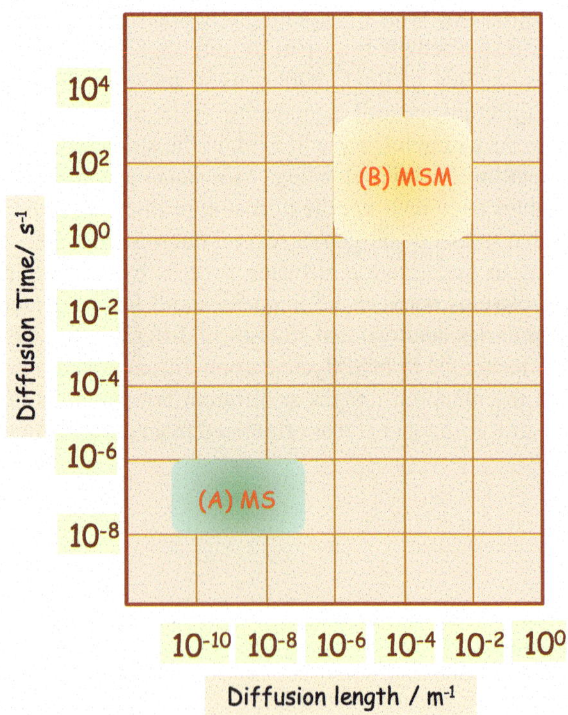

Fig. 9.1 Characteristic diffusion time and length accessible by ^{57}Fe nuclear probe through (**A**) line-broadening and/or hyperfine relaxation of Mössbauer Spectrum (MS) and (**B**) the Mössbauer Spectroscopic Microscope (MSM) imaging

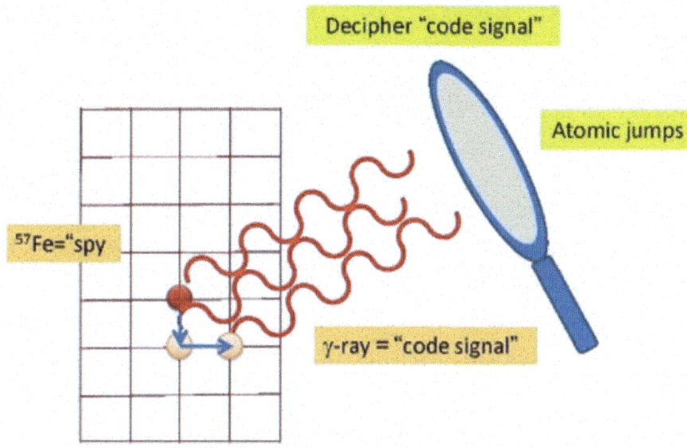

Fig. 9.2 Atomic motion influences the shape of Mössbauer spectrum, providing direct information of elementary jump process, such as jump distance and directions

lifetime of 14.4 keV excited state of ^{57}Fe nucleus. To measure the second region of (**B**), on the other hand, we use a new imaging technique in Mössbauer spectroscopy, which has opened a unique possibility to investigate the diffusion properties through imaging directly the different spectral components after annealing at different temperatures. This method provides the diffusion profiles separately for the various ^{57}Fe chemical states, which is in contrast to a standard diffusion measurement yielding the total concentration profile of the diffusion element only. This new methodology will make it possible to challenge ***"in-situ observations on the atomic diffusion" without disturbing jump processes under different circumstances*** because the imaging technique uses the Mössbauer effect, i.e., the recoil-free γ-ray resonant absorption and emission [1].

In this chapter, we are going to explain typical diffusion studies; iron diffusion in pure iron and silicon crystals as well as in iron steel and multi-crystalline silicon solar cells. In the next Sect. 9.2, we will shortly summarise the tracer diffusion studies and ^{57}Fe hyperfine interactions of Mössbauer spectroscopy. In Sects. 9.3 and 9.4, the readers will find the Mössbauer measuring techniques specially developed for the direct observations of the diffusion and the point defects in pure iron, such as the in-situ techniques for "High-Temperature Mössbauer spectroscopy" and "Accelerator Beam Mössbauer Spectroscopy". In Sect. 9.5, we will explain the imaging techniques in Mössbauer spectroscopy realized in our group. In Sect. 9.6, we describe some examples of diffusion studies by applying imaging techniques.

Since the discovery in 1958, there have been many textbooks and conference proceedings published on the Mössbauer effect and its applications for physics, chemistry, biology, materials science, earth science, and other research fields [2–14].

9.2 Historical Development of Diffusion Studies

9.2.1 *Fick' Principles and Brownian Motion Theory*

Historically, there appear to be two dominant directions for diffusion studies in materials. In 1855, A. Fick contributed to define the diffusion coefficient and to measure the diffusion of salt in water [15]. Further, G. K. von Hevesy first used radioisotopes to measure self-diffusion in water and lead [16]. Figure 9.3 schematically presents a concentration depth profile of tracer atoms after diffusing solute atoms at a specific temperature. The other direction originated from the study on Brownian motion, which was theoretically explained in term of the mean-square displacement of the particle by Einstein [17] and separately by Smoluchowski [18]. Many excellent textbooks were published so far [19–21] on diffusion in solids, which concerned mainly with a tracer diffusion experiment, providing us with a big data-base for diffusivities in wide varieties of materials [23, 24].

In a tracer diffusion experiment, typically, one measures a concentration profile of solute as a function of the depth of x from the surface after the annealing at different temperatures. In the case of Fig. 9.3, one puts the tracers on the surface as a thin film, the initial concentration being $c(x, 0) = M\delta(x)$. We solve the diffusion equation of (9.1) and obtain the concentration profile $c(x, t)$ as a Gaussian (9.2) for an annealing time of t. Accordingly, we can analyze the slope of $\ln c(x, t)$ as a function of x^2, yielding D (T) at the annealing temperature, T. Since the diffusion is a thermally activated process, the diffusivity of $D(T)$ depends on the Arrhenius formula with a pre-exponential factor D_0 and an activation enthalpy of ΔH, as is shown in (9.3).

$$\frac{\partial c(x,t)}{\partial t} = D\frac{\partial^2 c(x,t)}{\partial x^2} \tag{9.1}$$

$$c(x,t) = \frac{M}{\sqrt{\pi D t}} \cdot \exp\left(-\frac{x^2}{4Dt}\right). \tag{9.2}$$

$$D = D_0 \exp\left(-\frac{\Delta H}{k_B T}\right) \tag{9.3}$$

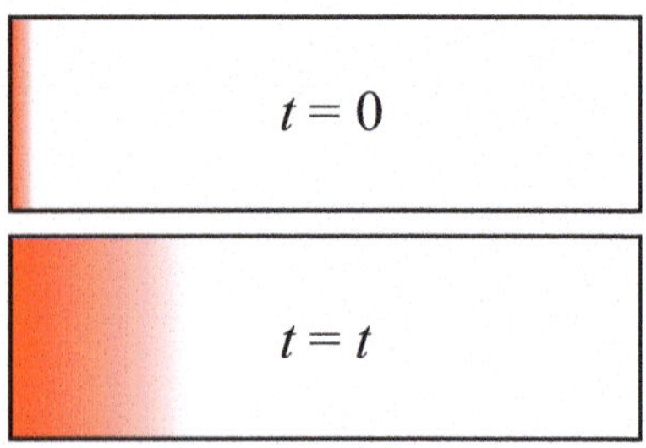

Fig. 9.3 A schematic drawing for diffusion experiment: the concentration profile of solute is measured as function of depth after annealing at different temperatures

Let us see the general trend of the solute diffusivities in metals as a function of temperature T/T_m. Here temperature T is divided by the melting point T_m. Figure 9.4 indicates that both values of D_0 and ΔH consist of three groups depending on the crystal structures and the diffusion mechanisms. From the lowest group, vacancy diffusion in fcc metals, and bcc metals are responsible as in Fig. 9.5. The highest group, on the other hand, corresponds to the fast-interstitial diffusion of the light

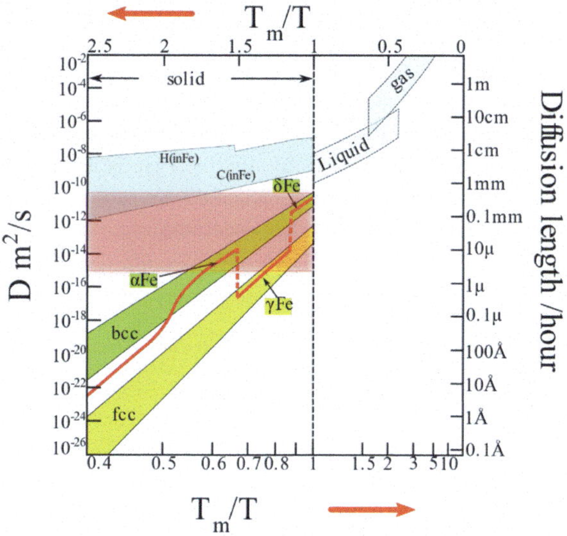

Fig. 9.4 Typical diffusivities and diffusion length per hour for metals as functions of T/T_m where T_m is melting temperature Each metal

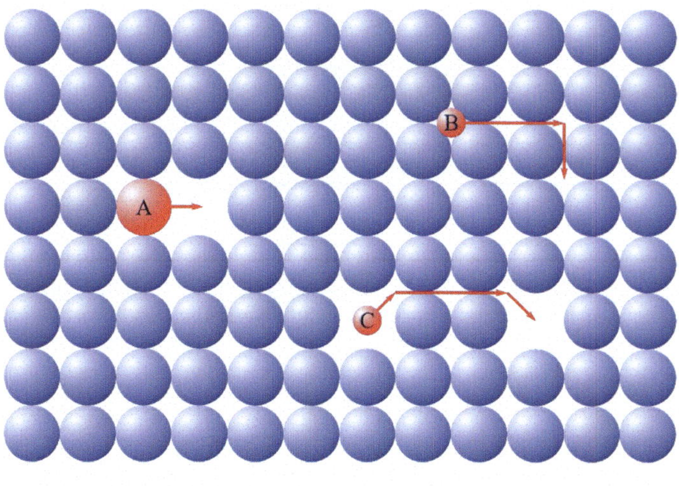

Fig. 9.5 Diffusion mechanisms via (**A**) vacancy, and (**B**) and (**C**) interstitials

elements such as H, C. One would get an idea of how fast the atoms can move in metallic systems on the right perpendicular axis in Fig. 9.4, i.e., the diffusion length after annealing for one hour.

As we will see in Sect. 9.2.3, one can detect the atomic diffusion as a line broadening of the Mössbauer spectrum for the diffusivities of ^{57}Fe atoms in pure Fe from $D = 10^{-15}$ to 10^{-11} m^2/s as is indicated by red zone in Fig. 9.4. We expect such a high diffusivity only at high temperatures close to the melting point. To achieve an in-situ observation of the atomic jumps by Mössbauer spectroscopy, therefore, one has to develop an experimental set-up accessible for measuring a spectrum with a line broadening $\Delta\Gamma$, between 0.01 and 30 mms^{-1}, yielding us a challenging task to attack the experimental conditions.

In Sect. 9.4, the readers will find a series of challenges in how we observed the line broadenings in pure iron as well as in silicon crystals. In the next subsection, we explain the basics of ^{57}Fe Mössbauer spectroscopy, which will be necessary to understand the contents described in the later sections.

9.2.2 *Hyperfine Interactions of ^{57}Fe Nuclei and Mössbauer Experimental Set-Up*

Here we are going to describe how we measure a Mössbauer spectrum of ^{57}Fe using a conventional set-up for an absorber experiment, i.e., the sample to be measured is the absorber; the lower part of Fig. 9.6 shows a set-up schematically, where the γ-rays are emitted from the 14.4 keV nuclear level, shown in the upper part, with a half-lifetime of 98.1 ns following the Electron Capture (EC) decay from the ^{57}Co radioisotope with a half-lifetime of 271 days. The standard ^{57}Co source, which is commercially available as a sealed radioactive source, is produced by electroplated and diffused in

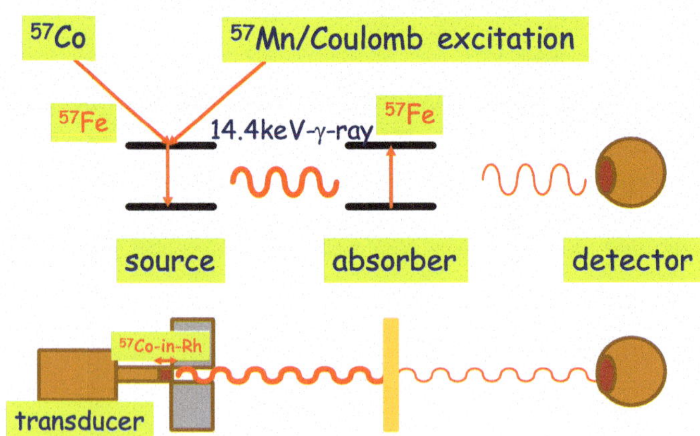

Fig. 9.6 Conventional set-up for Mössbauer absorber experiment

a Rh thin foil, and the strength will be selected from 0.37 to 3.7 GBq depending on the ^{57}Fe concentration contained in the sample. The 14.4 keV γ-rays are transmitted through the absorber with a thickness from 20 μm to 1 mm, depending on the sample. We should choose the specimen thickness carefully [25] because a too-thick sample would provide a spectrum shape strongly deviating from Lorentzian, and therefore, a fitting analysis would be complicated using a superposition of Lorentz functions. In the case of the iron sextet spectrum, an ideal thickness is 25 μm.

We count the transmitted γ-rays by using a gas proportional counter, or a scintillation counter, or a Si-PIN detector, and subsequently, record them on a multi-channel scaler with 1024 channels typically as a function of Doppler velocity of the ^{57}Co-in-Rh source. A transducer operates with a constant-acceleration mode, yielding a triangular-Doppler-velocity function between −10 and 10 mms^{-1}, for instance. In our particular case for a "Mössbauer spectroscopic microscope" described in Sect. 9.5, on the other hand, we use a linear actuator with a constant velocity operation to adjust a resonance condition, to obtain the image corresponding to a spectrum component.

We draw a typical single Lorentz spectrum in the upper part of Fig. 9.7. To measure such a very narrow absorption spectrum with a line width of an order of 10^{-8} eV for ^{57}Fe nuclear probe, we use the Doppler effect to modulate the γ-ray energy. Since the first-order Doppler shift, ΔE_D, is described by

$$\Delta E_D = \frac{v}{c} E_\gamma, \tag{9.4}$$

where $E\gamma$ is the γ-ray energy of 14.4 keV, c the light velocity, and v the Doppler velocity of the ^{57}Co source. Accordingly, $v = 1$ mms^{-1} corresponds to an energy shift of $\Delta E_D = 4.80766 \times 10^{-8}$ eV. One should notice that we cannot use a usual spectroscopic detector for analyzing the γ-ray energy down to 10^{-8} eV, but for counting after selecting the 14.4 keV γ-rays by a single-channel analyzer within a resolution from meV to keV range. We usually present the spectrum as a function of Doppler velocity in Mössbauer spectroscopy.

Furthermore, the velocity is not the absolute value against the laboratory system, but the relative one against the centre position of α-Fe sextet spectrum at room temperature, as in the lower part of Fig. 9.7. If one uses ^{57}Co-in-Rh source, the zero-velocity for Fe-in-Rh absorber ($\delta = +\ 0.109$ mms^{-1}) shifts to be $+\ 0.109$ mms^{-1} against the centre of Fe sextet spectrum. One should notice, however, in a source experiment, in which the source will be the sample, that the velocity scale is shown converted in comparison to an absorber experiment.

A spectrum shape depends on the hyperfine interactions between the ^{57}Fe nuclear probe and the electrons, i.e., electric monopole, electric quadrupole, and magnetic dipole interactions. The monopole interaction affects the position of resonance lines, leading to an "isomer shift (δ) in the Mössbauer spectrum, while the electric quadrupole and magnetic dipole interaction remove the degeneracy of nuclear levels, causing the electric quadrupole splitting ($\Delta \boldsymbol{E}_\mathrm{q}$) and the magnetic Zeeman splitting ($\Delta \boldsymbol{H}_\mathrm{m}$), respectively. The electric monopole interaction energy, i.e., the Coulomb interaction energy between the nucleus and the s-electrons penetrating, is given by

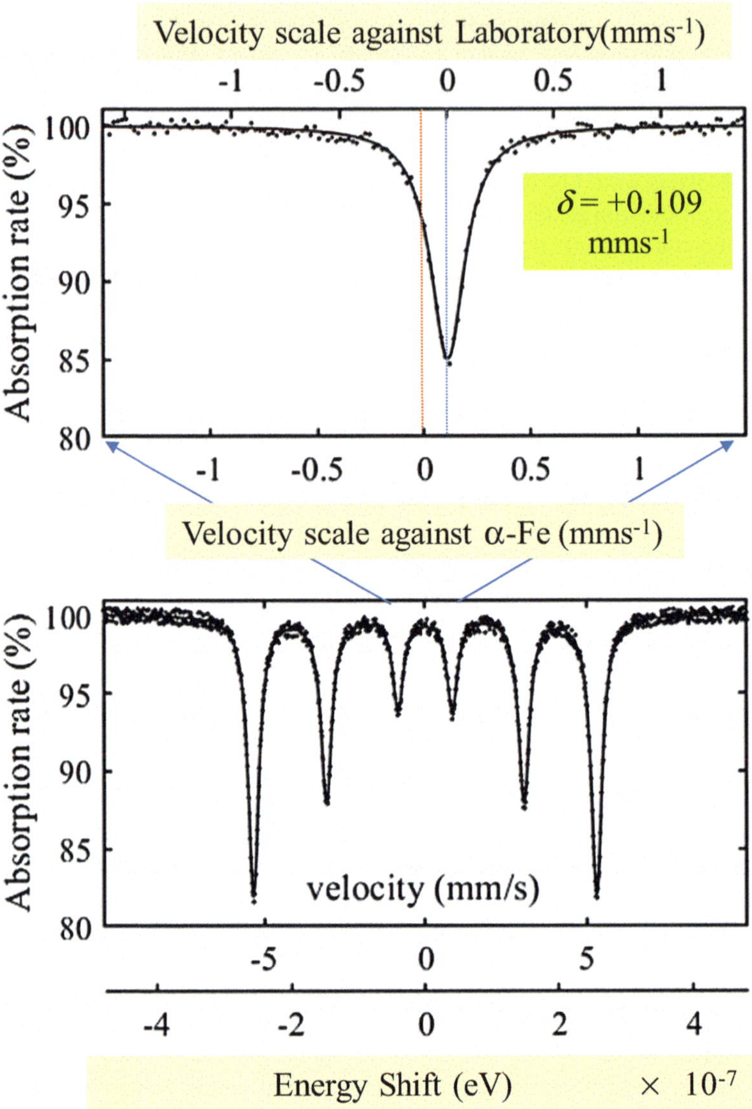

Fig. 9.7 Mössbauer spectrum is shown as a function of Doppler velocity of ^{57}Co-source, i.e., energy shift in 10^{-7} eV

$$E_{\mathrm{I}} = \frac{2\pi}{3}|\psi(0)|^2 Z_e \langle r_n^2 \rangle, \tag{9.5}$$

where $|\psi(0)|^2$ is the electron density at the nucleus, Z_e nuclear charge, and $\langle r_n^2 \rangle$ the expectation value of the square of the nuclear radius. In the Mössbauer spectrum, the isomer shift appears as.

$$\delta = C\left[|\psi(0)|_A^2 - |\psi(0)|_S^2\right], \tag{9.6}$$

i.e., the difference in the charge densities between the source (S) and absorber (A). As a reference of isomer shift, we use the centre position of the sextet Mössbauer spectrum of pure Fe at room temperature in this book. The isomer shift depends, for instance, on different charge states of ^{57}Fe atoms in Si crystal, as will be explained in Sects. 9.4 and 9.5.

When an electric field gradient (EFG) is present at the site of the ^{57}Fe nucleus in a paramagnetic phase, the electric quadrupole interaction causes a doublet with a splitting ΔE_Q, as in Fig. 9.8(**a**),

$$\Delta E_Q = \frac{1}{2} e^2 q Q \left(1 + \frac{\eta^2}{3}\right)^{1/2} \tag{9.7}$$

$$\eta = \frac{V_{xx} - V_{yy}}{V_{zz}}, \tag{9.8}$$

where eQ is the quadrupole moment, η the asymmetry parameter, V_{xx}, V_{yy}, V_{zz} the electric field gradient, for axial symmetry, $V_{xx} = V_{yy}$ leading to $\eta = 0$. By choosing the principal axes of EFG tensor such $V_{zz} \geq V_{xx} \geq V_{yy}$, the asymmetry parameter becomes $0 \leq \eta \leq 1$.

The nuclear states of ^{57}Fe consist of the ground state with $I = 1/2$ and the excited state with $I = 3/2$, and will be degenerated through magnetic dipole interaction (Zeeman effect), as shown in Fig. 9.8(**b**). When the ^{57}Fe probe exists in a magnetic field, the nuclear energy levels are given depending on the spin quantum numbers m_I by

$$E_M = -\mu \boldsymbol{B} m_I / I = -g_N \beta_N \boldsymbol{B} m_I, \tag{9.9}$$

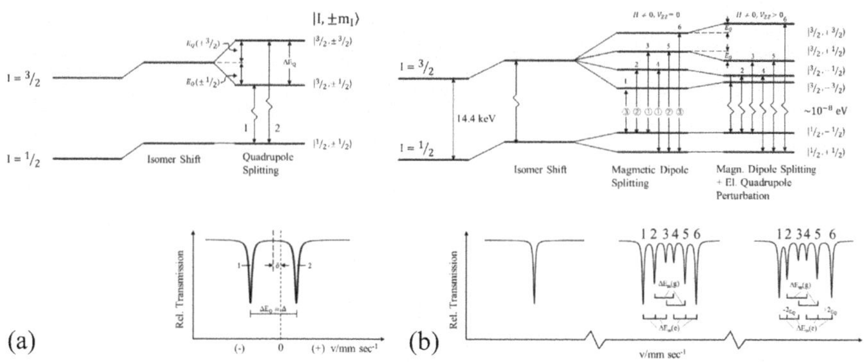

Fig. 9.8 Hyperfine interactions lead shifts of spectral positions and splitting

where μ is the magnetic dipole moment, g_N the nuclear Landè factor, and β_N the nuclear Bohr magneton. The selection rule for magnetic dipole transitions leads to the sextet as in Fig. 9.8(**b**). Furthermore, each sextet line will be shifted, as in the right-hand side in Fig. 9.8(**b**), when the magnetic dipole interaction is dominant in comparison with the electric dipole interaction. If the EFG tensor is axially symmetric and its principal axis makes an angle β, the nuclear energy is given by

$$\mathrm{E} = -g_N \beta_N \boldsymbol{B} m_I + (-1)^{|m_I|+1/2}(eQV_{zz}/8)\left(3\cos^2\beta - 1\right). \tag{9.10}$$

A typical example of (9.10) will appear in a spectrum of Fe–C steel in Sect. 9.6.

9.2.3 Principle of Mössbauer Study on Atomic Diffusion

In this section, we explain how atomic jumps affect the Mössbauer spectrum. Consider that a ^{57}Fe atom is jumping on an empty Bravais lattice while emitting γ-ray, as is schematically shown in Fig. 9.9 for the ^{57m}Fe nucleus on the 14.4 keV first excited state. This excited state can be occupied through different processes, such as electron capture (EC) decay from ^{57}Co, β-decay from ^{57}Mn, Coulomb excitation, ^{56}Fe (d, p)^{57m}Fe nuclear reaction, and Mössbauer absorption of 14.4 keV γ-ray. Subsequently, the 14.4 keV γ-ray will be emitted resonantly without recoil (the Mössbauer effect), while in Fig. 9.9, the ^{57m}Fe atom is jumping between different lattice sites with a typical frequency of 10^7 s^{-1}, which is roughly equal to the inverse of the lifetime, 140 ns. The ^{57}Fe is staying on the lattice positions, simultaneously vibrating with a phonon frequency of 10^{12} s^{-1}. When the atom occasionally obtains thermal energy enough to overcome the potential barrier between the different lattice

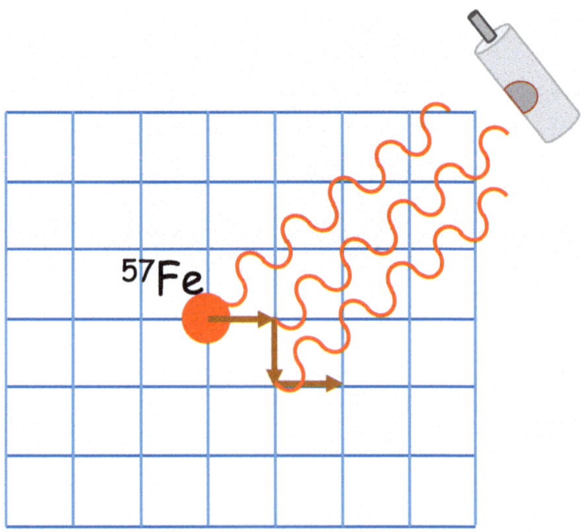

Fig. 9.9 Atomic jumps of ^{57}Fe in solid, leading wave shorter than the life-time, 140 ns. This cause a line broadening of the Mössbauer spectrum

sites, the ^{57}Fe can jump from a lattice site to another, leading to diffusion. The duration of one jump is about 10^{-12} s, which is orders magnitude shorter than that of the atom staying on a lattice site. It turns out that the ^{57}Fe performs a few jumps during emitting or absorbing γ-ray, and that the jump process breaks the coherency of the wave train. Consequently, we observe the lifetime of the 14.4 keV γ-ray apparently shorter than the natural lifetime. Considering of the Heisenberg time-energy uncertainty principle, $\Delta E\ \tau_M \approx \hbar$, a shorter lifetime provides a broader linewidth of the Mössbauer resonance. The line broadenings have been observed indeed in pure Fe using different experimental techniques, such as high temperatures UHV-furnace, Coulomb excitation and recoil-implantations and (d, p) reactions, as will be presented in Sect. 9.4. These techniques provide different point defects and their concentrations adjacent to ^{57m}Fe atoms.

In 1960, immediately after the discovery of the Mössbauer effect, Singwi and Sjölander [26] showed in a classical case that the Mössbauer emission spectrum $\sigma(\boldsymbol{k}, \omega)$ can be described as a space–time Fourier transformation of the self-correlation function $G_s(\boldsymbol{r}, t)$, which relates to a classical self-diffusion function, i.e., a probability to find ^{57m}Fe atom at a lattice site $\boldsymbol{r}$ at a time t, if the atom starts jumping from $\boldsymbol{r} = 0$ at $t = 0$.

$$\sigma(\boldsymbol{k}, \omega) = \frac{\sigma_0 \Gamma_0}{2} \cdot f \cdot \iint \exp\left[i\{\boldsymbol{k} \cdot \boldsymbol{r} - (\omega - \omega_0)t\} - \frac{\Gamma_0 |t|}{2\hbar} \right] G_s(\boldsymbol{r}, t) d\boldsymbol{r} dt. \tag{9.11}$$

Here $\boldsymbol{k}$ and ω are the wave vector and the angular frequency of the 14.4 keV γ-ray and σ_0 the resonance emission cross-section, Γ_0 the natural line width, and f the Debye–Waller factor (f-factor), respectively.

Based on the predictions [26], many groups tried to observe the broadenings and to interpret the experimental results by the jump models. At the early stage, however, there remained still a large discrepancy between the "tracer diffusivity" obtained from a radioactive tracer experiment and the "Mössbauer diffusivity" from a line broadening of Mössbauer spectrum. To learn the historical developments until early 1980′, we refer the following publications by Chudley and Elliot [27], Dibar-Ure and Flinn [28], Janot [29], Wolf [30], Mullen [31], Vogl and his group [32, 33], Nowik et al. [34], and Dattagupta [35]. The most successful challenges were the diffusion studies of iron in single-crystal aluminium by Mantl et al. [32], and also in single-crystal copper by Steinmetz et al. [33]. For the diffusion jumps on non-Bravais lattices, furthermore, where the unit cell consists of more than one atom, we have to consider different jump vectors with different jump frequencies. For more details, we recommend the readers to look up a review article on "Diffusion study by Mössbauer spectroscopy" by Vogl, in which he describes the challenging studies on the elementary jump processes in single crystals [10].

In the following, we will calculate the Mössbauer spectrum, as a simple example, when ^{57}Fe atom is jumping within the lifetime on an empty Bravais lattice such as schematically shown in Fig. 9.9. Here we assume that the self-correlation function, $G_s(\boldsymbol{r}, t)$ follows a kinetic equation based on a theory by Chudley and Elliot [27]:

$$\frac{\partial G_s(\boldsymbol{r},t)}{\partial t} = \frac{1}{\tau}\sum_{j=1}^{Z}\frac{1}{Z}G_s\left(\boldsymbol{r}-\boldsymbol{R}_j,t\right) - \frac{1}{\tau}G_s(\boldsymbol{r},t), \tag{9.12}$$

τ being the meantime for which ^{57}Fe atom stays on a given lattice site before jumping to a new lattice position, $\boldsymbol{R}_j$ a jump vector from one of the NN lattice sites $\boldsymbol{r}-\boldsymbol{R}_j$ to the site $\boldsymbol{r}$, and Z the number of the NN lattice sites; for instance, Z = 6 for the simple 3D-cubic lattice. Multiplying $e^{i\boldsymbol{k}\cdot\boldsymbol{r}}$ to both sides of (9.12), and integrating with $\boldsymbol{r}$, i.e., Fourier transformation in space, we obtain:

$$\frac{\partial I_s(\boldsymbol{k},t)}{\partial t} = -f(\boldsymbol{k})\cdot I_s(\boldsymbol{k},t), \tag{9.13}$$

Here $I_s(\boldsymbol{k},t) = \int G_s(\boldsymbol{r},t)e^{i\boldsymbol{k}\cdot\boldsymbol{r}}d\boldsymbol{r}$, and $f(\boldsymbol{k}) = \frac{1}{\tau}\left(1-\frac{1}{Z}\sum_{j=1}^{Z}e^{i\boldsymbol{k}\cdot\boldsymbol{R}_j}\right)$.

We can easily deduce the solution of (9.13),

$$I_s(\boldsymbol{k},t) = I_s(\boldsymbol{k},0)e^{-f(\boldsymbol{k})\cdot t}. \tag{9.14}$$

After inserting $I_s(\boldsymbol{k},t)$ in (9.11) and integrating in time further, we obtain

$$\sigma(\boldsymbol{k},\omega) = \frac{\sigma_0\Gamma_0}{2}\cdot f\cdot\int_0^\infty \exp\left[-i(\omega-\omega_0)t - \frac{\Gamma_0|t|}{2\hbar}\right]\cdot\exp[-f(\boldsymbol{k})\cdot t]dt. \tag{9.15}$$

Then, we take the real part of the integral, and finally, obtain the Mössbauer emission spectrum expressed as a Lorenz function [27],

$$\sigma(\boldsymbol{k},\omega) = \frac{\sigma_0\Gamma_0}{2}\cdot f\cdot\frac{\Gamma_0+\Delta\Gamma_D(\boldsymbol{k})}{(\hbar\omega-\hbar\omega_0)^2+[\Gamma_0+\Delta\Gamma_D(\boldsymbol{k})]^2/4}. \tag{9.16}$$

Here we define a line broadening,

$$\Delta\Gamma_D(\boldsymbol{k}) = 2\hbar f(\boldsymbol{k}). \tag{9.17}$$

Accordingly, we expect the "diffusion line broadenings $\Delta\Gamma_D$" to be anisotropic, and therefore, can measure as a function of the angle θ between the direction of the wave vector, $\boldsymbol{k}$ and the jump vectors, $\boldsymbol{R}_j$.

$$\Delta\Gamma_D = \frac{2\hbar}{\tau}\cdot\left(1-\frac{1}{Z}\sum_{j=1}^{Z}e^{i\boldsymbol{k}\cdot\boldsymbol{R}_j}\right) \tag{9.18}$$

Here we simulate the line broadenings $\Delta\Gamma_D$ and the Mössbauer spectra for the jump processes of ^{57}Fe in bcc Fe crystal with migration energy E_m of 0.3 eV as a function of temperature between 200 and 500 K. The jump frequency is

$$\frac{1}{\tau} = \frac{1}{\tau_0} \exp\left(-\frac{E_m}{k_B T}\right). \tag{9.19}$$

We consider two jump models on an empty bcc lattice: (A) the jump vectors from the body centre to the eight corner sites, as in Fig. 9.10(**a**), and (B) those from the body centre to the six octahedral sites, as in Fig. 9.10(**b**). We present the line broadening normalized by the natural linewidth as functions of the angle and temperature in Figs. 9.11(**a**) and 9.12(**a**), respectively, for the jump models of (A) and (B). The line broadenings depend strongly on the jump models. Also, the Mössbauer spectra in Figs. 9.11(**b**) and 9.12(**b**) show that the linewidths increase so quickly that the spectra disappear apparently within the temperature width of 100 K above 300 K. When we measure an angular dependence of the linewidth of the Mössbauer spectrum experimentally, we will be able to obtain the jump vectors in a crystal lattice.

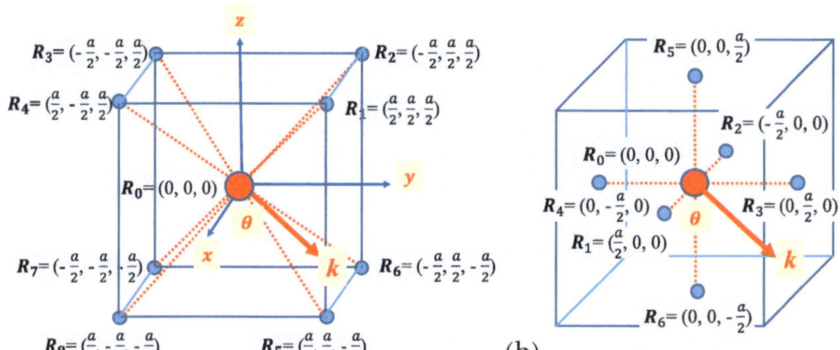

Fig. 9.10 (**a**) Jump vectors, $\boldsymbol{R}_j$ (j = 1, 8) from body centre to eight corner sites in bcc lattice and (**b**) those from body centre to octahedral sites, $\boldsymbol{R}_j$ (j = 1, 6). θ is the angle between [100] and wave vector $\boldsymbol{k}$ when the wave vector changed on x–y plain. The lattice constant of a is 0.2855 nm for α-Fe

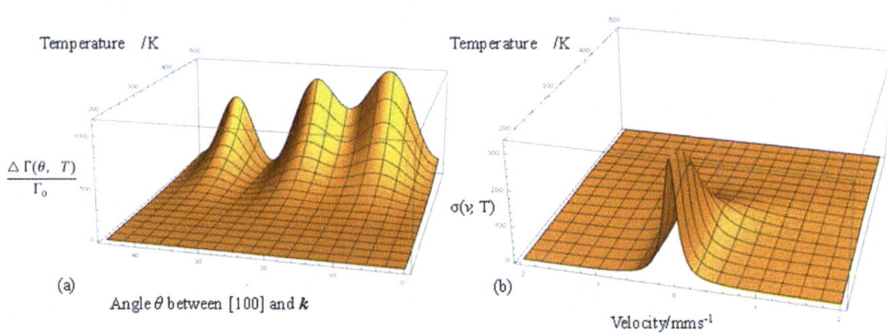

Fig. 9.11 (**a**) Line broadenings for different direction of wave vector $\boldsymbol{k}$ as function of temperature, and (**b**) Mössbauer spectra for [100] direction as function of temperature, which are simulated based on the jump model Fig. 9.10 (**a**)

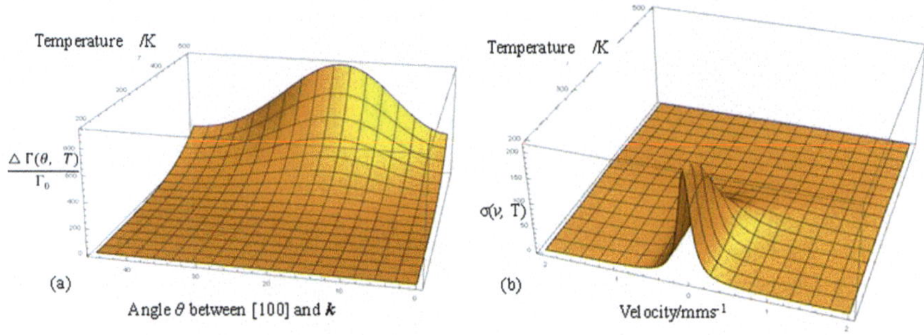

Fig. 9.12 (**a**) Line broadenings for different direction of wave vector $\boldsymbol{k}$ as function of temperature, and (**b**) Mössbauer spectra for [100] direction as function of temperature, which are simulated based on the jump model Fig. 9.10 (**b**)

We should mention that the diffusion strongly correlates with the point defects such as vacancies and interstitials with which the tracer atoms exchange their lattice positions, leading the atomic jumps, as are shown in Fig. 9.5. Accordingly, in a diffusion process on a crystal, τ in (9.19) must be considered to be a mean time of τ_{enc} between "encounters," when the Mössbauer experimental results are compared with those from tracer experiments (see, for instance, in Sect. 9.4.1). Wolf proposed an "encounter model" [30] to consider the correlations between the point defects and a marked atom such as the Mössbauer probe of ^{57}Fe atom. An "encounter" begins by the initial jump of ^{57}Fe atom due to an exchange with a point defect and ends when a different point defect approach to the ^{57}Fe atom. The successive encounters are uncorrelated because the concentration of the point defects is supposed to be less than 10^{-4} in solids. It means that one encounter yields practically one effective jump of ^{57}Fe atom, as is described in this section.

In the next Sect. 9.3, we will concentrate on **the experiments on pure iron**, which is the most straightforward system consisting of Fe atoms. However, there remain still basic questions on the dynamics of point defects such as vacancies and interstitials in Fe crystal after energetic particle irradiations.

9.3 Search for Point Defects in Pure Iron by Thermal Scanning Method

9.3.1 Thermal Scanning Study on C-doped Fe

Atomic diffusion is mediated by point defects in solids such as vacancies and interstitials, as shown in Fig. 9.5. Their concentration, however, is usually too low to observe directly as a defect component in the ^{57}Fe Mössbauer spectrum. To follow an elementary jump process of ^{57}Fe atoms in a crystal lattice, and to clarify the jump mechanism, *we would like to observe the point defects, which jump adjacent to the nuclear probes, and subsequently, exchange the lattice positions, and finally lead to the atomic jumps of ^{57}Fe.* However, the concentration of such point defects will be rather low on the thermal equilibrium; for instance, only below $10^{18}/cm^3$ in metals even just below the melting point. It may be possible, on the other hand, to create an experimental condition where the concentration around ^{57}Fe atoms could be orders of magnitude higher than that on the thermal equilibrium, i.e., under irradiations by highly energetic particles such as neutrons, electrons and heavy ions at low temperature.

In this section, we explain "a thermal scanning method" in Mössbauer spectroscopy, which enables us to detect the recovery stages of the point defects in pure Fe after low-temperature irradiations. Concerning the migration enthalpies of vacancies, interstitials, and also carbon impurities in pure Fe, there were intensive discussions among different research groups [36, 37] during the 1970s and 1980s. Accordingly, we developed a thermal scanning method and applied for the pure Fe samples carburized [38], or irradiated by energetic neutrons and electrons at low temperatures [39], to clarify the kinetics of the recovery stages of the point defects in pure iron.

In the **thermal scanning,** we use two different ^{57}Co sources deposited in pure Fe and Rh matrixes alternatively to measure ***the count rates accurately through Fe sample foil at zero Doppler velocity as a function of the isochronal annealing temperature.*** Figure 9.13 presents the experimental configuration schematically. The count rate at zero velocity strongly depends on the whole shape of the Mössbauer spectrum, being influenced by not only the defect lines but also the changes in the matrix lattice, such as the Debye–Waller factor and the hyperfine fields. Therefore, production, movements, and reactions of the defects can be detected and examined by utilizing all changes arising not only from the probe atom-defect complexes but from their environments. Also, this method can reduce the measurement-time remarkable and eliminate the mechanical errors in the transducer system and the unavoidable broadening in the spectrum measurement. Accordingly, we can examine the existence and movements of point defects even by tiny changes unrecognizable in the usual spectral measurements.

As the first step, we applied this thermal scan method to study the clustering and precipitation process of carbon in iron [38]. After quenching a Fe-7.8 at.%C alloy from 1100 °C into iced water, the Mössbauer spectrum consists of two components as in Fig. 9.14(**a**). A paramagnetic part arises from residual Austenite phase and a

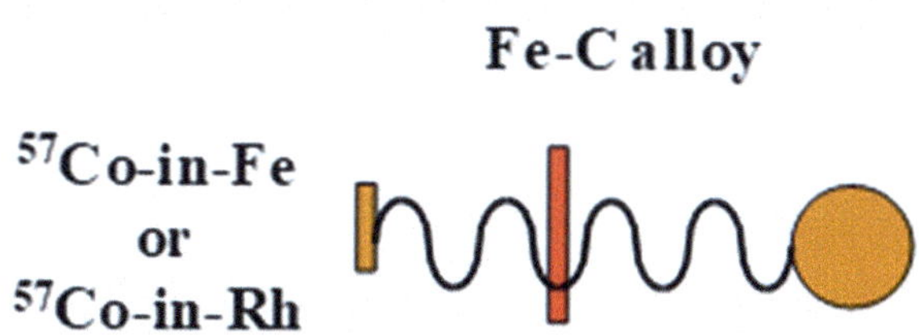

Fig. 9.13 Schematic set-up for thermal scan measurement

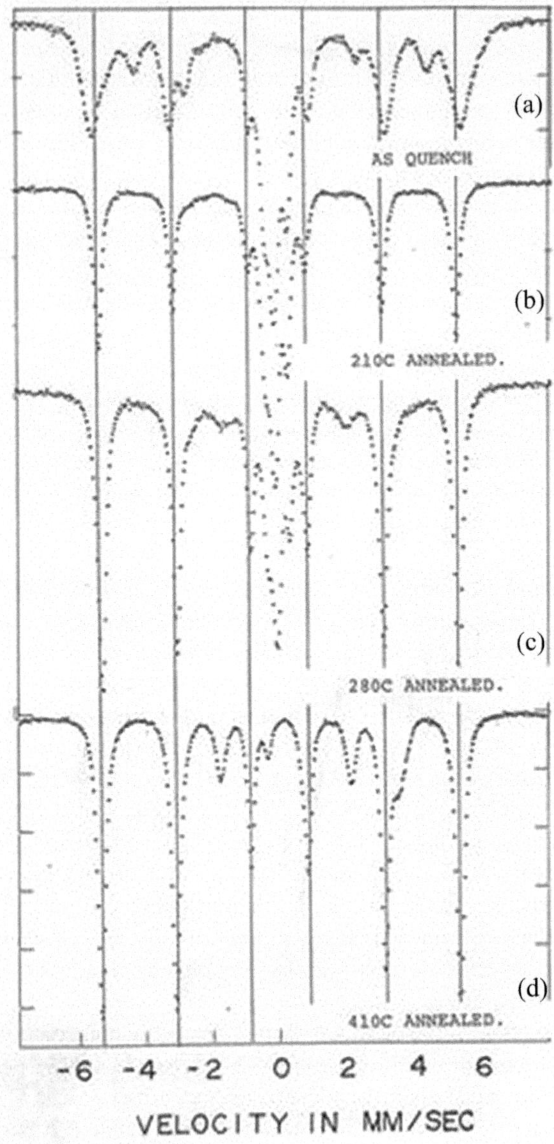

Fig. 9.14 Mössbauer spectra of Fe-7.8at%C: (**a**) as quench, (**b**) annealed at 210 °C for 10 min., (**c**) annealed at 280 °C, and (**d**) annealed at 410 °C. (From Y. Yoshida, Doctor thesis 1983, Osaka Univ.)

ferromagnetic component from Martensitic phase. Both phases contained the super-saturated interstitial carbon atoms. Therefore, the two components changed during the isochronal annealing due to the clustering and precipitation processes of carbon, as is seen in other spectra in Fig. 9.14(**b**)–(**d**).

These processes were observed as the recovery steps in the curve of thermal scan measurement, as shown in Fig. 9.15. Both the ^{57}Co-in-Fe six-line source (closed circles) and the ^{57}Co-in-Rh single-line source (open circles) were used alternatively for the zero-velocity counting, i.e., the thermal scanning. The small changes below 100 °C correspond to the clustering process of carbon in the Martensitic phase, and the remarkable differences between 100 and 250 °C are due to the precipitation process of carbon in that phase. Finally, those between 250 and 320 °C are the decomposing

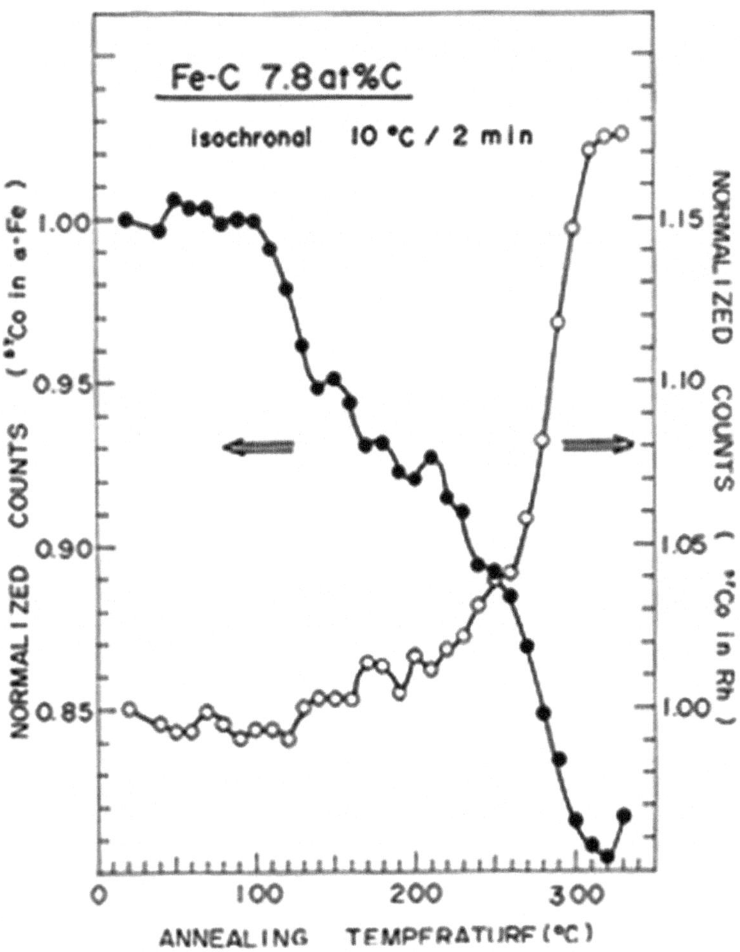

Fig. 9.15 Thermal scan technique: count rate as function of isochronal annealing temperature through transmission resonance absorbers. (From Y. Yoshida, Doctor thesis 1983, Osaka Univ.)

process of the residual Austenite phase. This method appears, therefore, to be very sensitive to detect the kinetics in Fe steels for low concentrations of carbon, as low as 0.5 at. %.

This thermal scanning shows that the zero-velocity counts of the 14.4 keV-γ-rays emitted from ^{57}Co-in-Fe source strongly depend on the number of Fe atoms in pure α-Fe, i.e., in a region without any perturbations from carbon impurities. The on-resonance absorption in the sample increases, and therefore the transmission counts decrease when the carbon clustering and the carbide formation processes gradually proceed during the isochronal annealing. On the other hand, the counts from ^{57}Co-in-Rh source increase when the number of Fe atoms in the Austenite phase decreases. Finally, the thermal scan technique convinced us that the zero velocity counts would be sensitive enough to detect small changes due to the defect's recovery stages in pure Fe.

9.3.2 Thermal Scanning Studies of ^{57}Co-doped-Fe Irradiated by Neutrons and Electrons at Low Temperature

We investigated the recovery process of the excess point defects introduced by energetic particle irradiation at low temperature, to clarify the kinetics of the point defects in pure Fe. Mössbauer spectroscopy appears to be ideal for identifying the "defect components" through the hyperfine interactions. Their concentrations are, however, expected to be too small to be detected as a "defect component" in a Mössbauer spectrum. The thermal scanning technique [38, 39] must, therefore, enable us to study the kinetics of the recovery process of point defects in pure Fe irradiated at low temperature as a function of isochronal annealing temperature [39].

An ultrahigh pure-iron foil with a resistivity ratio of $R_{300K}/R_{4.2\,K}$ (RRR_H), 5000, was irradiated at 30 K by fast neutrons with a total fluence of 5×10^{16}n/cm^2, and the Mössbauer spectra were measured at 77 K using a stainless-steel absorber. The left-hand side of Fig. 9.16 show the spectra as irradiated, annealed at 110 K, 150 K, 180 K, 230 K, and 280 K, respectively, while we plot their successive subtractions on the right-hand side. Although there appears to be no defect component as a well-resolved satellite in the spectrum after the neutron irradiation, we obtained the average hyperfine field of 35.7 T from a fitting analysis with one sextet, the value of which is considerably higher than that of pure iron, i.e., 33.8 T at 77 K. This change can be found in the subtraction between the spectrum after the annealing at 110 K and the spectrum as-irradiated. The red lines in the subtraction plot indicate the positions of the outermost sextet of Fe corresponding to the hyperfine splitting of 33.8 T. The result means that ***a defect component with a larger hyperfine field was formed presumably in the cascades by the neutron irradiation.*** The spectra and also the subtractions obtained from further annealing treatments indicate only small changes.

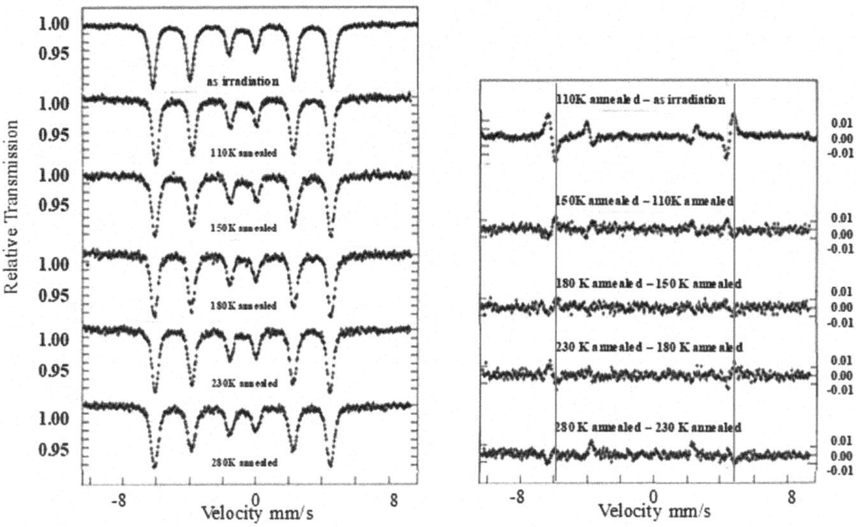

Fig. 9.16 Mössbauer emission spectra of super pure Fe doped with 74 MBq-^{57}Co during isochronal annealing and their successive subtractions. The low-temperature neutron irradiation and all the spectral measurements were done at the Research Reactor Institute of Kyoto University. (From Y. Yoshida, Doctor thesis 1983, Osaka Univ.)

Furthermore, we irradiated two ultrahigh pure-iron foils at 77 K by 28 MeV-electrons provided by an electron linear accelerator at Kyoto-University Research Reactor Institute, and the total fluence was 2×10^{18} and 4×10^{18} e^-/cm^2, respectively. Figure 9.17(**a**), (**b**) show the Mössbauer spectra and their subtractions for the fluence of 4×10^{18} e^-/cm^2. We measured the spectra at 77 K before and after the irradiation, and subsequently, after the annealing at 100 K, 110 K, 160 K, and 180 K, respectively. Unfortunately, there was an accident after the annealing at 180 K, and therefore, we could not measure the spectra above 200K. The average hyperfine field after the irradiation is 33.5 T at 77 K, the value of which is slightly smaller than that of pure Fe, 33.8 T. In the subtraction between the spectrum before and after irradiation on the right-hand side of Fig. 9.17, the positive peaks can be seen on the top, suggesting the appearance of a defect component with a hyperfine field lower than that of pure Fe, being in contrast to the neutron-irradiated specimen. Also, the same component with the lower hyperfine filed appeared when we annealed the sample at 100 K, the temperature of which is known to be close to the recovery stage I, 110 K. These results in the subtractions suggest that ***the ^{57}Co atoms on the undisturbed sites in the Fe matrix were trapped not only by the defects created by the electron irradiation but also by the mobile defects at 100 K, i.e., interstitial Fe atoms.***

In Fig. 9.18, we show the thermal scanning plots obtained from (**a**) the neutron-irradiated samples with a total fluence of 5×10^{16} n/cm^2, and the electron irradiated with the total fluence of (**b**) 2×10^{18} electrons/cm^2, respectively. In the figure, all the normalized counts are plotted as a function of the isochronal annealing temperature

around 110 K for pure Fe, the self-interstitials become mobile within the isochronal annealing time, i.e., five min., annihilating with the vacancies, and subsequently forming interstitial clusters. (**c**) At Stage II at around 150 K, the interstitial clusters grow at impurities such as carbons, and finally (**d**) at Stage III at about 220 K, the vacancies start migrating and disappearing at the interstitial clusters. There will be further recovery stages above room temperature. Above 500 K, all the defects and their agglomerates will disappear completely, as was observed by in-situ observation of High Voltage Electron Microscope [40].

The Electron Microscope provides a possibility to find directly such defect clusters and their creation and annihilation processes. Figure 9.19(**e**) shows an EM image observed after Fe-ion implantation with the energy of 100 keV at room temperature. M. Kiritani used a set of kinetic equations to analyze the growth kinetics of the defect-clusters found by the High Voltage Electron Microscope. They successfully obtained the mobility of vacancy in pure iron [40]: They measured the growth rate of the vacancy loops, which were supposed to play a sink for the mobile interstitials in pure Fe created by HVEM observations, i.e., in fact, electron irradiation with an energy of 2 MeV. They could compare the experimental results with the predictions simulated from the set of kinetic equations. Finally, they could deduce the activation enthalpies of the mobility of interstitials and vacancies as 0.3 eV and 0.6 eV, respectively. They reported in the second publication [40] that the carbon impurities influence the mobilities of vacancies and interstitials in the Fe matrix strongly so that the activation enthalpy of the vacancy changes from 1.2 to 0.6 eV as the carbon concentration decreases.

After neutron and electron irradiations of the ^{57}Co/^{57}Fe-doped super-pure Fe-foils, the Mössbauer emission spectra showed broad sextets without any isolated satellites, which could assign to a vacancy and an interstitial associated ^{57}Co/^{57}Fe probes. The neutron irradiations at 30 K, however, provided the spectrum with a different hyperfine field. The value was 35.0 (1) T higher than 33.8 (1) T due to regular site in pure Fe. After the electron irradiations at 77 K, on the other hand, another sextet appeared to possess the hyperfine field of 31.9 (1) T, being considerably lower than that of the regular site. Consequently, there exist ***three different ^{57m}Fe components due to substitutional, vacancy, and interstitial sites, corresponding to 33.8 T, 35.0 T, and 31.9 T,*** respectively. However, they appear not to be well-resolved from each other at the first insight.

Furthermore, the thermal scanning in Fig. 9.18(**a**) shows that the counts through Fe absorber at 77 K increases, as can be seen at around 110 K (Stage I) *in the neutron-irradiated specimen*, suggesting the number of ^{57}Co trapped by the defects increases. In contrast, the counts through Fe absorber at room temperature decreases due to the formation of ^{57}Co/^{57}Fe-interstitial clusters. *In the case of the electron irradiation* with 2×10^{18}electrons/cm^2 (Fig. 9.18(**b**)), on the other hand, the counts through Fe absorber at 77 K decrease due to the annihilation reaction of mobile self-interstitials with the vacancies, as is schematically shown in Fig. 9.18(**b**). At around 220 K (Stage III), the counts through Fe absorber at 77 K increase again, suggesting a formation of different defect clusters due to mobile vacancies. Finally, we found ***the recovery***

stages I and III in the thermal scanning, corresponding to the activation enthalpies of the mobility of interstitials and vacancies as 0.3 eV and 0.6 eV, respectively.

9.4 In-Situ Observations of Elementary Jump Processes in Iron and Silicon

Between 1983 and 2014, we developed unique measuring techniques in ^{57}Fe Mössbauer Spectroscopy as international joint ventures between the Hahn-Meitner Institut Berlin, Germany, and Institut für Festkörper Physik der Universität Wien, Austria and later the RIKEN, Japan. We applied these techniques for the in-situ observations of ^{57m}Fe atoms jumping within the lifetime in pure iron where different point defect concentrations and distributions are produced at around the nuclear probes of ^{57m}Fe. In this section from 9.4.1 to 9.4.3, we describe these experimental techniques to study the self-diffusion in pure Fe, and also the Fe impurity-diffusion in Si. First of all, in Fig. 9.20(**a**)–(**c**) we show the point defect distributions expected at ^{57m}Fe probes in pure Fe matrix: (**a**) the thermal vacancies at high temperature, (**b**) the excess vacancies and interstitials in the cascade damages created immediately after the implantation of Coulomb-excited and recoil-implanted ^{57m}Fe probes by 100 MeV Ar ion beams, and (**c**) the excess vacancies and interstitials introduced by ^{56}Fe (d, p) ^{57m}Fe reactions through 8 MeV deuteron injection into Fe matrix. Later in 9.4.4, we also explain the nuclear fragmentation and implantation technique to introduce ^{57}Mn/^{57}Fe nuclear probes into Si crystals.

In the last section, we have investigated the kinetics of the point defects in pure Fe for the isochronal annealing steps with 10° per 5 min after neutron and electron irradiations at low temperature. Consequently, we have found the recovery stages due to different point defects in pure Fe. The possible interpretation is that the recovery stages of I at 110 K and that of III at 200 K correspond to the migration energy of

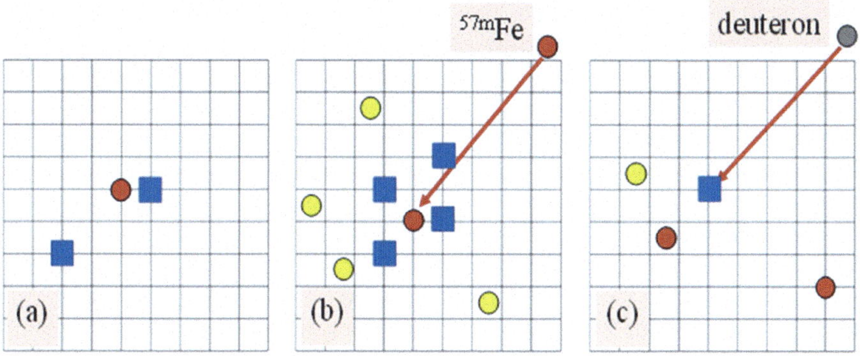

Fig. 9.20 Point defect distributions expected at around ^{57m}Fe probes (red circles): (**a**) thermal vacancies (blue squares) at high temperature, (**b**) cascade damage of vacancies (blue squares) and interstitials (yellow circles), and (**c**) vacancies and interstitials

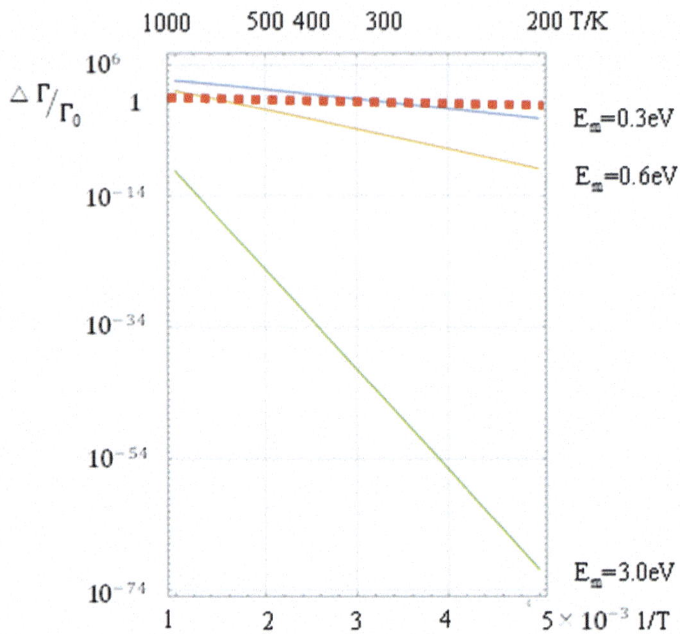

Fig. 9.21 Expected line broadening Γ/Γ_0 for the migration energy of 0.3, 0.6, and 2.8 eV

0.3 eV for self-interstitials, and 0.6 eV for vacancies, respectively. When we want to observe the ^{57}Fe jumps in terms of the line broadenings of Mössbauer spectrum due to interstitials and vacancies within the lifetime of ^{57m}Fe, i.e., 140 ns, the influences will appear at higher temperatures, as estimated roughly by the formula (refer to the formula 9.18). In Fig. 9.21, we present the Arrhenius plots of the line-broadenings due to the ^{57}Fe jumps with the migration energies of 0.3 eV and 0.6 eV only on ***empty bcc lattice***. The broadenings reach to the natural linewidth at around 300 K and 600 K, respectively. The broken red line in Fig. 9.21 indicates the line broadening equivalent to the natural linewidth. To observe the line broadenings in a real crystal of α-Fe, however, we have to create either self-interstitials or vacancies adjacent to the ^{57m}Fe atoms on the 14.4 keV excited state, leading the ^{57m}Fe jumps within the lifetime. As a reference, we show the line broadenings estimated from the self-diffusion with the activation energy of 2.8 eV in paramagnetic phases; the value of which is a sum of the formation energy, 2.2 eV and the migration energy of the vacancy, 0.64 eV [41]. By the three different techniques presented in Fig. 9.21, we expected to find the line broadenings due to the atomic jumps of ^{57m}Fe in pure bcc-Fe immediately after activating the excited state of ^{57m}Fe in Fe crystal by (**a**) 14.4 keV-γ-rays absorption, (**b**) Coulomb-excitation and recoil implantation and (**c**) (d, p) reaction. These techniques could provide high concentrations of point defects at the excited ^{57m}Fe probes (**a**) in γ- and δ-phases and (**b**) and (**c**) even in ferromagnetic α-Fe phase.

9.4.1 High-Temperature UHV-Furnace and Encapsulation Techniques

As the first step, we developed a "high-temperature UHV-furnace" at the Hahn-Meitner Institute Berlin, at University Wien, and later at Shizuoka Institute of Science and Technology (SIST) to achieve an in-situ observation of **self-diffusion of Fe** within the lifetime of 140 ns. It was not an easy task to measure pure Fe at high temperatures. Iron has three different solid phases: the α-phase (bcc) up to 1184 K, the γ-phase (fcc) up to 1665 K, and the δ-phase (bcc) up to the melting point of 1809 K. Especially challenging was to measure the Mössbauer Spectra in the δ-phase because of the high vapour pressure of Fe. In the δ-phase, we could not keep the sample foil even for several hours in the UHV furnace. Accordingly, it was necessary to develop a capsulation technique combining the Ta holder and the Be-O capsule [42]. Figure 9.22 shows the cross-section of the furnace. It consists of a water-cooled UHV chamber with Tantalum heat-shields, Zirconium foil-windows for 14.4 keV γ-rays, Alumina-heater core with Tungsten heater wire, and Pt- Pt Rh thermo-couples to measure the specimen temperature. The vacuum was 10^{-6} Pa using an ion pump system combined with a turbomolecular pump during all measurements at high temperatures. We could reach the highest temperature up to 1766 K. We measured by 3.7 GBq ^{57}Co-in-Rh source, and a gas proportional counter, as shown in Fig. 9.22.

Figure 9.23(**a**) and (**b**) show the spectra of γ- and δ-phases, respectively. The linewidths above 1656 K are much broader than those measured in γ-phase below

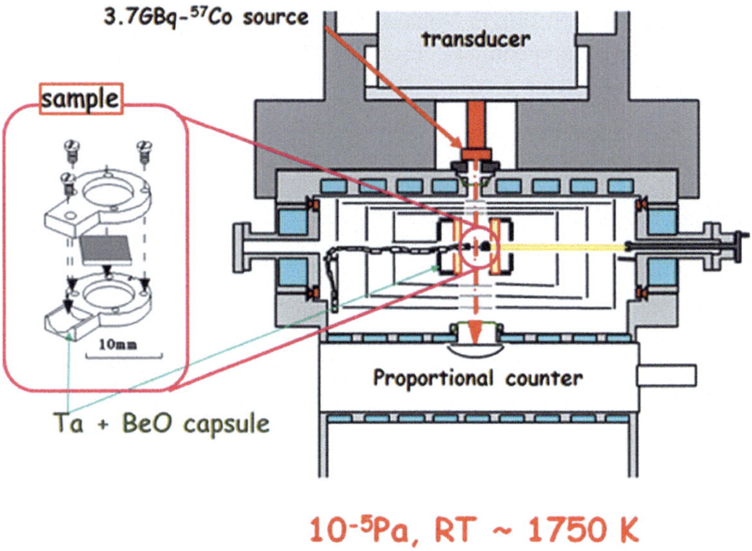

Fig. 9.22 High temperature Mössbauer UHV-furnace

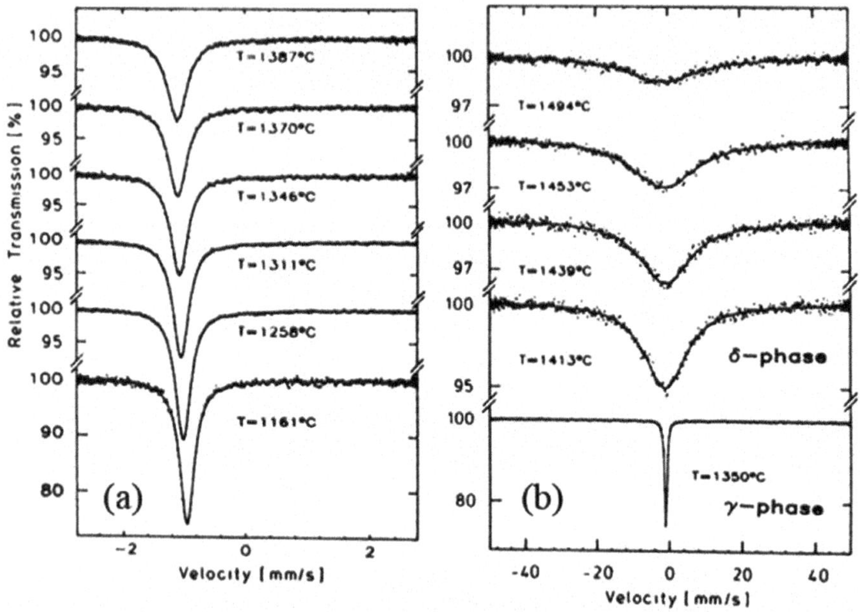

Fig. 9.23 High temperature Mössbauer Spectra in (**a**) γ-phase, and (**b**) δ-phase. Notice that the velocity scale for δ-phase is a factor of 20 larger than that of γ-phase (Reprinted figure with permission from [42], ©IOP publishing. Reproduced with permission. All right reserved)

1623 K. The line broadenings are due to the atomic jumps of ^{57}Fe atoms within the lifetime, i.e., self-diffusion in δ-phase. Since the line broadening, $\Delta\Gamma$, which is shown as a function of temperature in Fig. 9.24, is related to the jump frequency, $1/\tau$, we can deduce the diffusion coefficient, D, using the following formula [27]:

Fig. 9.24 Line broadening of Mössbauer resonance in Fe self-diffusion. Reprinted figure with permission from [42] ©IOP publishing. Reproduced with permission. All right re-served

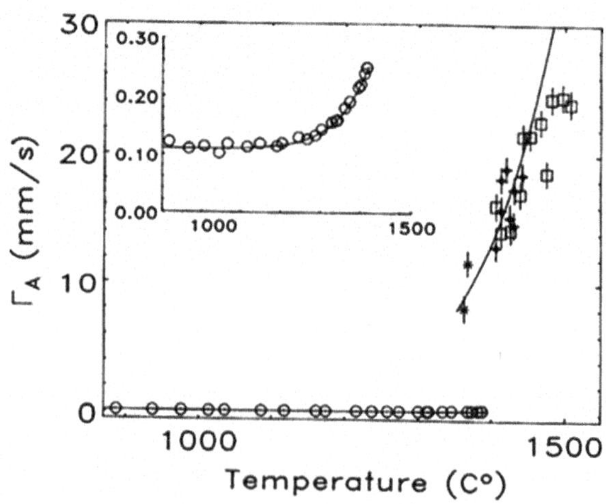

$$\Delta\Gamma = \frac{2\hbar}{\tau}\left(1 - \frac{1}{Z}\sum_j \exp(-i\boldsymbol{k}\cdot\mathbf{R}_j)\right) \tag{9.20}$$

For poly-crystals, we average the Eq. (9.20) overall possible jump directions for $|\boldsymbol{k}| \cdot |\boldsymbol{R}_n| \gg 1$,

$$\Delta\Gamma = \frac{2\hbar}{\tau} f_M. \tag{9.21}$$

Here f_{M} is called "Mössbauer correlation factor," Wolf calculated to be 0.64 [30] in the case of bcc lattice, which is related to a probability of the Mössbauer probe jumping to the vacancies. On the other hand, in the tracer diffusivity, we have to consider a tracer correlation factor, $f_{\mathrm{T}} = 0.73$, for self-diffusion in bcc lattice.

$$D = \frac{R^2}{6\tau} f_{\mathrm{T}} \tag{9.22}$$

Combining both equations of (9.21) and (9.22), we finally obtain a relation between the tracer diffusivity, D, and the line broadening $\Delta\Gamma$, as in equation of (9.23).

$$D = \frac{R^2}{12\hbar}\frac{f_{\mathrm{T}}}{f_{\mathrm{M}}}\Delta\Gamma \tag{9.23}$$

Assuming a vacancy mechanism with a jump distance of R for the self-diffusion in γ- and δ-phases of pure Fe, the self-diffusivity was estimated from the line broadening using the above equation. We plot the results as a function of $1/T$ in Fig. 9.25, in comparison with the tracer diffusivities for γ- and δ-phases [22]. The diffusivities in both phases are in good agreement, indicating that the diffusion theory in Mössbauer spectroscopy can be applied even for studying an unknown diffusion mechanism such as fast impurity diffusion in metals. This result was the first experimental proof that Mössbauer spectroscopy can provide the atomistic information on the long-range self-diffusion based on the Singwi and Sjölander theory [26].

We further applied the high-temperature technique for studying the Fe fast diffusion in β-Zr-Fe [43], β-ZrNb-Fe [44], α-Zr and α-Ti [45], β-Ti-Fe [46] and also the short-range ordering and the clustering properties in $\underline{\text{Au}}$-Fe and $\underline{\text{Cu}}$-Fe [47–50], and the Fe diffusion and segregations in Si single crystals and multi-crystals for Si solar cells [51–54].

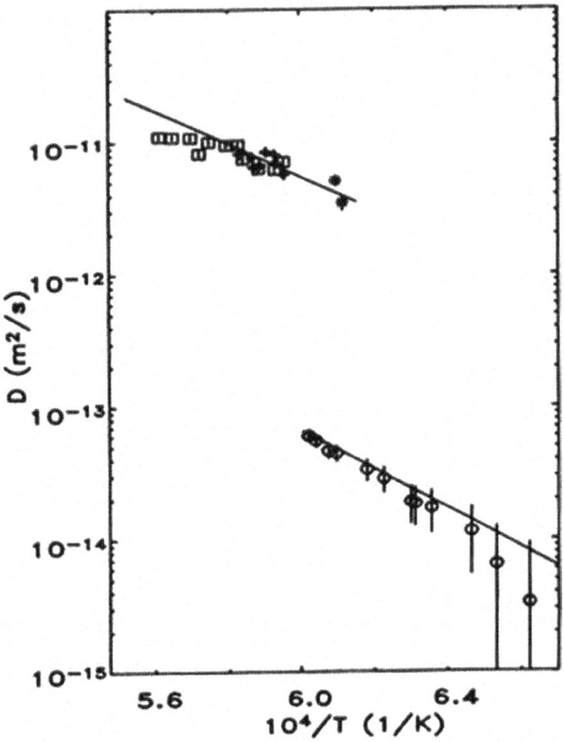

Fig. 9.25 Arrhenius plot of diffusivities both for γ- and δ-phases in comparison of tracer diffusivities. Reprinted figure with permission from [42], ©IOP publishing. Reproduced with permission. All right reserved

9.4.2 *In-Beam Technique Using Coulomb Excitation and Recoil-Implantation*

At a heavy ion accelerator (VICKSI), Hahn-Meitner-Institute Berlin (Fig. 9.26), an in-beam experimental set-up was built for Mössbauer Spectroscopy in the years between 1983 and 1986. Figure 9.27 sketches the arrangement of the in-beam Mössbauer experiment. A ^{40}Ar-ion beam at an energy of 100 meV strikes a thin target of ^{57}Fe, leading to a knock-out of ^{57}Fe atoms and their implantation into the desired host material. Simultaneously, the ^{57}Fe atoms are Coulomb-excited to the 14.4 keV Mössbauer state ^{57m}Fe. The angular distribution of the ejected ^{57m}Fe peaks between about 20° and 70° to the beam direction, which makes it easy to separate the ejected ^{57m}Fe atoms from the primary ^{40}Ar beam which is not allowed to strike the catcher material. The ^{57m}Fe atoms receive high recoil energies in the MeV range and penetrate the host material to depths up to several μm. Recoilless emission of the Mössbauer γ-ray occurs from the implanted probe within the lifetime $\tau = 141$ ns of the 14.4 keV state after it has come to rest. In contrast to a conventional Mössbauer experiment, we have to construct a unique measuring system for the in-beam Mössbauer experiment using the highly energetic Ar-beam, because a massive amount of the background radiations will completely mask the Mössbauer γ-rays. Accordingly, it was inevitable

Fig. 9.26 VICKSI-heavy ion accelerator

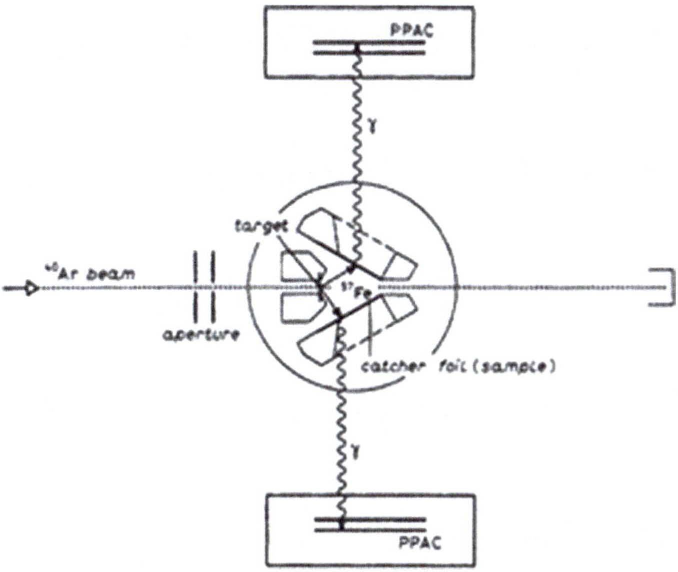

Fig. 9.27 Experimental set-up for Coulomb excitation and recoil-implantation technique

to develop a parallel-plate avalanche counter (***PPAC***), which is ***sensitive only for the Mössbauer γ-rays*** emitted from the sample. This detector counts the conversion electrons emitted from a ^{57}Fe-enriched stainless-steel absorber, which is fixed inside the PPAC directly mounted on a Mössbauer transducer.

As the first research project, we performed the in-beam experiments on a fast diffusion of ^{57}Fe in metals and semiconductors, where no practical solubility of Fe impurity exists to perform a conventional absorber experiment. We measured the in-beam Mössbauer spectrum as a function of temperature, which indicated a sudden area decrease accompanied with a quadrupole relaxation below 50 K, which enable us to interpret as a local jump process of ^{57}Fe atom (i.e., cage motion) on interstitial sites in Al [55] and α-Zr [56, 57]. Subsequently, the line broadenings due to a long-range interstitial diffusion were also found in the spectra of Fe in Si [58]. Furthermore, we obtained the in-beam spectra in Alkaline metals [59, 60], Pb [59–61] and Fe [59, 62]. Later, at RIKEN, we applied this technique for the diffusion studies of Fe in solid Argon [63, 64], and in graphite [65].

Here, we are going to explain another application of this novel technique for an in-situ experiment on ***a radiation-enhanced diffusion in pure iron* via *excess vacancies*** distributing in the cascades created by the recoil-implantation of Coulomb-excited ^{57m}Fe. In the cascades, there are various defects, such as self-interstitials and vacancies in the vicinity of the implanted ^{57m}Fe atoms. At higher temperatures, these defects become mobile even within our short measuring time (400 ns) just after the implantation. Consequently, the mobile defects could induce the diffusion of the implanted ^{57m}Fe atom, if ***the defects would encounter the implanted ^{57m}Fe atom and would trigger exchange jumps between the ^{57m}Fe atom and the defects.*** This process might be observable as a line broadening and a relaxation effect on hyperfine parameters. There was some evidence for such behaviour [59, 62].

Figures 9.28 shows a Mössbauer spectrum (**a**) in ^{56}Fe foil at 5 K, and (**b**) in natural Fe foil at 30 K, respectively, and (**c**) all the spectra measured in ^{56}Fe foil from 5 to 716 K after Coulomb-Excitation and Recoil-Implantation of ^{57m}Fe. The ^{56}Fe foil does not absorb the 14.4 keV-γ-ray resonantly, while the natural Fe foil absorbs the γ-rays emitting from ^{57m}Fe atoms on substitutional site. This explanation is possible because the latter contains 2.14% ^{57}Fe, distributed randomly on substitutional sites in the ^{56}Fe matrix. Accordingly, a defect component can appear more pronouncedly in the spectrum of the natural Fe foil. Indeed, the spectrum in Fig. 9.28(**b**) consists of two sextets due to a defect component with a hyperfine field of 35.04 (4) T in addition to the substitutional one with the hyperfine field of 33.8 T. The result coincides with the effects observed in the sample of ^{57}Co-doped-Fe after neutron irradiation at 30 K, which produces the cascade damages around ^{57}Co/^{57}Fe probes; we found a sextet component with a larger hyperfine field of 35.7 T.

In Fig. 9.29, we present all the fitting parameters as functions of the measuring temperature. We obtained those values from one sextet analysis on all the spectra of as-rolled ^{56}Fe sample between 5 and 716 K, shown in Fig. 9.28 (**c**). The resonance area decreases stepwise at around 300 K and at 650 K, which deviates from the expected change predicted by the Debye model with a Debye temperature of 470 K, as in Fig. 9.29(**a**). The hyperfine fields below 400 K are slightly higher than those of the laboratory experiment (Fig. 9.29(**b**)), as can be seen in the difference plot (**c**). The difference tends to zero at around 400 K, and increases and finally decreases again at about 700 K. The centre shift follows the second-order Doppler shift (**d**). It

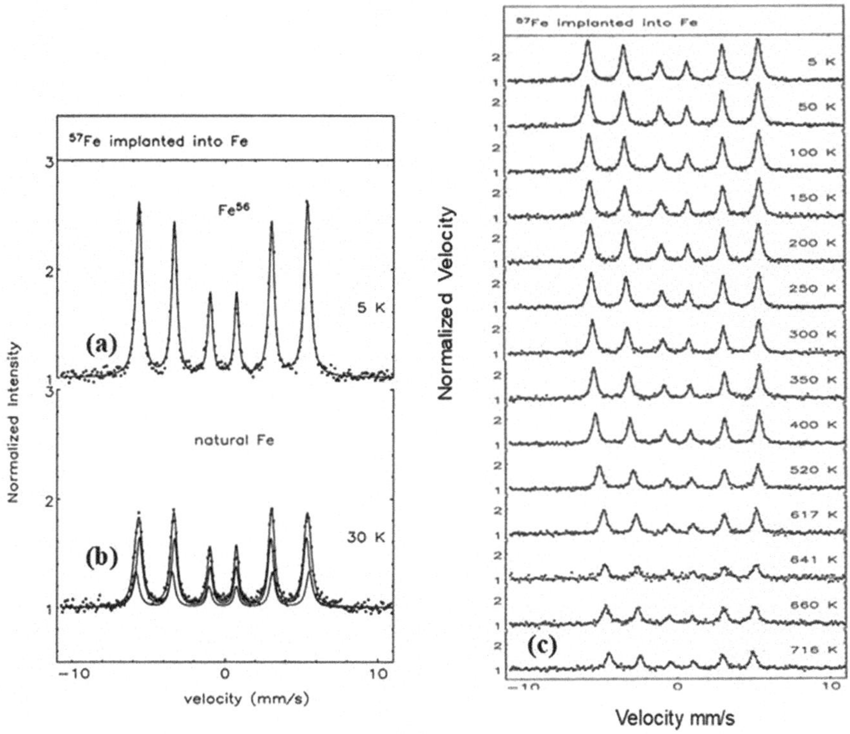

Fig. 9.28 (**a**) ^{57m}Fe In-beam Mössbauer spectrum in ^{56}Fe at 5 K after Coulomb-Excitation and Recoil-Implantation. (**b**) "Defect line" of natural Fe observed after Coulomb-excitation and recoil-implantation at 30 K. The hyperfine field of "defect site" is 35.04 (4) T, while 33.8 T for substitutional site of α-Fe. (**c**) ^{57m}Fe Mössbauer spectra in ^{56}Fe foil for the measuring temperature from 5 to 716 K [from Ref. 62]

is notably that the line width depends anomalously on the temperature, as shown in Fig. 9.29(**e**).

Before interpreting the above-mentioned experimental results, one should notice that all the Mössbauer spectra correspond to the measurement within 400 ns immediately after the implantation of the Coulomb-excited ^{57m}Fe probe, enabling us to perform "one-to-one measurement" without overlapping cascades. The recovery stages due to the migrations of interstitials and vacancies produced by the implantation appear to be shifted with a factor of about three because the "annealing time" is practically supposed to be orders of magnitude shorter in the in-beam experiment than that of several minutes in the conventional annealing experiments. Accordingly, the stage I at 110 K in the isochronal annealing experiments with the annealing time of 5 min. is expected to shift to at around 300 K, and the stage II of 150 K to 450 K, and the stage III of 220 K to 650 K, respectively. We observed the recovery stages in the thermal scans (Fig. 9.18), i.e., the isochronal annealing experiments, on pure iron after low-temperature irradiations [39].

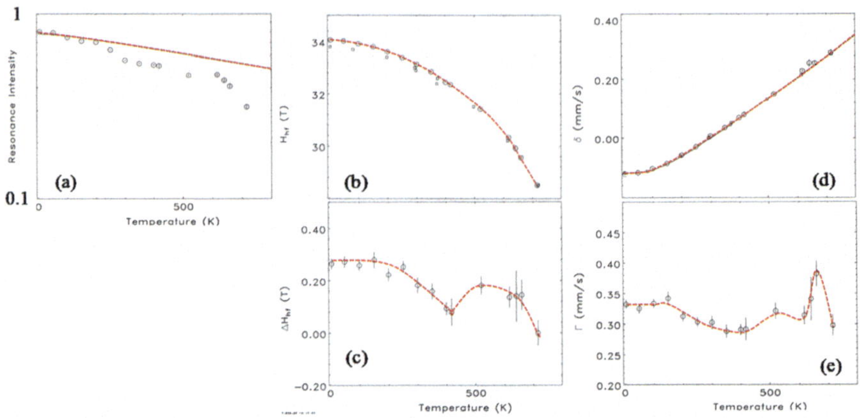

Fig. 9.29 Mössbauer fit parameters as functions of temperature: (**a**) Total resonance area, (**b**) hyperfine field, and (**c**) the difference of hyperfine field between in-beam and absorber experiments, (**d**) centre shift, and (**e**) Full width at half maximum. The red dotted lines are only for eye-guide [from Ref. 62]

The total resonance area, the hyperfine fields, and the linewidth change at around 300 K, and also at about 650 K (Fig. 9.29), which coincide well with the stage I and III, respectively, in our time scale. **At Stage I (300 K)**, the self-interstitials start migrating and reacting with the vacancies at the core of the cascade, where the implanted ^{57m}Fe probe stops on substitutional and interstitial sites, although we cannot distinguish the two spectral components. We interpret the gradual decreases of both the hyperfine field and the linewidth are due to the defect annihilations. The 20% of area disappears at around 250 K, which could be due to ***local jumpsas a precursor and subsequently to long-range jumpsof interstitial*** ^{57m}***Fe atom above 300 K.*** The local jumps, such as "cage jumps" [66, 67], are expected to cause a sudden decrease of the area before long-range diffusion. Accordingly, these local jumps seem to relate with the "***crowdion motion***" as was proposed by Seeger et al. [36, 68], and also by Derlet et al. [41]. **At Stage III (650 K)**, we find the second stepwise decrease of the area in Fig. 9.29(**a**), which is accompanied by the ***sharp increase of the linewidth at around 660 K.*** Figure 9.30(**a**) shows the spectra between 520 and 714 K, and Fig. 9.30(**b**) the Arrhenius plot of the jump frequency, ν, which provides us with an activation enthalpy of 0.6 eV, suggesting ***an enhanced diffusionof substitutional*** ^{57m}***Fe atom* via *the excess vacancies*** created by the implantation of the ^{57m}Fe atom.

We obtained these experimental results from a ***polycrystal-***^{56}***Fe-foil*** as-received. In a well-annealed Fe sample, however, we found a different temperature dependence both on the area and the line width at around 600 K. Accordingly, the local concentration of the point defects at the ^{57m}Fe atom appears to depend on the concentration of dislocations, which can be a sink to the point defects produced in each implantation of ^{57m}Fe. In the next section, we are going to explain an in-beam experiment on a ***single crystal α-Fe*** to measure ^{57m}Fe jumps via self-interstitials and vacancies as a function of crystal orientation.

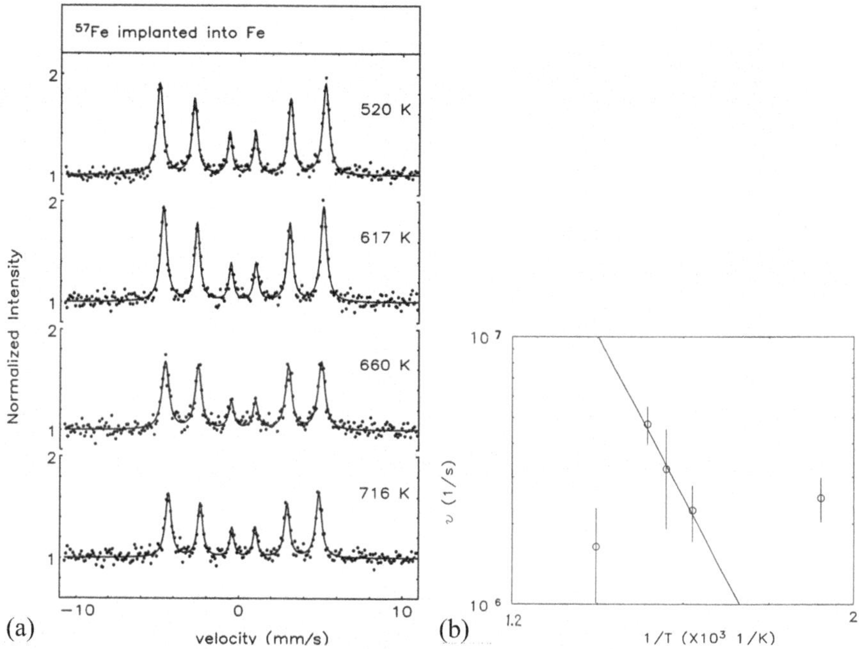

Fig. 9.30 (**a**) The Mössbauer observed between 520 and 716 K, and (**b**) the Arrhenius plot of the jump frequency, ν, yielding a thermal activation enthalpy of 0.6 eV [from Ref. 62]

9.4.3 *In-Beam Technique Using ^{56}Fe (d, p) ^{57m}Fe Nuclear Reaction*

An energetic deuteron beam offers a possibility to excite the ^{57m}Fe nuclear level (lifetime $\tau_{\text{Möss}} = 140\ 10^{-9}$ s) for In-beam Mössbauer spectroscopy [69], which may be applied to study the movement of self-interstitials in iron crystals [68]. The ^{56}Fe (d, p) ^{57m}Fe reaction creates the ^{57m}Fe probes on the 14.4 keV level in natural Fe, and simultaneously they are ejected from their lattice sites approximately in the beam direction with kinetic energies up to about 1 MeV (i.e., for elastic head-on collision). A certain fraction of the ejected ^{57m}Fe probes end up as self-interstitials in the Fe matrix. We obtain information on the jump vectors of the interstitial ^{57m}Fe [70] by measuring the angular dependence of the Mössbauer spectrum at different crystal orientations.

In an experimental set-up for the in-beam Mössbauer spectroscopy (Fig. 9.31(**a**)) [70] at the RIKEN AVF cyclotron (K = 70 MeV), a single-bunch 8 MeV deuteron beam with repetition period 1.2 μs, pulse width two ns, and cross-section 9.3 mm × 10 mm was used to implant the ^{57m}Fe nuclei into Fe single crystal. The time window was set at 100–500 ns, thus suppressing background counts due to short-lived reaction products. Apart from small bunching peaks every 80 ns, the background-to-signal ratio stayed constant throughout the time window, so that the area under

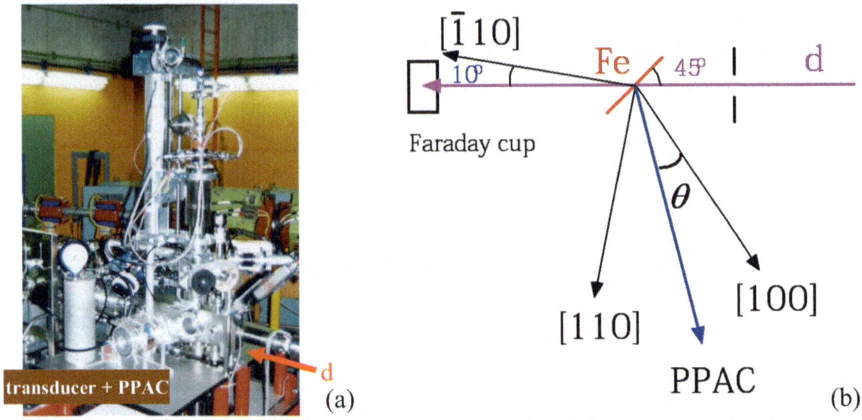

Fig. 9.31 (**a**) In-beam Mössbauer set-up beam line of AVF cyclotron, RIKEN, (**b**) Schematic view of the at the experimental arrangement. Reprinted figure from [42] with permission, ©Trans Tech Publications, Ltd. Reproduced with permission. All right reserved

the Mössbauer spectra could be determined accurately. The beam current was 75 or 110 nA, corresponding to fluxes of 4.6×10^{11} or 6.9×10^{11} d m^{-2} s^{-1}. High-purity Fe single crystals of 5 mm × 12 mm × 1 mm prepared at the Max-Planck-Institut für Metallforschung by strain annealing were chemically etched down to thicknesses of 50 μm and finally purified by annealing for 24 h at 1123 K under wet H_2 gas and 24 h at 923 K under dry H_2 gas. We selected the thickness of the samples so that no deuterons stopped in the sample. Thermal stress on the samples was avoided by fixing the crystals between two iron-plate frames with a rectangular window of 5 mm × 12 mm. The side of the iron frame facing the beam was covered by a 1 mm thick Ta plate to prevent ^{56}Fe (d, p) ^{57}Fe reactions in the frame. We checked the orientation of the crystal foil mounted on the sample holder by the Laue method.

Figure 9.31(**b**) shows the measuring geometry against the crystal orientation. The deuteron beam direction was 10° from [$\bar{1}$10]. A liquid-He flow cryostat was used for the measurements between 10 and 400 K. At the same time, an infrared lamp heater for those between 300 and 700 K. The temperature was measured on the Ta plate close to the sample, below room temperature by a Si diode and above room temperature by a Chromel/Alumel thermocouple (Type K). We kept the vacuum in the chamber at about 10^{-5} Pa during all measurements. The detector (PPAC) counts the conversion electrons emitted from an enriched stainless-steel absorber, which is fixed inside the PPAC directly mounted on a Mössbauer transducer. We used the different distances of 0.07 or 0.11 m between the sample and PPAC, respectively, for the temperature variation of the Mössbauer spectrum or its dependence on the emission direction. At the detector distance, 0.07 m typical 14.4 keV γ-ray count rates were about 12 s^{-1} for currents of 75 nA. At a distance 0.11 m, on the other hand, the counts were about 5 s^{-1} with an angular resolution of 5°.

We show typical in-beam spectra measured at different temperatures between 150 and 700 K in the [100] direction in Fig. 9.32. Figure 9.33 plots the fit parameters as functions of temperature: (**a**) total area $\boldsymbol{A}$, (**b**) hyperfine field $\boldsymbol{B}_{\mathrm{hf}}$, (**c**) centre shift $\boldsymbol{\delta}$, and (**d**) full width at half maximum Γ (full red symbols). We compare the results

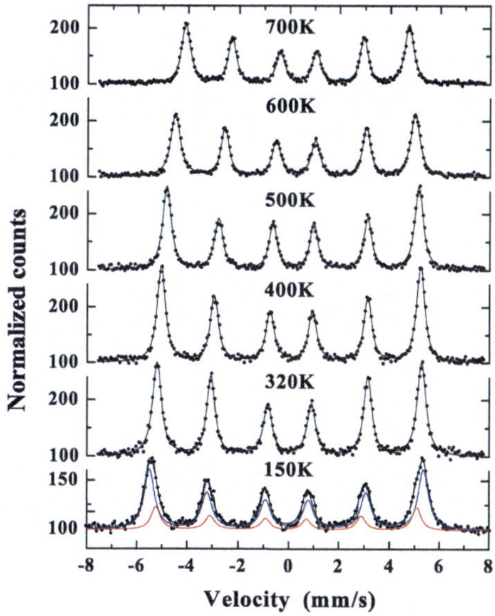

Fig. 9.32 In-beam Mössbauer spectra of α-Fe between 150 and 700 K measured in [100] at a detector distance of 0.07 m. The 150 K spectrum was obtained in an earlier experiment under otherwise similar conditions. Reprinted figure from [42] with permission, ©Trans Tech Publications, Ltd. Reproduced with permission. All right reserved

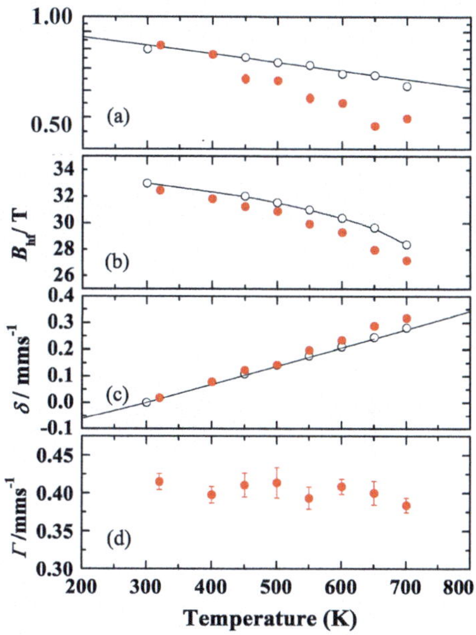

Fig. 9.33 Temperature dependence of the fit parameters above 300 K; (**a**) Area A, (**b**) hyperfine field B_{hf}, (**c**) centre shift δ, and (**d**) full width at half maximum Γ. Open symbols refer to the control experiment by an absorber experiment in laboratory. Reprinted figure from [42] with permission, ©Trans Tech Publications, Ltd. Reproduced with permission. All right reserved

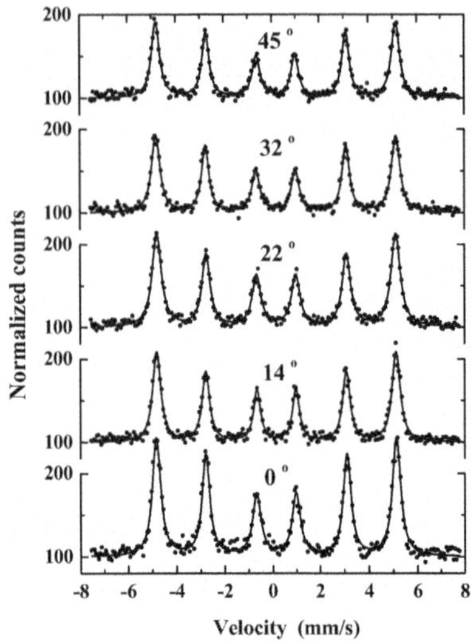

Fig. 9.34 In beam Mössbauer spectra at 500 K at different measuring directions from [100] (0°) to [110] (45°). The angular resolution was 5° at the detector distance 0.11 m. The deuteron beam current was 75 nA. Reprinted figure from [42] with permission, ©Trans Tech Publications, Ltd. Reproduced with permission. All right reserved

in Fig. 9.33 with those obtained in a conventional laboratory experiment on pure Fe in ultra-high-vacuum (open symbols). None of the spectra of Fig. 9.32 show well-resolved satellite lines. At first glance, the spectra appear to be very similar to those of pure α-Fe. The temperature dependence of the fit parameters is, however, quite different, as may be seen in Fig. 9.33: (**a**) The area $\boldsymbol{A}$ decreases much faster than expected from a Debye model with the Debye temperature of α-Fe, $\Theta_D = 470$ K. (**b**) The hyperfine fields $\boldsymbol{B}_{hf}$ are lower than those of α-Fe. (**c**) Above 550 K, the centre shift δ starts deviating from the second-order Doppler shift corresponding to substitutional Fe. (**d**) The full width Γ at half maximum stays at a constant of about 0.40 mms^{-1}.

Figure 9.34 shows the spectra at 500 K along with the measuring directions of 0° (= [100]), 14°, 22°, 32° and 45° (= [110]) in the (001) plane. The fit parameters change with the directions at 500 K and at 700 K (Fig. 9.35). Both $\boldsymbol{B}_{hf}$ and δ depend on the emission directions, suggesting that the spectra must arise from the superposition of several sextets. If the spectra consisted of the substitutional Fe component only, the area would change as a simple "cos θ effect" due to Mössbauer self-absorption. Nevertheless, the data analysis of $\boldsymbol{A}$, $\boldsymbol{B}_{hf,}$ and $\boldsymbol{\delta}$ was based on one sextet only; hence the numerical values of Figs. 9.33 and 9.35 must be considered as those motional-averaged over different contributions to the spectra.

Let us consider more details on the measuring process of the in-beam Mössbauer spectra of the pure Fe single crystal. It is known that the slowing down of the ^{57m}Fe after ejection from their lattice sites with kinetic energies of the order of magnitude 1 MeV is at first almost entirely due to electronic excitations. Only towards the ends of the ^{57m}Fe paths, there exists the cross-section for the atomic displacements accompanied by the productions of Frenkel pairs (self-interstitial-vacancy pairs). As a result, each ^{57m}Fe comes to rest in a displacement cascade with a diameter of a

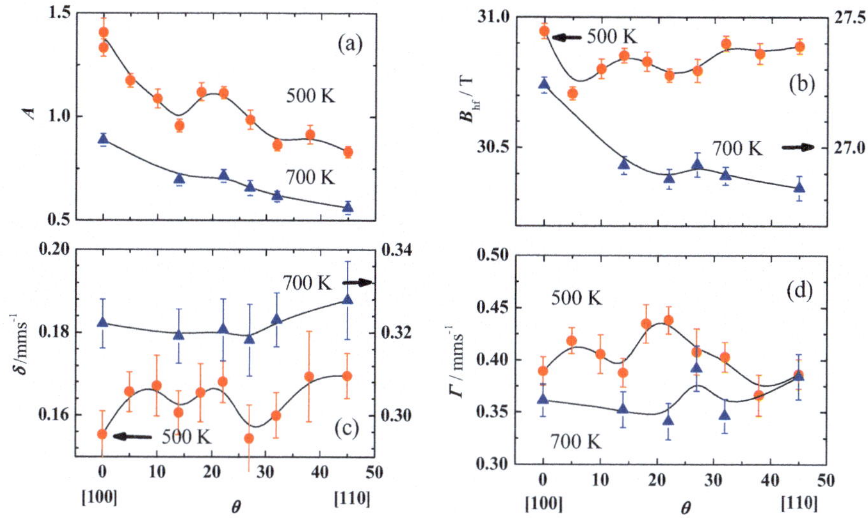

Fig. 9.35 Angular dependence of the fit parameters at 500 and 700 K: (**a**) Total Area *A*, (**b**) hyperfine field B_{hf}, (**c**) centre shift *δ*, and (**d**) full width at half maximum Γ. The lines are guides to the eye. Reprinted figure from [42] with permission, ©Trans Tech Publications, Ltd. Reproduced with permission. All right reserved

few nm containing about 10 2 Frenkel pairs. The defect reactions with ^{57m}Fe, such as interstitial-interstitial ^{57m}Fe and vacancy-vacancy clustering, as well as vacancy-interstitial ^{57m}Fe annihilation, will take place at elevated temperature even within 100 ns, the lifetime of ^{57m}Fe. At first sight, the differences between the in-beam and the control measurements increase with the increasing temperature. The trends may appear surprising since at high enough temperatures the irradiation damage produced by the ^{57m}Fe projectiles should anneal out in times that are short compared with $\tau_{Möss}$; hence almost all ^{57m}Fe should decay on substitutional sites.

At 150 K, we observe two sextets, as is shown in Fig. 9.32. Based on the experimental results on the electron-irradiated ^{57}Co-doped-Fe explained in Sect. 9.3.2, we interpret one component with the larger $\boldsymbol{B}_{hf}$ to ^{57}Fe at substitutional sites, and the other to ^{57}Fe interstitial sites. The fitting analysis of the 150 K data gives us 4:1 for the ratio of substitutional to interstitial sites. In Sect. 9.3.2, we simulated the spectrum as functions of temperature and crystal orientation for the different jump models, as in Figs. 9.11 and 9.12. The models suggest that the interstitial ^{57m}Fe component must be too broad to be seen apparently above 300 K if ^{57m}Fe atom could freely migrate on bcc Fe lattice with migration energy of 0.3 eV. **Above 300 K**, however, the temperature variations of the fit parameters in Fig. 9.33 deviate considerably from that of pure Fe, so that the spectra must be ***motional-averaged sextet between substitutional and the other excess defect components such as*** ^{57m}***Fe interstitial-interstitial and*** ^{57m}***Fe substitutional-vacancy clusters.***

The angular dependences of the parameters in Fig. 9.35 provide complete information on both the self-correlation function and the hyperfine relaxations [35] which

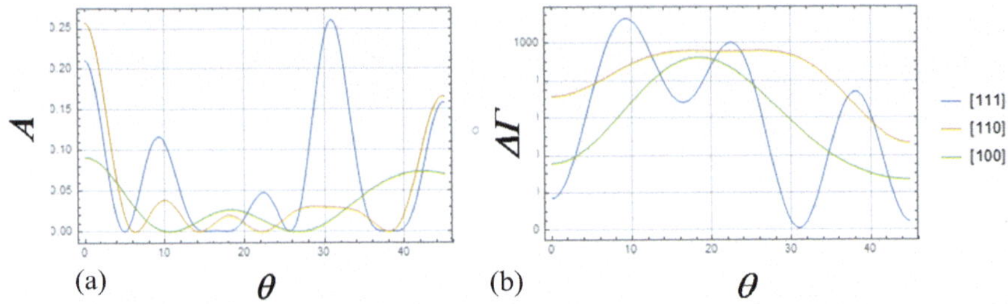

Fig. 9.36 Angular dependence of (**a**) area $\boldsymbol{A}$ and (**b**) line broadening $\Delta\Gamma$ for [111], [110] and [100] jumps with jump distances of $\sqrt{3}/2\boldsymbol{a}$, $\sqrt{2}\boldsymbol{a}$ and $\boldsymbol{a}$, respectively

requires the comparison of the experimental results with the defect kinetics for the conceivable jump processes via interstitials and vacancies. Within 100 ns, only the limited numbers of the defects created within the cascade approach to the ^{57m}Fe atom, and subsequently, trigger the exchange jumps and finally lead the ^{57m}Fe atomic jumps. That is the "defect induced diffusion of ^{57m}Fe atom", which causes the angular dependence of the Mössbauer parameters, such as area $\boldsymbol{A}$, line broadening $\Delta\Gamma$, internal magnetic field $\boldsymbol{B}_{\mathrm{hf}}$ and isomer shift $\boldsymbol{\delta}$. Here we show simulated results on the angular relationship of $\boldsymbol{A}$ and $\Delta\Gamma$ in Fig. 9.36(**a**) and (**b**) by a jump model for [111], [110] and [100] jumps with jump distances of $\sqrt{3}/2\boldsymbol{a}$, $\sqrt{2}\boldsymbol{a}$ and $\boldsymbol{a}$, respectively. So far, we cannot reproduce the experimental findings by each jump model, expecting, however, that an average of the different jumps with the corresponding jump probabilities could produce a similar change in the line broadening $\Delta\Gamma$ at 500 K, as in Fig. 9.35(**d**). It seems to be beyond the present review to get a complete set of the simulation parameters. So far, we have only finished such a kinetic model for an on-line experiment on ^{57}Mn/^{57m}Fe implantation in Si crystals, where substitutional and interstitial ^{57m}Fe components well separate each other.

9.4.4 On-Line Implantation of ^{57}Mn/^{57}Fe into Si Using Projectile Fragment Separator

Iron impurities in Si crystals have been intensively investigated more than half century, because they are inevitably incorporated into the devices during the production processes as one of the most active trapping centres for the carrier, leading to the degradation of the Si devices [71]. Especially serious are the problems of the Fe contamination in the case of Si solar cells. We have been challenging a series of the experimental investigations on the lattice sites and the charge states using ^{57}Fe Mössbauer spectroscopy, i.e., the Mössbauer experiments on ^{57}Fe deposited Si wafers at high temperatures [51–54, 72], on ^{57}Mn/^{57}Fe implanted Si crystals [72–76], on ^{57}Mn/^{57}Fe implanted Si solar cells [77] as well as on Si and Si solar cells under uni-axial stress [78], under light illumination [79], and carrier injections [80–82].

Furthermore, we have developed a mapping technique, which we will describe later, to study ^{57}Fe diffusion and segregation in Si materials [83–90]. These studies have clarified the solid solution of Fe impurities in Si [91–93]: The Fe impurities occupy not only interstitial sites but also substitutional sites. Besides, the carrier trapping cross sections for the interstitial components with different charge states, Fe_i^+ and Fe_i^{2+}, can be successfully obtained by evaluating the dynamical charge fluctuations within a time scale of 100 ns between Fe_i^+ and Fe_i^{2+} which were observed directly in the Mössbauer spectra of ^{57}Fe doped mc-Si solar cells [82].

After GeV-$^{57}Mn/^{57}Fe$ implantation into Si wafers, we found ^{57m}Fe atoms to occupy interstitial and substitutional sites [72–77]. The results appear to coincide those obtained at CERN ISOLDE using low energy implantation of ^{57}Mn into Si [93]. We use the assignments based on the isomer shift values, which relate to the charge density at the ^{57}Fe nucleus. We refer the first-principle theoretical calculations [94, 95], but there remains still discrepancies when we discuss the charge states of interstitial Fe in Si materials. Figure 9.37(**a**) shows Mössbauer spectra of ^{57}Fe in n-type FZ-Si in the temperature region between 330 and 1200 K [77]. The spectra from 330 to 700 K can be fitted by two Lorentzian singlets, while those from 800 K up to 1200 K can be analyzed only by a broad singlet. We assigned the two singlets at 330 K at the left, and the right-hand side has been to Fe_{int}^+ and Fe_{sub}^0 atoms on interstitial and substitutional sites in the Si matrix, respectively. The spectra change

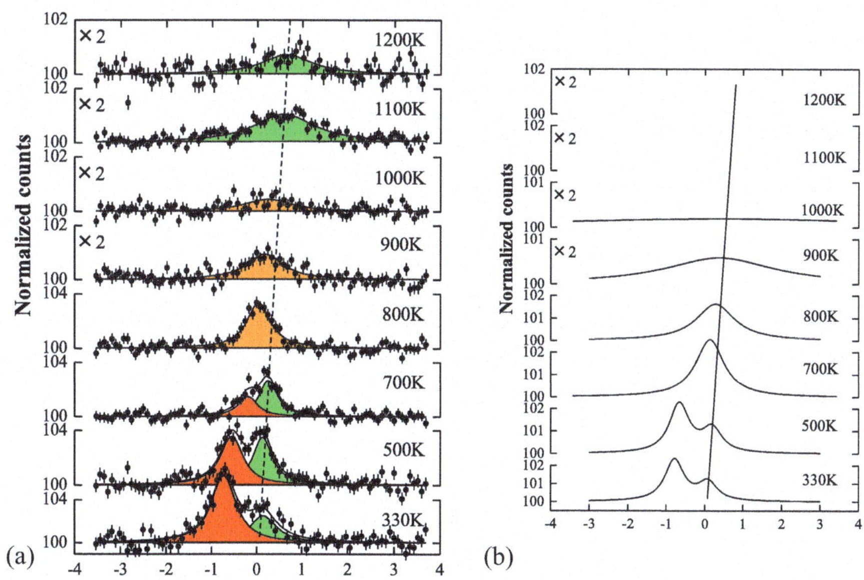

Fig. 9.37 (**a**) Mössbauer spectra of ^{57}Fe in n-type FZ-Si measured between 330 and 1200 K. Notice that the vertical scale for the region above 900 K is blown up with a factor of two, in order to show the small and broad resonance effects more clearly. (**b**) The model reproduces the overall temperature dependence of the Mössbauer spectra of ^{57}Fe in n-type FZ-Si measured between 330 and 1000 K. (From Ref. [77])

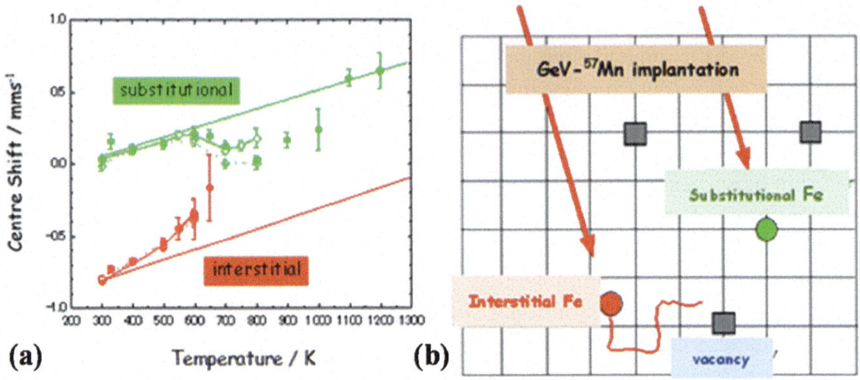

Fig. 9.38 (**a**) Centre shifts of substitutional and interstitial lines, (**b**) A model describes the relaxation behavior. (From Ref.[77])

anomalously with elevating temperature. The resonance area of the interstitial ^{57}Fe component at the left-hand side decreases sharply above 500 K. As can be seen in Fig. 9.38(**a**), the centre shifts of both components deviate clearly from the usual temperature dependence of the SOD shift. The centre shift of the interstitial component differs from the SOD shift, moving continuously to the substitutional position with increasing temperature. This anomaly accompanies by a sharp decrease in the resonance area. Furthermore, the centre shifts of the singlet between 800 and 1000 K are different from the extrapolation of the centre shifts of both components. Above 1100 K, the centre shift goes back to the position of the substitutional site again.

These findings suggest a model presented in Fig. 9.38 (b): interstitial ^{57}Fe atoms start jumping already above 300 K within the lifetime of ^{57m}Fe nuclei, and the simultaneous relaxation effects on the centre shift indicate that the charge state of interstitial Fe atoms changes with increasing temperature. Substitutional Fe atoms form via a kinetic reaction between interstitial Fe and excess vacancies above 600 K. The process must be related to the recovery processes to the equilibrium states of the Si lattice around the nuclear probes. Above 1100 K substitutional Fe atoms start migrating within the lifetime of ^{57m}Fe, leading to long-range diffusion via vacancies in the Si matrix.

$$\frac{dC_I^{Fe}}{dt} = K1(1 - C_I^{Fe}) - K2C_I^{Fe}C_V$$
$$\frac{dC_V}{dt} = K1(1 - C_I^{Fe}) - K2C_I^{Fe}C_V + K3(1 - C_V) - K4C_VC_I$$
$$\frac{dC_I}{dt} = K3(1 - C_V) - K4C_VC_I$$

The relaxation behaviour observed in this experiment can be interpreted in terms of a diffusion–reaction process of interstitial Fe atoms with vacancies within the time scale of 100 ns, leading to the formation of substitutional Fe atoms in the Si matrix.

This model corresponds to a set of the reaction kinetic equations described above. We now try to reproduce the whole changes observed in the spectra up to 1000 K. We simulate the kinetic equations of the interstitial ${C_I}^{Fe}$ and the substitutional ${C_S}^{Fe} = (1-{C_I}^{Fe})$ reacting with vacancies C_V and self-interstitials C_I which arises from the above mentioned "diffusion–reaction". Here, K1, K2, K3, and K4 are the reaction coefficients.

$$\Phi(\omega) = \frac{const}{\left[\Gamma + \Delta\Gamma_s + i(\omega - \varpi - \alpha \cdot \Delta\omega) + \frac{(\Delta\omega)^2(1-\alpha^2)}{\Gamma+\Delta\Gamma_I+i(\omega-\varpi+\alpha\cdot\Delta\omega)+\omega_R}\right]}$$

To take account of the relaxation behaviours, we use a theoretical model proposed by Dattagupta [35], which initially applied for a system with a charge fluctuation. Also, we have to incorporate the effects, $\Delta\Gamma_S$ and $\Delta\Gamma_I$, from the diffusion broadenings of substitutional and interstitial Fe to the spectra $\Phi(\omega)$, respectively. Here ω_R is a relaxation rate, $\alpha = p_S - p_I$ where p_S and p_I are the substitutional and the interstitial fractions, respectively. $\varpi = 1/2\ (\delta_S - \delta_I)$ and $\Delta\omega = 1/2\ (\delta_S + \delta_I)$ relate to the isomer shifts, δ_S and δ_I for the substitutional and the interstitial components, respectively. $\Delta\Gamma_S$ and $\Delta\Gamma_I$ are the line broadenings due to diffusion. We simulate the values of p_S and p_I from the kinetic equations, mainly due to the interstitial jumps within 100 ns [77]. In this model, an interstitial Fe atom is jumping with an activation enthalpy of 0.67 eV [71] within the lifetime of ^{57}Fe, 100 ns, and subsequently finding a vacancy, which results in the formation of the substitutional Fe atoms. The number of vacancies related to one interstitial ^{57}Fe is a parameter. Figure 9.37b shows the simulated spectra up to the temperature at 1000 K. The spectra above 1100 K corresponds to the substitutional ^{57}Fe, and the broadening appears to originate from the long-range diffusion via vacancies. Consequently, the general tendency of the temperature dependence of the Mössbauer spectra, as is shown in Fig. 9.37(**a**), is reproducible by the simulation.

9.5 Mössbauer Spectroscopic Camera

Since the discovery of the Mössbauer effect in 1958, the Mössbauer spectrum has been measured in wide varieties of materials to clarify the physical and chemical properties under different circumstances. Recently, imaging techniques are becoming more and more essential for material evaluations because there always exist microstructures in the samples. Therefore, highly desirable must be a ***"Mössbauer spectroscopic camera,"*** which could take ***"photographs"*** every minute separately for different spectral components with a spatial resolution of several micrometres or even sub-micrometres. So, we could achieve ***"Operand measurement,"****following the dynamical atomic motions and the chemical reactions in the vital industrial materials* such as iron steel, energy, chemical, and electronics functional materials. There appear to be two different methods among presently available techniques to realize such ***"camera"***: (1) A mapping technique by focusing the γ-rays down to several micrometres, and subsequently by measuring either the transmitted γ-rays or the reemitted conversion and Auger electrons as well as the 6 keV-X-rays as functions of the positions of the focused 14.4 keV γ-rays on the sample. (2) A direct imaging technique using a CMOS camera for X-ray imaging (Hamamatsu). In this section, we are going to explain both methods applied for the diffusion studies on Fe in multi-crystalline silicon wafers for silicon solar cells as well as for carbon segregation process in Fe steel.

9.5.1 Mapping Technique for Mössbauer Spectroscopic Microscope (MSM)

Recently, a set-up of "3D-Mössbauer Spectroscopic Microscope (MSM) (Fig. 9.39(**a, b**))" has been built to investigate the iron contamination in silicon solar cells in our group [83–90]. The MSM system combines with different evaluation techniques such as a scanning electron microscope (FE-SEM, APCO), an electron beam induced current (EBIC, APCO), an energy dispersive X-ray spectrometer (EDS, BRUCKER), an electron backscatter diffraction (EBSD, BRUCKER), and a cylindrical sector electron analyzer (HV-CSA, FOCUS). We show the whole configuration in Fig. 9.39 (**b**). Furthermore, Fig. 9.40 explains the principle of the MSM: We use a multi-capillary X-ray lens (MCX, HAMAMATSU) to focus 14.4 keV γ-rays on a sample mounted on a precision XYZ-θ stage (KOHZU). The electrons and transmitted γ-rays are measured as functions of the sample position by a funnel type Micro-Channel-Plate (MCP, HAMAMATSU) and a Si-PIN diode detector (AMPTEK), respectively. We use a 3.7GBq-^{57}Co-in-Rh source with an active area of 4 mmϕ in diameter (RITVERC) mounted on a moving-coil linear actuator (SMAC) operated with a constant Doppler velocity corresponding to the isomer shift of a Mössbauer spectral component. Finally, two different mapping images with 80 × 80 pixels are created typically for an area of 400 μm × 400 μm or 4 mm × 4 mm corresponding to each Fe component selectively with a spatial step of 5 or 50 μm, respectively.

In the following, we explain the essential elemental techniques combined in the Mössbauer Spectroscopic Microscopy (MSM). The whole system can operate by a LabVIEW program developed within our project. It is crucial to tune a resonance condition of the γ-ray energy emitted from a ^{57}Co-in-Rh source to the isomer shift corresponding to a Mössbauer component to be mapped. First of all, therefore, we need a precision linear actuator to realize a constant Doppler-velocity operation with an error of 0.005 mms^{-1} for the case of ^{57}Fe Mössbauer microscope. We use a ***moving-coil actuator*** coupled with a 100 nm-resolution linear encoder (SMAC). This actuator enables us to operate with either a constant acceleration mode or a constant velocity mode, as shown in Figs. 9.41(**a**) and (**b**), respectively. We obtained the linewidth error of 0.007 mms^{-1} which is sufficiently small for the mapping

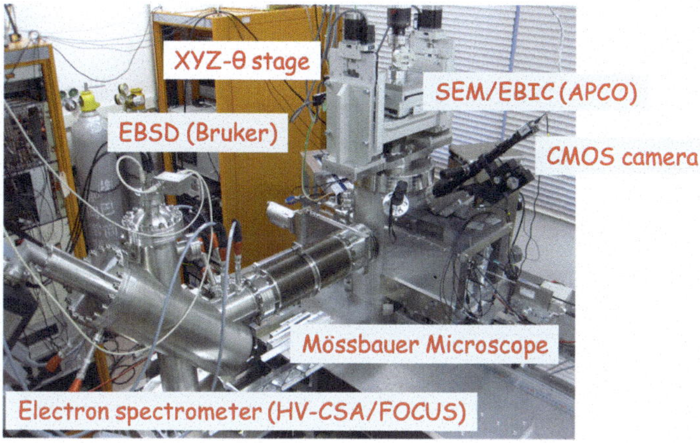

(a)

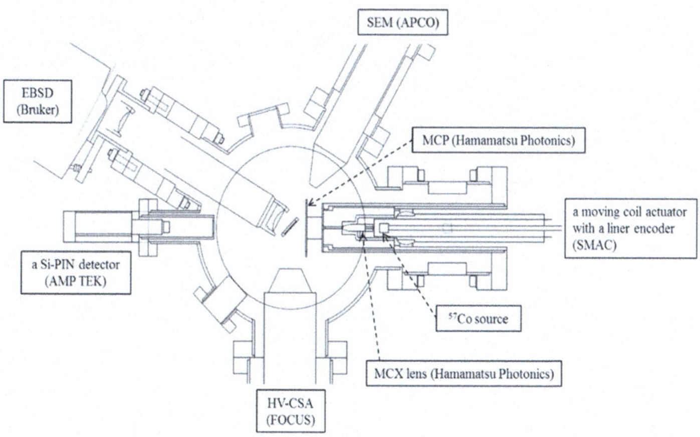

(b)

Fig. 9.39 (**a**) Photograph of Mössbauer Spectroscopic Microscope (MSM) combined with Scanning Electron Microscope (SEM) with Electron Beam Induced Current (EBIC), Electron Back Scattered Diffraction (EBSD), Electron Spectrometer, (**b**) Cross section of the whole system showing the arrangement of the optional systems. Reprinted by permission from Springer Nature: Springer, Hyperfine Interactions, 3D-Mössbauer spectroscopic microscope for mc-Si solar cell evaluation, Ino, Y., Soejima, H., Hayakawa, K. Yukihira, et al., © Springer International Publishing Switzerland 2016

measurements. Figure 9.41(**c**) shows the Mössbauer spectra of α-Fe foil with a thickness of 28 μm. They were measured either by a conventional Mössbauer driver or by the moving- coil actuator.

The second technique is the focusing of the 14.4 keV γ-rays. We use a multi-capillary X-ray lens ***MCX lens,*** developed initially for X-ray micro-beam analysis. The MCX lens provides a spot size of 75 μmϕ in diameter and a focal distance of 50 mm (HAMAMATSU). The spot size of γ-rays was measured by mapping the transmitted 14.4 keV-γ-rays through a Ta-pinhole collimator of 200 μmϕ in diameter. In Fig. 9.42, we show the image and the line profile through the spot centre

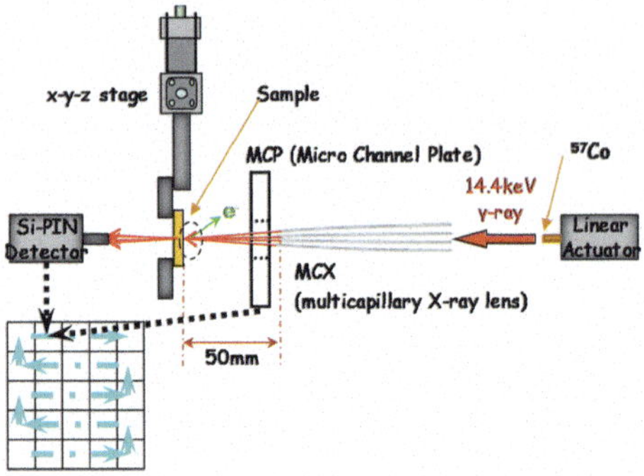

Fig. 9.40 Mapping technique consisting of linear actuator and multi-capillary X-ray lens, precision XYZ-θ stage, and micro channel plate for electron detection, and Si-PIN detector for transmitted γ-ray detection

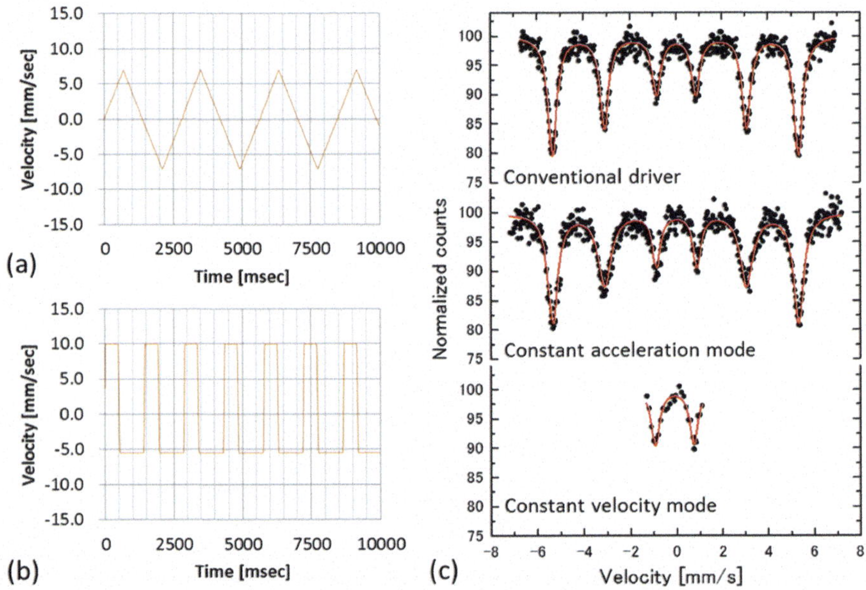

Fig. 9.41 Doppler velocity of the programmable actuator (SMAC) as functions of time: (**a**) the constant acceleration mode (acceleration = ± 10 mm/s^2) and (**b**) the constant velocity mode (velocity = + 10 and -5.42 mm/s), respectively. (**c**) The Mössbauer spectra of α-Fe foil measured either by a conventional Mössbauer driver (upper), or by the moving-coil actuator with the constant acceleration mode (middle) and with the constant velocity mode (lower). Reprinted by permission from Springer Nature: Springer, Hyperfine Interactions, 3D-Mössbauer spectroscopic microscope for mc-Si solar cell evaluation, Ino, Y., Soejima, H., Hayakawa, K. Yukihira, et al., © Springer International Publishing Switzerland 2016

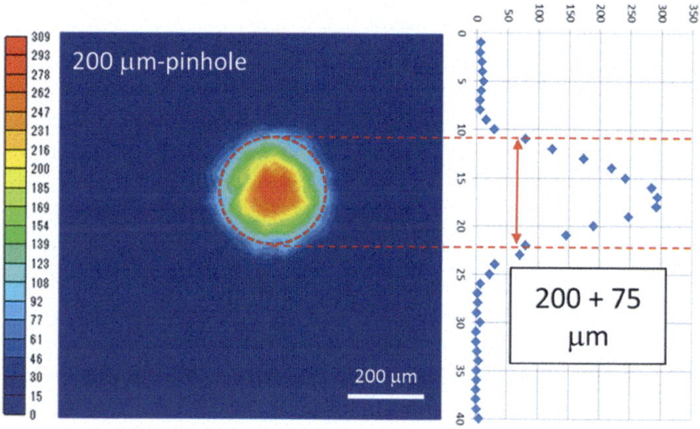

Fig. 9.42 Mapping image of a γ-ray spot focused with MCX lens (HAMAMATSU). The photograph is shown in Fig. 9.37(**b**). It also shows line profile through the spot centre. Reprinted by permission from Springer Nature: Springer, Hyperfine Interactions, 3D-Mössbauer spectroscopic microscope for mc-Si solar cell evaluation, Ino, Y., Soejima, H., Hayakawa, K. Yukihira, et al., © Springer International Publishing Switzerland 2016

as a function of the position on the right-hand side. The FWHM of the spot was about 275 μm. Thus, the spot size could be estimated to be 75 μmϕ in diameter. This value is small enough to distinguish the grains in mc-Si, for instance. We measured further the Mössbauer spectra of α-Fe-foil shown in Fig. 9.43 for the different set-ups

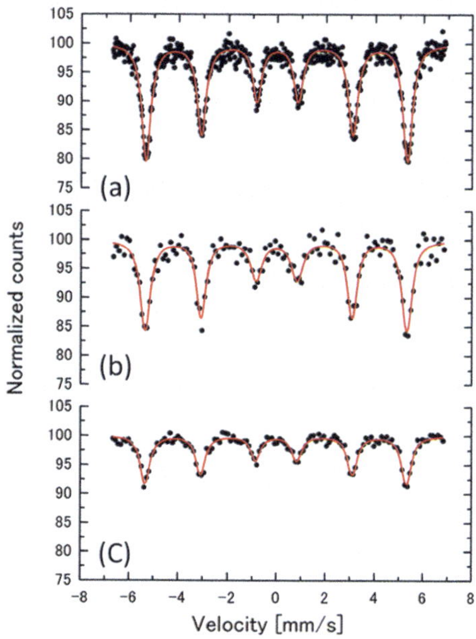

Fig. 9.43 Mössbauer spectra for 28 μm-thick α-Fe-foil (**a**) without MCX lens (γ-rays through a Pb collimator with a diameter of 2 mmϕ), (**b**) with MCX lens (the spot size of 265 μmϕ and the focus length 100 mm), and (**c**) with MCX lens (75 μmϕ and 50 mm). Reprinted by permission from Springer Nature: Springer, Hyperfine Interactions, 3D-Mössbauer spectroscopic microscope for mc-Si solar cell evaluation, Ino, Y., Soejima, H., Hayakawa, K. Yukihira, et al., © Springer International Publishing Switzerland 2016

with/without the MCX lens: (**a**) without MCX lens, i.e., we collimated the γ-rays only through a pin-hole of a lead plate with a diameter of 2 mmϕ, (**b**) with different MCX lens of spot size of 250 μmϕ and a focus length 100 mm, and (**c**) 75 μmϕ and 50 mm, respectively. It is evident in Fig. 9.43 that the absorption area decreases with decreasing the spot sizes, affecting the S/N for the Mössbauer-effect image contrast seriously. This tendency appears to be because the number of total reflections inside the capillary channels must increase with decreasing the spot size. Since we use the "total reflections" of the γ-rays inside of the capillaries with a diameter of 5 μmϕ, there must exist an energy loss during the focusing process by the MCX lens. For observing the Mössbauer effect, the energy loss must be less than 10^{-9} eV, which is entirely different from the case for X-ray micro-beam analysis. Conclusively, we can use this focusing technique for mapping an image by selecting the resonance condition on a spectral component. The spatial resolution can reach about 10 μm by overlapping the mapping points, as will be shown in the examples.

Finally, the third essential technique is the ***noise-free detection system*** for the transmitted γ-rays and the electrons after the Mössbauer resonant absorption of 14.4 keV γ-beams. Therefore, we must carefully consider the measuring geometry to avoid background radiations. Figure 9.44 presents (**a**) the cross-section and (**b**) the MCX lens, respectively. The set-up makes us possible to adjust the γ-ray axis of the MCX lens, so that the focused γ-rays will not touch the walls along the γ-ray path before illuminating the specimen surface except for a vacuum Be-window adjacent to the centre hole of the MCP.

In addition to the measuring geometry, we have a ***MgF₂coated Funnel-type MCP*** for detecting the conversion and Auger electrons (Fig. 9.45(**a**) and (**b**)), which was specially designed for our system by Hamamatsu. This detector yields an excellent detection efficiency with a factor of three higher than that of our former system. Figure 9.45(**c**) shows the pulse height spectra of the MCP output pulses obtained from a pure Fe foil both for ON resonance (blue) and OFF resonance (red) conditions.

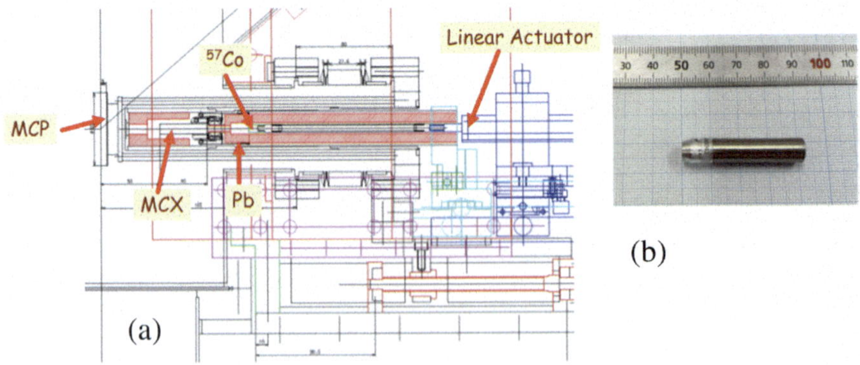

Fig. 9.44 (**a**) Schematic cross section of Mössbauer Spectroscopic Microscope (MSM) along the γ-ray path from ^{57}Co source to MCP through lead collimators and MCX lens, and (**b**) a photograph of MCX

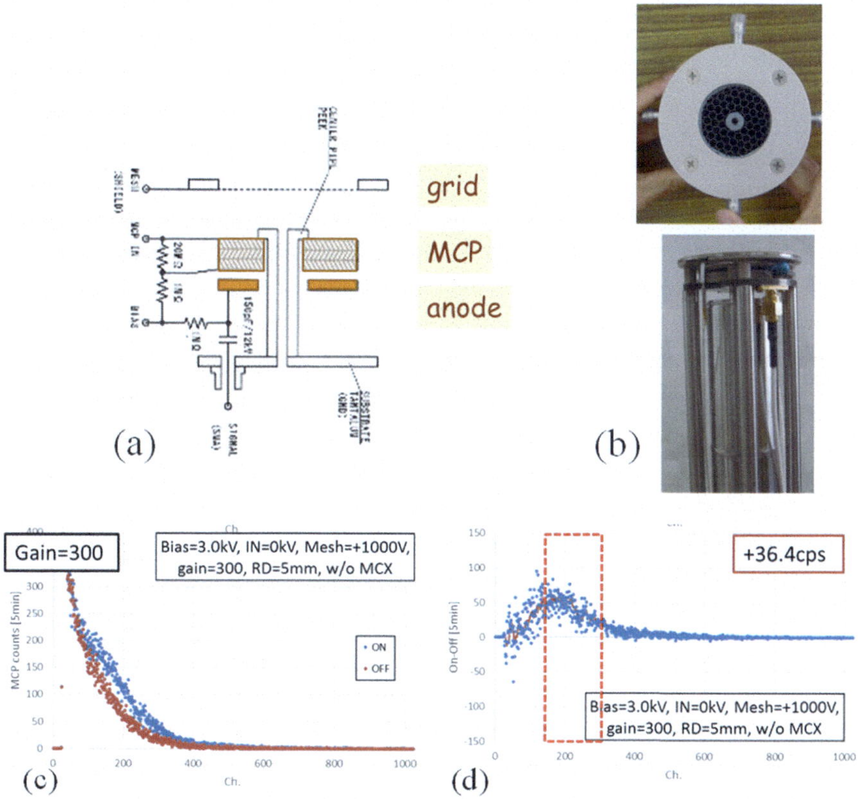

Fig. 9.45 (**a**) A cross section of MgF_2 coated Funnel-type MCP (Hamamatsu) mounted on a CF114-Flange, (**b**) Photographs, (**c**) the pulse height spectra of MCP output signals from pure Fe foil corresponding to the on-resonance (blue) and off-resonance (red) conditions, (**d**) the difference of the signals between ON and OFF resonance conditions

Subsequently, we make the difference between ON resonance and OFF resonance signals plotted in Fig. 9.45(**d**), which must, therefore, arise only from the Mössbauer effect. Such a subtraction procedure is beneficial to map a low-concentration of Fe contamination in Si solar cells, as will be shown later. While the 14.4 keV γ-rays transmitted through a sample are detected by a SI-PIN detector (AMPTEK), also providing a good S/N. We mount the specimen on a holder connected with a precision XYZ-θ stage (KOHZU). Accordingly, we can obtain a mapping image by counting both the electrons and transmitted γ-rays simultaneously as functions of the sample position.

9.5.2 MSM Mappings and Complemental Techniques Installed in Our Set-Up

In this section, we present typical examples of the MSM images. Figure 9.46(**a**) shows a Mössbauer spectrum of Fe-C 2.73 at.% quenched from 1200 °C, consisting of Austenite and Martensite phases. We measured two images of (**b**) and (**c**) related to Austenite and Martensite, respectively. To get the pictures, we tune the velocities to the isomer shifts of the singlet (red) and the most outer lines of the Ferro-magnetic sextet (green), respectively. The mapping area is 400 × 400 μm in a square with the step numbers of 80 × 80, resulting in a mapping step of 5 × 5 μm. We use the two sets of 80 × 80 matrix data for drawing the contour maps shown in Fig. 9.46(**b**) and (**c**). We notice that the Austenite phase with a size of about 20 μmϕ can be distinguished in the image of (**b**), although the focused spot size is 75 μmϕ. We will describe the details of the annealing behaviour in Sect. 9.6.

As the second example, Fig. 9.47 shows the MSM mapping data of Fe impurities in mc-Si solar cells before the image processing, which will be explained later in the next section. Looking at the energy level of Fe interstitial impurity in the bandgap of Si pn-junction in Fig. 9.47(**a**), we expect different charge states of the Fe interstitials, depending on the depth from the surface, as is schematically presented in (**b**). The photo (**c**) is the solar cell measured before antireflection coating, and the finger electrodes and the backside electrode can be seen on the left- and on the right-hand sides, respectively. After the deposition of 2 nm-^{57}Fe on the front side of the Si solar cell at room temperature in a vacuum of 10^{-5} Pa, we first measured the spectrum at room temperature in the dark before the electron irradiation. As the first step toward the mappings, we usually start measuring the Mössbauer spectra to determine the Mössbauer parameters for the spectral components, which are necessary to tune the resonance conditions. We present three spectra of a 2 nm-^{57}Fe deposited mc-Si solar cell in Fig. 9.47(**d**)–(**f**) under different circumstances, i.e., with/without the carrier injections. The spectra (**d**) and (**f**) correspond to the dark condition, while

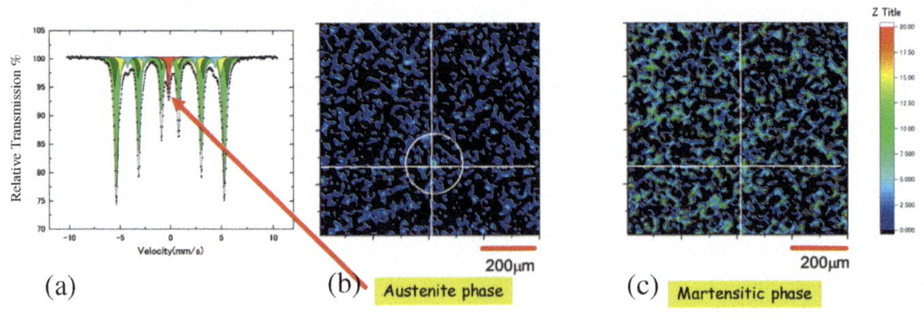

Fig. 9.46 (**a**) Mössbauer absorber spectrum of Fe-2.73 at. % C as quenched, (**b**) MSM mapping for paramagnetic Austenite phase, and (**c**) one for ferromagnetic Martensitic phase for the area of 400 × 400 μm. The former shows that the austenite phase with a size less than 20 μm may be identified by the mapping using an MCX lens of the spot size of 75 μmϕ for 80 × 80 mapping points

the spectrum (**e**) under 5 keV electron irradiations. Based on our assignment model [91], the spectrum consists of substitutional Fe_s^0 (green), interstitial Fe_i^0 with neutral charge state (yellow), and interstitial Fe_i^{2+} state adjacent to the defects (brown) such as dislocations and grain boundaries. During 5 keV-electron irradiation, however, Fe_i^{2+} state changes to Fe_i^+ (red) by a reaction between the injected electrons and Fe_i^{2+}, forming Fe_i^+ state. After irradiation, the spectrum in the dark becomes again more or less the same as before irradiation. The linewidth of Fe_i^{2+} seems broader than the others, which could arise from a dynamical fluctuation due to electron trapping and de-trapping at Fe_i^{2+} in the dark condition. Accordingly, it becomes clear that we should measure four Fe components of Fe_s^0, Fe_i^0, Fe_i^+ and Fe_i^{2+} for the mapping experiment.

Based on the above-mentioned spectral information, we obtained the ***MSM mapping images of a mc-Si solar cell*** without any intentional ^{57}Fe deposition. Figures 9.48 corresponds to the square area of 4 × 4 mm, comparing with the observations in Fig. 9.49 obtained by the complemental evaluation techniques installed in our MSM set-up in Fig. 9.39. Figure 9.48(**a**) is the photograph corresponding to the mapping area of the MSM images of the mc-Si solar cell, which were measured by counting (**b**) the transmitted 14.4-keV-γ-rays and (**c**) the conversion and Auger electrons, respectively. The MSM images measure with a constant velocity mode which generates reciprocal oscillations in two different speeds at +0.30 (ON-resonance) and −10 mms^{-1} (OFF-resonance) corresponding to the spectrum component of a neutral interstitial Fe_{int}^0 in the n-region of the solar cell (pn-junction), and to the background, respectively. We perform the mapping at 80 × 80 points by using an MCX lens with

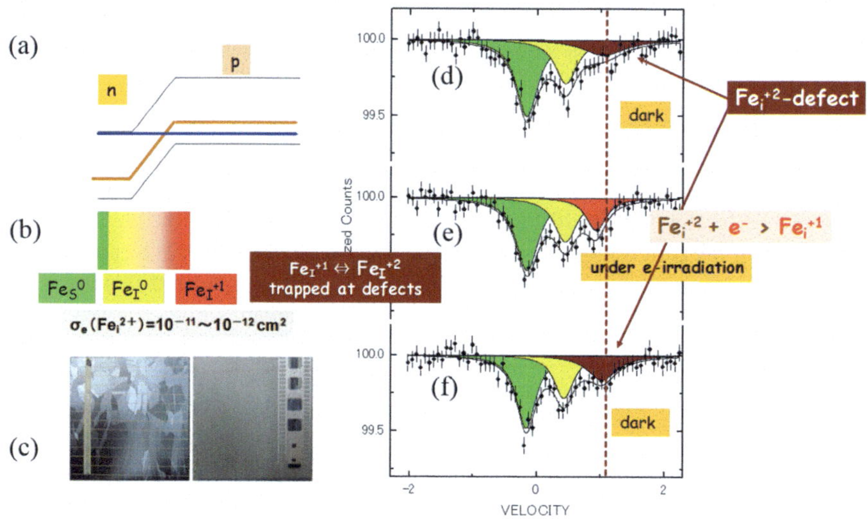

Fig. 9.47 (**a**) A schematic band structure of the pn-junction with the Fermi level (blue) comparing with the energy level of Fe interstitial impurity (brown), (**b**) an expected charge states of Fe interstitial based on this model, (**c**) photos of the mc-Si solar cell, (**d**) the spectrum in dark, (**e**) the spectrum under electron irradiation, and (**f**) the spectrum in dark after the electron irradiation

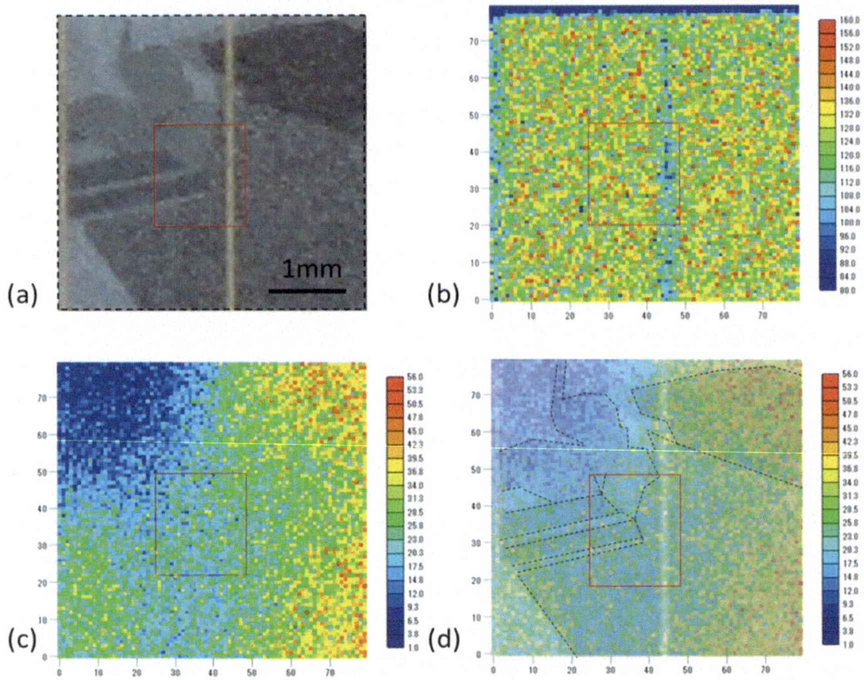

Fig. 9.48 (**a**) Photograph, (**b** and **c**) MSM mapping images with 80 × 80 points obtained by: (**b**) transmitted 14.4-keV γ-rays, (**c**) conversion and Auger electrons, and (**d**) a superposition of (**a**) and (**c**). The red area was observed by FE-SEM, EBIC, and EBSD, the results are shown in Fig. 9.49. Reprinted by permission from Springer Nature: Springer, Hyperfine Interactions, 3D-Mössbauer spectroscopic microscope for mc-Si solar cell evaluation, Y. Ino, H. Soejima, K. Hayakawa, K. Yukihira, et al., © Springer International Publishing Switzerland 2016

a spot size of 265 μmϕ. In Fig. 9.48(**b**), we can see the Ag electrodes with a width of 100 μm and a specimen holder edge, indicating that this image has a spatial resolution much better than that of 100 μm. The intensity distribution of the electron map is different in the different grains, as can be seen in Fig. 9.48(**c**). We can see this effect more evident in (**d**) where we superimpose the electron map and also the grain boundaries on the photograph. We find the electron intensities higher at the upper right and the lower left regions of the map, suggesting different Fe concentrations in the different grains. Fe impurities in mc-Si are incorporated from the feedstock as well as from the crucible wall during melting. The concentrations are expected up to 10^{15} cm^{-3} [71]. Although we do not measure the Fe concentration directly in this mc-Si solar cell in this measurement, the sample could contain Fe impurities with rather high levels in the solar cell.

Figure 9.49 show (**a**) a photograph for the square area shown in Fig. 9.48(**a**), and also the observations related with (**b**) SEM (**c**) EBIC, (**d**), and (**e**) EBSD, respectively, for the same area. The SEM image of Fig. 9.49(**b**) shows a vertical white line of the Ag electrode with a width of 100 μm on the lower right-hand side, while we can

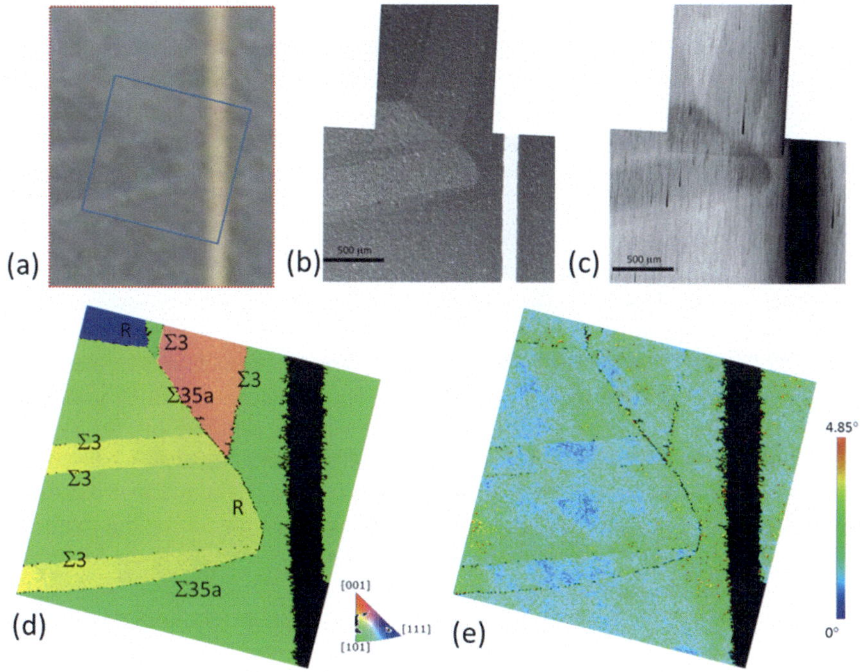

Fig. 9.49 Complementary observations of the mc-Si solar cell for the red area shown in Fig. 9.48: (**a**) photograph, (**b**) FE-SEM, (**c**) EBIC, (**d**) inversed pole figure map of EBSD, and (**e**) local misorientation map by EBSD, being related to an internal stress distribution. Reprinted by permission from Springer Nature: Springer, Hyperfine Interactions, 3D-Mössbauer spectroscopic microscope for mc-Si solar cell evaluation, Y. Ino, H. Soejima, K. Hayakawa, K. Yukihira, et al., © Springer International Publishing Switzerland 2016

identify the characteristic grain structures in the middle region of (**a**). We obtain the EBIC contrast of (**c**) by measuring the current flow from the electron-injected surface to the backside electrode. Accordingly, the contrast becomes dark at the carrier recombination centre, such as electrically active dislocations or grain boundaries, both of which must directly connect to the solar cell degradations. Also, near the Ag electrode, the current appears to flow preferentially into the electrode, so that the EBIC image yields the Ag electrode broader than the SEM image, as is observed in Fig. 9.49(**c**). Thus, ***a carrier recombination centre in mc-Si solar cells*** can detect by the EBIC technique, the same area of which can be investigated directly with the MSM images due to the Fe impurities. Accordingly, the correlations between the electro-active defects and the Fe impurities can be clarified.

Furthermore, the EBSD technique gives us a possibility not only to identify ***the crystal orientation of each grain*** of mc-Si solar cells as in Fig. 9.49(**d**) but also to investigate ***the internal residual strain distributions*** in the grains shown in (**e**). Both information must be essential to understand the Fe contamination processes, i.e., the diffusion of Fe impurities into mc-Si wafers and solar cells. Figure 9.49(**d**) is an inversed pole figure map along a normal direction of EBSD. The crystal orientation refers to as the colour distribution in the unit triangle pole figure inserted at a lower right-hand side of the Figure. In this region, there are the twin boundaries of $\Sigma 3$, the coincidence boundaries of high Σ such as $\Sigma 35$a, and the random boundaries (R). Figure 9.49(**e**) shows a local misorientation map relating to the residual strains. This image indicates large strains at a particular region in the grain and at a location near the electrode.

Our set-up provides a possibility to get a 3D-mapping image by combining the 2D-mapping with the depth profiles, which can be measured by an ***Electron Spectrometer*** (HV-CSA, FOCUS). The conversion and Auger electrons are emitted from the ^{57}Fe nucleus distributing at a different depth from the surface. Subsequently, the electrons lose their energy due to the inelastic collisions, which can be measured by the spectrometer. Therefore, it is possible to obtain a depth distribution of ^{57}Fe atoms by selecting the energies of the electrons, yielding a possibility for a ***depth selective mapping.*** This technique is called a depth selective conversion Mössbauer spectroscopy (DSCEMS) with the probing depth of about 100 nm, and the depth resolution of 5–10 nm for ^{57}Fe. Previously, we measured for the sample with the Ag layer of different thicknesses of 20, 50, 70 nm on ^{57}Fe layer using the MCP, and detected different electron counts depending on the Ag layer thickness, as shown in Fig. 9.50. These results indicate that we can obtain a depth profile for ^{57}Fe down to a depth of 100 nm with the depth resolution of about 20 nm.

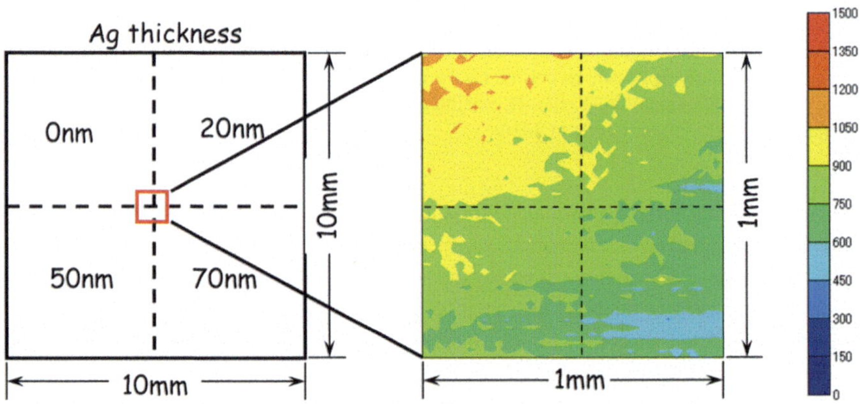

Fig. 9.50 MSM electron mapping of 50 nm-^{57}Fe layer on Si wafer after the deposition of different thicknesses of Ag with 0, 20, 50 and 70 nm

9.5.3 MSM Image Data Processing to Deduce the Concentration Distributions

It is fundamentally important to realize a background-free measuring system to determine the concentration distributions for the Fe components. In our practice, we measure the electrons and the transmitted γ-rays simultaneously under the ON-resonance and the OFF-resonance conditions. Then, we subtract the latter from the former counts at each mapping point. Accordingly, we can produce the electron and γ-ray intensity distributions originating only from the Mössbauer effect. Furthermore, we create the image data consisting of typically an 80 × 80 matrix finally converted to a contour map corresponding to one spectral component. As an example, we show the electron image data processing in the four columns of Fig. 9.51: the most upper one corresponding to the four original images of substitutional $Fe_s{}^0$, interstitial $Fe_i{}^0$, $Fe_i{}^+$, and $Fe_i{}^{2+}$ states in a Si-solar cell, respectively, from left to right, which measures with the ON-resonance conditions, and in the second column the OFF-resonance image, i.e., the mapping at the Doppler velocity of $\pm$ 10mms^{-1}. In the third column,

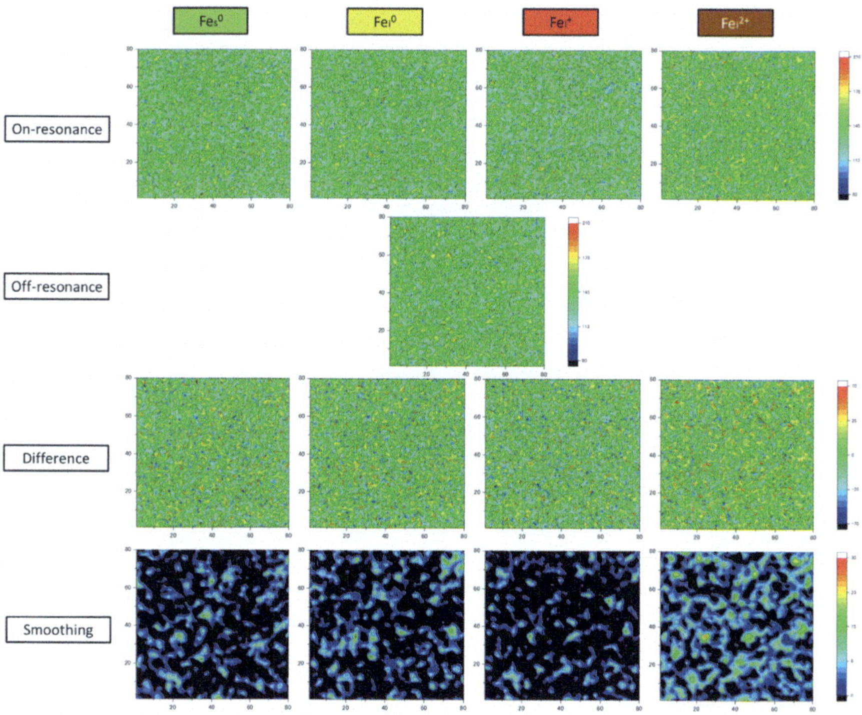

Fig. 9.51 The typical mapping image for each spectral component is produced by subtracting the OFF-resonance image from the ON-resonance one. Subsequently, the difference images are smoothed by taking into account of the spot size of the MCX lens on the sample: In this case we used an MCX lens with the spot size of 265 μmϕ

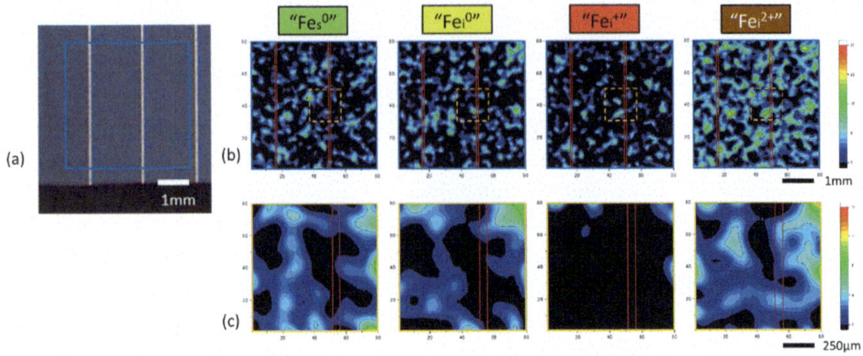

Fig. 9.52 MSM electron images of the emitter region (n-region) of a Si solar cell corresponding to the different Fe components of substitutional $Fe_S{}^0$, interstitial $Fe_i{}^0$, $Fe_i{}^+$ and $Fe_i{}^{2+}$ states: (**a**) photograph, (**b**) MSM images for the area of 4 × 4 mm, (**c**) those for 1 × 1 mm. The focal spot size of the MCX lens is 265 μmϕ [from Ref. 96]

the OFF-resonance images are subtracted from the ON-resonance images in the first column, yielding the background-free images. After the subtractions, indeed, we can distinguish more clearly some delicate structures in all the contour maps. Finally, in the fourth column, we further perform smoothing of the difference maps using by the focal size of the MCX lens, 265 μmϕ.

This mapping technique enables us, for instance, to investigate the origins of the degradation of the energy conversion efficiency of a Si-solar cell [96]. For such evaluations, it is crucial to deduce the absolute numbers of the carrier trapping centres as well as the cross-sections based on the mapping data. In Sect. 9.5.2, we have mentioned that both $Fe_i{}^+$ and $Fe_i{}^{2+}$ impurities in the Si matrix play a crucial role as the carrier trapping centres in addition to the other electro-active defects such as dislocations, grain boundaries in the mc-Si matrix. We show the MSM electron images of the emitter region (n-region) of a Si solar cell in Fig. 9.52: (**a**) the photograph of the solar cell, and the mapping images corresponding to the different Fe components of substitutional Fe_S, interstitial $Fe_i{}^0$, $Fe_i{}^+$, and $Fe_i{}^{2+}$ states for the observation areas of (**b**) 4 × 4 mm and (**c**) 1 × 1 mm, respectively. Considering of the integrated electron intensity obtained from these image data as well as of the cross-section of the recoil-free emission for the conversion electrons after the 14.4 keV-γ-ray absorption, the image data-processing described in Fig. 9.51 allows us to estimate the number of Fe impurities to be 4×10^{16} ^{57}Fe/cm^2. Assuming that the Fe impurities distribute within the observed depth of about 500 nm, we obtain the average concentration of 10^{21} ^{57}Fe/cm^3. This value means that this solar cell contains surprisingly high Fe contamination in the emitter region, i.e., 400 nm from the surface. To confirm our evaluation, we performed ICP luminescence and obtained the Fe concentration of 0.050 ppm, which coincides with that deduced by our MSM estimation. Accordingly, our method can provide the absolute value of the Fe concentration.

9.5.4 *Imaging Technique Using sCMOS-Camera*

Here we show shortly the second technique under development [97] (Fig. 9.53), i.e., Mössbauer spectroscopic camera based on an X-ray sCMOS camera (C12849-101U, HAMAMATSU) combining with a capillary plate, i.e., aggregate lead-glass collimators with a thickness of 1 mm and each capillary hole-size of 6 μm in diameter. This camera can connect with the Mössbauer spectroscopic mapping system mentioned above. The new camera achieves 4×10^6 pixels and 5 μm resolution. Figures 9.53(**b**) and (**c**) show the transmitted γ-ray images of Fe-2.73 at% C foil consisting of ferrite and austenite phases for the areas of 400 × 400 μm. The space resolution appears to be less than 10 μm. The maximum exposure time of this camera is only 30 s, and, therefore, we create the contour pictures by the sum of all the images accumulated for several hours. To improve signal to noise ratio, also, each one is made by subtraction between the OFF-resonance and ON-resonance images. Finally, smoothing is performed, considering the capillary hole-diameter, 6 μm. With this technology,

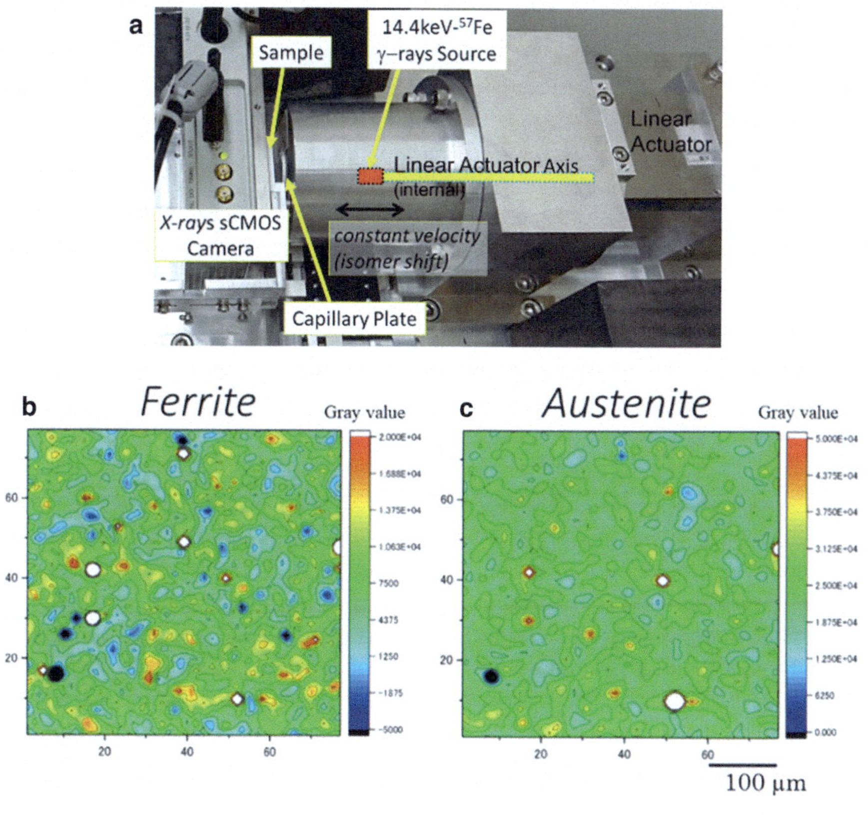

Fig. 9.53 (**a**) MSC setup [97]. Figure 9.53 (**b**) and (**c**) MSC images of Fe-2.73 at% C foil [From Ref. 97]

measuring the time necessary for one picture becomes drastically shorter than that of the mapping technique. We can apply these imaging technologies for a wide variety of Fe containing materials, such as silicon solar cells, steels, and even biomaterials soon.

9.6 Diffusion and Segregation Studies by Mössbauer Spectroscopic Camera

Recently, the imaging techniques in Mössbauer spectroscopy are opening a new possibility to study the diffusion process not only in homogeneous but also inhomogeneous materials for a diffusion-length from μm to mm. The materials provide different spectral components due to their atomic structure as well as complex micro-structures. For instance, a Si-solar cell is often contaminated with Fe impurities, leading to severe degradation in the light-energy conversion efficiency [71]. To clarify the origin of the Fe contaminations, however, one has to investigate the Fe diffusion and segregation processes in the p–n junction, which contains electrodes, anti-reflection coating layer, crystal grains, grain boundaries, dislocations, and strain distributions. In such systems, the atoms are moving and gathering on different lattice sites with other chemical states under operation.

In the next two paragraphs, we are going to explain the diffusion studies of Fe in Si wafer [87] and also in a multi-crystalline Si wafer [90]. The former system is supposed to be homogeneous, and the latter inhomogeneous. The MSM technique provides unique information on the diffusion process, i.e., on the diffusion profiles belonging to the spectral components selectively.

9.6.1 Fe Impurity Diffusion in Single-Crystalline Si Wafer [87]

We deposited a 2 nm-thick ^{57}Fe layer in an area of 10 mm × 10 mm on a CZ-Si wafer (10^{15}B/cm^3) with 30 mm × 30 mm × 500 μm shown in Fig. 9.54(**a**) and subsequently annealed at 430 °C for 1 h in a vacuum furnace. A Mössbauer spectrum at room temperature consists of three components corresponding to substitutional Fe_s^0, and interstitial Fe_i^0 and Fe_i^+, which are shown in Fig. 9.55. Subsequently, 1/3 of the sample was polished with an angle of 6° to the surface, as is seen in Fig. 9.54(**a**). Three MSM electron mapping images were obtained corresponding to the three spectral components. The focal size of the lens is 250 μmϕ. We evaluate the diffusivity of Fe in Si in two mapping areas of (A) and (B) as indicated in Fig. 9.54(**b**): (A) along the grinded surface, and (B) along the original wafer surface crossing through the ^{57}Fe-deposition boundary. By integrating the mapping intensity, we can evaluate the diffusivity of each Fe component along with two different directions by

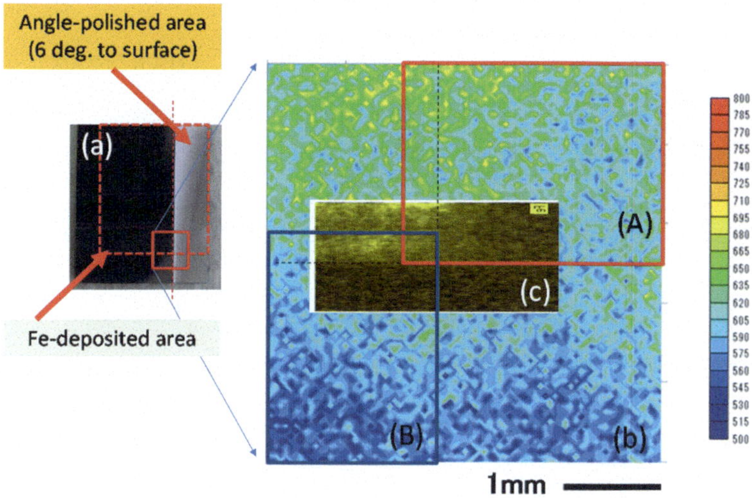

Fig. 9.54 (**a**) ^{57}Fe-deposited Si wafer, and ^{57}Fe-deposited area is shown by red broken-line rectangular, (**b**) MSM mapping for red-line square area in (**a**), (**c**) EDS mapping of Fe–$L_{\alpha 1}$ X-ray. The mapping areas of (**A**) and (**B**), which are shown in (**b**), are evaluated for the ^{57}Fe diffusivities along the angle-polished area and along the surface crossing through the ^{57}Fe-deposition boundary, respectively. Reprinted by permission from Springer Nature: Springer, Feasibility study to investigate diffusion of Fe in Si using a Mössbauer spectroscopic microscope, Y. Yoshida, Y. Ino, K. Matsumuro, T. Watanabe, et al., © Springer International Publishing Switzerland 2016.), © Springer International Publishing Switzerland 2016

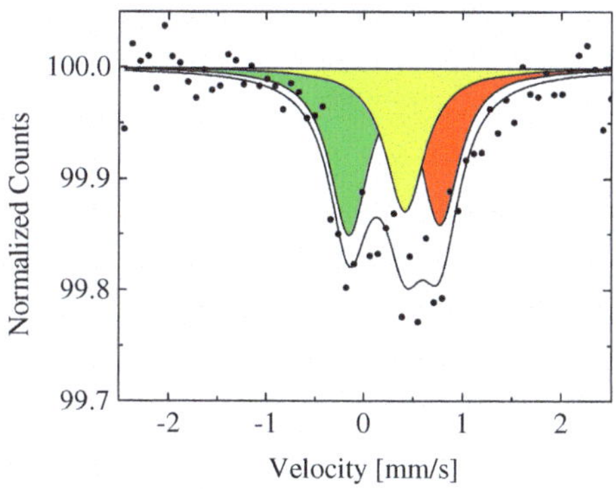

Fig. 9.55 Mössbauer Spectrum of ^{57}Fe deposited Si wafer, which was annealed at 430 °C. Reprinted by permission from Springer Nature: Springer, Feasibility study to investigate diffusion of Fe in Si using a Mössbauer spectroscopic microscope, Y. Yoshida, Y. Ino, K. Matsumuro, T. Watanabe, et al., © Springer International Publishing Switzerland 2016

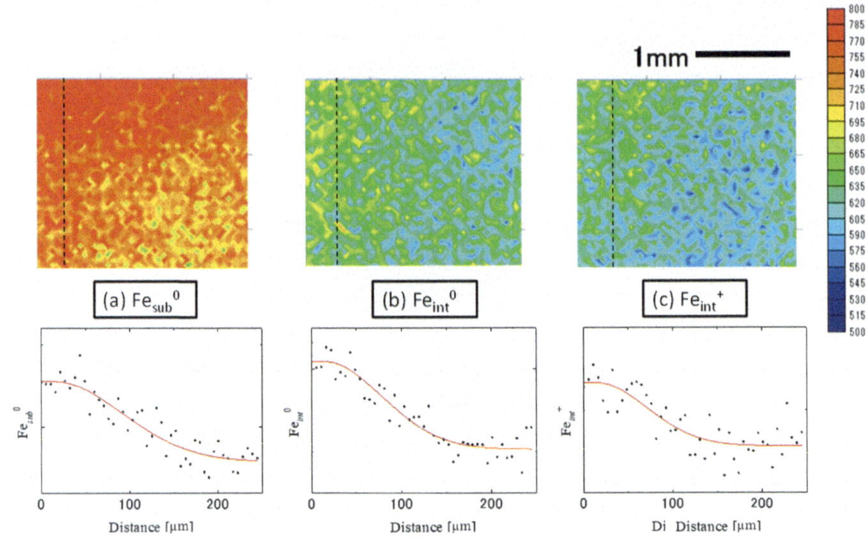

Fig. 9.56 Three mapping images of (A) along *angle-grinded surface*, corresponding to different spectral components, i.e., (**a**) substitutional $Fe_{sub}{}^{0}$, and (**b**) interstitials $Fe_{int}{}^{0}$ and (**c**) interstitial $Fe_{int}{}^{+}$, respectively. The diffusion profiles are plotted as function of depth, which is deduced by correcting the angle of 6°. Reprinted by permission from Springer Nature: Springer, Feasibility study to investigate diffusion of Fe in Si using a Mössbauer spectroscopic microscope, Y. Yoshida, Y. Ino, K. Matsumuro, T. Watanabe, et al., © Springer International Publishing Switzerland 2016

Gaussian fitting. We compare these diffusivities further with the total Fe diffusivity in Si, which is measured by EDS mapping of Fe–$L_{\alpha 1}$ X-ray for the area presented in Fig. 9.54(**c**).

Figure 9.56(**a**)–(**c**) present three mapping images of (A) along the angle-grinded surface are presented, corresponding to (**a**) substitutional $Fe_{sub}{}^{0}$, and (**b**) interstitials $Fe_{int}{}^{0}$ and (**c**) interstitial $Fe_{int}{}^{+}$, respectively. We use the same colouring scale for all the images between 500 and 800 counts/mapping points. The broken line on the left-hand side of each mapping indicates the position of the surface deposited with ^{57}Fe. After the diffusion annealing at 430 °C for one hour, the ^{57}Fe diffusion profiles appear to be different from each other, suggesting different diffusivities for different Fe states in Si. The mapping intensity is substantially proportional to the number of the Fe atoms which distribute down to about 100 nm from the surface in the case of Si. The average counts of Fig. 9.56(**a**) is higher than those of (**b**) and (**c**). This effect might be due to a different signal to noise ratio during the mapping measurement on the substitutional $Fe_{sub}{}^{0}$. For simplicity, however, we analyze only the relative changes of the integrated mapping intensity as a function of the distance from the surface. These 1D-diffusion profiles can be obtained by summing up the counts along the direction perpendicular to the depth. By considering of 6°, i.e., the angle between the grinding plane and the surface, we estimate the diffusion depth. Accordingly, we can deduce the diffusivity for each Fe component as a function of depth from the

Table 9.1 Fe Diffusivities in Si evaluated from the MSM mapping images

Methods	Diffusion profiles	Fe states	D/cm^2s^{-1}	
EDS	Depth	Total	8.0×10^{-9}	
	Parallel to surface		1.1×10^{-9}	
MSM	Depth	Total	5.7×10^{-9}	
		Substitutional Fe_{sub}^0	8.3×10^{-9}	
		Interstitial Fe_i^0	5.6×10^{-9}	
		Interstitial Fe_i^+	4.5×10^{-9}	
	Parallel to surface	Total	6.1×10^{-6}	
		Substitutional Fe_{sub}^0	7.4×10^{-6}	
		Interstitial Fe_i^0	4.4×10^{-6}	
		Interstitial Fe_i^+	7.0×10^{-6}	
Reference	[71]	Total	1.7×10^{-8}	at 430 °C

surface by assuming a Gaussian distribution. We evaluate the total diffusivity of all Fe components to be as 5.7×10^{-9} cm^2/s. The value is close to those obtained from the EDS mapping image, 8.0×10^{-9}, and 1.7×10^{-8} cm^2/s from the reference [71]. We summarise the results in Table 9.1.

On the other hand, we show the mapping images of (B) along the surface and their diffusion profiles in Fig. 9.57(**a**)–(**c**), respectively. The broken line on the left-hand side indicates the ^{57}Fe-deposition boundary. Surprisingly, the diffusivity along the surface is three orders of magnitude larger than that of along the depth, although the EDS image shows only 1.1×10^{-9} cm^2/s. This significant discrepancy may be due to the detection range: 100 nm for MSM and 300 nm for EDS. We notice that there is oxygen distribution, which coincides with the ^{57}Fe-deposition area and may have an influence on the diffusivities.

We conclude that the MSM mapping method makes it possible to measure directly ***the diffusion profiles and their diffusivities separately for different Mössbauer components,*** which correspond to the various lattice sites and the chemical states in the homogeneous system. The present results, therefore, suggest strongly that the Fe impurities deposited on the surface diffuse into Si crystal not only via interstitial Fe_i^+, but also via both interstitial Fe_i^0 and substitutional Fe_i^0, and that the diffusion must proceed with the strong correlations between all the components. The diffusion mechanism of Fe in Si appears to be completely different from that generally accepted in the semiconductor community, i.e., a simple interstitial diffusion [71].

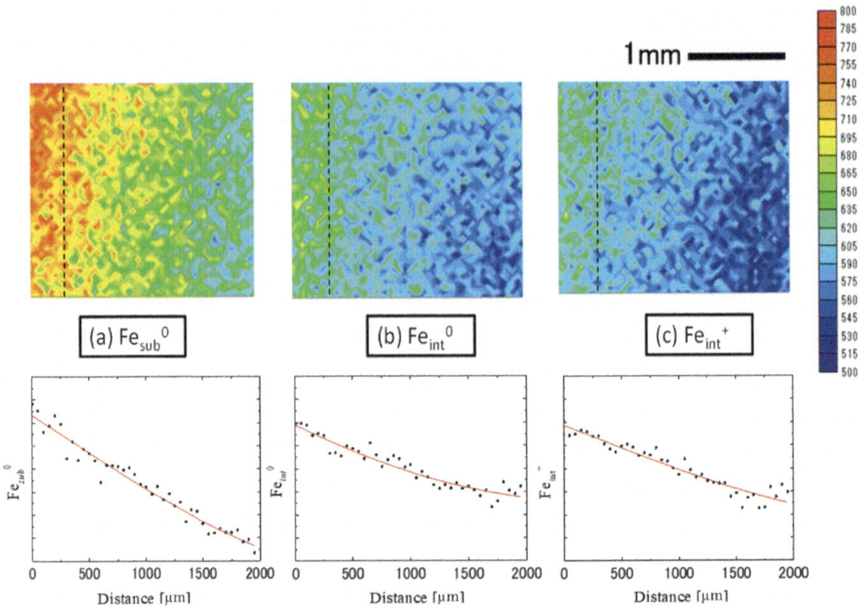

Fig. 9.57 Three mapping images of (B) *along the surface*, corresponding to different spectral components, i.e., (**a**) substitutional Fe_{sub}^{0}, and (**b**) interstitials Fe_{int}^{0} and (**c**) interstitial Fe_{int}^{+}, respectively. The diffusion profiles are plotted as function of surface distance from the ^{57}Fe-deposited boundary. Reprinted by permission from Springer Nature: Springer, Feasibility study to investigate diffusion of Fe in Si using a Mössbauer spectroscopic microscope, Y. Yoshida, Y. Ino, K. Matsumuro, T. Watanabe, et al., © Springer International Publishing Switzerland 2016

9.6.2 Fe Impurity Diffusion in Multi-crystalline Si Wafer [90]

The MSM images can be directly compared with the Raman and Photo-Luminescence (PL) mappings, which provide information on the crystal grains and the electroactive defect distributions, respectively. The concentrations of ^{57}Fe impurities and the PL intensities must correlate each other. We cut a Multi-Crystalline-Si ingot (10^{16} B/cm^3) into a square plate of 30 mm × 30 mm × 1.3 mm. After mirror polishing and chemical etching by 5%-HF, we performed Raman and PL mappings for 1 mm × 1 mm areas. A micro-PL spectrometer (Photon Design Co., Ltd.) uses a wavelength of 532 nm line emitted from a diode-pumped solid-state (DPSS) laser as an excitation source.

Figure 9.58(**a**) and (**b**) show Raman and PL images, respectively, for the areas of 23 mm × 23 mm and 1 mm × 1 mm. In the figures, small squares indicate the MSM mapping areas where we deposited ^{57}Fe impurities and subsequently annealed at 200 °C, and finally at 800 °C for 1 h. In the Raman map, (**a**), we find three grains and their boundaries. In the PL images, (**b**), on the other hand, there are "blue regions" containing the carrier traps along the grain boundaries. However, the electro-active defects exist also in the upper grain. After the mappings, we chemically etched the

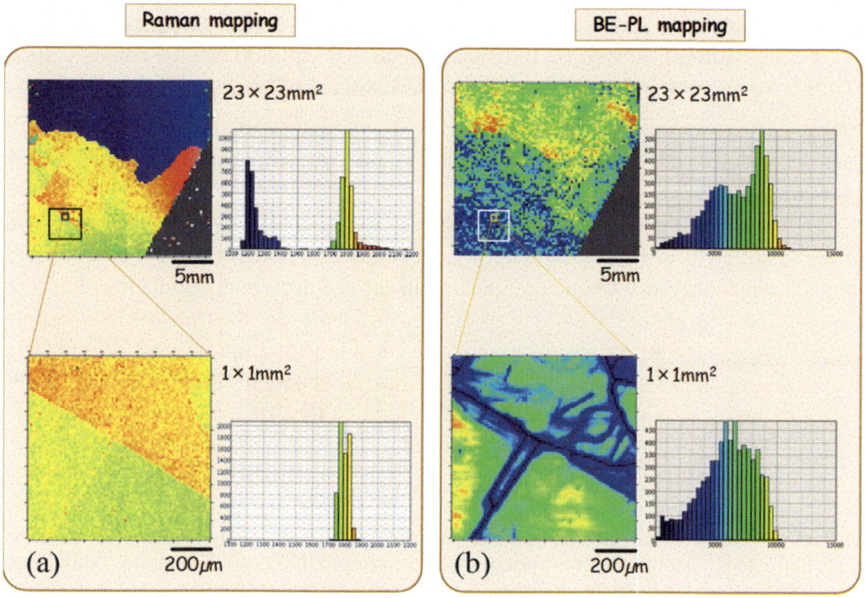

Fig. 9.58 (**a**) Raman and (**b**) Photo-Luminescence mappings of mc-Si wafer are measured after mirror polishing and chemical etching by 5%-HF. Two different areas are mapped on the wafer using a micro-PL spectrometer, their intensity histograms are also shown on the right-hand side of each mapping. The MSM mappings are performed in the same areas which are shown in the figures, after 2 nm-^{57}Fe deposition followed by the annealing at 200 and 800 °C for 1 h. Reprinted by permission from Springer Nature: Springer, Hyperfine Interactions, Mössbauer spectroscopic microscope study on diffusion and segregation of Fe impurities in mc-Si wafer, Y. Yoshida, T. Watanabe, Y. Ino, M. Kobayashi, et al., © Springer International Publishing Switzerland 2020

wafer to get a fresh surface again and deposited 2 nm-thick ^{57}Fe by an electron beam gun on an area of 30 mm × 30 mm of the wafer. We subsequently annealed the wafer at 200 °C for 1 h and measured a Mössbauer spectrum in a vacuum of 10^{-5} Pa at room temperature. The spectrum consists of a substitutional $Fe_{sub}{}^{0}$, and interstitials $Fe_{int}{}^{0}$ and $Fe_{int}{}^{+2}$ components, as is shown in Fig. 9.59. To obtain the MSM images for 1 mm × 1 mm area, we accumulated the electrons and the γ-rays simultaneously for 35 s at 80 × 80 points with a mapping step of 50 μm. The focal distance and size of the Multi-Capillary X-ray lens were 50 mm and 75 μmϕ, respectively, for the whole experiments. After the mappings, we annealed it at 800 °C for 1 h in a vacuum furnace of 10^{-5} Pa. Then, the mapping measurements were done with the same conditions again.

The MSM electron and γ-ray images are shown in Figs. 9.60 and 9.61, respectively, to the areas of 1 mm × 1 mm. We draw the contour maps by cutting a part of 20 × 20 points from the whole 80 × 80 points. Subsequently, we compare the Raman and the PL mappings with the MSM mappings for the same area. In the MSM images, we depict the positions of the grain boundaries and the defects in the grains by the red dotted lines. Seen is that all the Fe components appear to localize adjacent to the grain boundaries and the defects. Since the electro-active defects exist before the ^{57}Fe deposition followed by the annealing at 200 and 800 °C, we consider such defects have a strong influence on the diffusion process of Fe impurities in the mc-Si matrix.

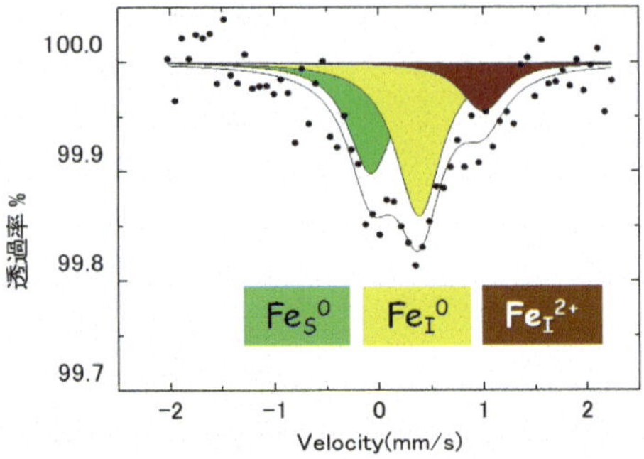

Fig. 9.59 Mössbauer Spectrum of 2 nm-^{57}Fe-deposited mc-Si wafer after the annealing at 200 °C for 1 h. Reprinted by permission from Springer Nature: Spring-er, Hyperfine Interactions, Mössbauer spectroscopic micro-scope study on diffusion and segregation of Fe impurities in mc-Si wafer, Y. Yoshida, T. Watanabe, Y. Ino, M. Koba-yashi, et al., © Springer International Publishing Switzer-land 2020

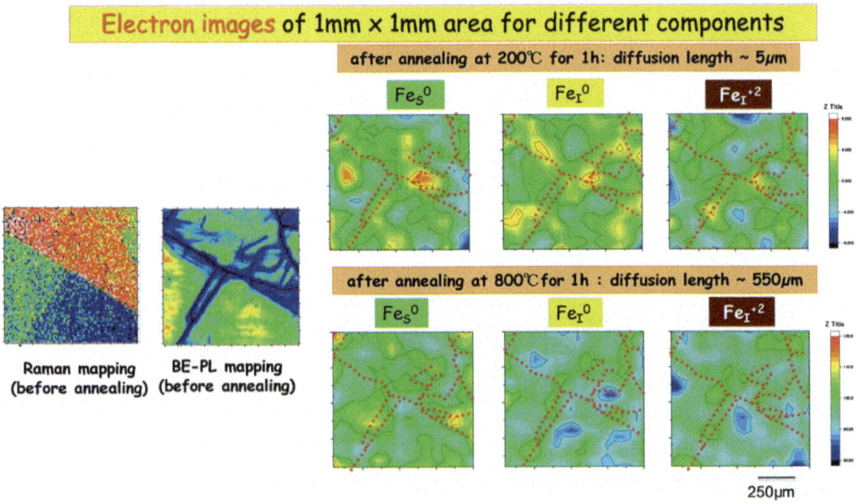

Fig. 9.60 MSM electron images for 1 mm × 1 mm area, which correspond to the Fe spectral components of substitutional Fe^0_{Sub}, and interstitials Fe^0_{int} and Fe^{+2}_{int}, are compared with the Raman and PL mappings for the same area. Reprinted by permission from Springer Nature: Springer, Hyperfine Interactions, Mössbauer spectroscopic microscope study on diffusion and segregation of Fe impurities in mc-Si wafer, Y. Yoshida, T. Watanabe, Y. Ino, M. Kobayashi, et al., © Springer International Publishing Switzerland 2020

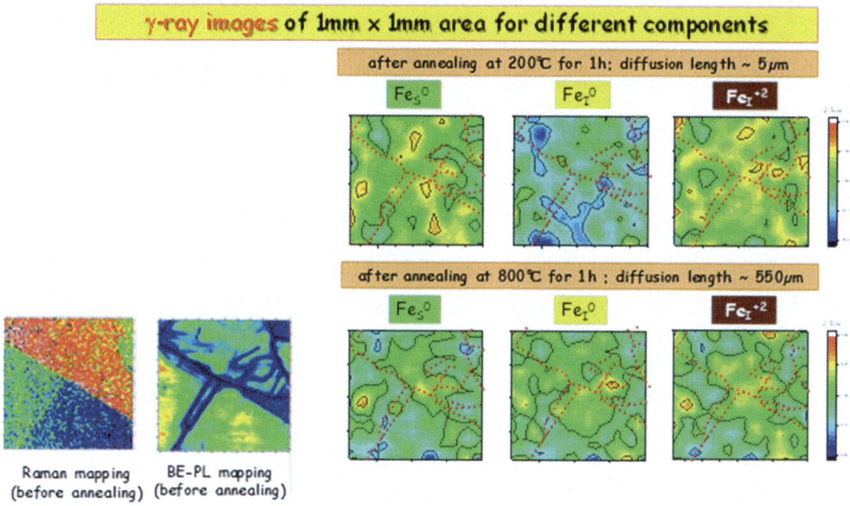

Fig. 9.61 MSM γ-ray images for 1 mm × 1 mm area, which correspond to the Fe spectral components of substitutional $Fe_{sub}{}^{0}$, and interstitials $Fe_{int}{}^{0}$ and $Fe_{int}{}^{+2}$, are compared with the Raman and PL mappings for the same area. Reprinted by permission from Springer Nature: Springer, Hyperfine Interactions, Mössbauer spectroscopic microscope study on diffusion and segregation of Fe impurities in mc-Si wafer, Y. Yoshida, T. Watanabe, Y. Ino, M. Kobayashi, et al., © Springer International Publishing Switzerland 2020

Moreover, we deposited 2 nm-thick-^{57}Fe on the present p-type Si wafer with 1 × 10^{16} B/cm^3. After the annealing at 200 and 800 °C for one hour, ^{57}Fe concentrations are 3 × 10^{19} and 3 × 10^{17} ^{57}Fe/cm^3, respectively, if we estimate the diffusion depth using the former diffusivity data yielding 5 and 550 μm in the Si crystal [87]. We think that B dopants distribute homogeneously as the charge state of -1 in contrast to that of + 1 of interstitial Fe, which we expect from the Fe_i energy level of 0.4 eV in the Si bandgap. Accordingly, one assumes that Fe impurities are at least forming Fe-B pairs partly. In the Mössbauer spectrum in Fig. 9.59 there exists a "yellow component" assigned to interstitial $Fe_i{}^0$. We reported [81] that the isomer shift of the pair to be quite similar to that of $Fe_i{}^0$. Besides, there must be strong correlations between the PL intensity and the MSM electron and γ-ray intensities of substitutional $Fe_{sub}{}^{0}$, and interstitials $Fe_{int}{}^{0}$ and $Fe_{int}{}^{+2}$ components, but the present statistic and the space resolution appear to be good enough to find the correlations. There seem to be different correlations between the PL suggesting these components are associating with the electro-active defects existing be intensity and the MSM electron and γ-ray images. The former originates from the depth regions close to the surface and in bulk, respectively: After the annealing at 200 °C, the MSM electron images of interstitials $Fe_{int}{}^{0}$ and $Fe_{int}{}^{+2}$ yield small negative correlations, for the ^{57}Fe deposition. After the annealing at 800 °C, on the other hand, a negative correlation of substitutional $Fe_{sub}{}^{0}$ is found only close to the surface, while no correlations of all the Fe components in bulk. These results appear to be reasonable because the Fe impurities are trapped in the defects at low temperature, while they can diffuse overcoming the trapping barriers at high temperature. The MSM image seems to be quite helpful in studying such diffusion and segregation processes in such an inhomogeneous matrix.

9.6.3 Carbon Diffusion in Fe Steel

Here we are going to show another potential of the MSM imaging technique, i.e., direct observation of the carbon diffusion and segregation processes in Fe steel, which is a vital material in our modern world. The system contains a complex microstructure consisting of Martensitic and Austenite phases, which will change drastically as a function of the annealing temperature. Finally, it will be followed by the phase separation to ferrite and cementite phases. These phenomena are indeed a long-standing research theme for Mössbauer Spectroscopy: Already in 1971, F. E. Fujita reported his pioneering works on Fe carbon alloys in a review book edited by Gonser [4]. In Sect. 9.3, we have already shown the experimental results on Fe-7.8 at. % C alloy using the thermal scanning combined with the spectrum measurements.

After half-century, low- to medium-carbon martensitic steels have been one of the essential subjects due to their potential for saving the weight as well as for improving the crash resistance of structural parts in automobiles. It is, therefore, crucial to understand not only the phase evolution during quenching and tempering but also the relationship between the microstructure and mechanical properties. The strength of martensite dominates by its carbon contents, distributing and even diffusing in the solid. Fe steel is a mixture consisting of carbides, retained austenite, carbon at dislocations and grain boundaries in martensite. It is, however, still not fully understood how all the phases and the lattice defects contribute to the mechanical properties. The main reason could be due to the lack of an evaluation technique to determine the concentration and their distribution of carbon in martensite under operation. Especially crucial is to obtain atomistic information. We expect the MSM imaging technique enables us to follow the development of the microstructure at high temperatures. As a first step toward an operand measurement on the carbon diffusion and segregation processes at high temperature, we measured the Mössbauer spectra and their mapping images of Fe-1Mn-2.7C foils [at%] with a martensitic-transformation start-temperature (Ms) of 210 °C after different heat treatments [98].

We first annealed the samples at 1100 °C, and water-quenched to room temperature with a cooling rate of 500–2000 °C/s, and subsequently put into liquid nitrogen. We tempered five foils separately at various temperatures from 100 to 700 °C. The Mössbauer spectra were measured at room temperature with a 1.85 GBq ^{57}Co-in-Rh source. Figure 9.62(**a**) and (**b**) show the Mössbauer spectra and their successive subtractions, respectively. After quenching, there are four components for martensite phase and one singlet for retained austenite [4]: The green one corresponds to α-Fe (bcc), the light blue to the Fe atom on the 1st Nearest Neighbour of Octahedral-site carbon atom, the dark green to Fe atom on the 2nd NN of O-site carbon, and the yellow to the 1st NN of Mn, and the red singlet to Austenite phase. The successive

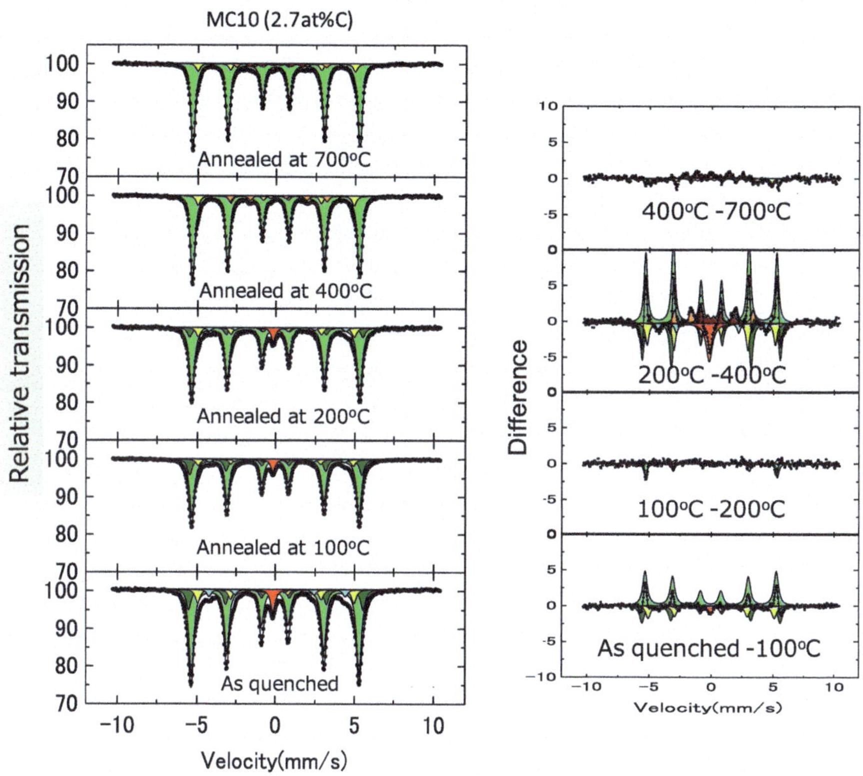

Fig. 9.62 (**a**) Mössbauer spectra after the annealing from 100 to 700 °C, and (**b**) their successive subtractions [from Ref. 98]

subtractions indicate clearly that the first segregation of carbon atoms appears to start after the annealing 100 °C. The overall phase separation from martensite to ferrite + cementite occurs at 400 °C, which is accompanied by the transformation from "retained austenite" to "ferrite" phases.

Figure 9.63 summarizes the MSM electron and γ-ray images of the Fe steel for the annealing at 100, 200, 400 and 700 °C as well as the spectral components of Ferrite, Austenite, 1st NN Fe of O-site Carbon in Ferrite, and Cementite. Since we measured five different samples after quenching and annealing at high temperatures, you could compare the contour maps for the different components. Although the present sample contains many components, we obtained the mapping images separately for the different components by tuning to the 1–6 lines among all the sextet. In a ferromagnetic sample, however, we have to consider magnetic domains by measuring all six lines to get the concentration distributions of the different phases. To study the carbon diffusion in the microstructure of martensite and austenite phases, we have to investigate one sample either after isochronal annealing at high temperatures, or

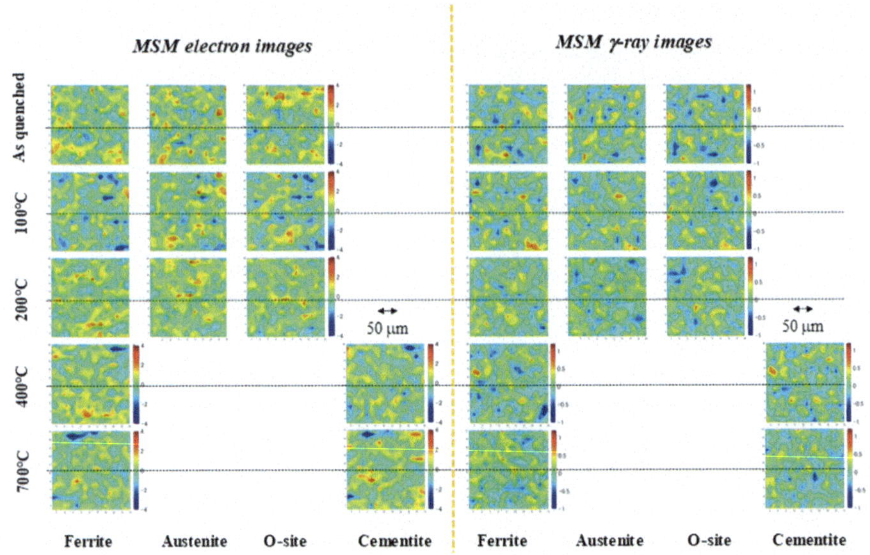

Fig. 9.63 MSM electron and γ-ray images measured after the annealing at 100, 200, 400 and 700 °C, corresponding to the spectral components of Ferrite, Austenite, 1st NN of O-site carbon, and Cementite, respectively. We used an MCX lens with a γ-ray spot size of 75 μmϕ [from Ref. 98]

after isothermal annealing. Finally, I only mention that the MSM imaging technique was also applied to study the microstructure of $BiFeO_3$ Thin Films [99–102].

9.7 Summary

This chapter has demonstrated several new methods in ^{57}Fe Mössbauer spectroscopy specially designed and constructed to study diffusion and point defects on an atomistic scale in metals and semiconductors. We performed various international joint ventures from 1980 to 2019 with different research groups in Japan, Germany, Austria, and Belgium. In-beam and on-line Mössbauer techniques were developed using heavy-ion accelerator facilities at the Hahn Meitner Institute Berlin, and later at RIKEN. The experimental set-ups provided unique possibilities for in-situ observations on atomic jump processes of ^{57m}Fe in homogeneous materials for the diffusion lengths to a few nanometers and the specific times from 10^{-6} to 10^{-8} s, respectively. Secondary, we have shown both mapping and imaging techniques of Mössbauer spectroscopy, accessible to the studies on Fe diffusion and segregation selectively for different chemical and physical states in inhomogeneous solid such as Si-solar cells and Fe steel. Soon, the Mössbauer camera will play a vital role to clarify unsolved problems in solid-state physics and material science. We have started to perform ***"operand observations"*** on a fracture process of metallic mold, on magnetic domains in magnetic materials, as well as a pattern formation during Belousov–Zhabotinsky reaction.

Acknowledgements I am entirely indebted to my mentors, Emeritus Profs. F. E. Fujita and S. Nasu, Osaka University, for their creative ideas and kind support to the point defect studies. I thank Emeritus Profs. Y. Maeda and H. Yoshida, KUR Kyoto University, for their warm support to low-temperature irradiation works. Since 1983, I had a lot of exciting opportunities to join the international joint ventures. I sincerely appreciate to Emeritus Prof. G. Vogl, who invited me to Hahn Meitner Institute Berlin and Universität Wien and gave me the best period to perform the Mössbauer diffusion studies by developing Accelerator Spectroscopy. I worked with excellent scientists and acknowledged the collaborations with Dr R. Sielemann HMI, Prof. P. Fratzl MPI of Colloids and Interfaces, Emeritus Prof. A. Seeger MPI Stuttgart, Emeritus Prof. G. Langouche Uni. Leuven, and Dr Y. Kobayashi RIKEN. For JST project for MSM development, I appreciate Dr Y. Ino, Dr K. Tanaka, Mr K. Hayakawa, and Mr K. Yukihira, and many students in my laboratory for intensive supports. Finally, I am deeply appreciated to warm help from Emeritus Profs. G. Langouche, Uni. Leuven and N. Kojima Tokyo univ., for organizing international conferences such as ICAME2011 Kobe and IWMS2018 Nagoya.

The project for developing Mössbauer Spectroscopic Microscopy (MSM) was supported by the "Development of Systems and Technologies for Advanced Measurement and Analysis" Program of Japan Science and Technology Agency (JST). I acknowledge intensive technical supports from APCO Ltd and Hamamatsu Photonics K.K.

References

1. R.L. Mössbauer, Fluorescent nuclear resonance of γ-radiation in Iridium-191. Z. Physik **151**, 124–143 (1958); R.L. Mössbauer, Kernresonanz-absorption von gammastrahlung in Ir^{191}. Naturwissenschaften **45**, 538–539 (1958); R.L. Mössbauer, Recoiless nuclear resonance absorption of γ-radiation. Science **137**, 731–738 (1962); R.L. Mössbauer, The discovery of the Mössbauer effect. Hyperfine Interact. **126**, 1–12 (2000)
2. V.I. Goldanskii, R. Herber (eds.), *Chemical Applications of Mössbauer Spectroscopy* (Academic Press, New York, 1968)
3. N.N. Greenwood, T.C. Gibb, *Mössbauer Spectroscopy* (Chapman and Hall, London, 1971)
4. U. Gonser (ed.), *Mössbauer Spectroscopy, in Topics in Applied Physics*, vol. 5 (Springer, Berlin, 1975)
5. P. Gütlich, R. Link, A.X. Trautwein, Mössbauer *Spectroscopy and Transition Metal Chemistry. Inorganic Chemistry Concepts Series*, vol. 3, 1st ed. (Springer, Berlin, 1978)
6. G.J. Long (ed.), *Mössbauer Spectroscopy Applied to Inorganic Chemistry, Modern Inorganic Chemistry Series*, vol. 1. (Plenum, New York, 1984)
7. G.J. Long (ed.), *Mössbauer Spectroscopy Applied to Inorganic Chemistry, Modern Inorganic Chemistry Series*, vol. 2. (Plenum, New York, 1989)
8. G.J. Long, F. Grandjean (eds.), *Mössbauer Spectroscopy Applied to Inorganic Chemistry, Modern Inorganic Chemistry Series*, vol. 3. (Plenum, New York, 1989)
9. G.J. Long, F. Grandjean (eds.), *Mössbauer Spectroscopy Applied to Magnetism and Materials Science*, vol. 1. (Plenum, New York, 1993)
10. G.J. Long, F. Grandjean, (eds.), *Mössbauer Spectroscopy Applied to Magnetism and Materials Science*, vol. 2. (Plenum, 1996)
11. P. Gütlich, E. Bill, A.X. Trautwein, *Mössbauer Spectroscopy and Transition Metal Chemistry* (Springer, Berlin, 2011)
12. Y. Yoshida, G. Langouche (eds.), *Mössbauer Spectroscopy-Tutorial Book* (Springer, Berlin, 2012)

13. Y. Yoshida, G. Langouche (eds.), *Defects and Impurities in Silicon Materials: An Introduction to atomic-level Silicon Defect Engineering* (Springer, Berlin, 2015)
14. Y. Yoshida, K. Mibu (eds.), *Proceedings of the 65th Yamada Conference and the 31st International Conference on the Applications of the Mössbauer Effect (ICAME2011, Kobe), 25–30 September 2011*, (Springer, New York, London, 2011)
15. A.E. Fick, Annalen der Physik und Chemie, **94**, 59 (1855); Phil. Mag. **10**, 30 (1855)
16. J. Groh, G. von Hevesy, Annalen der Physik **63**, 85 (1920); Annalen der Physik **65**, 216 (1921)
17. A. Einstein, Annalen der Physik, **17**, 549 (1905); Annalen der Physik, **19**, 371 (1906)
18. M. van Smoluchowski, Ann. Phys. **21**, 756 (1906)
19. P.G. Shewmon, *Diffusion in Solids, McGraw-Hill Book Company, New York/San Francisco/Tronto/London*, (Springer, Berlin, 1963)
20. J. Philibert, Atom movements: Diffusion and mass transport in solids, Les Editions de Physique (1991)
21. H. Mehrer, *Diffusion in Solids: Fundamentals, Methods, Materials, Diffusion-Controlled Processes* (Springer, Berlin, 2007)
22. H. Mehrer (Vol. ed.), *Diffusion in Solid Metals and Alloys, Landort-Bornstein, New Series, Group III*, vol. 26, (Springer, Berlin, 1990)
23. D.L. Beke (Vol. Ed.), *Diffusion in Semiconductors and Non-Metallic Solids, Sub-volume A, Diffusion in Semiconductors, Landort-Bornstein, New Series, Group III*, Vol.33 (Springer, Berlin, 1998)
24. D.L. Beke (Vol. Ed.), *Diffusion in Semiconductors and Non-Metallic Solids, Sub-volume B1, Diffusion in Nonmetallic Solids (part 1), Landort-Bornstein, New Series, Group III*, vol. 33, (Springer, Berlin, 1999)
25. S. Margulies, J.R. Ehrman, Nucl. Instr. Methods **12**, 131 (1961)
26. K.S. Singwi, A. Sjölander, Phys. Rev. **120**, 1093 (1960)
27. C.T. Chudley, R.J. Elliot, Proc. Phys. Soc. (London) **77**, 353 (1961)
28. M.C. Dibar-Ure, P.A. Flinn, Appl. Phys. Lett. **23**, 587 (1973)
29. Ch. Janot, J. De Physique **37**, 253 (1976)
30. D. Wolf, Appl. Phys. Lett., **30**, 617, (1977); Phil. Mag. A, **47**, 147, (1983).
31. J.G. Mullen, Phys. Lett. **79A**, 457, (1980); in *Proceeding of ICAME1981* (Jaipur, 29, 1982)
32. S. Mantl, W. Petry, K. Schroeder, G. Vogl, Phys. Rev. **27**, 5313 (1983)
33. K.H. Steinmetz, G. Vogl, W. Petry, K. Schroeder, Phys. Rev. B **34**, 107 (1986)
34. I. Nowik, S.G. Cohen, E.R. Bauminger, S. Ofer, Phys. Rev. Lett, **50**, 152, (1983); E.R. Bauminger, I. Nowik, p. 219, (1986), in *Mössbauer Spectroscopy*, eds. by D.P.E. Dickson, F.J. Berry, (Cambridge University Press)
35. S. Dattagupta, in *Mössbauer Spectroscopy*, eds.by D.P.E. Dickson, F.J. Berry (Cambridge University Press, 1986), p. 198
36. A. Seeger, H. Mehrer, in *Vacancies and Interstitials in Metals*, eds. by A. Seeger, D. Schumacher, W. Schilling, J. Diehl (North-Holland Publishing Co., Amsterdam, 1970). p. 1
37. W. Schilling, P. Erhart, K. Sonnenberg, *Interpretation of Defect Reactions in Irradiated Metals by One Interstitial Model in Fundamental Aspects of Radiation Damages in Metals*, eds. by M. T. Robinson and F. W. Young, jr., U.S.E.R.D.A.Conf. 751006, (1976), p. 470.
38. S. Nasu, Y. Yoshida, F.E. Fujita, in *Proceeding of ICAME1981*, (Jaipur, 1982), p. 212
39. Y. Yoshida, S. Nasu, F.E. Fujita, ^{57}Fe Mössbauer Study on Point Defects in Pure Iron, in *Proceeding of International Conference of Yamada Conference V* (1982), pp. 199-202; Y. Yoshida, S. Nasu, F.E. Fujita, Y. Maeda, H. Yoshida, J. Magn. Magn. Mater. **34-35**, 753 (1983); Y. Yoshida, Doctor thesis, *Mössbauer Effect Study of Point Defects in Pure Iron*, (Osaka University, Graduate school of engineering science, Department of Material Science, Toyonaka, Osaka, 1983)
40. M. Kiritani, H. Takata, K. Moriyama, F.E. Fujita, Mobility of lattice vacancies in iron. Phil. Mag. **40**, 779–802 (1979); M. Kiritani, Nature of point defects and their interactions revealed by electron-microscope observation of their clusters, in *Proceeding of International Conference of Yamada Conference V*, (1982), pp. 59–66

41. P.M. Derlet, D. Nguyen-Manh, S.L. Dudarev, Multiscale modelling of crowdion and vacancy defects in body-centred-cubic transition metals. Phys. Rev. B **76**, 054107 (2007)
42. A. Heiming, K.H. Steinmetz, G. Vogl, Y. Yoshida, Mössbauer studies on self-diffusion in pure iron. J. Phys. F: Metal Phys. **18**, 1491 (1988)
43. Y. Yoshida, W. Miekeley, W. Petry, W. Stehr, K.H. Steinmetz, G. Vogl, Anomalously fast diffusion of Fe in -Zr-Fe Alloys, A Mössbauer study. Mat. Sci. Forum **15–18**, 487–492 (1987)
44. Y. Yoshida, P. Fratzl, W. Miekeley, G. Vogl, Anomalously fast diffusion of Fe in Zr-Nb alloys. Defect Diffus Forum **66–69**, 353 (1990)
45. Y. Yoshida, M. Sugimoto, D. Tuppinger, G. Vogl, Search for anomalously fast diffusion of Fe in -Ti and -Zr with Mössbauer spectroscopy. Defect Diffus Forum **66–69**, 347 (1990)
46. Y. Yoshida, Y. Masuda, H. Häßlein, F.E. Fujita, H. Nakajima, High temperature Mössbauer study on the fast diffusion of Fe in -TiFe alloys. Defect Diffus Forum **143–147**, 461–466 (1996)
47. Y. Yoshida, P. Fratzl, F. Langmayr, G. Vogl, Short-range-order in Au-Fe alloys studied by high-temperature Mössbauer spectroscopy. Phys. Rev. B **39**, 6395 (1989)
48. P. Fratzl, Y. Yoshida, F. Langmayr, G. Vogl, Defect-mediated nucleation of α-Fe in Au-Fe alloy. Phys. Rev. B **44**, 4192–4199 (1991)
49. P. Fratzl, Y. Yoshida, G. Vogl, H.G. Haubold, Phase-separation kinetics of dilute Cu-Fe alloys studied by anomalous small-angle X-ray scattering and Mössbauer spectroscopy. Phys. Rev. B **46**, 11323–11331 (1992)
50. Y. Yoshida, P. Fratzl, G. Vogl, H. Hofer, F. Dworschak, Radiation-induced Segregation in Proton-irradiated AuFe Studied by Mössbauer Spectroscopy. J. Phys. Condens. Mat. **4**, 2415–2428 (1992)
51. Y. Yoshida, F. Shimura, In-Situ Observation of Diffusion and Segregation of Fe atoms in Si Crystals at High Temperature by Mössbauer spectroscopy, in *Electrochemical Society Proceedings*, vol. 98–1, (1998), pp. 984–996, (The 8th International Symposium on Silicon Materials Science and Technology, San Diego May 1998).
52. Y. Yoshida, S. Ogawa, K. Arikawa, Direct observation of substitutional Fe Atoms in Si and SOI wafers at 1273 K. Phys. B **340–342**, 605–608 (2003)
53. Y. Yoshida, S. Horie, K. Niira, K. Fukui, K. Shirasawa, In-situ observation of iron atoms in multi-crystalline silicon at 1273 K and 300 K by Mössbauer spectroscopy. Phys. B **376–377**, 227–230 (2006)
54. Y. Yoshida, S. Aoki, K. Sakata, Y. Suzuki, M. Adachi, K. Suzuki, Iron impurities in multi-crystalline silicon studied by Mössbauer spectroscopy. Phys. B **401–402**, 119–122 (2007)
55. M. Menningen, R. Sielemann, G. Vogl, Y. Yoshida, K. Bonde-Nielsen, G. Weyer, Interstitial Implantation of Iron into aluminum. Europhy. Lett. **3**, 3–8 (1987)
56. Y. Yoshida, M. Menningen, R. Sielemann, G. Vogl, G. Weyer, K. Schroeder, Local atomic jump process of iron in -Zirconium. Phys. Rev. Lett. **61**, 195–198 (1988)
57. Y. Yoshida, In-Beam Mössbauer study of atomic jump processes in metals. Hyperfine Interact. **47**, 95–113 (1989)
58. P. Schwalbach, S. Laubach, M. Hartick, E. Kankeleit, B. Keck, M. Menningen, R. Sielemann, Diffusion and Isomer shift of Interstitial Iron in Silicon observed via in-beam Mössbauer Spectroscopy. Phys. Rev. Lett. **64**, 1274 (1990)
59. R. Sielemann, Y. Yoshida, In-beam Mössbauer spectroscopy at heavy-ion accelerators. Hyperfine Interact. **68**, 119–130 (1991)
60. B. Keck, R. Sielemann, Y. Yoshida, Iron on substitutional and interstitial lattice sites in alkali metals and isomer shift systematics for interstitial iron in elemental metals. Phys. Rev. Lett. **71**, 4178–4181 (1993)
61. J. Kapoor, D. Riegel, Y. Li, C. Polaczyk, J. Andres, F. Mezei, R. Sielemann, Y. Yoshida, W.D. Brewer, L. A. de Mello, S. Frota-Pessoa, Observation of Magnetism of Fe at an Interstitial Site in a Metal Host. Phys. Rev. Lett., **78**, 1279–1282 (1997)
62. Y. Yoshida, D. Tuppinger, G. Vogl, B. Keck, R. Sielemann, In-beam Mössbauer Spectroscopy on pure iron, in *HMI-Bericht Berlin*, vol. 490, eds. by J. Eichler, W. Von Oertzen, H. Lindenberger, H. Homeyer, R. Michaelsen (1991), pp. 112–113

63. Y. Yoshida, Y. Kobayashi, K. Hayakawa, K. Yukihira, J. Nakamura, S. Nasu, E. Yagi, F. Ambe, In-beam Mössbauer study on ^{57}Fe jumps in solid Ar and Solid Xe after coulomb-excitation and recoil-implantation. RIKEN Rev. **16**, 19–20 (1997)
64. Y. Yoshida, Mössbauer spectroscopy to investigate atomistic jump processes on an atomistic scale. Hyperfine Interact **113**, 183–198 (1998)
65. R. Sielemann, Y. Kobayashi, Y. Yoshida, H.P. Gunnlaugsson, G. Weyer, Magnetism at single isolated iron atoms implanted in graphite. Phys. Rev. Lett. **101**, 137206 (2008)
66. G. Vogl, W. Mansel, P.H. Dederichs, Plys. Rev. Lett. **36**, 1497 (1976)
67. W. Petry, G. Vogl, W. Mansel, Phys. Rev. Lett. **45**, 1862 (1980)
68. H.J. Blythe, H. Kronmuller, A. Seeger, F. Walz, Phys. Stat. Sol. (a) **181**, 233, (2000)
69. J. Christiansen, E. Recknagel, G. Weyer, Phys. Rev. Lett. **20**, 46 (1966)
70. Y. Yoshida, Y. Kobayashi, F. Ambe, E. Yagi, A. Seeger et al., Direct observation of self-interstitial motion in pure iron by ^{56}Fe (d, p)^{5}7Fe In-beam mössbauer spectroscopy. Defect Diffus Forum **194–199**, 29–34 (2001)
71. A.A. Istratov, H. Hieslmair, E.R Weber, Iron and its complexes in silicon. Appl. Phys. A **69** (1999) 13–44; A.A. Istratov, H. Hieslmair, E.R Weber, Iron contamination in silicon technology Appl. Phys. A **70**, 489–534 (2000)
72. Y. Yoshida, *Direct Observation of Substitutional and Interstitial Fe atoms in Si by high-temperature and In-beam Mössbauer Spectroscopy, ALTECH 2003, ECS*, (Salt Lake City, US, 479–482, 2003)
73. Y. Yoshida, Y. Kobayashi, K. Hayakawa, K. Yukihira, F. Shimura, A. Yoshida, X. Diao, H. Ogawa, Y. Yano, F. Ambe, In-beam Mössbauer study on interstitial and substitutional ^{57}Mn/^{57}Fe jumps in Si. Defect Diffus Forum **194–199**, 611–616 (2001)
74. Y. Yoshida, Y. Kobayashi, A. Yoshida, X. Diao, S. Ogawa, K. Hayakawa, K. Yukihira, F. Shimura, F. Ambe, In-Beam Mössbauer spectroscopy after GeV-ion implantation at an on-line projectile-fragments separator. Hyperfine Interact **141–142**, 157–162 (2002)
75. Y. Yoshida, Y. Kobayashi, K. Hayakawa, K. Yukihira, A. Yoshida, H. Ueno, F. Shimura, F. Ambe, In-situ observation of substitutional and interstitial Fe atoms in Si after GeV-implantation: an in-beam Mössbauer study. Phys. B **376–377**, 69–72 (2006)
76. Y. Yoshida, Y. Kobayashi, K. Yukihira, K. Hayakawa, K. Suzuki, A. Yoshida, H. Ueno, A. Yoshimi, K. Shimada, D. Nagae, K. Asahi, G. Langouche, ^{57}Fe diffusion in n-type Si after GeV-implantation of ^{57}Mn. Phys. B **401–402**, 101–104 (2007)
77. Y. Yoshida, K. Suzuki, Y. Kobayashi, T. Nagatomo, Y. Akiyama, K. Yukihira, K. Hayakawa, H. Ueno, A. Yoshimi, D. Nagae, K. Asahi, G. Langouche, ^{57}Fe charge state in mc-Si solar cells under light illumination After GeV- implantation of ^{57}Mn. Hyperfine Interact **204**, 133–137 (2012)
78. K. Suzuki, Y. Yoshida, K. Hayakawa, K. Yukihira, M. Ichino, K. Asahi, Observation of iron impurities diffusion in silicon under bending stress by Mössbauer spectroscopy. Hyperfine Interact **197**, 213–217 (2010)
79. M. Ichino, Y. Yoshida, In-situ observation of iron impurities in n-type silicon under UV light illumination by Mössbauer spectroscopy. in *Proceeding of The Forum on the Science and Technology of Silicon Materials*, vol. 357 (2010)
80. K. Tanaka, Y. Akiyama, K. Hayakawa, K. Yukihira, Y. Yoshida, Mapping analyses of Fe-diffused mc-Si using Mössbauer microscope and photoluminescence. Hyperfine Interact **206**, 75–78 (2012)
81. K. Tanaka, T. Watanabe, T. Uenoyama, Y. Ino, Y. Yoshida, Search for Fe-B pairs in highly B-doped Si wafers by Mössbauer spectroscopy. in *Proceedings of Si-Forum 2014, Hamamatsu,* (2014), pp. 107–111
82. Y. Yoshida, Y. Tsukamoto, M. Ichino, K. Tanaka, Direct observation of carrier trapping processes on Fe impurities in mc-si solar cell. Solid State Phenom. **205–206**, 40–46 (2014)
83. Y. Yoshida, K. Suzuki, K. Hayakawa, K. Yukihira, H. Soejima, Mössbauer spectroscopic microscope. Hyperfine Interact. **188**, 121–126 (2009)
84. Y. Yoshida, K. Hayakawa, K. Yukihira, M. Ichino, Y. Akiyama, H. Kumabe, H. Soejima, Development and applications of "Mössbauer cameras." Hyperfine Interact. **198**, 23–29 (2010)

85. K. Hayakawa, Y. Tsukamoto, Y. Akiyama, M. Kurata, K. Yukihira, H. Soejima, Y. Yoshida, Deployment of system and technology for Mössbauer spectroscopic microscope. Hyperfine Interact **206**, 79–82 (2012)
86. Y. Ino, H. Soejima, K. Hayakawa, K. Yukihira, K. Tanaka, H. Fujita, T. Watanabe, K. Ogai, K. Moriguchi, Y. Harada, Y. Yoshida, 3D-Mössbauer spectroscopic microscope for mc-Si solar cell evaluation. Hyperfine Interact. **237**, 13 (2016)
87. Y. Yoshida, Y. Ino, K. Matsumuro, T. Watanabe, H. Fujita, K. Hayakawa, K. Yukihira, K. Ogai, K. Moriguchi, Y. Harada, H. Soejima, Feasibility study to investigate the diffusion of Fe in Si using a Mössbauer spectroscopic microscope. Hyperfine Interact. **237**, 130 (2016)
88. Y. Ino, T. Watanabe, K. Hayakawa, K. Yukihira, K. Matsumuro, H. Fujita, K. Ogai, K. Moriguchi, Y. Harada, H., Soejima, Y. Yoshida, A new set-up of Mössbauer spectroscopic microscope to study the correlation between Fe impurities and lattice defects in Si crystals. J. Cryst Growth. https://doi.org/10.1016/j.jcrysgro.2016.12.109
89. Y. Ino, H. Soejima, K. Hayakawa, K. Yukihira, K. Tanaka, H. Fujita, T. Watanabe, K. Ogai, K. Moriguchi, Y. Harada, and Y. Yoshida, 3D-mössbauer spectroscopic microscope for mc-Si solar cell evaluation. Hyperfine Interactions. **237**, 13 (2016).
90. Y. Yoshida, T. Watanabe, Y. Ino, M. Kobayashi, I. Takahashi, N. Usami, Mössbauer spectroscopic microscopy study on diffusion and segregation of Fe impurities in mc-si wafer. Hyperfine Interact. **240**, 114–121 (2019)
91. G. Langouche, Y. Yoshida, Ion Implantation, in *Mössbauer Spectroscopy-Tutorial Book*, eds. by Y. Yoshida, G. Langouche (Springer, 2013). https://doi.org/10.1007/978-3-642-32220-4
92. G. Langouche, Y. Yoshida, Nuclear Methods to study defects and impurities in Si materials. 10.1007/978-4-431-55800-2, pp. 373, Chapter 8, in *Defects and Impurities in Silicon Materials- An Introduction to Atomic-Level Silicon Engineering*, eds. by Y. Yoshida, G. Langouche (eds.), in *Lecture Note in Physics*, (Springer, Berlin, 2016)
93. H.P. Gunnlaugsson, G. Weyer, M. Dietrich, M. Fanciulli, K. Bharuth-Ram, R. Sielemann, ISOLDE collaboration. Appl. Phys. Lett. **80**, 2657–2659 (2002)
94. J. Kübler, A.E. Kumm, H. Overhof, P. Schwalbach, M. Hartick, E. Kankeleit, B. Keck, L. Wende, R. Sielemann, Isomer-shift of interstitial and substitutional iron in silicon and germanium. Z. Phys. B **92**, 155–162 (1993)
95. E. Wright, J. Coutinho, S. Oberg, V.J.B. Torres, J. Appl. Phys. **119**, 181509 (2016). https://doi.org/10.1063/1.14948243
96. Y. Ino, K. Hayakawa, K. Yukihira, K. Moriguchi, K. Ogai, K. Shirasawa, H. Takato, Y. Yoshida, Iron contamination near surface of silicon solar cells observed by Mössbauer spectroscopic microscope, under preparation for publication, (2020)
97. M. Kobayashi, Y. Yoshida, in *Abstract book for The 9th TOYOTA RIKEN International Workshop on New Developments and Prospects for the Future of Mössbauer Spectroscopy (IWMS2018)*
98. N. Maruyama, H. Fujita, M. Kobayashi, Y. Yoshida, H. Kubo, Mössbauer and atom probe tomography characterization of low and medium carbon martensitic steels, in *Abstract book for The 9th TOYOTA RIKEN International Workshop on New Developments and Prospects for the Future of Mössbauer Spectroscopy (IWMS2018)*
99. K. Tanaka, Y. Tsukamoto, K. Hayakawa, Y. Yoshida, Characterization of ^{57}Fe-Enriched $BiFeO_3$ thin films by mössbauer spectroscopy. Jpn. J. Appl. Phys. **51**, (2012)
100. K. Tanaka, Y. Tsukamoto, S. Okamura, Y. Yoshida, Observation of ^{57}Fe-Enriched $BiFeO_3$ thin films by Mössbauer spectroscopic microscope. Jpn. J. Appl. Phys. **52,** 09KB02 (2013)
101. K. Tanaka, Y. Tsukamoto, S. Okamura, Y. Yoshida, Mössbauer spectra of ^{57}Fe-Enriched $BiFeO_3$ thin films fabricated on SiO_2/Si substrates by chemical solution deposition process. Key Eng. Mater. **582**, 63–66 (2014)
102. K. Tanaka, Y. Fujita, S. Okamura, Y. Yoshida, Mössbauer spectra and electric properties of ^{57}Fe-enriched $BiFeO_3$ Thin Films. Jpn. J. Appl. Phys. **53** (2014)

Index

Y. Yoshida and G. Langouche (eds.), *Modern Mössbauer Spectroscopy*,
Topics in Applied Physics 137, https://doi.org/10.1007/978-981-15-9422-9

The manufacturer's authorised representative in the EU is Springer Nature Customer Service Centre GmbH, Europaplatz 3, 69115 Heidelberg, Germany. If you have any concerns regarding our products, please contact ProductSafety@springernature.com

Printed and bound by CPI Group (UK) Ltd, Croydon, CR0 4YY

30/09/2026

02225240-0012